AF657816

3rd International Conference on

Thermo-mechanical Simulation and Processing of Steels

3rd International Conference on

Thermo-mechanical Simulation and Processing of Steels 12-14 December 2012 : Ranchi, India

CONFERENCE PROCEEDINGS

Editors

Dr. B.K. Jha
S. Srikanth
Dr. Anjana Deva
S. Mitra Mazumdar
D. Karmakar
S. Varadarajan

3rd *International Conference on*

Thermo-mechanical Simulation and Processing of Steels

Editors

B.K. Jha, S. Srikanth, Anjana Deva, S. Mitra Mazumdar, D. Karmakar, S. Varadarajan

Organised by

R&D Centre for Iron & Steel
Steel Authority of India Limited
Ranchi - 834 002, India

Indian Institute of Metals
Ranchi Chapter
Ranchi - 834 002, India

The Minerals Metals and
Materials Society
USA

Computer Society of India
Ranchi Chapter
Ranchi - 834 002, India

Platinum Sponsors

Poestenkill NY 12140, USA

São Paulo-SP, Brazil

Gold Sponsors

Pennsylvania 15205-5005, USA

Mumbai 400 002 India

Silver Sponsors

Thermo-Calc Software AB

Stockholm SWEDEN

ICON Analytical

Mumbai 400 018 India

New Delhi | Mumbai | Chennai | Kolkata | Bengaluru | Hyderabad | Kochi | Guwahati

Publisher's note

Every possible effort has been made to ensure that the information contained in this book is accurate at the time of going to press, and the publisher and author cannot accept responsibility for any errors or omissions, however caused. No responsibility for loss or damage occasioned to any person acting, or refraining from action, as a result of the material in this publication can be accepted by the editor, the publisher or the author.

Every effort has been made to trace the owners of copyright material used in this book. The author and the publisher will be grateful for any omission brought to their notice for acknowledgement in the future editions of the book.

First Published 2013

Viva Books Private Limited

- 4737/23, Ansari Road, Daryaganj, New Delhi 110 002
  Tel. 42242200, 23258325, 23283121, Email: vivadelhi@vivagroupindia.net
- 76, Service Industries, Shirvane, Sector 1, Nerul, Navi Mumbai 400 706
  Tel. 27721273, 27721274, Email: vivamumbai@vivagroupindia.net
- Jamals Fazal Chambers, 26 Greams Road, Chennai 600 006
  Tel. 28294241, 28290304, Email: vivachennai@vivagroupindia.net
- B-103, Jindal Towers, 21/1A/3 Darga Road, Kolkata 700 017
  Tel. 22816713, Email: vivakolkata@vivagroupindia.net
- 7, GF, Sovereign Park Aptts., 56-58, K. R. Road, Basavanagudi, Bengaluru 560 004
  Tel. 26607409, Email: vivabangalore@vivagroupindia.net
- 101-102, Moghal Marc Apartments, 3-4-637 to 641, Narayanguda, Hyderabad 500 029
  Tel. 27564481, Email: vivahyderabad@vivagroupindia.net
- First Floor, Beevi Towers, SRM Road, Kaloor, Kochi 682 018
  Tel. 0484-2403055, 2403056, Email: vivakochi@vivagroupindia.net
- 232, GNB Road, Beside UCO Bank, Silpukhuri, Guwahati 781 003
  Tel. 0361-2666386, Email: vivaguwahati@vivagroupindia.net

www.vivagroupindia.com

ISBN: 978-81-309-2301-7

Published by Vinod Vasishtha for Viva Books Private Limited , 4737/23 Ansari Road, Daryaganj, New Delhi 110 002

Printed and bound by Syndicate Binders, Noida.

E0000

Contents

Section V: Emerging Trends in Steel Products

Section VI: Process Simulation

Section VIII: Advances in Materials Characterization

Section IX: Process Optimization

Preface

Steel research is chiefly driven by the desire to produce steels with tailor-made microstructure and properties suited to diverse applications. Significant progress has been made in understanding the influence of alloying elements, hot deformation of austenite and its controlled decomposition during cooling, cold rolling and annealing to generate a wide array of microstructures and mechanical properties in hot and cold rolled products. Thermo-mechanical simulations of varied purposes are resorted to, as cost effective means, in processing of steels to achieve aimed attributes and functions in the finished products. Based on the results of painstaking research in this important area, a number of new processing technologies have been introduced in the steel industry and are now being gainfully utilized in developing new market oriented products.

In this edition, more than sixty selected research papers, contributed by leading experts from India and abroad, covering diverse areas of research ranging from microstructural engineering, process simulation, thermo-mechanical processing, modeling, continuous casting, metal joining and recent advances in high strength, line pipe and formable quality steels have been organized as reference publication. This compilation of proceedings of SimPro'12 is third in series of international conferences on "Thermo-mechanical Simulation and Processing of Steels", which is organized every four years by the Research & Development Centre for Iron & Steel, Ranchi, India, a unit of Steel Authority of India Limited.

We sincerely hope that this special volume will help steel researchers in shaping their future course of research and development activities in these exciting fields of material science.

12th December, 2012

Editors

SECTION I

Novel Insights into Material Behaviour through Simulation and Modeling

Predicting the Occurrence of Dynamic Transformation and Rolling Mill Loads Drops by Hot Torsion and Compression Testing

John J. Jonas, Chiradeep Ghosh, Xavier Quelennec and Vladimir V. Basabe
Materials Engineering, McGill University,
3610 University St., Montreal, H3A 2B2, Canada
Email: john.jonas@mcgill.ca

ABSTRACT

Flow curves were determined in hot compression and hot torsion at a series of temperatures on eight C steels of increasing C concentration. The critical strains for the initiation of dynamic transformation (DT) as well as of dynamic recrystallization were determined by the double differentiation method. It is shown that dynamic transformation is initiated well before dynamic recrystallization under industrial conditions of rolling. The mean flow stresses (MFS's) pertaining to each experimental condition were then calculated from the flow curves by integration. These are plotted against inverse absolute temperature in the form of Boratto diagrams. The stress drop temperatures, normally defined as the upper critical temperature applicable to rolling A_{r3}*, were determined from these diagrams. These are shown to be about 40°C above the paraequilibrium and about 20-30°C above the orthoequilibrium upper critical transformation temperatures. This type of behavior is ascribed to the occurrence of dynamic transformation during deformation. The general characteristics of the dynamic transformation of austenite to ferrite are reviewed. It is suggested that some of the unexpectedly low rolling loads, as well as the load drops that have been reported to take place above the Ae_3 temperature in strip mills, may be attributable to this phenomenon.

KEYWORDS: *Rolling Load, Mean Flow Stress, Dynamic Transformation, Dynamic Recrystallization, Critical Strain, Mechanical Testing.*

INTRODUCTION

The dynamic transformation (DT) of austenite to ferrite during rolling above the Ae_3 was first demonstrated by Yada and co-workers in the 1980's [1,2]. They were able to produce this kind of transformation as much as 166°C above the orthoequilibrium Ae_3. On holding after deformation, the DT ferrite was shown to re-transform to austenite, as required by equilibrium thermodynamics. Later, the occurrence of this phenomenon was verified by in-situ x-ray methods [3] as well as by laser dilatometry [4,5] and considerable modeling work has also been carried out [6,7]. More recently, the double differentiation method has been employed to determine the critical strains for DT [8,9]. These have been shown to be well below the critical strains required for the initiation of dynamic recrystallization, so that DT is generally underway well before DRX under the usual conditions of industrial rolling.

Although DT is generally initiated during finish rolling, it does not produce readily detectable effects for a number of reasons. One is that the grains produced in this way are less than a micron in size and so are not readily observable using optical microscopy [10-15]. Another is that the rate of production of DT ferrite is quite low, ½ to 1% volume fraction per 1% strain [16], so that its initiation does not produce dramatic effects or sudden decreases in rolling load. However, because DT ferrite is softer than the austenite from which it has formed (at a given temperature), at the stage in rolling where significant volume fractions of DT ferrite have formed, they can have a measurable softening influence on the load. This is an effect that can be detected under appropriate conditions.

One way of predicting the occurrence and magnitude of the load drops due to DT is by carrying out hot compression and hot torsion experiments at the appropriate temperatures and then calculating the mean flow stress (MFS) by integration of the area under the curve. Such calculations can also be used to determine the effect of the accumulated strain below the T_{nr} on the magnitude of the load drop. Experiments of this type are described below and it is shown that load drops of 10 to 50 MPa can be readily produced in this way. It therefore appears that one of the possible causes of unexpected reductions in rolling load in the austenite phase field as well as rolling loads that are lower than predicted in the absence of strain accumulation may well be dynamic transformation. The calculations also explain why the values of the dynamic upper critical Ar_3* temperature derived using the Boratto [17] method can actually be higher than the equilibrium Ae_3 temperature.

EXPERIMENTAL PROCEDURE

Materials

The flow behaviors of an ultralow 0.019%C, a 0.11%C-0.038%Nb microalloyed, and a 0.22%C-0.045%Nb steel were determined by means of hot compression testing. A further five steels, containing 0.06%C, 0.09%C-0.036%Nb, 0.21%C, 0.45%C, and 0.79%C were tested in hot torsion. Their detailed compositions and equilibrium transformation temperatures are presented in Table 1.

Test Procedure

The hot compression tests were carried out on an MTS test frame with the aid of a Research Incorporated radiation furnace and superalloy tooling. The TestStar software was employed for test programming and data acquisition and conversion. Details of the heating and cooling procedure followed are given in Refs. 18-20. The torsion tests were performed on an MTS-based hydraulic torsion machine, again using a Research Incorporated radiation furnace and superalloy tooling. Test programming and data acquisition and conversion were again accomplished using the TestStar software. Details regarding the austenite conditioning as well as the heating, cooling and quench schedule are provided in Refs. 10-15. Metallographic examinations of the quenched samples were carried out on a Philips XL-30 FE-SEM.

Table 1: Chemical composition (wt %) and Ae_3 equilibrium transformation temperatures (°C).

Steel	Composition (wt%)	Ae_3 (°C) Orthoequilibrium	Ae_3 (°C) Paraequilibrium
1	0.019 C; 0.2 Si; 1.5 Mn; 0.017 Ti; 0.008 S; 0.005 P; 0.04 Al; 0.0089 N; 0.0005 B	872	860
2	0.11 C; 0.26 Si; 1.1 Mn; 0.038 Nb; 0.003 S; 0.004 P; 0.03 Al; 0.003 N	848	836
3	0.22 C; 1.56 Si; 1.56 Mn; 0.045 Nb; 0.017 Ti; 0.004 S; 0.006 P; 0.031 Al; 0.004 N	860	850
4	0.06 C; 0.01 Si; 0.30 Mn	877	870
5	0.09 C; 0.02 Si; 1.30 Mn; 0.036 Nb	836	820
6	0.21 C; 0.24 Si; 1.30 Mn	822	808
7	0.45 C; 0.24 Si; 0.70 Mn	772	753
8	0.79 C; 0.24 Si; 0.65 Mn	733	709

RESULTS

Flow Curves

Some typical flow curves determined in compression are presented in Fig. 1. In the ultralow C steel, Fig. 1a), flow softening took place under all the experimental conditions, while in the Nb-modified plain C steel, Fig. 1b), there was net work hardening at the lower temperatures and flow softening at the higher temperatures. In the Nb-modified TRIP steel, the flow stress nearly saturates at the lowest temperature, while there is clear evidence of flow softening, usually attributed to dynamic recrystallization (DRX) at the higher temperatures. Some typical flow curves determined in torsion are illustrated in Fig. 2. Only the low C (0.06%C) steel exhibits the conventional 'single peak' behavior, while the four remaining steels display work hardening, leading to saturation in three of the steels.

Critical Strains

The double derivative procedure [21] was employed In order to determine the critical strains for the initiation of dynamic transformation (DT) as well as of DRX. This required the fitting of polynomials to the rising portion of each stress-strain curve. Polynomials of order 8 to 10 were used, as described in more detail in Ref. 22. Examples of the second-derivative plots determined on the 0.09%C-0.036%Nb steel are provided in Fig. 3, where θ is the slope of the stress-strain curve (first derivative) and $\partial\theta/\partial\sigma$ is the second derivative. Here it can be seen that there are two local minima on each plot, the first of which is associated with the initiation of DT and the second with the start of DRX. Similar results were obtained on the other steels [15].

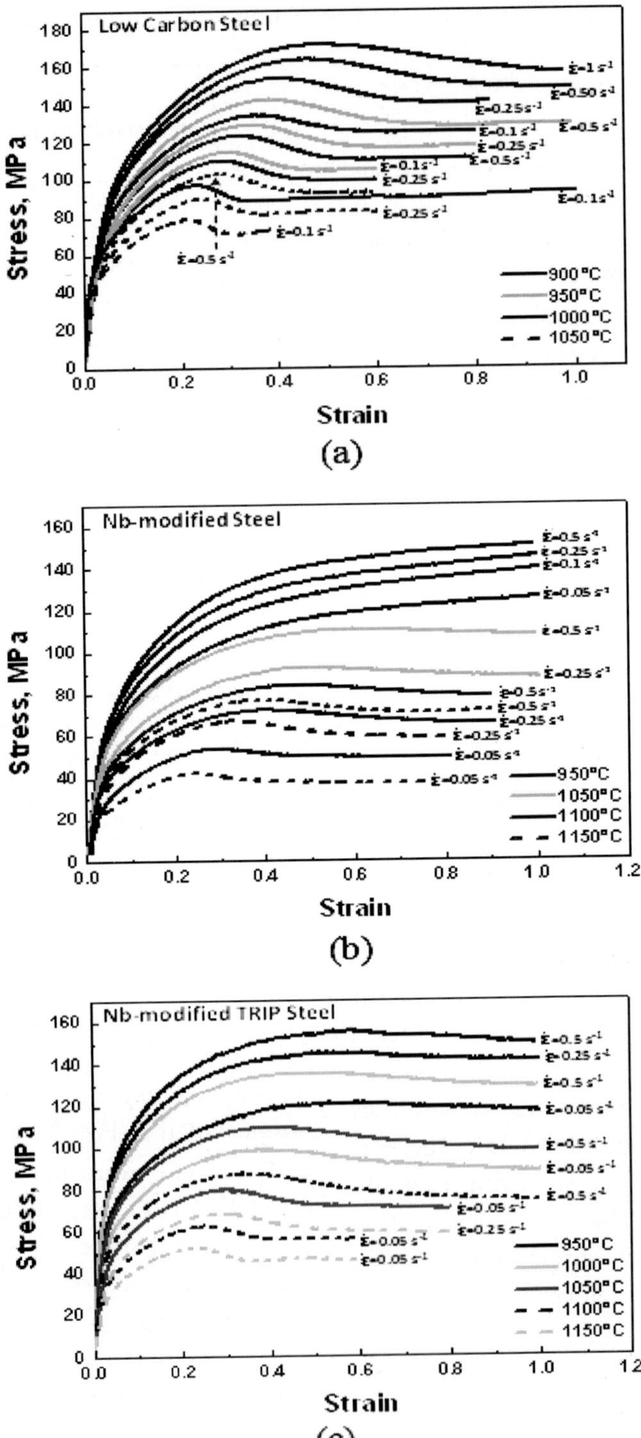

Fig. 1: Hot compression flow curves determined on the: a) 0.019%C ultralow C; b) 0.11%C Nb-modified; and c) Nb- modified TRIP steels [19].

The critical strains determined on four of the steels tested in torsion are illustrated in Fig. 4. Here it can be seen that the critical strains for the initiation of DT are quite low (about 0.1 to 0.25) and relatively insensitive to temperature and strain rate. They are always lower than those associated with DRX. The latter decrease with increasing temperature and decreasing strain rate in the manner normally associated with DRX [23]. The DT critical strains are in good agreement with the metallographic observations of ferrite formation during testing [10-15]. Similar results were obtained on the materials tested in compression [22], although the compression critical strains were always somewhat lower than those determined in torsion under comparable conditions.

Mean Flow Stresses

The mean flow stresses (MFS values) were determined from the flow curves by simple integration. These calculations were carried out using the Excel™ and Origin™ software programs. The results obtained were then plotted against inverse absolute temperature following the Boratto [17] procedure. The dependences observed are shown in Fig. 5 for the five steels tested in torsion.

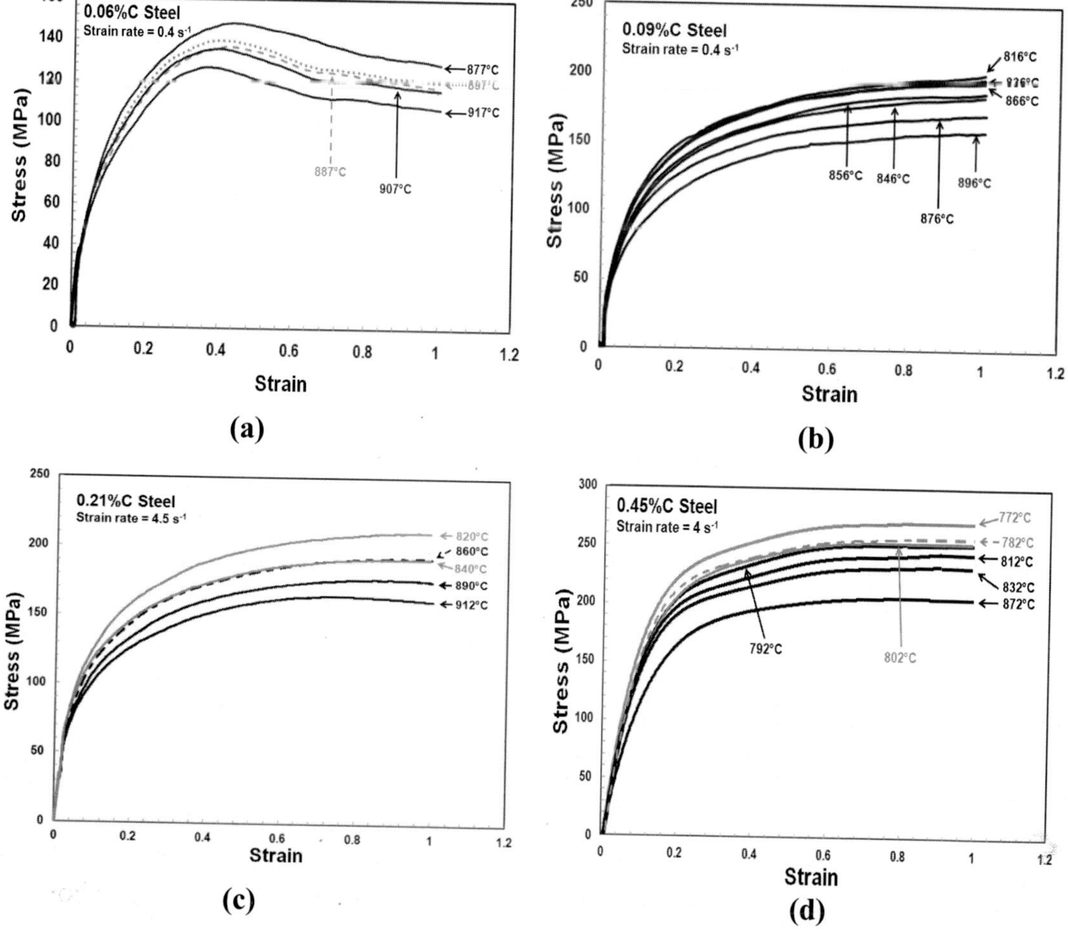

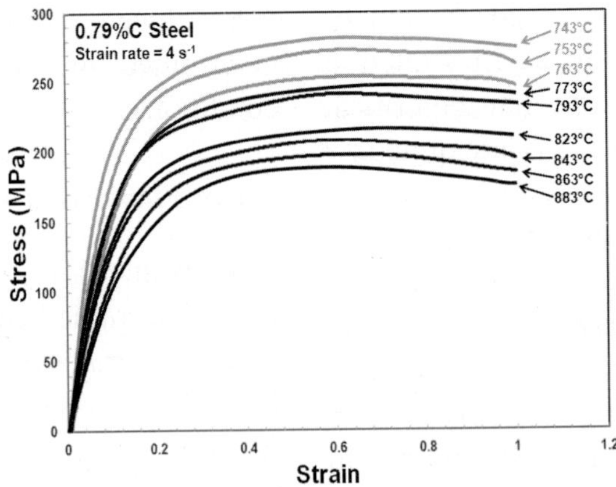

Fig. 2: Hot torsion flow curves determined on the: a) 0.06%C [10]; b) 0.09%C-0.036%Nb [11]; c) 0.21%C [15]; d) 0.45%C [16] and e) 0.79%C [15] steels.

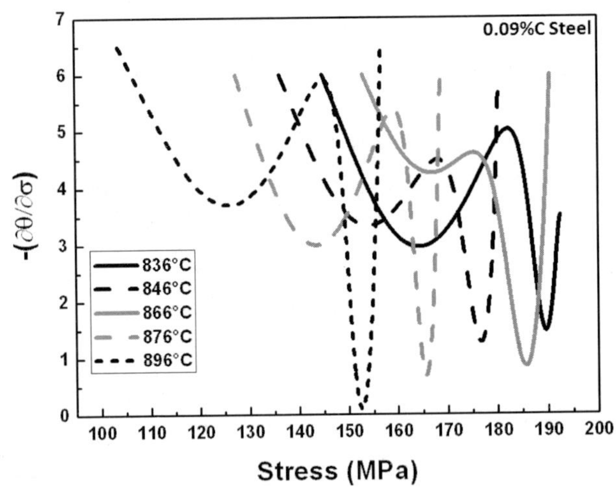

Fig. 3. -(∂θ/∂σ) vs. σ plots derived from the flow curves of the 0.09%C-0.036%Nb steel [15].

DISCUSSION

Mean Flow Stresses

Several features of these plots are of interest. First of all, it should be noted that all of the plots exhibit an Ar_3*, i.e. a point at which the MFS suddenly drops, even though the temperature is *decreasing*, so that the MFS is expected to rise. This is because the ferrite being formed by DT is softer than austenite at the temperature of interest. Of particular interest is the observation that in every case, the drop temperature *is well above* the equilibrium upper critical Ae_3 temperature normally associated with phase transformation. Both the orthoequilibrium as well as paraequilibrium temperatures are indicated on the diagrams, from which it can be seen that the dynamic Ar_3*'s are something like 20-40 degrees above the ortho Ae_3 and as much as 30-60 degrees above the para Ae_3.

This unusual behavior can be attributed to the considerable increase in free energy of the austenite associated with straining [24]. The dislocations introduced raise the free energy of the deformed austenite to well above that of the undeformed ferrite, so that

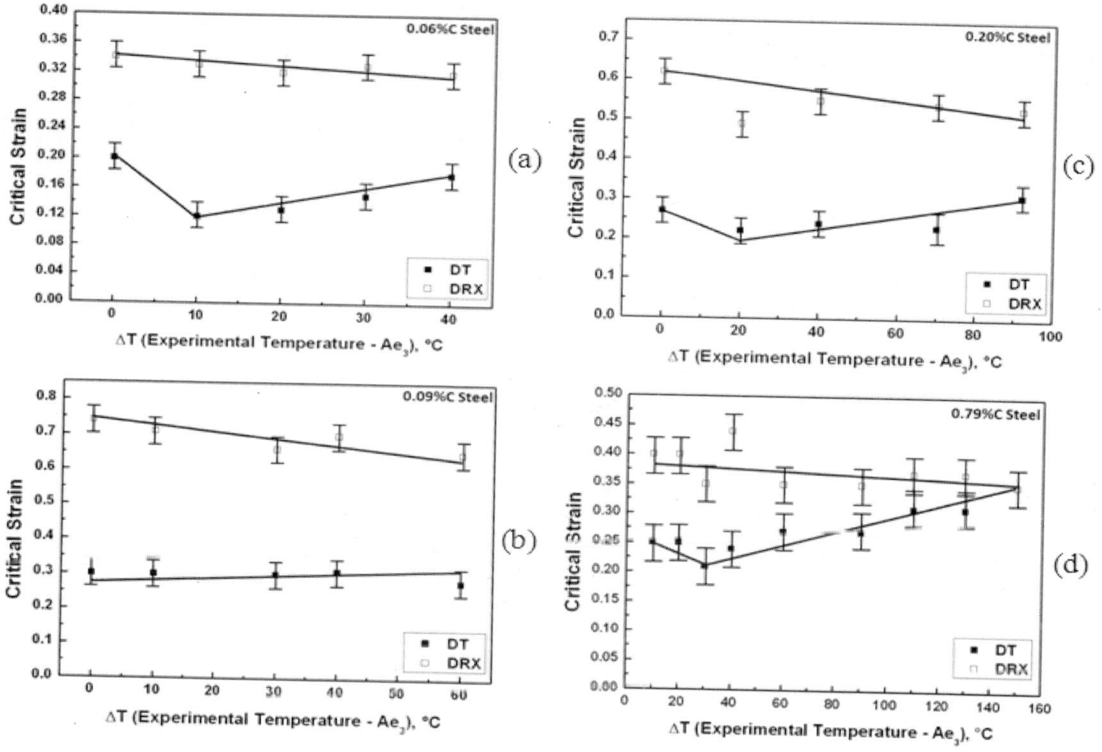

Fig. 4: Critical strain determined by means of torsion testing: a) 0.06%C steel; b) 0.09%C steel; c) 0.21%C steel; and d) 0.79%C steel [15].

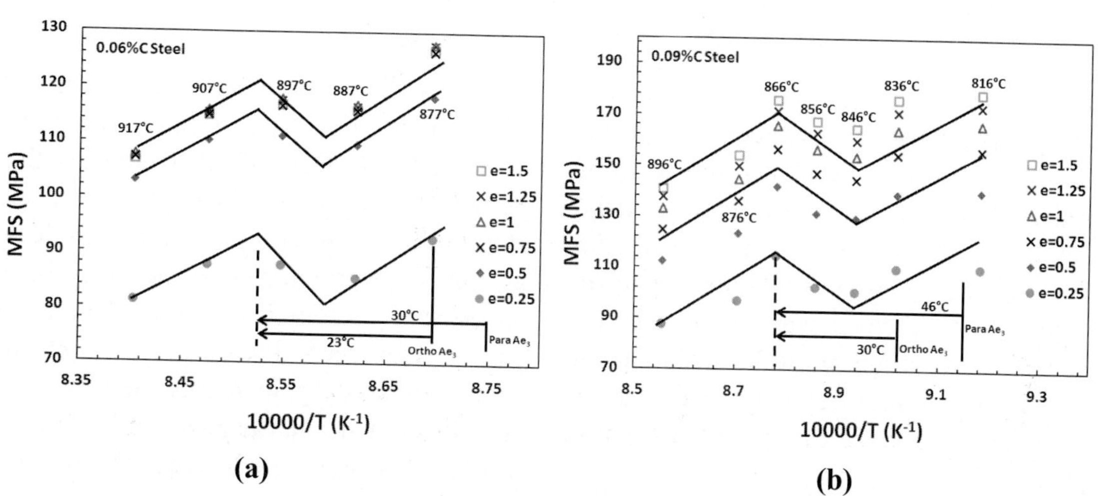

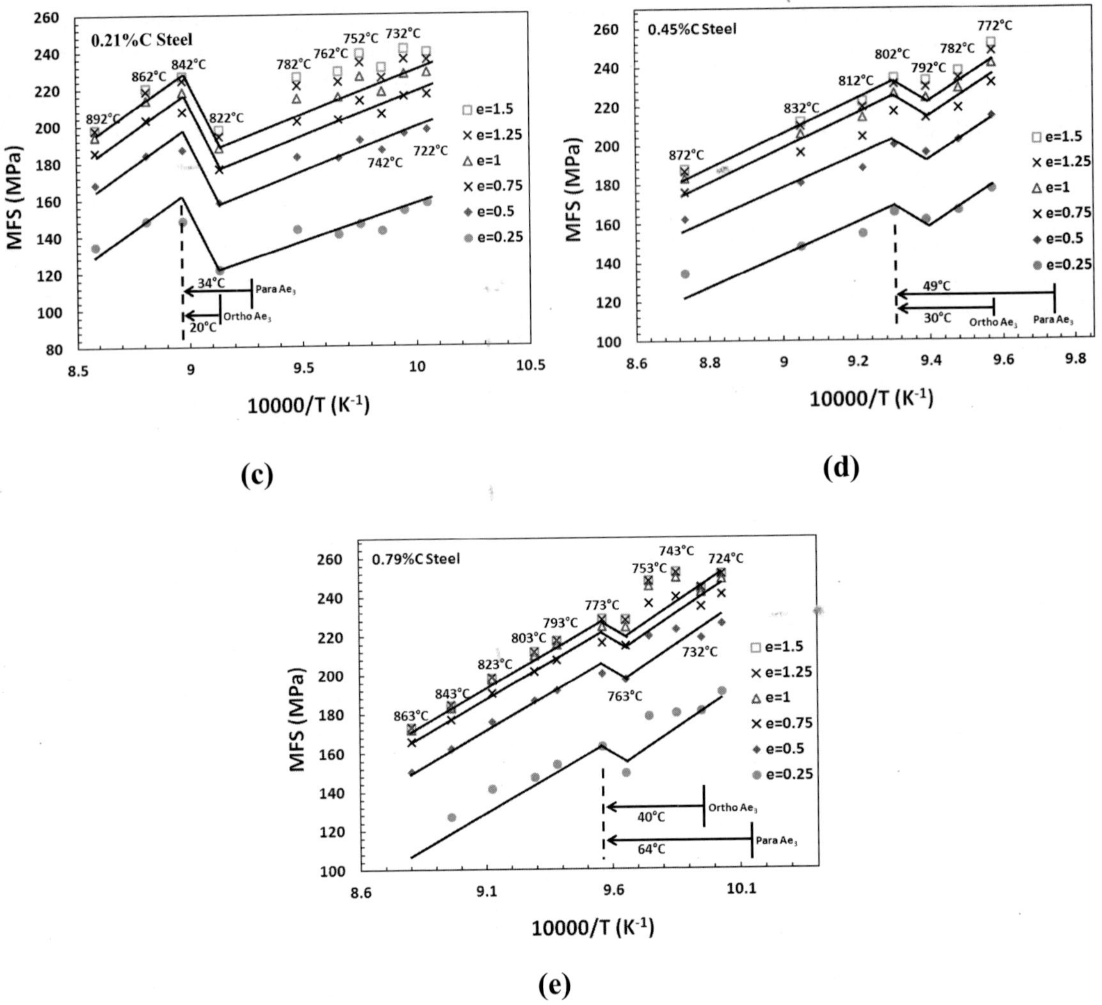

(c)

(d)

(e)

Fig. 5: Mean flow stresses (MFS's) determined from the torsion testing flow curves.

the effective transformation temperature can be increased by as much as 100-150 degrees. However, such DT ferrite forms quite slowly during deformation, being limited to about ½ to 1% volume fraction per 1% strain. Thus, even though DT ferrite begins to be formed well above the Ae_3, considerable strain must be accumulated before the volume fraction transformed is sufficient to produce a noticeable load drop [16].

Another feature of interest concerns the effect of accumulated strain on the stress drop behavior. First of all, it appears that increasing the strain does not affect the stress drop *temperature* or Ar_3*, at least as can be determined from the fairly large temperature intervals employed here. It can also be seen that, while the *amount* of the stress drop increases slightly with accumulated strain, the *percentage* drop is nearly constant in each material over the strain range investigated. The percentage drops appear to decrease with increasing C concentration, ranging from about 35% for the 0.06%C steel, to 22%, 19%, 5%, and 6% for the 0.09%C, 0.21%C, 0.45%C, and 0.79%C steels, respectively.

From a practical point of view, there are several important conclusions to be drawn from MFS diagrams such as those of Fig. 5. The first is that the effective Ar_3*'s determined from laboratory measurements and interpreted using Boratto diagrams can frequently be *higher than* the Ae_3's determined from the chemical composition using ThermoCalc. The second is that, due to DT, rolling loads can frequently be unexpectedly *lower* than estimates based on the increasing flow stress behavior of austenite when cooled down to the conventional Ae_3. This is because appreciable volume fractions of the softer ferrite can be present at a given temperature even though the presence of such ferrite is not called for by the conventional phase diagram. Finally, the presence of a dynamic Ar_3* that is higher than the Ae_3 has the practical effect of decreasing the expected temperature interval over which pancaking strains can be applied during controlled rolling.

Critical Strains and Softening Mechanisms

The conventional 'single peak' austenite flow curves of Fig. 1a) and 2a) are usually associated with the occurrence of DRX; their iconic shapes are thus generally attributed to the motion of the recrystallizing austenite grain boundaries, which remove dislocations and cause flow softening. However, it is clear from the critical strains presented in Fig. 4 that *two* flow softening mechanisms are acting concurrently during the high temperature deformation of austenite (in addition to dynamic recovery). That is, that the flow softening of austenite is actually caused by dynamic transformation as well as by dynamic recrystallization. This effect of DT has not been recognized to date for a number of reasons: i) one is that the DT ferrite plates produced in this way are so fine (under a micron in size) that they are not readily observable by optical microscopy; ii) the second is that the polygonal grains resulting from the coalescence and growth of the plates are usually considered to have formed *statically*, i.e. during cooling to room temperature (and not *dynamically*); and iii) finally the rate of formation of such DT ferrite is quite low, so that it does not generally produce any dramatic effects.

CONCLUSIONS

1. Hot torsion and hot compression testing can be used to determine the critical strains for dynamic transformation as well as the mean flow stresses applicable to rolling.
2. The double differentiation method is readily adaptable to calculating the critical strain for DT.
3. The DT critical strains are well below those associated with the initiation of DRX and indicate that, up to fairly high temperatures, the flow softening taking place during the high temperature deformation of austenite is generally caused by these two flow softening mechanisms occurring simultaneously.

4. As a result of DT, the Ar_3* temperatures determined by the Boratto method are generally *higher than* the ortho or para equilibrium Ae_3's.
5. Ferrite formation at such high temperatures is brought about by the increase in free energy of the deformed austenite associated with the presence of the dislocations introduced by straining.

ACKNOWLEDGEMENTS

The authors are indebted to the Natural Sciences and Engineering Research Council of Canada for financial support.

REFERENCES

[1] H. Yada, Y. Matsumura and T. Senuma, Proc. of the Int. Conf. on Martensitic Transformation, JIM, 1986, pp. 515 - 520.

[2] H. Yada, Y. Matsumura and T. Senuma, Proc. 1st Conf. on Physical Metallurgy of Thermomechanical Processing of Steels and Other Metals (THERMEC-88), Ed. I. Tamura, ISIJ, Tokyo, 1988, pp. 200 - 207.

[3] H. Yada, C. M. Li and H. Yamagata, ISIJ International, Vol. 40, 2000, p. 200.

[4] Y. Chen and Q. Chen, J Iron Steel Research International, Vol. 10, 2003, p. 46.

[5] X. Sun, H. Luo, H. Dong, Q. Liu and Y. Weng, ISIJ International, Vol. 48, 2008, p. 994.

[6] M. Tong, J. Ni, Y. Zhang, D. Li and Y. Li, Metall. Mater. Trans. A, Vol. 35, 2004, p. 1565.

[7] N. Xiao, M. Tong Y. Lan, D. Li and Y. Li, Acta Mater., Vol. 54, 2006, p. 1265.

[8] X. Sun, H. Luo, H. Dong, Q. Liu and Y. Weng, ISIJ International, Vol. 48, 2008, p. 994.

[9] J. K. Choi, D. H. Seo, J. S. Lee, K. K. Um and W. Y. Choo, ISIJ International, Vol. 43, 2003, p. 746.

[10] V. V. Basabe, J. J. Jonas and H. Mahjoubi, ISIJ International, Vol. 51, 2011, p. 612.

[11] V. V. Basabe and J. J. Jonas, ISIJ International, Vol. 50, 2010, p. 1185.

[12] V. V. Basabe, J. J. Jonas and C. Ghosh, Adv. Mater. Res., Vol. 409, 2012, p. 829.

[13] V. V. Basabe, J. J. Jonas and C. Ghosh, Steel Res. International, submitted.

[14] C. Ghosh, V. V. Basabe and J. J. Jonas, ISIJ International, submitted.

[15] C. Ghosh, V. V. Basabe and J. J. Jonas, Steel Res. International, submitted.

[16] J. J. Jonas, C. Ghosh and V. V. Basabe, Steel Res. International, In press.

[17] F. Boratto, R. Barbosa, S. Yue and J.J. Jonas, Proc. of the 1st Conf. on Physical Metallurgy of Thermomechanical Processing of Steels and Other Metals (THERMEC-88), Ed. I. Tamura I, ISIJ, Tokyo, 1988, pp. 383 - 390.

[18] J. J. Jonas, X. Quelennec and Lan Jiang, Materials Science Forum, Vol. 715-716, 2012, p. 81.

[19] X. Quelennec and J. J. Jonas, ISIJ International, Vol. 52, 2012, p. 1145.

[20] X. Quelennec, N. Bozzolo, J. J. Jonas and R. Loge, ISIJ International, Vol. 51, 2011, p. 945.

[21] E. Poliak and J. J. Jonas, Acta Mater., Vol. 44, 1996, p. 127.

[22] J. J. Jonas, C. Ghosh, X. Quelennec and V. V. Basabe, ISIJ International, in press.

[23] T. Sakai and J. J. Jonas, Acta Metall., Vol. 32, 1984, p. 189.

[24] C. Ghosh, V. V. Basabe, J. J. Jonas, Y. M. Kim, I. H. Jung and S. Yue: Acta Mater., submitted.

Microstructural Simulations via Thermal Processing of Roll Bonded Steel Laminates

David K. Matlock, Ralph J. Johnson, Emmanuel De Moor and John G. Speer
Advanced Steel Processing and Products Research Center
Colorado School of Mines
Golden, COUSA 80401
Email: dmatlock@mines.edu

ABSTRACT

Laboratory processed laminated composites, designed to simulate compositional gradients present in cast and hot rolled industrial microstructures, wereused to assess microstructural evolution during heat treating and the effects of specific microstructural variables on mechanical propertiesin steels. In this paper, two microstructural systems are presented to illustrate the broad applicability of the simulation technique. In one example, artificial compositional segregation of Mn was introduced to simulate microstructural banding observed in most industrial cast and hot rolled steels. Laminates were produced from two nominally 0.4 wtpct C steels with either 0.82 or 1.83 wtpctMn. The effects of composition band width and heat treatment, designed to alter the sharpness of the Mn gradients between layers, on microstructures andtensile properties are presented. In a second example, multiphase microstructures characteristic of the next generation advanced high strength sheet steels (AHSS) were simulated by roll bonding alternating layers of a steel with 2 wtpctMn and another steel with 18 wtpctMn. The resulting mechanical properties correlate to predictions of a composite model when sufficient interfacial strength was achieved to prevent tunnel crack formation.

KEYWORDS: *Laminated Composites, AHSS, Simulated Microstructures, Banding*

INTRODUCTION

Multi-scale modeling analyses which lead to property predictions of steel microstructures with multiple constituents (e.g. ferrite, bainite, martensite, austenite, carbide precipitates, inclusions, etc.) have become important components of research programs designed to identify, and eventually produce, new steels with enhanced properties for critical applications [1-4]. Essential to all models, are expressions that describe the deformation characteristics (yielding, strain hardening, etc.) of the individual constituents. The various modeling approaches differ with respect to the method used to integrate properties of individual phases and constituents and to incorporate interfacial properties and interactions between constituents. While most models concentrate on deformation behavior, some incorporate critical fracture criteria [5].

In steels with stable microstructures (i.e. without deformation induced transformation of austenite to martensite) interrelationships between microstructural constituents, strength, strain hardening, and ductility in tension are considered in terms of equations that relate operable strengthening mechanisms to strain-dependent expressions [1,2] and take the form shown in Eq. 1

$$\sigma(\varepsilon) = \sigma_{\mathrm{i}} + \sigma_{\mathrm{SS}} + \sigma_{\mathrm{D}} + \sigma_{\mathrm{PPTN}} + k_y d^{-1/2} \quad (1)$$

where $\sigma(\varepsilon)$ is the flow stress, σ_i is the friction stress of pure iron, σ_{SS} is the sum of solid solution strengthening contributions, σ_D is the strain-dependent strengthening from dislocation substructure, σ_{PPTN} is the contribution from precipitation hardening, ky is the Hall-Petch slope, and d is the grain diameter. Alternately, Ashby showed that for selected cases, the strength and strain hardening behavior of a multiphase material can be described by a "one-parameter" theory where strain-dependent strength depends only on the strain dependence of the average dislocation density, ρ_{AVG} [6].

Descriptions of constituent interactions primarily involve micromechanical models based on finite element analyses which incorporate predictions utilizing representative volume elements (RVE) [3,4,7] and composite models which describe modified rule-of-mixtures [1,2,8]. Both approaches recognize microstructural inhomogeneitiesand assess position-dependent variations in phases, constituents, and properties. In each case, experimental measurements are required to provide data to validate model predictions. Materials utilized in validation experiments may include commercially produced steels selected to assess microstructural variablesor specially processed model alloys designed and processed to produce controlled phase distributions. Production of roll bonded laminate composites, the subject of this paper, offers a unique method to produce model materials with systematically controlled microstructural variables. The potential importance of layered materials has been recognized by the Ministry of Education, Culture, Sports, Science and Technology of Japan through funding of the multi-site Layer-Integrated Steels and Metals Project [9].

In this paper the procedure to produce roll bonded laminated composites is summarized and a brief review of selected studies of laboratory-produced materials designed to evaluate specific variables in non-uniform microstructuresis presented [10-13]. This background information provides the basis for a study to physically simulate microstructural variables in potential 3rd generation advanced high strength steels (AHSS) and evaluate predictions of a composite model that has provided insight into new AHSS microstructures [1,2].

LABORATORY PRODUCTION OF SIMULATED MICROSTRUCTURES

A roll bonding procedure was developedto processa variety of alloy combinations on a laboratory rolling mill with 133 mm diameter by 203 mm face work rolls [10-13]. Typically two component materials were evaluated; the procedure could be easily modified toincorporate more than two primary constituents. Prior to fabrication of the final roll bonded compact, each experimental material was hot rolled, and in some cases cold rolled, to a final thickness of nominally 0.5 mm. To facilitate bonding and to eliminate the presence of undesirable inclusions at bond interfaces, prior to roll bonding, all sheet surfaces were pickled to remove oxides and mechanically cleaned with a commercial abrasive cloth.

Figure 1 shows the procedure for fabricating the roll bonded laminated compacts. For each two-component material of interest,alternating sheet layers, nominally 0.5 mm thick with widths and lengths of 90 and 125 mm, respectively, were encased in a welded stainless steel box. Figure 2 shows an example welded compact that included a vacuum attachment tube welded to one end. During TIG welding the layered stack and formed stainless steel box parts were compressed in a hydraulic press to minimize space between layers. The relative volume fractions of each constituent were varied by controlling sheet thickness.

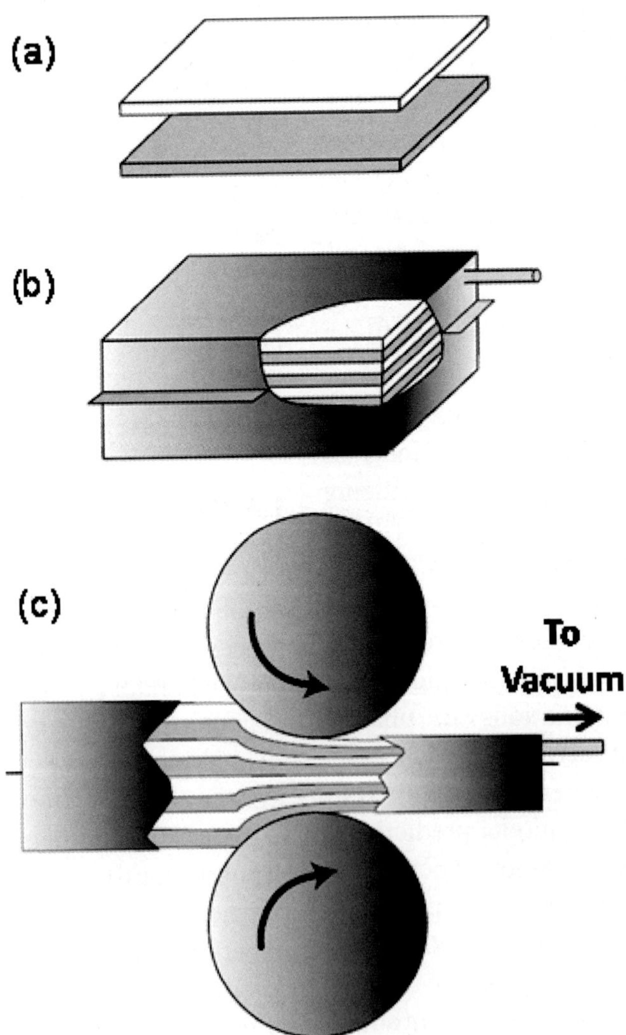

Fig. 1: Schematic of production steps for roll bonded laminated composites (a) alternately stacked sheets; (b) sheet compacts sealed in stainless steel boxes; and (c) boxes attached to an active vacuum system and hot rolled [13].

The welded boxes were connected to an active vacuum system and preheated to the desired rolling temperature selected based on the specific alloys of interest. The compacts were hot rolled to the final sheet thicknesses selected to produce specific band thicknesses based on the initial sheet thicknesses of the individual layers. To ensure bonding, the initial pass was the maximum achievable on the laboratory rolling mill. Typically primary bonding occurred in the first one or two passes and after three to four passes the vacuum tube separated from the compact. Subsequent rolling was completed in air, and intermediate reheating was used as necessary.

Fig. 2: Example TIG welded stainless steel box prior to hot rolling [12].

REVIEW OF MODEL SYSTEMS

Roll bonded laminate composites have been used to model properties in steels and other alloys with multi-component microstructures and have been used to produce new materials with enhanced properties [14,15]. Selected example systems and fundamental principles that have been considered include the following: fabrication and properties of Damascus steels [14]; fabrication and fracture assessment of high energy absorbing materials [16,17]; alloy homogenization during annealing with results applicable to cored castings and weldmentsand the effects of cored microstructures on the high temperature mechanical properties of weldments [18,19]; homogenization of thin thermally-sprayed coatings on substrates at elevated temperatures [20]; neutron diffraction studies of stress partitioning during deformation of multi-phase steels [21]; fundamental analyses of the Kirkendall effect and diffusion-induced porosity formation [22]; and banding in steels [11-13].In the latter studies, manganese banding and its potential effects on phase transformations and heat treat responses of steels were considered and selected results are presented here.

Banded 0.4 C steels with alternating equal-thickness layers of 0.8 or 1.8 Mn (wtpct) were processed with band widths between 10 and 320 μm to simulate banding observed in rolled and/or forged SAE 5140 steel and used to evaluate the effects of band width on phase transformations and tensile properties [11-13]. Results assessed the effects of heating and of cooling rates (83 °C/s to 8.3×10^{-3} °C/s) on phase transformations and layer interactions critical to the development of the final microstructures. To illustrate the importance of banding on layer-dependent properties, Fig. 3 shows engineering stress-strain curves for samples with various band widths cooled at 0.6 °C/s (Fig. 3a) and 1.7×10^{-2}°C/s (Fig. 3b) from 850 °C [11]. The tensile properties for samples cooled at 0.6 °C/s depend sensitively on band width, which, as discussed below, reflects the development of diffusion controlled interfacial layers of constant thickness. In contrast, the properties for samples cooled at a much lower rate (Fig. 3b) are essentially independent of band thickness.

The cooling-rate dependence of the effects of layer thickness were clarified via light optical microstructure analyses [11,12] and selected micrographs for samples cooled at 0.6 oC/s from 850 oC are shown in Figs. 4a and 4b for 160 and 20 μm samples,

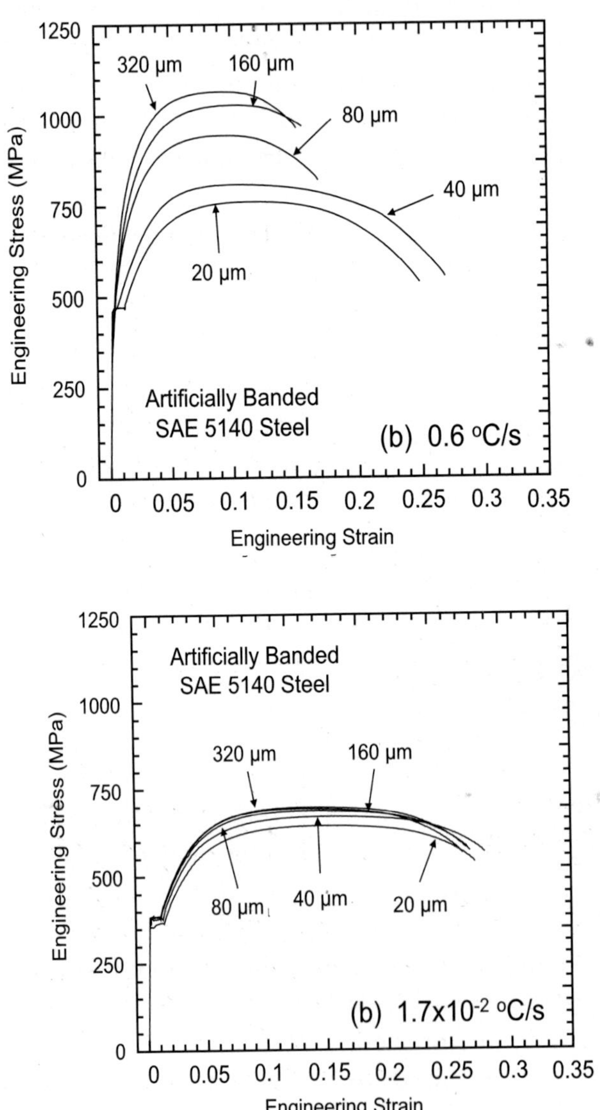

Fig. 3: Room temperature engineering stress strain curves for artificially banded SAE 5140 steel processed with band widths between 20 and 320 μm and cooled at (a) 0.6 °C/s and (b) 1.7x10-2°C/s [11].

respectively. At this intermediate cooling rate, a 20 μm interfacial band of pearlite [12] developed in addition to the ferrite pearlite band in the low Mn layer and a high strength bainitic band in the high Mn layer. In contrast, the pearlitic interface band is shown to completely encompass the high Mn layer in the 20 μm thick banded condition. Thus, with an increase in layer thickness, the volume fraction of the high strength bainitic band increases with band width leading to the strength effects shown in Fig. 3a. At the lower cooling rate ($1.7x10^{-2}$ °C/s), alternating ferrite-pearlite bands developed and the tensile properties were independent of band width (Fig. 3b). Related studies have shown that as a result of composition gradients in banded steels, displacements associated with the differential transformation response between layers and response to non-uniform heating, as imposed during induction hardening, depend on orientation with respect to the banded structure [12,23].

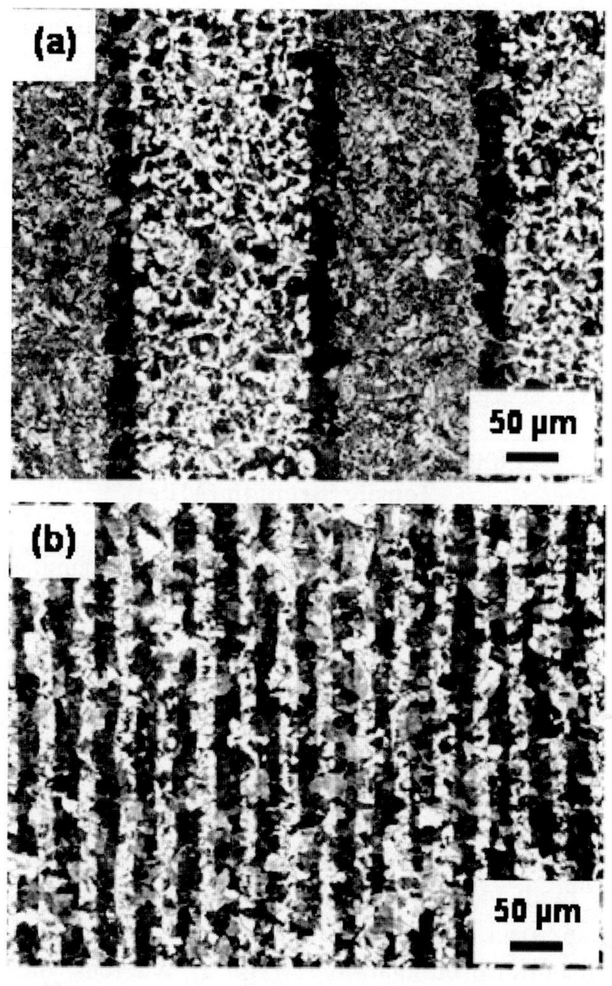

Fig. 4: Light optical micrographs of artificially banded SAE 5140 steel cooled at 0.6 °C/s with band widths of (a) 160 μm and (b) 20 μm (2 pctnital etch) [11].

THIRD GENERATION ADVANCED HIGH STRENGTH STEELS (AHSS)

Recently there has been considerable interest to develop and evaluate models to predict properties characteristic of the new class of sheet steels referred to as the "Third Generation of Advanced High Strength Steels (AHSS)" [24,25]. One approach, based on the application of a multi-component iso-strain composite model [1,2,26] to multi-phase steels, has provided insight into microstructural constituents that will be required for Third Generation AHSS products. Predictions of tensile instability (i.e. combinations of uniform elongation and ultimate tensile strength) based on assumed mixtures of ferrite + martensite or (stable) austenite + martensite are overlaid in Fig. 5 on a property map developed by AISI [24]. The property comparisons show that mixtures of ferrite and higher strength martensite (or "ultra-fine" ferrite) are attractive and represent a broad range of ferrite-based steels while austenite + martensite combinations generate properties well within the targeted Third Generation property band.

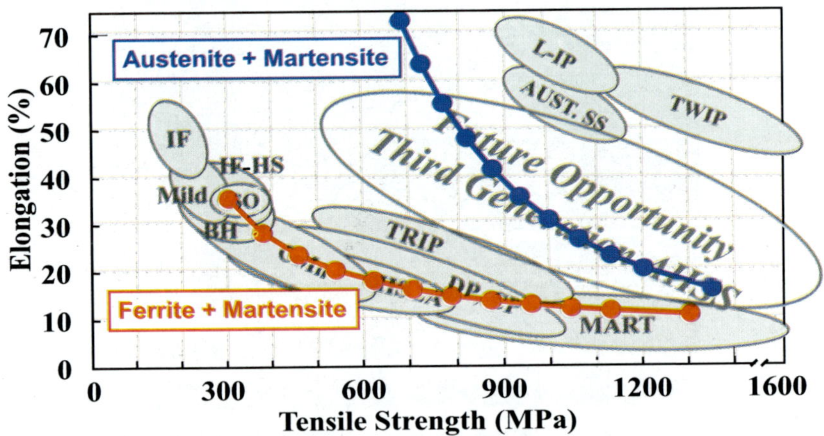

Fig. 5: Tensile elongation/strength combinations for various steel families and 3rd Generation AHSS overlaid with property combinations predicted for various combinations of martensite and ferrite or (stable) austenite [1].

The composite analysis shows significant volume fractions of austenite, having a high capacity for work hardening, are probably necessary for the next-generation AHSS grades. The importance of austenite stability was also predicted with the composite modeling approach [2] and Fig. 6 illustrates predicted uniform elongation/ tensile strength combinations for ferrite +austenite mixtures, where the mechanical stability of the austenite is varied according to the hypothetical behaviors shown in the inset. This analysis highlights the importance of controlling austenite stability as the very unstable material D exhibits properties essentially equivalent to the ferrite + martensite materials shown in Fig. 5, whereas the greater stability in material A provides exceptional property combinations.

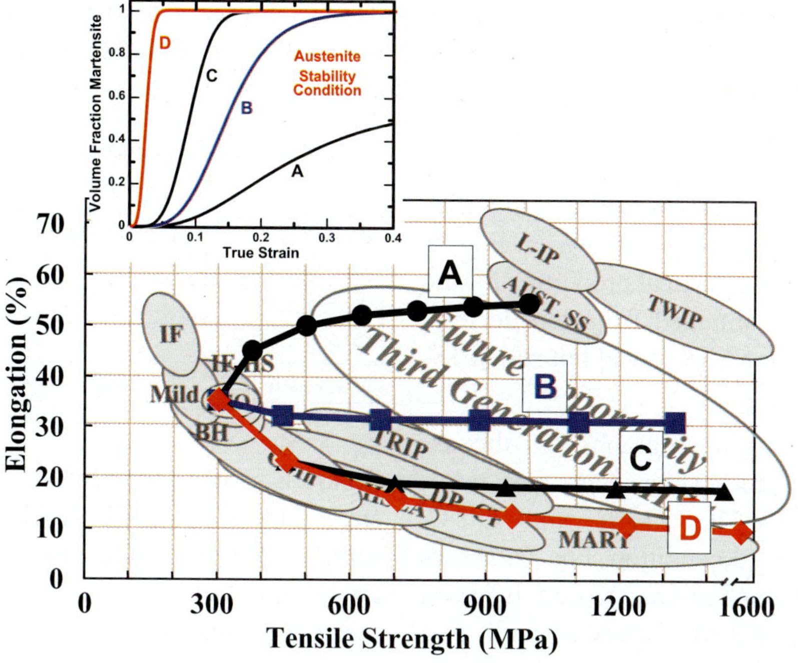

Fig. 6: Tensile ductility/strength combinations for mixtures of martensite and metastable austenite, with austenite mechanical stability varying as shown in inset [2].

To further understand the applicability of the simple composite model used to develop the predictions in Figs. 5 and 6, the following study to assess alternating roll-bonded layers of a stable austenitic steel and a dual-phase ferritic steel, in which the volume fractions of martensite were systematically varied by heat treatment, was undertaken. Roll bonded laminate composites consisting of equal thickness layers of a Twinning Induced Plasticity (TWIP) steel (18-Mn: in wtpct - 0.6C, 18Mn, 0.31Si, 0.38Cr, 0.04Nb, 0.04V, 1.5Al, 0.046S, 0.16P) and a dual-phase steel grade (2-Mn:in wtpct - 0.08C, 2.04Mn, 0.03Si, 0.27Cr, 0.28Mo, 0.06Al,0.01N, 0.01S, 0.02P) were hot rolled to 1.3 mm. Sub-sized ASTM E8 tensile samples were machined transverse to the rolling direction and contained a reduced gage section of 25 mm by 6 mm. These samples were then annealed in salt pots at 685 °C, 730 °C, 790 °C, and 865 °C for 10 min and immediately water quenched. Temperatures were selected to produce systematic variations in the volume fraction of martensite in the dual-phase steel layers as summarized in Table 1.A minimum of three tensile tests for each heat treat condition was performed at room temperature with an imposed engineering strain rate of 3.4×10^{-4}.

Table 1: Constituent fractionsfor annealed and quenched samples.

Phase	Volume Fractions for Annealed Samples			
	685°C	730°C	790°C	865°C
Austenite	0.5	0.5	0.5	0.5
Ferrite	0.3	0.2	0	0
Martensite	0.2	0.3	0.5	0.5

Table 2 summarizesthe assumed constituent propertiesincorporatedinasimpleHollomon power-law equation relating stress to plastic strain [1] used in the model predictions discussed below. The parameters for ferrite and martensite represent typical properties within the property bands shown in Figs. 5 and 6 [1]. The mechanical properties for the austenite were obtained from tensile tests performed on the as received hot rolled 18-Mn sheet [27].

Figure 7 shows micrographs for the as-rolled composite (Fig. 7a) and after annealing at 685°C, 730°C, and 790°C. For the hot rolled condition, Fig. 6a shows a thin diffusional mixing region between the 18-Mn and 2-Mn layers as well as a darker ferritic based microstructure in the 2 Mn layer. Samples annealed at 685 °C (Fig. 6b) and 730 °C (Fig. 6c) contain four thin dark bands of martensite, andlayers of ferrite in the 2-Mn layers. Figure 1d, for the sampleannealed at 790 °C indicates that the 2-Mn layer was fully martensitic, and the same is true for the sample annealed at 865 °C, not shown here. It should be noted that due to the significant compositional differences some etching artifacts (e.g. dark spots) were apparent but assumed to not represent significant microstructural constituents.

Table 2: Model parameters for composite analysis.

Phase	UTS (MPa)	Uniform True Strain
Austenite	1000	0.55
Ferrite	300	0.3
Martensite	1600	0.08

Figure 8 shows representative engineering stress strain curves for each heat treated sample. With an increase in annealing temperature the volume fraction of martensite increased leading to the observed increase in strength and decrease in ductility. The sample annealed at 685 °C was the only sample to exhibit significant post uniform strain.To correlate with predictions based on the simplified composite model leading to the predictions in Fig. 5, the constituent parameters in Table 2 were used along with phase fractions based on light optical microscopy of deformed microstructures and calculated by measured line lengths through the thickness for each phase.

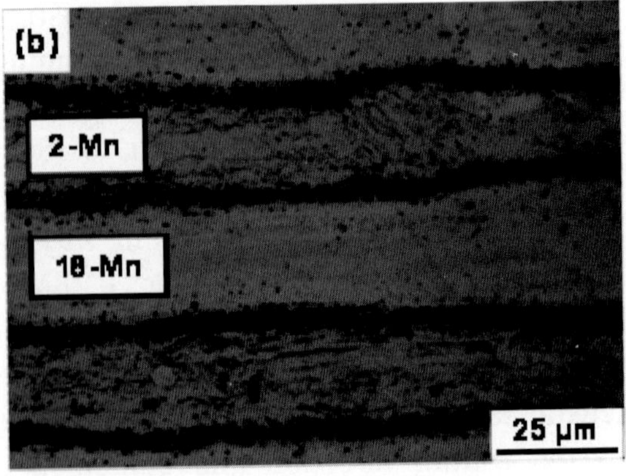

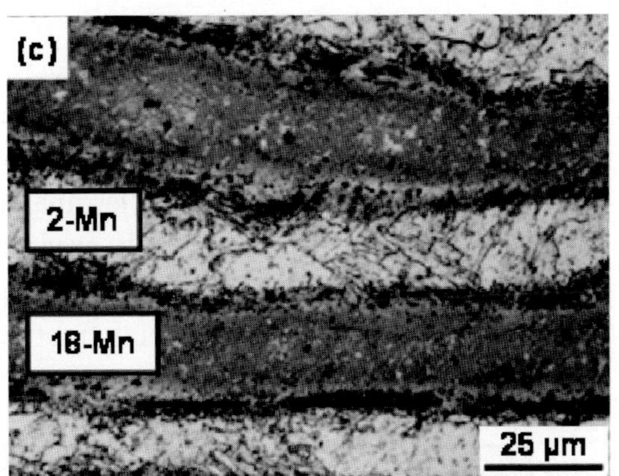

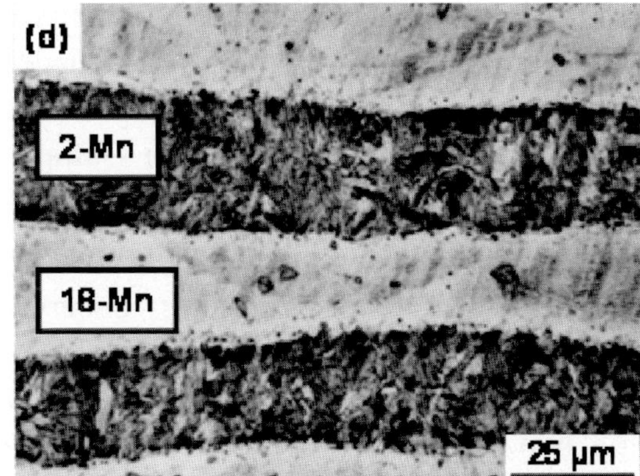

Fig. 7: Light optical micrographs of the 2-Mn/18-Mn laminate composite in the (a) as hot rolled condition or after annealing at (b) 685°C, (c) 730°C, and (d) 790°C (etched with 10 pctnital).

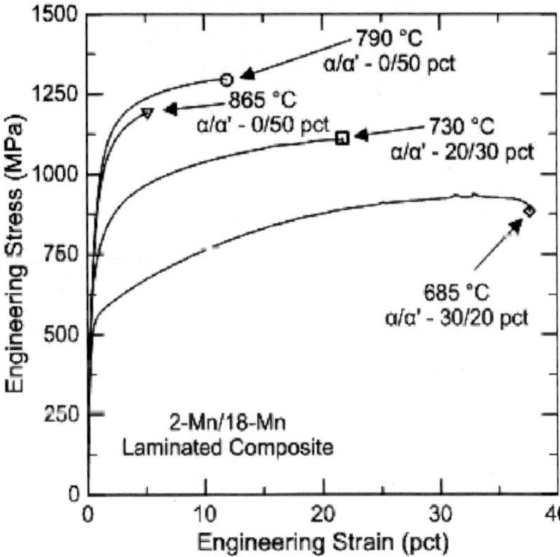

Fig. 8: Representative engineering stress-strain curves for the 2-Mn/18-Mn laminate composite after quenching from the indicated temperatures.

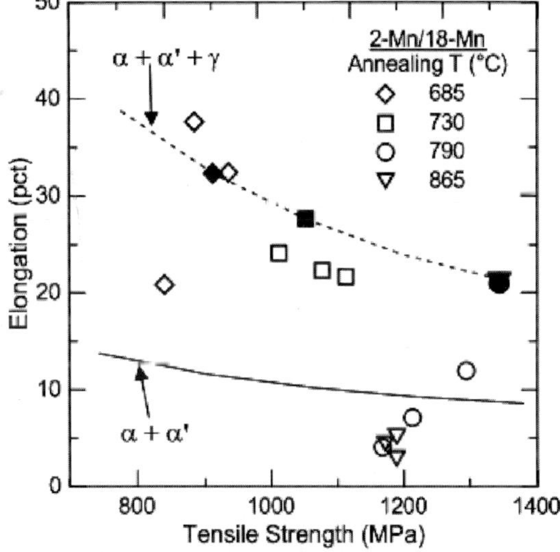

Fig. 9: A comparison of the predictions (closed symbols) for the 2-Mn/18-Mn laminate composite model analysis with experimentally measured tensile properties (open symbols).

Figure 9 compares modeled and actual tensile data. For reference, the solid line represents composites consisting of ferrite (α) and martensite (α') and the dashed line represents composites of three phases consisting of 50 percent austenite (γ), and systematically varied ferrite (α) and martensite (α') contents. For the calculations here, the effects of martensite carbon content on martensite properties were not considered but could be incorporated in future studies. The mechanical properties predicted by the model are plotted as solid symbols for each annealing condition.

The open symbols are the experimentally measured values. The samples annealed at 685 °C and 730 °C exhibit good agreement with the model, however the samples annealed at higher temperatures exhibit ductilities significantly less than predicted.

To resolve the discrepancies observed between measured and predicted properties for the samples annealed at the two higher temperatures, the fracture behavior of each sample was evaluated with light optical microscopy of cross sections of failedsamples. Figure 10 shows micrographs for samples annealed at 730 °C and 790 °C, which arealso representative of the behavior of samples annealed at 685 °C and 865 °C, respectively. Samples annealed at the lower temperatures (e.g. Fig. 10a) exhibited ductile fracture of both layers and sufficient plasticity in the interfacial martensite layer to maintain cohesion between layers. In contrast, Fig. 10b shows that for the higher annealing temperatures the martensitic 2-Mn layers failed in a brittle manner without localized necking and with significant interfacial crack formation. As a consequence of the brittle fracture, a critical assumption of the simplified isostraincomposite model, i.e. continuous bonding leading to enhanced ductility of the high strength layers [1,26], was not met. As a result, the measured ductilities for samples annealed at the two higher temperatures were less than predicted. It was also interpreted that with an increase in the volume fraction of martensite, in the presence of the P added to the 18-Mn steel, P diffusion to the interface may have decreased the critical stress required to grow tunnel cracks [28], further decreasing the laminate ductility by removing constraint on the high strength martensitic layers. This study illustrates that the laminated composite model [1,2] successfully predicts third generation AHSS tensile properties for those conditions where interfacial fracture was suppressed.

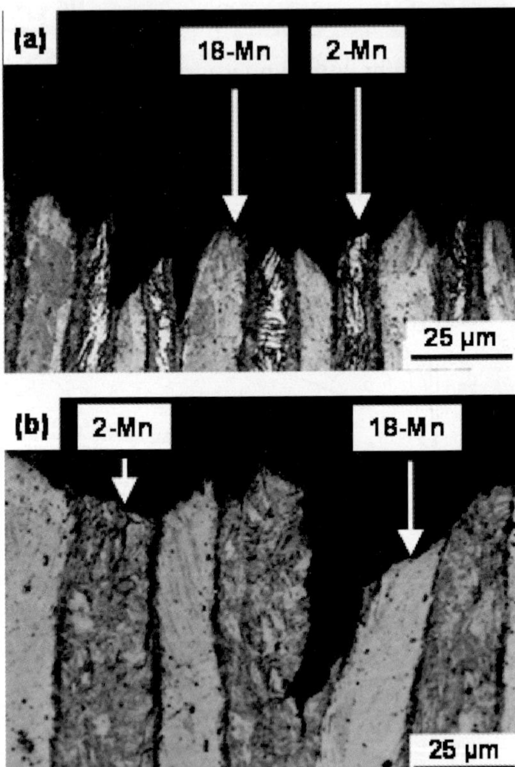

Fig. 10: Light optical micrographs of fractured tensile samples of 2-Mn/18-Mn laminatequenched from (a) 730 °C and (b) 790 °C.

SUMMARY

A laboratory technique to produce simulated complex microstructures with roll bonded laminated composites has been shown to provide a unique method for evaluating the effects of heat treating on microstructures and properties of multi-phase steel alloys. Insight provided by the simple composite model analysis which successfully predicted tensile properties in a microstructural system designed to simulate potential Third Generation AHSS products provides guidance for the development of future AHSS microstructures and properties. The results also illustrate the potential importance of interdiffusion between layers on microstructure development.

ACKNOWLEDGEMENTS

The authors gratefully acknowledge the support of the National Science Foundation for support under award CMMI-0729114 and the sponsors of the Advanced Steel Processing and ProductsResearchCenter, an industry/university cooperative research center at the Colorado School of Mines.

REFERENCES

[1] D.K. Matlock and J.G. Speer, Proc. of the 3rd Int. Conf. Structural Steels, Ed. H.C. Lee, The Korean Institute of Metals and Materials, Seoul, Korea, 2006, pp. 774-781.

[2] D.K. Matlock and J.G. Speer, Microstructure and Texture in Steels and Other Materials, Ed. A. Haldar, S. Suwas, and D. Bhattacharjee, Springer, London, 2009, pp. 185-205.

[3] K.S. Choi, W. Liu, X. Sun, and M.A. Khaleel, Metall. and Mat. Trans. A, Vol. 40, No. 4, 2009, pp. 796-809.

[4] C. Krempaszky, E. Werner, T. Hebesberger and A. Pichler, Proc. Int. Conf. on New Developments in Advanced High-Strength Sheet Steels, Ed. J.G. Speer, B. Nelson, and R. Pradhan, AIST, Warrendale, PA, 2008, pp. 305-314.

[5] F.J. Franklin, J.E. Garnham, D.I. Fletcher, C.L. Davis, A. Kapoora, Wear, Vol. 265, 2008, pp. 1332-1341.

[6] M.F. Ashby, Strengthening Methods in Crystals, Ed A. Kelly and R.B. Nicholson, Elsevier Pub., Amsterdam, 1971, pp. 137-192.

[7] K.S. Choi, A. Soulami, W.N. Liu, X. Sun. M.A. Khaleel, Computational Mat. Sci., Vol. 50, 2010, pp. 720–730.

[8] R.G. Davies, Metall. Trans. A, Vol. 9A, 1978, pp. 41-52.

[9] S. Nambu, M. Michiuchi, Y. Ishimoto, K. Asakura, J. Inoue, and T. Koseki, Scripta Mat., Vol. 60, 2009, pp. 221-224.

[10] I.D. Choi, D.K. Matlock, and D.L. Olson, Metall. Trans. A, Vol. 21A, 1990, pp. 2513-2520.

[11] T.F. Majka, D.K. Matlock, G. Krauss, and M.T. Lusk, Proc. of the 42nd Mech. Working and Steel Proc. Conf., October 2000, ISS AIME, Warrendale, PA, 2000, pp. 75-87.

[12] T.F. Majka, D.K. Matlock, G. Krauss, Metall. and Mat. Trans., Vol. 33A, 2002, pp. 1627-1637.

[13] R.A. Jaramillo, M.T. Lusk, and M.C. Mataya, Acta Mat., Vol. 52, 2004, pp. 851-858.

[14] D. R. Lesuer, C. K. Syn, O. D. Sherby, J. Wadsworth, J. J. Lewandowski, and W. H. Hunt, Jr, ," Int. Mat. Rev., Vol. 41, 1966, pp. 169-197.

[15] S. Nambu, M. Michiuchi, J. Inoue, and T. Koseki, Comp. Sci. and Tech., Vol. 69, 2009, pp. 1939-1941.

[16] F. Carreño, J. Chao, M. Pozuelo, O.A. Ruano, Scripta Mat., Vol. 48, 2003, pp. 1135-1140.

[17] C.K. Syn, D.R. Lesuer, J. Wolfenstine, and O.D. Sherby, Metall. Trans. A, Vol. 24A, 1993, pp. 1647-1653.

[18] I.D. Choi, D.K. Matlock, and D.L. Olson, Recent Trends in Welding Sci.e and Tech., Ed. by S.A. David and J.M. Vitek, ASM, Metals Park, OH, 1990, pp. 787 791.

[19] S.H. Wang, D.K. Matlock, and D.L. Olson, ScriptaMet.and Mat., Vol. 27, 1992, pp. 35-40.

[20] B. Klusemann and B. Svendsen, Int. J. of Solids and Structures, Vol. 49, 2012, pp. 1828–1838.

[21] M. Ojima, J. Inoue, S. Nambu, P. Xu, K. Akita, H. Suzuki, and T. Koseki, Scripta Mat., Vol. 66, 2012, pp. 139-142.

[22] I.D. Choi, D.K. Matlock, and D.L. Olson, Mat. Sci. and Engr., Vol. A124, 1990, pp. L15-L18.

[23] P.I. Anderson, D.K. Matlock and J.G. Speer, Proc. of Int. Conf. on New Dev. in Long and Forged Products: Metallurgy and Applications, Ed. J.G. Speer, E.B. Damm and C.V. Darragh, AIST, Warrendale, PA, USA, 2006, pp. 107-116.

[24] "Third Generation Advanced High Strength Steel (AHSS)," Research and Development Solicitation, AISI, Washington, DC, 2006.

[25] E. De Moor, P. J. Gibbs, J. G. Speer, D.K. Matlock, and J.G. Schroth, AIST Trans., Iron & Steel Tech., Vol. 7, No. 11, 2010, pp. 133-144.

[26] S.T Mileiko, J. of Mat. Sci., Vol. 4, 1969, pp. 974-977.

[27] J.A. Ronevich, J.G. Speer, and D.K. Matlock, SAE Int. J. Mater. Manuf., vol. 3, no. 1, 2010, pp. 255-267.

[28] S. Ho and Z. Suo, Trans. of the ASME, Vol. 60, 1993, pp. 890-894.

Process Modeling of Niobium Microalloyed Line Pipe Steels

S.V. Subramanian[1, a], Md Kashif Rehman[1], H. Zurob[1] and J.M. Gray[2]
[1]Department of Mat. Sci. and Engg., McMaster University, Hamilton, Canada
[2]Microalloyed Steel Institute, 5100 Westheimer, St. 540, Houston, Texas 77056, USA
[a]Email: Subraman@mcmaster.ca

ABSTRACT

Quantitative modeling of (a) strain induced precipitation of NbC and its interaction with recovery and recrystallization, (b) solute drag effect of Nb on boundary mobility have contributed to understanding the evolution of microstructure during thermo-mechanical rolling and accelerated cooling of plates or strips. While a fully integrated model is yet to be developed based on multi-variants associated with time evolution of local chemistry, deformation effects on precipitation and recrystallization and their mutual interactions influencing the microstructure, a modular approach has been developed to control the microstructural evolution at different stages of processing, based on current metallurgical understanding the underlying phenomena controlling the structure at each stage. The key concept is based on integrating modules of upstream austenite conditioning with downstream austenite transformation to control the morphological microstructure, and density and dispersion of high angle boundaries associated with crystallographic structure to which strength and fracture properties of the steel are related. The microstructural control at each of the various stages of hierarchical evolution is captured and integrated in "Through Process Modeling" based on modular approach. The application of process modeling of microstructure evolution based on modular approach for successful development of higher grade line steels based on high niobium (0.1wt%) is the focus of this paper.

KEYWORDS: *Niobium Microalloying, Process Modeling, Recrystallization, Line Pipe Steel*

QUANTITATIVE MODELING OF SOLIDIFICATION BEHAVIOUR OF MICROALLOYED STEELS (MODULE-1)

The precipitation behavior of microalloyed steel is influenced by solute redistribution from dendritic solidification of steel in the caster and the post-solidification cooling schedule.

Dendritic solidification of the steel causes a concentration variation on a local scale, which is of the order of the interdendritic spacing. The interdendritric spacing ranges typically from ten to a few hundred μm, depending on the local solidification rates occurring in industrial slabs. During the solidification of steels, the dendrite spacing is empirically related to the local solidification time (the resident time between the liquidus and solidus) raised to an exponent ranging between 0.5 to 0.33; the faster the cooling rate, the finer the interdendritic spacing. If the steady state dendritic solidification is disturbed by fluid flow, the solute enriched liquid may be transported over a large distance, resulting in macro-segregation, notably centreline segregation. While macrosegregation can be prevented by the control of fluid flow in the caster, microsegregation is inherent in the mechanism of dendritic solidification.

A quantitative model has been developed to predict the solute redistribution from dendritic solidification for multi-component micro-alloyed steels [1]. A finite difference method is used to predict the microsegregation at a given location (node point) in the slab. The diffusion equation is represented by the Crank-Nicholson implicit method of a finite difference equation. Local equilibrium is assumed at the solid–liquid interface. Multi-component phase equilibria are rigorously calculated from thermodynamic data for the case of Fe-Mn-C because of the large Mn concentration in the steel. The effects of other alloying elements are taken into account as a first order correction to this base. The slab thermal history used in the analysis is typical of an industrial caster. The algorithm results in the computation of the interdendritic microsegregation during freezing (including the effects of solid state diffusion during freezing) at stage-I and modifies this result by accounting for post-solidification homogenisation in the delta ferrite (stage-II), in the delta and gamma field (stage-III) and in the single phase gamma austenite region (stage-IV), see Fig. 1. Quantitative results of microsegregation in a typical high Nb, Ti bearing line pipe steel (0.03 C, 0.003 N, 0.09 Nb, 0.014Ti in wt%) solidifying at a location 5 mm from the slab surface with an interdendritic spacing of 70 μm are analyzed for the typical cooling schedule of the industrial caster. The solute redistribution of C and N exhibits a negligible variation in the concentration in the solid as a function of fraction solidified. The interstitial solutes exhibit negligible micro-segrgation, suggesting that the diffusion of the interstitial solutes is able to keep pace with the rate of advance of the solid-liquid interface during solidification. In contrast, the substitutional solutes give rise to significant microsegregation that persists to lower temperatures. Fig. 2 and 3 respectively show the redistribution of solute Nb and Ti as a function of interdendritic spacing. The reduction in the intensity of microsegregation as defined by $C_{maximum}/C_{minimum}$ from Stage-I to Stage-II, corresponding to the delta phase field (BCC structure) is far more significant than those of stage-III and stage-IV. The diffusion coefficients of Nb and Ti in the delta BCC phase are about 20 times greater than in those in the gamma phase (FCC structure). An increase in the temperature window of the delta phase field, corresponding to a decrease in the carbon content, brings about a significant homogenisation of the substitutional solutes. By comparison, the degree of homogenisation is not significant in austenite at a lower but wider temperature range corresponding to stage-IV. The effect of decreasing the carbon content in a given base chemistry was found to decrease the degree of microsegregation of Nb and Ti. The effect of lowering carbon is to increase the temperature window in the delta phase field, where significant diffusional homogenisation of substitutional solute occurs. Thus, a distinct advantage of low carbon design is to reduce the intensity of the microsegregation of the substitutional solutes, thereby promoting a more homogeneous microstructure. The reduced tendency of microsegregation in microalloyed steels with low carbon has been confirmed by Taira et al. [2] using mill data, which is consistent with the foregoing analysis. The microsegregation and precipitation models are used to design the base chemistry and temperature window required to prevent surface cracking during the bending and straightening of the slab. The base chemistry is designed to reduce the intensity of microsegregation using a low interstitial design. Ti additions are made to the stoichiometric requirement to tie up all the nitrogen at high temperatures as TiN. This is possible only if the nitrogen content of the steel is controlled to low levels (<40 ppm). A significant thermodynamic potential for NbCN in the temperature window corresponding to bend and unbend regions of caster is known to cause surface cracks. A low carbon design ensures less intensity of microsegregation of Nb, which allows high Nb design in the base chemistry.

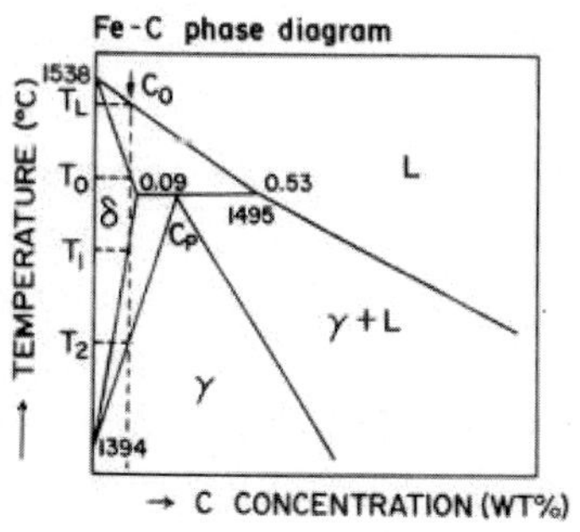

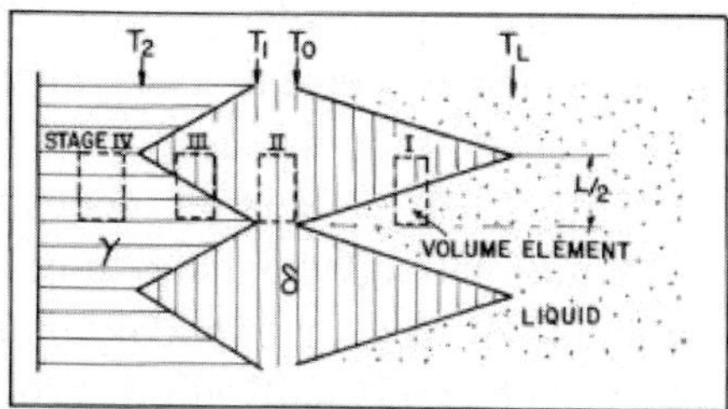

Fig. 1: A schematic diagram representing the various stages of microsegregation during solidification (Stage I) and post-solidification homogenization (Stages II, III and IV) [1].

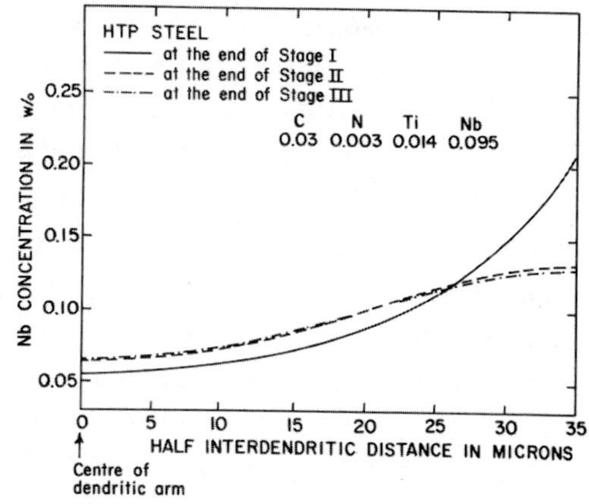

Fig. 2: The distribution of niobium as a function of interdendritic distance at the end of Stage I, II and III respectively [1].

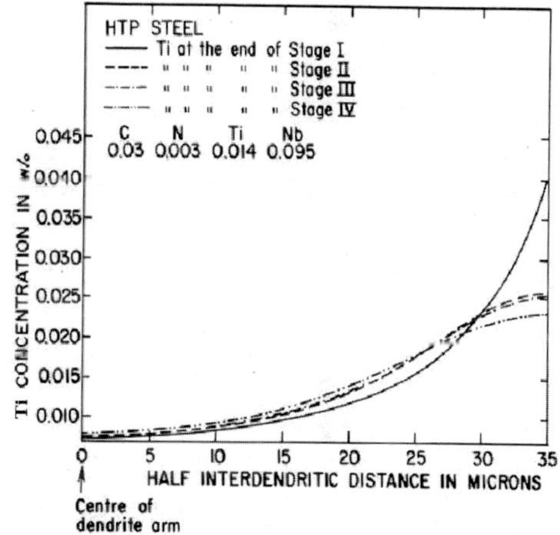

Fig. 3: The distribution of titanium as a function of interdendritic distance at the end of Stage I, II, III and IV (1200°C) respectively [1].

Thus, quantitative analysis of both the solidification and precipitation behaviour of a Ti-Nb microalloyed steel has underscored the scientific merits of low nitrogen as well as low carbon design in order to obtain sound slabs without surface cracking, which can be thermo-mechanically processed to obtain the required microstructure with enhanced strength and fracture toughness at low temperature.

Summary: *Quantitative modeling of solidification processing of slab shows that in order to obtain a homogeneous microstructure in the slab without surface cracks, the design of base chemistry warrants interstitial control. A low nitrogen design (30- 40ppm) is required coupled with stoichiometric Ti addition to tie up all nitrogen as TiN upstream (0.015%). A low carbon design (0.03 to 0.04%) is required to minimize microsegregation of Nb and Ti. A low carbon (0.03) design is coupled with high Nb (0.09%) addition which allows thermo-mechanical processing in high temperature window to gain increased productivity.*

THERMODYNAMIC MODELING TO PREDICT MOLE FRACTION AND COMPOSITION OF PRECIPITATES IN MICROALLOYED STEELS (MODULE-2)

Quantitative modeling based on Hillert and Staffanson's regular solution model is used for predicting the equilibrium mole fraction and the microchemistry of the precipitates of the compound (Ti_xNb_{1-x}) (C_yN_{1-y}) as a function of base chemistry and temperature [3]. TEM characterization of precipitates of Ti-Nb microalloyed steel has confirmed that the precipitates are mixed Ti-Nb carbo-nitrides.

These precipitates are described by sub-regular solution, in which the substitutional sites in cubic lattice are interchangeably taken by Ti and Nb, the interstitial sites are taken by N and C, with a certain amount of vacancy concentration as quantified by statistical thermodynamic approach. Thermodynamic model predictions

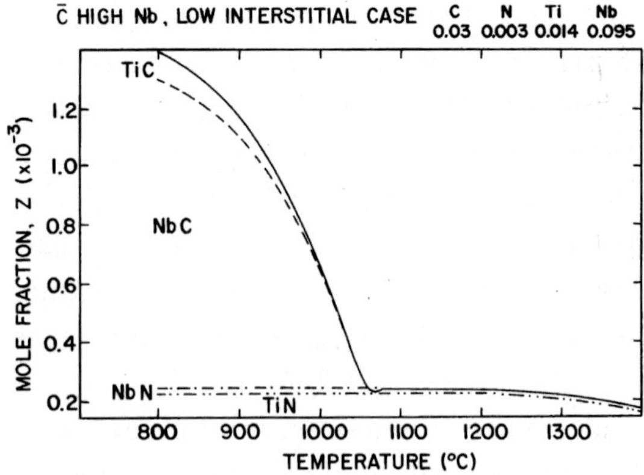

Fig. 4: Precipitate evolution curve for the high niobium low interstitial steel HTP. Thermodynamic potential for the precipitation of mixed carbides starts at 1060ºC.

were validated in HTP steel by TEM characterization of precipitates of samples equilibrated at different temperatures and quenched. Fig. 4 shows a typical plot of thermo-dynamic potential for precipitation of mixed carbo-nitrides of Ti and Nb occurring in low interstitial high niobium base chemistry (HTP). The high temperature precipitates are essentially mixed nitrides, enriched in TiN. The low temperature precipitates are essentially mixed carbides enriched in NbC. It is well known that while the kinetics of nitride precipitates are relatively rapid, the nucleation of NbC precipitates is sluggish and requires large undercooling even when their nucleation is aided by dislocations (strain induced). A quantitative understanding of the relative contribution of solute niobium, coherent and in-coherent precipitation of NbC on boundary mobility, recovery and recrystallization processes during hot rolling of niobium microalloyed steel is essential for microstructure control in thermo-mechanical rolling. This can be gained by quantitative modeling.

Summary: Mixed carbo-nitride formation is predicted in Ti-Nb microalloyed steel. Mixed nitrides enriched in TiN are promoted that are stable at high temperature during high temperature excursion in HAZ region in high heat input welding. Strain induced precipitation of NbC is promoted during thermo-mechanical rolling to retard static recrystallization in finish rolling. Refinement of austenite grains before pancaking should be carried out by rough rolling in high temperature window where there is little thermodynamic potential for precipitation of mixed carbides enriched in NbC.

REFINEMENT OF AUSTENITE GRAIN SIZE BY RECRYSTALLIZATION DURING ROUGH ROLLING AND PREVENTION OF COARSENING OF GRAINS BETWEEN ROUGHING AND FINISH ROLLING (MODULE-3)

A typical rolling schedule with softening kinetics pass by pass during hot rolling of high niobium microalloyed steel (HTP) is given in Fig. 5. The model is based on quantitative database generated by Lutz Meyer and coworkers on softening kinetics based on total niobium. Rough rolling (the first 5 passes in Fig. 5) is carried out in a regime where there is no potential for thermodynamic precipitation of NbC or NbCN. Refinement of austenite grains by breaking down the as-cast structure is brought about by static recrystallization [4]. While the models for recrystallization process tend to focus on percentage softening, the critical information required is number density and grain size distribution of austenite grains, for which one still has to rely on actual experimental data-base. This requires metallographic techniques to reveal prior austenite grains. Modeling of recrystallization nucleation kinetics of austenite is still at a stage of infancy.

The recent trend is to introduce another intermediate processing step in between roughing and finish rolling, where austenite grain size can be significantly refined by accelerating the kinetics of recrystallization nucleation. Though the modeling is yet to capture this aspect fully, experimental results on austenite grains are presented that demonstrate this concept [5].

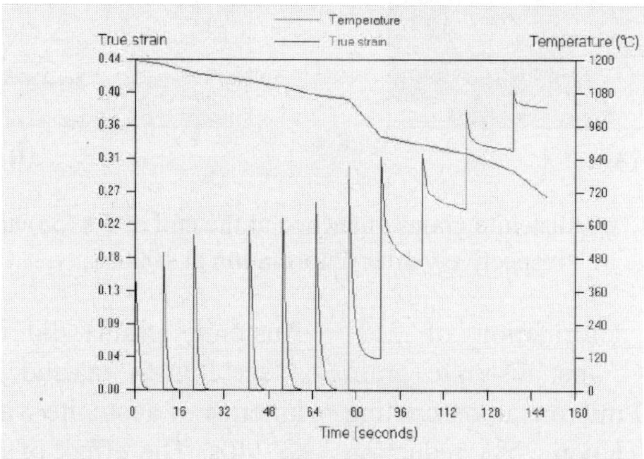

Fig. 5: Softening kinetics of high niobium microalloyed steel for a typical hot rolling schedule comprising of roughing (first six passes) and finish rolling (last 5 passes).

A metallographic technique was developed to reveal the prior austenite grain size in low interstitial- high Nb microalloyed steel sample interrupted by quenching at different stages during static recrystallization. This allowed time evolution of austenite grain size and distribution to be determined after single and multi-pass deformation. Conventional double compression test, stress relaxation test, plane strain experiments were also used along with the new etching technique of prior austenite grain to investigate static (SRX) and dynamic (DRX) recrystallization behavior of low interstitial, high Mn (>1.5%) and high Nb (0.1wt%) pipeline steel [6,7].

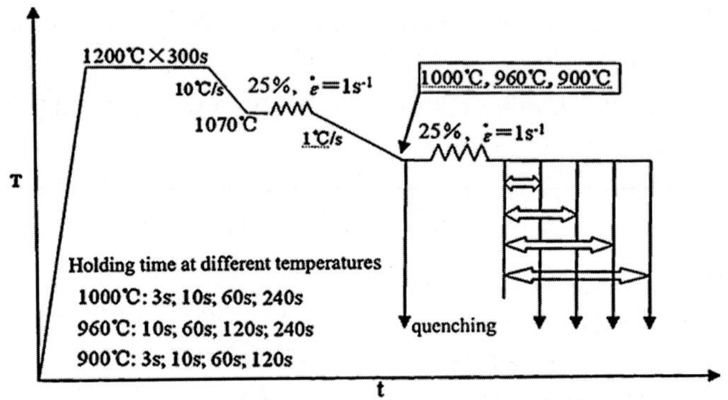

Fig. 6: A schematic of experimental rolling simulation investigations on niobium microalloyed steel.

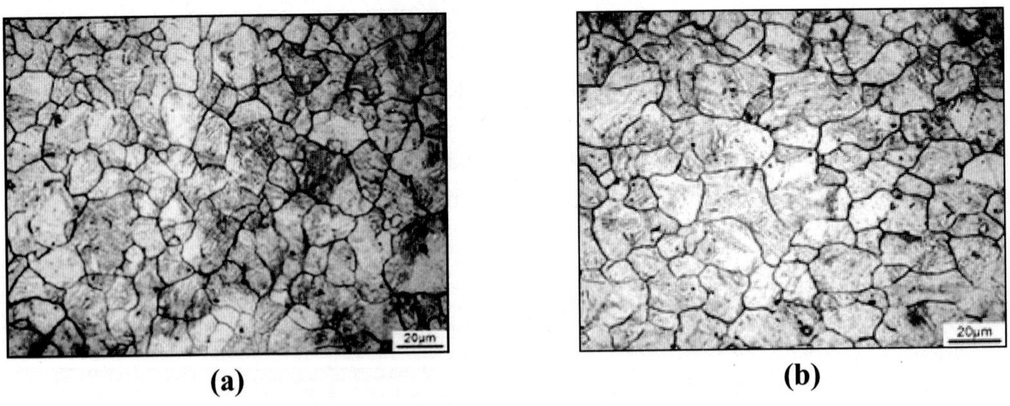

Fig. 7a and 7b: Austenite grains quenched at the end of 3 s (5a) and 60 s (2b) respectively after deformation at 1000°C.

After single pass deformation of 25% reduction at 1070°C, the austenite grain size coarsened to 80 microns upon cooling at 1K/s to 1000°C. Upon 25% reduction at 1000°C, steel exhibited uniform grain size of 13 microns at the end of 3 seconds. Austenite grains did not coarsen for 60 s, see Fig. 6, 7a, and 7b.. The coarsening kinetics of austenite was relatively slow up to 240s. The effect of prior deformation is to accelerate recrystallization nucleation kinetics, which can be related to the number density of austenite grains. More importantly, solute niobium was effective in retarding the grain coarsening of austenite at 1000°C compared to 1070°C. These two findings turned out to be the key to introduce an intermediate processing step in optimizing the rolling schedule to refine the austenite grains before pancaking in optimized rolling of HTP for X-80 referred to as OHTP process in a modern mill to roll 5 meter wide plate of thickness up to 25 mm [5]. Analysis of basic science aspects of recrystallization nucleation kinetics shows that the dislocation density is the driving force for recrystallization nucleation, which increases with applied strain. In consequence, the critical radius for recrystallization decreases, which, in turn, decreases the incubation time for recrystallization nucleation.

Summary: *Refinement of austenite grain size by static recrystallization is the key to refining austenite grain size before pancaking.*

The as-cast austenite grain which is between 800 to 1500 microns can be refined to less than 25 microns by static recrystallization.

A large pass reduction at the end of roughing will increase the nucleation sites for static recrystallization, thereby reducing the austenite grain size in the range of 10-15 microns.

PREVENTION OF GRAIN COARSENING OF AUSTENITE BEFORE PANCAKING IN FINISH ROLLING (MODULE-4)

The fine austenite grains will coarsen by the driving force, determined by reduction in surface energy accompanying coarsening of grains. The coarsening kinetics of grain refined austenite is a significant problem under industrial processing, which is examined in further detail by using quantitative modeling.

Modeling of coarsening kinetics of austenite

The driving force for grain coarsening is the reduction in surface energy associated with grain boundary, expressed as:

$$G = \frac{2\gamma}{R} \tag{1}$$

The boundary mobility for plain C-Mn steel as determined by Zhou et al. [8] is:

$$M_{C-Mn} = \frac{0.192}{T(t)} \, exp\left(\frac{-20837.14}{T(t)}\right) \tag{2}$$

This is modified to incorporate solute drag effect which was modeled using Cahn's equation [9]

$$M(t) = \left(\frac{1}{M_{C-Mn}} + \alpha C_{Nb}\right)^{-1} \tag{3}$$

where,

$$\alpha = \frac{\delta N_v (k_B T)^2}{E_b D}\left(sinh\left(\frac{E_b}{k_B T}\right) - \frac{E_b}{k_B T}\right) \tag{4}$$

$M_{C-Mn}M_{C-Mn}$ and $C_{Nb}C_{Nb}$ refer to the intrinsic grain boundary mobility of the C-Mn steel and the concentration of Nb in solution respectively, $\delta\delta$ is the grain boundary width $(\sim 1\text{ nm}),(\sim 1\text{ nm})$, N_v N_v is the number of atoms per unit volume, $E_b E_b$ is the solute-boundary binding energy and DD is the trans-interface boundary diffusion which is equal to twice the bulk diffusion coefficient of Nb in austenite [10].

$$D = 2 * [\, 0.000083 \exp(-266{,}500/RT)].$$

Finally the grain coarsening is expressed as:

$$R_f^2 = R_i^2 + 4\gamma M_{gb}\left(t_f - t_i\right) \quad (5)$$

where R_iR_i and R_fR_f are the initial and final grain size corresponding to initial and final time t_it_i and t_ft_f respectively. According to model predictions, the effect of adding 0.09 wt% Nb in retarding the coarsening kinetics of plain Fe-Mn-C steel (solute drag effect of Nb) is more pronounced at 1000°C compared with any higher temperature (>1100°C), which is consistent with the experimental observations. Even without Nb, diffusion is too low at low temperature (<900°C) to cause any significant coarsening. The effect of Nb addition in retarding the coarsening kinetics of austenite is plotted for a starting austenite grain size of 15 microns in Fig. 8. There is significant potential for grain coarsening of austenite between the end of roughing and start of finish rolling, which can be controlled by combining high solute Nb with rapid cooling between roughing and finish rolling.

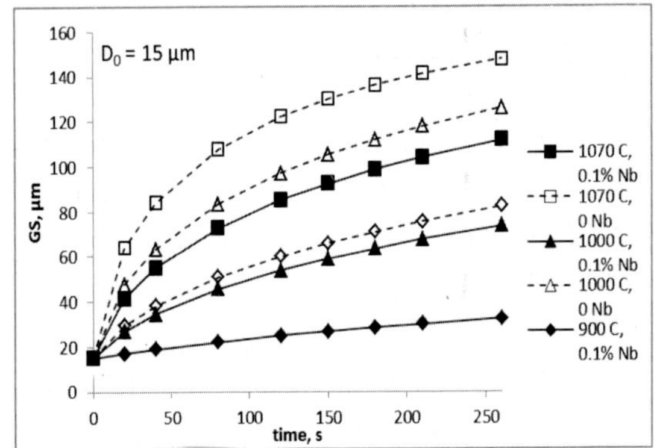

Fig. 8: Effect of 0.1%Nb on grain coarsening at different temperatures and time.

The model predictions of austenite grain size at entry to finish rolling are summarized in Fig. 9 for typical case histories. In order to achieve a target austenite grain size of 25 microns, high solute niobium and rapid cooling rate have to be coupled for a typical temperature window of 1000-910°C for a typical hot strip mill processing.

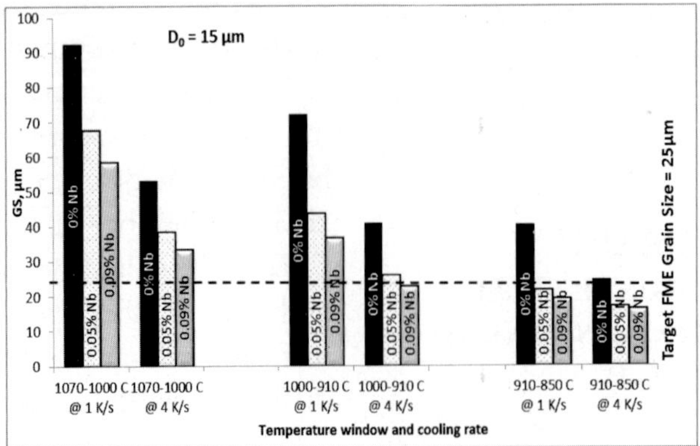

Fig. 9: Effect of cooling rate, temperature window and %Nb on grain coarsening kinetics. The starting GS= 15 µm.

Summary: *High solute niobium and rapid cooling rate between roughing and finishing (1050-910 °C) are shown to the key parameters to minimize grain coarsening of austenite. At low temperature (<900°C), grain coarsening is not a significant problem, which is used to advantage during pancaking of austenite in finish rolling of strips, where there is inadequate time for Zener pinning of precipitates.*

RECRYSTALLIZATION CONTROL FOR STRAIN ACCUMULATION IN FINISH ROLLING (MODULE-5)

Quantitative model for time evolution of strain induced precipitation of NbC was developed by Dutta, Palmiere and Sellars [11] and strain induced precipitate interaction with recovery and recrystallization after single pass deformation was quantitatively modeled by Zurob et al.[12]. Attempts have been made to refine the site saturation assumption for recrystallization nucleation in single pass deformation. The model is being refined to capture fresh continuous nucleation of strain induced precipitation with each new pass in multi-pass rolling. Zurob's model captures time evolution of recovery, which decreases the driving force for recrystallization. Precipitates nucleating on dislocations pin the dislocations which impedes the recovery until coarsening unpins the dislocations. Zurob's analysis [12] has confirmed that it is only at relatively large time when the precipitate volume fraction is large, Zener effect is effective in pinning the grain boundary, and the recrystallization is effectively retarded.

Summary: *Time evolution of strain induced precipitation occurs within seconds of deformation pass but time evolution of growth of precipitates takes tens of seconds before the precipitate volume fraction is adequate to exert significant pinning pressure on boundary to retard recrystallization.*

Solute drag due solute niobium is essential to prevent boundary break away and prevent potential for abnormal grain growth.

Prevention of mixed grain size due to rolling in partial static recrystallization regime

Physically based models are developed based on time evolution of strain induced precipitation and interaction of Zener drag and solute drag with recrystallization. Experimental results have confirmed the window of partial static recrystallization is not fixed. Temperature of no recrystallization (T_{NR}) is sensitive to both chemistry and processing parameters, which means that there is no fixed T_{NR} for HTP but it is also sensitive to mill processing parameters [4]. Pass No-6 in Fig. 5 exhibits partial recrystallization. The experimental results on the effect of strain on partial static recrystallization window are given in Fig. 10. It is imperative to avoid rolling in partial static recrystallization in order to avoid mixed grain size.

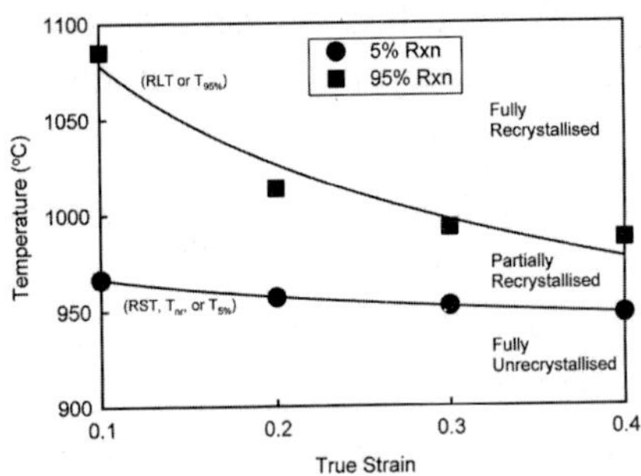

Fig.10: Map showing the start and finish of recrystallization as a function of temperature and strain [4].

Summary: *Rolling should be avoided in partial static recrystallization zone. Otherwise, mixed grain size would occur, which will impair toughness as measured by DWTT results.*

Roughing should be done well above partial static recrystallization regime for a given chemistry and austenite conditioning. Finish rolling should start below temperature of partial static recrystallization region in the region of no recrystallization. This window is sensitive to both chemistry and mill rolling parameters!

Prevention of mixed grain size due to partial dynamic recrystallization

In finish rolling, large strain accumulation is required without any partial dynamic recrystallization at any stage [13]. Modeling is used to predict the strain accumulation from pass to pass without exceeding the critical strain for dynamic recrystallization.

In strip rolling of HTP, there is inadequate time for adequate Zener drag to develop, which implies that there is significant potential for partial dynamic recrystallization to occur unless the temperature window is adequately lowered [12]. Experimental results have confirmed that mean flow stress can used to monitor and control the time-temperature-deformation schedule of finish rolling to prevent partial dynamic recrystallization, see Fig 11 and 12.

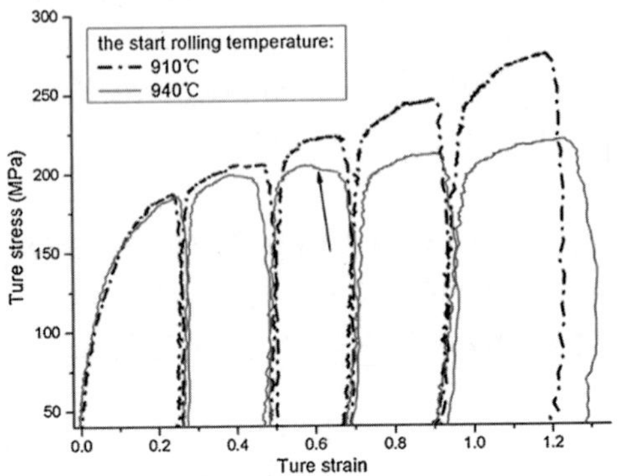

Fig. 11: Map showing the drop in mean flow stress due to recrystallization [6].

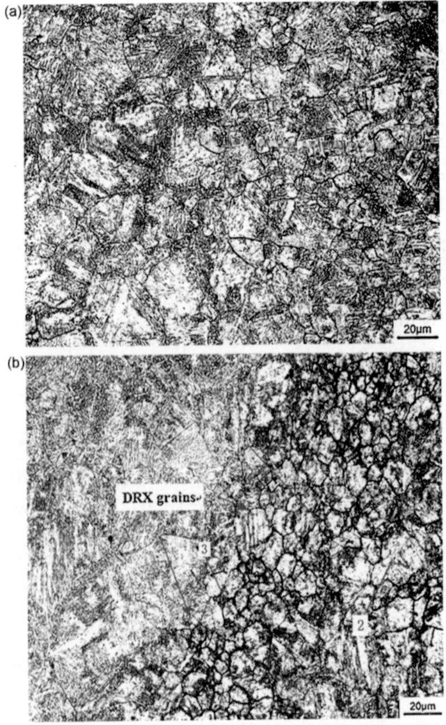

Fig. 12: Micrographs of the two cases shown in Fig. 11 demonstrating that the drop in flow stress is due to dynamic recrystallization [6].

Summary: *Partial dynamic recrystallization will give mixed grain size and therefore should be avoided. Increase in mean flow stress from pass to pass can be used for on-line monitoring to prevent partial dynamic softening during mill rolling.*

STRUCTURE-PROPERTY CORRELATION STUDIES (MODULE-6)

While the morphological microstructure control is used to enhance toughness associated with resistance to ductile fracture, the competition from brittle fracture as manifested by decrease in percentage shear area in DWTT is controlled by density and dispersion of crystallographic high angle boundaries [14], see Fig. 13

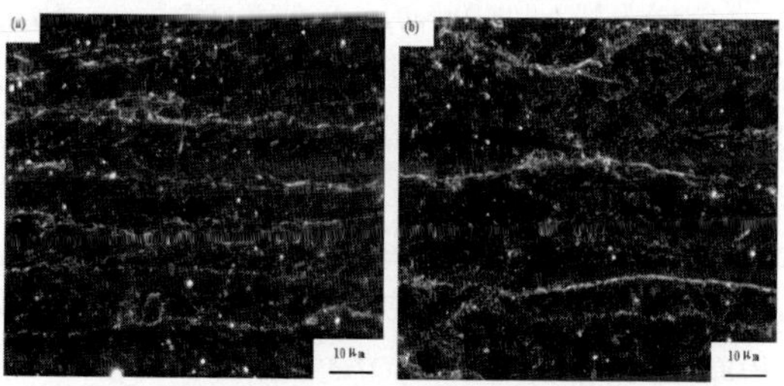

DWTT property for two processes

Heat	$R_{t0.5}$ MPa	R_m MPa	A %	$R_{t0.5}/R_m$	DWTT（-15°C），SA%		
					1	2	平均
(a)	535	705	44	0.76	100	95	98
(b)	548	662	50	0.83	45	60	55

---WJ Nie, XB Zhang, CJ Shang

Fig. 13: Micrographs showing the effect of Sv on DWTT.

Control of competition from brittle fracture

Studies based on EBSD characterization of crystallographic high angle boundaries in higher grade line pipe steels have confirmed that control of density and dispersion of crystallographic high angle boundaries is essential to control ductile to brittle transition temperature [15]. Crystallographic high angle boundaries act as microcrack arresters that prevent brittle crack initiation. Acicular ferrite structure with high density and dispersion of high angle boundaries can be promoted in niobium microalloyed steel. The domain size is determined by high density (number count) of high angle boundaries. This, in turn, is found to be a strong function of prior austenite grain size before pancaking as well as the degree of pancaking. A large strain accumulation in a relatively fine grained austenite when accelerated cooled through the transformation to promote coherent transformation to form acicular ferrite, and interrupted and held well above martensite start temperature to complete the transformation is shown to be the key to obtain the required domain size in higher grade line pipe steels. A coarse austenite grain exhibits a low density of high angle boundaries within the austenite grain. The large interspacing between high angle boundaries is weak zone to resist brittle fracture initiation, see Fig. 14

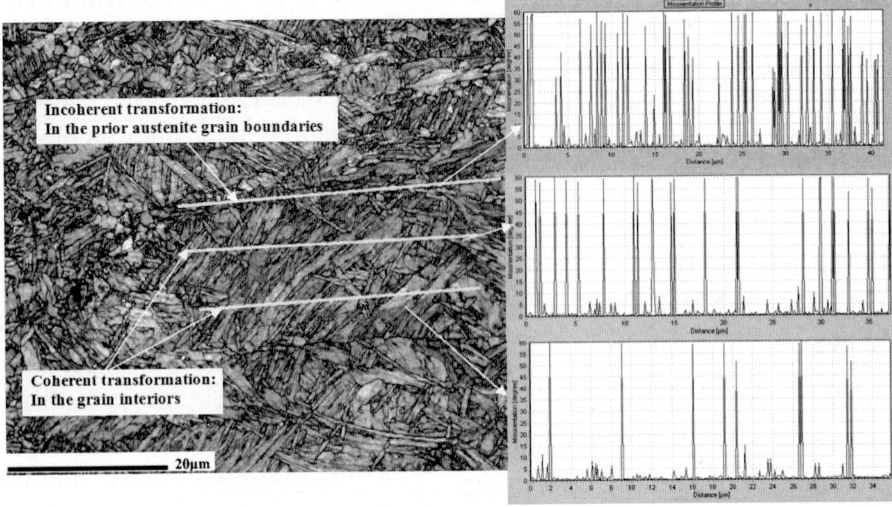

Fig. 14: Micrographs showing the density and dispersion of high angle grain boundaries.

The effect of prior austenite grain size on the density of high angle boundaries in austenite is illustrated in Fig. 15 X-120 exhibits typically a pancaked austenite thickness of 6 microns, which implies that for 75% total reduction the austenite grain size before pancaking is less than 20 microns.

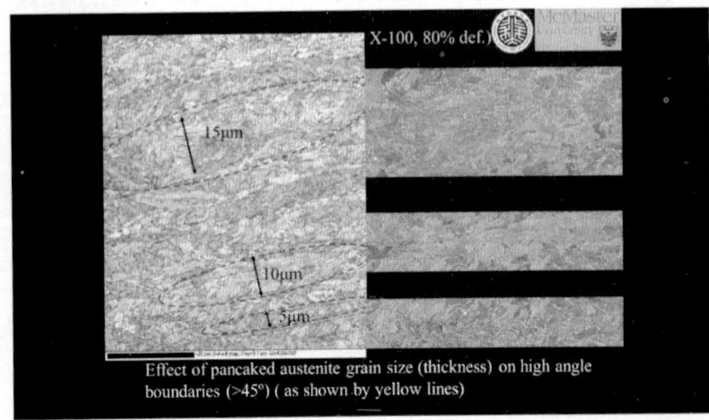

Fig. 15: Effect of pancaked austenite grain size on high angle grain boundaries.

High niobium steel has turned out to be a cost effective option over V-Nb microalloyed grade with high Mo to produce X100.

Summary: *Control of density and dispersion of high angle boundary is essential to prevent competition from brittle fracture, that lowers ductile to brittle transition temperature. Acicular ferrite structure is promoted by high niobium, which gives high density and dispersion of high angle boundaries. Refinement of austenite grains before pancaking, large strain accumulation and high cooling rate are key parameters that influence density and dispersion of high angle boundaries. High strength austenite transformed at low temperature increase the density and dispersion of crystallographic high angle boundaries.*

CONCLUSIONS

A modular approach is used to control upstream processing of austenite with downstream transformation product, to which strength and fracture properties are related in high niobium microalloyed steel of HTP design. The evolution of HTP as the workhorse for line pipe steel owes its success to the concept of product-process integration by "through process modeling" approach.

ACKNOWLEDGEMENTS

The authors wish to thank CBMM- Brazil, NSERC- Canada and Chinese Government Scholarship agencies for funding support. Research contributions from Prof. C.S. Shang, Dr. Miao Chengliang and You Yang of USTB, China are gratefully acknowledged.

REFERENCES

[1] S.V.Subramanian et al., Technology and Applications of HSLA Steels (Beijing), (Materials Park: ASM International, 1985), p. 151.

[2] T.Taira et al., Technology and Applications of HSLA Steels (Philadelphia), (Metals Park: ASM, 1983), p. 723.

[3] S.V.Subramanian, J.D.Embury and G.R.Purdy, "Solidification and Precipitation evolution analyses for Ti-Nb bearing line pipe steels" Proc. Int. Conf. Pipe Technology, Rome Nov 1987 AIM/CEC Publication, p. 153-170.

[4] E.J.Palmiere, C.M.Sellars, and S.V.Subramanian, "Modeling of thermomechanical rolling", The International Symposium Niobium 2001, AIME-TMS, Dec. 2-5, 2001, p. 501- 526, Orlando, U.S.A

[5] Nie, W. et al., Enhancing the Toughness of Heavy Thick X80 Pipeline Steel Plates by Microstructure Control, Adavanced Materials Research Vol, 194-196 (2011) p 1183.

[6] Miao, C.L., et al. Materials Science and Engineering: A, 2010. 527(18–19): p. 4985.

[7] Shang, C. and C. Miao, Refinement of Prior Austenite Grain in Advanced Pipeline Steel, Advanced Steels, 2011, Springer Berlin Heidelberg. p. 341.

[8] Zhou, T., R.J. O'Malley, and H.S. Zurob. Metallurgical and Materials Transactions A, 2010. 41(8): p. 2112.

[9] Cahn, J.W. Acta Metall Mater, 1962. 10(9): p. 789.

[10] Zurob, H.S., et al. Acta Mater, 2002. 50: p. 3075.

[11] Dutta, B., E.J. Palmiere, and C.M. Sellars, Acta Mater, 2001. 49: p. 785.

[12] Zurob, H.S., et al. ISIJ Int, 2005. 45(5): p. 713.

[13] S.V.Subramanian and H. Zou, Processing, Microstructure and Properties of Microalloyed and Other Modern High Strength Low Alloy Steels (Pittsburgh), (Warrendale: ISS-AIME, 1991), p. 23.

[14] S.V.Subramanian et al. Microalloyed Bar and Forging Steels (Hamilton), M. Finn Ed., (Montreal: CIM, 1990), p. 120.

[15] S.V.Subramanian et al, Microstructural development for brittle fracture control in Nb microalloyed line pipe steel. HSLA Beijing 2011, june 1st.

SECTION II

Advances in Thermo-mechanical Controlled Processing

Development of a New Direct Quenched Abrasion Resistant Steel

Esko Kinnunen[1], Ilkka Miettunen[1], Mahesh C. Somani[1], David A. Porter[1], L. Pentti Karjalainen[1], Ilari Alamattila[2], Anu Kemppainen[2], Tommi Liimatainen[2] and Vilma Ratia[3]
[1]University of Oulu, Department of Mechanical Engineering, Centre for Advanced Steels Research, POB 4200, 90014 Oulu, Finland, Email: david.porter@oulu.fi
[2]Rautaruukki Oyj, Ruukki Metals, POB 93, FI-90101 Raahe, Finland
[3]Department of Materials Science, Tampere University of Technology, POB 589, FI-33101 Tampere, Finland

ABSTRACT

A new way of producing abrasion resistant steel with a hardness substantially greater than 500 HV via direct quenching has been investigated. To avoid the ductility problems associated with fully martensitic microstructures, we investigated the possibility of achieving the desired hardness by incorporating a small amount of soft ferrite in a martensite matrix. The approach used was to study the potential suitability of several candidate compositions with varying levels of aluminium and manganese using physical simulation and laboratory rolling combined with metallography and abrasive wear testing. Physical simulation included the determination of CCT diagrams for strained and unstrained austenite on a Gleeble thermo-mechanical simulator. The CCT diagrams quantified 1) the increase in ferrite formation temperatures as a function of aluminium content, 2) the level of alloying required to obtain fully martensitic microstructures at appropriate cooling rates, and 3) the influence of prior austenite straining on the ferrite and bainite transformation kinetics. On the basis of the simulation results, laboratory rolling trials were made. They showed that a hardness even above 600 HV could be obtained with the desired martensitic-ferritic microstructure by designing a suitable composition and selecting appropriate finish rolling and quench start temperatures. Abrasive wear testing was used to compare the wear resistance of the laboratory rolled material with that of commercial direct quenched abrasion resistant grades.

KEYWORDS: *Steel, Abrasive Wear, Multiphase Microstructure, Physical Simulation*

INTRODUCTION

Abrasive wear is a severe problem in the mining, earth moving and transportation industry, and huge savings could be made if suitably priced and easily fabricated materials with improved resistance to abrasive wear were available [1]. Steel is the most commonly used abrasion resistant material, and since the life of abrasion resistant steel is largely dependent on hardness, there is a desire to find ways of making and using steels with as high hardness as possible. Commercially available abrasion resistant steel grades are often fully martensitic with a hardness of 400, 450 or 500 HB (420 to 527 HV). During recent years, there has been an increasing interest in the use of abrasion resistant steel produced via direct quenching where austenite is quenched to martensite immediately following hot rolling without the need for intermediate cooling and reaustenitization.

With hardnesses in the range 400 – 500 HB, relatively good combinations of abrasion resistance, toughness, cutability, bendability, weldability and resistance to hydrogen induced cracking can be obtained. However, as the hardness increases, these technological properties deteriorate and the production and use of the steel become ever more demanding. Consequently, it is difficult to profitably produce and utilize martensitic steels with a hardness higher than 500 HB (527 HV, 51 HRC).

It has been postulated that the low ductility and toughness associated with overall hardness levels above 500 HB might be improved by changing the microstructure from 100% martensite to a mixture of ferrite and martensite thereby enabling more novel abrasion resistant solutions to be achieved [2,3]. The objective of the work described in this paper was to explore the practicality of such an approach when using direct quenching. This has been done by studying the potential suitability of candidate compositions with varying levels of aluminium and manganese using physical simulation, laboratory hot rolling and direct quenching together with metallography and abrasive wear testing. Physical simulation included the determination of CCT diagrams for strained and unstrained austenite on a Gleeble thermo-mechanical simulator for designing composition and laboratory hot rolling schedules.

EXPERIMENTAL

Materials

The chemical composition of the five steels used are presented in Table 1. In order to promote ferrite formation, high levels of aluminium, silicon and vanadium were included. While V(CN) precipitates are known to be effective nucleation sites for ferrite [4,5,6], aluminium increases both the A_3 and A_1 [7,8,9] and allows ferrite to form at higher temperatures and lower cooling rates [8]. Aluminium also supresses the formation of cementite [10]. Titanium was added to all compositions to form fine TiN precipitates to suppress austenite grain growth and possibly provide nucleation sites for ferrite [11,12]. Different manganese levels were included due to its effect on ferrite formation and the hardenability of the austenite. Mn slows down the decomposition of austenite and lowers the M_s temperature. MnS inclusions also can act as nucleation sites for proeutectoid ferrite [4,5,6], but sulphur levels were kept low due to the detrimental effect of MnS inclusions on impact toughness. In some cases, chromium was also added to adjust hardenability.

The experimental steels were cast as 60 kg ingots (450 x 300 x 55 mm), which were cut into (225 x 75 x 55 mm) pieces for hot rolling experiments. The starting material for the Gleeble simulations was made by hot rolling to a thickness of 11 mm and subsequent air cooling.

Dilatometry and CCT Diagrams

CCT diagrams (*Continuous cooling transformation*) were constructed for all test material compositions using dilatometric data obtained with a Gleeble 1500 thermo-mechanical simulator starting with both recrystallized and deformed unrecrystallized austenite. Test samples were 6 mm diameter cylinders machined with their axis transverse to the rolling direction.

In the Gleeble, the specimens were heated to 1100°C at 20°C/s, held at temperature for 4 min and then cooled at 2°C/s. When no deformation was applied, cooling was stopped at 1000°C where the specimens were held for 2 min. In the case of deformed austenite, cooling was stopped at 850°C for10 s for compressive deformation.

Table 1: Compositions in wt.% of test materials, balance Fe (S≤ 50 ppm and P≤ 60 ppm).

Cast	C	Si	Mn	Al	V	Ti	Cr	N
0.4C-0.65Mn-0.08Al-V-Cr	0.38	0.18	0.65	0.08	0.20	0.019	0.37	35 ppm
0.4C-1.49Mn-0.08Al-V	0.39	0.22	1.49	0.08	0.21	0.017	-	38 ppm
0.4C-1.15Mn-1.23Al-V	0.39	0.30	1.15	1.23	0.20	0.021	-	35 ppm
0.4C-1.50Mn-2.10Al-V	0.43	0.26	1.50	2.10	0.20	0.019	-	38 ppm
0.5C-0.77Mn-1.90Al-V-Cr	0.47	0.56	0.77	1.90	0.192	0.020	0.24	40 ppm

Deformation was applied in three passes each with a strain of 0.2 and a strain rate of 1 s^{-1}. The interval between each compression was 25 s and the specimens were kept at 850°C for 10 s after the third compression. Both undeformed and deformed specimens were finally cooled to room temperature at different cooling rates ranging from 1.5°C/s to 80°C/s. The diameter of the cylinders during cooling was measured using a strain gauge. For metallographic examination and hardness testing, the specimens were cut in half along the compression axis. Microstructural characterization and hardness measurements were made as explained below.

Laboratory Rolling Experiments

The aim of the laboratory rolling experiments was to find rolling process parameters suitable for obtaining a dual-phase microstructure with ferrite and martensite. On the basis of the CCT diagrams, 0.4C-1.50Mn-2.10Al-V and 0.5C-0.77Mn-1.90Al-V-Cr where chosen as the suitable materials for the rolling experiments.

All the ingot pieces were annealed at 1200°C for 2 h before hot rolling to a final thickness of 12 mm which means a total reduction 78%. Plates were rolled with all hot rolling passes completed above

1000°C, except the last one. Like other passes, the final pass too had a reduction of 0.2 strain. The temperature of the final pass (*FRT*) was 1000°C or lower (950°C, 900°C or 850°C) to complete rolling in the no-recrystallization regime. The rolled plates were finally quenched in water (*WQ*). The starting temperature of quenching was about 20-30°C lower than the FRT and the mean cooling rate between 800°C and 300°C was estimated to be about 75°C/s. For presentational purposes, the FRT is used to identify the different test plates.

Material Characterization

Microstructural characterization was made using a light optical microscope and a field emission scanning electron microscope (*FESEM*) Zeiss Ultra Plus on samples that were polished and etched with 5% Nital. Ferrite fractions were estimated from the etched samples using the point counting method with a total of 663 points from three micrographs per test material.

Hardnesses of the mounted and polished samples were measured with a Struers Duramin A-300 hardness tester. A load of 10 kg was used for rolled and quenched material and 5 kg was used for CCT specimens. The values presented are means of 10 measurements

Wear Testing

Crushing pin-on-disk testing was employed to study the resistance against abrasive wear of the most interesting test materials. Tests were carried out at the Tampere Wear Center of the Technical University of Tampere. The pin-on-disk test principle is described in the ASTM G 99 standard. The only differences were the use of a free abrasive between the pin and the disk and the presence of confining walls that kept the abrasive on the disk. The disk is 160 mm in diameter and the confining walls have an inner diameter of 150 mm. The pin, which in this case is made of the material tested, is 36 mm in diameter. The flat end area of 1000 mm^2 is pressed close to the disk with a pneumatic cylinder. The pin and the disk are set not to come into contact with each other, leaving a gap of ca. 1 mm between the pin and the disk. In the test, the pin is repeatedly pressed against a gravel bed for 5 s, followed by a 2.5 s hold in the lifted position. This procedure allows recycling of the abrasive between the pin and the disk [13]. The pin is weighed before and after the test. The mass loss of the pin is used as an indication of resistance against wear. In these tests, 500 g of uneven sized (2-10 mm) granite was used as the abrasive, the test time was 30 min and the pin pressure was 90 kPa. The disk used was varied so that the hardness ratio between the pin and the disk was kept in the range (1.00-1.16) i.e. almost constant. The mass losses of the test materials were compared to those of the commercial wear-resistant steels Raex 400 and Raex 450 provided by Ruukki Metals.

RESULTS AND DISCUSSION

CCT Diagrams

In general, prior deformation of austenite accelerated the formation of ferrite so that some ferrite formed even with high cooling rates when deformation was applied leading to lower levels of hardness at a given cooling rate.

The effect of aluminium on the formation of ferrite becomes apparent when comparing the compositions 0.4C-1.49Mn-0.08Al-V and 0.4C-1.50Mn-2.10Al-V that only differ in their aluminium content (Figure 1 and Figure 2, respectively). The ferrite transformation of compositions with 2.1% Al started at temperatures just below 800°C at cooling rates from 1.5°C/s to 3°C/s (Figure 2) whereas with 0.08% Al (Figure 1) ferrite transformation started slightly above 700°C at the same cooling rates when austenite was not strained.

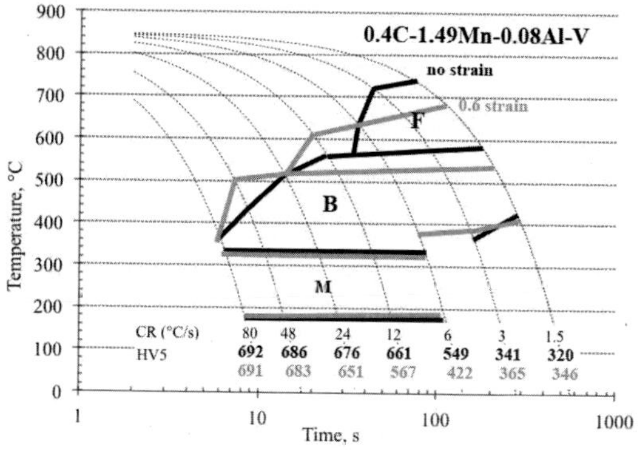

Fig. 1: CCT diagram; 0.4C-1.49Mn-0.08Al-V.

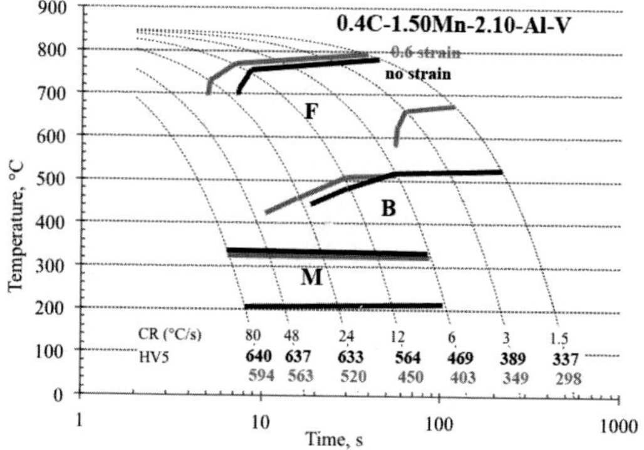

Fig. 2: CCT diagram; 0.4C-1.50Mn-2.10Al-V.

The effect of composition to formation of ferrite can also be seen from the hardnesses obtained with high cooling rates. For compositions 0.4C-1.49Mn-0.08Al-V and 0.4C-0.65Mn-0.08Al-V-Cr (Figure 1 and Figure 3, respectively), the hardness obtained at 80°C/s cooling rate is the same for both specimens tested with and without deformation of austenite. For other compositions, the hardness is lower when austenite is deformed, as discussed earlier.

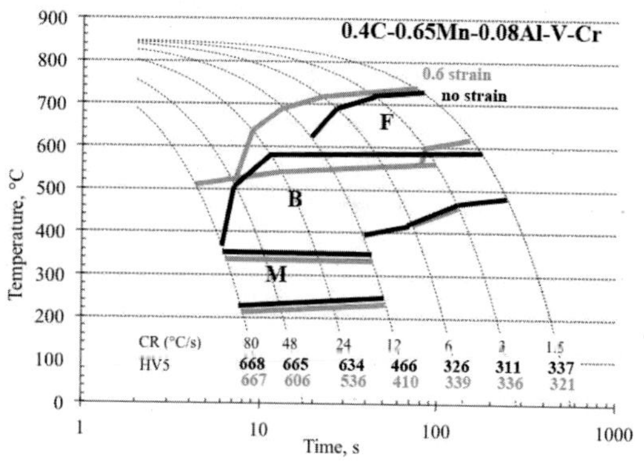

Fig. 3: CCT diagram, 0.4C-0.65Mn-0.08Al-V-Cr.

The effect of manganese on the phase transformation kinetics is evident when comparing the CCT diagrams of 0.4C-0.65Mn-0.08Al-V-Cr and 0.4C-1.49Mn-0.08Al-V (Figure 3 and Figure 1, respectively). Ferrite formation starts at lower temperatures in the composition with 1.49% Mn than in the one with 0.65% Mn, even though it also has 0.37% Cr, which is known to have a similar effect on hardenability as Mn. The effect is also clear in the critical cooling rates for ferrite formation: higher Mn content promotes martensitic (or bainitic) microstructure to form at lower cooling rates.

Compositions 0.4C-1.50Mn-2.10Al-V and 0.5C-0.77Mn-1.90Al-V-Cr were chosen as test materials for the rolling experiments, due to their relatively high ferrite formation temperatures combined with high hardness levels (Figure 2 and Figure 5, respectively).

Hardness of over 600 HV was obtained with both these compositions. With the carbon content of 0.5%, a martensitic microstructure containing a small amount of ferrite and a hardness of over 750 HV was attained from deformed austenite.

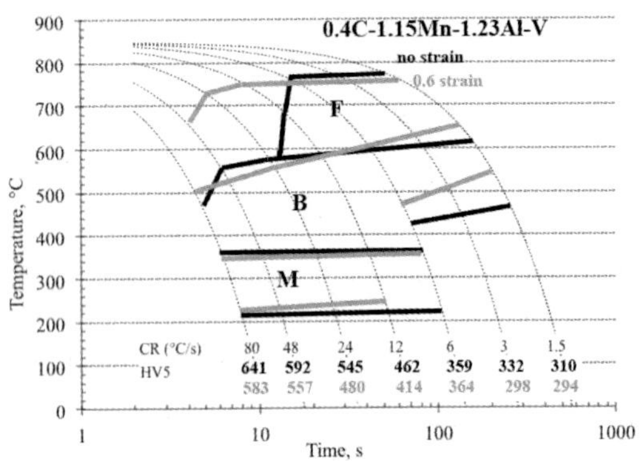

Fig. 4: CCT diagram; 0.4C-1.15Mn-1.23Al-V.

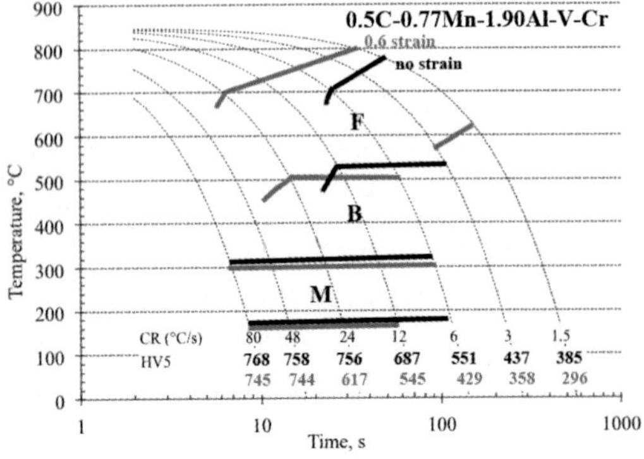

Fig. 5: CCT diagram; 0.5C-0.77Mn-1.90Al-V-Cr.

Laboratory Rolled Plates

Measured hardness values and corresponding proportions of ferrite in the rolled and direct quenched plates are presented in Figure 6. Plates with hardness from 483 to 734 HV were obtained. Highest hardness values were obtained with 0.5C-0.77Mn-1.90Al-V-Cr, as expected. The plates had low amounts of ferrite being almost totally martensitic. The measured hardnesses were close to the theoretical hardness of martensite, i.e. 744 HV (Table 2). In the Gleeble simulations, a marginally higher hardness of 768 HV was obtained at the cooling rate of 80°C/s (Figure 5), though the difference could, of course, simply be due to scatter.

With 0.4C-1.50Mn-2.10Al-V, relatively higher amounts of ferrite were obtained with ferrite content increasing with decreasing FRT. The highest hardness for 0.4C-1.50Mn-2.10Al-V (630 HV) was obtained with a FRT of 1000°C. Still it was significantly softer than the theoretical hardness of martensite calculated for that composition (Table 2). However, the hardness was close to the hardness obtained at 80°C/s from recrystallized austenite in the CCT tests (Figure 2). The lower hardness than the theoretical maximum is caused by a small amount of bainite in the microstructures of both the rolled plate (Figure 7c) and the CCT specimens.

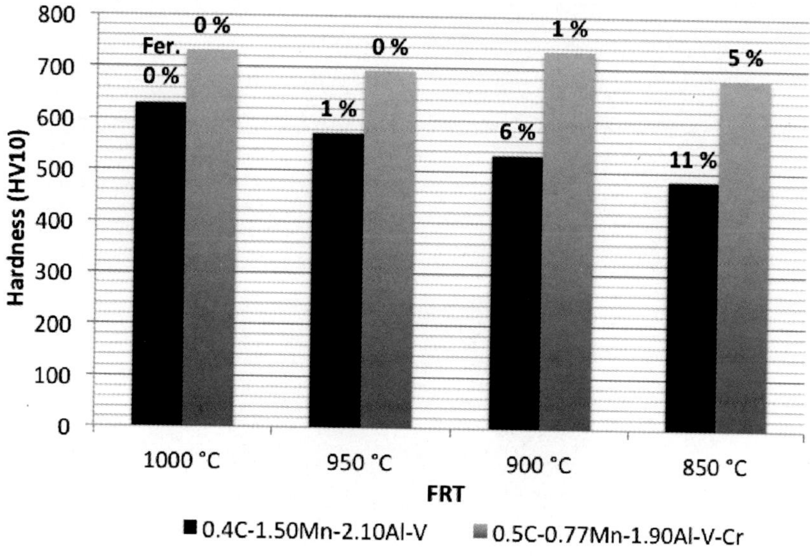

Fig. 6: Hardness of the rolled and quenched plates with the amount of ferrite as a percentage.

Table 2: Carbon content and theoretical hardness of fully martensitic microstructure.

	C %	100%M HV
0.4C-1.50Mn-2.10Al-V	0.43	705
0.5C-0.77Mn-1.90Al-V-Cr	0.47	744

The effect of composition and rolling schedule on the microstructure can be seen from the micrographs presented in Figure 7 and Figure 8. From the deformed microstructure in Figure 7a, it is clear that the FRT of 850°C is below the recrystallization stop temperature (RST) for 0.4C-1.50Mn-2.10Al-V. The structure clearly shows grain boundary ferrite (about 11%). With an increase in FRT to 900°C, the microstructure reveals the formation of some bainite as well as traces of ferrite (<6%) and mild pancaking (Figure 7b). Likewise, the structure of the 0.5C-0.77Mn-1.90Al-V-Cr steel showed the formation of some ferrite (5%) at grain boundaries for specimens finish rolled at 850°C (Figure 8a). Evidence of mild pancaking in the specimen finish rolled at 900°C (Figure 8b) suggests a higher RST, at least above 900°C, for this steel. Traces of ferrite (about 1%) seen in this sample suggests that strain induced transformation has happened before cooling.

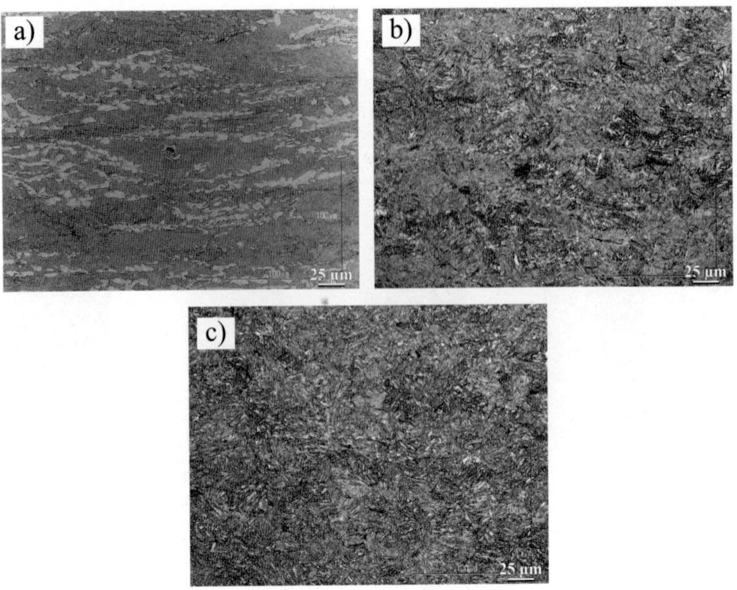

Fig. 7: 0.4C-1.50Mn-2.10Al-V; a) FRT 850°C, b) FRT 900°C, c) FRT 1000°C.

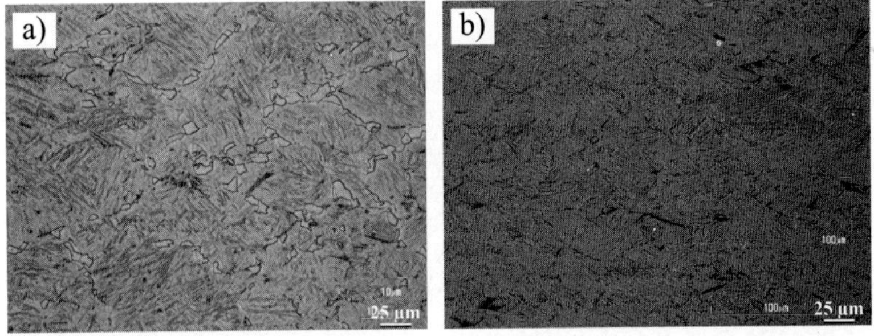

Fig. 8: 0.5C-0.77Mn-1.90Al-V-Cr; a) FRT 850°C, b) FRT 900°C.

Wear Properties

The results of the crushing pin-on-disk tests done for the rolled and quenched test materials and two commercial reference materials are presented in Figure 9. It is clear that higher hardness resulted in lower average mass loss. When comparing the results of the reference material Raex 450 and the laboratory steels with the smallest mass loss, it can be concluded that increasing the hardness from the 457 HV of Raex 450 to 570 HV or more can result in over 35% savings in mass loss.

It can also be seen that at a hardness of about 450 HV, the multiphase laboratory steel with 11% ferrite (0.4C-1.50Mn-2.10Al-V-850°C) suffered more mass loss than the fully martensitic reference material Raex 450 (457 HV). This suggests that for a given hardness, a ferritic-martensitic microstructure is not as abrasive resistant as a totally martensitic microstructure.

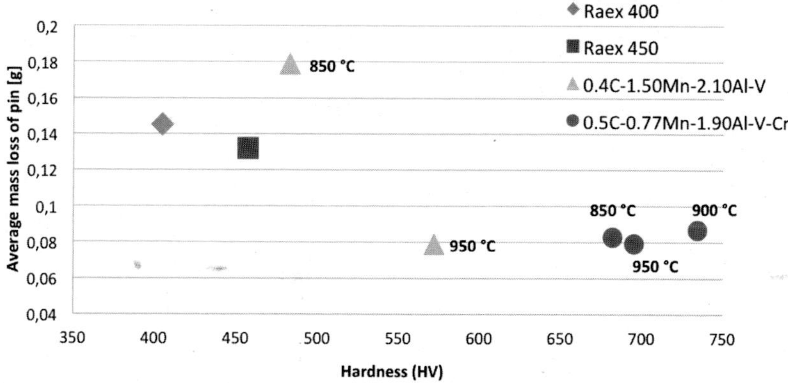

Fig. 9: Results of the crushing pin-on-disk tests: mass loss vs. hardness.

The huge difference in mass loss between the specimens with 11 and 1% ferrite made from 0.4C-1.50Mn-2.10Al-V also suggests that the dual-phase microstructure is not beneficial for abrasion resistance. This is contrary to the results reported by Jha et al. [3], who found that for a specific hardness level and composition, a ferritic-martensitic microstructure had better abrasion resistance than that of totally martensitic steel. They argue that dual-phase microstructure gives a good combination of hardness and toughness/ductility, which is more beneficial to abrasion resistance than merely a high hardness. Their studies concerned steels with lower hardness levels (252-451 HV), however. There were also major differences in testing methods between this work and the one used by Jha et al., which may explain the different observations. They used crushed silica sand (quartz) with particle size of 212-300 μm as abrasive. Quartz is much harder than granite and the particle size was much smaller than used in this work. They have also used the rubber wheel abrasion test described in ASTM G65-81. Hylkilä [14] has stated that when testing specimens hard enough using the rubber wheel abrasion test, the abrasive starts to penetrate into the rubber wheel, which causes the wear rate to increase with increasing hardness up to a certain level. This changes the wear from three-body abrasion to more like two-body abrasion, where the abrasive particles are rigidly attached to the counterbody [15]. Two-body abrasion tests are known to produce higher wear rates than three-body abrasion under comparable loading [15,16]. This could explain the lower wear rate obtained at a lower hardness reported by Jha et al. [3].

Based on the results presented in Figure 9, the abrasion resistance increases with the hardness as expected. However, further increase in hardness beyond about 600 HV does not show further improvement in abrasion resistance for the conditions studied here.

SUMMARY AND CONCLUSIONS

Physical simulation in form of Gleeble CCT diagrams was used for material selection and designing rolling experiments for obtaining dual-phase steels with hardness over 500 HV. Rolling experiments with direct quenching from varying final rolling temperatures were carried out. Abrasive wear resistance was determined using crushing pin-on-disc tests with granite and the results were compared to those from two commercially available reference materials. Martensitic and dual-phase steels with varying proportions of ferrite and martensite having hardness from 480 to 734 HV10 were obtained. The results indicated that increasing hardness from 450 HV of a fully martensitic steel to 570 HV or more can result in over 35% reduction in mass loss. However, under the present test conditions, the dual-phase steel with ferrite and martensite had poorer abrasion resistance than that of the totally martensitic reference material with the same hardness.

ACKNOWLEDGEMENT

This work was performed under the Demanding Applications program of FIMECC Ltd, financed by The Finnish Funding Agency for Technology and Innovation that is acknowledged with gratitude.

REFERENCES

[1] J.H. Tylcak, Abrasive Wear in ASM Handbook, vol. 18: Friction, Lubrication and Wear Technology, ASM Inter-national, 1992, pp. 336-350

[2] T. Liimatainen, personal communication, 2009.

[3] A.K. Jha, B.K. Prasad, O.P. Modi, S. Das and A.H. Yegneswaran, Wear, Vol. 254, 2003, pp. 120-128

[4] T. Furuhara, J, Yamaguchi, N. Sugita, G. Miyamoto and T. Maki, ISIJ International, Vol. 43, 2003, pp. 1630-1639

[5] T. Furuhara, T. Shinyoshi, G. Miyamoto, J, Yamaguchi, N. Sugita, N. Kimura, N. Takemura and T. Maki, ISIJ International, Vol. 43, 2003, pp. 2028-2037

[6] J.H. Shim, Y.J Oh, J.Y. Suh, Y.W. Cho, J.D. Shim, J.S. Byun and D.N. Lee, Acta Materialia, Vol. 49, 2001, pp. 2115-2122

[7] J.E. Garcia-Gonzalez, Doctoral Thesis, University of Pittsburgh, 2005

[8] A. Mein, G. Fourlaris, D. Crowther and P.J. Evans, Materials Characterization, Vol. 64, 2012, pp. 69-78

[9] M. Gomez, C.I. Garcia, D.M. Haezebrouck and A.J. DeArdo, ISIJ International, Vol. 49, 2009, pp. 302-311

[10] W.C. Leslie, G.C. Rauch, Metallurgical and Material Transactions A, Vol. 9, 1978, pp. 343-349

[11] C. Garcia de Andrés, C. Capdevila, D. San Martin and F.G. Caballero, Materials Science and Engineering A, Vol. 328, 2002, pp. 156–160

[12] J.H. Shim, Y.W. Cho, S.H. Chung, J.D. Shim and D.N. Lee, Acta Materialia, Vol. 47, 1999, pp. 2751-2760

[13] J. Terva, T. Teeri, V-T. Kuokkala, P. Siitonen and J. Liimatainen, Wear, Vol. 267, 2009, pp. 1821-1831

[14] A. Hylkilä, Master's Thesis, University of Oulu, 2008.

[15] J.D. Gates, Wear, Vol. 214, 1998, pp. 139-146

[16] E. Rabinowicz, L.A. Dunn, P.G. Russell, Wear, Vol. 4, 1961, pp. 345-355.

Phase-Field Simulation of the Thermomechanical Processing of Steels

Rongshan Qin[1] and Xu Tan[2]

[1]Department of Materials, Imperial College London, London, SW7 2AZ, United Kingdom

[2]Graduate Institute of Ferrous Technology, Pohang, 790-784, Republic of Korea

Email: r.qin@imperial.ac.uk

ABSTRACT

A scheme has been developed to integrate the phase-field model with the homogeneous deformation. This has been applied to the simulation of the microstructure evolution during warm rolling of steels. The phase transition is characterized by the evolution of a phase-field order parameter. The rolling of the steels is calculated by the displacement of vectors in materials by the deformation matrix. The governing equation for phase-field order parameter is solved numerically in the irregular and changing lattices. Significant discrepancies of crystal morphologies are observed in crystal growth before deformation, deformation of materials after crystal growth and deformation of materials while crytal growing. Those three thermomechanical processing conditions can be considerred as the cases of hot rolling, cold rolling and warm rolling, respectively. It has been found that the interface anisotropy plays an important role for the observed microstructural descrepacies. Other mechanisms affecting the grain morphologies are discussed for various thermomechanical processing of steels.

KEYWORDS: *Steel, Warm Rolling, Phase-Field, Homogeneous Deformation, Microstructure*

INTRODUCTION

Thermomechanical processing implements heat and stress in tailoring the microstructure of steels. Hot rolling, cold rolling and warm rolling are characterized by the sequences of austenite-ferrite phase transformations regarding to the mechanical deformation, i.e., the hot rolling deforms steels before austenite-ferrite phase transition, cold rolling deforms steels after this phase transition, and warming rolling deforms steels during the phase transition. In practice, steel processing may use more than one rolling scheme, e.g. starting from the hot rolling but completing at the warm rolling temperature. Warm rolling has attracted significant attention due to its complexity. Experimental observations such as the evolution of microstructure [1] and texture [2] in warm rolling of low carbon steels have been reported. Numerical simulations including finite element calculation are carried out to the investigation of the warm rolling [3-4]. This work aims to study the microstructure evolution in warm rolling by coupling the phase-field model with homogeneous deformation theory.

The deformation of materials can be described by either the equal-stress or equal-strain approximations. The homogeneous deformation is an equal-strain method, and is suitable particularly for those deformations without the formation of cracks and other defects. The fundamental principle of this method is represented by following equation

$$\begin{pmatrix} S_{11} & S_{12} & S_{13} \\ S_{21} & S_{22} & S_{23} \\ S_{31} & S_{32} & S_{33} \end{pmatrix} \begin{pmatrix} U_1 \\ U_2 \\ U_3 \end{pmatrix} = \begin{pmatrix} V_1 \\ V_2 \\ V_3 \end{pmatrix} \quad (1)$$

where S is called the deformation matrix. $\vec{U}$ and $\vec{V}$ are the vectors before and after the deformation, respectively. The elements of the matrix S take different values to represent various types of homogeneous deformations. Zhu et al has applied the homogeneous deformation method to the study of the metallography of a tetrakaidecahedron grain [5]. Chae et al applied the method to the deformation of non-uniform grains [6]. The computation of the microstructure evolution in mechanical deformation can be achieved by applying Eq. (1) to every pixel of the micrograph images.

The microstructure evolution in phase transition, including both displacive transformation and reconstructive transformation, can be calculated in the framework of a phase-field model [7]. The model introduces an order parameter (ϕ) to represent the physical state of the materials. ϕ takes one value from one phase to another value in another phase and changes smoothly from one value to another across the interface. The governing equation for the phase-field order parameter can be derived explicitly from the second law of the thermodynamics. The parameters in the governing equation can be obtained by manipulating the governing equation to the simplest conditions. For a case where the phase transition is driven by a constant driving force and the solute and temperature heterogeneous are ignored completely, the governing equation for the phase-field order parameter is [8]

$$\frac{\partial \phi}{\partial t} = M_{\phi}\left(\varepsilon^2 \nabla^2 \phi - \frac{\partial g(\phi)}{\partial \phi}\right) \quad (2)$$

where M_{ϕ} is called the phase-field mobility related to the interface kinetics. ε is called the gradient energy coefficient reflected the interface energy and the interface anisotropy. $g(\phi)$ is the free energy density of the material and is a double well potential in non-equilibrium state. The calculation of the phase transition is achieved by solving Eq. (2) with time iteration. Application of the phase-field models in various materials have been summarized in a number of review articles, as listed in ref. [9].

In the present work, the phase-field model has been coupled with the homogeneous deformation theory. Application of the coupled model to the thermomechanical processing of steels has been demonstrated. The microstructural evolution in various processing conditions is calculated. The mechanisms responsible for the characterized microstructures will be investigated.

THEORETICAL CONSIDERATION

Warm rolling has two coupled physical processes, namely the phase transition and materials deformation. Phase-field model can be used to address the microstructure evolution in phase transition. Homogeneous deformation is to be used to address the mechanical deformation. Warm rolling is taking place in relative high temperature. This enables the ignorance of the residual stress in materials which otherwise will affect the free energy density and interface energy. The formation of various defects such as dislocations, voids and cracks are ignored. It is also assumed that the time of each rolling pass is short enough in comparison with the overall phase transition time. This is reasonable because the phase transition takes place in a wide temperature range but the temperature dropping during each rolling pass is small. With such consideration, we can ignore the phase transition during the rolling pass, i.e., phase transition takes place in many time steps but each rolling pass can complete within a single time step.

The double well potential for the free energy density can be represented as

$$g(\phi) = \frac{1}{4\omega}\phi^2(1-\phi)^2 + h(\phi)g_1 + [1-h(\phi)]g_0$$

where $h(\varphi) = \varphi^3(6\varphi^2 - 15\varphi + 10)$. ω is related to the barrier height and can be determined by the interface thickness (2λ) and interface energy (σ) via $\omega = \lambda/(26.4\sigma)$. g_0 and g_1 are the free energy density of bulk phases represented by $\phi = 0$ and $\phi = 1$, respectively. $h(\phi)$ is associated with the volume fraction of the phases. This gives the second term in the right-hand-side of Eq. (2) as

$$\frac{\partial g}{\partial \phi} = \frac{1}{2\omega}\phi(1-\phi)(1-2\phi) + h'(\phi)\Re \qquad (3)$$

where $h'(\varphi) = 30\varphi^2(1-\varphi)^2$. $\Re = (g_1 - g_0)$ is the driving force for phase transition. The gradient energy coefficient ε in Eq. (2) is related to the interface energy and its anisotropy. For cubic crystal, the anisotropic interface energy can be represented as [8]

$$\varepsilon(\hat{n}) = \varepsilon_0 + \varepsilon_1\left(n_x^2 n_y^2 + n_y^2 n_z^2 + n_z^2 n_x^2\right) + \varepsilon_2 n_x^2 n_y^2 n_z^2 + \varepsilon_3\left(n_x^2 n_y^2 + n_y^2 n_z^2 + n_z^2 n_x^2\right)^2 \qquad (4)$$

where $\hat{n} = n_x\hat{x} + n_y\hat{y} + n_z\hat{z}$ is the normal vector to interface and is determined in phase-field calculation by $\hat{n} = \nabla\phi / |\nabla\phi|$. The coefficients ε_0, ε_1, ε_2 and ε_0 are determined by

$$\varepsilon_0 = \lambda\sqrt{k_0} \qquad (5.1)$$

$$\varepsilon_1 = \lambda k_1 / (2\sqrt{k_0}) \qquad (5.2)$$

$$\varepsilon_2 = \lambda k_2 / (2\sqrt{k_0}) \qquad (5.3)$$

$$\varepsilon_3 = \lambda k_3 / (2\sqrt{k_0}) - \lambda k_1^2 / (8k_0\sqrt{k_0}) \qquad (5.4)$$

The coefficients k_0, k_1, k_2 and k_0 are determined according to the interface anisotropy. A method to specify those coefficients for various metals was presented in our earlier works [8, 10].

In phase transition, Eq. (2) is calculated in space represented by discrete lattices that can be arranged as regular or irregular patterns. Crystal growth is obtained by iteration of Eq. (2) in a series discrete time steps. Each rolling pass is calculated within one time step according to Eq. (1) and the calculation is performed throughout the discrete lattices in space. The different values of ϕ is plotted with different colours so that the crystal morphology can be examined.

NUMERICAL CALCULATION

The materials are initially represented by rectangular lattices with lattice distance in all three dimensions equivalent to Δh. The finite difference method gives the first order differentiation of phase field order parameter regarding to the space coordinate parameters as following forward step approximation

$$\frac{\partial \phi_{ijk}}{\partial x} \approx \frac{\phi_{i+1\,jk} - \phi_{i\,jk}}{\Delta h} \tag{6}$$

where the meaning of ϕ_{ijk} is illustrated in Fig. 1(a). After the shape deformation, the rectangular lattices are not able to retain its original shape in all three dimensions. In some cases, the lattices are arranged in a shape similar to that illustrated in Fig. 1(b). The partial difference regarding to space coordinate should use more generic equation of followings [11]

$$\frac{\partial \varphi_{ijk}}{\partial x} = \lim_{\Omega_y \to 0} \frac{\oint \varphi_{ijk}\, dy}{\Omega_y} = \lim_{\Omega_z \to 0} \frac{\oint \varphi_{ijk}\, dz}{\Omega_z} \tag{7}$$

In discrete format as that illustrated in Fig. 1(b), it has

$$\oint \varphi_{ijk}\, dy = \frac{\varphi_{i+1,j,k} - \varphi_{i,j+1,k}}{2}\left(y_{i+1,j+1,k} - y_{i,j,k}\right) + \frac{\phi_{i+1\,j+1,k} - \phi_{i,j,k}}{2}\left(y_{i,j+1,k} - y_{i+1,j,k}\right) \tag{8}$$

To reduce the mathematical error, following calculation is applied in the calculation

$$\frac{\partial \varphi_{ijk}}{\partial x} = \frac{1}{2}\left(\lim_{\Omega_y \to 0} \frac{\oint \varphi_{ijk}\, dy}{\Omega_y} + \lim_{\Omega_z \to 0} \frac{\oint \varphi_{ijk}\, dz}{\Omega_z} \right) \tag{9}$$

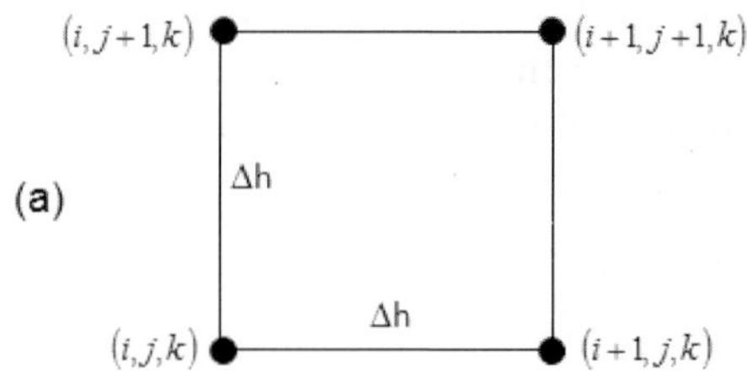

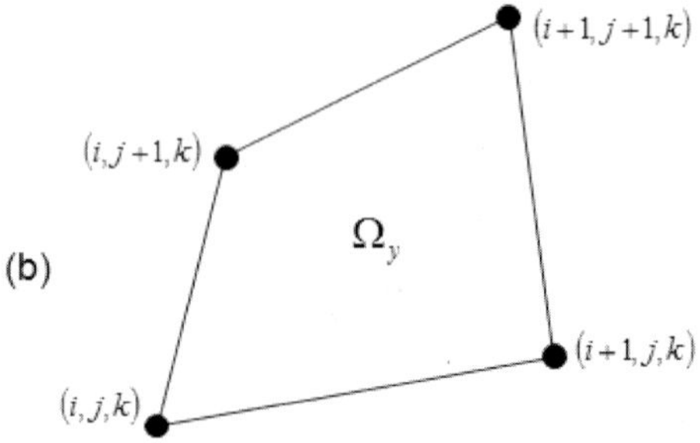

Fig. 1: Schematic diagrams of (a) rectangular and (b) non-rectangular lattices

The important condition for using Eq. (8) and Eq. (9) is that all the four lattice points should be in the same plane. The differentiation regarding to time is represented by following explicit discrete approximation in a time step of Δt

$$\frac{\partial \phi_{ijk}^{t}}{\partial t} \approx \frac{\phi_{ijk}^{t+\Delta t} - \phi_{ijk}^{t}}{\Delta t} \tag{10}$$

In the present simulation, a rectangular lattices of 160 × 160 × 120 are defined initially. The initial condition for the phase field order parameter is specified according to following conditions

$$\begin{cases} \varphi(r, t=0) = 1 & for\ r < \Delta h \\ \varphi(r, t=0) = \dfrac{2}{1+e^{r-\Delta h}} & for\ \Delta h < r < 4\Delta h \\ \varphi(r, t=0) = 0 & for\ r \geq 4\Delta h \end{cases}$$

The constant driving force is selected as $\Re = -3.686 \times 10^{8} J/m^{3}$. The experimental measurement gives the average interface energy between face centred cubic and body centred cubic crystals of steel as k_0= 0.8 J/ m^2. Because the actual anisotropic property between austenite and ferrite phases is not known, one uses a general expression of the anisotropic interface energy for the cubic metals and chooses following arbitrary values for the calculation:

$$\varepsilon_1 / \varepsilon_0 = -0.863$$
$$\varepsilon_2 / \varepsilon_0 = 0.395 \qquad (11)$$
$$\varepsilon_3 / \varepsilon_0 = 0.0238$$

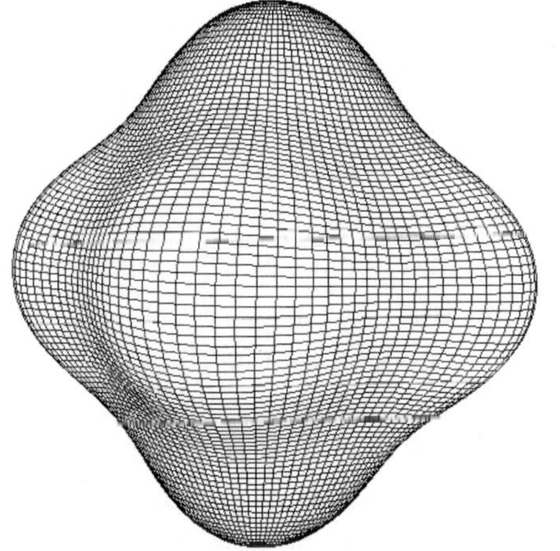

Fig. 2: Polar diagram of anisotropic interface energy represented by Eq. (4) with coefficients value in Eq. (11) [12].

The corresponding anisotropic interface energy in its polar diagram is illustrated in Fig. 2. Other parameters include M_ϕ = *100* and λ = 14.3 nm. This gives $\omega = 1.35 \leftrightarrow 10^{-9}\ m^3/J$ and $\varepsilon_0 = 2.488 \leftrightarrow 10^{-4}\ J^{1/2}/m^{1/2}$. Without consideration of deformation, the crystal morphology at t = 12000 time steps is demonstrated in Fig. 3. This is a typical anisotropic crystal. Due to the assumption of constant driving force and ignorance of heat and mass transfer and also the ignorance of fluctuation, there is no equilibrium shape of the crystal can be obtained.

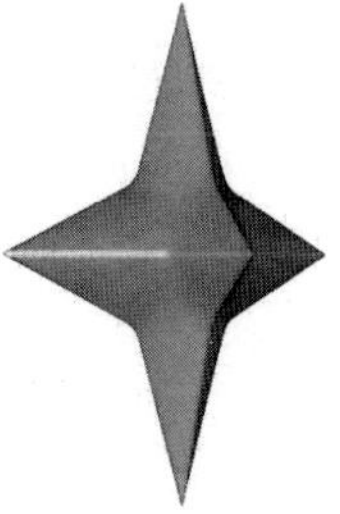

Fig. 3: The crystal morphology without deformation at 12000 time steps.

Application of shear deformation to the crystal once after 12000 steps of crystal growth gives the crystal morphology demonstrated in Fig. 4. The deformation matrix for shear deformation is

$$\begin{pmatrix} 1 & 0 & 0.3 \\ 0 & 1 & 0 \\ 0 & 0 & 1 \end{pmatrix}$$

After one rolling at the 12000 time steps, the morphology of crystal changes into that demonstrated in Fig. 4. The topological geometry of the crystal is not changed but the morphology is homogeneously sheared alone z-direction.

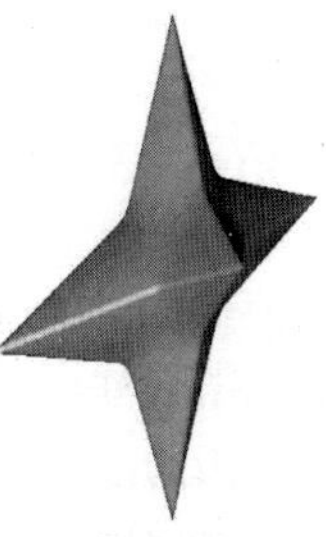

Fig. 4: The crystal morphology after the shear deformation at 12000 time steps.

Another case is to have the shear deformation after every 6000 time steps of crystal growth. The crystal morphological evolutions are demonstrated in Fig. 5. I t can be that the crystal morphology is completely different from that of Fig. 3 and Fig. 4.

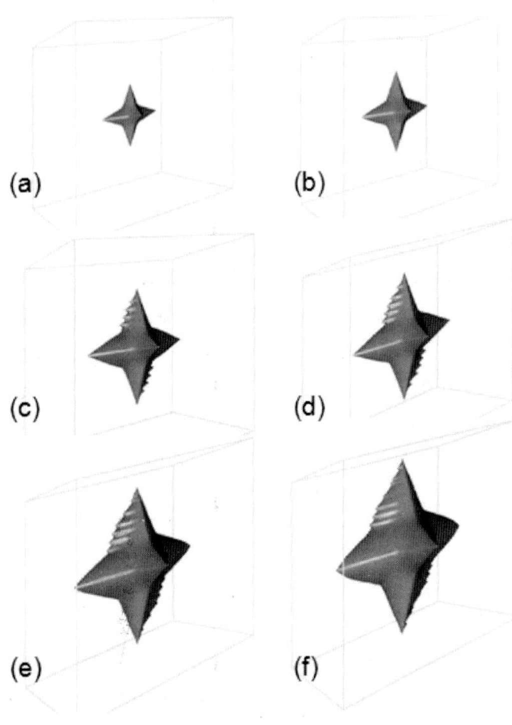

Fig. 5: Crystal shape at (a) 6000, (b) 8000, (c) 10000, (d) 12000, (e) 14000 and (f) 16000 time steps. The shear deformation applied after every 6000 time steps.

In summary, Fig. 3 corresponding a crystal growth without deformation. This is similar to the microstructure evolution in hot rolling. Fig. 4 corresponding to a mechanical deformation after crystal growth. This is similar to the case of cold rolling. Fig. 5 gives microstructure evolution with phase-transition and mechanical deformation taking place in a same period, and is corresponding to warm rolling. Warm rolling generates significant different microstructure in comparison with hot rolling and cold rolling.

DISCUSSION

It can be seen from Fig. 5 that the major difference of the crystal morphology is at the crystal arms facing and parallel to the shear direction. The interface at this direction is getting more steeper after shear deformation. This makes its interface energy bigger than before deformation. To reduce the interface energy, crystal growth along this direction is accelerated so that there are less amount of interface will be faced to this direction and hence the total surface energy and total free energy of the system can be reduced. This can be clearly seen in Fig. 6. The same situation appears to the interfaces at the plane diagonal direction. It can be seen that the similar crystal morphology evolution are appeared in that crystal arm. On the contradictory, the interface parallel and along the shear direction is getting less steep and more close to an interface normal direction which possesses smaller interface energy. The crystal growth at those areas is not encouraged in order to maintain its smallest interface energy and hence the total free energy. This is clearly seen in Fig. 5. The shear deformation does not change the orientation of other four crystal arms significantly. Therefore, the crystal morphology at those places are not altered obviously.

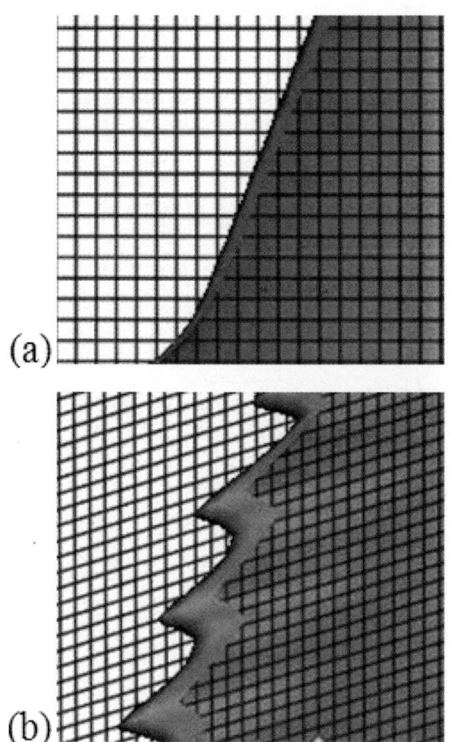

Fig. 6: Lattices and crystal interfaces after shear deformation and crystal growth

In summary, the combination of deformation and interface anisotropy are the major reason to cause the unusual microstructure evolution. Quantitative and more general description of this effect requires more systematic investigations.

CONCLUSION

A phase-field model has been coupled to the homogeneous deformation method. The integrated scheme is able to simulate the thermomechanical processing of steels. The microstructure evolution due to phase transition can be handled by the phase-field model. The microstructure evolution due to mechanical deformation can be calculated by the deformation theory. The coupled scheme has been applied to the crystal growth in warm rolling. The drastic microstructure alteration in this case is considered due to the deformation of the interface and the anisotropic properties of the interface.

ACKNOWLEDGEMENT

RQ is grateful to Professor HKDH Bhadeshia from University of Cambridge for invaluable discussion and to TATA Steel and the Royal Academy of Engineering for financial support of this work.

REFERENCES

[1] G.H. Akbari, C.M. Sellars and J.A. Whiteman, Acta Mater, Vol. 45, No. 12, 1997, p. 5047.

[2] A. Haldar and R.K. Ray, Mater. Sci. Eng. A, Vol. 391, No. 1-2, 2005, p. 402.

[3] J. Majta and A.K. Zurek, Int. J. Plasticity, Vol. 19, No. 5, 2003, p. 707.

[4] C.A. Greene and S. Ankem, Mater. Sci. Eng. A, Vol. 202, No. 1-2, 1995, p. 103.

[5] Q. Zhu, C.M. Sellars and H.K.D.H. Bhadeshia, Mater. Sci. Tech., Vol. 23, No. 7, 2007, p. 757.

[6] J.Y. Chae, R.S. Qin and H.K.D.H. Bhadeshia, ISIJ international, Vol. 49, No. 1, 2009, p.115.

[7] R.S. Qin and H.K.D.H. Bhadeshia, Mater. Sci. Tech., Vol. 26, No. 7, 2010, p.803.

[8] R.S. Qin and H.K.D.H. Bhadeshia, Acta Mater., Vol. 57, No. 7, 2009, p. 2210.

[9] R.S. Qin and H.K.D.H. Bhadeshia, Curr. Opin. Solid State Mater. Sci. Vol. 15, No.3, 2011, p.81.

[10] R.S. Qin and H.K.D.H. Bhadeshia, Acta Mater., Vol. 57, No. 11, 2009, p.3382.

[11] M. Shashkov and S. Steinberg, Conservative Finite-Difference Methods on General Grids, 1st Edn., CRC Press, 1996, pp. 26-45.

[12] R.S. Qin, Mater. Manufac. Proc., Vol. 26, No. 1, 2011, p. 132.

Thermo-Mechanical Behavior in Steels: Transformation Plasticity

Heung Nam Han[1], Yi-Gil Cho[2], Hoon-Hwe Cho[1], Moon-Jo Kim[1], Dong-Wan Kim[1] and Yang-Hoo Kim[1]

[1]Department of Materials Science and Engineering and Center for Iron & Steel Research, RIAM, Seoul National University, Seoul 151-744, Republic of Korea

[2]Korea Institute of Science and Technology (KIST), Seoul, 136-791, Republic of Korea

Email: hnhan@snu.ac.kr

ABSTRACT

Permanent deformation that originates from transformation plasticity has favorable aspects for steels with improved strength and ductility. However, it also causes undesirable deformation of products or specimens, leading to their degradation. This article reviews recent investigations of transformation plasticity. A combination of newly suggested models, numerical analyses, and novel experiments has attempted to reveal the mechanism. Since the nature of the transformation plasticity is still unclear, there are significant challenges still to be solved. Fundamental understanding of transformation plasticity will be essential for the development of advanced steels.

KEYWORDS: *Steels, Transformation Induced Plasticity (TRIP), Accelerated Creep, Mechanically Induced Martensite Transformation, Dilatometry, FEM.*

INTRODUCTION

Transformation plasticity has been believed to be a mechanism that causes permanent deformation during phase transformations of polycrystalline materials even under their yield stresses. As usage of the transformation plasticity in the contexts of strength and ductility [1,2], so-called transformation-induced plasticity (TRIP) steels [3], have attracted great interest in the automotive industry. However, sometimes the deformation due to transformation plasticity is undesirable for products or specimens, and that it degrades material qualities. For example, the situations listed below describe degradation in processing or experiments.

(a) Hot-rolled coils sometimes contract asymmetrically during steel making/ storing processes. Transformation plasticity is believed to be the primary reason [4].

(b) Permanent deformations and residual stresses occur in welding processes [5]. Considerable portions of deformation might be caused by transformation plasticity.

(c) Dilatation curves that contain the contribution of transformation plastic deformation may cause significant errors in dilatometry [6,7].

Since super-plastic like deformation due to transformation plasticity was reported, many researchers have begun to investigate its origin and nature extensively. Many theories and models have been proposed, which have been categorized into two main groups. For diffusive transformation, Greenwood and Johnson [8] derived an analytical solution for transformation plastic deformation on ideally plastic materials. They assumed that plastic deformation occurs in a weaker phase to accommodate external and internal stresses by volume mismatches between two solid phases. This model has been modified, extended and reproduced by other researchers. For example, Leblond et al. demonstrated the interaction between classical plasticity and TRIP [9,10], Taleb and coauthors re-evaluated the Leblond model by providing various experimental grounds [11-13], Fischer et al. quantified the effect of the orientation on the deformation of shape memory alloys [14-16]; Mahnken et al. utilized a unit-cell RVE model to find the mechanical behavior of macroscopic austenite/martensite composite, which combines the effect of classical plasticity and transformation induced plasticity [17]. On the other hand, Magee paid attention to the displacive martensitic transformation by which transformation plastic deformation is also induced [18]. The preference of a specific variant under a certain stress field was adopted to describe the deformation. Based on this theory, several models were proposed with extended or more generalized constitutive relations. Han et al. [1] proposed a nucleation-controlled kinetics based on Olson and Cohen's approach [3,19] in order to compensate for the disadvantage of kinetics in the strain-induced martensitic transformation [20-23]. Other approaches frequently adopted for the representation of the macroscopic behavior of phase transformed materials are used to introduce the micromechanical or multi-scale modeling techniques [14,24-31]. Recently, Levitas and Ozsoy developed a micromechanical approach by modeling the universal thermodynamic driving force for the interface reorientation to derive the stress- or strain-induced martensite transformation. In the application part of their micromechanical model, various types of representative volumes were built to validate the effect of the athermal threshold, the martensite variants and an interface orientation under three dimensional thermo-mechanical loading [27,28,32].

In this article, recent investigations of transformation plasticity, which attempted to reveal its mechanism and nature, are reviewed[33]. These investigations include newly developed theories, numerical approaches, and novel experiments.

REPRESENTATIVE THEORIES

Accelerated creep model [34]

During the diffusive phase transformation, the phase interface migrates through the movement of atoms across the interface. Generally, it can be assumed that the overall atomic flux across the interface will be perpendicular to the interface and that the migrating atoms will rearrange at the nearest atomic site in the transformed phase. However, when an external stress is applied, the migrating atoms will move to positions where they can release the applied stress field, and this can give rise to an atomic flux along the phase interface. This phenomenon may be similar to the mechanism of Coble creep.

The model of Greenwood and Johnson has been modified to include the temperature-dependence of the strain rate caused by the transformation plasticity [35,36]. These models could simulate the thermally activated behavior of the transformation plasticity in a relatively high temperature range. All of these models were based on the volume mismatch between the two phases and the creep deformation of the weaker phase during the phase transformation under externally applied stress. Recently, however, several sets of experimental results have been reported, which were difficult to explain by the Greenwood and Johnson's consideration. One of these is that an externally applied stress, even much lower than the yield stress of the material, induces considerable permanent strain during the recrystallization and growth in extra low-carbon steels [37]. Since there is little volume mismatch between the unrecrystallized and recrystallized region, the evolution of permanent strain during recrystallization is difficult to interpret by the volume mismatch induced internal stress model. Besides, it was found that the deviation angle from the Kurdjumov-Sachs (KS) orientation relationship considerably increased with the uniaxially applied stress during the austenite-to-ferrite transformation of steel [38,39], which implies that the sliding of the migrating transformation interface possibly occurs when the external stress is applied during phase transformation. From the above experimental observations, the transformation plasticity behavior of steel during the phase transformation under externally applied stress is modeled on the basis of a migrating interface diffusion mechanism, which is described as an accelerated Coble creep. The mathematical form of the model can be expressed as follows:

$$\dot{\varepsilon} = \frac{1}{3}\frac{d_0}{\delta}\dot{X}\frac{\sigma\Omega}{k_B T}c_{v0}\exp\left(-\frac{Q_f}{k_B T}\right) + \frac{B_c}{\pi}\frac{\delta D_I^s \sigma\Omega}{d_0^{\,3} k_B T} \qquad (1)$$

where d_0, δ, and Ω represent the initial grain size of the parent phase, the effective thickness of the interface and the volume of the vacancy, respectively. The Boltzmann constant, k_B, has a value of 1.38×10^{-23} J/°C. c_{v0} is a dimensionless constant determined by the change in thermal entropy associated with the formation of the vacancy and Q_f is the formation enthalpy of the vacancy at the interface. σ is the applied stress, and the X with a dot above it is the transformation rate.

δD_I^s denotes the effective diffusion coefficient at the stationary interface. B_c, is known to be 148 for Coble creep [40]. The first term on the right side of Equation (1) indicates the component of the creep strain rate during the interface migration, which can be attributed to the progress of the phase transformation. When no phase transformation occurs, the first term on the right side of Equation (1) becomes zero and equation (1) becomes the typical equation for Coble creep.

Fig. 1 shows the relationship between the transformation shape strain after the phase transformation and the applied stress. The transformation shape strain is proportional to the applied stress. This linear relationship is in agreement with the many experimental observations [8,20,41,42] of the transformation plasticity.

Martensitic transformation [1]

A one-dimensional model for the kinetics of strain-induced martensitic transformation was developed by Olson and Cohen [19]. Stringfellow et al. derived a constitutive model for the kinetics of mechanically-induced martensitic transformation that considered the external stress state in terms of the hydrostatic pressure and the equivalent shear stress [20].

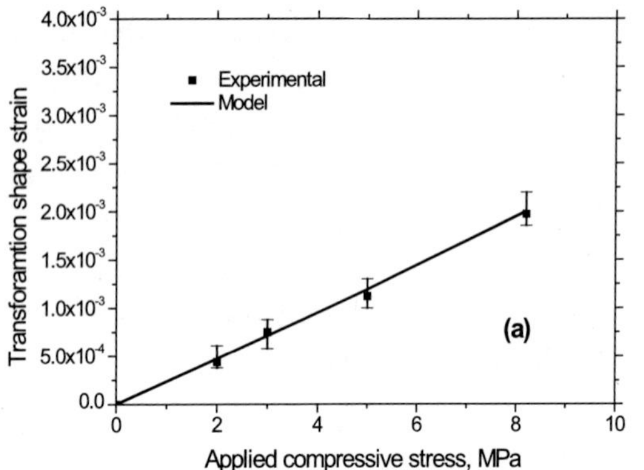

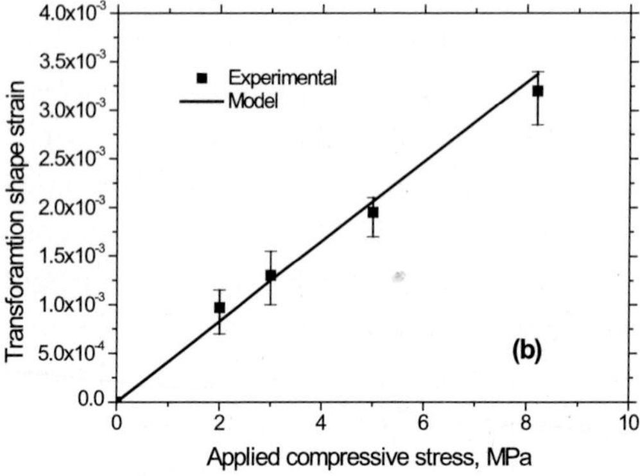

Fig. 1: Transformation shape strain after transformation as a function of applied stress during (a) cooling and (b) heating by 2 °C/sec[34].

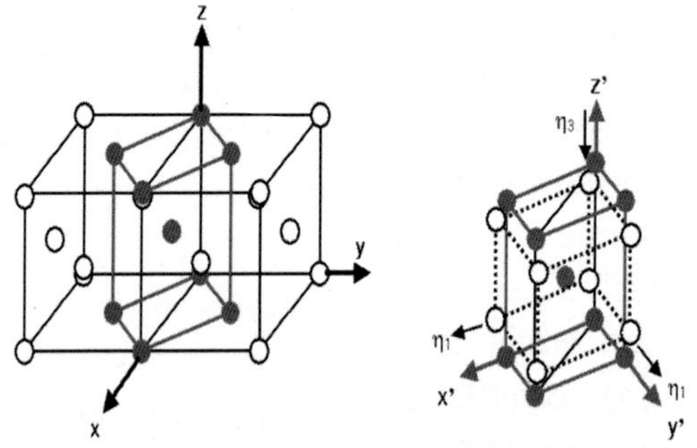

Fig. 2: Lattice correspondence between FCC and BCT and the Bain deformation accompanying with the transformation[1].

Tomita et al. modified Stringfellow's model to account for the experimental finding that the deformation behavior is controlled by the shear band mode as the strain rate increases [21-23]. However, the previous models were derived only for the kinetics of strain-induced martensitic transformation and have a disadvantage in that the athermal martensitic transformation cannot be analyzed. In addition, the models dealt with the TRIP strain as a strain softening without respect to the microstructural change, which is caused by the preferential selection of variants under the external stress state.

Martensitic transformation kinetics was assumed as a nucleation-controlled phenomenon on the basis of Olson and Cohen's approach [19,20]. The probability that a nucleation site would really act was derived for each variant as a function of the interaction energy between the externally applied stress state and the lattice deformation based on the Kurdjumov-Sachs (K-S) relationship. The TRIP strain was also evaluated by assessing the difference of the nucleation rate of 24 K-S variants. Each of the 24 variants of the K-S relation has one compressive axis and two tensile axes for the martensitic transformation, as can be seen in Fig. 2; this is called the Bain distortion. To reflect the athermal martensitic transformation kinetics as well, Koistinen and Marburger's empirical formula [43] was included in the probability equation.

REPRESENTATIVE RESEARCHES

Dilatometric non-isotropy [6]

Generally, it is known that non-isotropic volume changes in dilatometry have also been observed during the phase transformation, even in steel with an isotropic microstructure [44]. The contribution of the non-isotropic volume change to the dilatation data was recently quantified and this change was incorporated into a dilatometric analysis model [44].

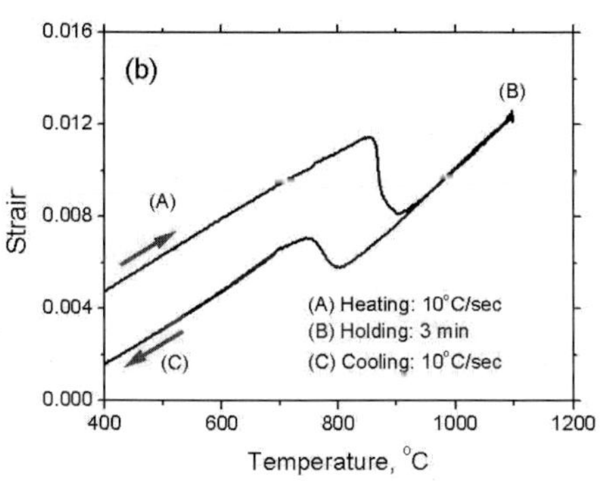

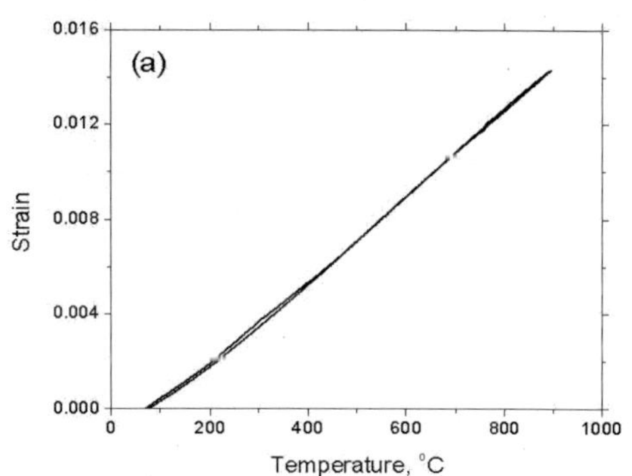

Fig. 3: Measured dilatation curves for (a) ultra-low carbon steel (steel ULC) during a thermal cycle with 1 K/s and (b) low carbon steel (steel LC) during a thermal cycle with 10 K/s[6].

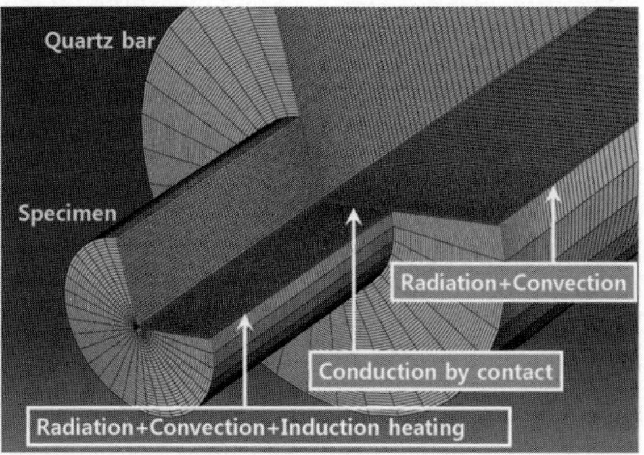

Fig. 4: FE meshes and boundary conditions for heat transfer in the dilatometric system[6].

However, up to now, there have been no studies that have attempted to pinpoint the cause of the non-isotropic dilatation in specimens with isotropic microstructure or to predict the non-isotropic dilatation of this kind of specimen.

Fig. 3a and 3b show typical examples of the dilatation curves for an ultra-low carbon (0.0003wt.%C-1.0wt.%Mn, ULC) and low carbon (0.02wt.%C-1.0wt.%Mn, LC) steel during cyclic heat treatments up to 900 and 1000 °C, respectively. In the case of the ultra-low carbon steel, since the steel underwent no transformation in the temperature range up to 900 °C, only thermal expansion and contraction could be observed. On the other hand, in the case of the low carbon steel, length change during the phase transformation could be found. Note that a mismatch at the starting and ending points of the dilation curve was observed in the case of the low carbon steel, which implies that the volume change accompanying the phase transformation has non-isotropic characteristics.

In order to describe this non-isotropic dilatometric behavior during the phase transformation in steel, the concept of transformation plasticity was used. The constitutive equation for the transformation plasticity was incorporated into a finite-element (FE) model (Fig. 4), which was adopted to describe the non-isotropic dilatometric behavior during the phase transformation in the steel without any specific microstructural directionality. An implicit numerical solution procedure to calculate the deformation during the dilatometric experiment was incorporated into the general purpose implicit FE program.

Besides the thermo-elasto-plastic constitutive equations, the phase transformation kinetics was characterized by a Johnson-Mehl-Avrami-Kolmorgorov (JMAK) type equation. For the FE formulation, the stress increment was defined as:

$$d\sigma = C^e : (d\varepsilon^T - d\varepsilon^v - d\varepsilon^P - d\varepsilon^p) \qquad (2)$$

where C^e and $d\varepsilon^e$ are the elastic stiffness tensor and the elastic strain increment, respectively. $d\varepsilon^v$, $d\varepsilon^P$ and $d\varepsilon^p$ are the volumetric strain increment due to the phase transformation and temperature change, the transformation plasticity strain increment associated with the phase transformation and the plastic strain increment, respectively.

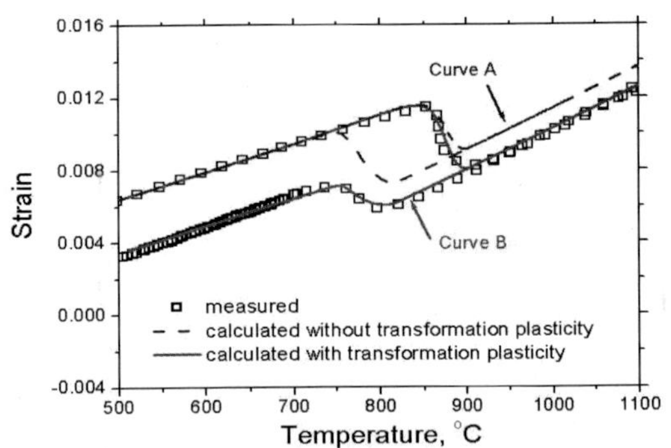

Fig: 5: Comparison between calculated (lines) and measured (symbols) dilatation curves. Curves A and B represent the dilatometric curve calculated without and with considering transformation plasticity, respectively[6].

In order to estimate precisely the interfacial heat transfer rate on the surface of the dilatometric specimen, including the induction heating, an inverse heat transfer technique was used. The dilatometric behaviors of both conventional low carbon and ultra-low carbon steels were simulated by using the FE model. To validate the suggested FE model, the measurement of the dilatation curves was carried out for these low carbon and ultra-low carbon steels and we compared the simulation results with the experimental data (Fig. 5). From the simulation results, the origin of the non-isotropic dilatation in the specimens was discussed.

The transformation plasticity was caused by the small amount of stress that naturally developed in the specimen during the dilatometric experiment. In the conventional low carbon steel, the stress in the specimen formed mainly due to the very small external force supplied to support it during the dilatometric experiment. As regards the ultra-low carbon steel, whose phase transformation occurs within an extraordinarily narrow temperature range, the inhomogeneous phase transformation due to the thermal gradient along the radial direction in the specimen was mainly responsible for the stress field in the specimen during the dilatometric experiment.

Microstructurally banded steel [7]

Microstructural bands in steels are alternating layers of ferrite and pearlite. They originate from the alignment of segregated regions with substitutional elements during casting and subsequent hot-rolling, which eventually produces layers with different microstructural constituents. A number of studies [45-48] have considered the correlation between the microstructural directionality and the dimensional non-isotropy, which is associated with non-isotropic volume changes during phase transformations.

The orientation dependent dilatometric behavior in microstructurally banded steel was simulated using the finite element analysis combining the thermal, elastic, and conventional plasticity as well as the transformation plasticity (Fig. 6). To examine the effect of transformation plasticity on the non-isotropic dilatations, numerical calculations were carried out in two different ways: (a) not considering the transformation plasticity so that only the elastic, conventional plastic, and volumetric deformations were taken into account, and (b) considering the transformation plasticity as well. Figure 7. Comparisons between the calculated and measured dilatation curves: (a) not considering transformation plasticity;

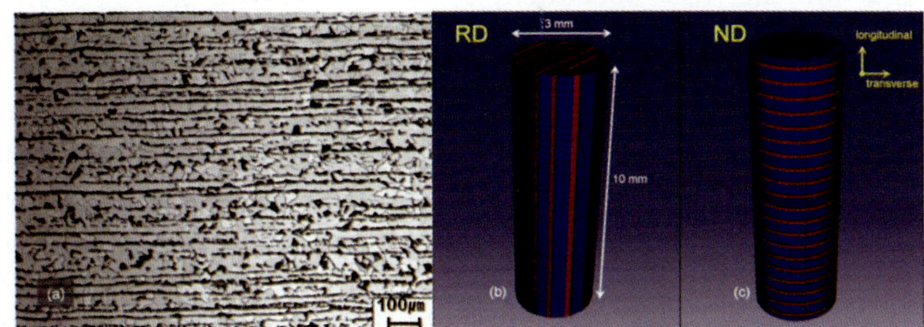

Fig. 6: (a) Microstructure of the hot-rolled steel. (b and c) Finite-element mesh for the dilatometric specimens: (b) RD specimen showing the longitudinal direction parallel to the rolling direction, and (c) ND specimen showing the longitudinal direction perpendicular to the rolling direction[7].

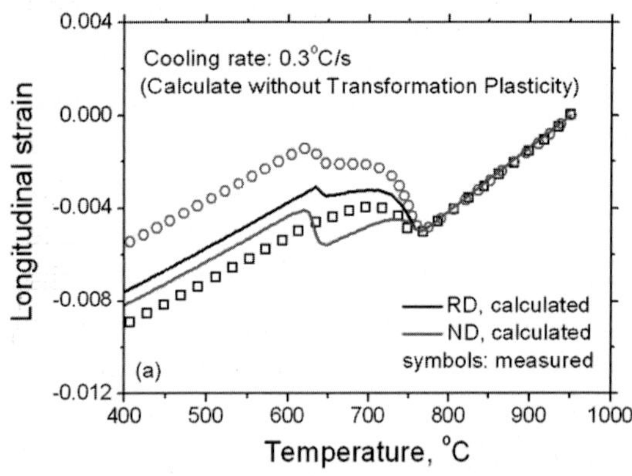

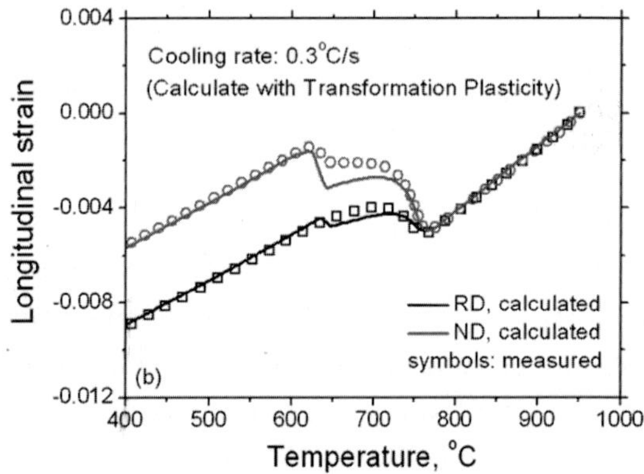

Fig. 7: compares the calculated and measured dilatation curves of the RD and ND (indicated in the figure) specimens[7].

(b) considering transformation plasticity. In the calculation not considering the transformation plasticity (Fig. 7a), the change in length of the RD specimen during cooling was larger than that of the ND specimen, which is inconsistent with the measured results, which show a smaller length change in the RD specimen. In the case of taking the transformation plasticity into account (Fig. 7b), the calculated changes in length in both directions captured the experimentally observed dimensional non-isotropy well. The results suggest that transformation plasticity plays a major role in generating the characteristic dilatometric behavior, derived from dimensional non-isotropy during transformations.

Asymmetric contraction of coil [4]

A hot-rolled steel strip is generally stocked in the form of a hollow cylindrical coil after the hot rolling process. The hot coil is cooled from 500-700 °C to room temperature over a 4-5-day period under natural air cooling conditions [49,50]. In most hot strip rolling processes, the phase transformation of the steel is finished on the run-out table (ROT) before coiling [51], and the hot coil is normally cooled down but maintained in a cylindrical shape [52]. However, asymmetric contraction occurs in an actual mill during cooling after the coiling of hot-rolled steel, which has significantly high hardenability due to its high content of carbon or other alloying elements and shows incomplete phase transformation prior to coiling, as can be seen in Fig. 8. This asymmetric contraction behavior is closely related to the phase transformation that occurs after coiling, and cannot be described by conventional creep behavior. This shape change in the hot coil causes acute problems in industrial applications, such as serious scratching on the strip surface during uncoiling. The constitutive equation for transformation plasticity was incorporated into a general purpose implicit FE program. In addition to the thermo-elasto-plastic constitutive equations, the phase transformation kinetics was characterized by a JMAK type equation.

Fig. 8: An example of asymmetric contraction of hot coil after coiling[4].

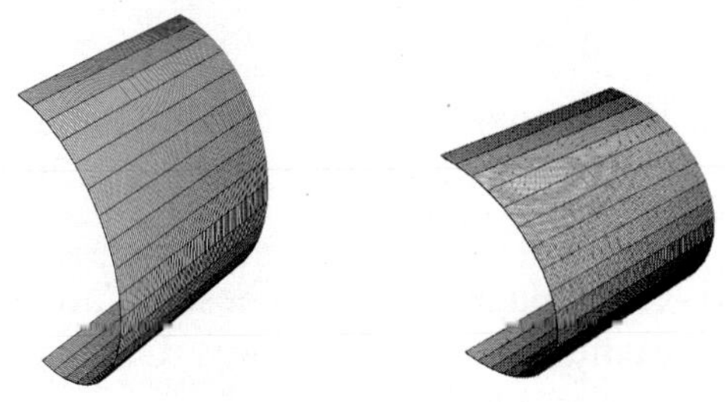

Fig. 9: Deformed shapes of the hot coil after cooling to room temperature, which are calculated by one layer FE modeling without (a) and with (b) the consideration of the transformation plasticity, respectively. Displacement was enlarged by 1.5 times for both cases[4].

The validity of the proposed model was examined by reproducing the asymmetric contraction behavior of the coil. The effect of some selected process variables on the asymmetric contraction was investigated through a series of process simulations.

Considering the transformation plasticity, the asymmetric contraction behavior of the coil during cooling could be reproduced successfully using FE simulation (Fig. 9). It was confirmed that the asymmetric contraction was caused by the small stress that develops naturally in the hot-rolled coil due to gravity. The FE simulations showed that the extent of the phase transformation before coiling, the tension force during coiling, and the steel weight per unit area of the inner layer in the coil are controllable process variables that can reduce the asymmetric contraction during cooling after the coiling of hot-rolled steel.

CHALLENGES AND FURTHER STUDIES

Although this article starts by looking at the undesirable aspects of transformation plasticity, the primary concern is not the degradation by itself, but rather its mechanism and nature. Fundamental understanding of transformation plasticity will be a key to future research. Many previously published studies have attempted to take advantage of transformation plasticity. The representative materials are TRIP aided steels [3], which have engineering importance in the automotive industry. Twinning-induced plastic deformation is also attracting considerable interest for the possibility that it might lead to remarkable improvement in both strength and ductility. However, there is still a lack of fundamental investigations, possibly due to obstacles that derive from the requirements of nano-scale engineering. Transformation plasticity can hardly be observed or measured on the nano/micro-scale. Despite recent study that have adopted the phase field model (PFM) into the FE model in order to consider micro-scaled transformation plasticity [53], the mechanism and nature of transformation plasticity are not clear. Obviously, the requirements to understanding this uncertainty will increase, and many opportunities will arise. Future works will have to be carried out as close collaborations between nano/micro-scale experiments and observations, along with numerical analysis.

REFERENCES

[1] H.N. Han, C.G. Lee, C.-S. Oh, T.-H. Lee and S.-J. Kim, Acta Mater., Vol. 52, 2004, pp. 5203 - 5214.

[2] K. Tao, H. Choo, H. Li, B. Clausen, J.-E. Jin and Y.-K. Lee, Appl. Phys. Lett., Vol. 90, 2007, p. 101911.

[3] G.B. Olson and M. Cohen, Metall. Trans. A, Vol. 13, 1982, pp. 1907 - 1914.

[4] H.-H. Cho, Y.-G. Cho, Y.-R. Im, J.K. Lee, J.-H. Kwak and H.N. Han, J. Mater. Proc. Tech., Vol. 210, 2010, pp. 907 - 913.

[5] T. Wu, M. Coret and A. Combescure, Inter. J. Solids & Structures, Vol. 45, 2008, pp. 2973 - 2989.

[6] Y.-G. Cho, Y.-R. Im, J.K. Lee, D.-W. Suh, S.-J. Kim and H.N. Han, Metall. Mater. Trans. A, Vol. 42, 2011, pp. 2094 - 2106.

[7] Y.-G. Cho, D.-W. Suh, J.K. Lee and H.N. Han, Scripta Mater., Vol. 65, 2011, pp. 569 - 572.

[8] G.W. Greenwood and R.H. Johnson, Proc. R. Soc. (Lond) A, Vol. 283, 1965, pp. 403 - 422.

[9] J.B. Leblond, G. Mottet and J.C. Devaux, J. Mech. Phys. Solids, Vol. 34, 1986, pp. 395 - 409.

[10] J.B. Leblond, Int. J. Plast., Vol. 5,1989, pp. 573 - 591.

[11] L. Taleb, N. Cavallo and F. Waeckel, Int. J. Plast., Vol. 17, 2001, pp. 1 - 20.

[12] L. Taleb and F. Sidoroff, Int J. Plast., Vol. 19, 2003, pp. 1821 - 1842.

[13] L. Taleb and S. Petit, Int. J. Plast., Vol. 22, 2006, pp. 110 - 130.

[14] F.D. Fischer and G. Reisner, Acta Mater., Vol. 46, 1998, pp. 2095 - 2102.

[15] F.D. Fischer, T. Antretter, F. Azzouz, G. Cailletaud, A. Pineau, K. Tanaka and K. Nagayama, Arch. Mech., Vol. 52, 2000, pp. 569 - 588.

[16] F.D. Fischer, G. Reisner, E. Werner and K. Tanaka, Int. J. Plast., Vol. 16, 2000, pp. 723 - 748.

[17] R. Mahnken, A. Schneidt and T. Antretter, Int. J. Plast., Vol. 25, 2009, pp. 183 - 204.

[18] C.L. Magee, Ph.D. Thesis, Carnegie Institute of Technology University, Pittsburgh, PA, USA, 1966.

[19] G.B. Olson and M. Cohen, Metall. Trans. A, Vol. 6, 1975, pp. 791 - 795.

[20] R.G. Stringfellow, D.M. Parks and G.B. Olson, Acta Metall. Mater., Vol. 40, 1992, pp. 1703 - 1716.

[21] Y. Tomita and T. Iwamoto, Int. J. Mech. Sci., Vol. 37, 1995, pp. 1295 - 1305.

[22] Y. Tomita and T. Iwamoto, Int. J. Mech. Sci., Vol. 43, 2001, pp. 2017 - 2034.

[23] Y. Tomita and Y. Shibutani, Int. J. Plast., Vol. 16, 2000, pp. 769 - 789.

[24] F. Marketz and F.D. Fischer, Metall. Mater. Trans. A, Vol. 26, 1995, pp. 267 - 278.

[25] A.V. Idesman, V.I. Levitas and E. Stein, Comput. Methods Appl. Mech., Vol. 173, 1999, pp. 71 - 98.

[26] V.I. Levitas, A.V. Idesman and D. Preston, Phys. Rev. Lett., Vol. 93, 2004, pp. 105701 - 1105704.

[27] V.I. Levitas and I.B. Ozsoy, Int. J. Plast., Vol. 25, 2009, pp. 239 - 280.

[28] V.I. Levitas and I.B. Ozsoy, Int. J. Plast., Vol. 25, 2009, pp. 546 - 583.

[29] S. Turteltaub, and A.S.J. Suiker, J. Mech. Phys. Solids, Vol. 53, 2005, pp. 1747 - 1788.

[30] V.G. Kouznetsova and M.G.D. Geers, Mech. Mater., Vol. 40, 2008, pp. 641 - 657.

[31] P.J. Jacques, Q. Furnemont, F. Lani, T. Pardoen and F. Delannay, Acta Mater., Vol. 55, 2007, pp. 3681 - 3693.

[32] V.I. Levitas and D.-W. Lee, Phys. Rev. Lett., Vol. 99, 2007, pp. 245701-245704.

[33] Y.-G. Cho, H. N. Han, J. Kor. Ceram. Soc., Vol. 49, 2012, pp.1 – 10.

[34] H.N. Han, J.K. Lee, D.-W. Suh and S.-J. Kim, Phil. Mag., Vol. 87, 2007, pp. 159 - 176.

[35] K. Kitazono, E. Sato and K. Kuribayashi, Acta Mater., Vol. 47, 1999, pp. 1653 - 1660.

[36] C. Schuh and D.C. Dunand, Acta Mater., Vol. 49, 2001, pp. 199 - 210.

[37] H.N. Han, K.J. Lee and S.-J. Kim, Mater. Lett., Vol. 59, 2005, pp. 158 - 161.

[38] J.-H. Kang, D.-W. Suh, J.-Y. Cho, K.H. Oh and H.-C. Lee, Scripta Mater., Vol. 48, 2003, pp. 91 - 95.

[39] D.-W. Suh, S.-J. Kim and H.N. Han, CAMP-ISIJ, Vol. 18, 2005, pp. 1331 - 1334.

[40] R.L. Coble, J. Appl. Phys., Vol. 32, 1961, pp. 793 - 799.

[41] P. Zwigl and D.C. Dunand, Acta Mater., Vol. 45,1997, pp. 5285 - 5294.

[42] H.N. Han and D.-W. Suh, Acta Mater., Vol. 51, 2003, pp. 4907 - 4917.

[43] D.P. Koistinen, R.E. Marburger, Acta Metall., Vol. 7, 1959, pp. 59 - 60.

[44] D.-W. Suh, C.-S. Oh, H.N. Han and S.-J. Kim, Acta Mater., Vol. 55, 2007, pp. 2659 - 2669.

[45] T.A. Kop, J. Sietsma and S. Van der Zwaag, J Mater. Sci., Vol. 36, 2001, pp. 519 - 526.

[46] R.A. Jaramillo, M.T. Lusk and M.C. Mataya, Acta Mater., Vol. 52, 2004, pp. 851 - 858.

[47] R.A. Jaramillo and M.T. Lusk, Acta Mater., Vol. 52, 2004, pp. 859 - 867.

[48] D.-W. Suh, C.-S. Oh, H.N. Han and S.-J. Kim, Metall. Mater. Trans. A, Vol. 38, 2007, pp. 2963 - 2973.

[49] S.-J. Park, B.-H. Hong, S.C. Baik and K.H. Oh, ISIJ Int., Vol. 38, 1998, pp. 1262 - 1269.

[50] A. Saboonchi and S. Hassanpour, J. Mater. Process. Technol., Vol. 182, 2007, pp. 101 - 106.

[51] H.N. Han, J.K. Lee, H.J. Kim and Y.-S. Jin, J. Mater. Process. Technol., Vol. 128, 2002, pp. 216 - 225.

[52] C.M. Park, J.T. Choi, H.K. Moon and G.J. Park, J. Mater. Process. Technol., Vol. 209, 2009, pp. 3714 - 3723.

[53] Y.-G. Cho, J.-Y. Kim, P.-R. Cha, D.-W. Suh, J.K. Lee and H.N. Han, *PLoS ONE*, Vol. 7, 2012, p. e35987.

Influence of Boron on the Hardenability of Unalloyed and Low Alloyed Steel

Anjana Deva, Saikat K. De, Vinod Kumar, M. Deepa and B.K. Jha
R&D Centre for Iron and Steel, Steel Authority of India Limited, Ranchi-834002, India
Email: anjana@sail-rdcis.com

ABSTRACT

Boron exerts a large influence on mechanical properties of steel through microstructural control. It increases the hardenability of steel by retarding the heterogeneous nucleation of ferrite at the austenite grain surfaces and the decomposition kinetics of austenite to ferrite transformation are governed by its location and chemical state. To understand the effect of boron (~ 25 ppm) on microstructural evolution and change in continuous cooling transformation (CCT) diagram , a systematic study has been carried out using Gleeble thermo-mechanical simulator in unalloyed (C: 0.05 wt%; Mn: 0.2 wt%) and low alloyed (C: 0.2 wt%; Mn: 1.2 wt%: Cr: 0.15 wt%) steels. CCT diagrams, plotted for the onset and end of pearlitic, bainitic and martensitic reactions, consisted essentially of two C-curves and a remarkable difference was observed on comparing the results for both steels. It is interesting to note that bainite and martensite are completely absent even at higher cooling rate of 70 °C/sec in the unalloyed steel with boron addition. In contrast, the addition of boron was observed to promote significant bainite formation even at a slower cooling rate of only 20 °C/sec in the low alloyed steel. These contradicting effects on hardenability can be explained by the effect of boron in shifting only the upper C curve to the right for reconstructive transformation.

KEYWORDS: *Boron, Cooling Rate, CCT, Unalloyed, Low Alloyed, Hardenability*

INTRODUCTION

The increase in grain boundary area per unit volume of austenite provides an increase in the number of potential nucleation sites for the subsequent austenite to ferrite transformation, which accelerates the transformation rate [1]. For the same cooling condition, a bigger austenite grain size results in decrease in the Ar3 temperature leading to decrease in the volume percentage of ferrite fraction, and the effect is manifested in terms of increased hardenability. It is well established that boron increases hardenability of steel by retarding the heterogeneous nucleation of ferrite at the austenite grain surfaces [2-4]. It is probable that this effect is due to the reduction in interfacial energy as boron segregates to the grain boundaries. This in turn makes grain boundary less effective as heterogeneous sites. However, boron effect is entirely different in low and high carbon steels, plain and alloyed steels, with low and high soaking temperatures, and more significantly with low and high cooling rates.

Addition of boron increases the hardenability of quenched and tempered steels. However, the extent of hardening depends on transformation characteristics of austenite to ferrite and the presence of other alloying elements. A systematic study through thermo-mechanical simulator has been carried out in unalloyed and low alloyed steel to understand the boron effect on microstructural evolution and thereby change in continuous cooling transformation (CCT) diagram. Role of boron as a hardener under different soaking temperatures and cooling rates has also been examined.

EXPERIMENTAL

Chemical composition of aluminium killed boron added low C-Mn steel used in the present study is shown in Table 1.

Table 1: Chemistry (wt. %) of unalloyed and low alloyed steels

Steel	C	Mn	P	S	Si	Al	Ti	Cr	B
Unalloyed	0.05	0.20	0.012	0.025	0.028	0.040	-	-	25 ppm
Low Alloyed	0.20	1.25	0.015	0.008	0.28	0.044	0.023	0.15	25 ppm

Suitable test specimens were prepared from the crop ends of 32 mm transfer bar plates, collected after rough rolling of slabs. Specially designed tests were performed on the prepared specimens using Gleeble-3500C with fully integrated digital closed loop control thermal and mechanical system for understanding of processing conditions of hot rolling. Dilation tests aimed at determination of phase transformation temperatures. Easy to use Windows based computer software, combined with an array of powerful processors, provided an extremely user-friendly interface to create, run and analyze data from thermal-mechanical tests and physical simulation programs. Standards metallographic procedures were followed for material characterization which included optical microscopy. Scanning / transmission electron microscopy was used for characterizing low temperature phases and precipitates.

Dilatometric Study

The austenite to ferrite transformation is accompanied by increase in atomic volume of the transformed-phase; resulting dilation in the test specimen is linked with the kinetics of transformation [5,6]. A number of dilation plots were generated by exposing prepared tubular specimens (85 mm long with 10 mm diameter) of unalloyed steel at different austenitising temperatures followed by cooling them through varied rates as shown schematically in Fig.1. While austenitising, specimens were heated at the rate of 3 °C/sec and soaked in the range of 950 to 1200 °C for 120 seconds to facilitate homogenization of chemistry and microstructure. These specimens were then cooled at different rates, ranging from 1 to 70 °C/sec. The change in diameter of these specimens during cooling was recorded using an ultra-sensitive dilatometer provided with the gleeble 3500 C simulator and the output of the dilatometer i.e. "dilatation plot" was used in identifying start and finish temperatures of transformed phase.

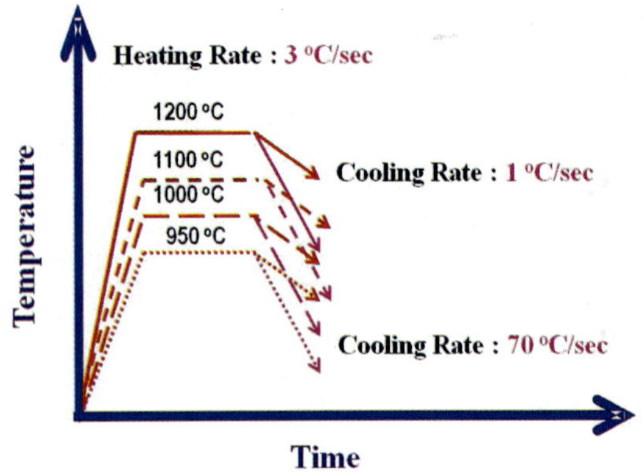

Fig. 1: Schematic depiction of thermal cycles of dilatation study (unalloyed steel)

The steel samples of low alloyed steel were first soaked Ac3 temperatures (900oC) in the simulator for 120s, presuming that this will be adequate for chemical and structural homogenization. They were subsequently cooled to room temperature by employing cooling rates ranging from 0.2oC/sec to 36oC/sec.

RESULTS AND DISCUSSION

Fig. 2 show a typical continuous cooling transformation (CCT) diagram of unalloyed boron added steel, constructed based on data generated on time taken for start to finish transformation for different cooling rates. It can be noted that with the increase in cooling rate, the onset of ferrite transformation shifted to a lower temperature.

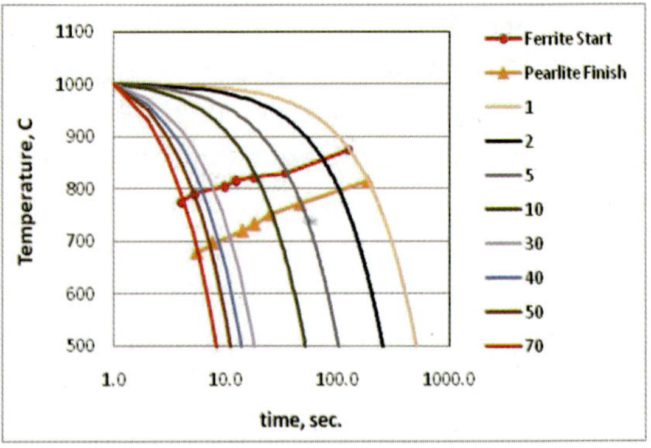

Fig. 2: Continuous cooling transformation diagram of unalloyed steel

Sample heated to 1100 °C and cooled with a rate of 70 °C/sec was subjected to transmission electron microscopy. Figs. 3 (a-d) show the bright and dark field images alongwith selected area diffraction pattern and the indexed key. It may be noted that there is no evidence of martenisitic laths.

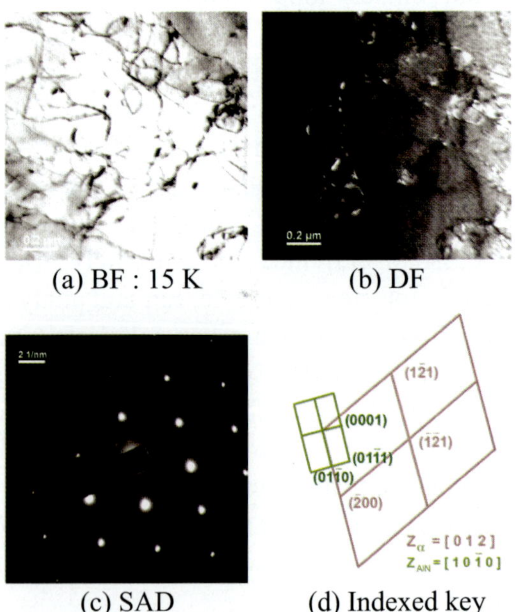

(a) BF : 15 K (b) DF

(c) SAD (d) Indexed key

Fig. 3: Transmission Electron Micrographs depicting AlN precipitates

Dislocations entangled with the precipitates can be clearly seen here (Fig. 4a). From Fig. 4d, which shows the key of SAD, it is clear that the precipitates observed in dark field image shown in Fig. 4b are those of AlN with Z = [1 0 $\bar{1}$ 0]. Ferrite diffation pattern pertains to Z = [01 2].

To observe the microstructural evolution with much higher cooling rate, steel sample was heated to austenitisising temperature of 1200 °C and subjected to very high cooling rate (~230 °C/sec) in thermo-mechanical simulator and found that martensitic laths are present in steel (Fig. 4).

It clearly indicates that there must be a critical cooling rate, above which bainite or martensitic transformation starts in low carbon unalloyed steel also which will depend on soluble boron content, austenitising temperature and cooling rate.

It is interesting to note that the Ar3 temperatures of both the steels, when subjected to varied cooling rates, showed almost similar trend with the increase in cooling rate.

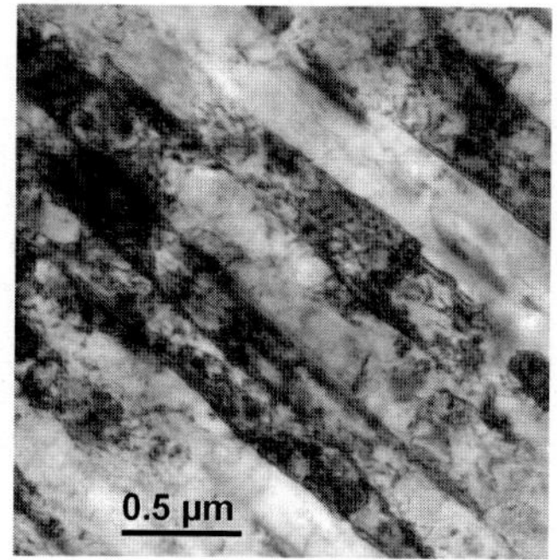

Fig. 4: Bright field image (8K) showing martensitic laths at higher cooling rate

This observation authenticates that austenitisation at low temperature (950 °C) causes boron to precipitate as boron nitride leading to no dissolved boron in lattice. As a result, there is no contribution to hardenability of steel by boron and there is rather some amount of softening, owing to removal of free nitrogen through scavenging effect. Contrary to this phenomenon, a different kind of situation emergesd when boron treated steel was austenitised at higher temperatures (up to 1200 °C) followed by cooling at varied rates (Fig. 5).

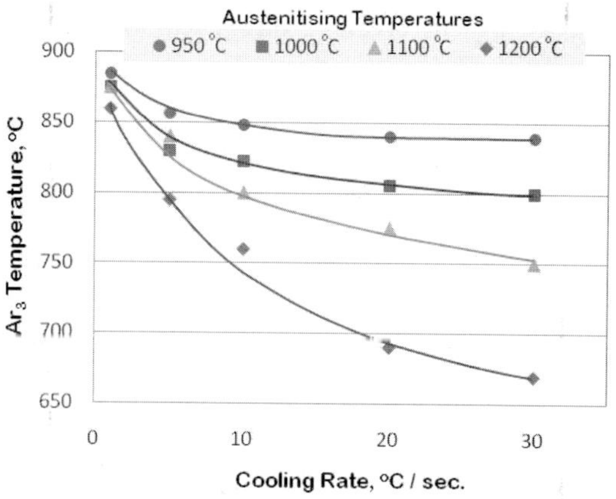

Fig. 5: Variation in Ar3 temperature of B-added steel with cooling rate for different austenitising temperatures

The results show that increase in cooling rate leads to significant decrease in Ar3 temperature at higher austenitising temperatures (> 950 °C). This effect is particularly greater at 1200 °C.

Fig. 6 show a typical continuous cooling transformation (CCT) diagram of low alloyed boron added steel. Against each cooling rate curve Ar3, Ps, and Pf corresponding to proeutectoid ferrite start, pearlite start and pearlite finish temperatures respectively have been marked.

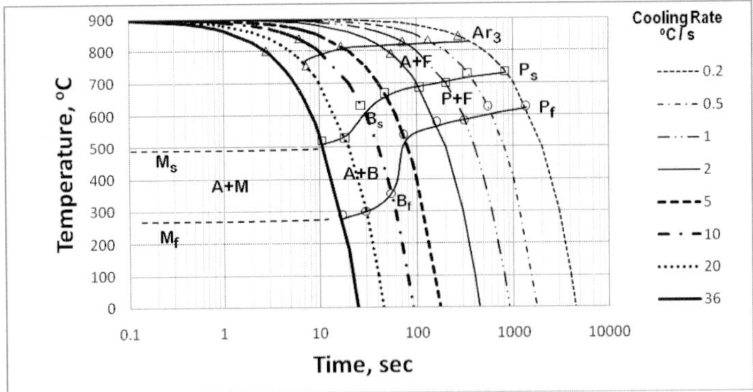

Fig. 6: Continuous cooling transformation diagram of low alloyed steel

It can be inferred from Fig. 6 that at all the curves associated with different transformation temperatures have diminishing tendencies as the cooling rates are increased. However, decrease is gradual upto cooling rate of 5°C/s. The proclivity towards decline becomes quite sharp at and above cooling rates of 10°C/s, which continues till it again levels off at cooling rates of 20°C/s and beyond. The CCT diagram can be interpreted in terms of first formation of proeutectoid ferrite from austenite labeled Ar3 line in the figure. There is not much variations in this temperature with cooling rates. The region between Ar3 line and the Ps line gets widened upon increase in cooling rates. The lines representing Ps and Pf remain somewhat parallel upto cooling rate of 5oC/s. Marked change in width of separation occurs after this cooling rate and area get widened progressively as it approaches towards higher cooling rates. Therefore, cooling rate of 5oC/s is critical and seems to signal departure from formation of mere pearlite and ferrite phase mixture. This may also be the cooling rate above which bainite may form.

Fig. 7 show the scanning micrographs from samples cooled at different rates as indicated in the figure. The microstructures of Fig. 7 (a) and (b) from samples cooled at slower rates of 0.5°C/s and 2°C/s is completely dominated by ferrite and pearlite phase mixture.

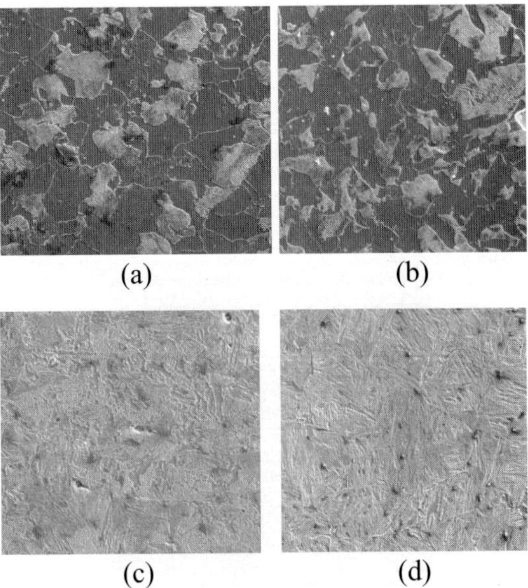

Fig. 7: Scanning electron micrograph from samples (a) 0.2°C/s (b) 1°C/s (c) 10°C/s (d) 20°C/s

However in the case of latter coarse peralite colony size is slightly bigger. The formation of proeutectoid ferrite has occurred along the prior austenite grain boundary. From this an estimate of average prior austenite grain size can be made, which works out to be ~130μm. The first indication of formation of bainite can be perceived in Fig. 7 (c). The bainite volume fraction has increased as can be discerned from Fig. 7 (d) obtained from sample cooled at the rate of 20°C/s. The above microscopic examinations corroborates the surmises gained from CCT diagram.

TTT diagrams are plotted for the onset and completion of reactions to form pearlite, bainite and martensite, consist essentially two C Curves, the one at higher temperature representing reconstructive transformation to ferrite or pearlite, and that at lower temperature where atomic mobility is diminished, representing displacive transformation such as Widmanstatten ferrite, bainite and acicular ferrite (Fig. 8).

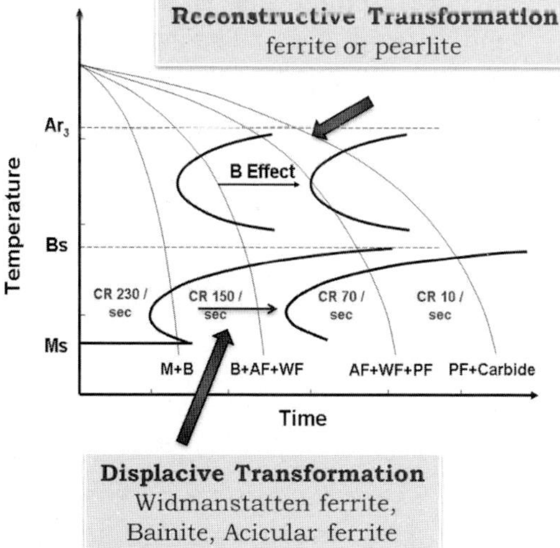

Fig. 8: Schematic representation of boron on pronounced effect of upper C curve

Boron has the largest effect on the C-curve for reconstructive transformation, leaving the lower C-curve hardly changed. In Fig. 5, the effect of cooling rate on the evolution of microstructure has been shown schematically which further collaborates the pronounced effect of boron on upper C curve only.

The increase in incubation period before reconstructive transformation, as a function of soluble boron concentration has been investigated quantitatively by Pickering [7], and it was found, that there is proportional increase in the incubation period until a concentration of about 20 ppm is reached, after which the function becomes insensitive to boron and overall effect becomes unreliable. In contrast, for medium carbon (0.4 wt.%) high manganese (1.5 wt.%) steel, addition of boron (40 ppm) was observed to enable some bainite and martensite formation at a cooling rate of only 0.6 °C/sec [8]. Even slow cooling rate, as low as < 1 °C/sec has suppressed the formation of pearlite. In this case, the dramatic effect on hardenability can be attributed to shifting upper C curve to right by boron and shifting both upper and lower C curves to right by C and Mn.

CONCLUSION

- Both austenitising temperature and cooling rate are critical in influencing the hardenability of boron added steel.
- In unalloyed boron containing steel, no bainite/ martensite was formed even at cooling rate of 70°C/sec, whereas formation of theses phases

occurred at much lower cooling rate of 20°C/sec in low alloyed boron added steel.

- It is not boron alone but its synergistic effect in presence of carbon, manganese, chromium has resulted in significant change in hardening behaviour of steel.

ACKNOWLEDGEMENT

The authors thankfully acknowledge the guidance and support extended by the management of Research & Development Centre for Iron and Steel during the course of this work. Contribution by Ms Smriti Ojha, National Institute of Foundry and Forge Technology, Hatia, Ranchi during carrying out experiment is also acknowledged with thanks.

REFERENCES

[1] Ohmori Y. and Yamanaka K., Conference proceeding on Boron in Steel, edited by Banerjee S. K. and Morral J. E., The Metallurgical Society of AIME, Warrendale, PA, 1979, p.44.

[2] Shoenberger L. R., Trans AIME, 1958, Vol.212, p.402

[3] Cho Y.R. and Kim S.I., Iron and Steel Technology, May 2004, p.46.

[4] Tsuji, N., Matsubara, Y., Sakai, T., and Saito, Y., ISIJ International, Vol. 37, No 8, (1997), p.797.

[5] Grange R.A. and Mickel J.B.,Trans ASM, 1961, Vol. 53, p.157.

[6] Pandi, R., Militzer, M., Hawbolt, E.B. and Meadowcroft, T.R., 37th Mechanical Working and Steel Processing, ISS, (1996), p.635.

[7] Pickering F.B., Physical Metallurgy and the design of steel, Applied Science Publisher, London, 1978, p.50.

[8] Ohmori Y. and Yamanaka K., Conference proceeding on Boron in Steel, edited by Banerjee S. K. and Morral J. E., The Metallurgical Society of AIME, Warrendale, PA, 1979, p.44.

Refinement of Ferrite Grain Size Near to the Ultra-Fine Range by Single-Pass and Multi-Pass Thermo-Mechanical Compression

D. Chakrabarti[1], S. Patra[1], A. Haldar[2] and Vinod Kumar[3]
[1]Department of Metallurgical and Materials Engineering, Indian Institute of Technology (I.I.T.), Kharagpur, West Bengal 721 302, India.
[2]Steel Metallurgy Division, Swinden Technology Centre, Tata Steel Europe, Moorgate Rotherham, UK.
[3]R&D Centre for Iron and Steel, RDCIS, SAIL, Ranchi.
Email: debalay@metal.iitkgp.ernet.in

ABSTRACT

Single-pass and multi-pass deformation schedules have been applied to low-C microalloyed steel containing Nb, Ti and V. Heavy deformation (~80%) in a single pass, just above or below the austenite-to-ferrite transformation temperature (A_{r3}), followed by slow-cooling (@1°C/s) of the samples to room temperature, resulted in the formation of ultra-fine ferrite grain structures with average grain size of ~2 to 3 μm. Ferrite grain size variation observed in the ultra-fine grained steel can be explained in view of several metallurgical phenomena such as, static and dynamic strain-induced austenite (γ) → ferrite (α) transformation, dynamic recovery, dynamic recrystallisation, and grain-growth of ferrite. Multi-pass deformation schedule can also refine the ferrite-grain size close to the ultra-fine range (~3 μm) provided more than 50% reduction is applied in the finishing stage, between A_{e3} and A_{r3} temperatures.

KEYWORDS: *Ultra-Fine Grain, Single-Pass, Multi-Pass Deformation, Plane-Strain Compression, Ferrite Grain Size, Grain Size Variation, Micro-Texture.*

INTRODUCTION

Beneficial effect of ferrite grain refinement on improving the strength and toughness of low-carbon steel is well established and that was the drive behind the industrial development of thermo-mechanical controlled processing (TMCP) of microalloyed steels, containing Nb, Ti and V [**1**]. Application of advanced TMCP refined the ferrite grain size down to the ultra-fine range (≤ 3 μm) [**2-4**]. The effects of prior austenite (γ) grain size, deformation temperature (T_{def}), true strain (ε), strain rate ($\dot{\varepsilon}$), and cooling rate (CR) on the formation of ultra-fine ferrite (UFF) grains, in low-carbon steels, have been studied extensively [**2-9**]. Use of finer γ-grain sizes (14-20 μm), large strain deformation ($\varepsilon > 0.5$) just above the A_{r3} temperature (750-850°C) at a high strain rate ($\dot{\varepsilon} > 0.1$/s), followed by rapid cooling (CR > 30ºC/s) of the samples can achieve homogeneous distribution of UFF grains through dynamic strain induced austenite (γ) → ferrite (α) transformation, DSIT [**2-9**]. In an industrial rolling mill, the application of single and heavy deformation pass at a low temperature, with high $\dot{\varepsilon}$ and fast CR, is however, difficult due to the requirement of excessive rolling load (resulted from the high material flow-stress), precise temperature control, and water-cooling unit [**2**]. Therefore, single-pass, as well as multi-pass deformation schedules has been performed in the present study, keeping the total deformation same, and the samples were cooled down at a slow-rate (1°C/s). The average ferrite grain size, grain size distribution, nature and distribution of the second phase and texture developed in both the schedules have been compared. Fine-scale grain size variation observed in the ultra-fine grain structures has also been studied as this aspect is yet to be understood fully.

EXPERIMENTAL DETAILS

Plane-strain compression testing has been carried out on the samples (10×15×20 mm rectangular blocks) of a Nb-Ti-V microalloyed steel, Table 1, using the Gleeble®3500® simulator. Single-pass and multi-pass deformation schedules, having different amount of roughing, intermediate and finishing deformation passes have been applied at different temperatures, varying between 1050°C and 650°C. Schematic of a typical single pass deformation schedule is shown in **Fig. 1,** where the deformed samples were soaked at 1200°C, cooled at 1°C/s to the deformation temperature and either water-quenched or continued to cool down at a slow-rate (1°C/s) after the compression. Total amount of reduction (~80%) remained the same. Samples have been prepared for the metallographic studies and investigated by optical microscope, scanning electron microscope (EBSD), transmission electron microscope (TEM) and electron backscattered diffraction analysis (EBSD).

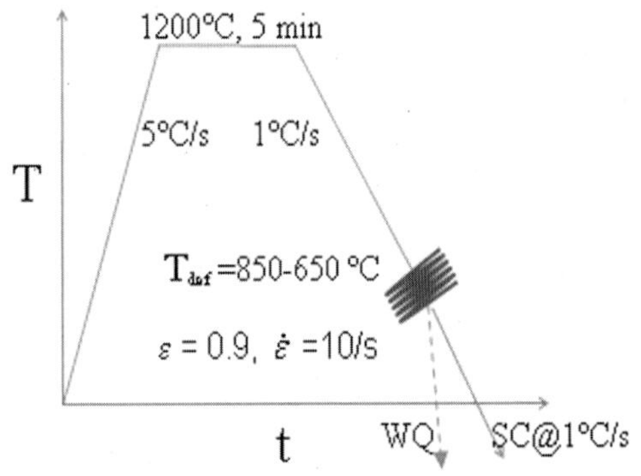

Fig. 1: A single-pass, thermo-mechanical processing schedule performed in Gleeble®3500.

Table 1: Chemical composition of the industrial slab (in wt %):

C	Si	Mn	P	S	Al	Nb	Ti	V	N
0.09	0.33	1.42	0.010	0.003	0.035	0.050	0.019	0.050	0.007

RESULTS AND DISCUSSION

Applying heavy deformation in a single pass, just above the austenite-to-ferrite transformation temperature (A_{r3}), followed by slow-cooling (@1°C/s) to the room temperature resulted in the formation of UFF grain structures, **Fig. 1**, with average α-grain size between 2 and 3 μm and the largest grain sizes extending up to ~10 to 12 μm. Water quenching just after deformation prevented the coarsening of UFF grains, offered finer average grain size (<2 μm), and restricted the largest grain sizes to under 6 μm [**10**]. Although the ferrite grain structures appeared homogeneous in slowly cooled samples, careful observation revealed the presence of alternate bands of coarse- (5 to 10 μm) and fine-α grains (<1 to 3 μm) [**10**].

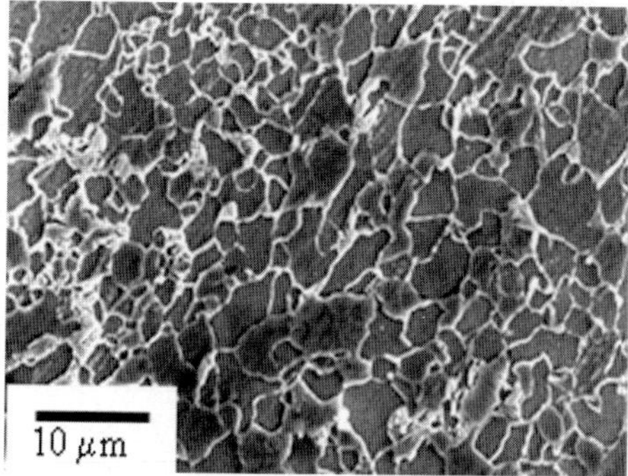

Fig. 2: SEM micrograph of the sample deformed at 800°C showing the ultra-fine ferrite grain structure.

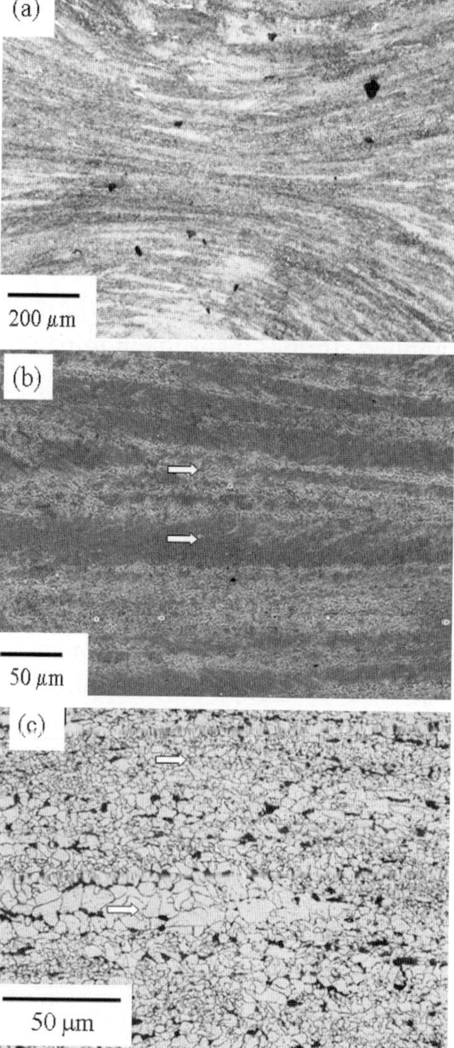

Fig. 3: (a and b) Deformed region of the compression tested sample showing the formation of UFF grains along the prior-γ grain boundaries; (c) After slow cooling, prior-γ grain centre turned into lens-shaped coarse-α grain regions, which were surrounded by the UFF grains formed along the γ-grain-boundary regions. The prior-γ grain-boundary and grain-centre regions are indicated by arrows.

The final α-grain size distributions can be explained in view of the starting γ-grain size variation, static and dynamic strain-induced γ-to-α transformation (DSIT), dynamic recovery, recrystallisation (DRX), and grain growth of α during slow cooling. For coarse γ-grains, higher strain accumulation near the γ-grain boundaries resulted in DSIT at those regions at a lower applied strain than at the γ-grain centre regions, **Fig. 3(a, b)**. Three-dimensional impingement (i.e. impingement from all sides) of the α-grains nucleated at the γ-grain-boundaries prevented their growth and restricted those grains to below ~2 μm, even in case of slow cooling after deformation [**9**]. Upon further straining, at the later stage of deformation, DSIT can start on intra-granular defects at the γ-grain interior. Lack of impingement between those α-grains can lead to grain-growth and the formation of closely spaced, parallel planar arrays (known as "rafts") at the γ-grain centre regions. Due to the two-dimensional impingement, the grains in the raft can grow along the direction of the raft. Slow cooling of the samples subsequent to deformation, promotes the formation of carbide banding (perpendicular to the direction of compression) utilizing the carbon rejected during transformation [**9, 10**]. Carbide bands can also restrict the lateral growth of α-grains and allows the α-grains to grow along the direction of the raft. If the applied strain is insufficient to penetrate DSIT within the γ-grain interior, γ-grain centre regions may also transform statically, forming coarser α-grains, surrounded by UFF grains developed at the γ-grain boundary regions. Statically transformed coarse α-region is indicated in **Fig. 3c** and the ferrite-raft is indicated in **Fig. 4**. The distribution of α-grain sizes becomes more uniform with the increase in applied strain and the decrease in prior γ-grain size, which resulted in the occurrence of DSIT in a more uniform fashion and distribute the carbide particles more uniformly throughout the structure [**10**]. Uniform distribution of carbide particles is beneficial for the grain

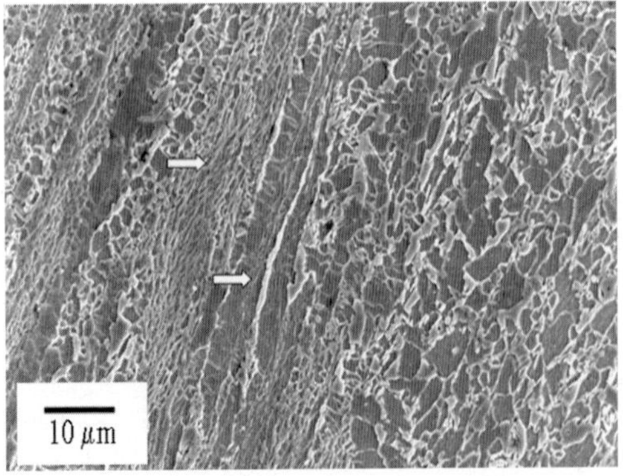

Fig. 4: Ferrite grain size variation observed in the inter-critically deformed (at 700°C) and slowly cooled sample. Coarse-grain raft and fine-grain region is arrowed and the arrows are along the direction of compression of the sample.

size uniformity, as carbides prevent the α-grain growth by pinning down the grain boundaries [**4-10**].

Grain size variation may also arise after inter-critical deformation due to the recovery / recrytsllisation of the deformed, pro-eutectoid ferrite and the dynamic / static transformation of the austenite. As ferrite and austenite co-deform, being the softer phase α is deformed preferentially. Due to its high stacking fault energy ferrite has a tendency to recover dynamically, which promotes further strain accumulation in ferrite. Strain partitioning between γ and α can influence the dynamic softening behaviour of those phases and the resultant grain size variation. Microalloy precipitates also play a vital role in this aspect by retarding the dynamic recrystallisation of γ and α, which alters the strain-partitioning and the resultant, dynamic softening.

Heavy finishing deformation (true strain ~ 0.8-1.0) between A_{e3} and A_{r3} using multiple-passes (for 10 s inter-pass time) effectively refined the ferrite (α) grain sizes (4.1-3.2 μm) nearly to the ultra-fine range [**11**]. Increasing the finishing deformation and the decrease in the number of passes and inter-pass time were found to be beneficial for the ferrite grain refinement. Besides grain refinement, heavy finishing deformation reduced the intensity of microstructural banding, pearlite fraction in the microstructure, and the fraction of harmful 'cube' texture component (to below 5%). Simultaneously, the fraction of small Fe_3C particles (50-600 nm size), high-angle (>15° misorientation) boundaries (75-80%), beneficial γ-fibre ((ND//<1 1 1>) components and {332}<113> and {554}<225> texture components (40-45%)

and the hardness of the deformed samples (184-192 HV) have been increased [**11**].

CONCLUSIONS

Major conclusions derived from the present investigation are summarized below:

- Single pass or multi-pass deformation refined the ferrite grain size nearly to the ultra-fine range (~3 μm or below), when heavy deformation (true strain, $\varepsilon > 0.8$) has been applied between A_{e3} and A_{r1} temperatures.
- Fine-scale ferrite grain size variation remains even in the ultra-fine grain structures due to the dynamic $\gamma \rightarrow \alpha$ transformation (DSIT) preferentially along the γ grain boundaries, whilst, γ-grain centre regions transform statically.
- Different rate of α-nucleation and α-grain growth between γ-grain boundary (high nucleation and stronger impingement) and γ-grain centre may also create a grain size variation after DSIT.
- Inter-critical deformation can also create ferrite grain size variation due to the extended dynamic recovery of ferrite grains and static transformation of austenite.

ACKNOWLEDGEMENTS

Authors acknowledge Department of Science and Technology, New Delhi and Tata Steel, Jamshedpur, for the research funding and I.I.T. Kharagpur, Tata Steel and RDCIS, SAIL for the provision of research facilities.

REFERENCES

[1] T. Gladman: 'The Physical Metallurgy of Microalloyed Steels'; Book 615, The Institute of Materials, London, 1997, pp. 80-115.

[2] R. Song, D. Ponge, D. Raabe, J.G. Speer, and D.K. Matlock: Mat. Sci. and Engg. A, 2006, 441, pp. 1–17.

[3] R. Priestner and A. K. Ibraheem: Mater. Sci. Technol., 2000, vol. 16, pp. 1267-1272.

[4] P. D. Hodgson, M.R. Hickson, and R.K. Gibbs: Scripta Mater., 1999, vol. 40, pp. 1179-1184.

[5] H. Beladi, G.L. Kelly, and P.D. Hodgson: Metall. Mater. Trans. A, 2007, vol. 38A, pp. 450–63.

[6] K. Mukherjee, S.S. Hazra, and M. Militzer: Metall. Mater. Trans. A, 2009, vol. 40A, pp. 2145–59.

[7] P.R. Rios, I. de, S. Bott, D.B. Santos, T.M.F. de Melo, and J.L. Ferreira: Mater. Sci. Technol., 2007, vol. 23 (4), pp. 417–22.

[8] B. Eghbali and A. Abdollah-zadeh: Scripta Mater., 2006, vol. 54, pp. 1205–09.

[9] H. Beladi, G.L. Kelly, A. Shokouhi, and P.D. Hodgson: Mater. Sci. Eng. A, 2004, vol. 371, pp. 343–52.

[10] S. Patra, S. Roy, Vinod Kumar, A. Haldar, and D. Chakrabarti: Metall. Mater. Trans. A, 2011, vol. 42, 2575-2590.

[11] S. Patra, S. Neogi, Vinod Kumar, D. Chakrabarti, A. Haldar: Accepted in Metall. Mater. Trans. A, 2012.

An Empirical Prediction Model of the Incremental Strengthening of Ferrite/Pearlite Steels with Additions of Vanadium and Nitrogen, with Emphasis on the Effective Nitrogen Level

Robert J. Glodowski
Evraz - East Metals (NA)
Pittsburgh, PA, USA
Glodowski@na.eastmetals.biz

ABSTRACT

Several prediction models of V-N strengthening have been published and used over the course of many years of vanadium microalloying experience. The basis and origin of a proven strength model previously distributed is reviewed and explained. Inherent in the prediction accuracy of the model is the assumption of the amount of nitrogen actually available for VN precipitation. This paper discusses various situations where the amount of nitrogen available can be significantly altered by the presence of other alloy elements, particularly Ti, Nb, and Al. Of these, the effect of the Al content (and prior processing history) seem to be most widely overlooked when evaluating and predicting the expected strengthening of the VN precipitate in polygonal ferrite. The possible interactions of Al and N that can affect nitrogen availability are reviewed, and the unintended effects of specification Al requirements are explained.

KEYWORDS: *Steel, Vanadium, Effective Nitrogen, Precipitation, Strengthening Model*

INTRODUCTION

Since the advent of microalloyed steels it has been recognized that vanadium precipitation strengthening is enhanced by the presence of nitrogen. Perhaps this was most widely announced in several papers included in the now classic "Proceedings of an International Symposium on High-Strength, Low-Alloy Steels", perhaps better known as MA '75. [1] However, at that time many of the products were heat treated, either normalized or quenched and tempered. As a result, conclusions about the strengthening factors for vanadium and nitrogen tended to vary significantly. The relative effects of austenitizing temperature, nitrogen and aluminum contents that control the availability of V and N for precipitation strengthening during subsequent transformation were generally unrecognized.

For the as-rolled situation, the need to control austenite recrystallization and growth during the rolling process was just being recognized. While the advantages of control rolling with higher recrystallization stop temperatures were just being exploited, the benefits of recrystallization controlled rolling were not promoted to any degree. Grozier [2] presented an extensive review of results from the production of VAN-80, vanadium-nitrogen bearing hot rolled strip steel produced with controlled cooling first introduced in 1968. The graphical model used by Grozier to describe the effect of vanadium and nitrogen content on incremental yield strength was first published by Korchynsky and Stuart [3] and is reproduced again here as Figure 1. This information, confirmed by numerous researchers, has been widely referenced. Also widely used from Grozier's work is Figure 2, showing the effect of coiling temperature on the precipitation strengthening of 0.12% vanadium steel at increasing nitrogen levels.

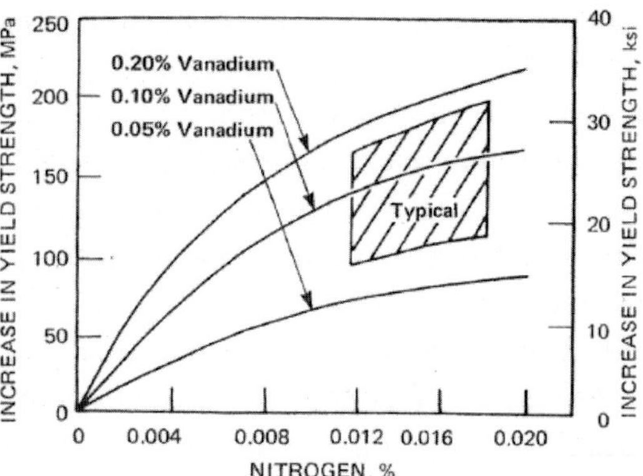

Fig. 1: Increase in yield strength from nitrogen and vanadium as a result of the precipitation of vanadium nitride. Control-cooled coil product, with a coiling temperature of 593 C. [3]

One problem with Figure 1 is that it predicts no strengthening of vanadium at zero nitrogen. This is not likely to be true except in the most extreme cases, as there should be some VC precipitation strengthening. Also, the slope of strengthening with increasing nitrogen increases with higher V contents. At sub-stoichiometric levels of nitrogen, the amount of VN precipitation is controlled by the nitrogen available, not the amount of V. Because of this, N strengthening slopes at different V levels are more likely to be parallel, but starting at different strengthening levels due to possible VC precipitation. The VC precipitation strengthening would logically be assumed to be relatively constant for constant processing conditions. Only the nitrogen level is changing. Thus, the strengthening curves at sub-stoichiometric levels of N would be expected to be slightly displaced by higher V contents, but parallel up to stoichiometric VN levels. As the nitrogen level exceeds stoichiometric V:N, the strengthening effect will be significantly reduced.

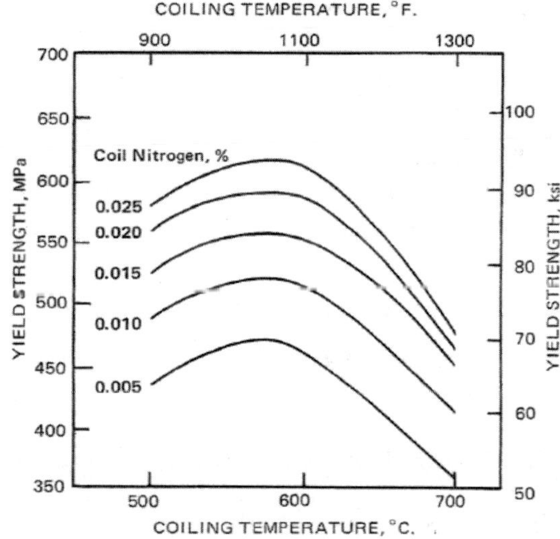

Fig. 2: Effect of coiling temperature and total nitrogen content on the yield strength of 6.3 mm strip with a FRT of 900 C and cooled at 17 C/sec. 0.13% C, 1.40% Mn, 0.5% Si, 0.12% V. [2]

Perhaps a good demonstration of this effect was published by Siwecki et.al. in 1981. [4] Figure 3 shows the contribution of V-N precipitation strengthening to a 0.12% C, 1.35% Mn, 0.09% V steel produced by either control rolling (CR) or recrystallization controlled rolling (RCR). For all processing conditions, the contribution of additional nitrogen was linear, consistently at 7 MPa for every additional10 ppm of nitrogen. For all nitrogen levels, the nitrogen is sub-stoichiometric to the vanadium level, so the linear nitrogen response is expected and consistent with the logic that the amount of addition VN precipitation is controlled by the additional N.

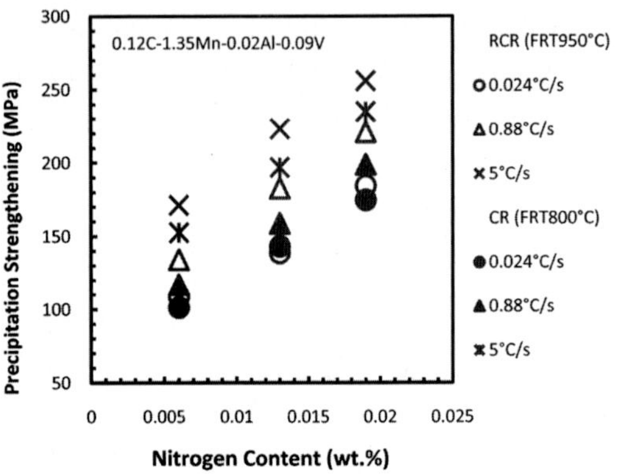

Fig. 3: Effect of nitrogen on precipitation strengthening [4]

DEVELOPING A DIFFERENT MODEL

To clarify some of the issues stated, a method was needed to generate a descriptive graphical model of V-N strengthening that would have a visual impact demonstrating the strengthening potential of the VN precipitation, and yet reflect the other strengthening processes that may be contributed by the presence of VC precipitation as well as interstitial nitrogen. As shown in Figure 3, there will always be the possibility of induced variation from processing variations. An empirical fit to a mechanical property data obtained from multiple production mills was chosen as the basis for the model. Next, for functionality and for ease of visual recognition, it was decided to use incremental yield strength as the Y axis, and vanadium content, the primary controlled variable, as the x axis. Different but fixed nitrogen levels would be represented by multiple curves on this graphical representation.

What is required to build this kind of graphical model is the strength contribution of three different situations. First, the effect of interstitial or free nitrogen is needed to predict the starting points. Second, the strengthening rate for V additions must be determined when there is a hyper-stoichiometric level of nitrogen available and VN precipitation is predominant. Third, the strengthening rate for V additions when no nitrogen is available (hyper-stoichiometric additions of V to N) must also be determined. It is presumed that VC will form at this time and be the primary precipitate for strengthening.

For the base data set, a previously published compilation of test results from three different thin-slab direct-rolled strip mills was used. [5] The as-rolled properties were used for the regression model, as the simulated coil-cooled samples may have had addition VC precipitation that may not happen in production coils. Given the minimum amount of data available, only linear regression was considered. Since the N strengthening coefficient would be very different depending on the presence of vanadium, only data from V containing heats were used. Also, almost all of those heats were sub-stoichiometric in N for the V level, meaning that sufficient nitrogen was available to tie up all V as VN. A number of different dependent variable configurations were investigated in linear regression analysis, finally settling on the simplified version as follows:

Predicted YS (MPa) = 226 + 40(%Mn) + 742(%V) + 8440(%N) Eq. 1

A brief review of the literature to estimate a value of the strengthening coefficient for interstitial or "free" nitrogen suggested that the effect is quite small. In MA'75, Pickering [6] reported a typical value of 350 MPa per 1% N, based on a survey of several different authors. Compared to the N coefficient in V steels, this value is insignificant, but not zero. This value is used to determine the starting (0% V) point of the different strength curves representing different nitrogen levels, providing separation in the initial portion of the graph. Since the practical levels of nitrogen variation in ferritic steels is only from 50 to 200 ppm, the strength increase for the additional 150 ppm of interstitial N would only be slightly over 5 MPa.

The initial section of each strengthening curve would represent the initial V addition, which would by necessity be nitrogen rich assuming a starting N level of at least 50 ppm. Since the data set was essentially a constant carbon level (.05% to .06%) a C coefficient could not be determined. As previously discussed, the interstitial N strengthening is considered to be relatively insignificant. The N coefficient is assumed to represent the N contribution to strength, therefore predicting the VN precipitation strengthening. In line with that assumption, the V coefficient is assumed to represent the V strengthening due to VC precipitation. While these assumptions are tenuous at best, the end results proved to be very credible.

Finally, we know that only precipitation in the ferrite during or after transformation provides strength. We also know that VN precipitation takes precedence over VC precipitation. Therefore, for the purposes of this model it is assumed that all VC precipitation takes place in the ferrite after transformation. The maximum solubility of carbon in ferrite immediately after transformation is considered to be about 0.02% C, and decreases rapidly as it cools further. That means the maximum VC precipitation that can be expected is that which forms from a maximum of 0.02% C, which stoichiometrically translates to 0.085% V. (V:C = 4.24:1) Vanadium added above the amount needed to form VN and VC from all of the N and C available will not contribute to precipitation strengthening.

The result is a strength prediction curve for increasing V additions consisting of three lines with three distinct slopes – the first line having a higher slope with N available for VN precipitation, and the second line representing VC precipitation having a smaller slope when V is added beyond the stoichiometric V:N ratio (3.64:1). After the available C in ferrite is consumed, presumed to be a maximum of 0.02% C, the strengthening slope for vanadium additions approaches zero. The first inflection point of the Strength-Vanadium curve is at a V level of 3.64 x N content. The second inflection point is .085 V above the first break point, where all C in the ferrite is consumed. The result is as shown in Figure 4.

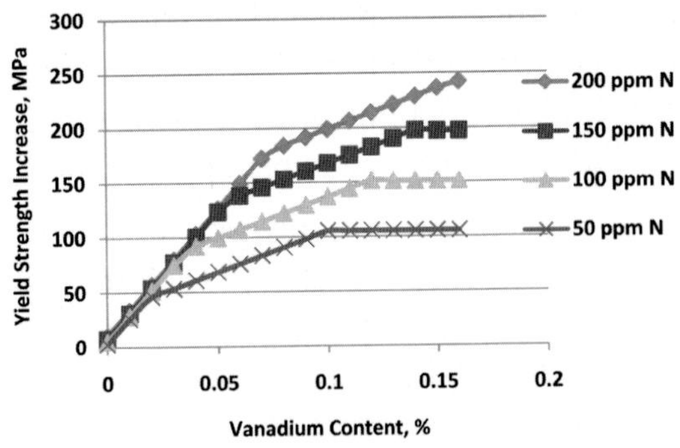

Fig. 4: Linear representation of strengthening model of ferritic steels with V and N

Since the assumption is that all available nitrogen and all available carbon in the ferrite are consumed, there are no interstitial elements left to form strengthening vanadium precipitates. Since the level of vanadium is low, less than 0.2%, no further alloying effect is expected. The microalloying (precipitation strengthening) effect of vanadium has been exhausted because of the consumption of the available interstitial elements in the ferrite.

REALITY CHECK

The relationship of yield strength increase with vanadium addition can be affected by a number of variables. First, all VC and VN (or V[C,N]) must be in solution in the austenite prior to the final transformation to ferrite. This usually limits this analysis to either as-rolled or as-forged products. Second, there should be no re-precipitation in the austenite during rolling or forging. At high levels of vanadium and nitrogen, this is likely to be some deformation-induced precipitation which, while it can contribute to intragranular ferrite formation, does not contribute directly to strength.

Secondly, most or all of the V should be precipitated in the ferrite, but we know that is unlikely to be true as well. The data used to derive this empirical description likely did not have much VN precipitated in the austenite, but it is likely that only a portion of the V was precipitated in the final as-rolled ferrite. This incomplete precipitation was demonstrated by the simulated coil-cooling strength increase that occurred in these materials and is reported elsewhere. [7] The strengthening shown in Figure 4 is not likely to contain all of the potential vanadium precipitation, but it does represent strength achieved in a variety of actual hot strip mill production processes. Since the VC precipitation is the last vanadium precipitation to occur and at a slower rate than the VN precipitation, it is subject to more dependence on process variation. Strengthening can be significantly less when a continuous cooling practice is used well beyond the austenite to ferrite transformation temperature. This reduction of V strengthening by rapid cooling to below 600°C is demonstrated by effect of lower coiling temperature as shown in Figure 2. For these situations where VC precipitation is limited, it is expected that the vanadium strengthening rate (slope of the line) in the intermediate region can be significantly less.

On the other hand, some authors(8) have proposed that the precipitation strengthening of VC can be enhanced at intermediate carbon levels by super-saturation of the ferrite. Deduced values of precipitation strengthening for isothermally transformed 0.12% V steels at 650°C indicated a strong carbon effect. However, the nitrogen effect on enhancing VN precipitation strengthening was equally strong at all carbon levels.

And finally, the rate of precipitation of VN and VC are controlled by various time-temperature parameters (solubility, diffusion and precipitation). As a result, different degrees of precipitation completion would be expected as process conditions changed. Besides affecting the actual amount of incremental strengthening, it would also cause difficulty in verifying with production data any of the inflection points suggested in Figure 4. Perhaps a better representation would be much more of a continuous curve. To reflect this reality, the model was modified as shown in Figure 5. Many years of experience has proven the model to be very useful in predicting the trade-off in V and N content in the final strength level. Because of the process dependence of the relationship, it is only an estimate of the actual strength achieved. So while quantitatively, the actual strength increments can be variable, there is always a qualitative relationship of the type shown in Figure 5.

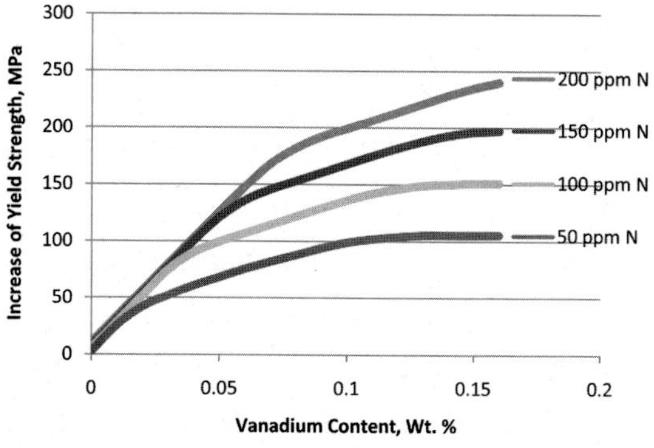

Fig. 5: Modified strengthening model of ferritic steel with vanadium and nitrogen

Even with the qualifications just mentioned, and the indirect derivation of the initial coefficients, this prediction of precipitation strengthening has proven to be very useful. In particular, an increase of yield strength by 7 MPa for each 10 ppm nitrogen added (when N is less than stoichiometric V:N) has been extremely consistent as reported by multiple investigators.

EFFECTIVE NITROGEN

With the understanding that the effectiveness of precipitation strengthening with vanadium is largely controlled by the available nitrogen, it is important to review the concept of "effective nitrogen". From the previous discussion, it is apparent the VN precipitation in ferrite will occur in preference to VC precipitation. There may be some co-precipitation of V(C,N), but most investigators have found that the first to form is VN, then followed by VC precipitation around the VN particle. So the effectiveness of the vanadium addition is primarily a function of the nitrogen available. The problem is that, even if the nitrogen is in the steel, it may not be available for VN precipitation in the ferrite. Any nitride former that has a thermodynamic "preference" for nitrogen is capable of reducing the beneficial strengthening effect of VN precipitation. The "Effective" nitrogen, as defined as the nitrogen available for formation of VN precipitates, may be much less than the total nitrogen reported by standard analysis.

The first, and most well know of the nitrogen scavengers, is titanium. TiN will form at very high temperatures, even before solidification, and will not typically dissolve to any significant extent during the reheat operation. So it can be assumed that for any Ti addition to the steel, a stoichiometric amount of nitrogen will be removed from availability. While Ti additions are desirable in some situations to provide grain refining precipitates for high temperatures grain growth control, the loss of N available for VN precipitation must be considered. For small amounts of Ti addition, additional nitrogen added can reduce the negative effects of effective N reduction.

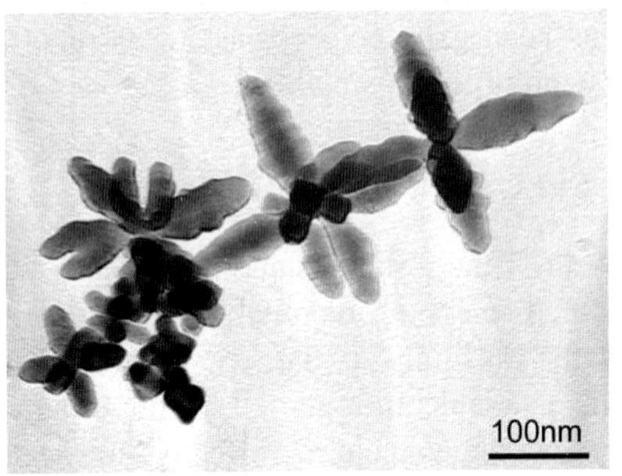

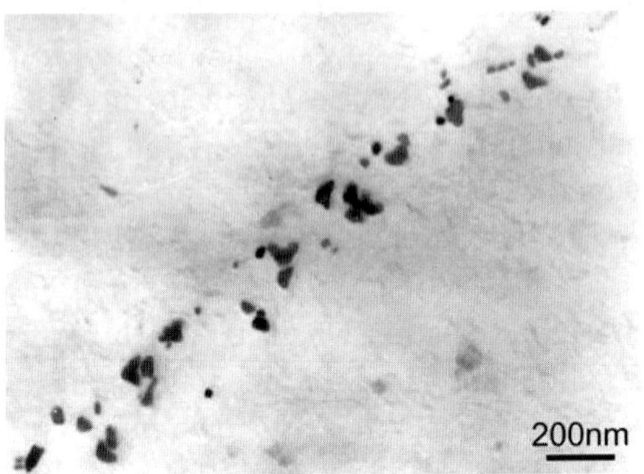

Fig. 6: TEM micrographs showing cruciform particles in a V-Ti-N steel (left), and cuboid particles in a V-Nb-Ti steel (right) after equalization at 1100°C.[8]

Niobium additions can complicate the issue. NbN is usually in solution during conventional reheating, but in CSP mills the reduced reheat temperatures may be insufficient to fully dissolve NbN. Also, because of the lower solubility of NbN vs. VN, the possibility of NbN precipitation during rolling is much higher than VN precipitation. Add to this the possibility that a Nb bearing steel is likely to be finish rolled at lower temperatures to take advantage of the higher recrystallization stop temperature. The combination creates a real problem in designing a V-Nb alloy system that takes advantage of the austenite conditioning properties of niobium while allowing the maximum VN precipitation. At this point there does not seem to be a desirable solution to this problem.

ALUMINUM CONTENT

This leaves one more nitride forming element that is often overlooked, but almost always present. Aluminum is routinely added to steels for deoxidation and for grain refinement. The soluble Al, that amount that is not oxidized, can form AlN. Aluminum does not form AlC. The solubility of AlN is lower than VN, making it a competitor to V for the available N. Grozier recognized this problem, and discussed it at length [2]. He pointed out that when vanadium-nitrogen heats are produced with high aluminum contents, adequate reheating temperatures must be maintained to avoid deterioration in strength properties. Numerous other authors have identified the same issues. Similar problems exist with the new thin-slab direct-roll mills that use tunnel furnaces with limited heating capability.

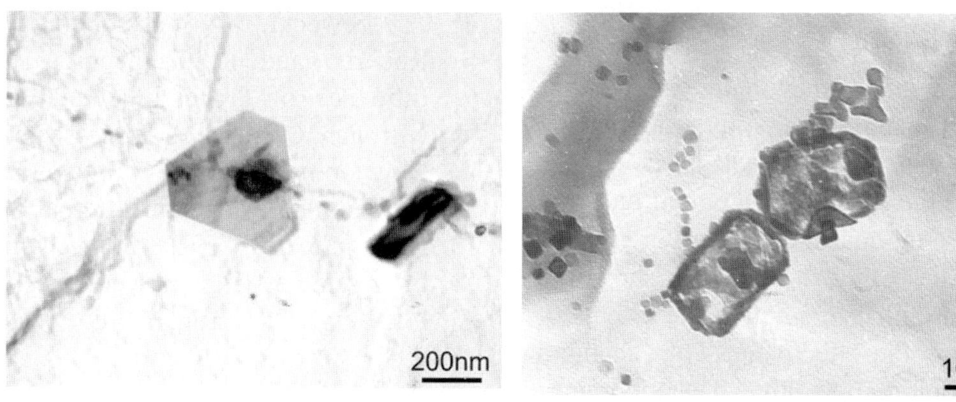

Fig. 7: TEM micrograph showing AlN + MnS in a V-N steel (left), and complex particles in a V-Ti-N steel (right) after equalization at 1050°C.(8)

Fortunately, even with the low solubility of AlN, it is kinetically very sluggish to precipitate during the cooling cycle. As long as the temperature is in a decreasing mode in normal production processing cycles, AlN precipitation is minimized. Any reheating, however, can cause rapid precipitation. Problem areas include the traditional slab/billet reheating as discussed. One place where this reheating can occur is at the surface during casting, where discontinuous but repeating water sprays can create repeated thermal cycling. These cool and reheat cycles are ideal for precipitating AlN. While these AlN precipitates may be re-dissolved during conventional reheat cycles, they can cause hot short issues at the caster, resulting in cracking problems. [9]

Very recent work reported by Rothleutner and Van Tyne[10] demonstrated the problem of Al effects in V-N microalloyed air-cooled forging steels. Again, the problem is getting the AlN in solution during the reheat cycle for forging. Air-cooled forging alloys rely heavily on VN precipitation, so the amount of effective nitrogen available is critical. His work demonstrated a loss of 50 MPa when reducing the forging reheat temperature from 1200 to 1100°C in a 0.37% C, 0.088% V, 0.031% Al, 153 ppm N steel.

Modern steelmaking practices are capable of adequately deoxidizing steels with substantially less Al than in the past. Clean steel practices using reducing slag and other techniques are now widely used. There is little need to maintain high Al contents to minimize steel reoxidation from high oxygen slags.

While fine ferrite grain size is a desirable trait for microalloyed steels, the method to achieve these fine grains in an as-rolled or as-forged microstructure is completely independent of the presence of aluminum. Rolling and cooling practices primarily determine the final grain size in microalloyed steels. Aluminum additions to meet specification grain size requirements are of no value in as-rolled steels. Neither is the reheat testing of these steels for subsequent austenite grain size. Since the as-rolled product will not be heat treated, the austenite grain size achieved on reheating is irrelevant to the performance of the steel in the as-rolled condition.

SUMMARY

A graphical model showing the interactions of V and N on the precipitation strengthening of polygonal ferrite steel is developed using a combination of empirical data and expected thermal-mechanical response. The resulting model has been shown to be very useful in predicting the effects of nitrogen on the effectiveness of vanadium precipitation strengthening. The importance of the effective nitrogen level is developed, considering the possible competing effects of the presence of alternate nitride forming elements. In particular, the importance of considering the competition of aluminum and vanadium for the nitrogen content is reviewed. When using the vanadium-nitrogen alloying system, it is highly recommended that the aluminum additions be limited to that necessary for the desired metallurgical function required of the aluminum addition. Since microalloyed steels were designed to be used in the as-rolled or as-forged condition, the presence of aluminum has no bearing on the grain size of the steel as used.

REFERENCES

[1] Micro Alloying 75 Proceedings, Oct. 1975, Washington, D.C. USA

[2] J.K. Grozier "Production of Microalloyed Strip and Plate by Controlled Cooling", MA'75, pp. 241-250

[3] M. Korchynsky, H. Stuart, Symposium-Low Ally High-Strength Steels, Nuremberg, May 1970, pp.17-27

[4] T. Siwecki, A. Sandberg, W. Roberts, and R. Lagneborg. "The Influence of Processing Route and Nitrogen Content on Microstructure Development and Precipitation Hardening in Vanadium-Microalloyed HSLA-Steels", Thermomechanical Processing of Microalloyed Austenite, Pittsburgh Pa. Aug. 1981. Pp. 163-194.

[5] R.J. Glodowski, "Effect of V and N on Processing and Properties of HSLA Strip Steels Produced by Thin Slab Casting", 42nd MWSP Conf. Proceedings, ISS, Vol. XXXVIII, 2000, pp. 441-449.

[6] F.B. Pickering "High-Strength, Low-Alloy Steels – A Decade of Progress", MA'75, 1975, pp 9-30.

[7] R.J. Glodowski, "Nitrogen Strain Aging in Microalloyed Steels", ISS Tech 2003 Conference Proceedings, pp.763-7711.

[8] Y. LI, J.A. Wilson, D.N. Crowther, P.S. Mitchell, A.J. Craven, T.N. Baker, "The effects of Vanadium, Niobium, Titanium and Zirconium on the Microstructure and Mechanical Properties of Thin Slab Cast Steels", Vanitec Publication, Westerham, Kent. UK. <www.vanitec.org>

[9] R.J. Glodowski, "Fine Grain Practice - Revisited" Proceedings, AISTech May 2010, Pittsburgh.

[10] L.M. Rothleutner and C.J. Van Tyne, "Mechanical Properties of a Vanadium Microalloyed Air-cooled Forging Steel with Various Aluminum Additions", Proceedings, MS&T'12, Oct. 7-12, 2012.

Development of Bulk Ultrafine-Grained Cold Reducible Grade Low Carbon Steel Produced by Equal Channel Angular Pressing

Rajpal Singh, Santosh Kumar#, N.K. Mukhopadhyay, G.V.S. Sastry and R. Manna
Centre of Advanced Study, Department of Metallurgical Engineering,
Indian Institute of Technology (Banaras Hindu University), Varanasi
Varanasi-221 005, India
#Research and Development Centre for Iron and Steel, Steel Authority of India Limited
Ispat Bhawan, Doranda, Ranchi-834 002, India
Emails: rmanna.met@itbhu.ac.in, rpmanna@rediffmail.com

ABSTRACT

Billets of cold reducible grade low carbon steel has been deformed at room temperature, up to an equivalent strain of 16.94, by equal-channel angular pressing (ECAP) to produce ultrafine grain (UFG) structure, adopting route Bc. Deformed microstructures are characterized by optical microscopy, transmission electron microscopy. Mechanical properties are evaluated by Vickers microhardness measurement and tensile testing. Mechanisms of grain refinement and strengthening are explored. Bulk ultrafine-grained microstructure could be produced by equal-channel angular pressing in low carbon steel. Grain refinement is strongly dependent on amount of strain. The refinement process involves several steps viz., elongation of grains by shear deformation, subdivision of grains to bands, subgrain formation in bands and fragmentation of bands into grains. Finally conversion of subgrains to grains takes place. Grain size is reduced from 68 μm to 0.20 μm after twenty eight passes or an equivalent strain of 16.8. The yield and tensile strengths increase rapidly at first few passes and then reach to a saturation level. The rate of strengthening as a function equivalent strain decreases to a minimum as deformation progressed. The yield strength increases marginally by 20% but tensile strength goes up by about 300%. The major cause of strengthening is attributed to grain refinement and work hardening. However, ductility of UFG low carbon steel is lower than that of their coarse grained counterpart.

KEYWORDS: *Low Carbon Steel, Equal-Channel Angular Pressing, Ultrafine Grains, Transmission Electron Microscopy, Microhardness.*

INTRODUCTION

Grain refinement is the major hardening technique that can be adopted to enhance strength in low carbon steel. Limited strengthening of steels of low carbon content can be achieved by thermomechanical treatment (TMT) due to limited refining upto 4-5 micron level [1, 2]. Ultra high strength can be achieved by grain refining to ultrafine (grain size <1 μm) level, through severe plastic deformation (SPD) method, which is an order of magnitude higher than that of TMT processes [2,3]. Heavy equipment (high tonnage) are required to process to final component size through TMT [2]. On the other hand, new processing technique called severe plastic deformation technique can deform metals and alloys repeatedly by simple shear without any change in cross-section [3-6]. Among these SPD techniques, equal-channel angular pressing can produce bulk ultrafine-grained alloys free from any contamination, fully dense and ultrafine in grain size [3,]. Steel being a major structural materials, recent research focuses on fabrication of bulk ultrafine-grained steel [1,7-22]. The grain size of low carbon steels processed by ECAP can be reduced to 0.2-0.5 μm at equivalent strain of 4 while ECAPed at 200-500°C i.e. 25 to 30 times lower than that can be achieved by TMT [1,7-9,20-21,23-24]. Grain size after ECAP at room temperature to an equivalent strain of 3 was obtained 0.2 μm [8]. The major works are either at low equivalent strain or at high temperature. Therefore, the present work focuses on producing bulk ultrafine-grained structure at high amount of equivalent strain and at room temperature.

MATERIALS AND EXPERIMENTAL PROCEDURE

Billets of 15 mm diameter and 60-120 mm long were prepared from cold reducible grade low carbon steel plates of size 330X220X38 mm^3. The plates were received from Research and Development Centre for Iron and Steel, Steel Authority of India Ltd., Ranchi, India. The steel sample was analysed by optical emission spectrophotometer (OES , ARL Model 3460). The chemical composition was (by wt%) C-0.075%, Mn-0.31%, Si-0.108%, S-0.014%, P-0.033%, Al-0.028%, Ti-trace, Nb-trace and Fe-balance. The billets of 60 mm length were deformed at room temperature by ECAP adopting route Bc [3] using a 30 ton hydraulic universal testing machine. The billets were coated with a lubricant mixture of high density paraffin and MoS_2 to reduce friction between billet and die. The ECAP die consisted of two equal-channels of 15 mm diameterintersected at an inner intersection angle (Φ) 120° and an outer arc angle (ψ) 60° that introduced an equivalent strain of ~0.6 in every passage of billet through the die. The equivalent strain ε was calculated for N number of passes by following equation [25].

$$\varepsilon = \frac{N}{\sqrt{3}}\left[2\cot\left(\frac{\Phi}{2}+\frac{\psi}{2}\right)+\psi\ \cos e\left(\frac{\Phi}{2}+\frac{\psi}{2}\right)\right]$$

The billets were deformed to a maximum equivalent strain of 16.94 or 28 passes at room temperature. Deformed samples were sectioned along X, Y and Z plane for microstructural investigation. X plane was the transverse plane perpendicular to extrusion direction. Y plane was the flow plane vertical to the extruded billet and Z plane was the horizontal but parallel to top surface along extrusion direction [3]. The polished samples were etched at 2% nital solution at room temperature and the microstructure were examined using Metalux-3 optical microscope. Thin foils of 3 mm diameter and 40 μm thickness were electropolished using an electrolyte of 10vol % perchloric acid and 90vol % methanol at 20V and at temperature at –40°C. Detailed microstructures are characterized by Tecnai20G^2 transmission electron microscope(TEM) at 200kV. Grain size is measured in micrographs by Heyn's linear intercept method. Hardness is measured by Shimadzu Microhardness 2T tester at varying load to find out minimum load for load independent hardness in X and Y plane for different type of samples. Based on the above result minimum load was selected as 980 mN for Vickers indenter. Tensile test was conducted on bar specimens of gauge length 16 mm and gauge diameter 4.5 mm at room temperature at a constant cross head speed of 0.05 cm/min and constant chart speed 2cm/min.

RESULTS AND DISCUSSION

Fig. 1 shows the appearance of ECAPed samples after various degree of deformation. The sample ends are wedge shape and the angle of wedge is the shearing angle of 60°. There are no cracks while samples were deformed upto twenty seven passes (Fig 1 (a)). However a minor crack could be observed on sample that had undergone twenty eight passes of ECAP (Fig 1(b)). The extra projections at die joining face are removed before subsequent pass. The loss in length is due to this problem in die.

Fig 2(a) depicts the optical microstructure of as-received low carbon steel. The microstructure of as-received sample consists of mainly ferrite grains (bright contrast) with a small amount of pearlite

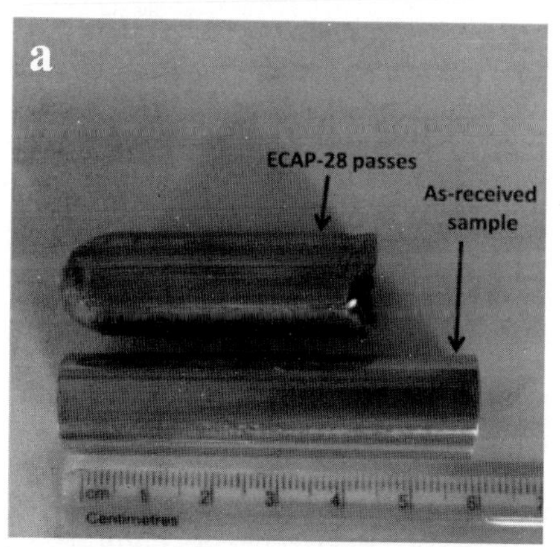

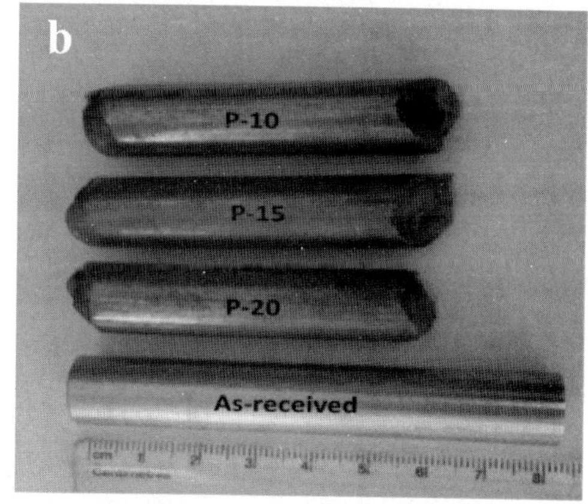

Fig. 1: Low carbon steel billet (a) as-received, after ECAP for ten passes (P-10), fifteen passes (P-15), twenty passes (P-20) and (b) twenty eight 28 passes

(dark contrast). The average grain size of ferrite is found to be 68 μm. Fig.2 (b)-(h) display the optical micrographs of low carbon steel of Y plane after ECAP. The grains are elongated in Y plane at initial stage of deformation (Fig 2 (b)-(c)). Average grain size decreases with amount of deformation (Fig.2 (b)-(f)). Grains subdivided to equiaxed grain/subgrain after fifteen passes. Structural modifications also took place to pearlite. Majority of lamellae were broken gradually. Cementite became spheroidised (Fig. 2(h)). However though grain size decreases with deformation but grains remains equiaxed in X plane. Bandings are not observed in X-plane.

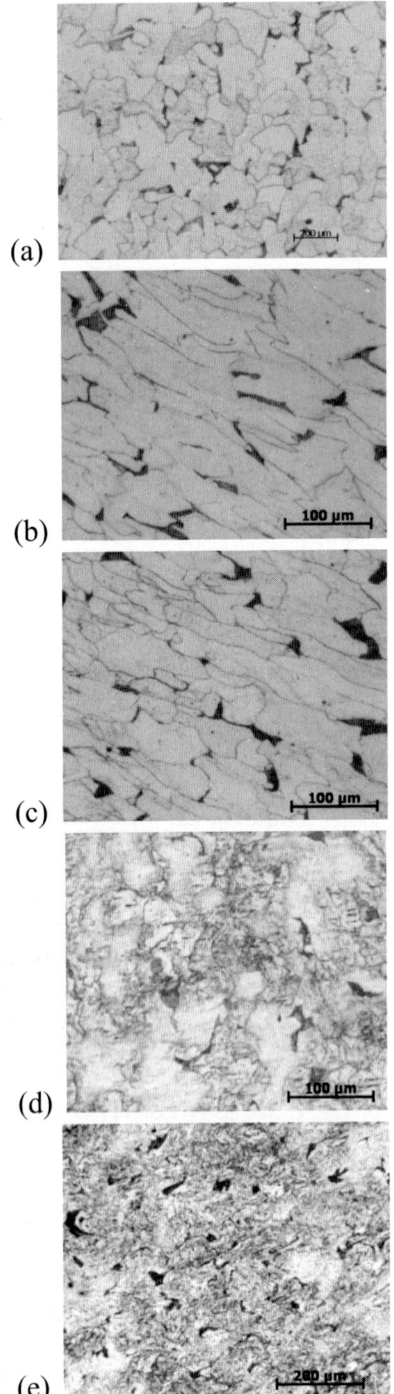

Fig. 2: Optical microstructure of low carbon steel (a) as-received, after ECAP for (b) one pass, (c) two paases, (d) five passes (e) ten passes, (f) fifteen passes

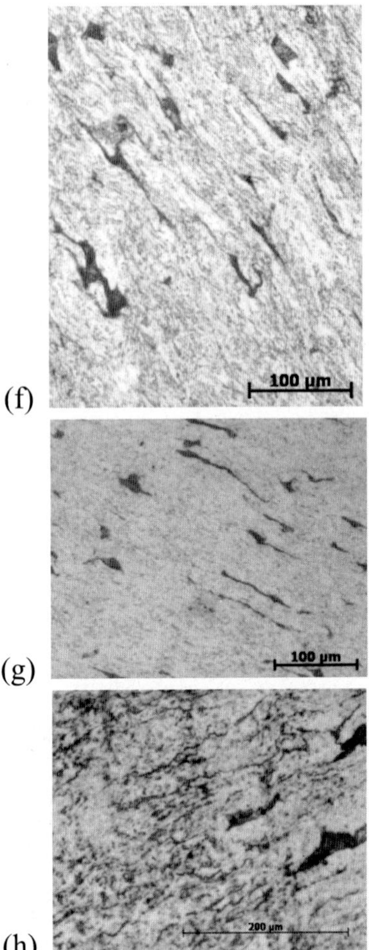

Fig. 2: Optical microstructure of low carbon steel after ECAP for (f) fifteen passes, (g) twenty passes and (h) twenty eight passes

Fig. 3 shows the bright field images of Y planes of low carbon steel after ECAP (a) two passes, (b) five passes, (c) ten passes, (d) fifteen passes, (e) twenty passes and (f) twenty eight passes. After two passes of ECAP the grains subdivided into bands (Fig. 3(a)). The microstructure contained elongated parallel bands with band width of approximate 350 nm. The band consists of elongated grains or subgrains. The average grain/subgrain size is 490±40nm. The extended parallel band boundaries are called lamellar-type boundaries (LBs). The grains contained high amount of dislocations. The transition bands are also observed after two passes of ECAP(Fig3.(b)). After five passes of ECAP the microstructure consists of elongated grains/subgrains with good amount of dislocations (Fig. 3(c)). The average grain/subgrain size is reduced to 379±32nm. The microstructure contains elongated bands of band width of 299nm. The pearlite colonies are broken. The lamellar cementite is converted to spherical isolated cementite particles (Fig.3(d)). After ten passes the microstructure consists of elongated bands of band width of 250 nm having average grain/subgrain size of 344±28 nm (Fig. 3(e)). After fifteen passes of ECAP microstructure displays ribbon grains of width 220nm. The average grain/subgrain size is 302±20nm (Fig. 3(f)). After twenty passes microstructure depicts lower density of dislocations inside ferrite grains and the average grain size is reduced to 247±18nm (Fig. 3(g)). After twenty eight passes the average grain is reduced to 204±17.5 nm.

Fig. 4 depicts bright field images of X planes of low carbon steel after ECAP (a) ten passes, (b) fifteen passes, (c) twenty passes, (d) twenty eight passes. Banded structure are not observed at X plane (Fig.4(a)-(d)). However, with increasing equivalent strain or number of passes grains have got refined continuously. At the initial stages of deformation the grain consists of subgrains. After ten passes the average grain/subgrain size is 346±27 nm (Fig. 4(a)). After fifteen passes the average grain/subgrain size is reduced to 290±15 nm (Fig. 4(b)). At this stage majority of the boundaries are high angle boundaries. After twenty passes the average grain became 270±12 nm (Fig. 4(c)). After twenty eight passes grain size reduces to 210±10 nm (Fig. 4(d)).

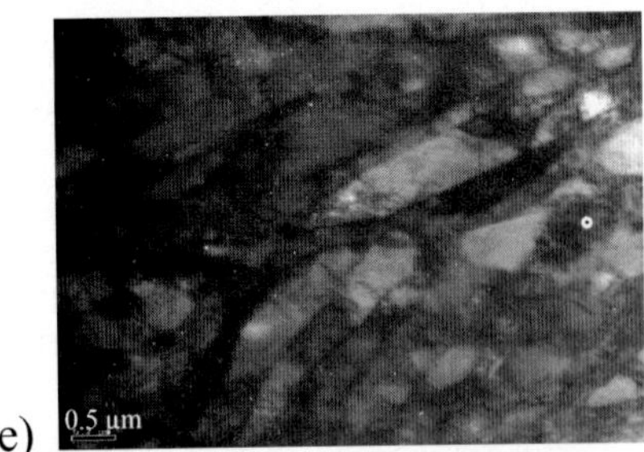

Fig. 3: Bright field images of Y plane of low carbon steel after ECAP for (a)-(b) two passes, (c)-(d) five passes (arrow indicates carbide particle, (e) ten passes.

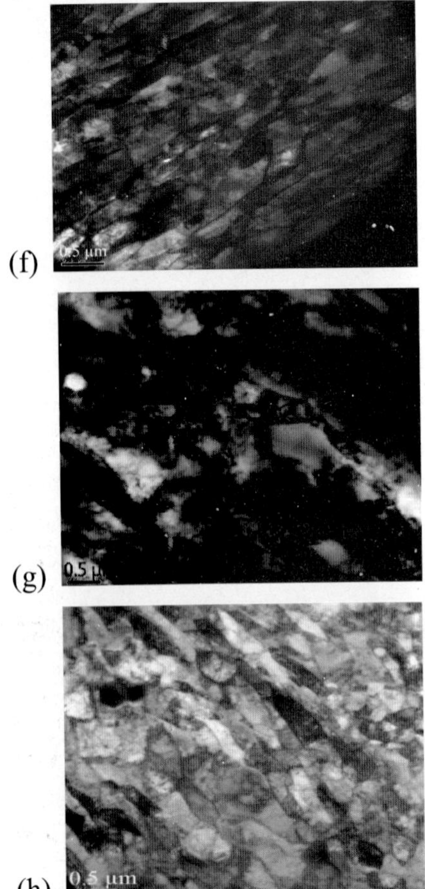

Fig. 3: Bright field images of Y plane of low carbon steel after ECAP for (f) fifteen passes, (g) twenty passes and (h) twenty eight passes.

Fig. 4: Bright field images of X plane of low carbon steel after ECAP for (a) ten passes.

Fig. 4: Bright field images of X plane of low carbon steel after ECAP for (b) fifteen passes, (c) twenty passes, (d) twenty eight passes.

Fig.5 displays variation of grain/subgrain sizes at X and Y planes. The grain size decreases to 0.5 μm just after two passes of ECAP which is two orders of magnitude of starting material (average grain size 68 μm). However most of the boundaries are of low angle boundary at this stage. Results suggests that grain/subgrain size strongly depended on equivalent strain. Grain size decreases with increasing strain gradually though it has reduced at slower rate at higher strain and finally it reaches to 0.2 μm at an equivalent strain of 16.94 (twenty eight passes). The rate of decrease in grain size with increase in equivalent strain in both X and Y planes were almost identical (Fig. 5).

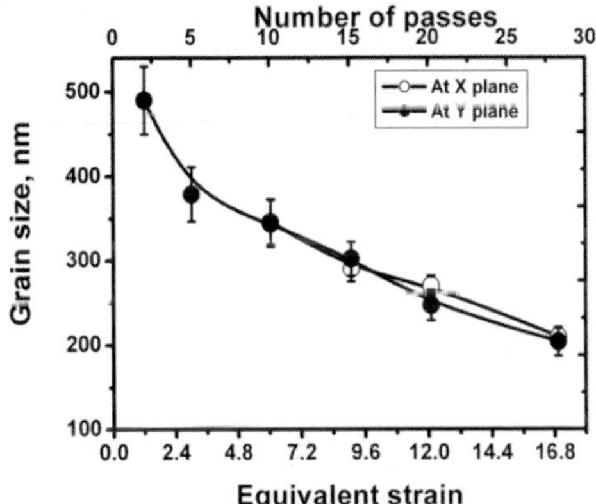

Fig. 5: Variation of grain size of low carbon steel with equivalent strain.

Fig. 6 shows the variation of ratio of grain/ sub-grain width to its length. The ratio approaches towards unity with increasing amount of equivalent strain. However, the rate at which this ratio increases are different in X plane than that in Y plane. For a given equivalent strain this ratio is higher for X plane than that of Y plane. Therefore, grains/subgrains are more equiaxed nature in X plane than in Y plane.

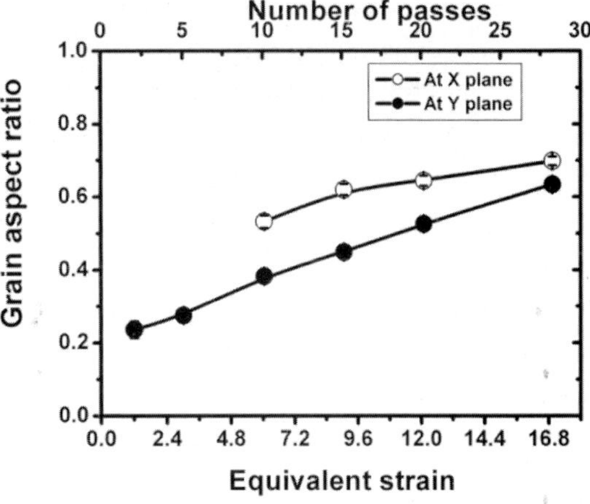

Fig. 6: Variation of width to length ratio of grain/subgrain in low carbon steel with equivalent strain or number of passes

Grain refinement mechanisms operative during ECAP are explored [26,28-29] earlier for high stacking fault energy materials. At initial passes banding takes place and the thickness of the bands decreases with the number of passes. The degree of grain size reduction is highest at low equivalent strain. Most of the boundaries at this stage are of low angle boundaries created by the rearrangement of dislocations. The grain refinement process involves (i) elongation of grains by simple shear deformation at low to intermediate equivalent strain levels, (ii) subdivision of grains to bands and subgrain formation in bands at low strain, (iii) fragmentation of bands into grains by intersection of bands due to rotation of the billet. Final microstructure contains grains with high dislocation density. This indicates that dynamic recrystallisation could not take place in low carbon steel due to continued straining by ECAP.

Fig.7 shows the engineering stress-strain curves for as-received as well as deformed samples. As-received low carbon steel yields at 225 MPa and uniformly deformed upto 27 % elongation and then neckes but got fractures with a total elongation of 42%. ECAPed material yields at higher stress but quickly reaches to necking condition with a total uniform elongation of only approximately 5.5-7.5%. Ultimate tensile strength of ECAPed samples are significantly higher than that of as-received material.

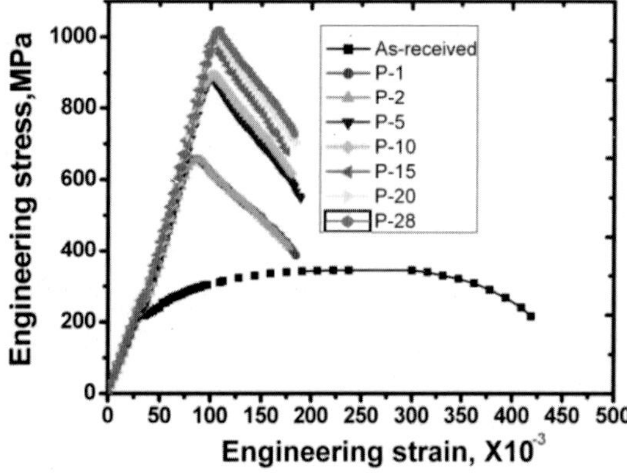

Fig.7: Engineering stress-strain curves of as-received low carbon steel and ECAPed samples.

Table –II exhibits the summary of tensile properties at room temperature for as-received as well as as-deformed samples.

Table II: Tensile properties of as-received and ECAPed low carbon steel at room temperature [Y.S.=yield strength, U.T.S.=ultimate tensile strength, T.E.=total elongation, U.E.= uniform elongation]

Equivalent Strain	Y.S. MPa	U.T.S. MPa	T. E. (%)	U.E. (%)
0	225	344	41.88	26.9
0.6	234	656	18.48	5.47
1.21	243	653	17.81	5.63
3.02	239	882	18.90	7.03
6.05	265	892	17.97	7.19
9.08	259	983	17.50	7.19
12.10	274	1011	18.44	7.19
16.94	271	1017	18.28	7.5

Fig.8 describes variation of yield strength, ultimate tensile strength, uniform elongation and total elongation with equivalent strain.

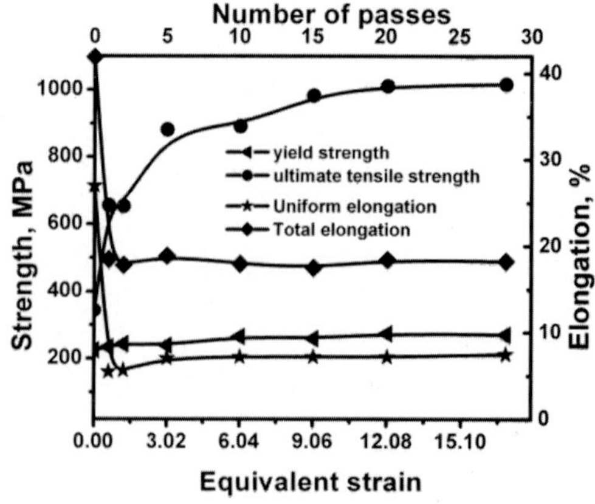

Fig. 8: Variation of yield strength, ultimate tensile strength and total elongation with equivalent strain or number of passes.

The yield strength of as-received low carbon steel is 225 MPa and just after one pass it increases marginally. It reaches to a saturation value of ~274 MPa. Ultimate tensile strength of ECAPed samples increases rapidly with increasing number of passes (upto 15 passes) at the initial stages of deformation or lower amount of equivalent strain and there after reaches to a saturation level at an equivalent strain of 9.1. The saturated ultimate tensile strength is about three times to that of as-received material. Fig.9 shows the variation of Vicker's microhardness (VHN) at 980 mN load with equivalent strain. The hardening behavior is very similar to the strengthening behavior. Hardness increases rapidly with increasing equivalent strain at initial stages of deformation and thereafter reaches to a saturation value which is about two and half times to that of as-received material. Hardness at X planes are very close to that at Y plane.

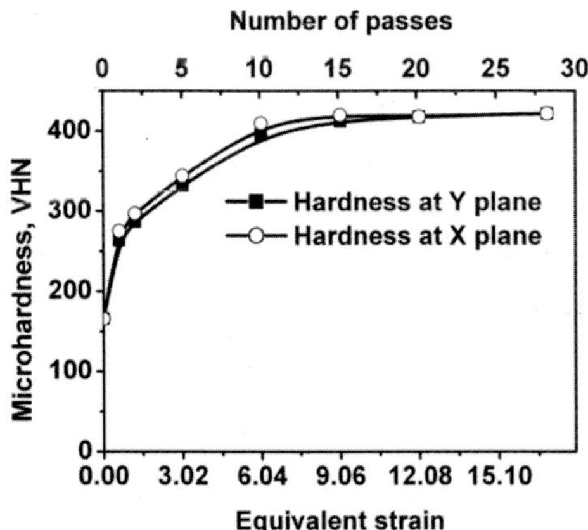

Fig. 9: Variation of microhardness measured at 980 mN in X and Y-plane of ECAPed samples with equivalent strain or number of passes.

The initial strengthening is due to work hardening (i.e. increase in dislocation density) and grain boundary strengthening. The rearrangement of these dislocations takes place at higher strain to form subgrain boundaries. These boundaries have got converted to high angle boundaries at large strain. Reduction in dislocation density reduces strength nevertheless formation of high angle boundaries enhances strength by grain boundary strengthening. Though strength increases at first rate with increasing equivalent strain after few passes of ECAP at initial stage, it decreases at intermediate level and reaches a saturation level at high strain [27-28].

Results describes that both total elongation and uniform elongation decrease drastically after initial pass and remain almost constant at low value though uniform elongation regains slightly with increasing number of passes. The loss of ductility is due to lack in work hardening ability of the material [27-28]. The decrease in work hardening is due to the absorption of dislocations into non-equilibrium grain boundaries of UFG metals [30].

CONCLUSIONS

Equal-channel angular pressing can be adopted to produce ultrafine-grained structure in low carbon steel of average grain size of 0.2 μm. Grain refinement is strongly dependent on amount of strain. The refinement process involves several steps viz., elongation of grains by shear deformation, subdivision of grains to bands, subgrain formation in bands and fragmentation of bands into grains. Finally conversion of subgrains to grains takes place. Ultimate tensile strength of ultrafine-grained low carbon steel could be enhanced to about 1000 MPa which is about three times to that of coarse grained counterpart. The yield and tensile strengths increase rapidly at first few passes and then reach to a saturation level. The rate of strengthening as a function equivalent strain decreases to a minimum as deformation progressed. The yield strength increases marginally by 20% but tensile strength goes up to by about 300%. The major cause of strengthening is due to grain refinement and work hardening. However, ductility of UFG low carbon steel is lower than that of their coarse grained counterpart.

ACKNOWLEDGMENTS

Authors wish to thank the Department of Science and Technology, Government of India for financial support under the project no. SR/S3/ME/0009/2010 (G).

REFERENCES

[1] R. Song, D. Ponge, D. Raabe, J.G. Speer, D.K. Matlock, Mater Sci Eng. A441 (2006) 1

[2] V. M. Segal, Mater Sci Eng. A386 (2004) 269

[3] R. Z. Valiev and T. G. Langdon, Prog. Mater. Sci. 51 (2006) 881

[4] R. Z. Valiev, R. K. Islamgaliev, I. V. Alexandrov, Prog. Mater. Sci. 45 (2000) 103

[5] M. Furukawa, Z. Horita, M. Nemoto, T. G. Langdon, J. Mater. Sci. 36 (2001) 2835

[6] R. Manna, Ph. D. Thesis "Design and Characterization of Ultrafine-grained Aluminum and Aluminum Alloys Processed by Equal-channel Angular Pressing", 2008, Banaras Hindu University, Varanasi, India.

[7] D. H. Shin, B. C. Kim, Y.-S. Kim, K.-T. Park, Acta Mater. 48 (2000) 2247

[8] Y. Fukuda, K. Oh-Ishi, Z. Horita, T.G. Langdon, Acta Mater. 50 (2002) 1359.

[9] D. H. Shin, K. T. Park, Mater Sci Eng. A410-411 (2005) 299

[10] S. Tamimi, M. Ketabchi, N. Parvin, Materials and Design 30 (2009) 2556

[11] Y. Okitsu, N. Takata, N. Tsuji J. Mater Sc 43 (2008) 7391

[12] W. Lei, T. S. Wang, Z. Li, X. J. Zhang, Q. F. Wang, F. C. Zhang, Mat. Sc. Eng A, (2010)

[13] M. A. Munoz-Moris, D. G. Morris, Scripta Mater. 63 (2010) 304

[14] K. –T. Park, Y. S. Kim, J. G. Lee, D. H. Shin, Mater. Sci. Eng. A293 (2000)165

[15] R. Saha, R. K. Ray, Mater. Sci. Eng. A459 (2007) 223

[16] R. Saha and R. K. Ray, Metal Metall. Trans, 40A (2009) 2160

[17] A. Sarkar, A. Bhowmik and S. Suwas, Applied Phys A, 94(2009)943

[18] A. Bhowmik, S. Biswas, S. Suwas, R. K. Ray, D. Bhattacharjee, Metal Metall. Trans, 40A (2009) 2729

[19] T. S. Wang, Z. Li, B. Zhang, X. J. Zhang, J. M. Deng, F. C. Zhang, Mat. Sc. Eng A, 527 (2010) 2798

[20] E. G. Astafurova, G. G. Zakharova, E. V. Naydenkin, S. V. Dobatkin and G. I. Raab, The physics of Metals and Metallography, 11 (2010) 260

[21] Y. G. Ko, C. W. Lee, S. Namgung, D. H. Shin, Jl. Alloys and Compounds, 504S (2010) 5452

[22] M. Calcagnotto, D. Ponge, D. Raabe, Mat. Sc. Eng A, 527 (2010) 7832

[23] Y. II. Son, Y. K. Lee, K. T. Park, C. S. Lee, D. H. Shin, Acta Mater 53 (2005) 3125

[24] K. –T. Park, S. Y. Han, B. D. Shin, Y. K. Lee, K. K. Um, Scripta Mater 51 (2004) 909

[25] Y. Iwahashi, J. Wang, Z. Horita, M. Nemoto, T.G. Langdon, Scripta Mater 35 (1996) 143.

[26] R. Manna, N. K. Mukhopadhyay, G. V. S. Sastry, *16th International Conference on Textures of Materials (ICOTOM 16)* December 12-17, 2011, Mumbai, India, *Mater Sci Forum* 702-703 (2012) 135.

[27] R. Manna, N. K. Mukhopadhyay, G. V. S. Sastry, *International Conference on Advances in Metallic Materials and Manufacturing Processes for Strategic Sectors (ICAMPS 2012),* January 19-21, 2012, Trivandrum, India, *Mater Sci Forum* 710 (2012) 241.

[28] R. Manna, N. K. Mukhopadhyay and G. V. S. Sastry, Metall. Mater. Trans. A, 39A (2008) 1525

[29] R. Manna, N. K. Mukhopadhyay and G. V. S. Sastry, Proc. 30thRisø Intl Symp on Mat. Sci: Nanostructured Metals – Fundamentals to Applications, edited by J.-C. Grivel et al., Risø National Laboratory for Sustainable Energy, T. U. D., Roskilde, Denmark, 7-11 Sept, 2009,P-245-252.

[30] R. Z. Valiev, I. V. Alexandrov, T. C. Lowe, Y. T. Zhu, J. Mater. Res., 17 (2002) 5

Evolution of Microstructure, Microtexture and Texture during Thermo Mechanical Processing Zirconium Alloy Structural Components

D. Srivastava[1], N. Keskar[1], K.V. ManiKrishna[1], S. Neogy[1], G.K. Dey[1], S.K. Jha[2] and N. Saibaba[2]
Materials Science Division
Bhabha Atomic Research Center, Mumbai,
Nuclear Fuel Complex, Hyderabad

ABSTRACT

The Indian pressurized heavy water reactors (PHWRs) use Zr-2.5% Nb pressure tubes and the in-reactor performance of these tubes, such as irradiation creep and growth behaviour, depends strongly on their microstructure and texture. The microstructure and textue of the tube essentially depends on the combination of different thermomechanical treatement imparted during manufacturing of these pressure tubes. In the present paper the microstructure evolution has been studied for pressure tubes produced by novel heat treated condition. The finished microstructure consisted of fine α lamalle with distribution of nanometer size β phase along the α/α interfaces having composition 75-80wt%Nb. Microstructure evolution in new optimized fabrication routes for manufacture of cold worked pressure tubes has been described here. The microstructure comprises of long α grain with higher aspect ratio and β being present as thin film having composition 15-20wt%Nb. This new fabrication route is proposed to be used to manufacture pressure tubes for all upcoming PHWR reactors. The paper also shows irradiated microstructure characterization at different location of the first pressure tube obtained from the KAPS2 reactor. The study shows substantial modification in β morphology.

KEYWORDS: *Zirconium Alloy, Thermo-Mechanical Processing, Microstructure, Texture*

INTRODUCTION

The present generation of Indian PHWRs use Zr-2.5%Nb pressure tubes. KAPS-2 is the lead reactor in Indian PHWR reactors using pilgered Zr-2.5% Nb PTs fabricated by Nuclear Fuel Complex, Hyderabad. The satisfactory performance and the life of the pressure tubes depend mainly upon the extent of degradation of the mechanical properties, corrosion under the influence of high temperature coolant and neutron flux and dimensional instability as a result of irradiation creep and growth in the reactor condition [1-3]. Irradiation growth and irradiation creep govern the performance of the cladding and the pressure tube material to a great extent. Since dimensional changes and mechanical property degradation can limit the life of PT, it is essential to study and develop an understanding of the microstructural evolution at different levels manufacturing and its exposure to neutron irradiation. The creep and growth are greatly influenced by the microstructural parameters such as shape, size and aspect ratio of the α-lamellae, the distribution and morphology of the β-phase, the interfacial structure, dislocation density and the crystallographic texture [1-3]. These aspects are sensitive to manufacturing process of the pressure tube.

The present paper describes the microstructure evolution in a new fabrication route for cold worked pressure tubes. This route has been developed after making industrial scale trial of nine different routes at NFC. Heat treated pressure tubes are known to show improved creep performance [4]. The present paper describes the microstructure assessment at different stages of fabrication of heat treated pressure tube. Samples taken from an irradiated PT in KAPS2 is evaluated and compared it with the microstructure of the unirradiated, as-fabricated material.

RESULTS AND DISCUSSION

(A) Microstructural Study of Zr-2.5Nb alloy Heat Treated Pressure Tube Fabrication

TEM micrographs after first extrusion showed dynamic recrystallization of the α grain. Refined and uniform microstructure suggested that cast structure was completely removed during hot extrusion process β phase was present at the α/α interface Fig.1(a)[5-6]. The billet was β quenched to obtain homogenous microstructure, texture and chemical homogeneity. β quenched microstructure showed martensitic microstructure consisting of coarse lath martensites (Fig.1(b)). The quenched billet was extruded to hollow blank. Microstructure after second hot extrusion exhibited that α grains become elongated in the axial direction of the tube and thin in the radial direction(Fig.1(c)). The longitudinal section exhibited α phase of lamellar morphology aligned along the extrusion direction. Transverse section exhibited α lamellae, whose width and thickness were not same. The average length, width and thickness of α lamellae were in the range of 15-20μm, 1-2.5μm and 0.3-0.5μm respectively. β phase was present as thin wafer between two α lamellae having average thickness of 0.02-0.05 μm. The extruded tube was further pilgered which was showing heavily deformed microstructure. Subsequent to pilgering α+β qunenching was carried out which is the most critical step in the fabrication of heat treated pressure tube. Quenching treatments were carried out in a controlled fashion, using quenching dilatometer, to study the evolution of microstructures as a function of different cooling rates. Systematic studies on transformation from α+β phase field as a function of cooling rate are far from being extensive in the open literature. In addition to cooling rate, effect of soaking temperature in the α+β phase field has been also examined in the present study. The samples cooled from 883°C at different cooling rates in argon showed primary α and transformed β phases. Samples cooled at the rate of 100°C/ sec and 50°C/sec showed predominantly internally twinned martensite and internally slipped martensite structure respectively. Below this cooling rate, transformed β product was observed to be Widmanstatten α +β. On the basis of these observations it could

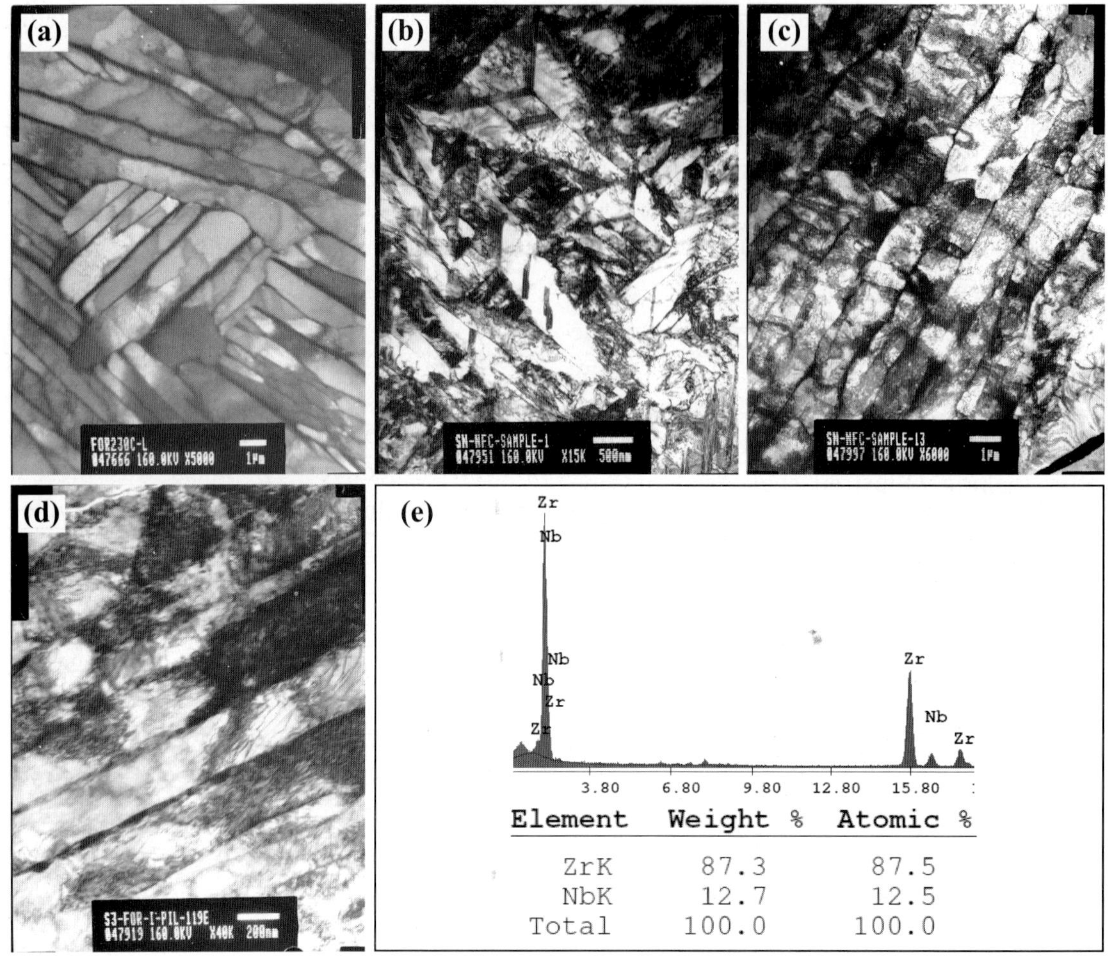

Fig. 1: TEM Microstructure evoluation at different stages of new fabrication route of cold worked Zr-2.5Nb alloy pressure tubes (a)Forged (b)β quenched (c) first pilgered (d) autoclaved

be concluded that in order to obtain primary α and martensitic microstructure after α+β quenching, a minimum cooling rate is required which is a function of soaking temperature. On the basis of the results obtained during controlled heat treatment performed in quenching dilatometer, 883°C was selected as soaking temperature and water as quenching medium for α+β quenching operation for large dimension tube (100 mm OD X 4.5 mm WT X 500 mm length). TEM micrographs revealed α+β quenched microstructure consisting of fine martensite phase along with 20-25% primary α volume fraction (Fig. 1 (d). The α+β quenched samples were cold deformed (by pilgering) to the extent of 23%.. Cold deformation of martensitic microstructure resulted in considerable increase in the dislocation density and dislocation substructure. (Fig.1(e)). These defects acted as nucleation sites for precipitation of β_{Nb} from the supersaturated martensitic phase during subsequent aging process. Ageing temperatures 540°C below the recrystallization temperature

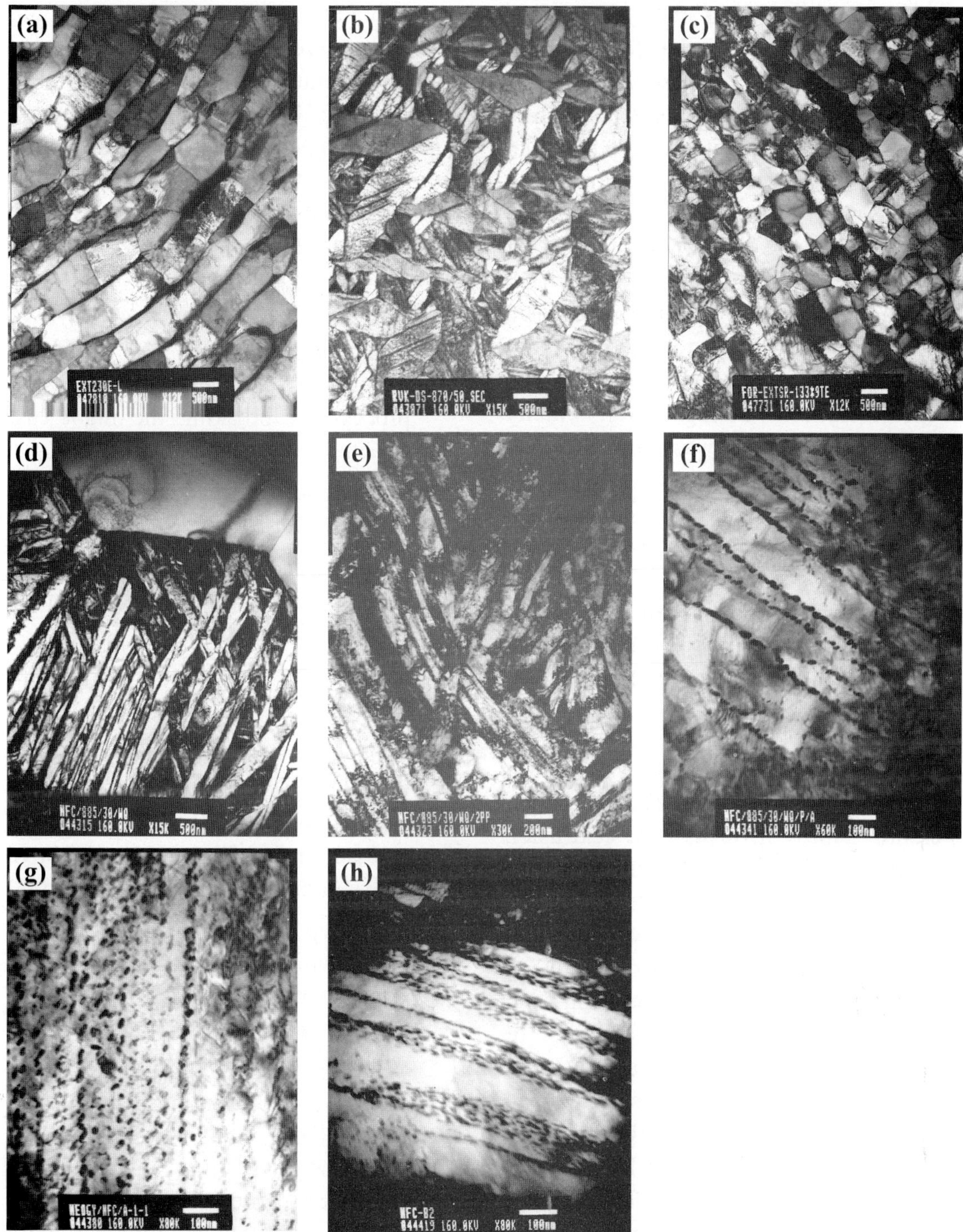

Fig. 2: TEM Microstructure evaluation at different stages of fabrication route of heat treated Zr-2.5Nb alloy pressure tubes (a)I-extrusion (b) β quenched (c) II extrusion) (d) α + β quenched (e) ii pilgered (f) and (g) aged(h) autoclaved

were used in the present study after the 2nd pilgering operation. Complete tempering of martensite at 540°C was observed whereas no recrystallization of the microstructure was observed. β precipitates (size less than 15nm) were observed only at the α/α lath interface (Fig. 1 (f-g)). Energy Dispersive Spectroscopy (EDS) analysis in TEM of samples aged at 540°C showed 75-80%Nb in β phase. This study thus shows that a temperature of 540°C is required to attain equilibrium concentration of Nb (85%) in β precipitates during aging. Finally autoclaving treatment was carried out at 290°C for 120 hrs, which did not modify the microstructure to any noticeable extent (Fig. 1(h).

(B) Microstructural Study of Zr-2.5Nb alloy Cold Worked Pressure Tube Fabrication by Optimized Fabrication Routes

The conventional method (CANDU) for fabrication of cold worked Zr-2.5Nb alloy pressure tubes involves hot extrusion followed by 25 pct cold drawing. Zr-2.5% Nb alloy pressure tubes in NFC fabrication involves hot extrusion (with lower extrusion ratio) and two-stage pilgering with an intermediate annealing treatment followed by autoclaving. The microstructure of the pressure tube fabricated at NFC has considerably different microstrucre than those reported for CANDU pressure tube. The α grains are much finer and aspect ratio is also small [7]. The β phase is mostly present in the globalised form unlike stringer in CANDU tube and it is substantially richer in Niobium. In order to examine the role of individual fabrication stages, in NFC a campaign was carried out to fabricate pressure tube with different fabrication parameters. In all pressure tubes were fabricated by nine different fabrication routes. The objective was to examine which route gives best desirable microstructural, textural and mechanical properties. The important variations in Zr-2.5Nb pressure tube fabricated with different route are given below

(i) Single stage forging and two step forging vs single stage extrusion process for breaking the ingot cast structure.

(ii) Comparison of tube produced with different extrusion ratio.

(iii) Comparison of the tube produced with two stage pilgering and with single stage pilgering

(iv) Role of stress relief treatment after extrusion

(v) Final 25% cold work by single stage drawing versus single stage pilgering.

The important microstructural observations in the pressure tubes fabricated with above routes are summarized below

1. Two step forging route has shown more homogenous microstructure. Grains were observed to be longer and aspect ratio was much higher, which is preferable.
2. There was variation in microstructure from leading end to trailing end in single stage forgiing and extrusion processes of breaking cast structure. In general trailing end microstructure was more lamellar and homogenous.
3. Higher extrusion ratio has resulted in still higher aspect ratio (desired).

The variation in microstructure from leading end to trailing end was much smaller in this case and microstructure appeared more uniform. Single pass always resulted in the continuous β along the α/α interface.

4. Intermediate annealing treatment resulted in the globulization of β phase in several regions the composition analysis has shown that these β were enriched to the level of 25 to 45 wt% of Nb.
5. Cold drawn process has shown that microstructure was very close to those observed in the tube produced by CANDU route
6. The α lamella in CANDU pressure tubes were much longer upto 20 micron length and β phase was continuous and the composition was always in the range of 15 to 20 wt%.
7. NFC produced Pressure Tubes: The α grains were not lamellar in all the regions; the grain size was much smaller. The β was not continuous in many regions and they had globulaized both at the α/α interface as well as within the α grains. The β phase composition varied from 15-45 %.

On the basis of this study described in detail in report [8] it has been proposed to fabricate Zr-2.5Nb alloy pressure by following fabrication route which involves two stages forging to billet followed by β quenching and extrusion to blank with higher extrusion ratio. The blank is then pilgered to final size and autoclave at 400C for 24 hours. The microstructure developed during each stage of this fabrication route is described in Fig.2. The forged microstructure showed widmanstatten α phase and β phase being present at α/α interface (Fig.2(a)). The grains were uniformly distributed from one end of the to other end of the forged billet. The microstructure was also uniform from center to the edge of the billet. The β quenched structure showed complete martensitic microstructure. The extrusion of billet with higher extrusion ratio resulted in much longer α phase (15-20μm) and the aspect ratio was in the range of 1:10:70. The β phase was present as thin film between α lamalle. The single pilgereing was done to obtain pressure tube in 25% cold work condition. Microstructure showed higher density of dislocations within α grain. Autoclaving treatment reduced the dislocation density to some extent. However, the finished tube exhibited lamellar α with stringer of β phase in between. No globulisation of β phase was seen in this route and EDS analysis have shown β composition was in the range of 15-wt20%Nb.

CONCLUSIONS

The present paper shows the microstructure at different stages of fabrication characterized by TEM. The microstructure is shown for the heat treated pressure fabricated first time at NFC Hyderabad. The finished microstructure consisted of fine α lamalle with distribution of nanometer size β phase along the α/α interfaces having composition 75-80wt%Nb. The microstructure developed during each stages of the new fabrication route for fabrication of pressure tube of 700MW PHWR reactor is also described. The final microstructure comprises of long α grain with higher aspect ratio and β being present as thin film having composition 15-20wt%Nb.

REFERENCE

[1] B. A. Cheadle, C. E. Coleman and H. Licht, Nucl Tech. , 57,1982, p. 413

[2] R. A. Holt, JNM 372 (2008) 182-214

[3] D.K. Rodgers, C.E. Coleman, M. Griffiths, G.A. Bickel , J.R. Theaker, I. Muir, A.A. Bahurmuz, S. St. Lawrence, M. Resta Levi JNM 383 (2008) 22–27

[4] M Koike and T. Asada JNM 159 (1988) 62-74

[5] N. Saibaba, S. K. Jha, S. Tonpe, Kumar Vaibhaw, V. Deshmukh, S. V. Ramana Rao, K.V. Mani Krishna, S. Neogy, D. Srivastava, G. K. Dey, R. V. Kulkarni, B.B. Rath, E. Ramadasan and S. A. Anantharaman, In Press, 2011 Zirconium ASTM-STP, China.

[6] N. Saibaba, S. K. Jha, S. Tonpe, V. Kumar, V. Deshmukh and Ramana Rao, D. Srivastava, S. Neogy, K.V. Mani Krishna, and G. K. Dey, E. Ramadasan, R. V. Kulkarni, B.B. Rath, and S. A. Anantharaman, BARC external report BARC/2010/E/010

[7] D. Srivastava, G. K. Dey, and S. Banerjee Metallurgical and Materials Transitions, **26A**(1995) 2707

[8] D. Srivastava, K.V. ManiKrishna, S. Neogy, G. K. Dey, S.K. Jha and N. Saibaba (to be published)

[9] D. Srivastava, R. Tewari, G. K. Dey, B. P. Sharma and D. N. Sah BARC Report, No. BARC/2005/E/023, 2005

[10] D. Srivastava, S. Neogy, Suparna Banerjee, E. Ramadasan, S. Anantharaman and G. K. Dey BARC external report BARC/2011/E/007

Thermomechanical Processing of 13/4 Martensitic Stainless Steel

Brij Kishor*, G.P. Chaudhari and S.K. Nath
Department of Metallurgical and Material Engineering,
Indian Institute of Technology Roorkee, India-247667
*Email: brijkdmt@gmail.com

ABSTRACT

13/4 martensitic stainless steel (MSS) finds extensive application in hydro turbine machinery, especially as turbine blades. The material is normally used in as- cast condition. This material suffers heavy damage during usage because of silt erosion. A number of efforts are being made to improve its erosion resistance, like development of coatings, and heat treatment cycles. However, these have limited effect. In the present investigation, thermomechanical working of 13/4 martensitic stainless steel has been carried out at different strain rates of 0.1, 1, and 10, at 1000 °C in Gleeble® 3800 simulator. Microstructural examination and mechanical characterization has been performed. It is observed that sample deformed with strain rate of 1 exhibit highest average hardness of 434 HV as compared to 300 HV obtained for as cast specimens. Slurry erosion behavior of the processed steels is performed and the results are correlated with microstructure and mechanical properties.

KEYWORDS: *Microstructure, 13/4 Martensitic Stainless Steel, Thermomechanical Processing, Slurry Erosion.*

INTRODUCTION

Hydro turbine components are subject to severe service conditions involving corrosive and erosive environment, accompanied with cavitation damage. This damage results in breakdowns, less efficiency of power generation and expensive repairs. Chauhan et al. [1] reported that erosion due to impact of hard and abrasive particles is a common phenomenon in hydro turbine under water parts and many other industrial situations, causing significant material loss (slurry erosion condition). Divakar et al. [2] report that the rate of erosion is influenced by the relative hardness of the target material and the particle that is impacting on the surface. Hence, proper material selection or optimization of surface and heat treatments of selected materials are necessary.

Martensitic stainless steel is used in hydroelectric power plants because of its high hardness and good corrosion resistance in severe aqueous environments. They are find wide application in the making of runner blades and vanes in hydro-turbines. They are capable of being heat treated in order to improve its hardness and strength levels also microstructure can be refined by thermo-mechanical treatment to provide better mechanical properties, and the surface of such processed specimen exhibits a good erosion resistance.

Limited literatures available on combined study of thermo-mechanical treatment and erosion behavior in hot compression test that reveals the microstructure features and effect of strain rates. Effect of microstructures on hardness [1, 3, 4] and erosion resistance [5, 6] of heat treated specimen has been studied by many researchers. They used plasma coatings of nitrides, laser techniques, high-velocity oxy-fuel (HVOF) spray process [7, 8, 9] and composites structures. Ebrahimi et al. [10] studied multistage thermo-mechanical treatment and reported that due to less available time at high strain rates, more strain energy is stored during deformation and the softening is attributed to recrystallization while the fraction of recovery is negligible. Krauss [11] investigated the deformation of martensite that is referred to as the lattice invariant deformation, and it produces a high density of dislocation or twins in martensite. Studies on martensite/ferrite duplex stainless steel revealed that various microstructural changes such as dynamic recovery and recrystallization can occur during hot rolling [5]. These changes can refine the grain size and alter their final microstructure and mechanical properties.

Table 1: Chemical composition of 13/4 MSS

C	Mn	Si	P	S	Cr	Ni	Mo	Cu	V	Al
0.065	0.73	0.62	0.027	0.008	12.30	3.78	0.45	0.22	0.04	0.005

In view of this, our study is mainly focused on investigating the microstructural features and variation of hardness and erosion behaviors for each thermomechanically processed specimen. Result show encouraging effect of different strain rates on the mechanical behavior of MSS composition. Effect of thermo-mechanical treatment on the microstructural evolution, hardness and erosion resistance of 13/4 martensitic stainless steel of as-received and thermomechanically processed specimens using different strain rates is presented in this work.

EXPERIMENTAL PROCEDURES

Material Selection

13/4 martensitic stainless steel was procured from Vaishnav Steel Pvt. Ltd. Muzaffarnagar India, conforming to the specification ASTM A743 CA-6NM is investigated in the present study. The chemical composition of this steel as obtained from spark emission spectrometry is shown in Table 1. The cylindrical samples of diameter 10 mm and length 15 mm were prepared from the given steel.

Thermomechanical Processing

The hot compression tests were carried out in thermo-mechanical simulator, Gleeble®3800 with Hydrawedge® attachment. Thermocouples were welded in the centre of the specimen and the end surfaces were lubricated with graphite. Tantalum foils was used to prevent sticking of specimen surface and anvil. The specimens were inserted in a vacuum chamber, where they were resistance heated. The schematic representation of the designed thermomechanical processing schedule is shown in (Fig.1).

The specimens were heated to 1100 °C using the heating rate 5°C/s held for 120 s at this temperature for homogenization. The specimen was then cooled at 1°C/s to the deformation temperature of 1000 °C at which it was held for 120 s to eliminate the thermal gradient before deformation and then deformed using strain rates of 0.1, 1, and 10 per second.

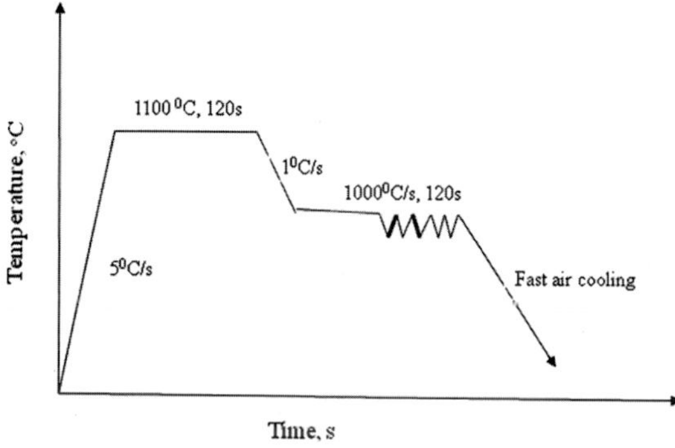

Fig. 1: Schematic diagram of the thermomechanical processing schedule

Metallographic Analysis

The deformed specimens were cut along the compression axis and the internal surfaces polished on emery papers followed by cloth polishing with emulsion of tap water and alumina powder until the surfaces were suitable for microscopic analysis. For optical microscopy, the specimens were etched by using Villella's reagent (100 ml ethanol+5 ml HCl+1g picric acid). A Leica DMI 5000M optical microscope equipped with digital imaging was used.

Hardness Measurement

The hardness of the 13/4 MSS as received and thermomechanically processed specimens was measured by using Vickers hardness tester FIE VM50 using 20 kg load with dwell time of 15 s. For hardness measurements, average of five readings at different points was taken.

Erosion Test

Erosion behavior of specimen was studied using the slurry erosion test apparatus and eroded surfaces were studied with the help of FE-SEM QUANTA 200 FEG scanning electron microscope.

Erosion Test Setup

Erosion test is carried out in the slurry pot erosion tester shown in (Fig. 2) Pot (container) assembled with tester contain 10 parts of water and 1 parts of sand particles, giving a slurry concentration of 10 wt%.

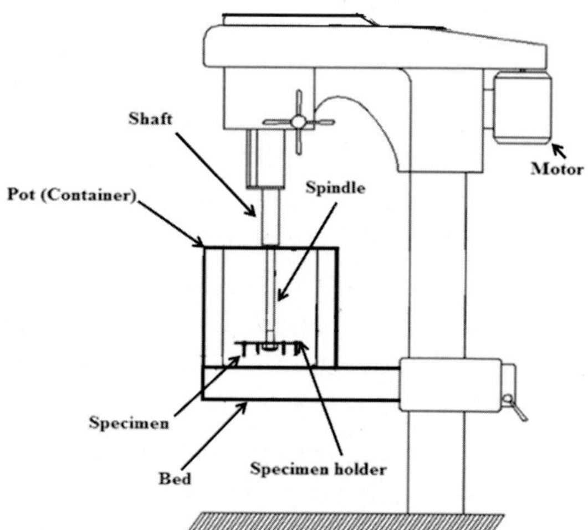

Fig. 2: Schematic diagram of the slurry pot test setup used for erosion testing.

The central shaft is fastened with spindle and sample holder was attached to bottom of central shaft on which samples were stuck with the help of an adhesive. The spindle is driven by an electric motor running at 500 rpm (linear velocity of 3.66 m/s).

The test conditions were tabulated in Table 2. Test is carried out for total of 24 h at room temperature. Weight loss was measured by Mettler electronic balance having an accuracy of 0.1 mg after every 3 h. The stuck specimens were removed by the acetone, cleaned, and dried before weight loss measurement.

Table 2: Test conditions for erosion test used in current investigation

Linear Velocity	Test duration	Weight loss measured	Sand particle size	Slurry wt%	Slurry medium
3.66 m/s	24 hours	every 3 hours	Between 210 mm and 354 mm	10 wt%	Tap water of pH 7.0

RESULT AND DISCUSSION

Microstructure Studies

Optical micrographs of 13/4 MSS in as-cast condition along compression axis and thermomechanically processed specimens are shown in (Fig. 3). From the microstructures studies it is found that as-received specimen (Fig. 3a) show lath martensite along with some retained austenite.

Processed specimens were heated up to 1100 °C at which according to the ternary phase diagram [11], ($\gamma + \delta$) phase field exists. At a temperature of 950 °C austenite is formed. The amount of austenite increases with the increasing temperature. Upon deformation at 1000 °C followed by fast air cooling, austenite phase transforms in to martensite. Thus, the resulting microstructure is predominantly martensite with small amount of ferrite.

Micrographs of specimen processed at strain rate of $0.1 s^{-1}$ are shown in (Fig. 3b). It shows the coarse-lath martensite with retained austenite. Microstructure of thermomechanically processed specimen at strain rate $1 s^{-1}$ is shown in (Fig. 3c). It shows the finer lath packets containing small amounts of retained austenite between the laths. Similar findings are reported earlier by Wen [5], that greater density of martensite packets were expected at finer structure. This may be Ebrahimi et al. [10] explain that as the strain rate increases, higher stored energy increases the driving force for recrystallization. They also found that as strain rates increases dislocation density increases and internal energy increases due to which hardness affected. Thus, at higher strain rates, partial dynamic recrystallization also occurred. Microstructure obtained at strain rate $10 s^{-1}$ is shown in (Fig. 3d). It shows decreased extent of lath packets of martensite. It also shows the coarsening of grains in a disperse way. Wang et al. [12] explain that as the strain rate increases, softening kinetics improves.

(a)

(b)

(c)

(d)

Fig. 3: Optical micrographs of 13/4 MSS specimens (a) As received specimen and specimens thermomechanically processed at strain rates of (b) 0.1 s^{-1} (c)1 s^{-1} and (d) 10 s^{-1}.

Hardness

Variation of Vickers hardness value of 13/4 MSS thermomechanically processed with different strain rates conditions are shown in (Fig. 4).

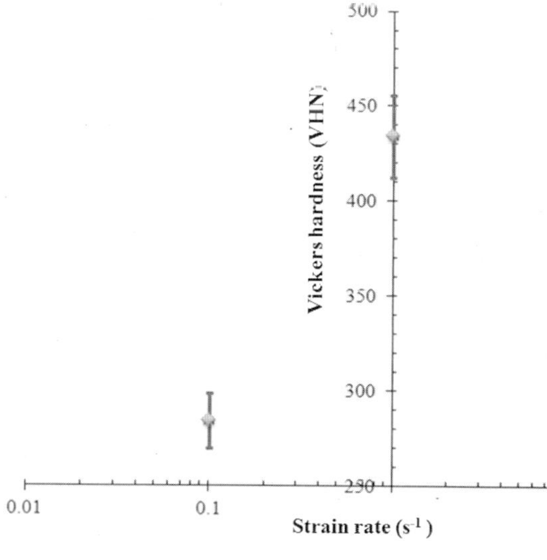

Fig. 4: Variation of hardness (VHN) for specimen thermomechanically processed with different strain (s^{-1}) rates.

The hardness of as-received specimen is 300 VHN. At 0.1 strain rate, it decreased to 284 VHN. There is significant increase in hardness values for specimens thermomechanically processed with strain rates 1 s^{-1} (434 VHN) and 10 s^{-1} (417 VHN) compared to as-received specimen. This correlates well with their refined martensitic microstructure. Thermomechanically processed specimen with strain rate 10 s^{-1} shows hardness value slightly lower than that of the specimens thermomechanically processed with strain rate 1 s^{-1}. Thus, at higher strain rate of 10 s^{-1}, softening occurred because of grain growth which clearly visible in (Fig. 3d).

Erosion Studies

Erosion tests were conducted for total duration of 24 h. (Fig. 5) shows the results of these studies. It can be seen that after 24 h of testing total weight loss is highest for as-received 13/4 MSS specimen.

Specimens thermomechanically processed with strain rates of 0.1s^{-1} and 1 s^{-1} show the minimum total weight loss. Specimen thermomechanically processed with strain rates of 10 s^{-1} shows higher total weight loss as compared to other thermomechanical processed specimens. Repetitive contact and striking of sand particles [13] suspended in water results in material removal leading to weight loss observed in the specimen. Initially higher material removal rate is observed from each specimen and as the test time increases the material removal rate decreased because of strain hardening of the MSS and rounding of slurry particles.

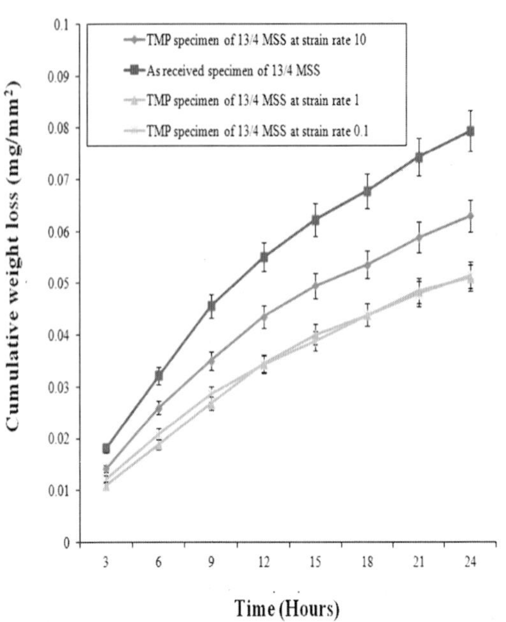

Fig. 5: Variation of cumulative weight loss with test time for as-received and thermomechanically processed (TMP) 13/4 MSS specimens at different strain rates.

Analysis of Eroded Specimens

SEM studies have been made on eroded surfaces of as received and thermo-mechanically processed specimens after a 24 h test. As-received specimen after erosion (Fig. 6a) shows a higher indentation with deepest and wider craters. Erosion is caused due to the sand particles suspended in water, by cutting and ploughing action along with plastic flow of metal [13]. Micrographs show the scratches, pits, craters and ploughing making the surface rougher.

Surface morphologies of thermo-mechanically processed with strain rates of 0.1, 1, 10 s^{-1} after erosion are shown respectively in (Fig. 6b), (Fig. 6c) and (Fig. 6d). Thermo-mechanically processed specimens exhibit the smoother surface as compared to as-received specimen after erosion. Micrographs reveal the effect of indentation of sand particles impacted during slurry erosion. Processing at strain rate 1 shows the minimum indentation and microcutting (Fig. 6c). At strain rate 0.1 (Fig. 6b) shows the similar morphologies as Fig. 6c. Specimen processed at strain rate 10 s^{-1} (Fig. 6d) shows the less erosion than as-received specimen.

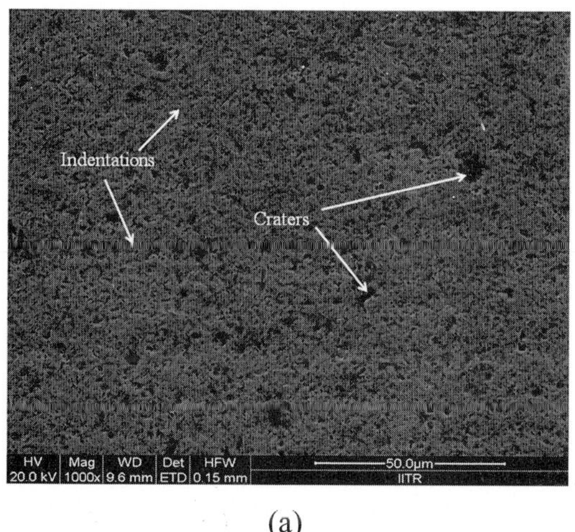

(a)

(b)

(c)

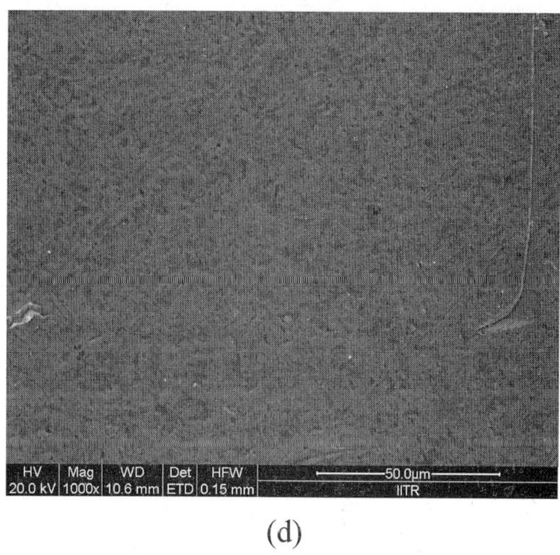

(d)

Fig. 6: FESEM micrographs of eroded surfaces 13/4 MSS (a) as-received, thermomechanically processed at strain rate of (b) $0.1s^{-1}$, (c) $1s^{-1}$, and (d) $10s^{-1}$.

CONCLUSION

Thermo-mechanical processing of 13/4 MSS resulted in refinement of microstructure. The lath martensite became finer. This is reflected in improvement in surface hardness with the best hardness obtained when processed the specimen with strain rates. Specimens, thermomechanically processed with strain rates of 0.1s^{-1} and 1 s^{-1} show the minimum slurry erosion. This is attributed to the refined martensite resulting from thermomechanical processing.

ACKNOWLEDGEMENT

Authors thank to Mr. Rajendra Sharma (lab technician) and Mr. Surya Deo Yadav (Research scholar), IIT Roorkee for their technical support during thermomechanical processing on Gleeble® 3800 simulator. Authors also want acknowledgement to Vaishnav Steel Pvt. Ltd. Muzaffarnagar India for providing the 13/4 martensitic stainless steel.

REFERENCES

[1] A. K. Chauhan, D. B. Goel, S. Prakash, Bull. Mater. Sci., Vol. 31, No. 2, 2008, p. 115.

[2] M. Divakar, V. K. Agarwal, S. N. Singh, Wear, Vol. 259, 2005, p. 110.

[3] X. Liu , K. Zhao, Y. Zhou, D. Ye, Jiang, Q. Yong, J. Su, Advanced Materials Research, Vol. 393-395, 2012, p.440.

[4] J. F. Santa, J. C. Baena, A. Toro, Wear, Vol. 263, 2007, p. 258.

[5] D. C. Wen Met. Mater. Int., Vol. 16, No. 1, 2010, p. 13.

[6] H. J. Amarendra, P. Kalhan,G. P. Chaudhari, S. K. Nath,S. Kumar, Materials Science Forum, Vol.710, 2012, p.500.

[7] K. H. Lo, F. T. Cheng, C. T. Kwok, H. C. Man, Materials Letters, Vol. 58, 2003, p.88.

[8] I. Alphonsa, A. Chainani, P. M. Raole, B. Ganguli, P.I. John, Surface and Coatings Technology, Vol. 150, 2002, p. 263.

[9] D. K. Goyal, H. Singh, H. Kumar, V. Sahni, Wear, Vol. 289, 2012, p. 46.

[10] G. R. Ebrahimi, H. Keshmiri, M. Mazinani, A. Maldar, M. Haghshenas, J. Material Sci. Engg., Vol.1, 2012, p.1.

[11] G. Krauss, Properties and Selection: Irons, Steels, and High Performance Alloys-ASM HandbookVol.-1, 1st Edn., ASM International, Ohio, 2005, pp. 211-23.

[12] B. Wang , W. Fu, Z. Lv, P. Jiang, W. Zhang, Y. Tian, Materials Science and Engineering A, Vol. 487, 2008, p.108.

[13] H. J. Amarendra, G. P. Chaudhari, S. K. Nath, Wear, Vol. 290-291, 2012, p. 25.

SECTION III

Microstructural Engineering

Deterministic Approach for Microstructurally Engineered Formable Steels

G. Anand[1], S. Datta[2] and P.P. Chattopadhyay[1]

[1]Department of Metallurgy and Materials Engineering, Bengal Engineering and Science University, Shibpur, Howrah, India

[2]Devbhoomi Institute of Technology, Dehradoon, Uttarakhand, India

Email: ppc@metal.becs.ac.in

ABSTRACT

Steel design is traditionally governed by the imprecise knowledge acquired from the experimental efforts occasionally invoking the empirical, physical and thermodynamic models supported by numerous assumptions and boundary conditions. The available frameworks are independently inadequate for complete and deterministic composition-process-microstructure-property (CPMP) correlation. The situation thus calls for judicious integration of the potential techniques with the capability of a *priori* prediction/optimization of the complete system. In view of the above the present article aims to provide a *bird's eye view* of the deterministic approach invoked so far in the traditional framework of design and development of formable steels. An attempt has also been made to propose the scheme for hybridization of the potential techniques for describing the complete CPMP correlation in the production schedule of hot strip coil.

KEYWORDS: *Steel, Microstructure, Property, Design, Technique Blends.*

INTRODUCTION

Progress in steel technology is primarily driven by the mounting performance demand in application areas, emergence of competitive materials and increasing stringency in the environmental regulations. Over the years, the experimental and theoretical capabilities have advanced from one paradigm to other as identified by compositional, thermo-mechanical, microstructural and deformation attributes.

At the initial phase, majority of compositionally designed steels were designated by single phase ferritic, pearlitic, austenitic or martensitic microstructures. Since 1970s, major emphasis has been on grain refinement by tailoring the thermo-mechanical schedule. The demand for weight reduction of the vehicle body structure evolved the concepts of composite microstructure coupling hard and soft phases as for example, in dual phase steel.

In the early 1990s, the ferritic microstructure containing bainitic phase with adequate amount of retained austenite was found attractive in respect of strength-ductity balance. Such microstructures were identified as TRanformation induced Plasticity (TRIP)-aided steels. A similar class of microstructures developed by subjecting the partitioning of alloying elements from martensite to austenite between martensite-start (M_s) and finish (M_f) temperatures are designated as "Quench and Partitioning Steels". Development of such TRIP-aided steels relies on the dynamic microstructure evolution during plastic deformation.

The response of austenite to strain and strain rate at different temperatures attracted major attention in the design of fully austenitic high manganese steel which offers most attractive strength-ductility-formability combination.

At this juncture, it may be worthy to take an account of the model capabilities and their integration with an aim to develop an integrated and deterministic design framework.

EMPIRICAL AND PHYSICAL APPROACH

Microstructural Evolution

Fig. 1 presents the mapping of the microstructures developed for high strength application of formable steels based on the popularly known *'banana'* diagram. The strain hardening and formability parameters are also incorporated in the figure. The diagram depicts the properties of three single phase microstructures namely, ferritic, martensitic and austenitic, and the evolution of composite microstructures obtained by combining two or three of the phases at different proportions. Among such multiphase varieties, strength of the DP and TRIP steels and constrained by the presence of intercritical ferrite (>60%), while the single phase austenitic steels requires addition of large amount alloying elements (15-25%). The austenite-bainite/martensitic verities of steel exhibit the possibility of achieving attractive strength-toughness formability balance and have drawn major attention in recent time. The empirical and physical frameworks available for understanding the microstructural evolution of established varieties of steels are summarized below in Fig. 1.

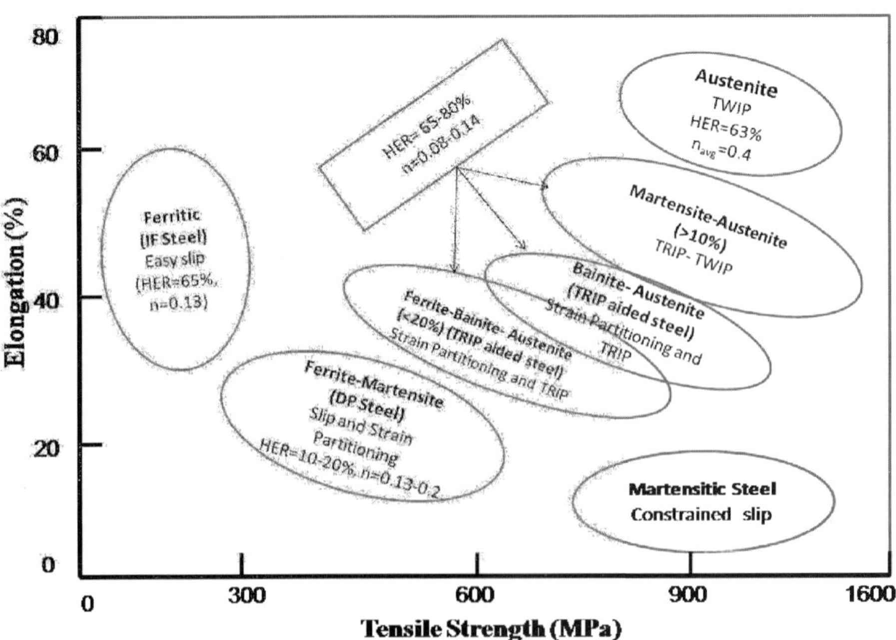

Fig. 1: Microstructure based mapping of formable steels

Ferritic Microstructure

Quantitative understanding of hot deformed/ thermo-mechanically processed low carbon low alloy ferritic steels concerns austenite recrystallization and austenite to ferrite transformation. The initial grain size of austenite determines the nucleation rate and recrystallization kinetics. Generally, recrystallization kinetics is expressed in terms of so called "Johnson-Mehl-Avarami-Kolmogorov (JMAK) theory [1-5]. Toloui et al [6] expressed JMAK equation for the time for 50% recrystallization ($t_{0.5}$) as:

$$X_v = 1 - \exp\left\{-0.693\left(\frac{t}{t_{0.5}}\right)^n\right\}$$

In this equation, $t_{0.5}$ *is* correlated to the processing history and may be calculated by empirical relationship as [6]:

$$t_{0.5} = \frac{A d_0^a Z^c \varepsilon^f}{E + F\varepsilon^b} \exp\left(\frac{Q_{rex}}{R_g T_{rex}}\right)$$

where *A, E, F, a, b* and *c* are material constants, d_0 the initial grain size, ε the strain, Q_{rex} the activation energy for recrystallization, R_g the universal gas constant, T_{rex} the recrystallization temperature and Z is the Zener–Hollomon parameter. The recrystallized fraction of austenite under non-isothermal condition is determined by incorporating the additivity rule. According to the Scheil's additivity rule [7], the reaction starts when the sum of the ratios of reaction starts time (τ_0) reaches unity, which may be written as:

$$\sum_{i=1}^{n} \frac{\Delta\tau_i}{\tau_0(T_i)} = 1$$

Where, $\Delta\tau_i$ represents the incremental hold time at temperature T_i.

The phase transformation from austenite to ferrite in the intercritical region has also been treated under the JMAK framework incorporating the Scheil's additivity rule. In the case of increased mole fraction of the substitutional solute, Onsager's extremel principle [8-9] has been applied. For estimation of the Gibbs energy (G) of the system as:

$$G = \frac{1}{\Omega} \sum_{i=1}^{n+m} \int_V x_i \mu_i dV$$

Where, Ω denotes the molar volume, *i* from 1 to n are the labels for the substitutional components and the interstitial components are labeled by i ranging from the $n+1$ to m. The volume of the system is V, and x_i and μ_i are the mole fraction and the chemical potential of the components.

Multiphase Microstructure

The common practice in development of the ferrite based multiphase microstructure of such steel is intercritical annealing (ICA) of the hot-rolled pearlitic steel with an aim to achieve ferritic and alloy enriched austenite. The intercritical austenite is subsequently transformed to martensite (Dual Phase steel) or bainite (TRIP steel). During inter critical annealing, the first step comprises the nucleation of austenite at ferrite-pearlite interface and growth of austenite into pearlite, until dissolution of pearlite is complete. This stage is followed by austenite growth controlled by Mn diffusion and final equilibrium with Mn diffusion in austenite. The Mn distribution in austenite and ferrite during intercritical annealing involves the following two processes [25]:

(i) Mn moves from ferrite to the ferrite-austenite grain boundary and forms Mn rich rim.

(ii) The Mn in Mn-rich rim at the ferrite-austenite boundary diffuses into austenite until final equilibrium is reached and Mn gradient in austenite is eliminated.

The diffusion analyses for these two processes combine with mass balance

$$\frac{\int_{S_\alpha} C_{Mn}^{\alpha} dS_\alpha}{S_\alpha + S_\gamma + S^i} + \frac{\int_{S_\gamma} C_{Mn}^{\alpha} dS_\gamma}{S_\alpha + S_\gamma + S^i} + \frac{S^i C_{Mn}^i}{S_\alpha + S_\gamma + S^i} = C_{Mn}^0$$

(S_α, S_γ *and* S^i represents the area occupied by ferrite, austenite and Mn-rich rim respectively)

The equation thus provides Mn redistribution and partitioning coefficient at any time during austenite grain growth.

Martensitic Transformation (Dual-phase Steels)

The martensite start or M_s temperature of the intercritical austenite may be calculated by least square fitting martensitic transformation data as determined from dilatometric experiments to the exponential Koistenin-Marburger (K-M) equation given as:

$$f = 1 - \exp\left[-0.011(M_s - T)\right]$$

Where, f and T are the martensite fraction and the absolute temperature, respectively [11]. Fraction transformed f_m is obtained from slope of dilatation (v) temperature before and after the Martensitic temperature is determined. The determined M_s temperature as a parameter of the linear fitting of $-ln(1-f_m)$ versus temperature, using exponential K-M equation expressed as $\ln(1-f_m) = a(M_s - T)$. From the measured M_s temperature, the carbon content of the intercritical austenite may be calculated by using the following empirical equation [12-13]:

$$Ms(K) = 539 - 423\%C - 30.4\%Mn \\ - 7.5\%Si + 30\%Al$$

The following exponential expression for M_s temperature has been proposed [14], as influence of carbon at higher concentration is less [15], leading to following modified equation [16]:

$$Ms\ (K) = 546 \times \exp(-1.362\%C) \\ -30.4\%Mn - 17.7\%\%Ni \\ -11\%Si - 12.1\%Cr$$

Bainite Transformation (TRIP-aided steel)

Bainite formation was envisaged to occur at a temperature just below T_0 line, which is the locus of transformation temperatures where both bainitic ferrite and austenite of similar composition have identical free energy [16].

The effect of substitutional solutes on reducing the T_0 temperature (ΔT_0) is given by [17]:

$$\Delta T_0 = \frac{\sum_i x_i \left(b_{NM}\Delta T_{NM} + b_M \Delta T_M\right)}{b_{NM} + b_M}$$

The T_0 curve for plain carbon steel is given by [18]:

$$T_0^{'}(K) = 970 - x_c$$

Where, x_c is the atomic fraction of carbon.

The volume fraction of austenite is also the function of transformation temperature. The maximum volume fraction of retained austenite may be given by following relationship:

$$a_r = 76.98\left(\%C + \frac{\%Si}{3} + \frac{\%Cr}{5}\right) + 0.3379T_{transf}\left({}^{0}C\right) - 211.5$$

Where, T_{transf} is bainite transformation temperature.

The transformation of austenite to bainite when fitted to JMAK equation shows two slopes in the plot of $\ln\{\ln(1/(1-f))\}$ versus $\ln t$. This indicates that the change in the transformation temperature

High Manganese austenitic steel

The transformation of austenite during straining is important depending upon the stacking fault energy (SFE) or γ_{SFE}. The transformation of austenite during deformation may take following routes:

1. For, $\gamma_{SFE} < 18mJ$, martensitic transformation occurs either by formation of ε martensite or even at lower SFE by formation of a'_{bcc} martensite following the $\gamma_{FCC} \rightarrow \varepsilon_{hcp} \rightarrow a'_{bcc}$ route [19].
2. For, $18mJ < \gamma_{SFE} < 45mJ$, twinning is the favorable phenomena [20-21].
3. For, $\gamma_{SFE} > 45mJ$, planer glide of dislocation determines the plasticity and strain hardening [19].

The important aspect of the phase evolution of TWIP steel concerns the variation of stacking fault energy (SFE) with appropriate adjustment of alloying elements, particularly in respect of Mn, Al and Si. In this regard, Mn [20] and Al [21] decrease the SFE, while Si increases the SFE [22]. The measurement of SFE by thermodynamic approach was proposed by Olson and Cohen [23] as:

$$\gamma_{SFE} = 2\rho\Delta G^{\gamma\rightarrow a} + 2\sigma^{\gamma/\varepsilon}$$

Where, ρ is the molar density along {111} plane, $\Delta G^{\gamma\rightarrow a}$ represents the molar free energy of the austenite to ferrite transformation and $\sigma^{\gamma/\varepsilon}$ represents the surface energy of the austenite-ferrite interface. The model accounts for the change in the Gibbs energy of each element upon the ε-martensitic transformation. The $\Delta G^{\gamma\rightarrow a}$ term include the first order interaction between various elements and change in Gibbs energy due to the magnetic contributions to the Gibbs energy, including paramagnetic to anti-ferromagnetic transitions of the austenite and the ε-martensite.

DEFORMATION BEHAVIOR AND TENSILE PROPERTY

The yield strength σ_{YS} as a function of ferrite content f_a, ferrite grain size d_a, pearlite spacing S_P, and compositional details such as Mn content, Si content and free nitrogen content is expressed as [24]:

$$\sigma_{YS}(MPa) = f_a^{1/3}\left[35.4 + 58.5(\%Mn) + 17.4d_a^{-1/2}\right] + (1 - f_a^{1/3})\left[178.6 + 3.85S_P^{-1/2}\right] + 63.1(\%Si) + 42(\%N)^{1/2}$$

Similarly, the tensile strength σ_{UTS} is expressed as [24]:

$$\sigma_{YS}(MPa) = f_a^{1/3}\left[35.4 + 58.5(\%Mn) + 17.4d_a^{-1/2}\right] + (1 - f_a^{1/3})\left[178.6 + 3.85S_P^{-1/2}\right] + 63.1(\%Si) + 42(\%N)^{1/2}$$

The yielding and straining of high strength low alloy steel (HSLA) includes the additional influence of precipitation strengthening by microalloying elements. The yield and ultimate stress increment due to precipitation hardening σ_{ppm} in the case of V-microalloyed steels is represented as [24]:

$$\sigma_{ppn}(MPa) = 573\log R + 700V + 7800N + 19$$

The stress strain behavior of single phase microstructures may be approximated by Mecking-Kocks model [25-26], where the evolution of dislocation density depends upon the competitive hardening and softening (due to recovery).

According to Mecking-Kocks theory, the evolution of plasticity results from the competition between rate of production of dislocation $\left(d\rho / d\gamma\right)^{+}$ and rate of annihilation $\left(d\rho / d\gamma\right)^{-}$. The strain hardening due to accumulation of dislocation, is given as:

$$\left(\frac{d\rho}{d\gamma}\right)^{+} = \frac{1}{\lambda b} = \frac{1}{d} + k\sqrt{\rho}$$

Where, λ is the mean free path of the dislocation, b is the burger vector of the dislocation, d is the grain size, k is the constant and ρ represents the dislocation density. The dislocation annihilation term, corresponding to the recovery, is represented as:

$$\left(\frac{d\rho}{d\gamma}\right)^{-} = -f\rho$$

Where, f is the dislocation annihilation parameter. The expression for the evolution of dislocation density is obtained as:

$$\frac{d\rho}{Md\varepsilon} = \frac{1}{\lambda b} + \frac{k}{b}\sqrt{\rho} - f\rho$$

Where, M is the Taylor parameter, independent of grain size and ε represents strain.

Model for tensile behavior of dual phase microstructure was presented by Tomata *et. al.* [27], which is general in nature, as it considers the internal stress produced by the inhomogeneous deformation. It is applicable to the case where, second phase (hard) is randomly distributed as grains of ellipsoidal shape.

The stress-strain behavior of the each constituent phase of TRIP steel is also determined by considering the Mecking-Kocks model [28]. The model for M/A constituent is determined by calculating the strain-induced martensitic fraction, as the initial austenite island transforms into M/A constituent, the separate behavior of the martensite and austenite is taken into consideration and coupled to the strain dependent evolution of the phase ratio of the M/A constituents. The martensitic volume fraction $f_{a'}$ may be determined from the Olson-Cohen formulation [28] as:

$$f_{a'} = 1 - \exp\left(-\beta\left[1 - \exp(-\alpha\varepsilon)\right]^{n}\right)$$

where, α is related to the volume fraction of shear bands and β is related to the number of shear interactions per unit volume of the austenite and n is the constant having the value of 2, for TRIP-aided steels [28]. The grain size of the retained austenite d_γ is also the function of the volume fraction of the strain induced martensite, which may be given as:

$$d_\gamma(\varepsilon) = d_{\gamma init}(1 - f_{\alpha'})^{1/3}$$

Where, $d_{\gamma init}$ is the initial grain size of the austenite. The stress-strain behavior of the austenite may be modeled using Mecking-Kocks Model, when λ is considered to be equal to d_γ. Thus,

$$\frac{d\rho}{Md\varepsilon} = \frac{1}{b}\left(\frac{1}{d_{\gamma init}(1 - f_{\alpha'})^{1/3}}\right) + \frac{k}{b}\sqrt{\rho} - f\rho$$

The stress strain behavior of the martensite under static deformation condition has been developed by Rodriguez and Gutierrez [29]:

$$\Delta\sigma = \alpha\mu M\sqrt{b}\sqrt{\frac{1-\exp(-Mf\varepsilon)}{fL}}$$

Where, L is the martensite lath width.

Now, Gladman-type power law [30] is used to define the stress-strain behavior of the M/A constituent, as given by:

$$\sigma_{M/A} = \sigma_\gamma\left(1 - f_{\alpha'}^{n'}\right) + \sigma_{\alpha'} f_{\alpha'}^{n'}$$

Where $f_{\alpha'}$ is the voume fraction of martensite and σ_γ and $\sigma_{\alpha'}$ are the stress in austenite and martensite repectively. The value of *n'* is determined by fitting stress-strain curves to the experimental data. Fig.2 shows the schematic representation of this scheme.

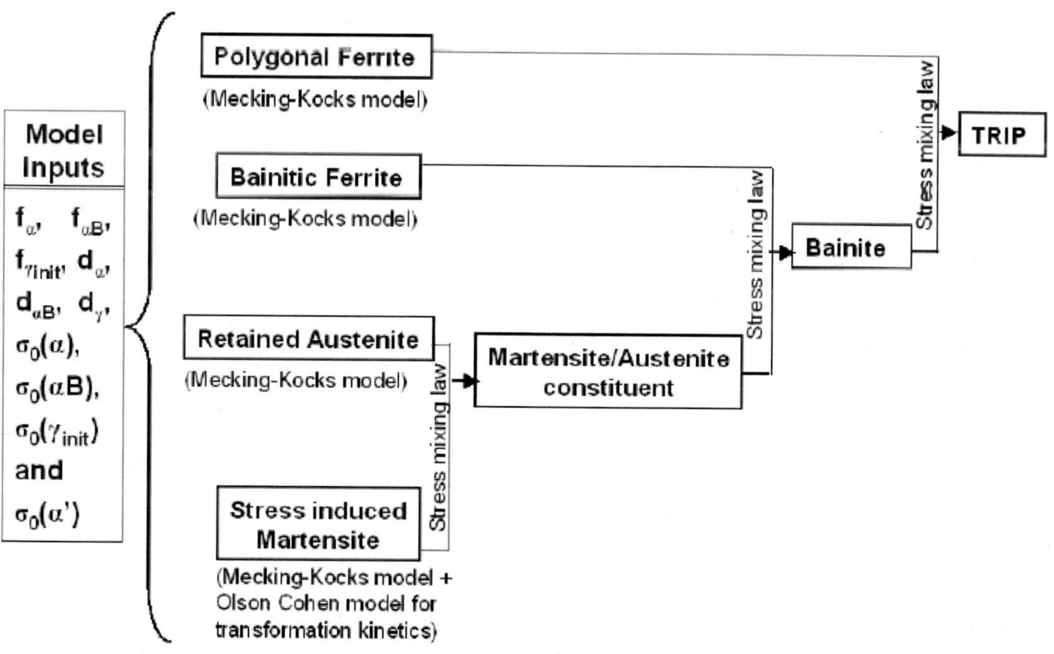

Fig. 2: Decomposition of multiphase TRIP steel and application of Mecking-Kocks model and Stress mixture law for determination of stress-strain curve.

Another important area concerns the twinning induced plasticity in TWIP steel, which presents attractive strength-ductilty combination [20] with impressive strain hardening [21]. The twinned crystal planes acts as the sub-boundary like obstacles to dislocation motion [31-32], leading to the Hall-Petch type strenghening of the material under progressing strain [33]. The role of twinning in strain hardening was proposed by Remy, considering the twin spacing to mean free distance of the dislocation, leading to the expression of flow stress as [34]:

$$\sigma = \sigma_m + k\frac{\mu \cdot b}{t}n$$

Where, σ_m is the flow stress of the untwinned austenite matrix, μ the shear modulus, b the burger vector, n the number of dislocation in the pile-up and t the mean distance between adjacent twin.

The second approch of the modelling concerns the dynamic reduction of mean free path of dislocation by the twin boundary [35-36], It also considers the Mecking-Kocks theory to relate the evolution of the statistically stored dislocation and it considers isotropic hardening behavior of the material. Here, the mean free distance has been modified and written as [35-36]:

$$\frac{1}{\lambda} = \frac{1}{d} + \frac{1}{t}$$

The average twin spacing is an inportant parameter for the deformation of TWIP steel. The same may be related to the twin volume fraction by the following relation [37]:

$$t = 2e \cdot \frac{1-F}{F}$$

Where, F is the twin volume fraction and e is the average twin thickness, which is independent of strain and of the order of one micrometer.The twinning kinetics may be derived from the Olson-Cohen' assumption [38], given as:

$$dF = (1-F) \cdot m d\varepsilon$$

Where, m is a function of SFE, which increases when SFE decreases. By integrating, the twin volume as a function of strain is obtained as:

$$F = 1 - \exp(-m\varepsilon)$$

Damage Behavior

The damage behavior of the formable multiphase steels has been successfully modeled by the Finite Element Analysis (FEA) [39] involving an adequate definition of the representative volume element (RVE). The RVE must be large enough to include the sufficient number of microvoids and microcracks, but at the same time small enough for stress and strain rate to be considered homogenous [40-41]. Simulations of dimple fracture mechanisms in TRIP steels have been carried out using GTN (Gurson-Tvergaard-Needleman) heory [42]. The GTN model considers that primary voids will grow after plastic deformation. Additionally, secondary voids emerges at characteristic equivalent strain.

In contrast, cohesive zone model (CZM), based upon separation law, has been used to describe the cleavage failure mechanism [43]. Moreover, the CZM is applied in the RVE between phases, to consider the interface de-bonding in the microstructure. When the normal and tangential components of separation reaches a critical values, and respectively, the continuum elements connected by cohesive elements becomes disconnected. The integral of traction Γ_0 as a function of the separation, up to the critical separation δ_0 provides the energy dissipated by the cohesive element at failure. The failure criterion for the cohesive zone is defined by the integration function, expressed as [44],

$$\Gamma_0 = \int_0^{\delta_0} T(\delta)d\delta$$

COMPUTATION APPROACH OF STEEL DESIGN

Thermodynamic Tools

Presently, several thermodynamic softwares are available for calculation of the thermodynamic parameters in steel. The salient features of the popularly used softwares like MTDATA [45], ThermoCalc [46] and DICTRA [47] are given below.

The MTDATA software allows equilibria to be calculated for the multicomponent systems from critically assesses data for their component binary and ternery subsystems. The application of MTDATA in steel research is useful in the areas of calculation of thermodynamic parameters of alloying elements, study of behaviors of particular alloying element under the influence of other alloying elements, approximation of solute partitioning effect, driving force for precipitation and phase composition calculation [48-49].

The ThermoCalc software has been developed for systems exhibiting complex heterogeneous interactions for strongly non-ideal solution phases [50-51]. Thermo-Calc allows explicit conditions on individual phase compositions or configuration while most of the similar software handles conditions on the overall composition only. The flexible way to set conditions is particularly useful when Thermo-Calc is used as a subroutine package in application programs, for example, in microstructure evolution or process simulations.

In parallel with the development of ThermoCalc, software for simulation of DIffusion Controlled TRAnsformations, DICTRA was also developed. It is closely linked with the ThermoCalc software, which provides all the necessary thermodynamic calculations. Some areas of application includes, homogenization of alloys [52], carburization and decarburization of steels [53], nitriding of steels [54], austenite/ferrite diffusional transformations in steels [55], calculation of TTT-diagrams [56] etc.

Microstructural Simulation Techniques

In the efforts on understanding the genesis of structure-property correlation, sharp interface simulation techniues like Cellular Automaton (CA) and Monte Carlo-Potts (MC) technique have found usefull application for prediction of morphology of the grains and to follow the nucleation and growth of the individual grains and their mutual interaction.

The microstructural simulation models concerning the kinetics of the evolution may be broadely calssified into two categories, Firstly sharp interface models and secondly diffused interface models [57].

In conventional modelling techniques, sharp interface is considered, with sudden change in the field variable or order parameter (when the state of the system is defined by the single variable defined as the order parameter). The difference may be conceptually visualized from the Fig. 3.

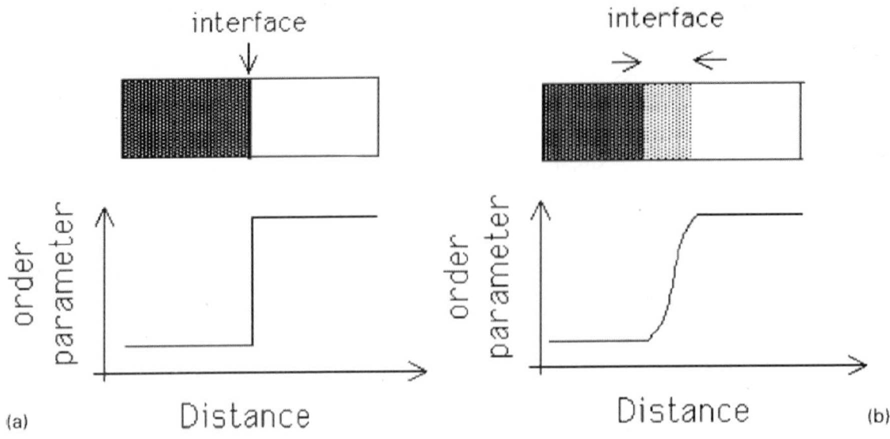

Fig. 3: (a) Sharp interface approach and (b) diffused interface approach [57].

The phase field simulation is basically a meso-scale simulation technique, based on diffused interface approach, considering a number of grains. The shape and mutual distributions of the grains are represented by the set of field variables, that are continuous in space and time [57]. Within a grain, the values of the phase field variables are constant, while at the interface between the two grains the phase field variables gradually vary between their values in neighboring grains. The equation for the evolution of phase field variables are derived on the basis of the chemical thermodynamics and kinetics. However, they do not deal with the behavior of individual atoms, so material properties must be introduced in the model through phenomenological parameters that are determined on the basis of the experimental and theoretical information.

In steel research, Phase field simulations has been applied to the wide variety of applications including the solidification [58], diffusive [59-60], and non-diffusive transformations [61-62].

Soft Computation Approach

In recent times, a group of soft computing techniques have been employed in the efforts of composition-process-microstructure-property correlation of steel based on the available information and imprecise knowledge. It is imperative to present here some of the potential techniques

Fuzzy Inference Systems

A fuzzy set is a set without a crisp or clearly defined boundary. The set contains elements with only a degree of membership, as in fuzzy logic the truth of any statement becomes a matter of degree. The membership value (or degree of membership) of each point in the input space is mapped to a value between 0 and 1 by a membership function (MF). Fuzzy linguistic descriptions are formal representations of systems made through fuzzy IF-THEN rules. Fuzzy IF-THEN rules are in the form of

IF (x_1 *is* A_1, x_2 *is* A_2, ……, x_n *is* A_n) **THEN** (y_1 *is* B_1, y_2 *is* B_2, ……, y_n *is* B_n)

where the linguistic variables x_i, y_j take the values of fuzzy sets A_i, B_j respectively. One of the most effective applications has been found in prediction of properties of TRIP-aided steels based only on the imprecise metallurgical knowledge [63].

Artificial Neural Network

An ANN model consists of a set of interconnected neurons having computing capabilities. The neurons interact with each other through connection strength, called weights. The process of adjusting connection weights with the objective of getting a better response is called learning or training.

The inputs X_i are multiplied by weights W_{ji} for a hidden node H_j; summation of all the $W_{ji}X_i$ is then added to a bias value θ_j and finally operated by a suitable transfer function (*f*). The operation may therefore be written as

$$H_j = f(\Sigma W_{ji} X_i + \theta_j) \quad .... (1)$$

Hidden layers contribute to the output nodes through a linear operation. The output Y can be written as:

$$Y = (\Sigma W_j H_j + \theta') \quad .... (2)$$

where W_j and θ' are new sets of weights and bias values. Fig. 1 presents the simplified structure of ANN technique. ANN has found extensive application in prediction of TTT diagram [64-65], CCT diagram [66], transformation temperature [67-68], prediction of mechanical properties of TMCP [69], dual phase [70] and TRIP aided steels [71].

Genetic Algorithm

Genetic algorithm (GA), is a evolutionary algorithm based stochastic global search method capable to mimic the principles of natural selection and natural genetics [72] to find the best solution for a specific problem. GA operates on a population of feasible solutions to produce better approximations to a solution applying the principle of survival of the fittest.

Technique Blends

The aforesaid techniques concerns particular transformation and/or microstructural state and are inadequate for complete description of composition-process-microstructure-property correlation under a given manufacturing schedule. The robustness as well as the inherent complexity and/or nonlinearity embedded in the process often limit the applicability of a single technique.

In the real world such limitations have been frequently circumvented by invoking the composite/hybrid approach. In this approach a given problem is judiciously segmented into different components and are solved by more than one techniques employed in parallel or sequential manner.

In view of the above, an attempt has been made here to present a hypothetical framework for composition-process-microstructure correlation in the case of continuously cooled hot strip coils by combining several models and computational techniques.

In the hot strip manufacturing, the first stage involves deformation of austenite followed by dynamic, metadynamic and static recrystallization of austenite grains. Here, the input variables are austenite grain size before deformation, temperature and amount of deformation (dislocation density) while the output variables are size of the recrystallized austenite grains and corresponding temperature.

Fig. 4 presents the scheme for modeling the dynamic, metadynamic and static recrystallization behavior. While the empirical correlations predict the grain size and volume fraction of austenite, the thermodynamic tools are capable to model the microstructures. In the case of low alloy addition and simple microstructure, CA and MCP methods are found effective when supported by appropriate thermodynamic conditions developed by taking the amount of deformation and temperature into account. The kinetic conditions may also be presented by considering the JMAK equation along with the additivity rule. For higher level of alloy additions, applicability of techniques like CA and MCP techniques are limited. In such occasions data driven techniques like ANN are found to be the appropriate tool. For the emerging alloys where data is sparse, rule based technique like fuzzy inference system may be considered.

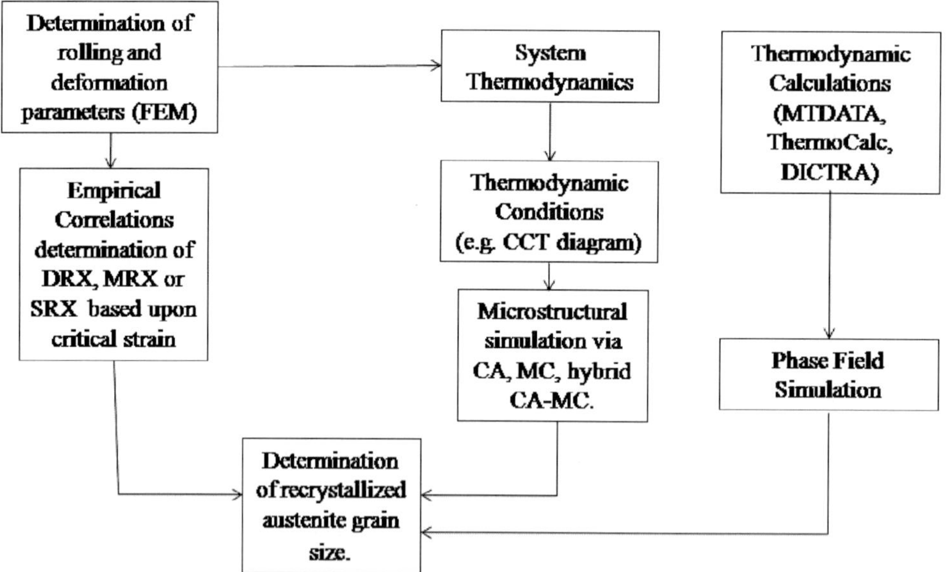

Fig. 4: Model for recrystallization during deformation in hot-strip mill

Output from the model presented in Figure 4, i.e., recrystallized volume and grain size of austenite and the corresponding temperature profile, is employed as the input variable for modeling the microstructure evolution during cooling on the run-out table followed by coiling, where ferritic and bainitic transformation of austenite is depicted by the concerned CCT diagram. Recently ANN technique has been successfully employed for prediction of CCT diagram for a given alloy [66]. Correspondingly, Thermal profile at this stage is determined by using the FEM. From the predicted CCT diagram transformation temperatures are determined for a given cooling profile, which provide the necessary thermodynamic conditions for simulation of microstructure by CA.

Phase evolution during coiling may effectively be solved in a similar manner to that of the run out condition. However, the role of contact pressure and curved profile of the coil make the heat dissipation model more complex than the run out table. In such cases controlled volume method perform more effectively than FEM.

Once the microstructural evolution of the hot strip coil predicted successfully for a given composition and process variables, the same can be subjected to estimation of mechanical properties by invoking the M-K models and/or rule of mixture as depicted in Fig. 2.

Hot strip production has been practiced in industries over the years and the information in the area is abundant. In such cases GA may be successfully employed to minimize the error between the experimental composition-process-property and predicted composition-process-microstructure correlation which make the prediction more reliable. Once the model is established with adequate reliability, the same can be used for multi objective optimization employing GA by invoking the compositional and mill constraints and desirability of various properties. In the effort to develop a comprehensive model for predicting mechanical properties from primary variables, viz. composition and rolling parameters, considering all the deformation and cooling schedule, deterministic models may fail to consider some particular features e.g., grain boundary precipitation/diffusion. Such issues can be mitigated by embedding uncertainty and imprecise knowledge through fuzzy logic.

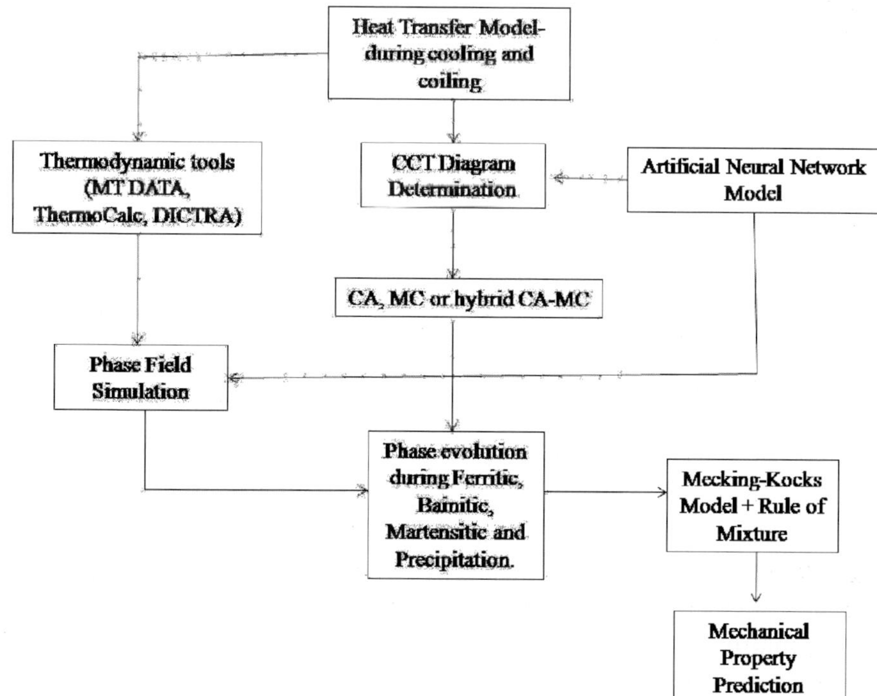

Fig. 5: Model for phase transformation during cooling and coiling in hot-strip mill

SUMMARY

The review of different models available for deterministic understanding of the path of microstructure and mechanical properties evolution in steel may be summarized as following:

1. Most of the model available so far stand in isolation and are independently inadequate to describe the complete spectrum of the composition-process-microstructure-property correlation in the time scale of the concerned process.
2. Large volume of information generated through industrial practice and research efforts over the years remain unutilized in the sense that significant amount of latent knowledge embedded in the volume of information remains unexplored.
3. Judicious hybridization of the physical and empirical models as well as the deterministic and fuzzy models may evolve the reliable framework for describing the complete path of microstructural evolution and correlating the same with mechanical properties.
4. Emergence of the potential simulation technique owing to the recent advancement in computing techniques and capability need to be effectively incorporated in the hybrid framework.
5. The successful hybridization of the potential techniques may develop the capability of *a priori* design of microstructurally engineered high performance steel to be manufactured under continuous cooling condition '*right at first time*'.

REFERENCES

[1] X. Li , M. Wang , F. Dua, Materials Science and Engineering A, Vol. 408, 2005, pp 33.

[2] A. N.Kolmogorov, Bull Acad. Sci. URSS (0 Sci. Math. Nat), Vol.3, 1937, pp 355.

[3] M. Avrami M., J Chern. Phys.. Vol 7, 1939, pp 1103.

[4] W. A. Johnson and R. F. Mehl , Trans. Arn. Inst Min. Engin. Vol 135, 1939, pp 416.

[5] M. Avrami M., J Chern. Phys., Vol. 8, 1940, pp 212.

[6] M. Toloui and S. Serajzadeh, J. Mater. Process. Technol., 2007, 184 (2), pp.345–353.

[7] E. Scheil, Arch. Eissenhuttenwes., Vol. 12, 1935, pp 565

[8] J. Svoboda, E. Gamsjager, F. D. Fisher and P. Fratzl, Acta Mat., Vol 52, 2004, pp 959.

[9] J. Svoboda, F. D. Fisher, P. Fratzl, A. Kroupa, Acta Mater., Vol 50, 2002, pp 1369.

[10] S. Sun and M. Pugh, Material Science and Engineering A, Vol 276, 2000, pp 167.

[11] D.P. Koistinen, R. E. Marburger. Acta Metall Mater, Vol. 7, 1959, pp 447.

[12] M. Victor Li, D.V. Niebuhr, L.L. Meekisho and D.G. Atteridge, Metall. Mater. Trans.B, Vol. 29, no.6, 1998, pp 661.

[13] B. Engl, Proc. Joint Workshop on Advanced hot rolling practice and products, Brussels, Belgium, October 2000, ECSC Steel/NEST.

[14] J. Wang and S. Van der Zwaag, Metall. Trans., Vol. 32, 2001, pp 1527.

[15] J. Zhao, Mater. Sci. Tech., Vol. 8, 1992, pp 997.

[16] C. Zener. Trans. AIME, Vol. 167, 1946, pp 550.

[17] R. W. K. Honeycombe and H. K. D. H. Bhadeshia. Steels, Microstructure and Properties. Edward Arnold, London, second edition, 1995.

[18] D. Quidort and O. Bouaziz. Can. Metall. Q., Vol. 25, 2004, pp 25.

[19] S. Curtze and V. T. Kuokkala, Acta Mater., Vol. 58, 2010, pp 5129.

[20] O. Bouziz, S. Allain, C.P. Scott, P. Cugy and D. Barbier, Current Opinion in Solid State and Materials Science, Vol 15, Issue 4, 2011, pp 148.

[21] K. Ishida and T. Nishikawa, Trans Jpn Inst Met, Vol 15, 1975, pp 225.

[22] R. E. Schramm and R. P. Reed, Metall. Trans. A, Vol 6, 1975, pp 1345.

[23] G. B. Olson and M. Cohen, Metall. Trans. A, Vol. 7, 1976, pp 1897.

[24] S. X. Zhou, . Mater. Proc. Tech., Vol. 134, 2003, pp 338.

[25] U. F. Kocks, J. Eng. Mater Technol (Trans ASME), Vol 98, 1976, pp 76.

[26] H. Mecking and U. F. Kocks, Acta Metall, Vol. 29, 1981, pp 1865.

[27] Y. Tomata, K. Kuroki, T. Mori, and I. Tamura, Mater. Sci. Eng., Vol. 24, 1976, pp 85.

[28] L. Samek, E. De Moor, J. Penning and B. C. De Coomen, Metall Trans. A, Vol 36, 2006.

[29] J. Bouquerel, K. Verbeken and B.C. De Coomen, Acta Mater., Vol 54, 2006, pp 1443.

[30] T. Gladman, I. D. McIvor, F. B. Pickering, ISIJ Int., Vol 210, 1972, pp 916.

[31] X. Zhang, A. Misra, H. Wang, M. Nastasi, J. D. Embury, T. E. Mitchell, R. G. Hoagland, and J. P. Hirth, Appl. Phys. Lett. Vol. 84, 2004, pp 1096.

[32] N. Tao, X. Wu, M. Sui, J. Lu and K. Lu, J. Mater. Res., Vol. 19, 2004.

[33] V. Randle, Acta Mater., Vol. 47, 1999, pp 4187.

[34] L. Rémy , Acta Metall, Vol. 26, 1978, pp 443.

[35] I. Karaman, H. Sehitoglu, A. J. Boudin, Y. I. Chumlyakov, H. J. Maier and C. N. Tome, Acta Mater., Vol. 48, 2000, pp 2031.

[36] O. Bouaziz, N. Guelton, Mater Sci Eng A, Vol. 246, 2001, pp 319.

[37] R. L. Fullman, Trans. AIME, Vol. 197, 1953, pp 447.

[38] G. B. Olson and M. Cohen, Met Trans. A, Vol. 6, 1974, pp 791.

[39] J. H. Song, H. W. Wang et. al., Computational Mecahnics, Vol. 42, no. 2, 2008, pp 239.

[40] S. Papaefthymiou, W. Bleck, U. Prahl, C. Acht, J. Sietsma, S. van der Zwaag, Proceedings of THERMEC 2003, International Conference on Processing & Manufacturing of Advanced Materials, 07–11 July 2003, Madrid, Spanien, Hrsg. T. Chandra et al., Part. 2. Materials Science Forum 426/432, Uetikon-Zu¨ rich: Trans Tech Publ., pp. 1355.

[41] W.M. Kaluza, M. Lake, L. Pesek, W. Bleck, Proceedings of Materials Week 2000, International Congress on Advanced Materials, their Processes and Applications, Munich, 25– 28 September 2000.

[42] V. Tvergaard, A. Needleman, Acta Metall. Vol. 32, no. 1, 1984, pp 157.

[43] X. P. Xu and A. Needleman, J. Mech. Phys. Solids, Vol 42, 1994, pp 1397.

[44] A. Cornec, I. Scheider and K. H. Schwalbe, Engng Fract Mech, Vol 70, 2003, pp 1963.

[45] R. H. Davies, A. T. Dinsdale, J. A. Gisby, J. A. J. Robinson, S. M. Martin, CALPHAD, Vol. 26, no. 2, 2002, pp. 229.

[46] M. Hillert, Phase Equilibria Phase Diagrams and Phase Transformations, (1999) Cambridge Univ. Press, ISDN o-521-56270-8.

[47] J. O. Anderson, T. Helander, L. Hoglund, P. Shi and B. Sundman, CALPHAD, Vol. 26, no. 2, 2002, pp 273.

[48] J. D. Robson and H. K. D. H. Bhadeshia, Materials Science and Technology, Vol.13, 1997, pp 631.

[49] You Fa Yin, Roy G. Faulkner Materials Science and Engineering A, Vol. 344, 2003, pp 92.

[50] B. Sundman, B. Jansson and J-O. Andersson, CALPHAD, 9 (1985) 153.

[51] B. Sundman, Anales de Fisica, Serie B, 86 (1990) 69.

[52] H. E. Lippard, C. E. Campbell, T. Bjorklind, U. Borggren, P. Kellgren, V. P. Dravid and G. B. Olson, Metall. Mater. Trans. B, Vol. 29, 1998, pp 205.

[53] L. Sproge and J. Agren, J. Heat Treat., Vol. 6, 1988, pp 9.

[54] H. Du and J. Agren, Z. Metallkd., Vol 86, 1995, pp 522.

[55] S. Crusius, L. Hoglund, G. Inden, U. Knoop and J.Agren, Z. Metallkd, Vol. 83, 1992, pp 729.

[56] Z.-K. Liu, Some aspects on theoretical calculation of TTTdiagram; Solid-Solid Phase Transformations, edited by W.C. Johnson, J.M. Howe, D.E. Laughlin and W.A. Soffa, The Minerals, Metals & Materials Society, 1994, pp. 39-44.

[57] R. Qin and H. K. D. H. Bhadeshia, Mater. Sci. Tech., Vol. 26, No. 7, 2010, pp 803.

[58] J. Tiaden, J Cryst Growth, Vol. 198/199, 1999,pp 1275–80.

[59] G. Pariser, P. Shaffnit, I. Steinbach, W. Bleck, Steel Res., Vol. 72, 2001, pp 354.

[60] I. Loginova, J. Ågren, G. Amberg, Acta Mater, Vol. 52, 2004, pp 4055.

[61] A.Artemev, Y.Jin, A.G.Khachaturyan, Acta Mater, Vol 49, 2001, pp 1165.

[62] O. U. Salman, A. Finel, R. Delville and D. Schryvers, J. Appl. Phys., Vol 111, 2012, pp 103517.

[63] S. Dey, S. Datta, P. P. Chattopadhyay and J. Sil, Computational Materials Science, Vol. 43, 2008, pp 501.

[64] S. Malinov, W. Sha and Z. Guo: Materials Science and Engineering A, 2000, 283, 1–10.

[65] M. Kundu, S. Ganguly, S. Datta, and P. P. Chattopadhyay: Materials and Manufacturing Processes, Vol. 24, 2009, pp 169.

[66] J. Wang, P. J. van der Wolk and S. van der Zwaag: ISIJ International, Vol. 39, 1999, pp 1038.

[67] L.A. Dobrzanski and J. Trzaska: Journal of Materials Processing Technology, Vol. 155-156, 2004, pp 1950.

[68] W. G. Vermeulen, P. F. Morris, A. P. de Weijer and S. Van der Zwaag: Ironmaking Steelmaking, Vol. 23, 1996, pp 433.

[69] M. Çöl, H.M. Ertunç and M. Yilmaz: Materials and Design, Vol. 28, 2007, pp 488.

[70] A. Bahrami, S. H. M. Anijdan and A. Ekrami: Journal of Alloys and Compounds, Vol. 392, 2005, pp 177.

[71] S. M. K. Hosseini, A. Zarei-Hanzaki, M. J. Yazdan Panah and S. Yue: ANN Materials Science and Engineering A, Vol. 374, 2004, pp 122.

[72] D. E. Goldberg, Genetic Algorithms in Search, Optimization and Machine Learning; Pearson-Education: New Delhi, 2002.

Characterisation of Residual Stress in Electron Beam Welded P91 Plates by Neutron Diffraction

Amrita Kundu[1], Peter John Bouchard[1], Santosh Kumar[2] and Gautam Kumar Dey[2]
[1]Materials Engineering, The Open University, Milton Keynes, MK7 6AA, UK
[2]Materials Science Division, BARC, Mumbai, India
Email: akundu05@gmail.com

ABSTRACT

This paper describes the characterisation of residual stress in electron beam welded P91 ferritic-martensitic steel plates (9 mm thick) by neutron diffraction. The novelty of the work lies in revealing the residual stress profile at a fine length scale associated with a ~ 1 mm wide fusion zone. A characteristic 'M' shaped distribution of stresses across the weld line is observed with high tensile peaks situated just beyond the HAZ/parent material boundary. Measured stresses close to the weld centre-line are significantly less tensile than the adjacent peaks owing to martensitic phase transformation during cool-down of the weld region. The effect of applying a second smoothing electron beam weld pass is shown to be undesirable because it increases the tensile magnitude and spread of residual stress.

KEYWORDS: *Electron Beam Weld, Residual Stress, Ferritic Martensitic Steel, Neutron Diffraction*

INTRODUCTION

Ferritic-martensitic steels containing 9-12 Cr are seen as offering the most promise for components such as headers and the main steam pipe in ultra-supercritical (USC) fossil-fired power plants, as well as for primary components in some of the design concepts for Generation IV nuclear reactors [1]. This is because 9-12 Cr steels offer superior creep strength to grades containing less chromium such as, for example, 2.25Cr-1Mo steels. Steels containing 9-12 Cr are also less expensive than austenitic stainless steels, and the fact that they have a higher thermal conductivity than austenitic steels means that they are less prone to the generation of thermal stresses when subjected to the thermal transients such as those that arise during plant outages and shut downs for scheduled maintenance.

While the advantages of 9-12 Cr steels are well recognized, the performance benefits often have not been realized in service operation due premature failures in the heat-affected zone (HAZ) of welded joints. The failure mechanism is type IV cracking [1], which refers to an enhanced rate of creep void formation in the fine-grained and intercritically-annealed HAZ of a weld. In recent years, significant efforts have been dedicated to understanding the influence that HAZ microstructures have in influencing the propensity for type IV failures in 9-12 Cr steels [1].

In this work we are focusing on modified P91 steel [2] as it is commercially available and currently in service. We wish to explore the influence that the choice of welding process might have in influencing the tendency for type IV failure. In particular, we have noted that there is some evidence to suggest that welds made with high-energy beam processes such as laser or electron beam welding may be more resistant to type IV cracking than welds made with conventional arc welding processes [1, 3]. It has been suggested that superior performance might be offered by laser- and electron-beam welds due to these processes generating a narrower HAZ than that for an arc welding process [1]. However, recent measurement work [1, 3] has also shown that when a P91 pipe girth weld was made with gas-tungsten arc (GTA) and submerged arc welding, substantial tensile residual stresses were generated in the type IV region. Moreover, these stresses were not entirely relieved by a conventional post-weld heat treatment (PWHT). It is possible, therefore, that residual stresses may also be contributing to type IV cracking in some way.

This article reports on the measurement of residual stresses in electron beam welds in a 9 mm thick P91 steel plate in the as-welded state. The purpose is to provide detailed characterisation of the residual stress field for validation of weld numerical models and to contribute to work investigating how the Electron Beam (EB) welding process influences the tendency for type IV cracking. Results for two plates are presented, with one containing a single pass weld and the other being made in two passes, that is a welding pass and a (cosmetic) smoothing pass. In this work, the residual stresses have been measured by neutron diffraction.

EXPERIMENTAL

The base material studied in the present investigation is P91 steel having the chemical composition given in the Table 1. As-received hot-rolled plate material was first normalised at 1050 °C for 1min/mm and then tempered at 770 °C for 3 min/mm, followed by an air cool. Several plate samples of dimensions 245 mm by 145 mm were machined from the tempered material which was nominally 10 mm thick. The samples were then surface ground on both sides in order to remove a slight waviness of the plates originating from the hot rolling fabrication process. The final thickness of the plate material ranged from 9.2 mm to 9.8 mm.

Table 1: Chemical composition of the steel (remaining Fe)

C	Mn	Zr	Si	P	S	Cr	Mo	Ni	Cu	Al	N	Nb	Ti	V
0.106	0.443	0.005	0.221	0.018	0.0008	8.965	0.901	0.212	0.045	0.010	0.0464	0.073	0.004	0.194

Two pairs of surface ground plates were joined by an EB welding process with a voltage of 60 kV, current of 70 mA and welding speed of 1 m/min. No pre-heat was applied and the plates were spot welded along the seam. The plates were held rigidly using pneumatically controlled clamps during welding. An additional smoothing pass was applied to the second plate with a voltage of 60 KV and defocused beam current of 30 mA, again at a welding speed of 1 m/min. The as-welded residual stresses in both of these plates were studied by neutron diffraction supported by metallographic characterisation, as described below.

Cross sectional samples from the ends of both EB welded plates by electro-discharge machining (EDM). The samples were first ground using 400 grade Emery paper and then metallographically polished to a 1 μm finish. The macrostructure of the weld fusion zone was revealed by etching with Villela agent following which a hardness survey was undertaken using a measurement pitch of 0.5 mm and a Vickers indenter with a load of 0.3 kg.

Neutron diffraction residual stress measurements were carried out on the ENGIN-X beam-line at the ISIS facility of the Rutherford Appleton Laboratory in the UK. At this facility a spallation source is used to produce a "white" beam of variable wavelength neutrons. Time of flight is used to determine the lattice spacing in a gauge volume defined by sets of incoming slits and the capture angle of the detectors. Because of the very narrow width of the EB weld fusion zone, 1 mm detectors were mounted and used for all the measurements. The incoming beam was controlled by slits to give a (1x1x1) mm^3 gauge size for strains measured in the longitudinal direction and an extended gauge length of 10 mm for measurements in the transverse and normal directions. In each welded plate measurements were made through the thickness at the weld centre-line and along a line positioned 1.5 mm below the top surface of the plate. In the case of the plate, which received an additional smoothing pass, this 1.5 mm below line is at the interface between the two passes. The Rietveld refinement method [4] and General Structure Analysis System software were used to analyse the measured spectra. It was ensured that the range of each spectrum selected for analysis contained at least 5 identifiable peaks from sets of crystallographic planes.

The lattice strain in a given direction is determined from the lattice spacing, *d*, measured at the location of interest, and the lattice spacing measured in representative (or ideally the same) material that is free from macroscopic stress; the latter quantity, the stress-free lattice parameter, is denoted as d_0. For direct strains in direction *i* the strain ε_{ii} is given by:

$$\varepsilon_{ii} = \frac{d_{ii} - d_{0ii}}{d_{0ii}}$$

If strains in three orthogonal directions are measured then the direct stresses, S_{ii}, can be determined using the following equation:

$$\sigma_{ii} = \frac{E}{(1+\nu)}\left[\varepsilon_{ii} + \frac{\nu}{(1-2\nu)}\left(\varepsilon_{ii} + \varepsilon_{jj} + \varepsilon_{kk}\right)\right]$$

Where *E* is the bulk Young's modulus and *ν* is Poisson's ratio. In the present case *E* = 218 GPa and *ν* = 0.3 were used for both the P91 base metal and fused zone [2].

A 2 mm thick strip was electro-discharge machined from the end of each plate and position dependent stress-free lattice parameter reference values, d_0, measured, relative to the weld centre-line and top surface, in order to take account of gradients in both the microstructure and strain history of the welded plates. For each point lattice parameters measurements were made in the longitudinal, transverse and the normal directions and an orientation independent lattice parameter determined assuming plane stress state conditions.

RESULTS AND DISCUSSION

Figure 1(a) shows the macrostructure of a single pass EB welded plate, with the line along which the cross-weld hardness profile at 1.5mm was determined superimposed. The width of the fused zone tapers from 1.2mm at the top of the plate to about 0.8mm at the bottom. The width of the HAZ appears to extend about 0.8mm beyond the fusion boundaries on both sides. The hardness profile, Figure 1(c), matches the visible extent of the HAZ boundaries, rising rapidly from the parent material level of 230HV (at ± 1.3mm) to a plateau of about 480HV magnitude that spans the fusion zone. The plateau hardness level is indicative of a full martensitic microstructure which is seen to extend to a width of 1.4mm (slightly wider than the fused width of 1.1mm). The effect of applying a second smoothing EB weld pass is evident in Figure 1(b). Here the fusion boundary associated with the second pass extends to a depth of about 1.5mm below the top surface. The maximum width of the second pass at the top surface is about 2.5mm. The hardness profile at 1.5mm depth is almost identical to the single EB welded case apart from being 0.2mm wider owing to the influence of the smoothing pass.

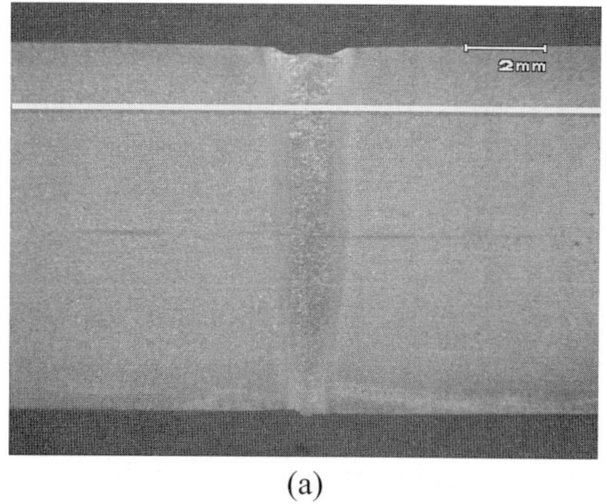

(a)

(b)

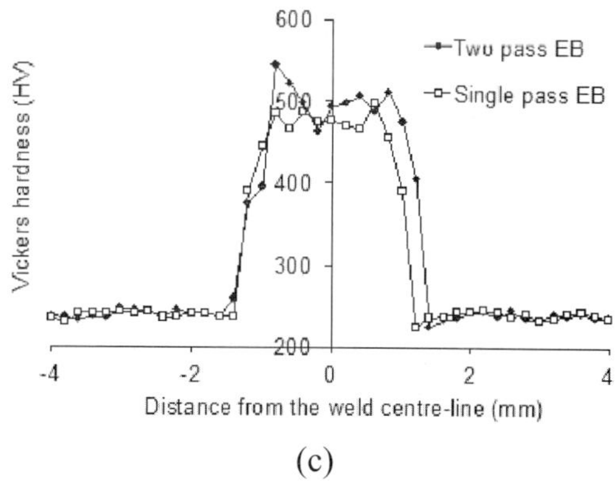

(c)

Fig. 1: Photographs showing the macrostructure with superimposed measurement lines at 1.5 mm below the top surface of the weld for (a) the single pass EB welded plate and (b) the two pass (smoothed) EB welded plate.

Figure 2 shows the variation in stress-free lattice parameter across both types of EB welded plate derived from neutron diffraction measurements on the 2mm thick strips (described earlier). No distinction was made between stress-free lattice parameter measurements from the two instrument detector banks or with measurement orientation as no systematic variation was observed. Despite a large degree of scatter in the results (Figure 2), a correlation with the martensitic zone and HAZ boundaries (± 0.8mm and ± 1.35mm respectively) is evident. The greater values of lattice parameter in the martensitic zone, and to a lesser extent in the HAZ, may be attributed to the lattice being supersaturated with carbon. The stress-free lattice spacing measurements at 1.5mm from the weld centre-line and beyond, show little scatter; these data represent parent material unaffected by the welding process. The idealised line in this region (see Figure 2) is based on the mean of all "stress-free" measurements made in parent material; that is in the 2mm strips, at the remote ends of the plates and in small parent cuboids. Stress-free lattice parameter measurements through the thickness in the fusion zones of the 2mm thick strips show a large scatter that appears to be random. For this reason all the lattice spacing measurements made with gauge volumes either fully immersed or partially within the martensitic transformed zone, that is within ±0.75mm of the weld centreline, have been averaged to give a single martensitic weld metal value for use in the strain and stress calculations. A linear variation in stress-free lattice parameter from the average weld to average parent values is assumed in the HAZ. This idealised variation in reference lattice parameter across the welds is marked in Figure 2.

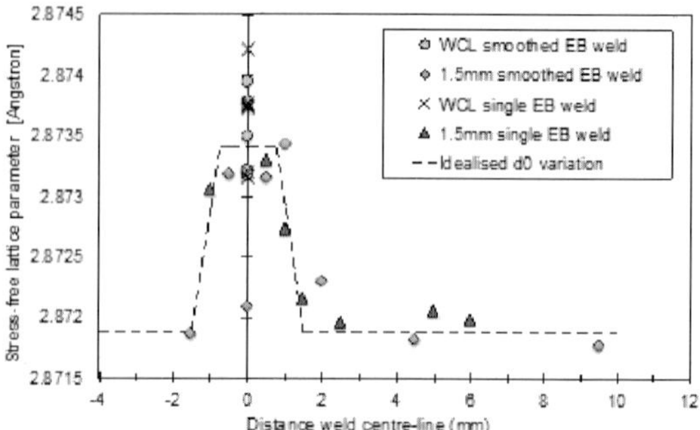

Fig. 2: Stress-free lattice parameter variation along the weld centre-line measured by neutron diffraction in both the single EB weld and two pass smoothed EB welds.

Figure 3(a) shows the distributions of residual stress acting in the longitudinal, transverse and normal directions measured by neutron diffraction across the single pass plate EB weld centre-line at 1.5 mm below the top surface. The longitudinal stress profile exhibits an 'M' shape having tensile peaks of about 475 MPa magnitude that are almost symmetric about the weld centre-line and roughly coincide with the maximum extent of the HAZ. Interestingly the stresses are much lower in a central zone 1.5 mm wide that spans the weld centre-line. The width of this zone correlates exactly with the weld martenisitically transformed region (plateau) revealed by the hardness profile of Figure 1(c). The normal component of residual stress has lower magnitude but follows the same profile as the longitudinal component, whereas the transverse stress varies smoothly across the weld apart from one point (at -1.0 mm) that may be an experimental anomaly.

The corresponding measured distributions of residual stress in the EB welded plate with the smoothing pass are shown in Figure 3(b). Again the longitudinal stress profile exhibits an 'M' shape with a tensile peak stress approaching 500 MPa. But in this test specimen the minimum stress appears to be displaced 1 mm from the notional centre-line of the first EB weld pass. This shift could be associated with a lateral offset in the position of the smoothing pass relative to the first pass, or measurement positioning error, or a combination of the two effects. Like the single EB weld the longitudinal stresses rise rapidly in the HAZ, but here the high tensile region is significantly (1.5 mm) wider. Both the normal and transverse components of residual stress have lower magnitude than the longitudinal component but follow the same shape.

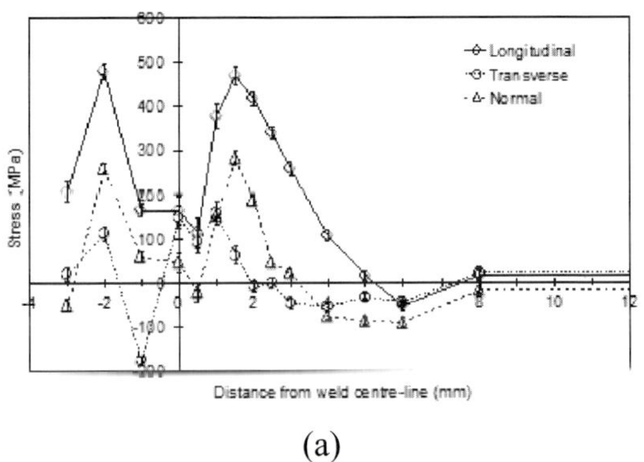

(a)

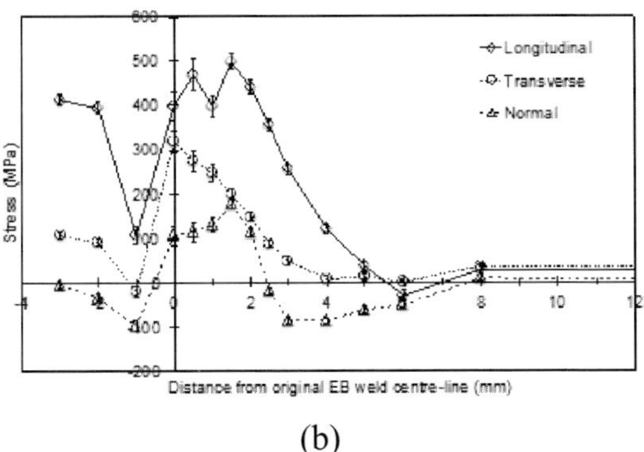

(b)

Fig. 3: Residual stress distributions in the as-welded plates at 1.5 mm below the top surface measured by neutron diffraction; (a) the single pass EB weld and (b) the two pass smoothed weld.

The stresses acting through the thickness of the single pass EB weld are shown in Figure 4(a). The most notable feature of these profiles is the low tensile magnitude in the longitudinal direction and the compressive nature of the normal stress profile. The corresponding results for the smooth weld test specimen are shown in Figure 4(b). The magnitudes of the stresses are higher for this case in the top 3 mm of the plate thickness; this is because stresses are being sampled in material adjacent to the martensitic zone of the smoothing pass. Interestingly the stresses profiles from 4 to 9 mm below the top surface from the two welds closely match each other. This evidence suggests that neutrons have sampled the lower part of the main EB weld fusion zone in both plates. Although the smooth variations in measured stress through the wall gives confidence in these results, the weld metal measurements have high uncertainty owing to the observed scatter in stress-free reference measurements (Figure 2).

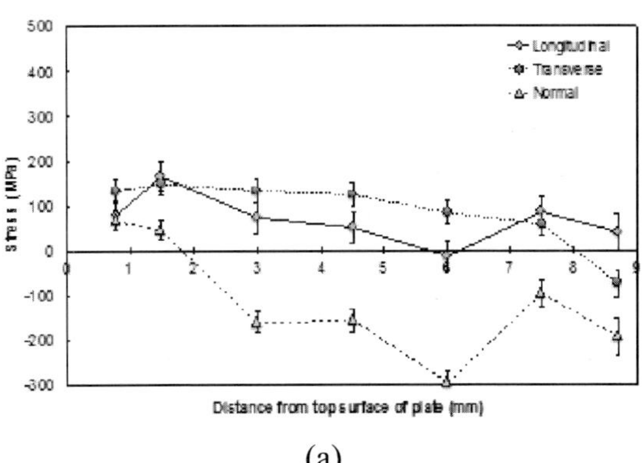

(a)

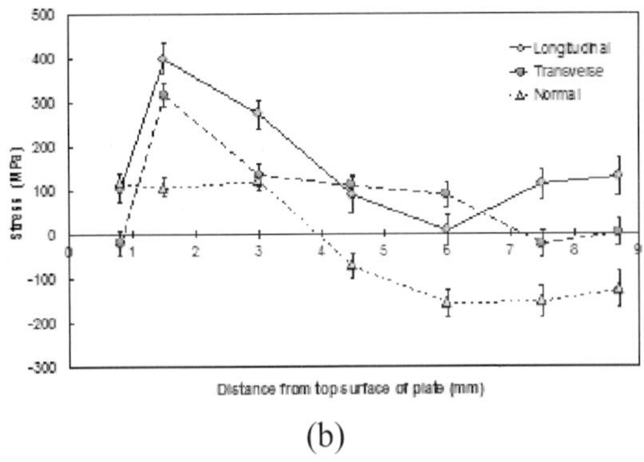

(b)

Fig. 4: Measured residual stress distribution through the thickness weld at the weld centre-line (a) the single pass EB weld and (b) the two pass smoothed weld.

As expected the highest tensile stresses across the EB welded joints are in the longitudinal orientation, since this is the direction that is exposed to the highest contraction strain as a consequence of the welding thermal cycle. The second highest tensile stresses are in the normal direction. This is unusual because the magnitude of the normal stress component is usually low in conventional fusion weldments. The normal stresses are greater than the transverse for an EB weld because the weld fusion width to depth aspect ratio is very small (≈ 0.1) whereas for a conventional weld bead this ratio is usually greater than unity.

The most striking feature of both M-shaped cross-weld stress profiles is that stresses close to the weld centre-line are significantly less tensile than at the HAZ boundary and in the parent plate material beyond. This feature can arise in steels that undergo a solid state phase transformation from austenite upon cooling to bainite and martensite [1, 5, 6]. It is notable that the low stresses coincide exactly with the extent of the high hardness plateau shown in Figures 1(a) for the single pass EB weld and Figure 1 (b) for the two pass (smoothed) EB weld, that characterises the extent of the martenisitic zone created during cool-down. Reduced levels of residual stress in martensitically transformed regions of multi-pass P91 weldments have been predicted by [5] and measured by Yaghi et al [7] using X-ray diffraction. More recently a zone of compression associated with the final weld bead deposit of a multi-pass weld in a P91 pipe has been measured by the contour method [3] and also by neutron diffraction [3] with the latter study specifically linking the compressive region to martensitic transformation. The current results are also broadly consistent with those of Francis et al. [1] for a single weld bead deposited on a 20 mm thick plate made of SA508 steel, and Dai et al [8] who interrogated low transformation temperature groove welds in 12 mm thick plates. In both of these cases the low stress region extended throughout the fused region into the HAZ and this appeared "to push" highest tensile stresses laterally outwards towards and beyond the HAZ/parent material boundary.

The potential for development of flaws during cooling after welding (e.g. hydrogen cracking), and the development of low toughness microstructures in either the weld metal or HAZ are of significant concern during fabrication and for safe life management. In P91 steel the toughness of the fine grained HAZ is a particular concern as this region is susceptible to type IV cracking [1, 3] when exposed to load at high temperature. The present measurements show that high tensile triaxial residual stresses are generated by the EB welding process in the Type IV region. Of course, P91 weldments are always post weld heat treated (PWHT) prior to engineering use for metallurgical reasons and this treatment can be effective in relieving high levels of residual stress introduced by welding. But Paddea et al. [3] has reported hydrostatic stresses of up to 100 MPa remaining in the Type IV zone of a conventional multi-pass pipe butt weld after PWHT.

Residual stress fields that vary rapidly over fine length-scales are difficult to measure with high reliability. Here special care was taken to use a small gauge volume ($1mm^3$) – at the expense of longer acquisition times, position the sample and reference strips to within tens of microns using theodolites and select appropriately spaced measurement locations. The measured results are suitable for validating finite element prediction of residual stresses in electron beam welded plates, although more detailed information from other techniques such as the Contour method [9] is desirable to confirm full field predictions.

CONCLUSIONS

Residual stresses in 9 mm thick EB welded plates have been measured using neutron diffraction. A characteristic 'M-shape' distribution of stresses across the weld line has been observed with high tensile peaks situated just beyond the HAZ/parent material boundary in both the single pass and two pass EB welded plates, that is immediately outside those regions that have either fully or partially austenitised during welding. The length-scale of this stress distribution is very short with the tensile peaks approximately 3 mm apart spanning a 1 mm wide fused zone. Measured stresses close to the weld centre-line are significantly less tensile than the adjacent peaks owing to martensitic solid state phase transformation during cool-down of fused weld metal and adjacent austenitised material within the HAZ. Residual stresses in the EB welded plate with a second smoothing pass have higher tensile magnitude and greater spread than the single pass case. Therefore from a residual stress and structural integrity point of view application of a smoothing (second) EB weld pass is undesirable.

REFERENCES

[1] Francis JA, Mazur W, Bhadeshia HKDH, Materials Science and Technology, Vol. 22, 2006, p. 1387.

[2] Haarmann JC, Vallant B, Vandenberghe W, Bendrick AA, The T91/P91 Book, 2nd edn. Vallourec & Mannesmann Tubes, 2002.

[3] Paddea S, Francis JA, Paradowska AM, Bouchard PJ, Shibli IA, Materials Science and Engineering A, 2012, Accepted.

[4] Von Dreele RB, Jorgensen J, Windsor CG, Journal of Applied Crystallography, 1982, p. 581.

[5] Yaghi AH, Hyde TH, Becker A, Sun W, Journal of Strain Analysis, Vol. 43, 2008, p. 275.

[6] Francis JA, Turski M, Withers PJ, Materials Science and Technology, Vol.25, 2009, p. 325.

[7] Yaghi AH, Hyde TH, Becker A, Sun W, Science and Technology of Welding and Joining, Vol. 16, 2011, p. 232.

[8] Dai H, Francis JA, Stone HJ, Bhadeshia HKDH, Withers PJ, Metallurgical and Materials Transactions A, Vol. 39A, 2008, p. 3070.

[9] Prime MB, Journal of Engineering Materials and Technology, Vol. 123, 2001, p. 162.

Thermo-mechanical Processing and Process Modeling of Power Plant Materials

A.K. Bhaduri, Dipti Samantaray and Sumantra Mandal
Materials Development and Technology Group, Indira Gandhi Centre for Atomic Research, Kalpakkam 603102, India
Email: bhaduri@igcar.gov.in

ABSTRACT

The paper discusses the details of process modelling for fabrication of components used in sodium-cooled fast reactors, in which 316LN austenitic stainless steel and modified 9Cr-1Mo steel are used as the major structural materials. For process modelling of thermo-mechanical processes, processing maps based on the dynamic material modelling approach have been developed for both the materials to visually represent their inherent deformation behaviour as a function of strain rate and temperature. Detailed microstructural investigations have also been carried out to validate the safe and unsafe domains of the processing maps. Based on these processing maps, microstructural studies and variation in activation energy, the parameters for the processing these materials have been optimised.

KEYWORDS: *316Ln Stainless Steel, Modified 9Cr-1Mo Steel, Hot Isothermal Compression, Flow Behaviour, Processing Map, Activation Energy*

INTRODUCTION

In power plants, the structural materials used define the upper temperature limit of the thermal power cycle and, hence, the efficiency of the power plant. In nuclear power plants, apart from elevated temperature thermal and mechanical properties, compatibility with coolant and corrosion behaviour, the irradiation properties are also taken into consideration for selection of suitable structural materials. In sodium-cooled fast reactors (SFRs), in which the material is exposed to high fast neutron flux, sodium environment and elevated temperatures, austenitic stainless steels are chosen as the structural material for the in-core, primary and secondary circuit components, while chromium-molybdenum ferritic steel is chosen for the steam generators [1, 2]. Proper fabricability to the required product forms is the key to optimum utilisation of the chosen materials. The components of the reactor and steam generators are fabricated through various thermo-mechanical processes, like extrusion, rolling and forging, etc. [2]. During these forming processes, the material undergoes various microstructural changes, which affect the mechanical properties of the final component and in turn their in-service behaviour [2]. To ensure prolonged service life of the components control of the microstructural quality and soundness of the product is of paramount importance. As the microstructural quality of materials is dependent on the forming process parameters, strain rate and temperature [2,

3], attention has to be paid towards their optimization. For this purpose, process modeling has proven to be a powerful tool. Dynamic materials model (DMM) [4] is one such tool, and DMM-based processing map, which gives a visual representation of the inherent material behavior in the strain rate and temperature domain, are used as a guide to choose the suitable domain for thermo mechanical processing. This paper discusses DMM-based thermo-mechanical process modeling of 316LN austenitic stainless steel and modified 9Cr-1Mo steel, which are used as materials of construction of in-reactor components and steam generators, respectively, in SFRs.

EXPERIMENTAL

The chemical composition (in wt-%) of 316LN stainless steel (SS) used in the present study is Fe-17.57Cr-12.15Ni-2.53Mo-0.025C- 0.14N- 1.74Mn- 0.20Si-0.0041S- 0.017P, while that for the modified 9Cr-1Mo steel is Fe-8.95Cr- 0.91Mo-0.102C- 0.059N- 0.37Mn- 0.23Si- 0.11Ni-0.03Al-0.182V-0.075Nb-0.003Ti-0.064Cu. Cylindrical specimens of 10 mm diameter and 15 mm length were fabricated from these steels. Uniaxial compression tests at constant true strain rates of 0.001, 0.01, 0.1, 1 and 10 s^{-1} were performed in the temperature range 1073-1423K for 316LN SS and 1123-1373K for modified 9Cr-1Mo steel. All the specimens were given 50% deformation after which they were quenched in water to freeze the microstructure. The experimental load-stroke data were converted to true stress–true plastic strain curves using the standard procedure [2, 3]. The water-quenched samples were cut along the longitudinal direction to prepare metallographic specimens. The diamond polished metallographic specimens of 316LN SS were electrolytically etched in 10% ammonium persulphate solution, while those of modified 9Cr-1Mo steel were etched with Villela's reagent. Optical microscopic examination was carried out to study the microstructures in the maximum deformation zone of the samples.

THE APPROACH

In process modeling, the most important pre-requisite is to understand and represent the flow behaviour of the material as a function of strain, strain rate and temperature [2, 3]. Depending on the objective to be accomplished through process modeling, many models have been developed to represent this complex behaviour. DMM is one such model which is intended to visually represent the inherent dynamic behaviour of the material in a two-dimensional temperature-strain rate frame at specified strain levels. According to DMM, the work-piece subjected to hot deformation is a nonlinear dissipater of power [4]. The power (P) input to the work-piece at any deformation level consists of two complementary parts, viz. power content (G) and power co-content (J) [2-4]. The power content represents the fraction of the power lost through temperature rise and the power co-content represents the fraction used for microstructural changes, with the partitioning between the two being decided by the strain rate sensitivity parameter (m) [2-4] given by:

$$m=\partial(ln\sigma)/\partial(ln\dot{\varepsilon})\approx\frac{\dot{\varepsilon}\sigma d(ln\sigma)}{\sigma\dot{\varepsilon}(ln\dot{\varepsilon})}=\frac{\dot{\varepsilon}d\sigma}{\sigma d\dot{\varepsilon}}=\frac{dJ}{dG}$$

where σ is flow stress and $\dot{\varepsilon}$ is strain rate. The J content is represented as:

$$J = \int_0^{\sigma} \dot{\varepsilon} d\sigma = \frac{\sigma \dot{\varepsilon} m}{m+1}$$

with J achieving a maximum value when m becomes 1. The dimensionless parameter, obtained by normalizing J with J_{max}, is known as the efficiency parameter η [2-4]. The plot showing the variation of η with temperature and strain rate is known as power dissipation map. This map exhibits the different domains that can be correlated to the microstructural mechanisms operative in that domain which are responsible for the deformation behaviour of the material. Usually, dynamic recrystallization (DRX), dynamic recovery and super-plasticity are considered as the favourable microstructural mechanisms for better deformability of the material, while localized slips, shear band formation, wedge cracking and void formation at hard particles, and dynamic strain aging are considered as the mechanisms that promote flow instability [5] that lead to poor deformability. However, the power dissipation map alone cannot be used to identify the exact domains of instability. Therefore, a continuum instability criterion based on the extremum principles of irreversible thermodynamics as applied to large plastic flow [4, 5] is used to identify the exact regimes of flow instabilities. This instability criterion parameter is expressed as:

$$\xi(\dot{\varepsilon}) = \frac{\partial ln[m/(m+1)]}{\partial ln \dot{\varepsilon}} + m < 0$$

The variation in the instability parameter is plotted in temperature and strain rate windows to obtain the instability map, and the contour with $\xi(\dot{\varepsilon}) = 0$ is used as the boundaries between the safe and unsafe domains. The instability map superimposed on the power dissipation map constitutes the processing map [2]. The processing map visually represents the stable and unstable dynamic behaviour of the material at various combinations of strain rate and temperature, and is therefore directly used for choosing the processing domain as per the requirement. The results obtained from the processing map can also be substantiated by studying the variation of activation energy in that domain. It has also been seen that in stable domains of the processing map the apparent activation energy Q remains almost constant [2, 3]. Hence, variation of Q is used as an additional indicator for the optimum processing domain [2, 3]. The apparent activation energy Q for the material is given by:

$$Q = \frac{k}{m} \frac{\partial ln\sigma}{\partial ln(1/T)}$$

RESULTS AND DISCUSSIONS

316LN Stainless Steel

The representative flow curves of 316LN SS obtained from compression tests are shown in Fig. 1, from which the effect of strain and temperature on the flow stress could be observed.

As the strain rate sensitivity of the material plays an important role in power distribution between the J-content and G-content, the value of m has been calculated at various sub-intervals of strain rate and temperature. Over this entire domain, m varies between

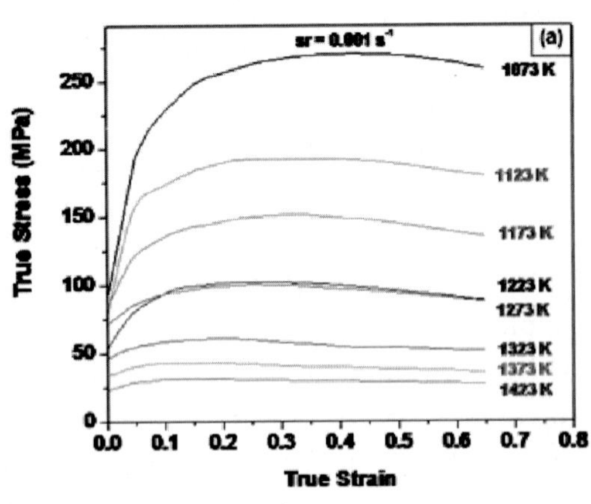

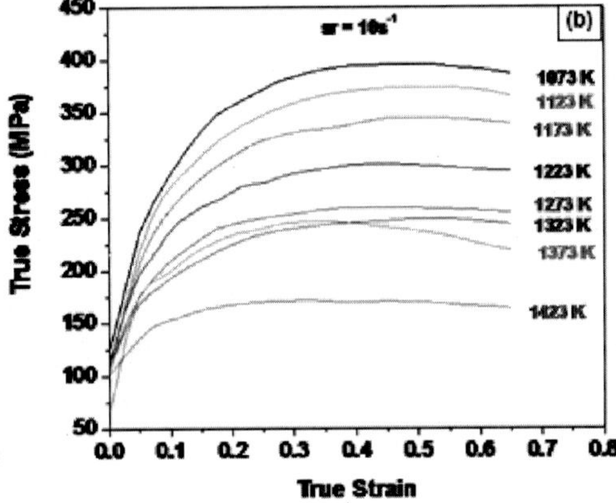

Fig. 1: Flow curves of 316LN SS at strain rates of (a) 0.001 s^{-1} and (b) 10 s^{-1}

–0.019 and +0.35. For this steel, *m* is positive at all conditions except in the region bounded by 1073–1123K and 0.1–1 s^{-1}. These values of *m* have been used for developing the processing map shown in Fig. 2. For ease of analysis, the map has been divided into six domains (a) to (f) as discussed below.

Domain (a) in occurs at about 0.01 s^{-1} and 1123K, with the peak η being 35% and *Q* being ~200 kJ/mol. The representative microstructure of this domain, as given in Fig. 2(a), shows initiation of DRX in the microstructure with the characteristic of necklace structure. Though this domain is a safe domain, it cannot be considered for the thermo-mechanical processing because of its proximity to the unstable domain in the processing map.

Domain (b) is spread over 0.001–0.01 s^{-1} and 1173–1273K, with the peak η being 45–30%. The representative microstructure of this domain (Fig. 2b) shows almost completely equiaxed DRX microstructure. Considering the microstructural quality this domain can be chosen for thermo-mechanical processing. However, the *Q* of this domain varies in the range of 100 to 500 kJ/mol. The rapid variation in both efficiency and activation energy indicates the sensitivity of grain size to the processing parameters. Therefore, this domain would not allow much flexibility for controlling the process parameters.

Domain (c) is another safe domain bounded within 0.001–0.05 s^{-1} and 1350–1423K with peak η of 50%. The representative microstructure of this domain is shown in Fig. 2(c). Like domain (b), the microstructure corresponding to this domain also shows completely equiaxed DRX grains that are comparatively larger in size. In this domain, *Q* varies in the range of ~100–200 kJ/mol.

Domain (d) occurs at about 0.1 s^{-1} and 1323K. This is the fourth safe domain with peak η of 45%.The representative microstructures of this domain (Fig. 2d) corresponding to this peak η shows evidence of DRX. However, the microstructure consists of equiaxed DRX grains and elongated deformed grains. The *Q* of this domain varies between 100-500 kJ/mole. This domain is not suitable for thermo-mechanical processing because of its proximity to the unsafe domain.

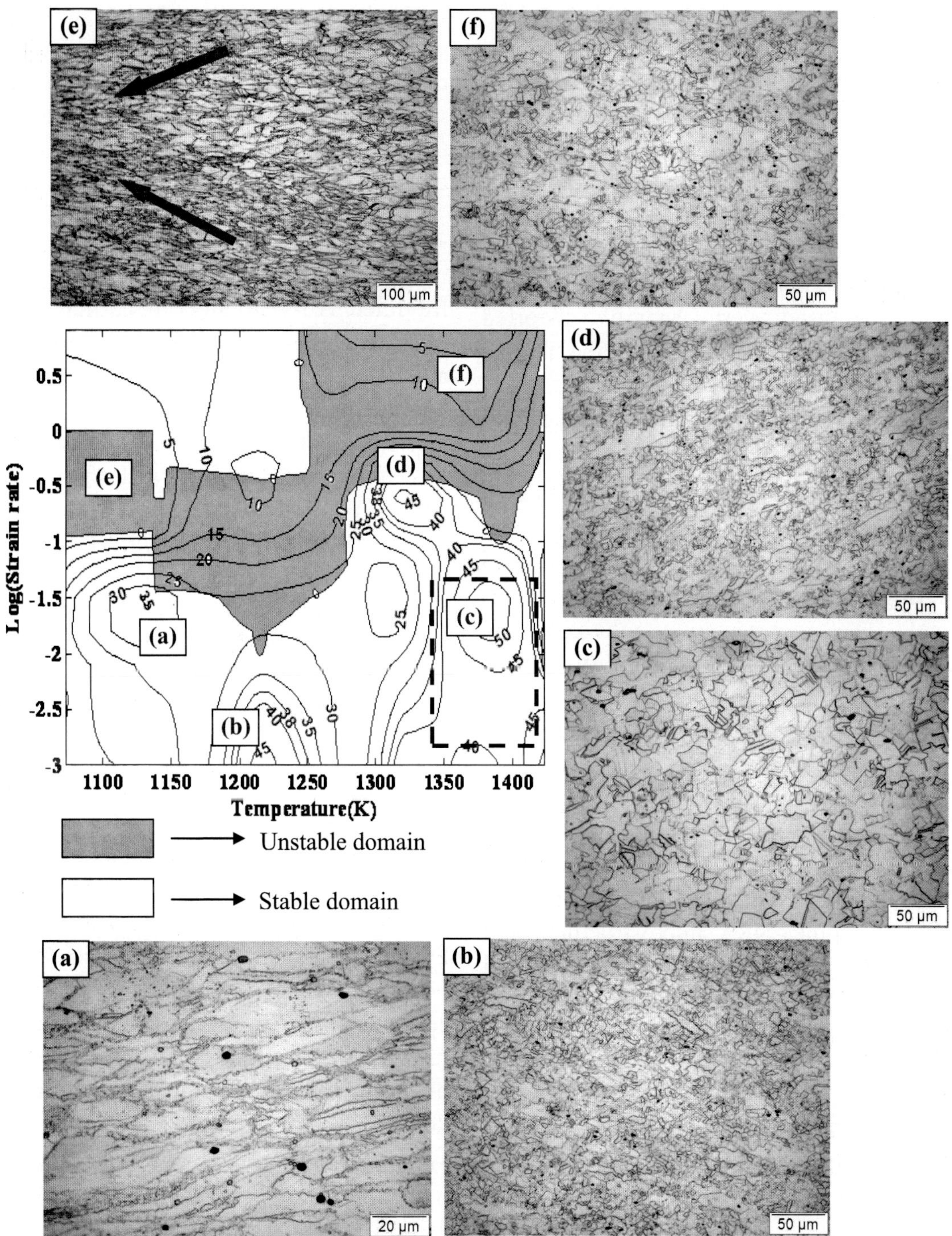

Fig. 2: Various domains of processing map of 316LN SS and their representative microstructures at: (a) 1123K, $0.01s^{-1}$; (b) 1223K, $0.001s^{-1}$; (c) 1373K, 0.01 s^{-1}; (d) 1323, $0.1s^{-1}$; (e) 1123K, $1.0s^{-1}$ (f) 1373K, $1s^{-1}$

Domain (e) is an unstable domain bounded within 0.1–1 s^{-1} and 1073–1123K. This domain matches with the local negative strain rate sensitivity (LNSRS) domain of the material as the activation energy map of this domain shows negative values of *Q*. The microstructure obtained in this LNSRS domain (Fig. 2e) shows presence of some moderately distorted grains along with localised very heavily distorted grains (marked by arrows in the figure) without any evidence of DRX. These are clear indications of the occurrence of inhomogeneous deformation in the matrix.

Domain (f) is the larger unsafe domain with positive *m* and η values in the range 5–30%. The *Q* in this domain also showed a large variation from ~200 to 500 kJ/mol. The representative microstructure corresponding to this instability region, given in Fig. 2(f), does not show any sign of the conventional instabilities like adiabatic shear band, localized flow or inter-crystalline cracking. The small nucleated grains in the microstructure indicates occurrence of DRX. However, the microstructure is significantly different from the other DRX microstructures and consists of more number of elongated (~54 μm) deformed grains along with a few equiaxed DRX grains. The random bimodal distribution of the grains can be taken as the signature of instability for this steel in this domain as this can lead to inferior mechanical properties in the material.

From the above analysis of the domains exhibited by the processing map and the microstructural observations, it can be concluded that the domain (c), which is bounded within the temperature range 1350–1423K and strain rate range 0.001–0.05 s^{-1} is the most suitable domain for hot working of this grade of 316LN SS with the variation in *Q* being moderate (average value of 150 kJ/mol) over a large domain (shown by a box in the map). In this domain, the efficiency variation is 45±5%.

Modified 9Cr-1Mo (Gr-91) Steel

Modified 9Cr-1Mo steel has a martensitic microstructure at room temperature and transforms to austenite over the temperature range of 1103–1143K [6]. The representative flow curves of the steel at different temperatures and strain rates, given in Fig. 3, shows that this material exhibits work hardening, thermal softening and strain rate hardening. The strain rate sensitivity parameter *m* for this steel varies in the range 0.04–0.3. The processing map for this material is shown in Fig. 4. The major domains exhibited in this map are discussed below.

Domain (a) occurs at about 0.1 s^{-1} and 1123K and is an unsafe zone with an average η of 20%. In this domain, *m* varies from 0.08 to 0.15. The lower strain rate sensitivity indicates that the material has lower ductility attributable to phase changes occurring in this region. The microstructure corresponding to this domain, given in Fig. 4(a), shows low intensity initiation of localized slip identified. The *Q* in this domain is ~100 kJ/mol.

Domain (b) occurs at low strain rates of 0.001–0.01s^{-1} and medium temperatures 1173–1223K, exhibiting a peak η of 38% with *Q* of ~200-500 kJ/mol. However, this domain is an unsafe' region as the instability parameter $\xi(\dot{\varepsilon})$ is negative. The representative microstructure (Fig. 4b) of

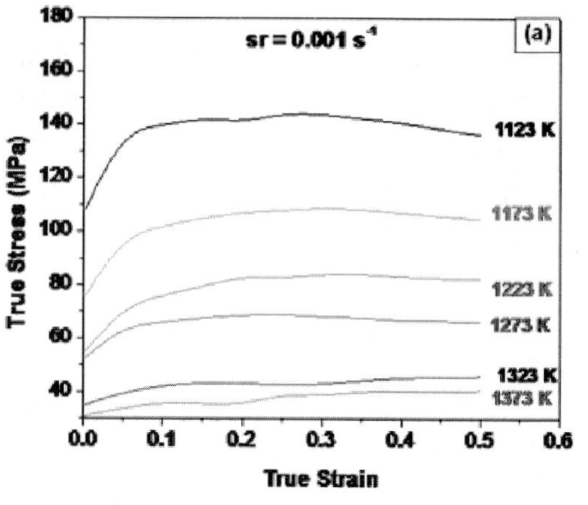

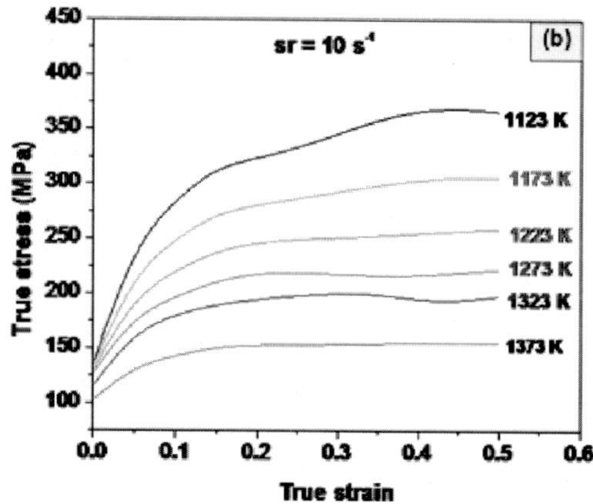

Fig. 3: Flow curves of Gr-91 steel at strain rates of: (a) 0.001s^{-1} and (b) 10 s^{-1}

confirms the flow instability of the material in this region.

Domain (c) occurs at low strain rate of 0.001–0.01 s^{-1} and highest temperature of 1373K with maximum η of 25% and *Q* of 200-300 kJ/mol. At higher temperatures and lower strain rates, some materials may exhibit super-plasticity that often results in abnormally elongated grains [2]. Although the representative microstructure of this domain (Fig. 4c) shows abnormally elongated grains and microstructural inhomogenity, the η and *m* values are not high enough to confirm super-plasticity.

Domain (d) is a relatively larger domain with very low *m* and low η occurring above 1 s^{-1} and at 1150–1373K. At higher strain rates, the material does not get enough time to dissipate the adiabatic heat generated during deformation [2]. Therefore, temperature rises locally causing local flow softening if DRX does not occur predominantly. Further application of load causes localised slip in the material, and when localised slip becomes intense the material may develop instabilities like shear band. Figure 4(d), which shows a representative microstructure obtained in this domain, indicates presence of such shear bands. Analysis of micrographs obtained in this domain leads to the conclusion that at higher strain rates, with increase in temperature, Gr-91 steel is prone to develop shear bands in spite of this domain having a *Q* of ~400 kJ/mol, which remains constant over a wide domain.

Domain (e) occurs at 1373K and 0.5 s^{-1} and exhibits maximum strain rate sensitivity of 0.3 and peak η of 45%, which decreases to 22% with increase in strain rate to 1 s^{-1}. Such a steep domain indicates the processes like cracking [2]. Though the microstructure of specimens tested in this domain shows equiaxed grain structure (Fig. 4e), this domain should be avoided because of its proximity to the unstable domain (d). In fact, half of this efficiency hill is part of the unsafe region (see Fig. 4). Thermo-mechanical processing in this domain would require very close control of strain rate and temperature, which may not always be possible in industrial-scale metal forming processes. *Q* in this domain is ~400 kJ/mol and it remains constant over a wide domain.

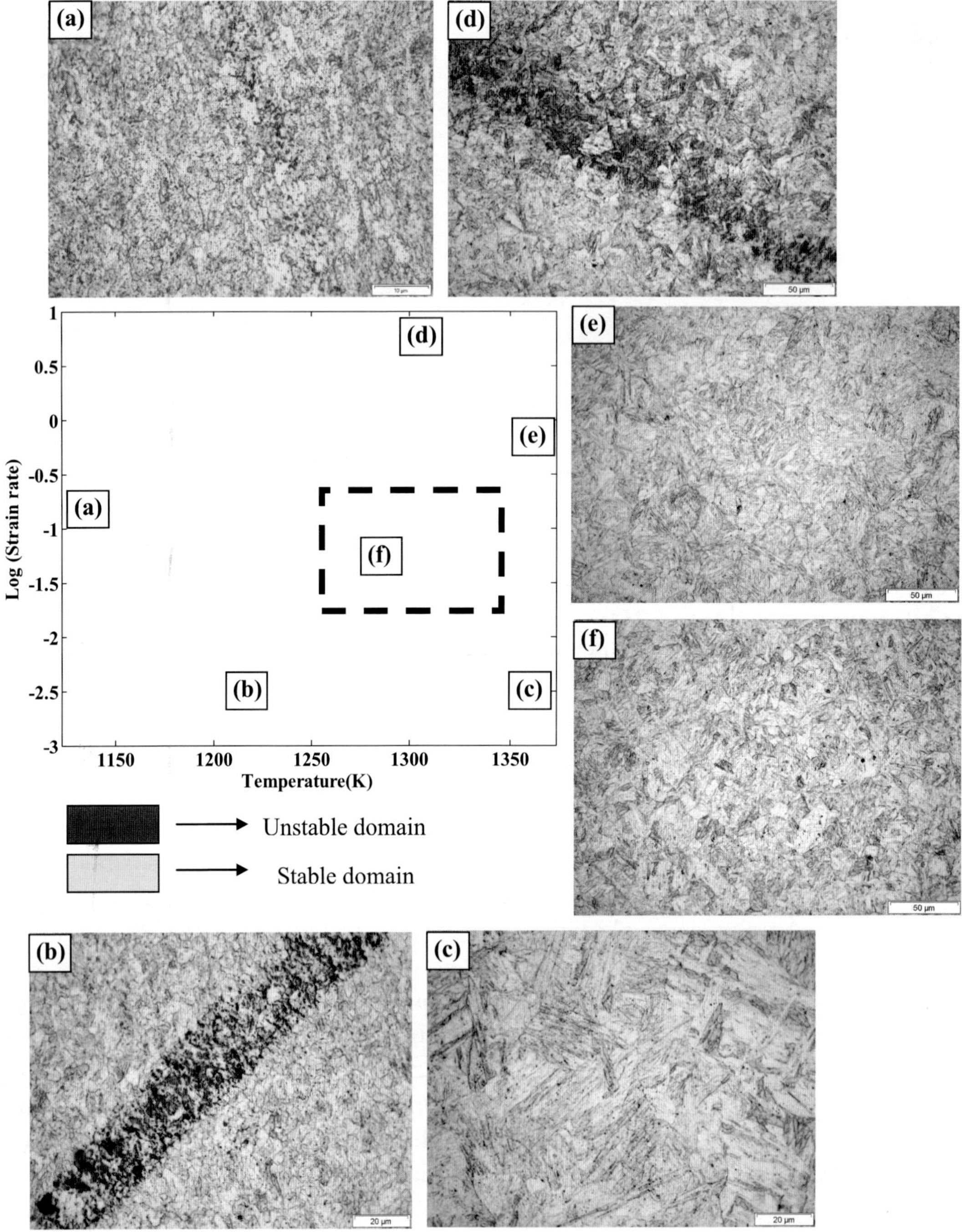

Fig. 4: Various domains of processing map for modified 9Cr-1Mo steel and their representative microstructures at: (a) 1123K, $0.1s^{-1}$; (b) 1173K, 0.001 s^{-1}; (c) 1373K, 0.001 s^{-1}; (d) 1323K and 1 s^{-1}; (e) 1373K, 1 s^{-1}; and (f) 1323K and $0.1s^{-1}$

Domain (f) is another safe domain of the processing map that occurs at about 1273K and 0.1 s^{-1}. The η hill of this domain is not a steep one, with η in the range 32–38% spread over wide range of temperatures (1250–1350K) and strain rates (0.01–1 s^{-1}). The high η of power dissipation and the flow curves at 1273–1373K at 0.001 s^{-1} (Fig. 3a) indicates DRX to be the key deformation mechanism in this domain. The representative microstructure of this domain (Fig. 4f) reveals presence of defect-free equiaxed grains. It is important to note that Gr-91 steel transforms into austenite on heating beyond 1123K and regains its ferritic/martensitic structure on cooling to room temperature by quenching [6].

The results obtained from the processing map and the microstructural investigation reveals that in the temperature range 1275–1325K and 0.015–0.1 s^{-1} strain rate, η and Q remains constant at ~34% and ~400 kJ/mol, respectively. Therefore, for thermo-mechanical processing of Gr-91 steel, the preferred processing parameters are 1300±25K and 0.015–0.1 s^{-1} strain rate. However, to provide more flexibility for controlling parameters during industrial-scale manufacturing of components, the practical processing parameters are 1250–1350K and 0.015–0.3 s^{-1}, as the effect of variation can be insignificant with η in this domain varying within 32–38%.

SUMMARY

Thermo-mechanical process modelling has been carried our for optimising the processing parameters for 316LN austenitic stainless steel and modified 9Cr-1Mo ferritic steel. Dynamic materials based processing maps were developed for both the steels using experimental data, and their safe and unsafe domains were validated by microstructural investigations of the specimens tested in each domain. Based on analysis of the processing map, microstructural investigations and activation energy variation, the optimum hot-working domain was identified for the 316LN SS to be 1350–1423K and 0.001–0.05 s^{-1} strain rate, and for modified 9Cr-1Mo steel to be 1250–1350K and 0.015–0.3 s^{-1} strain rate.

REFERENCES

[1] S.L. Mannan, S.C. Chetal, Baldev Raj, S.B. Bhoje, Transactions of Indian Institute of Metals, Vol. 56, No. 2, 2003, pp. 155–178.

[2] D.Samantaray, S.Mandal, A.K. Bhaduri, Materials Science and Engineering A, Vol. 528, 2011, pp. 5204–5211.

[3] D. Samantaray, S. Mandal, V. Kumar, S.K.Albert, A.K.Bhaduri, T. Jayakumar, Materials Science and Engineering A, Vol. 552, 2012, pp. 236–244

[4] Y.V.R.K. Prasad, T. Seshacharyulu, International Materials Review, Vol. 43, 1998, pp. 243–258.

[5] A.K.S.K.Kumar, Criteria for Predicting Metallurgical Instabilities in Processing Maps, MS thesis, Indian Institute of Science, Bangalore, 1987.

[6] S. Raju, B.J. Ganesh, A. Banerjee, E. Mohandas, Materials Science and Engineering A, Vol. 465, 2007, pp. 29–37.

Evolution of Microstructures and Mechanical Properties of Thermomechanically Processed Ultrahigh Strength Steels

S. Chatterjee* and S.K. Ghosh
Department of Metallurgy and Materials Engineering,
Bengal Engineering and Science University, Shibpur, Howrah - 711 103, India
*Email: schatterjee46@yahoo.com

ABSTRACT

The present study concerns the development of two low carbon microalloyed ultrahigh strength steels on a pilot scale. The dilatometric results of these steels yield flat top continuous cooling transformation (CCT) diagram along with predominantly bainitic microstructure at a lower cooling rate. The two experimentally determined CCT diagrams have further revealed that good hardenability is achievable in these steels even in air cooled condition (cooling rate ≈5°C/s). Based on the results of continuous cooling transformation, these steels were thermomechanically processed and finished at different finish rolling temperatures followed by air cooling. Variation in microstructure and mechanical properties at different finished rolling temperatures was studied. A mixture of granular bainite and bainitic ferrite along with microalloy carbide is the characteristic microstructural feature of these steels. The yield strength values in the range of 1040-1234 MPa with 1312-1439 MPa ultimate tensile strength along with 12-17% elongation have been obtained in the present investigated steels. The attainable high strength value is due to the accumulated contribution of fine grained pan-caked austenite, mixture of granular bainite and bainitic ferrite along with the presence of tiny precipitates of microalloy carbide and/or Cu rich precipitates. The steels reveal the impact toughness value in the range of 35-72J which exhibits nearly consistent fall with the lowering of testing temperature.

KEYWORDS: *Ultrahigh; Microalloyed; Steel; Microstructure; Mechanical Properties.*

INTRODUCTION

Ultrahigh strength steels can be subdivided into low alloy steel, precipitation hardened stainless steels, maraging steels, and secondary hardening steels [1]. Basically this class of steel comprises dislocated martensitic structure with fine precipitates. The alloy steels are heat-treated by austenitizing, a subsequent cooling to room temperature or below during which most of the austenite is transformed to martensite, and finally a tempering or aging treatment is carried out during which carbides and/or other hardening particles are precipitated. These steels are ordinarily restricted to aircraft applications as landing gears, arresting hooks, fasteners, jet engine shafts, and ballistic-tolerant components as well as non-aircraft heavy duty parts as high-strength bolts, automotive drive shafts and bicycle frames etc.

Low carbon steels containing 1-2% Cu have been already emerged as structural materials due to their excellent strength/ ductility balance [2, 3]. Microalloying elements i.e., Ti, Nb or V, which are strong carbide former and also act as grain refiner while in solution in austenite, are also added in these steels. On subsequent cooling, these microalloying elements precipitate as carbides and/or carbonitrides which are fine enough to offer hindrance to mobile dislocations leading to a significant increase in strength. An additional strengthening is obtained from fine ε-Cu particles on ageing [4, 5]. It is noteworthy that Cu in steels causes surface hot shortness during hot deformation [6] and addition of Ni by an amount half that of the Cu content is effective for the suppression of surface hot shortness [7].

In view of the above, considerable efforts have been made to develop two novel steels by thermomechanical controlled processing on a pilot scale. The low carbon content (wt%) (≈ 0.15) with high Mn (1.41-1.90) and marginally higher Ni (3.76-3.80) content instead of medium carbon (0.3-0.4) along with low Mn (0.5-0.8) and high Ni (2.7-3.3) containing commercial steels used earlier in the defence sector has been designed for achieving better solid solution strength together with the more than requisite through-hardening capability. This recent effort has been devoted towards the reduction of weight by achieving high strength to weight ratio together with improved weldability for the various prospective defense applications such as explosive ammunition, gun barrel, missile skins, light-weight military bridges etc. The continuous cooling transformation has been evaluated and the influence of various processing parameters on structure and properties of the steels have also been assessed.

EXPERIMENTAL

The steel was melted in an electric arc furnace of 2 ton capacity. The cast square ingot of 254 mm × 254 mm was carefully inspected and top end containing pipes was cropped off using electric saw. The cropped ingots of 6452 sq. mm cross section were subsequently hot forged to 95 mm × 95 mm cross section billets. Table 1 presents the chemical compositions of the two steels obtained by spectroscopic analysis, using Quantovac 80 and Laboratory Equipment Company Instruments.

Table 1: Chemical Composition of the Investigated Steels (wt%)

C	Mn	Si	S	P	Cr	Mo	V	Cu	Ni	Ti	Nb	O	N	CE*	Tnr(°C)#
0.15	1.41	0.21	0.014	0.011	0.89	0.66	0.02	0.33	3.76	0.036	0.067	0.003	0.009	0.40	1161
0.13	1.90	0.70	0.008	0.004	0.11	0.81	0.01	1.25	3.80	0.052	0.068	0.003	0.008	0.43	998

*CE [8] = C + (Si)/30 + (Mn+Cu+Cr)/20 + (Ni)/60 + (Mo)/15 + (V)/10 + 5B wt%

#Tnr [9] = 887 + 464C + (6445Nb - 644√Nb) + (732V - 230√V) + 890Ti + 363Al - 357Si (wt%)

The dilatometric study was carried out using Gleeble 1500 thermomechanical simulator. After soaking at 1200°C for 15 minutes, the samples were cooled at constant rates of 0.5, 1, 5, 10, 15, 25, 50 and 100°C/s. The experiments were carried out in argon atmosphere. Austenite transformation start and finish temperatures were determined from the dilatation curves and those were used for construction of the continuous cooling transformation (CCT) diagram.

The forged billets of 95 mm × 95 mm cross section were reheated at 1200°C for 3h and control rolled to 16 mm thick plates. The controlled rolling was made in three stages. In stage one (I), about 42% deformation was given in three multiple passes in the temperature range of 1100°C to 1020°C to reduce the thickness from 95 mm to 55 mm. In stage II, another 36% deformation was applied in the temperature range of 1020-920°C and in stage III, 54% deformation was given in similar three multiple passes in the temperature range of 920°C to finish rolling temperature (FRT). FRTs were selected as 850°C, 800°C and 750°C. Finally the rolled plates were subjected to air cooling. During the course of rolling, temperatures at different deformation stages were measured by infrared pyrometer.

Optical metallographic samples were prepared by conventional grinding and polishing techniques and subsequently etched with 2% nital solution for examination in a light microscope. Quantitative image analysis was also conducted using ImageJ software (version 1.42) [10] in order to quantify the relevant microstructural parameters. For transmission electron microscopy, 3 mm diameter discs were punched out from thin foils and subsequently subjected to twin jet electro-polishing using a mixture of electrolyte of 90% acetic acid and 10% perchloric acid. The thin electron transparent samples were examined in a transmission electron microscope (TEM) at 200 kV operating voltage. The chemical compositions of the precipitates were determined by Energy Dispersive X-ray (EDAX) analysis.

The tensile specimens were prepared from the rolled plates with their long axis parallel to the rolling direction as per ASTM specification (Vol.03.01 E8M-96) with 50 mm gauge length where as width and thickness measure 12.5 mm and 12 mm, respectively. Tensile testing was carried out in a 50 ton capacity universal tensile testing machine at a strain rate of $3.3 \times 10^{-4}\,s^{-1}$.

Charpy V notch impact testing of the standard samples as per the ASTM Standard (ASTM: Vol. 03.01: E23–96) with flat surface along the plane of rolling was conducted at room temperature 25°C, 0°C, -20°C, -40°C and -60°C. The average of three consistent test results was recorded as the impact value for the corresponding samples.

RESULTS

Continuous Cooling Transformation

Figure 1 depicts the CCT diagrams for the present investigated steels. The Ac_3 and Ac_1 temperatures of "no-Cu" steel have been recorded as 820°C and 700°C respectively; whereas the respective temperatures for Cu bearing steel are 825 and 706°C. The bainite start temperature (B_S) has been found to be as 436-440°C at the cooling rate of 0.5°C/s for "no-Cu" and Cu bearing steels respectively which fairly agrees with the estimated B_S temperature of 406-408°C, obtained from the following equation [11].

$$B_S = 830\text{-}270C\text{-}90Mn\text{-}37Ni\text{-}70Cr\text{-}83Mo \text{ (wt\%)} \quad (1)$$

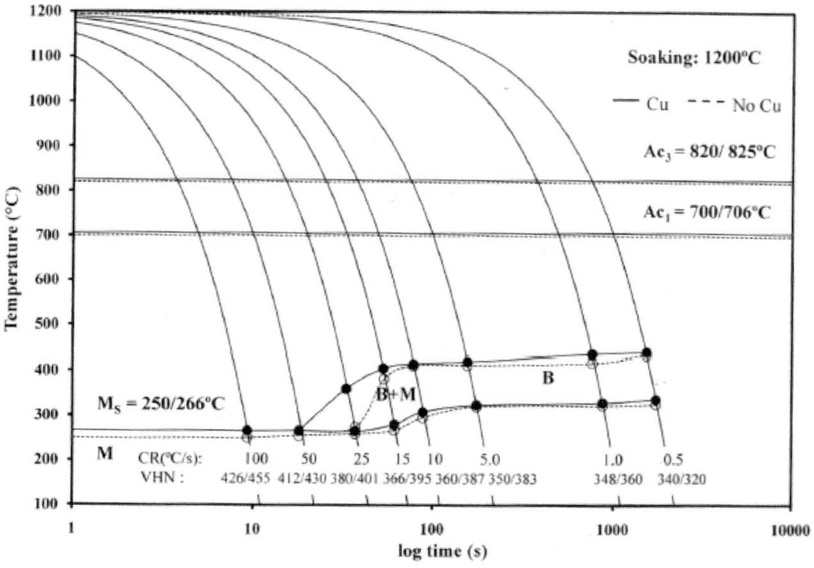

Fig. 1: The CCT diagrams for the present investigated steels.

The martensite start temperature (M_S) has been obtained in the range of 250-266°C for the above mentioned two steels and is represented as the horizontal straight line in the CCT diagram. It is imperative to mention that the transformation temperature range (Ar_3-Ar_1) of Cu bearing steel lies above to that of "no-Cu" steel. A flat top "C" curve with bainite and martensite structure has been obtained for both steels and the transformation temperatures do not vary much for a wide range of cooling rates. It is important to note that the variation in the hardness value as a function of the cooling rates has been appended in the concerned CCT diagram and the hardness value increases with the increasing cooling rate. At a given cooling rate the Cu bearing samples usually exhibit higher hardness than those of "no-Cu" steel.

The variation in hardness can be explained by the optical micrographs at various cooling rates as shown in Figs. 2(a)-(d). At a lower cooling rate (0.5°C/s), the microstructures (Figs. 2(a) and (b)) of both the steels shows granular bainite (α_g) and bainitic ferrite (α_h) within equiaxed prior austenite grains which yield lower hardness value in the range of 320-340 VHN. The granular bainite is usually revealed as small (10-20μm) equiaxed grains. The bainitic ferrite laths have been nucleated from the prior austenite grain boundaries and subsequently penetrated inside the grain body. The measured average prior austenite grain size lies in the range of 110-135 μm. With a further increase in cooling rate of 25°C/s, microstructures of both the steels become lath martensitic in packet form (Figs. 2(c) and (d)) which yields higher hardness value in the range of 380-401 VHN. However, the presence of some bainite cannot be ignored. The CCT diagrams have revealed that adequate hardenability is achievable in both steels usually at a cooling rate >5°C/s.

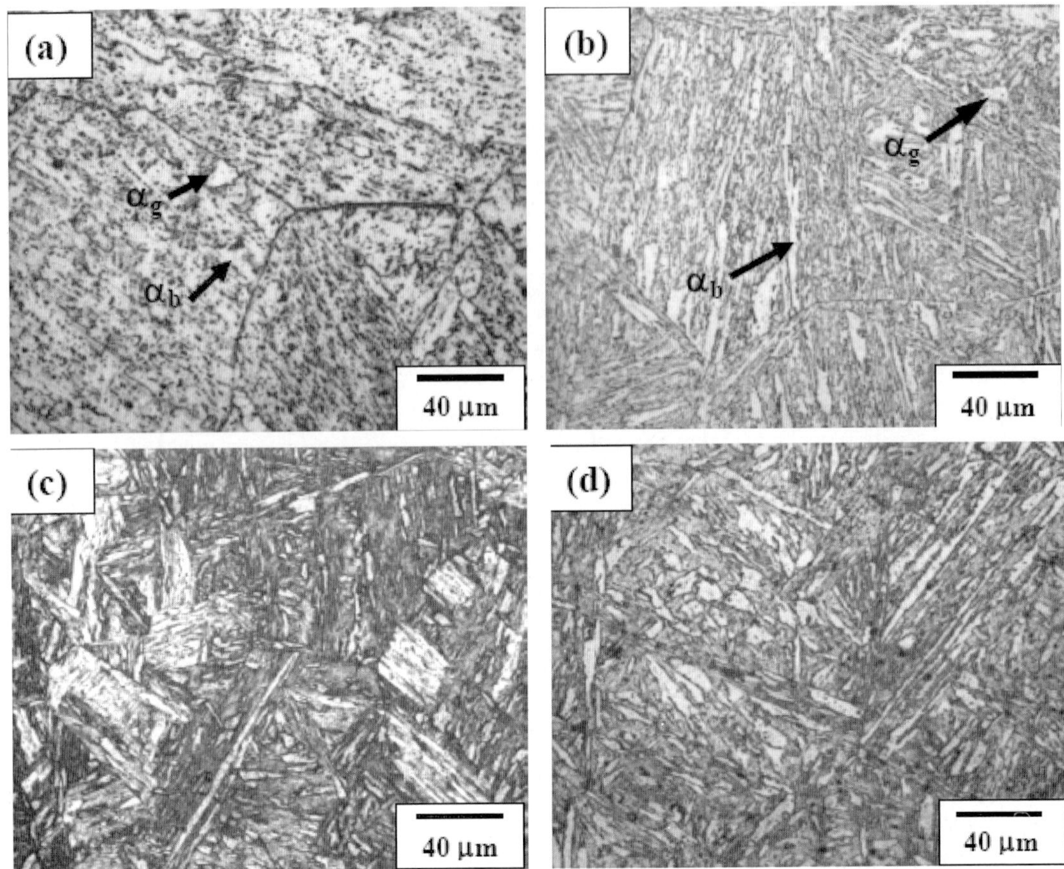

Figure 2. Optical micrographs of both (a) "no-Cu" steel and (b) Cu bearing steel specimens cooled at 0.5ºC/s, and both (c) "no-Cu" steel and (d) Cu bearing steel specimens cooled at 25ºC/s showing the evolution of microstructure under continuous cooling conditions.

Microstructural Evolution

Figures 3(a) and (b) show the bright field (BF) TEM images of air cooled "no-Cu" and Cu bearing steels, respectively which reveal mixture of granular bainite (α_g) and bainitic ferrite (α_b) along with high dislocation density in prior pan-caked austenite grains obtained at 750° FRT. The formation of fine parallel micro-twin substructure within the bainitic ferrite as well as interlath region parallel to the bainitic ferrite were confirmed as retained austenite(γ_R) in similar type of steel reported earlier [12].

Figure 4(a) presents the TEM BF micrograph of "no-Cu" steel obtained on air cooling from 850°C which reveals the dark precipitate (denoted by arrow) (≈20 nm) within the bainitic ferrite. The EDS spectra (Fig. 4(b)) from the particle denoted by arrow in (a) reveals that the particle is rich in Nb and C which indicates the formation of NbC precipitate.

Figure 5(a) is the BF TEM micrograph showing the fine precipitate of darker contrast (denoted by arrow) inside the bainitic ferrite. Figure 5(b) is the EDS spectra from the arrowed fine precipitate

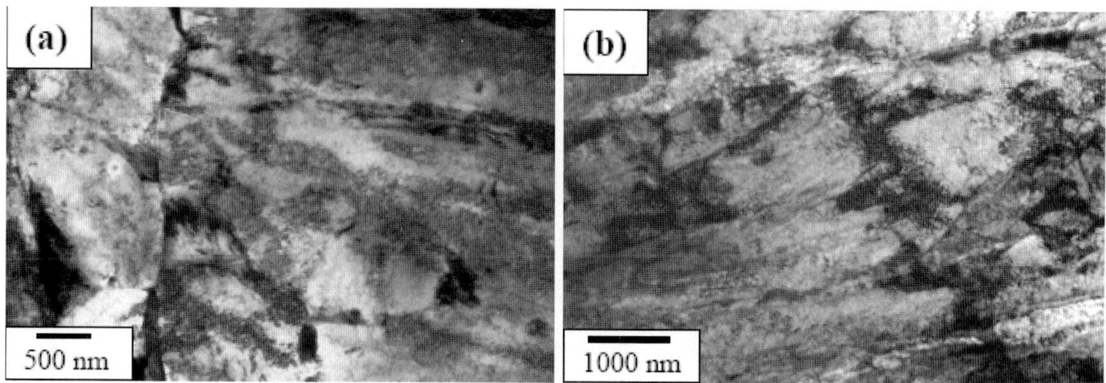

Fig. 3: TEM BF micrograph of (a) "no-Cu" steel and (b) Cu bearing steel showing mixture of granular bainite (αg) and bainitic ferrite (αb) in prior pan-caked austenite grains at 750°C FRT.

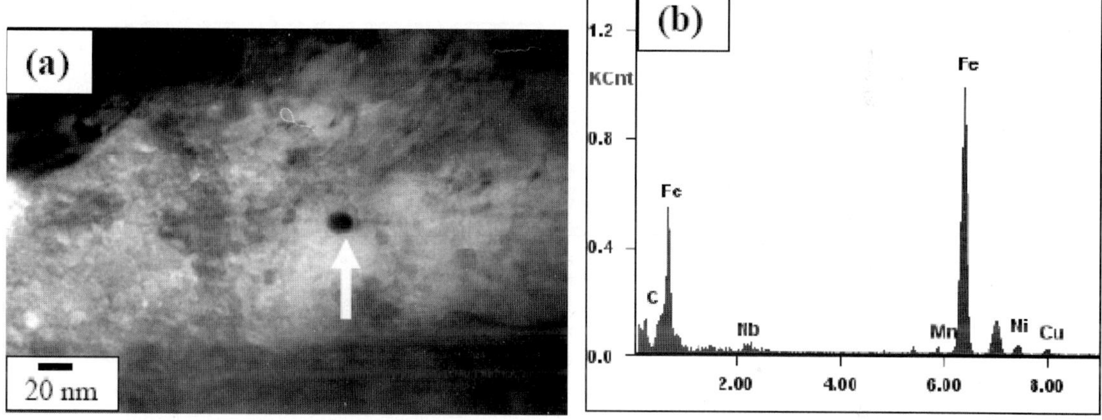

Fig. 4: (a) TEM BF micrograph showing precipitate particle within bainitic ferrite obtained at 850°C FRT in "no-Cu" steel and (b) EDS spectra from the arrowed precipitate particle in (a).

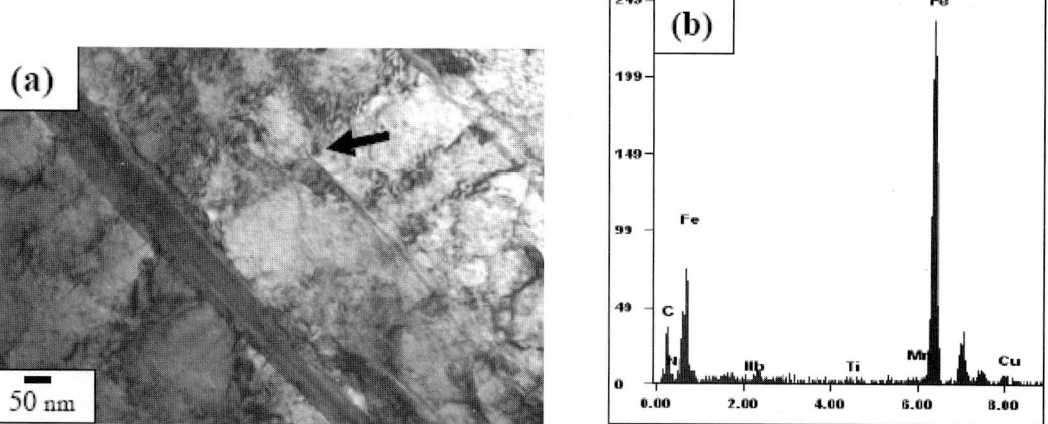

Fig. 5: (a) TEM BF micrograph showing dark precipitate (denoted by arrow) in Cu bearing steel and (b) EDS spectra from the arrowed precipitate particle in (a).

(<50 nm) in (a). The result shows that the precipitate is rich in Cu which indicates the formation of Cu precipitate due to the adequate Cu content (> 0.6 wt%) () [7] in the present steel. Such type of Cu precipitation during continuous cooling was reported earlier in commercial low alloy steel [13].

Mechanical Properties

Figure 6 shows the variation in yield strength (YS), ultimate tensile strength (UTS) and hardness (Hd) values of the investigated steels with the change in FRT. The minimum values of hardness (342 VHN), YS (1040 MPa) and UTS (1342 MPa) values were obtained in "no-Cu" steel at 850°C FRT. Similarly, the minimum values of hardness (434 VHN), YS (1182 MPa) and UTS (1312 MPa) values are obtained at same FRT (850°C) for Cu bearing steel. While the variation in hardness, YS and UTS of "no-Cu" steel processed in the range of 750°C – 850°C are 5%, 0.67% and 7% respectively; those values are 2.5%, 4% and 3% respectively for Cu bearing steel.

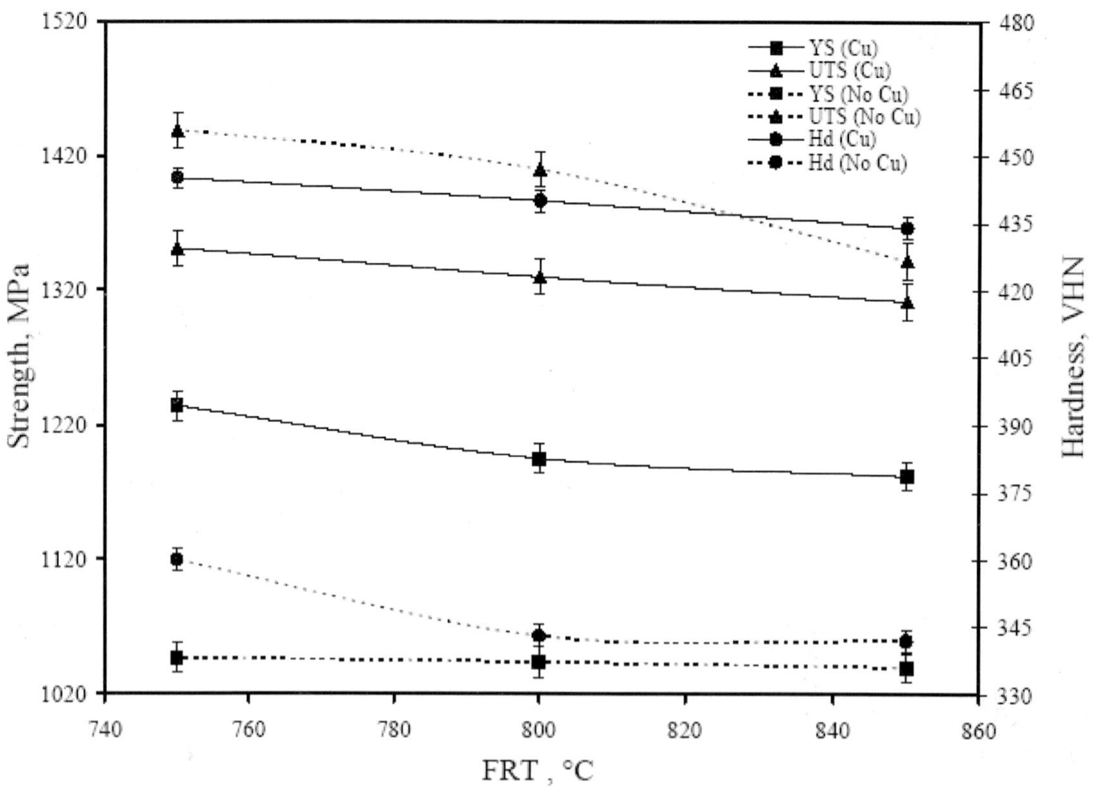

Fig. 6: Variation in yield strength, ultimate tensile strength and hardness values with the change in FRT.

Figure 7 reveals the variation in percent elongation (El.) and reduction of area (RA) with the change in FRT. It is observed that percent elongation increases with the increase in FRT in both the steels. The minimum elongation of 12 to 13%

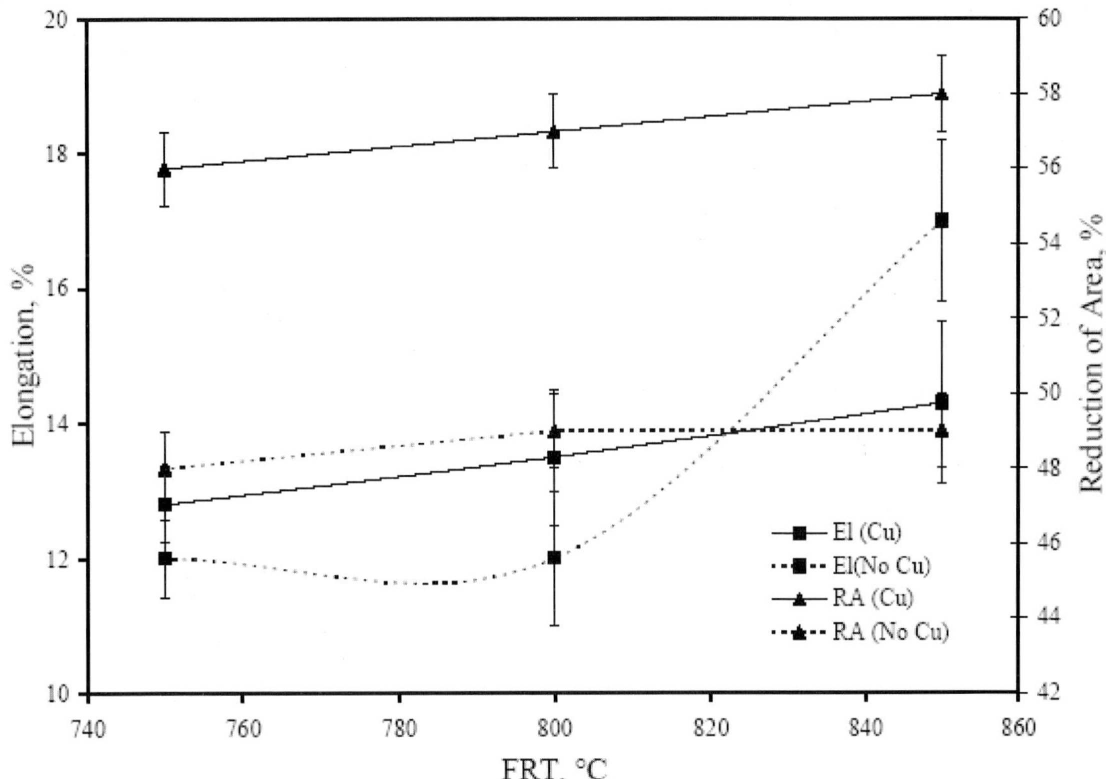

Fig. 7: Variation in percent elongation and reduction of area with the change in FRT.

is obtained at 750°C FRT for both steels and the same increases to 17% and 14% at 850°C FRT for "no-Cu" and Cu bearing steel, respectively. The percent RA for both the steels follow nearly the same trend as mentioned for percent elongation.

Figure 8 shows the variation in impact toughness values of both the steels processed at different FRTs with the various testing temperatures. It is noticed that generally impact toughness decreases with the decrease in testing temperature. The "no-Cu" steels record the impact toughness values in the range of 35-63J, whereas Cu bearing steels reveal in the range of 41-72J. It is evident that "no-Cu" steel has revealed lower impact properties in comparison with those of the Cu bearing steel within the range of testing temperature. While no distinct transition temperature range is recorded for the "no-Cu" steel, the Cu bearing steel samples exhibit a distinct transition temperature range from -20°C to -40°C. In the case of "no-Cu" steel samples, a gradual fall in impact toughness value is noticed with the lowering of testing temperature except for 850°C FRT. On the other hand, the upper and lower shelf energy values vary in a close range for Cu bearing steel.

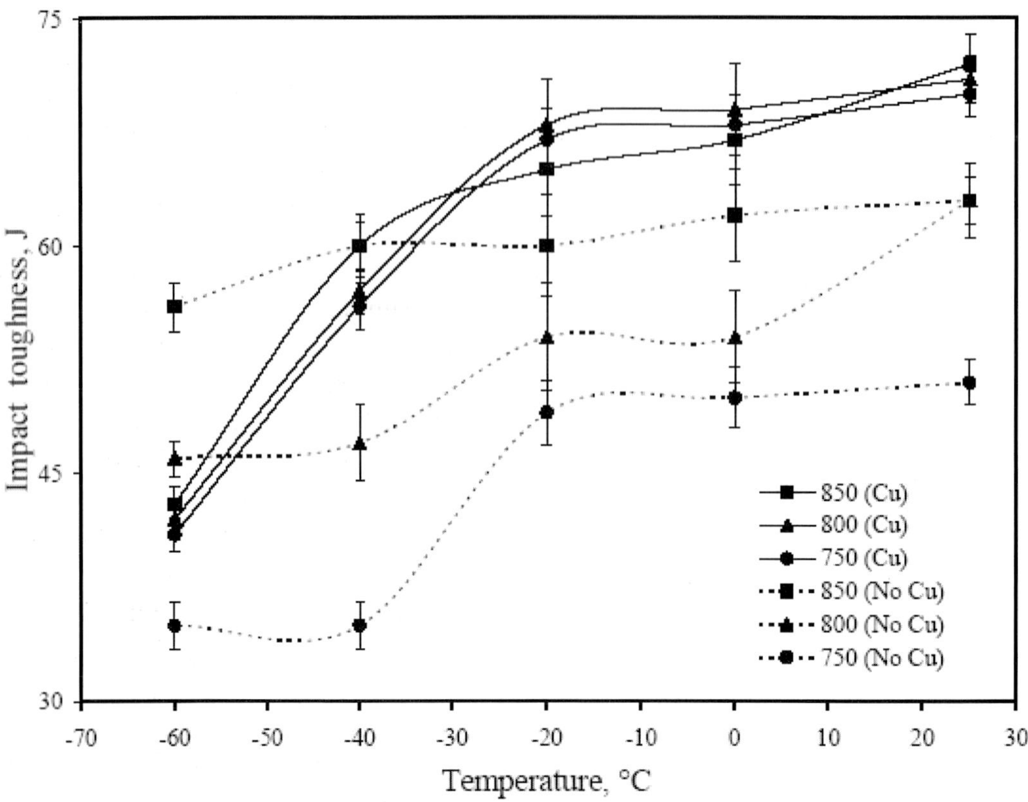

Fig. 8: Variation in impact toughness values with the testing temperature.

DISCUSSION

Austenite Conditioning

Austenite conditioning was performed by controlling the austenitising temperature or by the amount of deformation in the recrystallisation or non-recrystallisation region. In the present study, high soaking temperature (1200°C) and time (3h) were employed to ensure the maximum dissolution of microalloy carbides and carbonitrides in austenite. As the solubility limit of TiN calculated from the available references [14, 15] at 1200°C is 4.365×10^{-6} $(wt\%)^2$ which is very low as compared to the [Ti][N] solubility product obtained in the range of 3.24×10^{-4} $(wt\%)^2$ to 4.16×10^{-4} $(wt\%)^2$ for the present investigated steels, TiN/TiCN particles formed are likely to remain undissolved during reheating at 1200°C. However, the calculated solubility limits of TiC (9.9×10^{-3}) or NbC/NbCN (4.5×10^{-3}) are being rather high at this soaking temperature, it is expected that TiC, NbC or NbCN formed during cooling of the ingot will be dissolved again during slab reheating. In this context, it may be mentioned that as TiC has higher solubility than NbC or NbCN, it is expected that the latter will first come out from the solution.

Misra et al. [16] studied microstructure evolution in industrially processed hot rolled Nb-Ti microalloyed steel. The carbide forming elements Ti and Nb retard recrystallization during finish rolling and consequently a pan-caked microstructure is formed that promotes grain refinement during austenite to ferrite transformation. About 90% deformation in the non-recrystallization region (stage II and III; Tnr = 998 and 1161°C) results into the formation of pan-caked grains which restrict the size of bainite obtained on air cooling. It needs to be mentioned that for the carbon equivalent (CE) in the range of 0.40-0.43% (Table 1), there is a less potential for cracking in the heat affected zone (HAZ) on flame cut edges and welds of the present steels.

Multiphase Structure

On air cooling from the various FRTs, the deformed austenite has been transformed into granular bainite and bainitic ferrite with dislocated substructure (Fig. 3). The dislocation introduced into the bainite is a consequence of product of transformation. It is noteworthy, that the well deformed austenite prior to cooling leads to the formation of crystalline defects which contributes to the formation of fine bainite structure.

The microalloying elements i.e., Ti, Nb remain in solution at the soaking temperatures (1200°C) and upon cooling they combine with C and N to form carbide/carbonitride precipitates. With a further lowering in temperature the supersaturation increases and precipitation begins e.g., NbC precipitates (Fig. 4(a)) form during processing of the slab. It can be considered that the fine microalloy precipitate (≈20 nm) (Fig. 4(a)) and Cu rich precipitate (<50 nm) (Fig. 5(a)) formed after rolling in the non-recrystallization region. These precipitates are expected to be incoherent with the steel matrix. The precipitation hardening resulting from these nano-sized particles is significant in case of high strength steel.

Structure-Property Correlation

On air cooling, mixture of granular bainite and bainitic ferrite along with microalloy precipitate is obtained and tested to reveal higher level of hardness 434-445 VHN, 1182-1234 MPa YS and 1312–151 MPa UTS along with lower 13-14% elongation and comparable RA (56-58%) and higher level of impact toughness value (41-72J) as compared to those of the "no-Cu" steel. Highly dislocated bainite (Fig. 3) along with microalloy carbide (Fig. 4(a) increases the strength of the steel. A major strengthening comes from the contribution of Cu rich precipitates (Fig. 5(a)) in Cu bearing steel.

A modified version of the Hall-Petch equation (2) has been used to estimate the combined effect of dispersion strengthening (σ_p) conferred by incoherent particles and by dislocation strengthening (σ_d). The combination of $\sigma_p+\sigma_d$ is based on the strong interaction of dislocation and particles. Equation (2) shows that by subtracting the ferrite lattice friction stress and carbon + nitrogen in solution (σ_0), together with the solid solution (σ_s) and grain size strengthening (σ_g) components from the measured yield stress (σ_y), values of $\sigma_p+\sigma_d$ can be calculated. Cracknell and Petch [17] suggested that σ_0 (given in equation (3)) depends upon the frictional forces exerted by solute atoms and precipitates on the dislocation arrays that propagate the yield. The value of σ_0 increases when the concentration of carbon + nitrogen in solution is increased. Pickering and Gladman [18] and Leslie [19] extensively studied the substitutional solid solution strengthening effects in carbon steel and iron, respectively and suggested that σ_s can be estimated using equation (4). Morrison and Chapman [20] investigated the effect of grain size on yield stress in terms of equation (5).

$$\sigma_p+\sigma_d=\sigma_y-(\sigma_0+\sigma_s+\sigma_g) \quad (2)$$

$$\sigma_0=45 \text{ MPa} \quad (3)$$

$$\sigma_s=84(\text{Si})+32(\text{Mn})+38(\text{Cu})+43(\text{Ni}) \quad (4)$$

$\sigma_g=18.1L^{-1/2}$, where L is the average ferrite lath size in mm (5)

The average size of bainite (≈1μm) (Fig. 3(a)) provides σ_g of 572 MPa and the chemical composition yields σ_s ≈237 MPa. Then $\sigma_p+\sigma_d$ make a contribution of 193 MPa for the "no-Cu" steel (750°C FRT) with 1047 MPa yield strength. Similar average bainite size (≈1μm) (Fig. 3(b)) contributes σ_g of 572 MPa and thus $\sigma_p+\sigma_d$ make a higher contribution of 287 MPa for the Cu bearing steel (750°C FRT) with σ_s ≈ 330 MPa and 1234 MPa yield strength. Similar type of estimations of contributions of different strengthening components can be made for other FRT also.

CONCLUSIONS

1. Two microalloyed ultrahigh strength steels have been successfully produced on a pilot scale. The continuous cooling transformation behaviour of both steels has been studied. In both steels, flat top CCT diagrams along with bainite microstructure have been observed at a lower cooling rate.
2. "No-Cu" steel reveal mixture of granular bainite and bainitic ferrite along with the NbC precipitates. On the other hand, Cu bearing steel shows similar microstructural features like "no-Cu" steel and additional Cu rich precipitate.
3. Strength level of 1040-1047 MPa yield strength and 1342–1439 MPa ultimate tensile strength along with 12-17% total elongation have been obtained without Cu addition, whereas higher yield strength in the range of 1182-1234 MPa and lower range of 1312–1351 MPa ultimate tensile strength along with lower level of 13-14% total elongation have been obtained in Cu bearing steel.
4. "No-Cu" steel reveals the impact toughness value in the range of 35-63J which is lower compared to those of Cu bearing steel (41-72J) under similar processing condition.

REFERENCES

[1] Metals Handbook, Vol. 1, 10th Edn., ASM International, Ohio, 1990, pp. 430 - 448.

[2] M. Morita, K. Sato and Y. Hosoya, Tetsu to Hagane, Vol. 80, 1994, p. 48.

[3] N. Maruyama, M. Sugiyama, T. Hara and H. Tamehiro, Materials Transactions JIM, Vol. 40, No. 4, 1999, p. 268.

[4] M. Mujahid, A. K. Lis, C. I. Garcia and A. J. DeArdo, Journal of Materials Engineering and Performance, Vol. 7, No. 2, 1998, p. 247.

[5] S. S. Ghasemi Banadkouki, D. Yu and D. P. Dunne, ISIJ International, Vol. 36, No. 1, 1996, p. 61.

[6] N. Imai, N. Komatsubara and K. Kunishige, ISIJ International, Vol. 37, No. 3, 1997, p. 217.

[7] Metals Handbook: Properties and Selection: Iron; Steels and High Performance Alloys, Vol. 1, 10th Edn., ASM International, Materials Park, Ohio, 1995, pp. 389 - 423.

[8] Y. Ito and K. Bessyo, A Prediction of Welding Procedure to Avoid Heat Affected Zone Cracking, IIW Doc. IX, 1969, p. 631.

[9] F. Boratto, R. Barbosa, S. Yue and J. J. Jonas, Proc. Int. Conf. on THERMEC – 88, Iron and Steel Institute of Japan, Tokyo, 1988, pp. 383-390.

[10] M. Miller, ImageJ Software, 2011, URL: http://rsb.info.nih.gov/ij.

[11] F. B. Pickering, Proc. Int. Conf. on Microalloying '75, Union Carbide Corp., New York, 1977, pp. 9-31.

[12] P. S. Bandyopadhyay, S. K. Ghosh, S. Kundu and S. Chatterjee, Materials Chemistry and Physics, Article in Press, 2012.

[13] S. W. Thompson, D. J. Colvin and G. Krauss, Metallurgical and Materials Transactions A, Vol. 27, No. 6, 1996, p.1557.

[14] K. J. Irvine, F. B. Pickering and T. Gladman, Journal of Iron and Steel Institute, Vol. 205, No. 2, 1967, p.161.

[15] T. Gladman, D. Dulieu and I. D. McIvor, Proc. Int. Conf. on Microalloying '75, Union Carbide Corporation, New York, 1977, pp. 32-55.

[16] R. D. K. Misra, K. K. Tenneti, G. C. Weatherly and G. Tither, Metallurgical and Materials Transactions A, Vol. 34, No. 10, 2003, p. 2341.

[17] A. Cracknell and N. J. Petch, Acta Metallurgica, Vol. 3, No.2, 1955, p.186.

[18] F. B. Pickering and T. Gladman, Metallurgical Developments in Carbon Steels, ISI Spec. Rep. No. 81, Iron Steel Inst., London, 1963, p.10.

[19] W. C. Leslie, Metallurgical Transactions B, Vol. 3, No. 1, 1972, p. 5.

[20] W. B. Morrison and J. A. Chapman, Proc. Rosenhain Centenary Conf., The Royal Society, London, 1976, pp. 286- 303.

Role of Base Metal Microstructure on Tensile Properties and Weldability of Simulated Continuously Annealed Advanced High Strength Steels

Kumkum Banerjee
R&D Dept, Tata Steel Limited, Jamshedpur, India
Email: kumkum.banerjee@tatasteel.com

ABSTRACT

Advanced high strength steels (AHSS) are several families of steels (dual phase, transformation-induced plasticity, complex phase and martensitic steels) that constitute ferrite along with low temperature phases--martensite, bainite and/or retained austenite, in adequate quantities, to obtain desired mechanical properties. These steels find their applications in automobiles and are able to enhance mechanical properties with reduced weight that make them cost effective, fuel efficient and environment friendly. All AHSS are produced using proper chemistry design in association with controlled rate of cooling from the austenite–ferrite phase region, either on the run-out table in a hot rolling mill or in the cooling section of a continuous annealing line. This ensures high strength and other desired properties essentially through the low temperature phase transformation products. In the present investigation, the base metal microstructure had played a key role in dictating tensile properties and weldability of spot welded AHSS in simulated continuously annealed conditions. A few miocroalloyed AHSS with low carbon equivalent were produced in a laboratory air induction furnace and were subsequently hot forged and rolled in the laboratory. Continuous heating and cooling transformations were vividly studied in a Gleeble 1500D to determine intercritical temperature regimes and the transformation temperatures in association with cooling rates of the low temperature transformation products. The hot rolled samples after duly cold rolled were submitted to strip annealing simulation to mimic actual continuous annealing, employing a Gleeble 3500 thermomechanical simulator. Two types of annealing cycles were simulated to examine the effect of cooling rate and bainite alongside fine microalloying precipitates on tensile properties and weldability. The annealing cycles that resulted in ferrite, 37% bainite and martensite in association with nano-sized Mo-C were successful in delivering the satisfactory combination of tensile properties and weldability.

KEYWORDS: *Advanced High Strength Steel, Continuous Annealing, Microalloying Precipitates, Haz Softening, Tensile Strength, Ductility, Resistance Spot Welding, Weldability*

INTRODUCTION

In recent years, light-weight-fuel-efficient vehicles that ensure passenger safety have become a priority for automotive industries. Manufacturing of such vehicles has been possible by the development of high strength multiphase advanced high strength (AHSS) steels. AHSS is a term used to describe several families of steels (dual phase-DP, transformation induced plasticity—TRIP, complex phase–CP and martensitic steels) that contain ferrite, martensite, bainite and/or retained austenite in sufficient quantities to produce desired mechanical properties. Some types of AHSS have higher strain hardening capacity resulting in strength-ductility balance that is superior to conventional steels while other types possess ultrahigh yield and tensile strength in association with bake hardening behaviour [1]. All AHSS are produced by controlling the cooling rate from the austenite –ferrite phase, either on the run-out table of the hot rolling mill or in the cooling section of the continuous annealing line to ensure high strength levels through bainitic-martensitic transformation [2]. However, such transformations can not be achieved unless the cooling rate is over a critical threshold that depends on steel composition [3]. The critical cooling rate increases with decreasing alloying additions and as the cooling rates attainable by industries are limited, without the addition of alloying elements, achievement of high strength is not possible.

Problems in welding AHSS

Generally, sheet steels can be hardened by the addition of C, Si, Mn, P etc. However, increasing alloying elements, such as C, Si and Mn enhances quenching hardenability of steels that affects weldability adversely [2-5]. Weldability of steels depends upon carbon equivalent (CE) that gives the measure of hardenability of the steel and is a function of alloying elements present.

The following expression is widely used for expressing CE of AHSS [6].

$$CE = C + Si/30 + Mn/20 + 2P + 4S \text{ (mass\%)}$$

Resistance spot welding is a favoured method for joining advanced high strength steels [7]. The welding lobes of AHSS are narrower than mild and high strength steels. Further, AHSS tend to exhibit higher expulsion, higher electrode wear and most importantly, they fail during pull button test resulting in interfacial failure. The acceptable strength is achievable in these spot welded steels with full button pull, partial button pull or partial interfacial fracture.

Several researchers on AHSS have reported the welding related problems for the formation of matensite in the weld nugget [8, 9]. Martensite can also form in resistance spot welds, even at low carbon levels as cooling rate associated with spot welding [10, 11] is quite high. Similar observations were made while laser beam welding (LBW) was conducted on AHSS

[12, 13]. The hardness achieved in the steels is directly associated with the fast cooling in the welding processes that can achieve cooling rate of the order of 10^3-10^5 °C/s [5,14]. The hard martensite thus formed due to high cooling rate, provides a path for a crack to propagate. In addition, rapid cooling can cause porosities towards the edges of the weld where stress concentration is highest in a peel test [15] and also solidification cracking enhancing the chances of interfacial failure of a spot weld [14].

Shrinkage, solidification cracks and liquid metal embrittlement in DP 600 were also observed by Milititsky et al [16] in some welds, where weld current and hold time were found responsible for the formation of such imperfections. In another work [17], the detection and prediction of expulsion during spot welding of DP600 were reported. Several researchers have reported about the interfacial failures in DP600 steel while the weld size exceeded the minimum threshold size [189]. In addition, Wang et al. also found interfacial fracture in DP600 steel weld buttons for the thicker gages [19-21]. Such failures were attributed to solidification cracks and shrinkage voids [22].

It can be mentioned that HAZ softening may occur in the case of AHSS with significant amount of martensite [23, 24] in the base metal that can be attributed to martensite tempering at the points of the weld where the temperature approached A1 (subcritical HAZ). Further, as DP steel essentially derive its strength from the composite microstructure of ferrite and martensite and as martensite is a thermally unstable phase, it tends to decompose in the heat affected zone (HAZ) resulting in softening and thus causing hardness reduction. This phenomenon is termed as HAZ softening [25, 26]. HAZ softening is a very complex phenomenon that is affected by steel composition, base metal martensite content, prestrain and heat input [27-29] and it can have detrimental impact on weldment. HAZ softening was also reported by Dancette et al [30, 18] in DP980 due to martensite tempering, however, the presence of a low amount of martensite in the base metal of DP450, helped avoid HAZ softening.

In this article, a detailed discussion on the role of base metal microstructure on tensile properties and weldability of the steels is presented.

EXPERIMENTAL

Material

Two heats were made using a 25-kg in-house laboratory air induction furnace. Chemical composition (in wt%) of the heats (Heat Nos. 9 and 14) along with the carbon equivalents are presented in Table I.

Table I: Chemical composition and carbon equivalent of the developed steels

Heat No.	C	Mn	Si	Al	Mo	Ti	Nb	Cr	P	S	CE
9	0.12	1.52	0.37	0.15	0.29	0.009	0.004	0.021	0.008	0.013	0.27
14	0.09	1.55	0.41	0.27	0.21	0.031	0.009	0.036	0.008	0.008	0.24

Hot Rolling

Equilibrium transformation temperatures (Ae1 and Ae3) for the steels were calculated using Thermo-Calc (TCFE3) so that finish rolling temperatures remained 30-40°C above the A3 temperatures. The ingots were forged at 1200°C for 2 hrs. and subsequently hot rolled by soaking at 1200°C for 1-hr in a furnace at RDCIS SAIL, Ranchi. The Ae3 temperature of the steels is 847-850°C, while the finish rolling temperatures (FRT) of the steels 9, 10 and 14 were 875, 862 and 885°C, respectively.

Dilatometry

For designing continuous annealing cycles, AC_1 and AC_3 temperatures were determined by continuous heating and

Bs/Ms temperatures through continuous cooling transformation (CHT & CCT) tests on Gleeble 1500D thermomechanical simulator using 80mm long-3mm-dia samples from hot rolled materials.

For CHT, samples were heated at the rate of 1-100°C/s to 1050°C, followed by air cooling. While, for CCT, samples were heated to 1050°C for 2-mins. at the rate of 5°C/s, followed by cooling at the rate of 0.5°C/s-100°C/s.

For microstructural analysis of the CCT samples, all samples were cut in half, approximately at the position of the thermocouple and were prepared for metallographic examination following standard sample preparation technique, while using 2% nital to reveal microstructures.

Continuous Annealing Simulation

The hot rolled samples were cold rolled to ~ 0.8-1 mm using the laboratory cold rolling set-up at NML Jamshedpur for subsequent continuous annealing (CAL) simulation. To simulate continuous annealing cycles, strip annealing facility of RDCIS, SAIL Ranchi was utilized. The strip samples of 260mmX50mm were used for the annealing simulation.

The simulation was carried out in the intercritical temperature range ($\alpha+\gamma$) that was determined by CHT tests on Gleeble 1500D. Two types of continuous annealing cycles were designed for strip annealing simulation on Gleeble 3500 to examine the effect of cooling rate and bainite on tensile properties and weldability.

Effect of cooling rate

Cold rolled samples of Heat-14 were used to examine the effects of cooling rate in the simulated annealing cycles. The parameters used in the annealing cycles along with their thermal cycles are given in Table II.

Table II: Annealing parameters for cooling rate effect

Anneal. cycle	Heating rate, °C/sec	Soaking temp., °C	Soaking time, sec	Slow cooling rate, °C/sec	Fast cooling temp., °C	Fast cooling rate, °C/s
14-CR-1	10	830	120	2	800	100
14-CR-2	10	830	120	2	800	30

Effect of bainite

Cold rolled samples of Heat-9 were used to examine the effects bainite in the simulated annealing cycles. The parameters used in the annealing cycles along with their thermal cycles are given in Table III.

Welding of annealed samples

A DC MF-90 Spot Welder was used to spot weld the annealed samples using a Cu-Cr-Zr alloyed electrode of 5-mm truncated tip diameter. The squeeze force, weld current and weld time were 22daN, 5.8-6kN and 180-200s, respectively. After welding, the

Table III: Annealing parameters for bainite effect

Anneal. cycle	Heating rate,°C/s	Soaking temp., °C	Soaking time, sec	Slow cooling rate, °C/s	Fast cooling temp, °C	Fast cooling rate, °C/s	Bainite temp., °C	Bainite time, sec	Cooling rate,°C/s
9-B-1	10	780	60	2	760	40	470	60	70
9-B-2	10	780	120	2	760	20	470	60	10

samples were submitted for peel-testing using a laboratory vice.

Microscopy, hardness measurements and tensile testing (hot bands, annealed samples and welded samples)

Cross sections of samples after hot rolling, cold-rolled-annealing and welding were ground and polished following standard metallographic procedures. Thereafter, the samples were examined by optical (OM), high resolution scanning electron microscopy (FEG-SEM) and transmission electron microscopy (TEM) equipped with energy dispersive spectroscopes (EDS). For OM and FEG-SEM, the polished specimens were etched with 2% nital. For TEM, thin foils of 3-mm diameter were punched from mechanically thinned ~60 mm slices and were electro-polished in a solution of methanol and 10% perchloric acid at -40°C using a twin-jet electropolisher. Thin foils were examined using a Phillips CM12 transmission electron microscope operated at 120kV.

Microhardness measurements of various phases present in the hot bands and the CCT samples were conducted using LECO LM 247AT Vickers microhradness tester, while macro hardness for the hot bands and resistance spot weldments were measured using a LECO LV700AT Vickers Hardness tester.

The annealed samples were further submitted to tensile testing following ASTME-08 sample dimension on Instron 5582 tensile tester to record the tensile data.

RESULTS AND DISCUSSION

Hot band microstructure

The microstructure of hot band-9 (Heat-9) consisted of ferrite and pearlite (Fig 1a) in association with nano-sized Mo-C carbides (Fig. 1b, c). On the other hand, hot band-14 consisted of ferrite, pearlite, bainite and martensite (Fig.2a). Further, complex precipitates (Ti-Nb-Mo) C (Fig. 2b, c), rich in Ti were also observed in hot band-14 as Ti content was higher (0.031) in this band than that of the hot band-9 (0.009). The ferrite grain size of the hot bands 9 and 14 were 4.1 and 3.7 μm, while ferrite area fractions were 69 and 79, respectively.

Continuous cooling transformation

Heat-9

From continuous cooling transformation it was observed that below 5°C/s cooling rate, ferrite, pearlite and bainite were

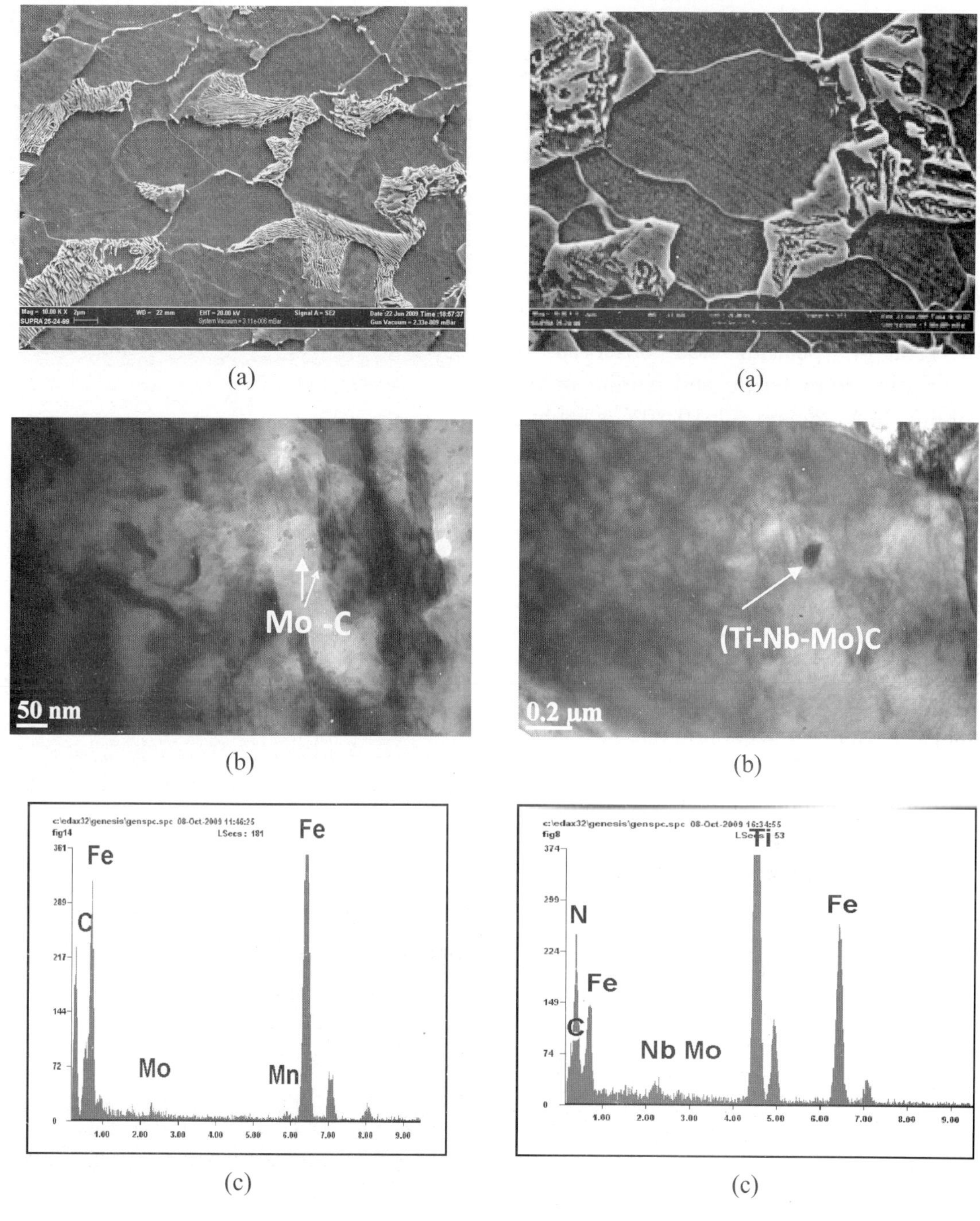

Fig. 1: (a) FEG-SEM micrographs showing ferritic-pearlitic structure, **(b)**TEM micrograph showing nano-sized Mo-C precipitates and **(c)** EDS of the precipitates in the hot band of Heat-9.

Fig. 2:(a) FEG-SEM micrographs showing ferritic-banitic-martensitic structure **(b)** TEM micrograph showing (Ti-Nb-Mo)C precipitates **(c)** EDS of the precipitates in the hot band of Heat-14.

present, while between 5°C/s and 20°C/s, the co-existence of ferrite and bainite were noticed. Further increase in cooling rate till 70°C/s showed the presence of martensite in association with ferrite and bainite, while above 70°C/s, bainite disappeared and only ferrite and martensite were present.

Heat- 14

In Heat-14, pearlite was observed in association with ferrite and bainite in the cooling rate range of 10-20°C/s, while till 10°C/s ferrite structure was dominant. While from 20 to 100°C/s, ferrite, bainite and martensite were present.

For designing continuous annealing cycles, fast cooling rate was maintained to avoid pearlite formation as the presence of pearlite develops yield point elongation and which in turn causes surface problems. Thus, from the CCT studies, the cooling rate for Heat-9 was kept above 5°C/s, while for Heat-14, the cooling rate maintained was above 20°C/s.

Continuously annealed samples

Effect of cooling rate—Annealing cycles 14-CR-1 and 14-CR-2

a) Microstructure

Samples that underwent cycles 14-CR-1: 10°C/s—830°C (120s)-2°C/s-800°C-100°C/s and 14-CR-2: 10°C/s—850°C (120s)-2°C/s-800°C-30°C/s, experienced cooling rates of 100 and 30°C/s, respectively. In both the samples ferritic-martensitic microstructure was obtained with 40% and 31% martensite fraction, respectively (Figs. 3a, b). Due to the fast cooling rate of 100 °C/s, no new precipitates could be

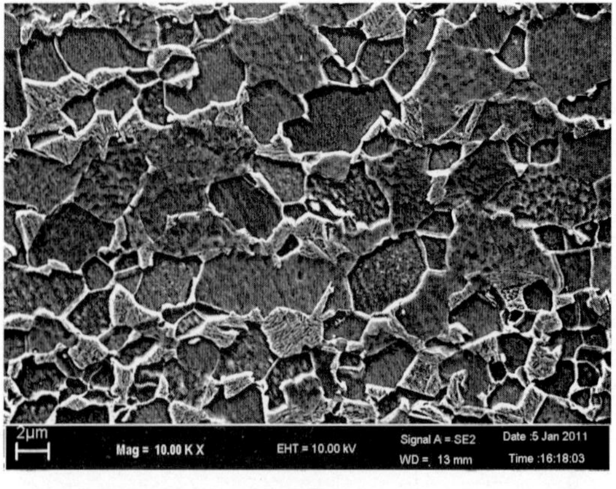

(a)

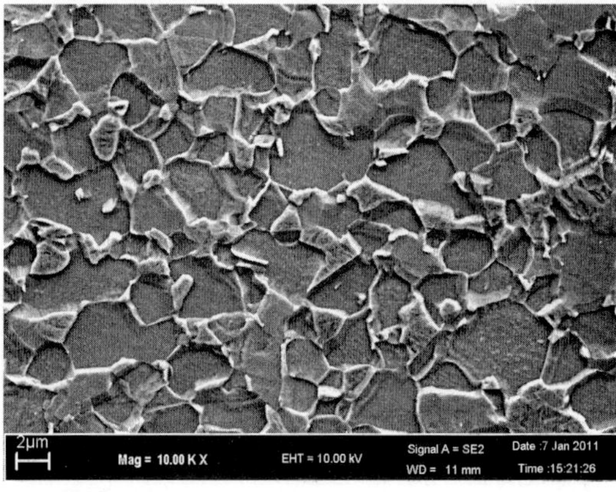

(b)

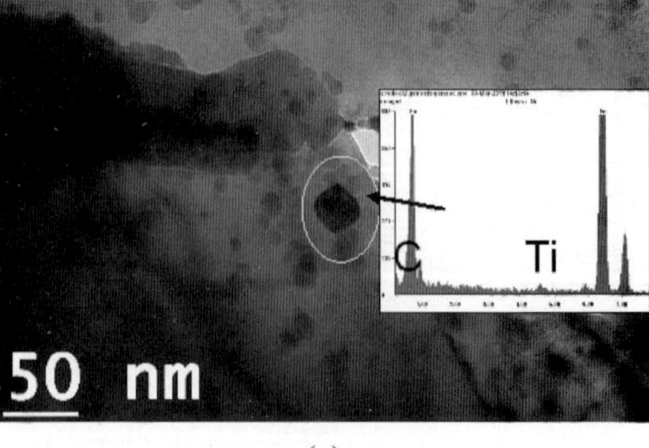

(c)

Fig. 3: (a,b) FEG-SEM micrographs showing ferritic-martensitic structures for annealing cycles 14-CR-1 and 2, respectively and (c) showing the EDS of TiC precipitate in 14-CR-2 sample.

formed during the cooling cycle of 14-CR-1, however, as the cooling rate in 14-CR-2 was slower (30°C/s), some TiC precipitates were also formed (Fig. 3c). Further, the presence of (Ti-Nb-Mo) CN precipitates was also observed in 14-CR-2.

b) Tensile properties

The YS and UTS of cold-rolled and continuously annealed 14-CR-1 were registered as 405 MPa (Table IV) and 750 MPa, respectively, with % UEL (%Uniform Elongation) of 13% and %TEL (%Total Elongation) of 19% for 100°C/s cooling rate The precipitates helped increase the YS to 494 MPa compared to 405 MPa in 14-CR-1, by grain refinement and restricting dislocation movement. However, UTS was reduced to 690 MPa for 14-CR-2, as martensite content reduced, while %UEL and TEL were increased to 17% and 27%, respectively. The strain hardening exponents (n) for 14-CR-1 and 2 were satisfactory with 0.18 and 0.20, respectively.

Table IV: Microstructure, tensile data and weld quality of simulated continuously annealed samples

Sample	Microstructure	YS, MPa	UTS, MPa	% UEL	% TEL	n	Weld quality
Cycle-14-CR-1	F+40% M	405	750	13	19	0.18	Porosities in weld zone and cracks in HAZ
Cycle-14-CR-2	F+31%M and TiC precipitate	494	690	17	27	0.20	No porosities and cracks in weldment
Cycle-9-B-1	F+ 25%(M+ B)	600	702	16	23	0.23	HAZ cracks
Cycle-9-B-2	F+37%(B+M) and Mo-C precipitate	449	753	16	22	0.18	No porosities and cracks in weldment

c) Weldability

The welded sample of 14-CR-1 showed plug type nugget failure with predominantly martensitic in association with bainitic welded structure (Fig.4a,b--representative photographs are presented as in all the cases failure was plug type with predominantly martensitic weld structure). The weld zone was examined by FEG-SEM and porosities were observed. Further, the cracks initiated in a zone where the alloy remained in the solidus-liquidus temperature range during welding. The dendrite structure inside open cracks (Fig. 4c) indicates that the liquid was present at the moment of crack formation and therefore, it is liquation cracking according to the classification of Hemsworth et al [31]. In addition, cracking in the heat affected zone also was initiated by the presence of MnS stringers (Fig. 4c). This can be attributed to the triaxial stresses at the sharp corners of the stringers during cooling cycle of the welding. This helped lower the cohesive force at the MnS-matrix interface and thus initiated a crack. While hardness values for the weld, HAZ and base metal were 309, 256 and 290 VPN, respectively. Thus, hardness of the HAZ was

(a)

(b)

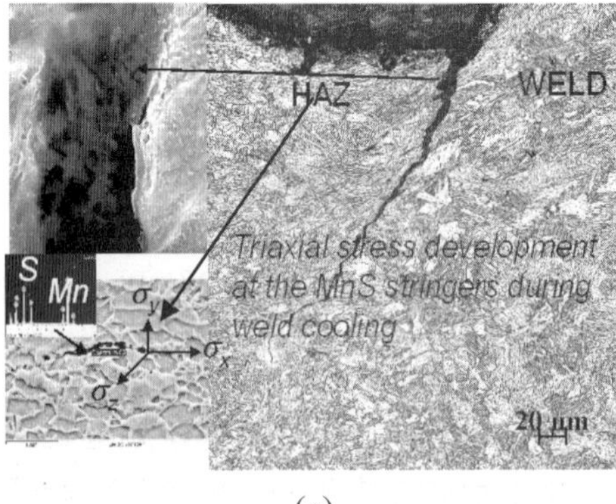

(c)

Fig. 4: (a) Showing a representative photograph of pullout button failure, **(b)** showing OM micrograph of predominantly martensitic structure in weld zone **(c)** OM micrograph showing liquation crack in the fusion line [FEG-SEM magnified view in the top-left inset] along with HAZ crack at the Mn-S stringers [bottom-left inset] for annealing cycle 14-CR-1.

quite lower than that of the weld region and base metal that is indicative of the fact that the HAZ softening was the possible reason for the HAZ cracking and poor weldability. The HAZ softening occurred in this case while significant amount of martensite (40%) was present in the base metal, which was in good agreement with the conclusion made in the literature [23, 24]. In HAZ softening, martensite tempering takes place at the points of the weld where the temperate approaches A1 (subcritical HAZ). This is because a hard weld metal can not deform plastically to release welding stresses and they are transferred to the softened HAZ, causing it to crack [32]. This is the reason why softer base metal consistently suffers a smaller amount of HAZ cracking than the harder consisting of large amount of martensite.

However, the weldment of 14-CR-2 exhibited better welding performance than 14-CR-1. A plug type nugget failure was obtained and the microstructure of the welded zone was again bainitic martensitic. However, no imperfections were in the weldment (Fig. 5). Further, hardness of the HAZ (390 VPN) was higher than both the

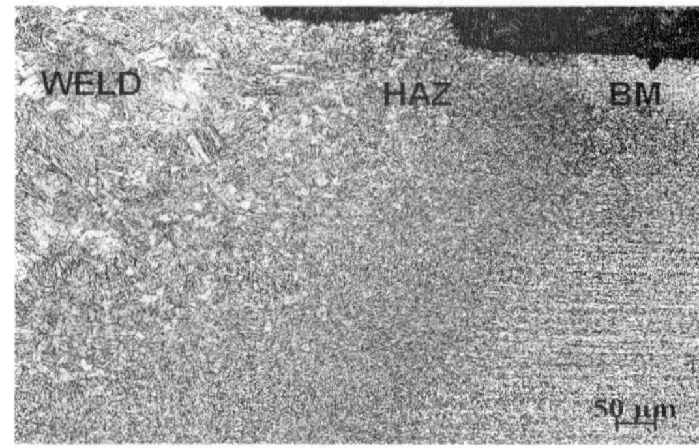

Fig. 5: OM micrograph showing defect free weld, heat affected zone (HAZ) and base metal for annealing cycle 14-CR-2.

welded (353 VPN) and base metal (250 VPN). Thus, there was no HAZ softening and subsequent cracking. This is possibly due to the presence of (Ti-Nb-Mo) CN precipitates that helped control the austenite grain growth in the HAZ, which subsequently resulted in reduced transformed grain size in the HAZ and avoided HAZ softening. Further, the presence of TiC (Fig. 3c) in the HAZ possibly also played a role in controlling the HAZ softening.

Effect of Bainite--Annealing cycles 9-B-1 and 9-B-2:

a) Microstructure

The base metal microstructure of the cold rolled and continuously annealed sample of cycle 9-B-1 (10°C/s—780°C (120s)-2°C/s-760°C-40°C/s—470°C (60s)-70°C/s) was ferrite with 25% martensite and bainite (Fig. 6a, b). In this case, bainite was quite nominal compared to martensite content. Another annealing cycle, 9-B-2: 10°C/s—780°C (120s)-2°C/s-760°C-20°C/s—470°C (60s)-10°C/s resulted in ferritic-bainitic-martensitic microstructure with bainite and martensite contents of 37%, in which the amount of bainite was relatively

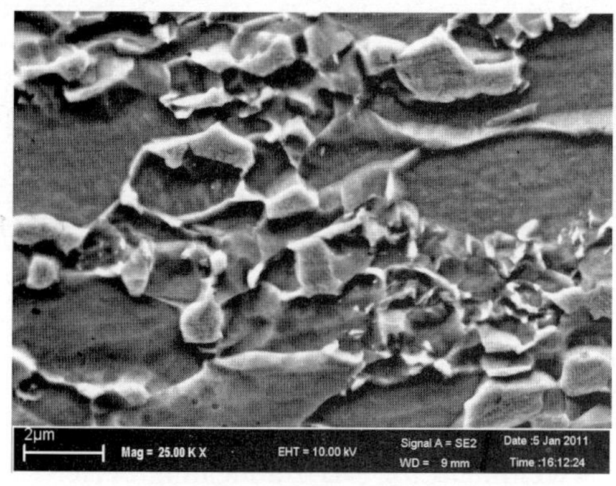

(a)

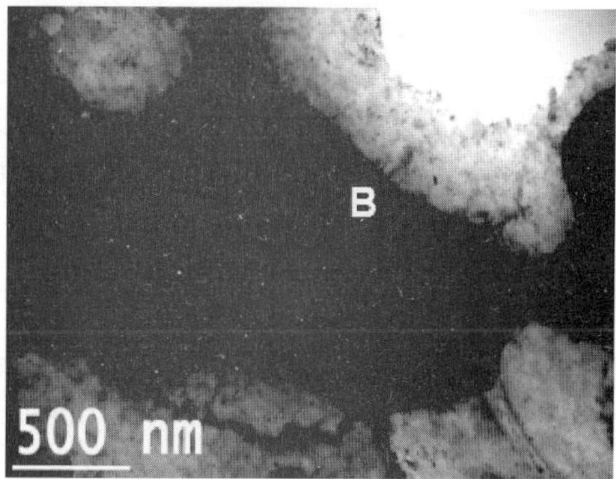

(b)

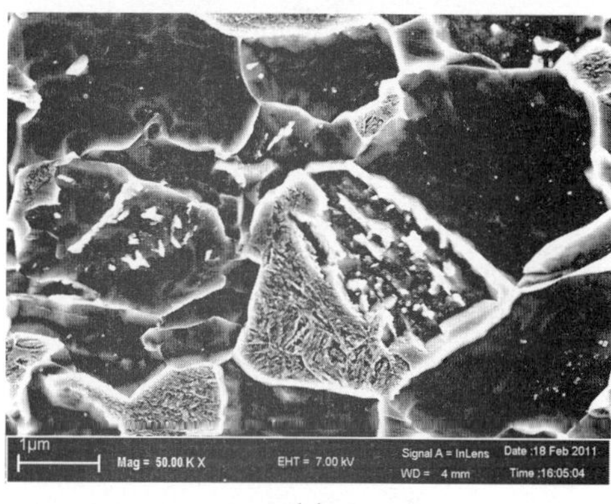

(c)

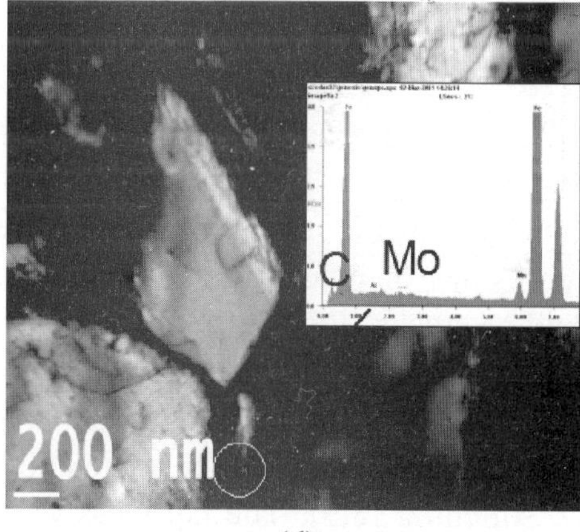

(d)

Fig 6: (a) FEG-SEM showing ferritic-martensitic-bainiticstructure and **(b)** showing bainite in TEM for annealing cycle 9-2-1, while **(c)** showing ferritic-martensitic-bainitic structure in FEG-SEM and **(d)** and intergranular Mo-C precipitate in TEM for annealing cycle 9-B-2.

more (Fig. 6c). Further, intergranular precipitation of Mo-C was also observed because of slower cooling rate of 20°C/s from quenching temperature to the bainitic temperature (Fig 6d).

b) Tensile properties

The YS and TS with 25% M+B in the ferrite matrix of 9-B-1 were obtained as 600 and 702

In this case bainite amount was quite less. The %EL in terms of %UEL and %TEL, was found to be 16 and 23%, respectively, while 'n' value was quite satisfactory with 0.23. The cycle 9-B-2 offered one of the best annealing cycles in the present study with YS and UTS of 449 and 753 MPa with % UEL, %TEL and 'n' as 16%, 22% and 0.18 (Table IV), respectively. This can be attributed to the presence of relatively substantial amount of bainite in association with nano-sized Mo-C precipitates that helped achieve the above properties.

c) Weldability

The 9-B-1 welded sample also resulted in a plug type failure. The welded zone with bainitic-martensitic structure though had no porosities or cracks the HAZ was not devoid of cracks (Fig. 7a). Short cracks in the HAZ were again found to have initiated and propagated at the MnS stringers. Crack generation in the HAZ can be attributed by HAZ softening [8,9]. The hardness of the HAZ was recorded to be 250 VPN while the hardness values of the weld and base metal were 330 and 300 VPN, respectively. On the other hand, 9-B-2 sample was welded successfully with pull-out nugget failure without any weld imperfection and HAZ cracking (Fig. 7b). The successful welding can be attributed to the presence of relatively substantial amount of bainite in the base metal microstructure that helped avoid HAZ softening (HAZ—360 VPN, Weld 355 VPN, base metal—244 VPN) by providing a formable base metal to accommodate the plastic residual stresses during weld cooling. Further, the presence of the carbide forming alloying element like Mo (Mo-C) was possibly able to increase the resistance to HAZ softening [31].

From the investigation it is clear that the base metal microstructure played an

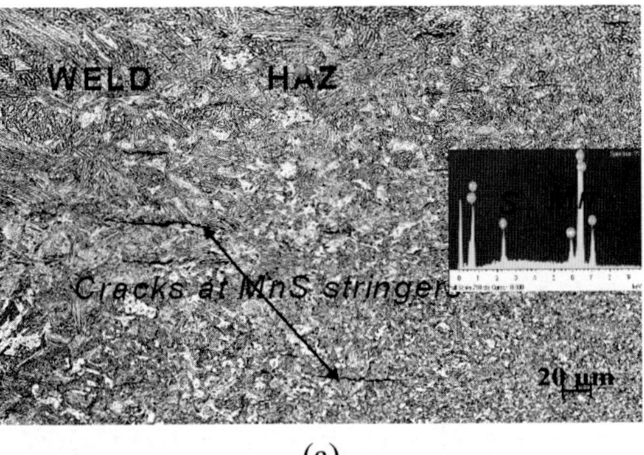

(a)

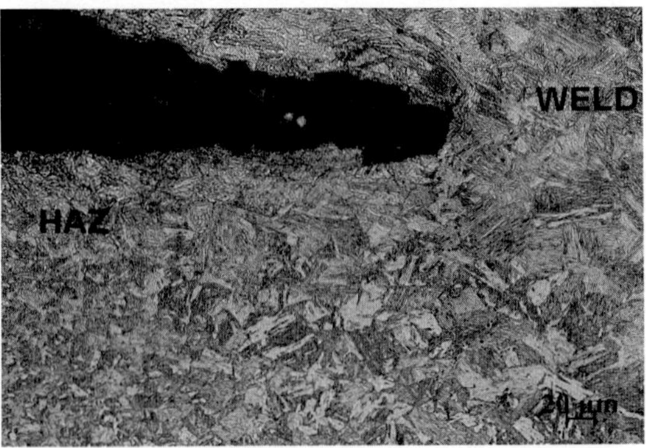

(b)

Fig. 7: (a) OM micrograph showing HAZ crack in the weldment of 9-B-1 annealing cycle while **(b)** showing defect free weldment for 9-B-2 annealing cycle.

important role in affecting tensile properties and weldability of the steels. It was inferred that weld HAZ hardness needed to be higher than the weld metal (i. e.$H_{HAZ} > H_{WM}$) while the same requires to be stronger than the base metal also to avoid HAZ softening and thus HAZ cracking.

CONCLUSIONS

- The steels developed were in the range of ~690-753 MPa ultimate tensile strength and 19-27% total elongation.
- In all the annealing conditions the desirable plug type nugget failure was observed
- Microstructure containing martensite >~ 30% in the base metal and cooling rate of >30°C/s from intercritical annealing temp., (without bainite effect) adversely affected weldability.
- Introduction of optimum amount of bainite in the ferrite and martensite structure resulted in satisfactory weldability behaviour.
- Nano-sized precipitates like TiC, Mo-C and (Ti-Nb-Mo)CN helped obtain desirable tensile and weld properties.
- HAZ softening was found to be the main cause for HAZ cracking. HAZ softening was associated with high martensite (> 30% content) in the base metal.
- In weldments where $H_{HAZ} < H_{WELD}$ and H_{BM} (HAZ softening), liquation cracks and cracks at MnS stringers in HAZ/HAZ-WM were observed.
- Cycle 9-B-2 (10°C/s—780°C (120s)-2°C/s-760°C-20°C/s—470°C (60s)-10°C/s) with ferrite, 37% bainite and martensite along with nano-sized precipitates of Mo-C resulted in the best combination of tensile properties: YS-449 MPa , UTS-753 MPa, %UEL-16%, %TEL-22% and n-0.18, and weldability with plug type nugget failure.
- Stronger HAZ than weld metal in association with formable base metal resulted in favourable combination of stength-ductility-weldability.

ACKNOWLEDGEMENT

The author is thankful to the management of Tata Steel for permission to publish the work. Special thanks are due to Dr. Vinod Kumar, RDCIS SAIL Ranchi and IIT Madras, for strip annealing tests and TEM examination, respectively.

REFERENCES

[1] V. Cuddy et al, "Manufacturing guidelines when using ultrahigh strength steels in automotive applications, EU report-ECSC R585, 2004.

[2] V. H. Lopez-Cortez and F. A. Reyes-Valdes, Welding Journal, Dec 2008, 36-40.

[3] C. Elgohen, A. De Paepe, A. Lucas and Y Hardy, Materials Sci Forum, Vols 539-543, 2007, p. 4405.

[4] B Girvin, W Peterson and J. Gould, Report on Development of appropriate spot welding practice for Advanced High Strength steels, Project No.

47819GTH, EWI, 1250 Arthur E Adams Drive Columbus, OF 43221, Nov 10, 2004.

[5] J. E. Gould, S. P. Kurana and T. Li, Weld J, May 2006, p.111.

[6] K. Osawa, Y. Suzuki and S. Tanaka, Kawasaki Steel Technical Report, No. 48, March 2003, p.9.

[7] Committee on Automotive applications: "AHSS-application guidelines", 2006, Washington, DC, Int. Iron and Steel Institute.

[8] J. E. Gould, L. R. Lehman and S. Homes, A. Design of experiments evaluation of factors affecting the resistance spot weldability of high strength steels. Proc Sheet Metal Welding Conf. VII, 1996, AWS Detroit Section.

[9] J. E. Gould and D. Workman, Fracture morphologies of resistance spot welds exhibiting hold time sensitivity behaviour, Proc. Sheet Metal Weld Conf VII, 1998, AWS Detroit Section.

[10] M. V. Li, P. Dong and M. Kimchi, 1997, Modelling and analysis of microstructure development in resistance spot welds of high strength steels, 981BECC-8.

[11] Z. Feng, S. S. Babu, B. W. Riemer, M. L. Santella, J. E. Gould, and M. Kimchi, "Modeling of resistance spot welds: Process and Performance." 2001, Welding in the World, 45 p.18.

[12] H. Shao, 2000, Blank welding high strength steels, Ph.D. dissertation, Columbus, Ohio, The Ohio State Univ.

[13] S. L. Ream, 2003, Remote laser weld performance in advanced high strength steels. Proc. ALAW 2003, Univ. of Michigan.

[14] W. L. Chuko and J. E. Gould, Weld J., 2002, p.1s.

[15] M. Milititsky, E. Pakalnins, C. Jiang, A. K. Thompson, On Characteristics of DP600 resistance spot welds, SAE Report, Paper No. 01-0520, Detroit, 2003.

[16] C. Ma, S. D. Bhole, D. L. Chen, A. Lee, E. Biro, G. Boudreau, Sci. Technol, Weld Joining, 11, 2005, p.296.

[17] M. I. Khan, M. M. L. Kuntz, P. Su, A. Gerlich, T. North and Y. Zhou, Sci. Technol. Weld Joining, 2007, 12, p.175.

[18] G. Wang & M. E. Barkety, Weld. J., 85, 2006, p.84s.

[19] M. Maryas and X. Q. Gayden, Weld. J., 84, 2005, p.172s.

[20] A. Joaquin, A. Elliot, C. Jiang, Weld. J., 2007, Vol. 86, No. 2, p. 24s.

[21] M. Marya, and X. Q. Gayden, Weld J, 2005, 84, No 11, p.172.

[22] V. H. Baltazar Hernandez, M. L. Kuntz, M. I. Khan and Y. Zhou, Sci. and Technol. of Welding & Joining, 13, No. 8, 2008, p.769.

[23] M. I. Khan, M. L. Kuntz and Y. Zhou, Sci. Technol. Weld Joining, 2008, 13, p.49.

[24] S. Hashimoto, S. Kanbe and M. Sudo, Trans. ISIJ, 1981, 21, p.B-497.

[25] N. Yamauchi, T. Taka, K. Kunishige and N. Nagao, Trans ISIJ, 1982, 22, B-107.

[26] E. Biro and A. Lee, Sheet Metal Welding Conf XII, AWS, Livonia, MI, 2006.

[27] M. Xia, E. Biro, Z. Tian and Y. N. Zhou, ISIJ, 2008, 48, p.809.

[28] S. Dancette, V. Massardier-Jourdan, D. Fabregue, J. Merlin, T. Dupuy and M. Bouzekri, Isij Int, 51, No 1, 2011, p.99.

[29] B. Hemsworth, T. Boniszewski and N. F. Eaton, Classification and definition of high temperature welding cracks in alloys. Metallurgy, Feb 1969, p.5.

[30] K. Banerjee, N.L. Richards and M.C. Chaturvedi, Metall. and Mater. Trans. A, Vol 36A, July, 2005, p.1881.

[31] E. Biro, J.R. McDermid, J.D. Embury, and Y.Zhou, Metall. and Mater. Trans. A, Vol. 41A, Sept, 2010, p.2348.

Effect of Strain, Strain Rate and Temperature of Hot Deformation on Microstructural Evolution of Ti and V Microalloyed Steel

Deepak Kundalkar and Asim Tewari
Department of Mechanical Engineering, IIT Bombay
Powai, Mumbai-400076

ABSTRACT

High-strength low-alloy steels (HSLA) have better mechanical and corrosion properties than carbon steels and lower manufacturing cost due to fewer heat-treatment steps. The superior properties of HSLA are attributed to the combined effect of multi-phase microstructure, grain size strengthening and precipitate hardening which are produced in situ during hot deformation. The final properties are a combined effect of solutionization/precipitation, austenitic grain growth/ DRX/recovery and kinetics of ferrite, pearlite, martensite and bainite formation. Hence it is of scientific and technological importance to understand the microstructural evolution of HSLA under High Temperature Deformation (HTD). In this work the effect of deformation rate and temperature on the competition among grain-growth/DRX/recovery kinetics of austenite phase is studied using analytical and numerical thermodynamic models. It is seen that for a given temperature there exists a critical deformation rate above which DRX is the dominant mechanism.

KEYWORDS: *Drx, Hot Deformation, Hsla Steel, Microstructure Evolution, Strain Energy Density*

INTRODUCTION

Structural steels with a ferrite-pearlite microstructure are widely used in civil, mechanical and chemical engineering applications. Higher working loads and increasing demand of reliability requirements have led to an increase in the requirement of high strength and toughness in these steels. In addition the requirement of good weldability leads to decrease in carbon content of the steel, thereby decreasing the strength of the steel, due to a decrease in the amount of pearlite. This decrease in strength can be overcome by microalloying the steel with elements such as titanium, niobium and vanadium in small amounts (< 0.15 wt. %). These microalloying elements improve the strength of steels through mechanisms such as precipitation strengthening and grain size refinement [1-6].

In hot deformation process the mechanisms such as recovery, DRX and growth also play a very important role in determining the prior austenite grain size and also the final evolved microstructure of the steel. R. D. Doherty el. Al., in a review on DRX, have precisely defined these mechanisms as below. "Recrystallization is the formation of a new grain structure in a deformed material by the formation and migration of high angle grain boundaries driven by the stored energy of deformation". On this definition "Recovery consists of all annealing processes occurring in deformed materials that occur without the migration of a high angle grain boundary". Grain coarsening was defined as processes involving the migration of grain boundaries when the driving force for migration is solely the reduction of the grain boundary area itself [7]. Based on the type of deformation the mechanisms of recrystallization can be static or dynamic [8]. In the case of hot deformation recrystallization takes place dynamically.

The effect of hot deformation on Nb, Ti and V microalloyed steel has been studied and reported by many authors [9-13]. These include the effect of strain rate and temperature of hot deformation on the microstructure and properties of steel.

In this work an effort has been made to model the effect of hot deformation parameters on the microstructure through experiments and modeling.

EXPERIMENT

The HSLA steel consisting of Ti and V microalloying elements (commercially known as 38MnVS6) was chosen for this investigation. The material composition is as given in Table 1. The raw material was received in the form of rolled billets of ϕ 90 –110 mm diameter and 3000 mm length. This material was then processed as shown in Figure 1 to obtain samples of ϕ 10 mm diameter and 15 mm length.

Table 1: Chemical composition of 38MnVS6 HSLA steel

Chemical Composition	Carbon [C]	Manganese [Mn]	Silicon [Si]	Phosphorous [P]	Sulfur [S]	Chromium [Cr]
Wt. %	0.38	1.45	0.55	0.02	0.4	0.16
Chemical Composition	Nickel [Ni]	Molybdenum [Mo]	Aluminium [Al]	Copper [Cu]	Vanadium [V]	Titanium [Ti]
Wt. %	0.13	0.04	0.02	0.18	0.12	0.01

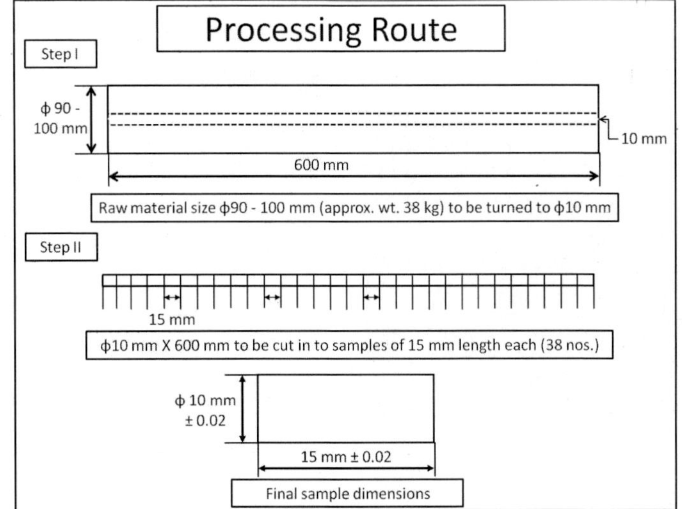

Fig. 1: Schematic of sample preparation from as-received material

Hot compression test were carried out at various temperatures and strain rates on GLEEBLE thermo-mechanical simulator.

The tests were conducted at true strain of 0.69 i.e. 50% compression of sample from its original length. Figure 2 represents the schematic of the hot compression testing process. Figure 3 shows the image of the GLEEBLE system with an enlarged view of sample mounting and testing chamber.

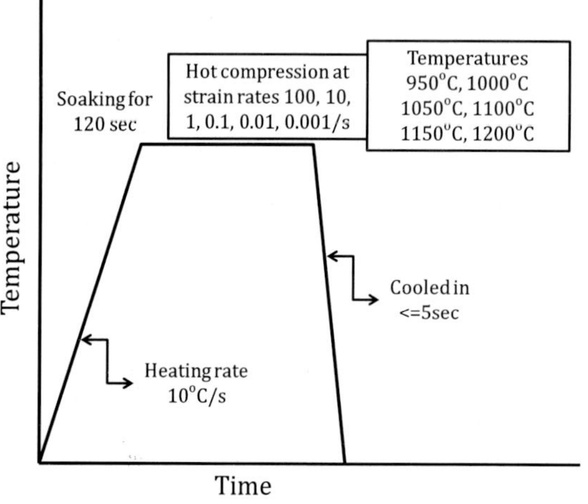

Fig. 2: Schematic of the hot compression testing process

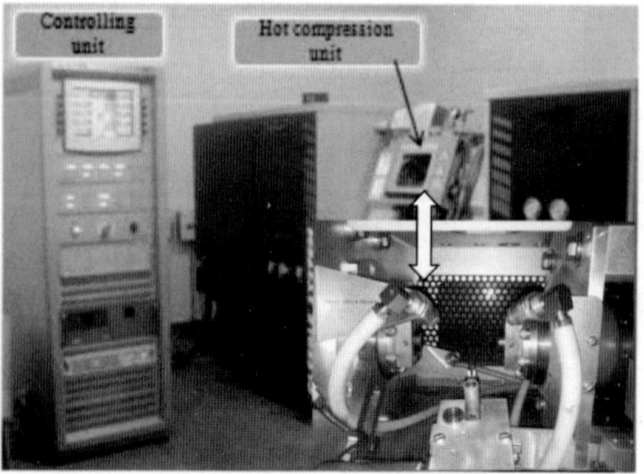

Fig. 3: GLEEBLE Thermo-mechanical simulator

MODELING

Constant strain rate hot compression model was build using Deform v.10 software. The geometry used for modeling was same as the sample geometry Figure 4. Material model was created using the experimental flow stress curves. Figure 5 shows the setup used for modeling the constant strain rate hot compression test. The contact friction between the dies and the work-piece was taken as 0.3, condition similar to the lubricated hot forging environment.

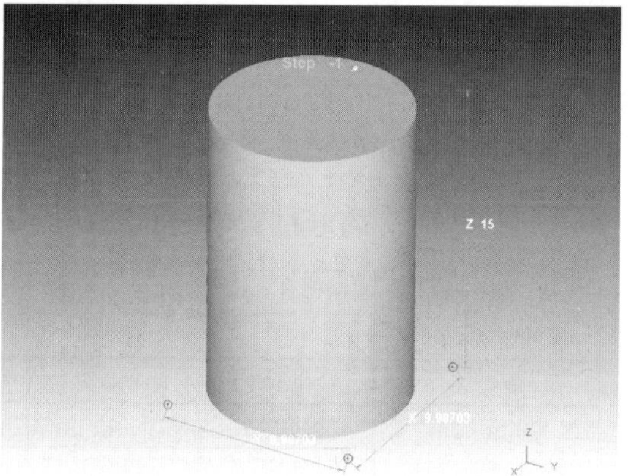

Fig. 4: Deform v.10 model geometry

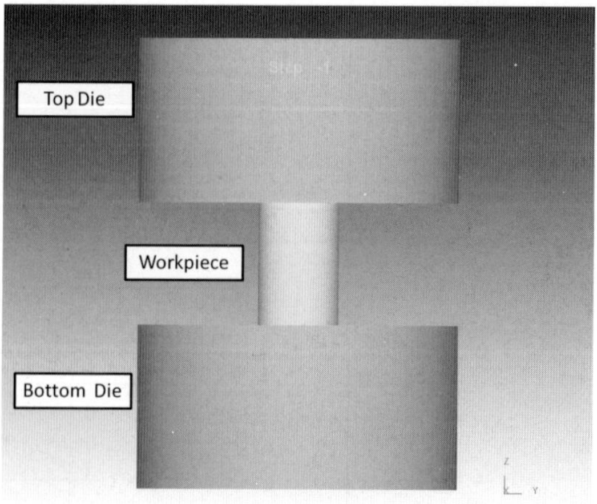

Fig. 5: Deform model hot compression setup

Adiabatic boundary condition was imposed on the circumference along the length of the sample to imitate the experimental conditions.

RESULTS & DISCUSSION

The as-received material was polished and etched using 2% nital etchant and observed under the Carl Ziess optical microscope. Figure 6 shows the optical micrograph of as received samples of 38MnVS6. The microstructure observed in the optical micrograph shows ferrite-pearlite structure with prior austenite grain size of approx. 40mm.

Figure 7a) shows the mean prior austenite grain size and Figure 7b) shows ferrite phase fraction along with 95% confidence interval for as-received sample of 38MnVS6 steel.

The sample of 38MnVS6 steel was initially soaked at 1300°C for 120 sec and then hot compressed at constant strain rate of 10/s and temperature of 1150°C in GLEEBLE thermo-mechanical simulator. Figure 8 shows the heating, soaking, hot deformation and the cooling temperatures as function of time collected from GLEEBLE data. The heating was very rapid at rate of 10°C/s and after deformation cooling was also very rapid (i.e. in less than 5 sec).

Fig. 6: Optical micrograph of as-received 38MnVS6 HSLA steel

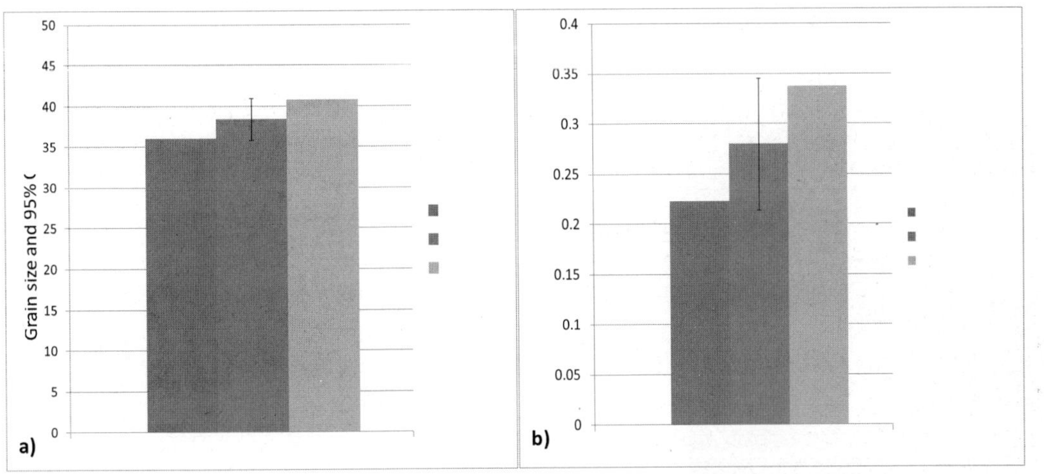

Fig. 7: a) Grain size and b) Ferrite phase fraction with 95% CI for as-received samples of 38MnVS6 HSLA steel

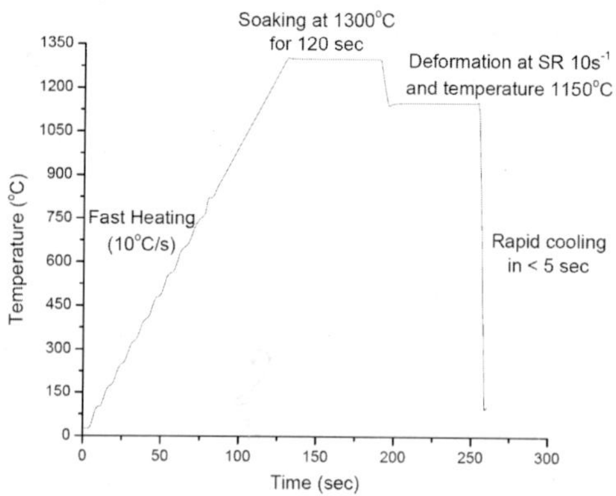

Fig. 8: Time temperature curve extracted from GLEEBLE data

The effect of this hot deformation process on the evolved microstructure is as shown in Figure 9. It is seen that the grain size at the cross-sectional surface is the largest and decreases towards the radial edge. Similarly the grain size reduces as we move from the surface towards the axial center. The clear distinction in the final grain sizes at these locations is shown in Figure 10.

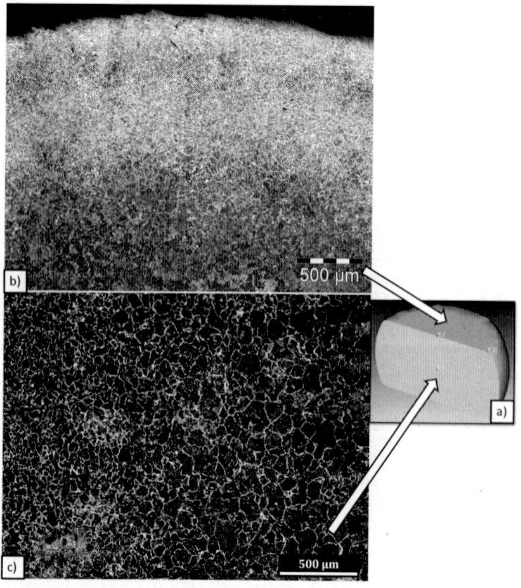

Fig. 9: Optical micrographs of a) sample at b) cross section and c) axial centre locations

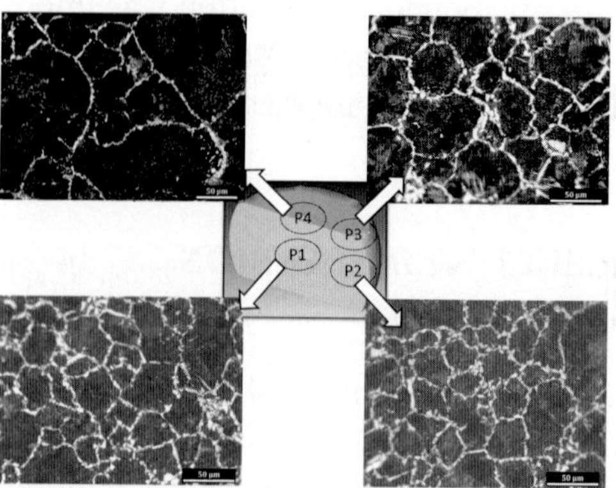

Fig. 10: Optical micrographs at various locations as marked on the sample

The variation in the prior austenite grain sizes at locations shown in Figure 10 on the sample along with the 95% CI values is shown in Figure 11. It is seen that the mean grain size at the surface (P4) is almost the twice of the mean grain size at the center (P1). The variation in the phase fraction of ferrite is shown in Figure 12. The locations with largest grain size have least fraction of ferrite whereas those with fine grain size have higher fraction of ferrite phase.

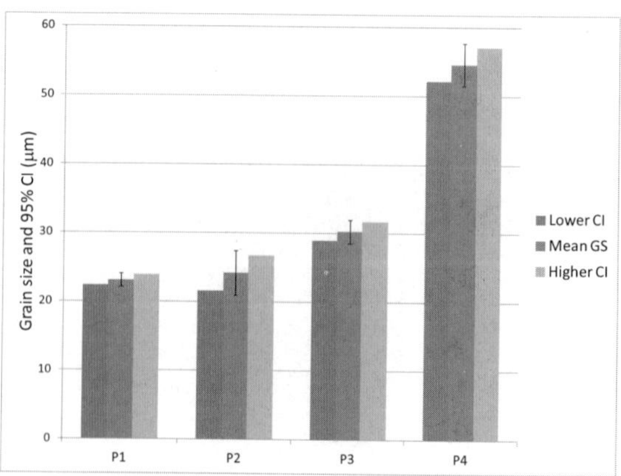

Fig. 11: Prior austenite grain size along with 95% CI at locations marked in Figure 10

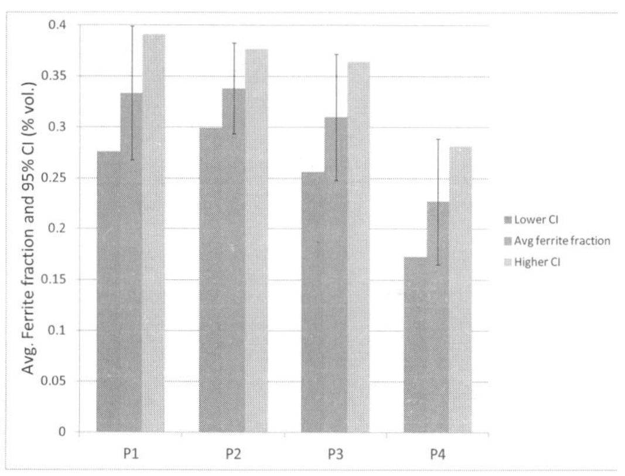

Fig. 12: Phase fraction of ferrite along with 95% CI at locations marked in Figure 10

Figure 13 shows the EBSD IPF and IQ maps at locations P1, P3 and P4, respectively.

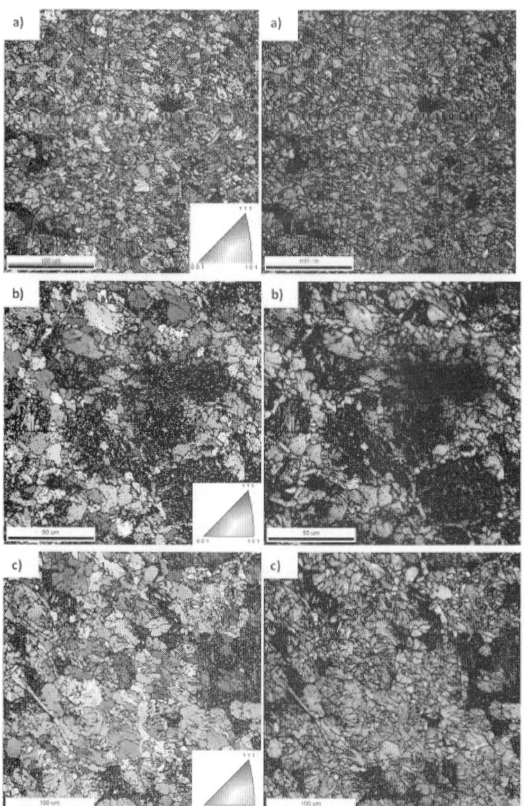

Fig. 13: EBSD IPF and IQ maps at locations a) P1, b) P3 and c) P4

The KAM values were measured at these locations using the EBSD results and are given in Table 2.

From Figure 11 and Table 2 it is seen that the mean grain size increases in the order P1 < P2 < P3 < P4, whereas the pre-eutectoid ferrite phase fraction decreases in the same order P1 > P2 > P3 > P4.

Table 2: Grain sizes and KAM values for various locations

Location	Grain size	KAM
P1	19 mm	1.20
P3	22 mm	1.05
P4	46 mm	0.9

KAM value provides the measure of post DRX deformation. As post DRX deformation increases KAM value also increases. Highest KAM value at location P1 indicates large deformation after final DRX, whereas lower KAM value at P4 indicates lesser deformation after final stage of DRX.

Hot compression model was prepared in Deform to understand the variation in the strain, strain rate and temperature at different locations within the sample. The data used for building the material model was based on the experimental flow stress data available in the literature and from own work. In the Deform model same four points were tracked and the effective strain rate for each of these four locations were plotted Figure 14. The equivalent stress vs. equivalent strain curves were also plotted for each of these locations as shown in Figure 15.

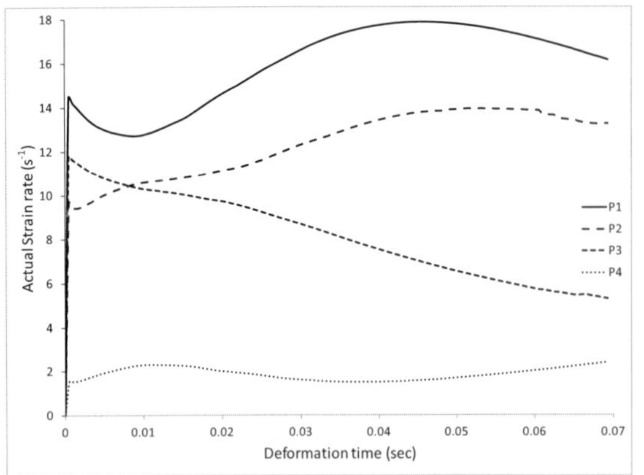

Fig. 14: Actual strain rates at locations P1 – P4 as a function of deformation time

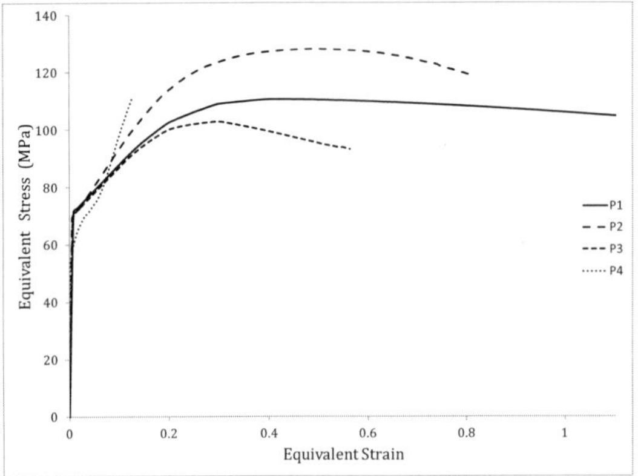

Fig. 15: Equivalent stress – strain plot for locations P1 – P4

Form Figure 14 it is seen that thought the nominal strain rate of deformation is constant $10s^{-1}$, but the local strain rates are seen to vary by an order of magnitude. This large difference in the stain rates at different locations results in a large difference in strain energy densities at these locations. As a result of these different strain energy densities the nucleation rate also tends to vary.

This stress strain data was further utilized for calculating the strain energy density at each location viz. P1, P2, P3 and P4. Figure 16 depicts the plots of strain energy density for each location as a function of true strain.

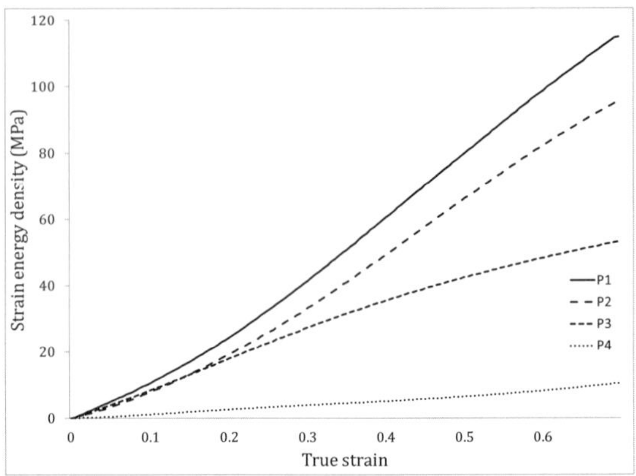

Fig. 16: Strain energy density at locations P1 – P4 as function of true strain

The experimental and modeling results indicate variation in the nucleation rates at various locations within the sample. Increase in the strain rate at location P1 resulted in increased strain energy density. The nucleation rate which is directly proportional to the strain energy density thus increases resulting in lower grain size. For a given interval of time, say Dt, the strain changes as a function of strain rate. From Figure 14 it is evident that point P1 would experience highest strain and point P4 would be least strained which is observed in Figure 15. This matches with observed values of KAM, which provide an indication of post DRX deformation.

Thus a qualitative model is proposed that with increase in strain-rate, the strain energy density would increase, which would trigger higher nucleation rate. Higher nucleation rate would result in lower grain size which is a direct measurable parameter and is seen to validate the above model. Moreover, higher strain-rate would mean larger post DRX deformation and thus larger KAM value. The measured KAM values show the above predicted trend thus further supporting the model. Finally, the inverse relation of pre-eutectoid ferrite with the grain size can be explained by the fact that the finer grains would have larger grain-boundary area (which is the site for pre-eutectoid ferrite nucleation) thus higher pre-eutectoid ferrite fraction.

CONCLUSION

1. It is seen that within a single hot compression sample, there exists large variations in grain size, and KAM values.
2. The variation of grain size is seen to be inversely proportional to the pre-eutectoid ferrite fraction of that location.
3. FEM simulations of the experiment also show that the local strain rates vary by an order of magnitude from one location to the other.
4. The strain energy density calculations through FEM results show that the region with the highest strain energy density also has the smallest grain size.
5. A model to explain all the observed results based on nucleation theory is proposed. This in mathematical form can be written as $\dot{\varepsilon} \uparrow \leftrightarrow E \uparrow \leftrightarrow N \uparrow \leftrightarrow G \downarrow \leftrightarrow f_f \uparrow$, where E is the strain energy density, N is the nucleation rate, G is the Grain size and f_f is the pre-eutectoid ferrite fraction.

REFERENCES

[1] A. Abdollah-Zadeh, B. Eghbali. "Mechanism of ferrite grain refinement during warm deformation of a low carbon Nb-microalloyed steel". Materials Science and Engineering A 457, 2007, p. 219–225.

[2] Majid Meysami. "Study on the behavior of medium carbon vanadium microalloyed steel by hot compression test". Materials Science and Engineering A 528, 2011, p. 3049–3055.

[3] B. Eghbali. "Microstructural development in a low carbon Ti-microalloyed steel during deformation within the ferrite region". Materials Science and Engineering A 480, 2008, p. 84–88

[4] D. K. Matlock, G. Krauss, J. G. Speer. "Microstructures and properties of direct-cooled microalloy forging steels".Journal of Materials Processing Technology 117, 2001, p. 324–328.

[5] A. Babakhani. "Investigation on the effects of hot forging parameters on the austenite grain size of vanadium microalloyed forging steel (30MnSV6)". Journal of Alloys and Compounds 490 , 2010, p. 572–575.

[6] Peter H. Wright, C. S. "High Strength Low Alloy Steel Forgings". In ASM Handbook, Volume 1, Properties and Selection: Irons, Steels, and High Performance Alloys. ASM International. 2005, pp. 578-593.

[7] R. D. Doherty, Hughes, J. J. Jonas, Humphreys. "Current issues in recrystallization: a review". Materials Science and Engineering A238, 1997, p. 219–274.

[8] B. Verlinden, J. Driver, I. Samajdar and R. D. Doherty. "Thermo-Mecanical Processing Of Metallic Metals". Elsevier publication, UK. 2007, p. 85–108.

[9] B. Eghbali. "Influence of deformation temperature on the ferrite grain refinement in a low carbon Nb-Ti microalloyed steel". Journal of Materials Processing Technology 180 , 2006, p. 44–48.

[10] B. K. Show. "Effect of vanadium and titanium modification on the microstructure and mechanical properties of a microalloyed HSLA steel". Materials Science and Engineering A 527 , 2010, p. 1595–1604.

[11] C. Garcıa-Mateo. "Influence of vanadium on static recrystallization in warm worked microalloyed steel". Scripta mater. 42 , 2000, p. 137–143.

[12] Hasan Karabulut. "Effect of vanadium content on dynamic strain ageing in microalloyed medium carbon steel". Materials and Design 25 , 2004, p. 521–527.

[13] Z. Wanga, X. Maoc, Z. Yanga. "Strain-induced precipitation in a Ti micro-alloyed HSLA steel" Materials Science and Engineering A 529, 2011, p. 459– 467.

Acoustic Emission during Tensile Plastic Deformation of 2.25Cr-1Mo Steel

A.B. Pattnaik and B.B. Jha
Surface Engineering Department
CSIR-Institute of Minerals and Materials Technology
Bhubaneswar, Odisha, India
Email: amulya@immt.res.in

ABSTRACT

In the present investigation, the effect of microstructural variation on AE signals obtained during tensile plastic deformation of 2.25Cr-1Mo steel has been studied. Tensile testing of normalised and thermally aged samples has been carried out at a strain rate of 6.66 x $10^{-4}s^{-1}$ with simultaneous monitoring of acoustic emission (AE) parameters. Results have shown that acoustic emission activities were higher for normalised condition and the same decreased as the severity of ageing was increased. Higher acoustic emission activities in normalised condition have been attributed to the presence of fine coherent M_3C and Mo_2C precipitates which results in strengthening of matrix by precipitation hardening. The reduction in acoustic emission activities in aged condition was found to be due to spheroidisation and coarsening of carbide precipitates, as a result of an increase in inter particle spacing. The microstructural characterisation has been done using optical microscopy and Field emission scanning electron microscope (FESEM).

KEYWORDS: *2.25Cr-1Mo Steel, Acoustic Emission, Tensile Deformation, Strain Rate*

INTRODUCTION

2.25Cr-1Mo steel has long been used as a structural material in power generating utilities and petrochemical industries for high temperature applications [1].This steel undergoes microstructural degradation when exposed to extremely high temperature and pressure due to precipitation and coarsening of carbides [2, 3]. It is also used as heat exchanger material in fast breeder nuclear reactors [4].

Acoustic Emission (AE) is a transient elastic stress wave generated as a result of deformation in materials which could be detected by a transducer [5]. Most of the previous studies on deformation of various materials have related the AE activities to the dislocation kinetics [6]. Scruby et al [7] during their investigation on AE during deformation of aluminium alloys with various microstructures have concluded that in case of age-hardened alloy the AE activities were due to shearing of Guinier-Preston (GP) zones. Heiple et al [8] studied the AE generated during deformation of precipitation strengthened alloys. They detected higher AE activities in solution treated condition and decrease in AE activities during precipitation.

Mukhopadyay et al. [9-12] systematically investigated the AE generated during deformation of austenitic stainless steels during notched and unnotched condition. They indicated that the AE generated in the micro plastic and yielding stage was mainly due to generation of dislocations from various sources like Frank-Read (FR) and grain boundaries. Jalaj Kumar et al. [13] have shown that acoustic emissions during deformation of near alpha titanium alloys for both notched and smooth specimens are due to slip and twinning. Other studies on Al,Mg alloys and low carbon steels [14-16] also showed that dislocation activities was the main source of acoustic emission.

Thermal ageing of 2.25Cr-1Mo steel at different temperature and time produces different types of carbides. We have studied the microstructural degradation in thermally aged 2.25Cr-1Mo steel and found that the carbide precipitation was in the sequence of M_3C, M_7C_3, $M_{23}C_6$, and M_6C [3]. Investigations have been done to study the deformation behaviour of various materials using acoustic emission technique [7-16]. However, no work has been reported so far to the best of our knowledge to correlate the deformation behaviour of thermally aged 2.25Cr-1Mo steel using acoustic emission technique. The present study dealing with this aspect has, therefore, been carried out to bring more clarity in our understanding towards strengthening mechanisms and the role of precipitates there on.

EXPERIMENTAL

The starting material used in this study was in the form of 12mm thick plate. The chemical compositions are given in Table 1.

Table 1: Chemical composition (wt %) of 2.25Cr-1Mo steel sample used in the present investigation.

C	Mn	Si	S	Cr	Mo	Ni	Fe
0.07	0.42	0.019	0.025	2.28	0.98	0.09	Balance

The plate was cut into strips of size 115mmx14mmx12mm using BAINCUT-XL cutting machine. These strips were normalised at 1223 K for 2 hrs and thermally aged at 873 K/10 hrs, 923 K/20 hrs, 973 K/40 hrs, and 973 K /80 hrs using a tubular furnace. Flat sub size tensile specimens of dimension 25 mm x 6 mm x 2 mm with rolling direction parallel to the tensile axis were prepared according to ASTM E8. Tensile tests of normalised and thermally aged samples were carried out on a Tinius Olsen make screw driven material testing machine at a strain rate of $6.66 \times 10^{-4} s^{-1}$ at ambient temperature (300 K). Microstructural characterisation was done using Leica DMI 3000 M optical microscope and ZEISS SUPRA 55 Field emission scanning electron microscope (FESEM).

Acoustic emission (AE) time domain parameters were studied using a four channel PCI-DiSP with an AEwin version E4.32 AE system (Physical Acoustic Corporation, USA). Two miniature sensors of resonant frequency 225 kHz and 325 kHz were mounted on the specimen using high vacuum silicone grease and tape as shown in Fig. 1. The sensors were linearly located at a distance of 40mm from each other and the wave velocity was maintained at 6.40 m/sec for 2.25Cr-1Mo steel. The signals were captured using a

40 dB gain preamplifier and a compatible filter (10 kHz-2 MHz). A threshold of 45 dB was set to eliminate the background noise by applying the Kaiser effect.

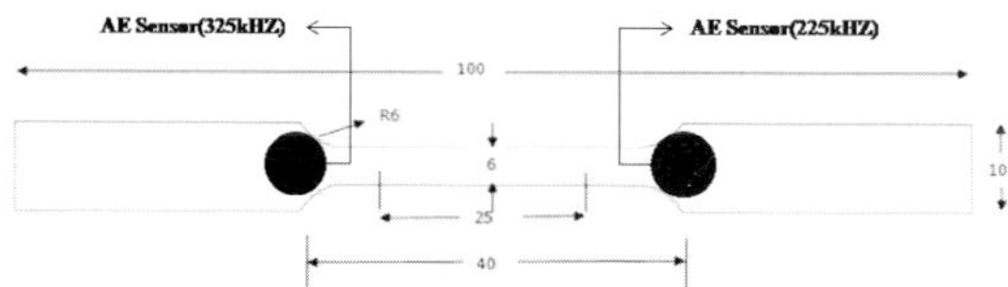

Fig. 1: Schematic of sub size tensile specimen made according to ASTM E8.

The dummy specimens were repeatedly loaded and unloaded below the yielding region to detect any noise. Fig. 2 shows the experimental set up for acoustic emission testing during tensile deformation. A total system gain of 100dB was used to capture the signals.

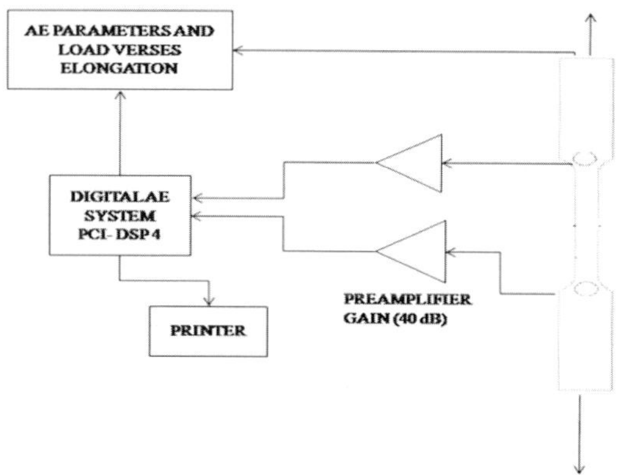

Fig. 2: Schematic of experimental set up.

RESULTS AND DISCUSSION

Microstructural degradation

Fig. 3 (a) (b) shows the optical and FESEM microstructure of normalised 2.25Cr-

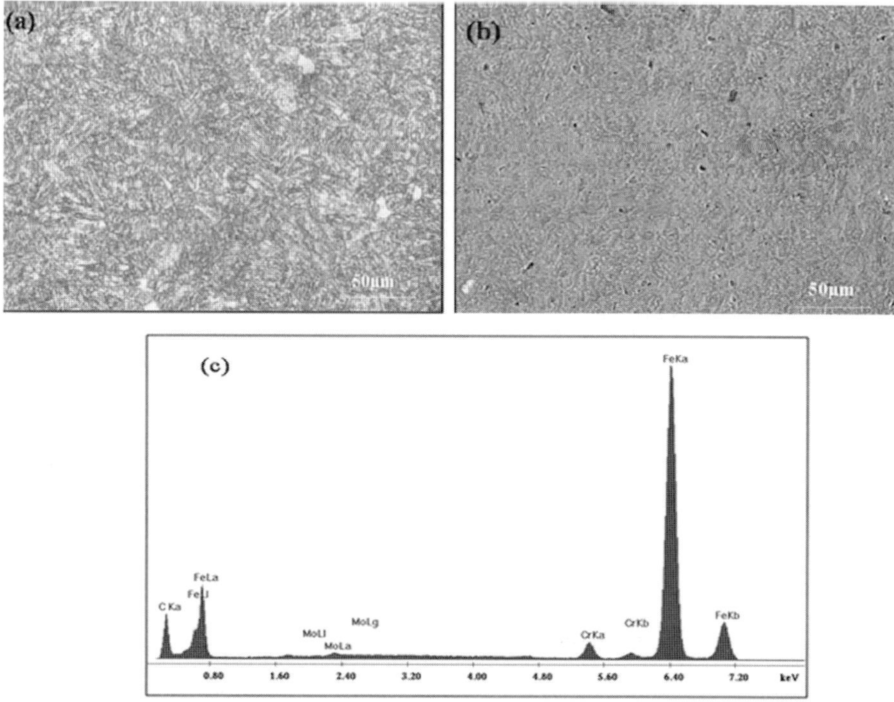

Fig 3: (a) Optical micrograph of normalised 2.25Cr-1Mo steel showing the bainitic phase (b) Scanning electron micrograph and (c) EDX of normalised sample showing the presence of Fe rich carbide.

1Mo steel. The initial microstructure of normalised steel is found to be consists of bainite and very little amounts of pro eutectoid ferrite containing iron carbides and fine M_2C carbide. The Energy dispersive X-Ray analysis (EDX) spectrum in Fig. 3 (c) for normalised condition shows that the carbides are Fe rich and are identified as M_3C based on TEM analysis [3].

The FESEM micrographs of thermally aged samples at 873K/10hrs, 923K/20hrs, 973K/40hrs, 973K/80hrs shown in Fig. 4 (a)-(d) clearly identify the different types of carbide precipitates and their coarsening which were obtained with the help of their corresponding EDX results attached to the FESEM and the selected area diffraction (SAD) pattern of these precipitates detailed in our work reported elsewhere [3].

The sequence of carbide precipitation in normalised and thermally aged samples was found to be M_3C, M_7C_3, $M_{23}C_6$ and M_6C where M is mostly molybdenum and chromium which is in agreement with other investigation [5].

It has been reported that thermal ageing at lower temperature and shorter time produces a microstructure consisting of Fe_3C, epsilon

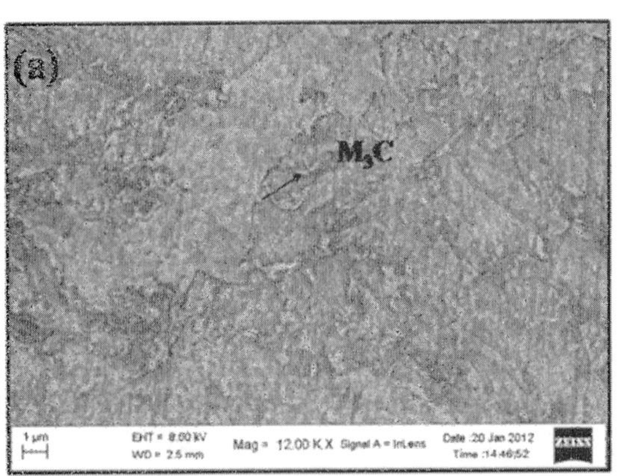

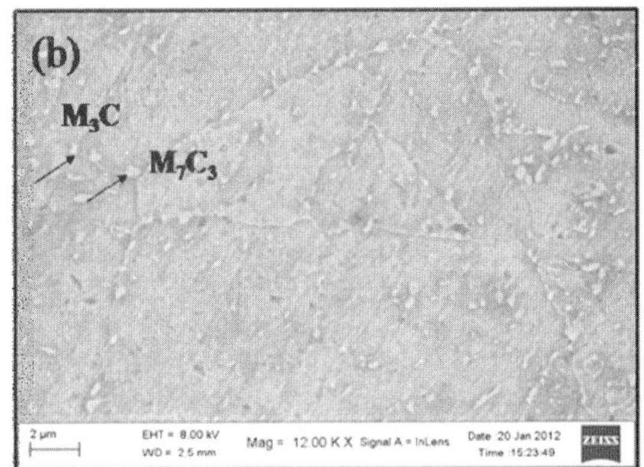

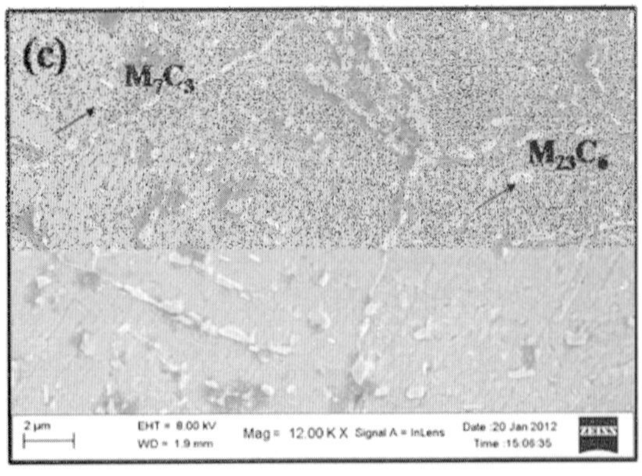

Fig. 4: FESEM micrograph of thermally aged samples at (a) 873K/10hrs (b) 923K/20hrs (c) 973K/40hrs (d) 973K/80hrs showing the coarsening of carbide precipitates.

carbides and fine Mo_2C carbides [5]. The Mo_2C forms fine needle like precipitates and contributes to strengthening by precipitation hardening [5]. As the degree of thermal ageing is increased, the fine coherent M_3C precipitates coalesce to form larger incoherent globular precipitates of $M_{23}C_6$ and M_6C ('M' is mostly Fe, Cr and Mo).

Effect of microstructural degradation on AE during tensile deformation

The tensile test results of normalised and thermally aged samples are shown in Table 2. It has been observed that offset yield strength and the ultimate tensile strength are higher for normalised sample as compared to thermally aged samples.

Table2: Tensile test results of normalised and thermally aged samples tested at a strain rate of $6.66x10^{-4}s^{-1}$.

Microstructural condition	0.2% offset YS (MPa)	UTS (MPa)	Total Elongation (%)
Normalised (virgin) (1223K/2hrs)	764	927	21.9
Thermally aged (873K/10hrs)	671	781	21.5
Thermally aged (923K/20hrs)	489	587	26.7
Thermally aged (973K/40hrs)	381	502	34
Thermally aged (973K/80hrs)	387	507	38

Upon ageing, the strength decreased considerably and the ductility increased.

The variation of AE counts and stress with nominal strain for normalised and aged samples is shown in Fig. 5(a)-(e). Significant increase in AE activities was detected in normalised condition as compared to the same of thermally aged condition. Most of the AE signals in normalised and thermally aged condition were found to be in the yielding region of the stress strain curve. This increase in AE activities in this region has been attributed to the sudden increase in number and velocity of mobile dislocations when the material transforms from elastic to plastic regime of deformation. Fig. 6(a)-(e) shows the variation of AE amplitude with strain for normalised and aged condition.

More AE events with amplitudes in the range of 45-100dB were observed in case of normalised condition. The AE amplitude considerably reduced as the severity of ageing increased. Higher tensile strength and larger AE activities in case of normalised condition has mainly been attributed to the presence of fine coherent Mo_2C precipitates providing a very small mean free path and creating obstacles to the dislocation motion. In this case probably the dislocation particle interaction is more frequent resulting in copious AE activities. In case of thermally aged condition the AE activities decreased with ageing time and temperature may be due to the increase in the mean free path of precipitates by spheroidisation and coarsening resulting in less frequent dislocation particle interaction.

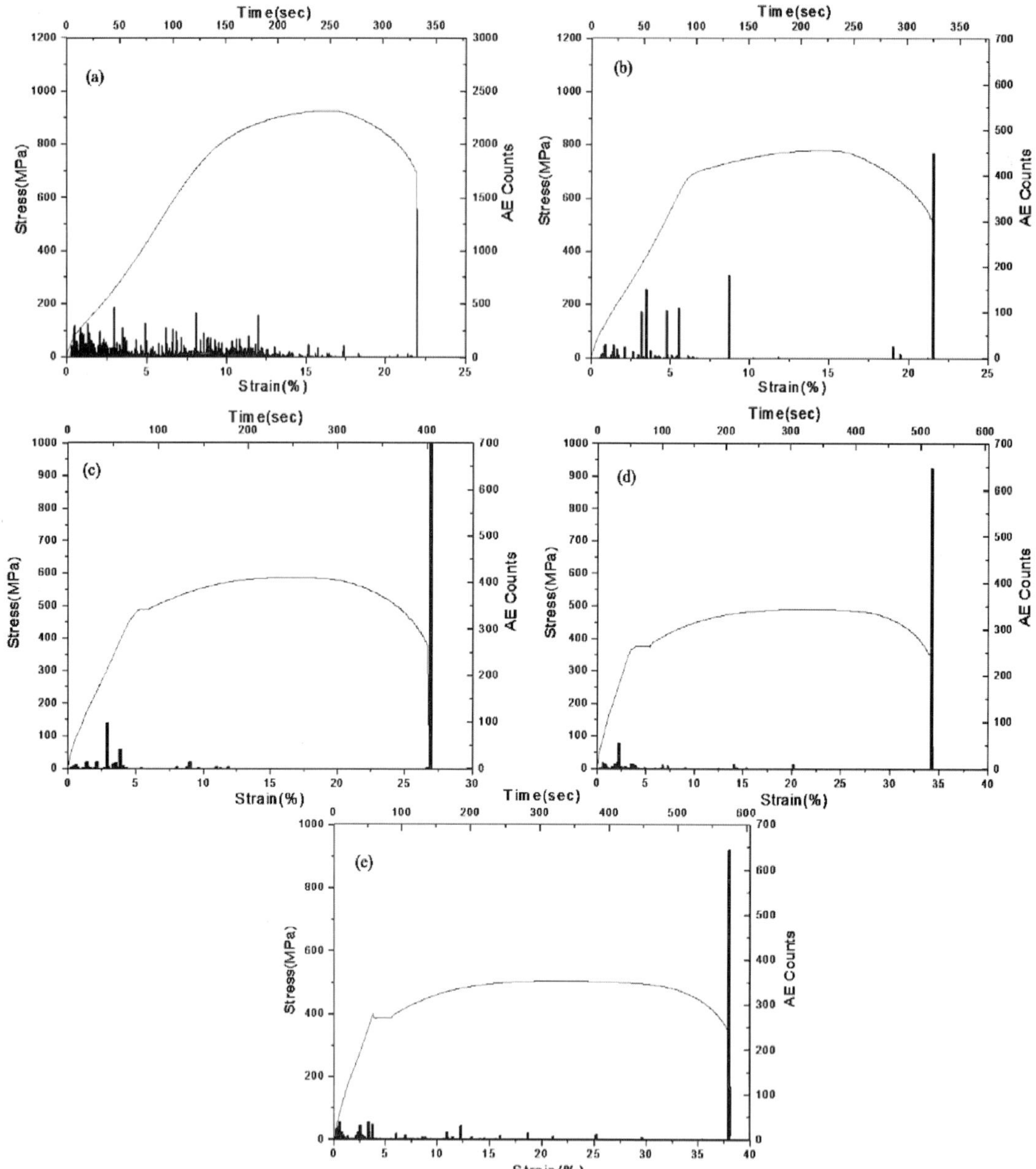

Fig. 5: Variation of engineering stress with AE Counts and engineering strain for normalised and thermally aged samples (a) Normalised (b) 873K/10hrs (c) 923K/20hrs (d) 973K/40hrs (e) 973K/80hrs.

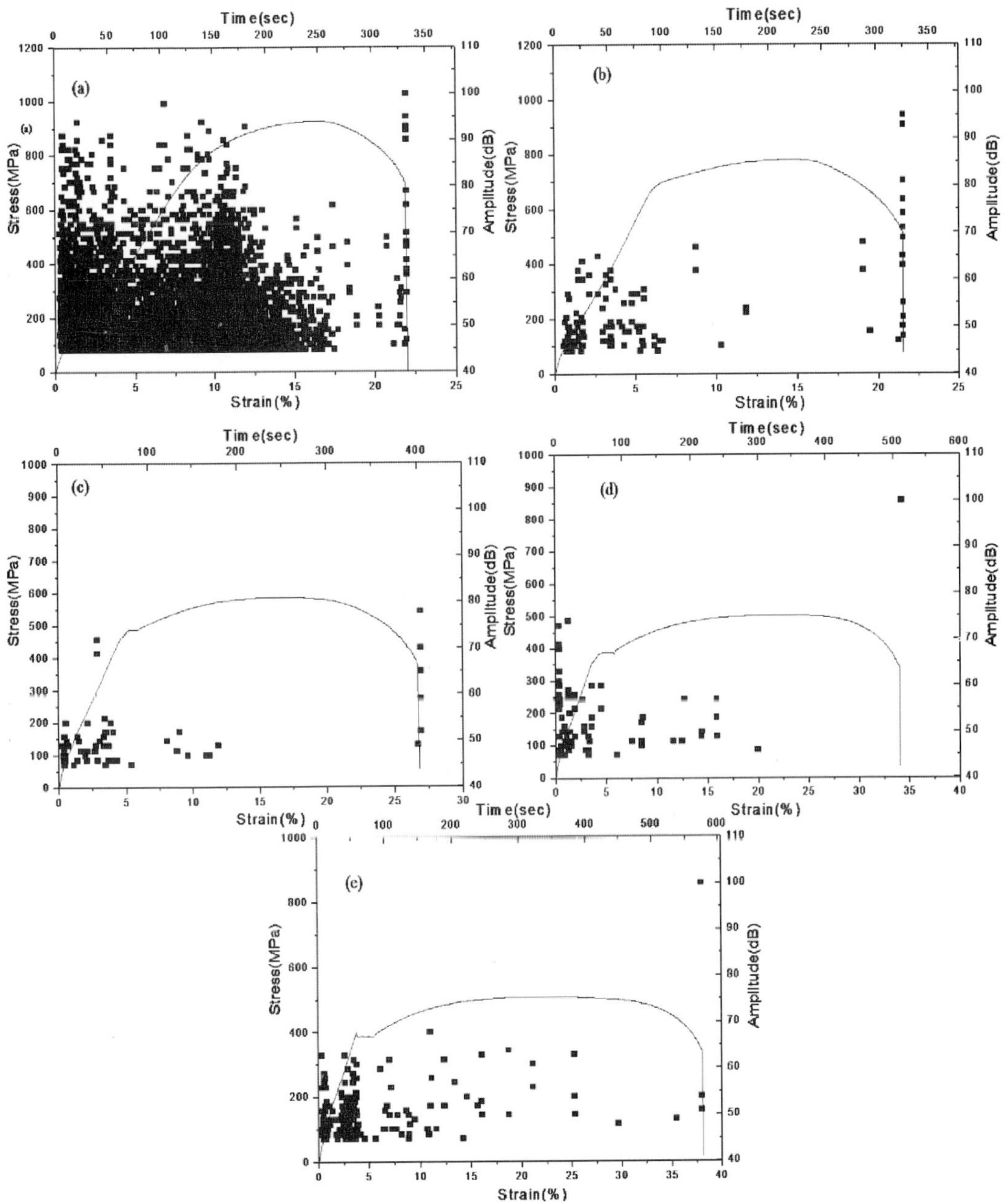

Fig. 6: Variation of engineering stress with AE amplitude and engineering strain for normalised and thermally aged samples (a) Normalised (b) 873K/10hrs (c) 923K/20hrs (d) 973K/40hrs (e) 973K/80hrs.

Fig. 7 shows a more quantitative result elucidating the effect of microstructural variation on acoustic emission cumulative counts. It shows the variation of AE cumulative counts with that of the ageing time and temperature. The AE cumulative counts has been found to be much higher for normalised samples of 2.25Cr-1Mo steel as compared to all the thermally aged samples which are in line with our earlier observations.

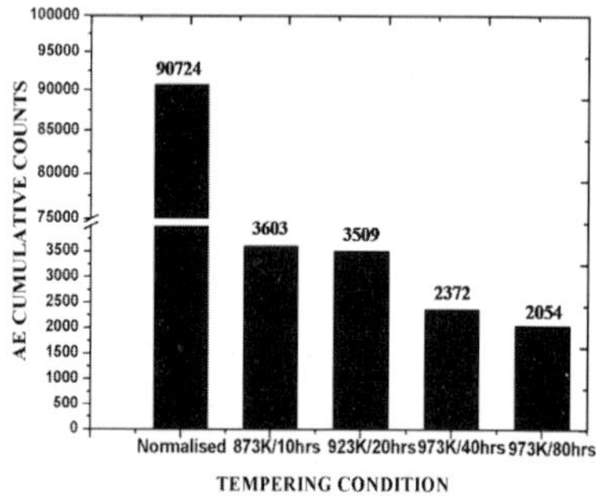

Fig. 7: Variation of AE cumulative counts with ageing time and temperatures.

CONCLUSIONS

The main conclusions that can be drawn are as follows.

1. Deformation behaviour of 2.25Cr-1Mo steel at room temperature could be studied using acoustic emission technique.
2. More intense AE has been found to be generated during the region of deformation where material transforms from elastic to plastic regime of deformation. This observation is independent of existing microstructural features.
3. Normalised 2.25Cr-1Mo steel has been found to generate more copious AE signals which has been attributed to the smaller mean free path provided by coherent precipitates of Mo_2C.
4. Thermally aged samples of 2.25Cr-1Mo steel have been found to generate much lesser AE activities in comparison to their normalised counterpart. This decrease could be attributed to spheroidisation and coarsening of carbide precipitates providing higher mean free path leading to less frequent dislocation particle interaction.

ACKNOWLEDGEMENT

The Authors would like to thank Prof. B.K. Mishra, Director, CSIR-Institute of Minerals and Materials Technology, Bhubaneswar, India for his constant encouragement and support. The Authors also thank Dr. Debdhyan Behera for his help during microstructural characterisation using FESEM.

REFERENCES

[1] J. Sanderson: 'Ferritic steels for high temperature applications' (ed. Ashoke Khare), 178, 1983 Proc. Conf. ASM.

[2] J.M. Race and H.K.D.K. Bhadeshia, Material Science and Technology, Vol. 8, 875-882, 1992.

[3] B.B. Jha, B.K. Mishra, B.K. Satpati, S.N. Ojha, Materials Science-Poland, Vol. 28, No1, 335-346, 2010.

[4] P. Parameswaran, S. Saroja, M. Vijayalakshmi, V.S. Raghunathan, Journal of Nuclear Materials, 232, 1996, 226-232.

[5] R.G. Baker and J. Nutting, Journal of Iron and Steel Institute, 192, 257 1959.

[6] 'Standard Terminology for Nondestructive Examinations', ASTM1316-11A, July 2011.

[7] C.B. Scruby, H.N.G. Wadley, K. Rusbridge, and D. Stockham-Jones, Metal Science, Vol.15, 599-608, 1981.

[8] C.R. Heiple, S.H. Carpenter and M.J. Carr, Metal Science, Vol.15, 587-598, 1981.

[9] C.K. Mukhopadyay, K.K. Ray, T. Jayakumar, Baldev Raj, Materials Science and Engineering A255, 98-106, 1998.

[10] C.K. Mukhopadyay, T. Jayakumar, Baldev Raj, K.K. Ray, Materials Science and Engineering A293, 137-145, 2000.

[11] B. Venkataraman, C.K. Mukhopadyay and Baldev Raj, Material Science and Technology, Vol.20, 1320-1316, 2004.

[12] C.K. Mukhopadyay, T. Jayakumar, Baldev Raj, K.K. Ray, Journal of Material Science, Vol.42, 5647-5656, 2007.

[13] C.K. Mukhopadyay, K.V Rajkumar, T. Jayakumar, Baldev Raj, Journal of Material Science, 45, 1371-1384, 2010.

[14] Jalaj Kumar, Sunny Punnose, C.K. Mukhopadyay, T. Jayakumar, Vikas Kumar, Research in Non-destructive Evaluation, 23:1, 17-31, 2012.

[15] F. Chemlik, E. Pink, J. Krol, J. Peickaand P. Lukac, Acta Metallugica Inc, Vol.46, No.12, 4435-4442, 1998.

[16] K. Mathis, F. Chemlik, Z. Trojanova, P. Lukac, J. Lendvai, Materials Science and Engineering A 387, 331-335, 2004.

[17] Meysam Akbari, Mehdi Ahmedi, Physics Procedia, 3, 795-801, 2010.

Aging Response of An Experimental Co-Ni Containing Secondary Hardening Ultra High Strength Steel

R. Veerababu[1, 2], K. Satya Prasad[1], R. Balamuralikrishnan[1],
S. Karthikeyan[2] and K. Muraleedharan[3, *]
[1]Defence Metallurgical Research Laboratory, Kanchanbagh, Hyderabad, India
[2]Department of Materials Engineering, Indian Institute of Science, Bangalore, India.
[3]Director, Directorate of Materials, Defence Research and Development Organization, DRDO Bhavan, New Delhi, India.
*Email: mularee@hqr.drdo.in

ABSTRACT

Typical secondary hardening ultra high strength steels contain C and Mo contents up to 0.25wt. %, and 2wt. % respectively. In the present work, aging response of secondary hardening ultra high strength steel, with relatively higher Mo content (3wt. %) has been investigated in the range 300 to 650°C for 5h aging time, through hardness measurements on the aged samples. This steel exhibited significant secondary hardening with a maximum hardness of 685VHN in the sample aged at 450°C. Microstructural investigation through optical microscopy, SEM and TEM has been carried out on select aged samples (350, 400, and 450°C) and also on the as-quenched and as-quenched and LN_2 treated samples. As-quenched sample shows the presence of significant amount of retained austenite, which has been transformed to martensite upon LN_2 treatment. SEM and TEM investigations on the aged samples revealed the presence of alloy cementite. In addition to the alloy cementite, 450°C aged sample also showed the presence of fine M_2C particles. These experimental results have been well correlated with the theoretical calculations using ThermoCalc.

KEYWORDS: *Aging, Cementite, M_2c Carbide, And Tem.*

INTRODUCTION

Secondary hardening ultra high strength and high toughness (SHUHS) steels are being used for niche applications such as aircraft landing gear, due to their outstanding combination of high strength, fracture toughness, K_{IC} and stress corrosion cracking resistance, K_{ISCC}. Some typical steels of this type are HY180, AF1410, and AerMet100 [1–4]. These are high Co-Ni containing steels with carbide forming elements such as Cr and Mo [5,6] These are essentially quench and temper steels and are believed to derive their excellent mechanical properties from a dispersion of fine M_2C carbides in a highly dislocated lath martensitic microstructure, formed during aging in the range 450–510°C [6,7]. The optimum heat treatment for these steels, corresponding to a slightly overaged condition, consists of austenitising in the range 825–900°C, followed by quenching to room temperature and LN_2 treatment and aging in the range 485–510°C for 5h [1,8-14].

Typically, SHUHS steels contain (in wt. %) 8-15Co, 10-11Ni, 2-3Cr, 1-2Mo and

0.1-0.25C. We are currently investigating a SHUHS steel with relatively higher Mo content (3wt. %). Our earlier work [15] on this high Mo steel has demonstrated that austenitizing temperatures of the order of 1150°C are desired for complete austenitization. In the present work, aging response of this high Mo steel austenitized at 1150°C is being reported using hardness, optical metallography, scanning and transmission electron microscopy and correlated with the theoretical predictions using ThermoCalc software. It is expected that Mo will have significant effect on the aging response of this class of steels and hence merits investigation.

EXPERIMENTAL PROCEDURE

The steel of desired composition was vacuum arc melted and cast in the form of 450gm pancake using commercially pure raw materials. The pancake was remelted at least 3-4times by turning upside down after each melting step to ensure homogeneity. The pancake was then soaked at 1100°C for 1h and hot rolled to 3-4mm thick plate. The chemical composition was determined using wet chemical analysis and LECO analyser for C and S content. Small samples (15x10mm) for heat treatment were cut from the rolled plates, austenitized at 1150°C for 30min and oil quenched to room temperature. The as-quenched samples were subsequently subjected to LN_2 treatment and aged in the range 300-650°C for 5h and oil quenched. Samples after each stage, i.e. after oil quenching, and after LN_2 treatment were preserved for reference hardness and microstructural examination. Samples for hardness testing, metallography and TEM examination were cut from the heat treated blanks using slow speed saw.

Hardness testing was performed using Vicker's hardness at 30kgf load. Samples for optical metallography and SEM were prepared using standard metallographic techniques and etched with waterless Kalling's reagent. Etched samples were examined using LEICA make optical microscope and Quanta 400 SEM equipped with energy dispersive spectrometer (EDAX make). Thin foils for TEM examination were prepared from ~200μm thick slices that were cut using a slow speed saw, mechanically ground to ~100μm thickness, and subsequently twin jet electropolished with mixed acids (by volume: 88 pct methanol, 10 pct lactic acid, 8 pct sulfuric acid, 3 pct nitric acid, and 2 pct hydrofluoric acid) electrolyte at -30°C. Carbon extraction replicas [18] were prepared following a standard technique using waterless Kalling's reagent for both initial light etching and final lift-off of the films. Thin replica films floated on the etchant surface were then collected and washed in a series of water and methanol mixture. These films were then collected on to the copper grids and then dried. Both thin foils and carbon extraction replicas were examined in a FEI Tecnai 20T transmission electron microscope operated at 200 kV and equipped with an EDAX EDS system. ThermoCalc software for Windows OS (TCW5) with TCFE6 (the database for iron and steel), was used to perform the thermodynamic calculations reported here.

RESULTS

Aging curve for the experimental steel between 300 and 650°C aging temperature is shown in Figure 1.

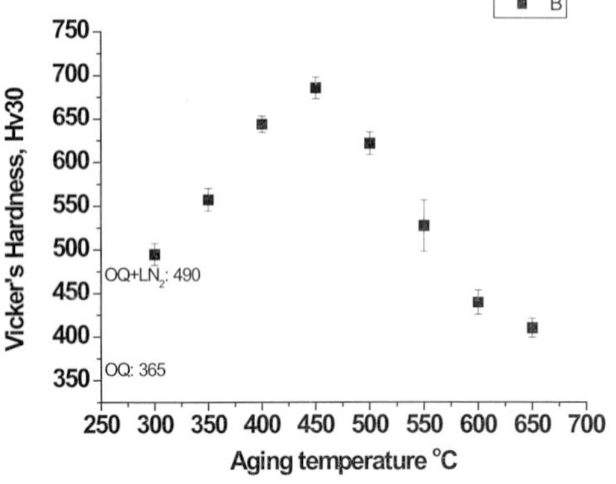

Fig. 1: Aging response curve for the experimental steel

The hardness of the as-quenched (OQ) as well as LN_2 treated ($OQ+LN_2$) samples are indicated. The data reveals that there is significant improvement in hardness (365 to 490VHN) from the OQ sample to the $OQ+LN_2$ sample. The hardness of 300°C aged sample is almost similar to that for the $OQ+LN_2$ sample and then started to increase rather sharply with aging temperature. The maximum hardness of 685 VHN (HRc 59.5) is observed upon aging at 450°C. The peak hardness temperature for the experimental steel is similar to the peak hardness temperature range (between 450 and 490°C) reported for AerMet100 and AF1410 steels [2,3]. This remarkably higher hardness can be attributed to the higher Mo and relatively higher C contents in the experimental steel. Increasing the aging temperature beyond 450°C resulted in a sharp decrease in hardness.

Optical micrographs of the samples after different heat treatments are shown in Figure 2. The as quenched sample (Fig. 2a) reveals significant amount of retained austenite (light etching phase) along with martensitic laths (dark etching needles). Theoretical calculations using ThermoCalc (Fig. 3) predict that the amount of retained austenite at 200°C is greater than 20 mole percent, which is consistent with the present experimental observations. The amount of retained austenite reduces significantly upon LN_2 treatment because of conversion to martensite (Fig. 2b). Samples aged at 350, 400 and 450°C (Figs 2c to e), exhibit similar tempered martensitic microstructure with no appreciable differences.

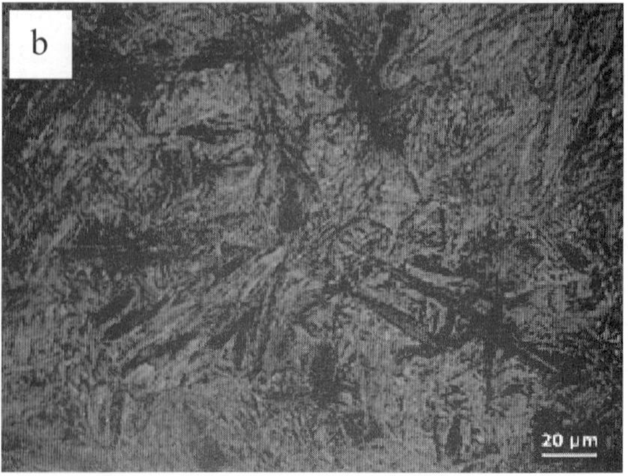

Fig. 2: Optical micrographs of the experimental steel; (a) as-quenched, (b) as- and LN_2 treated, (c) 350°C, (d) 400°C and (e) 450°C aged samples

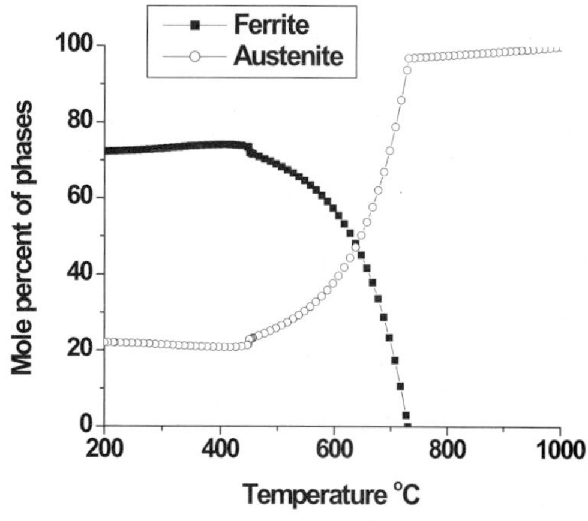

Fig. 3: Calculated mole percent of ferrite and austenite as a function of temperature in the experimental steel using ThermoCalc.

Figure 4 shows SEM micrographs of the experimental steel after different heat treatments. SEM micrographs for OQ, and OQ + LN_2 samples (Figs 4a and b) in general, reveal that there are no undissolved precipitates. Figures 4a and b further confirm the presence of retained austenite in the OQ sample that has been converted to martensite upon LN_2 treatment. SEM micrographs for aged samples (Figs 4c to e), show that there are significant differences in the microstructure from the OQ and OQ + LN_2 samples that could not be resolved

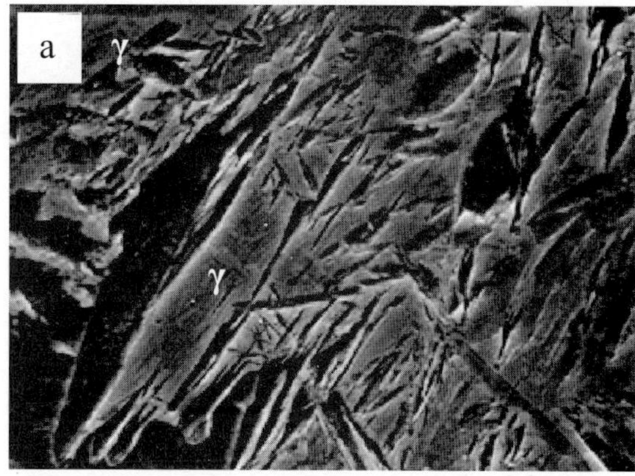

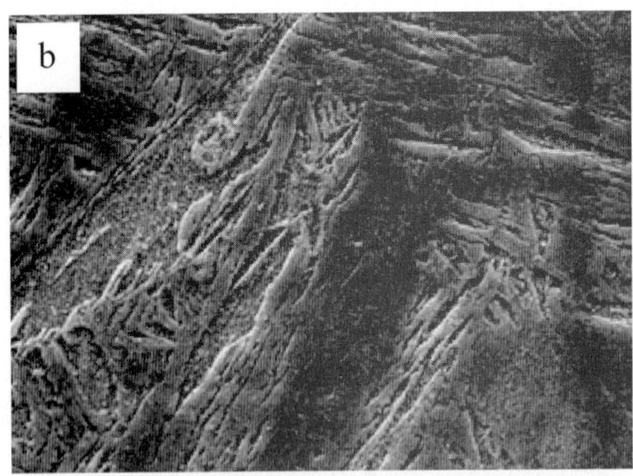

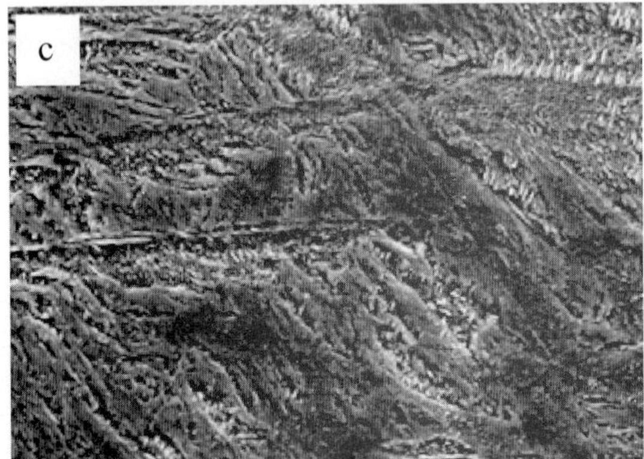

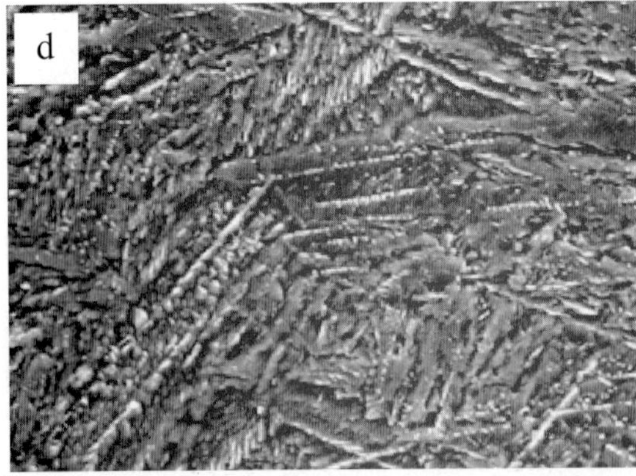

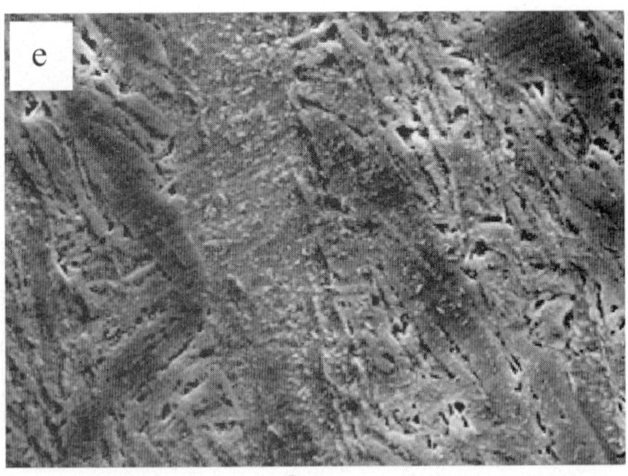

5μm

Fig. 4: SEM micrographs of the experimental steel; (a) as-quenched, (b) as-quenched and LN_2 treated, (c) 350°C, (d) 400°C and (e) 450°C aged samples. Retained austenite (γ) is indicated in Fig. (a)

through optical microscopy. Samples aged at 350°C and 400°C (Figs 4c and d) reveal the presence of both inter- and intra-lath precipitates. In 450°C aged sample (Fig. 4e), precipitates are coarsened relative to those in the 350°C and 400°C aged samples. The precipitates can be seen in greater detail in the high magnification SEM images (Fig. 5).

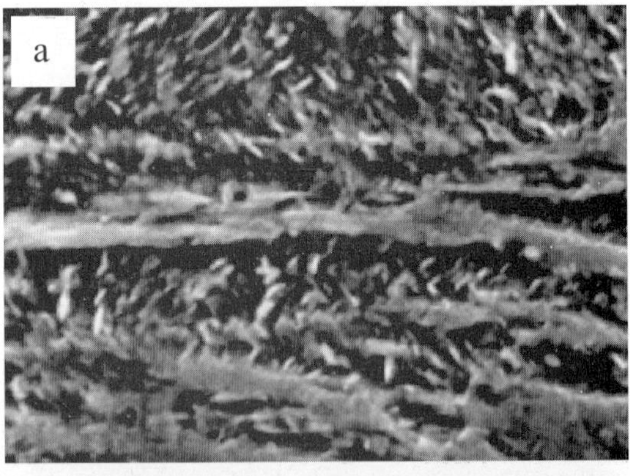

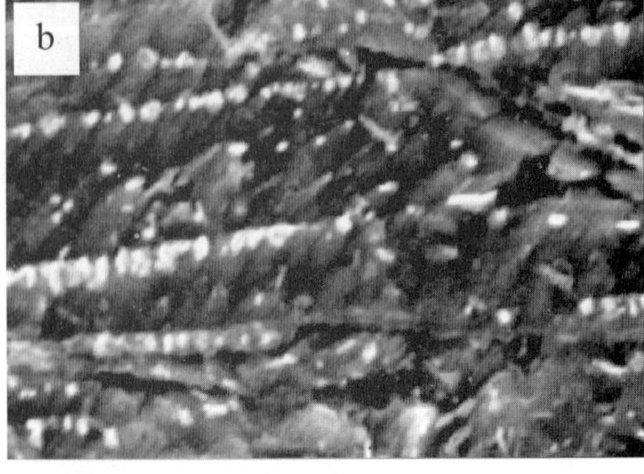

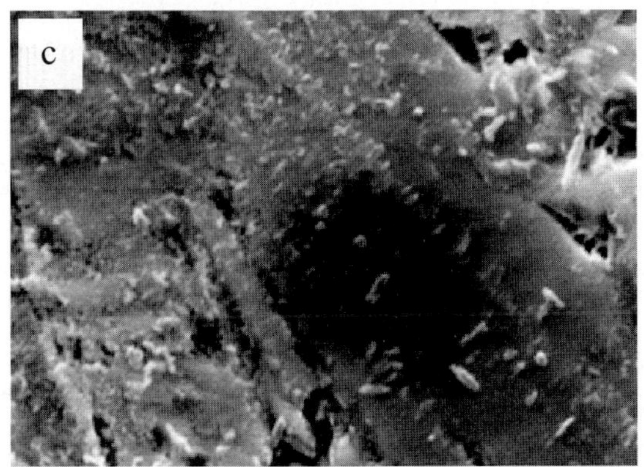

Fig. 5: High magnification SEM micrographs of the experimental steel; (a) 350°C, (b) 400°C and (c) 450°C aged samples

Bright field TEM image (Fig. 6) for the experimental steel in the oil quenched and LN_2 treated sample shows typical lath structure devoid of precipitates. Upon aging at 350°C, fine needle like precipitates are observed (Fig. 7). Indexing of the selected area diffraction (Fig. 7c) patterns indicates that these are likely to be cementite with orthorhombic structure. 400°C aged sample also reveals the presence of similar cementite precipitates (Fig. 8). However the sample aged at 450°C show fine particles (Figs 9a and b) present at both inter- and intra-lath

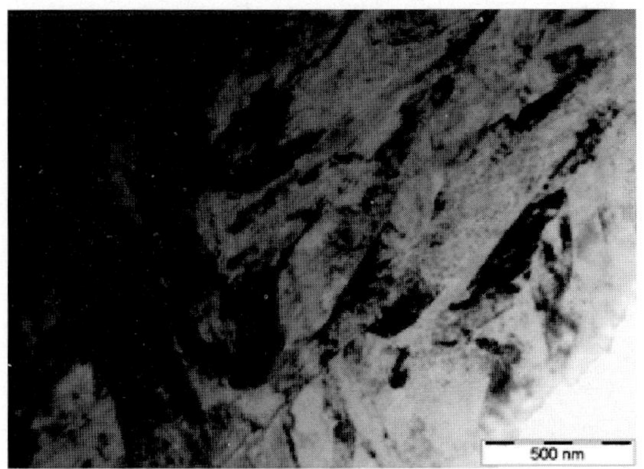

Fig. 6: Bright field TEM image of the oil quenched and LN_2 treated sample

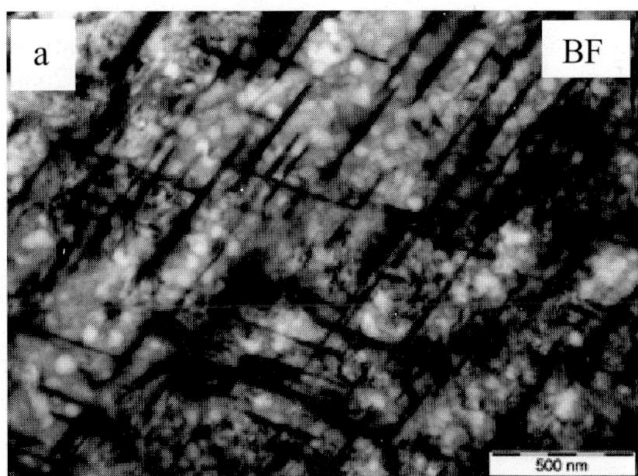

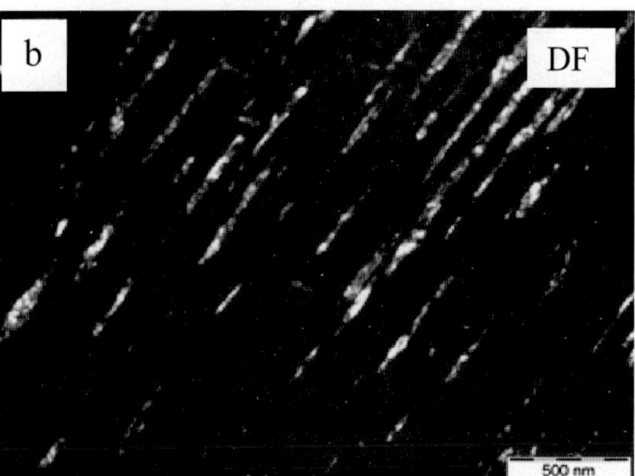

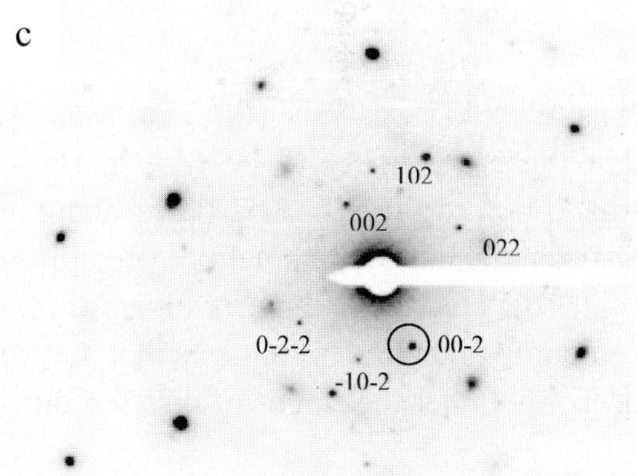

Fig. 7: TEM images of 350°C aged sample; (a) BF, (b) DF and (c) selected area diffraction pattern, some precipitate spots have been indexed. Diffraction spot used for obtaining the DF image is encircled in (c)

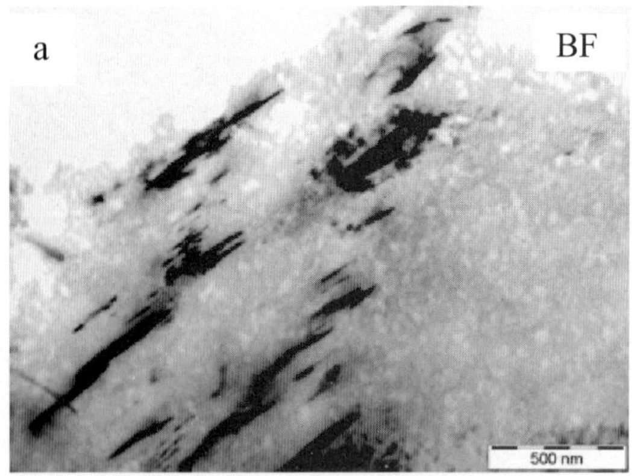

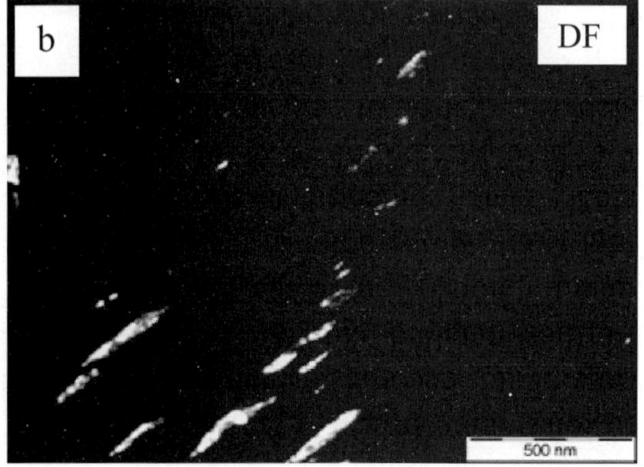

Fig. 8: (a) BF and (b) DF TEM images of 400°C aged sample

regions. SAD analysis (Fig. 9c) suggests that these are likely to be M_2C particles. In addition to these fine particles, there are also fine needle like cementite precipitates (Fig. 10). The size and morphology of these cementite precipitates suggests that they are probably under the stages of dissolution.

DISCUSSION

Isochronal aging curve for the experimental steel shows that the hardness of 300°C aged sample is almost similar to the OQ + LN_2 sample and started to increase from

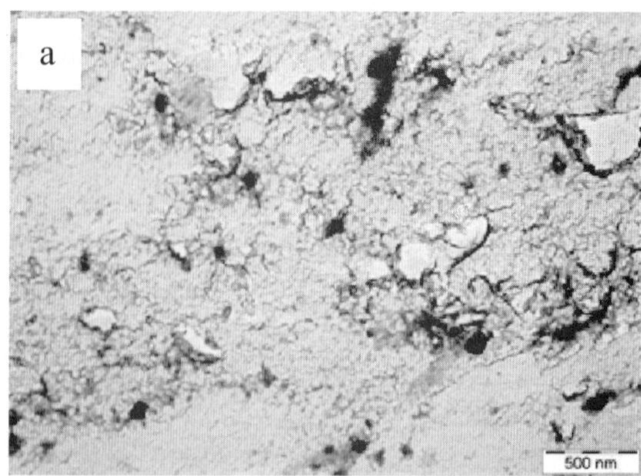

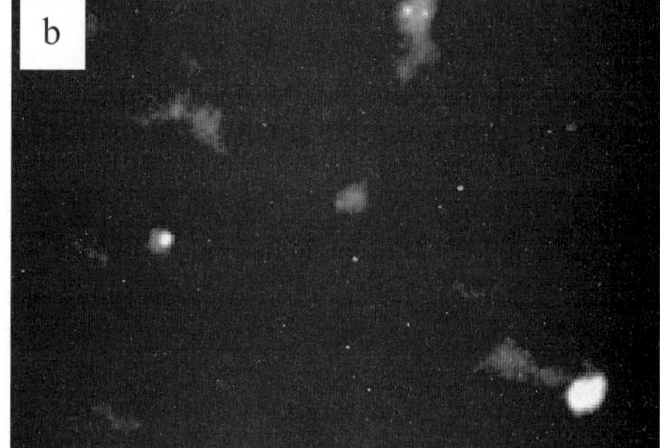

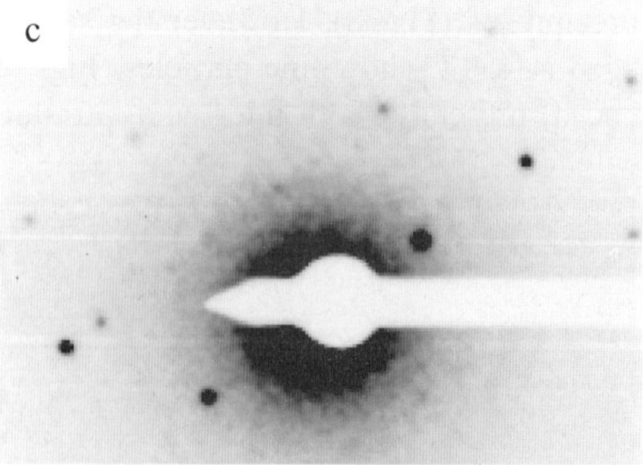

Fig. 9: TEM images of 450°C aged sample using carbon extraction replicas; (a) BF and (b) DF pair, and (c) SAD pattern.

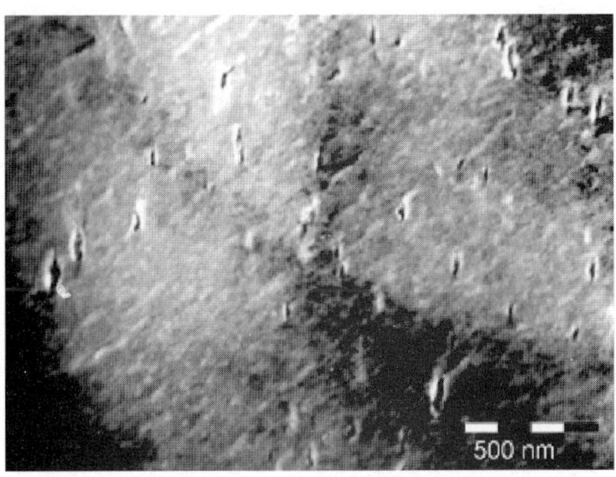

Fig. 10: Thin foil bright field TEM image of the experimental steel aged at 450°C.

300°C to a maximum at 450°C and then dropped sharply with increase in aging temperature. Microstructural observations at different length scales reveal that the OQ + LN_2 sample consists essentially of lath martensitic microstructure (Figs 2b, 4b and 6). Aged microstructures between 350 to 450°C reveal the presence of alloy cementite in the form of fine needles. These results suggest the formation of cementite would have initiated prior to 350°C aging. The increase in hardness within this aging regime could have resulted from the presence of fine cementite particles. The length of the cementite needles appears to decrease with increase in aging temperature (Figs 5, 7a &b, 8, and 10). The size of the cementite needles in the sample aged at 450°C (Fig. 10) suggests that that the cementite is under final stages of dissolution. Therefore, contribution of the cementite to peak hardness is less compared to its effect on hardening at other aging temperatures. However, another type of precipitate, i.e. M_2C carbides would have contributed more to the hardening at the peak aging temperature. It is difficult to prove the formation of M_2C carbide unambiguously at the normal aging treatment (either at 482°C or 510°C for 5h) of this class of steels because of small size of these precipitates [3]. Normally, these steels will be subjected to long aging times so that the size of the precipitates become amenable for the study. Further detailed studies are required to unambiguously prove the M_2C carbide at this aging temperature in the experimental steel. Thus, cementite is formed prior to the precipitation of M_2C in the present steel. Formation of cementite prior to M_2C precipitation in this class of steels has been reported earlier up to aging temperatures of 482°C [2,3] and 600°C [19]. The decrease in hardness, although not studied in this work, beyond 450°C aging would have resulted from the recovery of the martensitic laths and coarsening of the M_2C carbides. However detailed microstructural investigations on the samples aged at temperatures greater than 450°C need to be performed to completely understand the aging response of this steel.

CONCLUSIONS

In the present work, aging response of SHUHS steel with relatively higher Mo and C has been investigated using hardness, optical microscopy, scanning and transmission electron microscopy and correlated with the theoretical predictions using ThermoCalc. The main conclusions from the work are as follows:

1. Isochronal aging response of the experimental steel reveals that the hardness of the sample aged at 300°C is almost similar to that of the OQ+LN_2 sample. Hardness is increased sharply

to a maximum of 685VHN at 450°C and then decreased with increase in aging temperature.

2. Remarkable high hardness observed at 450°C aging can be correlated to the relatively higher C and Mo content.
3. Microstructural observations using optical metallography revealed that there is significant amount of austenite has been left over in the OQ condition, which was transformed to martensite upon LN_2 treatment.
4. Electron microscopy of aged samples shows the precipitation of alloy cementite, which gets coarsened upon aging at 450°C. In addition to the alloy cementite, 450°C aged sample also showed the presence of fine M_2C precipitates.
5. The maximum hardness observed at 450°C aging can be correlated to the formation of fine M_2C carbide following the dissolution of cementite.

Detailed microstructural studies of the samples aged at temperatures greater than 450°C are required to further investigate the aging response of this steel

ACKNOWLEDGEMENTS

This research was funded by the Defence Research and Development Organization (DRDO), India. The permission of the Director, Defence Metallurgical Research Laboratory (DMRL), Hyderabad, to publish the results is gratefully acknowledged. The authors wish to thank Dr. M. Srinivas, Scientist 'G' for fruitful discussions while performing this work. The authors are grateful to Mr. M. Rama Krishna and Dr. G. Ravi Chandra, Advanced Research Center International (ARCI) for Powder Metallurgy and New Materials, for giving permission to perform TEM experiments. The authors also wish to thank Mr. Deepak Kumar (DMRL) for his help in preparing thin foils.

REFERNCES

[1] G. R. Speich, D. S. Dabkowski, and L. F. Porter: *Metall. Trans.,* 4 (1), 1983, pp. 303–315.

[2] R. Ayer, and P. M. Machmeier: *Metall. Mater. Trans. A,* 28 (9), 1996, pp. 2510–2518.

[3] R. Ayer, and P. M. Machmeier: *Metall. Trans. A,* 28 (9), 1993, pp. 1943–1955.

[4] J.M. Dahl and P.M. Novotny: *Adv. Mater. Processes,* 1999, vol. 155 (3), pp. 23–25.

[5] R. Veerababu, R. Balamuralikrishnan, K. Muraleedharan, and M. Srinivas: Metall. Mater. Trans. A 39 (8), 2008, pp. 1486–1495.

[6] Z. Hu, X. Wu and C. Wang: J. Mater. Sci. Technol., 20 (4), 2004, pp. 425-428.

[7] H. Zhengfei, and W. Xingfang: Micron, 34, 2003, pp. 19–23.

[8] M. Grujicic, CALPHAD, 14 (1), 1990, pp. 49–59.

[9] L. E. Iorio: "The effect of increasing carbon level on titanium carbosulfides and their influence on toughness in ultra-high strength steels", Ph.D Thesis, Carnegie Mellon University, Pittsburgh, Pennsylvania, 2000.

[10] L. E. Iorio, J. L. Maloney, and W. M. Garrison, Jr.: 40th Mechanical Working and Steel Processing Conf. Proc., Vol. XXXVI, ISS-AIME, Warrendale, PA, 1999, pp. 901–20.

[11] H. R. Yang, K. B. Lee, and H. Kwon: Metall. Mater. Trans. A, 32 (9), 2001, pp. 2393–2396.

[12] K. S. Cho, J. H. Choi, H. S. Kang, S. H. Kim, K. B. Lee, H. R. Yang, and H. Kwon: Mater. Sci. and Engg A, 527, 2010, pp. 7286–7293.

[13] J.S. Montgomery and G.B. Olson: Proc. G.R. Speich Symp., G. Krauss and P.E. Repas, eds., ISS, Warrendale, PA, 1992, pp. 177–214.

[14] P. M. Machmeier, T. Matuszewski, R. Jones and R. Ayer: J. Mater. Engg Perf., 6 (3), 1997, pp. 279-288.

[15] R. Veerababu, R. Balamuralikrishnan, K. Satya Prasad and S. Karthikeyan, *unpublished research.*

[16] http://www.thermocalc.se/

[17] R. Veerababu, R. Balamuralikrishnan, and S. Karthikeyan, DMRL Tech. Report, DRDO-DMRL-EMG-018-2012, March 2012, pp. 1-16.

[18] D. B. Williams and C. B. Carter; Transmission Electron Microscopy, 2nd edition, Springer Science and Business Media, LLC, 233 Spring Street, New York, NY 10013, USA, p. 185.

[19] H. Kwon, C.M. Kim, K. B. Lee, H. R. Yang, and J. H. Lee: *Metall. Mater. Trans. A,* 28A (3), 1997, pp. 621–627.

Microstructure Modelling for Line Pipe Steels in Hot Strip Mill

Pradeep Agarwal, H.A. Lanjewar, R. Madhusudhan, S. Manjini and Ashish Chandra
JSW Steel Ltd., Vijayanagar Works
Email: Pradeep.agarwal@jsw.in

ABSTRACT

Microstructure and mechanical properties are the key parameters that need to be controlled during hot rolling process. The present study uses Hot Strip Mill Modelling (HSMM) interface to model hot rolling process. Two API line pipe grades equivalent to 'API X-70 & API X-80' have been modelled for its final microstructure and mechanical properties based on thermal, mechanical and microstructural changes occurring during hot rolling. The model has been validated with plant data and test results. Austenite recrystallisation model has been validated against calculated flow stresses from actual roll forces applied. Phase transformation model has been validated with the final microstructure and mechanical properties. The API X-70 grade shows strain accumulation (i.e., no recrystallisation) in the finish rolling stage only, while the X-80 grade carries the strain accumulation from the roughing passes also, resulting in very fine grain ferrite. It is due to high retained strain before the phase transformation on run out table. The predicted properties by the model are in close proximity for head, middle and tail end of the coil for both the grades.

KEYWORDS: *Microstructure, Recrystallisation, Phase Transformation and Mechanical Properties*

INTRODUCTION

Hot rolling of line pipe steel is a complex process in terms of microstructure modelling. Most of the reported works on recrystallisation kinetics and phase transformation relate to physical simulation or mathematical modelling but very limited information is available on mathematical modelling aspect of plant based trials on rolling [1]. The present study aims at to develop a model for predicting the microstructure and properties of grades equivalent to API X70 and X80 using HSMM interface. The study is carried out in three phases. First, developing the model using HSMM base values and literature based coefficients. Secondly, the validation and tuning of model based on plant data and test results. In third phase, the output of the model is extended to understand the effect of individual process variable on microstructure keeping the rest constant.

HSMM is a PC-based off-line model capable of predicting the thermal evolution, deformation behaviour, microstructure evolution and final mechanical properties of steel being processed in a hot strip mill [2].

MODEL INPUTS

Mill configuration: The location of each unit and its capabilities are fed to calculate heat transfer and temperatures at various section of rolling mill.

Chemical composition: In this particular model, the chemical composition is used to determine the type and amounts of the transformation products, and the mechanical properties.

Reheated (austenite) grain size: There have been many investigations concerning the effect of reheating temperature on grain size. The input does not have significant implication as this grain size is big enough for complete recrystallisation during roughing rolling.

Reheating temperature: The reheating temperature is not used to determine the reheated grain size but it is required to determine what elements are in solid solution and heat transfer calculation thereafter.

Number of passes: The roll speed and the roll diameter are used to calculate the average strain rate through the roll gap. The strain at the pass, or the relevant dimensional information is required to calculate the strain (the entry exit thickness is input and strain is calculated assuming plane strain deformation). Interpass time is calculated based on stand speed; it is to determine the extent of any transformations taking place in between passes. Cooling rate during the runout and the coiling temperature are calculated based on water flow inputs.

MODEL OUTPUTS

The HSMM uses a series of physical models to calculate thermal, mechanical and microstructure evolution. The model generates two types of data:

The microstructural evolution during hot rolling: Temperature of no recrystallization (T_{nr}), average recrystallized grain size before the Tnr, average retained strain in grains recrystallized before the Tnr, average non-recrystallized grain size and retained strain, Nb and other solutes in solution and fraction precipitated and the phase transformation kinetics.

Final microstructure and properties: Average ferrite grain size, phases present and volume fractions, yield strength, tensile strength (UTS) and elongation. Modeling the austenite decomposition on the run-out table and subsequent precipitation of carbides, nitrides, and/or carbonitrides in ferrite during coiling are of particular importance to determine the mechanical properties, which depend on the character of the transformation products (ferrite, pearlite, bainite, and martensite), the ferrite grain size, and the extent of precipitation and solid solution strengthening.

These parameters can be used in designing and optimizing the rolling schedules and cooling pattern.

VALIDATION AND TUNING OF MODEL

It is utmost important to validate the model for the desired chemistry and rolling schedule. The HSMM model was originally validated for seven different chemistries in steel [2]. For other chemistry it provides direct method of changing the equation coefficients for input models or separate user defined functions can also be used to calculate the recrystallisation or phase transformation behaviour.

Two chemistries in line pipe steels equivalent to API X70 and X80 are used as given in Table 1 to validate the model.

Multiple coils data or single coil multiple location data has been used. The chemistry shown is the nominal chemistry for which the results have been interpreted.

Table 1: Chemistry of steels used for validation

Grade	C	Mn	Si	S (max)	P (max)	Al	N	Micro-alloying Nb+V+Ti+Mo (max)
X-70	0.084	1.54	0.37	0.004	0.01	0.04	0.006	0.15
X-80	0.07	1.72	0.27	0.004	0.01	0.033	0.0045	0.30

Recrystallisation kinetics model validation for X-70 grade

First the slab (220 mm thickness) is heated in reheating furnace, after coming out of the furnace it passes through reversing roughing process (7 passes) then 6 stands finishing mill to attain the final thickness (10.5 mm). Table 2 shows the entry temperature and strain applied in all the reversing roughing (RR 1-7) passes and finishing mill passes (FM 1-6). Pyrometer finishing temperature (PY-FT) is given in the Table 2. The initial austenite grain size is assumed to be 500 μ. Strain applied is calculated based on thickness reduction in the stand and retained strain is the stored deformation after recrystallisation or recovery. During the roughing rolling, whatever strain applied gets annihilated due to complete recrystallisation, only the refinement of austenite grain size take place during roughing rolling. While in the finishing mill most of the strain applied gets accumulated and increase the flow stress significantly as shown in figure-1.

Table 2: Rolling parameters and calculated microstructure for X-70 type steel

API X-70 Recrystallisation Modeling	**Roughing Station**	**RR:1**	**RR:2**	**RR:3**	**RR:4**	**RR:5**	**RR:6**	**RR:7**
	Entry Temperature (°C)	1205	1188	1168	1147	1121	1094	1062
	Strain Applied	0.17	0.19	0.22	0.23	0.25	0.28	0.36
	Austenite Grain Size (μ)	500	332.2	277.58	226.9	206.1	170.6	150.7
	Retained Strain entry	0	0	0	0	0	0	0
	Finishing Station	**FM1**	**FM2**	**FM3**	**FM4**	**FM5**	**FM6**	**PY-FT**
	Entry Temperature (°C)	902	881	861	855	846	836	830
	Strain Applied	0.32	0.33	0.25	0.2	0.15	0.1	–
	Austenite Grain Size (μ)	118.1	100	100	100	100	100	100
	Retained Strain entry	0	0.31	0.68	0.97	1.2	1.37	1.48

To validate the recrystallisation kinetics model the actual roll force data has been used. As the mill configuration and rolling parameters are already fed into the model. The model calculates the actual flow stress based of rolling forces. The predicted flow stress is calculated based on recrystallisation pattern as shown in figure-1. The actual and predicted flow stress should match if the estimated recrystallisation kinetics is correct. The same exercise is repeated for multiple locations in the same coil or multiple coils and recrystallisation coefficients are tuned to the best satisfied level. The Figure-1 shows that the flow stresses for all the roughing and finishing passes is matching which means that the model can be used satisfactorily.

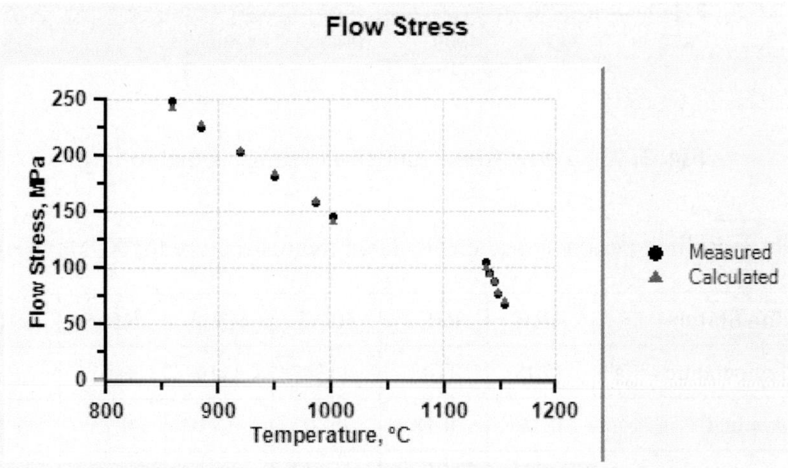

Fig. 1: X-70 flow stress validation against actual rolling

Recrystallisation kinetics model validation for X-80 grade

In the same line of validation for X-70 grade, the initial grain size is assumed to be 500 μ after reheating. The thickness reduction from 220 mm slab to 15 mm strip takes place in 7 reversing roughing passes and 6 finishing passes. Due to high microalloy, this grade has higher T_{nr}, and therefore complete recrystallisation did not occur during roughing passes. The strain applied in the 6th roughing pass gets accumulated for the next pass and thereafter total retained strain reaches to 2.04 before phase transformation on runout table.

The flow stress plot (Figure 2) reveals that the flow stress at all temperature is higher than the flow stress in X-70 grade (Figure 1). The recrystallisation pattern in roughing rolling is matching well with the actual flow stresses, while the flow stress pattern in the finishing mill is not matching well with the predicted values. The reason can be fluctuating rolling force during finish rolling. To satisfy multiple locations flow stress data the optimised weights are used for calculation of recrystallisation pattern of X-80 steel (Figure 2).

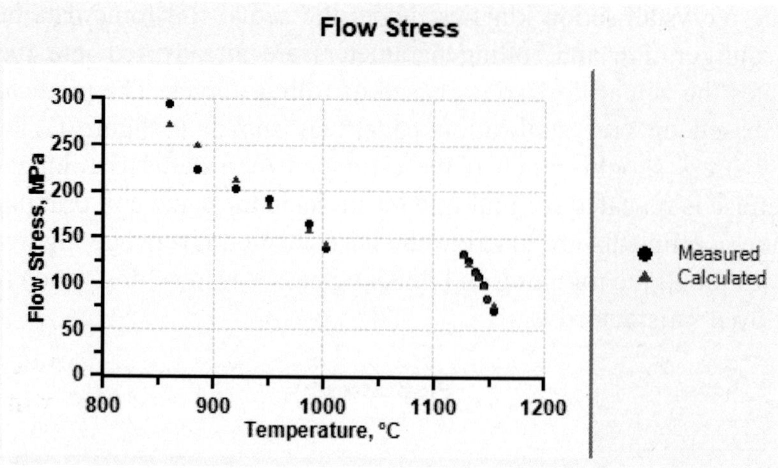

Fig. 2: X-80 flow stress validation against actual rolling

Table 3: Rolling parameters and calculated microstructure for X-80 type steel

	Roughing Station	**RR:1**	**RR:2**	**RR:3**	**RR:4**	**RR:5**	**RR:6**	**RR:7**
API X-80 Recrystallisation Modeling	Entry Temperature (°C)	1205	1148	1134	1119	1101	1082	1057
	Strain Applied	0.16	0.18	0.20	0.21	0.22	0.24	0.28
	Austenite Grain Size (μ)	500	442.99	213.37	154.86	122.48	83.9	83.9
	Retained Strain entry	0	0	0	0	0	0	0.27
	Finishing Station	**FM1**	**FM2**	**FM3**	**FM4**	**FM5**	**FM6**	**PY-FT**
	Entry Temperature (°C)	933	909	886	876	865	853	843
	Strain Applied	0.32	0.27	0.21	0.19	0.11	0.10	–
	Austenite Grain Size (μ)	83.9	83.9	83.9	83.9	83.9	83.9	83.9
	Retained Strain entry	0.63	1	1.31	1.56	1.78	1.96	2.04

Validation of phase transformation model

The phase transformation model calculates the austenite decomposition based on the final austenite grain size and retained strain in it [3, 4]. The grain size is normalised assuming zero retained strain using the formula; $d_{eff} = D_0 * \exp(-\varepsilon_r)$

Where d_{eff} is effective grain size, D_0 is initial grain size, and ε_r is retained strain in initial grains [5]. This effective grain size is then considered free of dislocations and nucleation of ferrite is assumed to take place on the austenite grain boundaries. For each time-step, the ferrite fraction and precipitation kinetics are calculated based on cooling rates and temperature drop.

To validate the results, predicted values (ferrite grain size and fraction of second phase) are compared with the actual microscopic results. Table 4 represents the validated result of phase transformation. For cross reference, continuous cooling transformation (CCT) plot is generated using the software JMatPro [8] as shown in Figure 3 and 4. The effective austenitic grain size for X-70 grade is calculated as 22.8 μ *and for X-80 steel* 10.9 μ. *CCT plot of X-70 steel*, as shown in Figure 3, is superimposed with actual cooling pattern of strip before and after coiling. Figure 4 shows the CCT plot for X-80 steel superimposed with cooling pattern. The cooling rates under the water headers are calculated based on water flow inputs and runout table configuration. The phase transformation behaviour is found to be matching well with the CCT plots and the final microstructure is similar to the predicted one as shown in Figure 5.

Table 4: Calculated runout table entry and exit microstructure

	Station	**PY-FT**	**PY-CT**
API X-70	Entry Temperature (°C)	830	565
	Austenite Grain Size (μ)	100.0	Dα=6.4μ α=90.8% P=9.2%
	Retained Strain entry	1.48	
API X-80	Entry Temperature (°C)	843	571
	Austenite Grain Size (μ)	83.9	Dα=2.2μ α=98.6% B=1.4%
	Retained Strain entry	2.04	

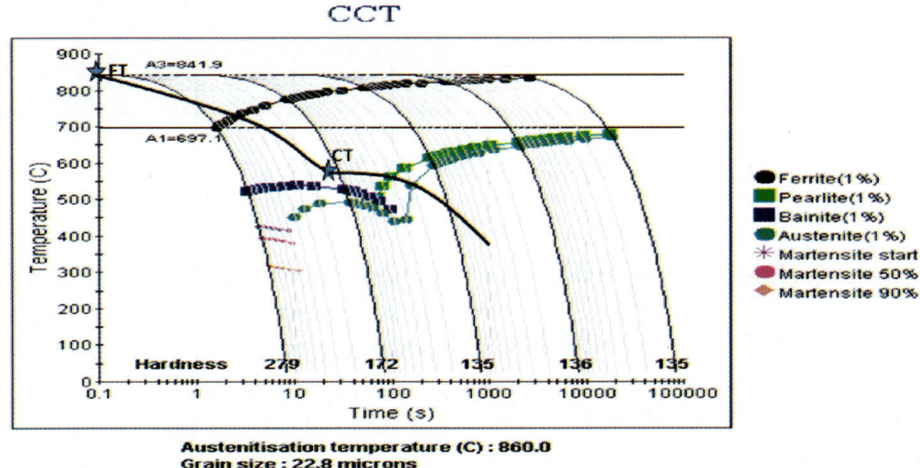

Fig. 3: CCT plot generated by software JMatPro for X-70 Grade

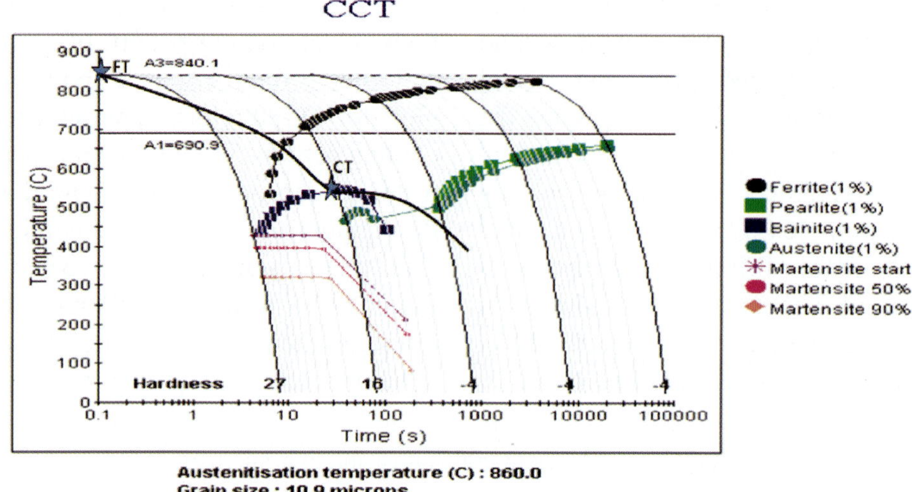

Fig. 4: CCT plot generated by software JMatPro for X-80 Grade

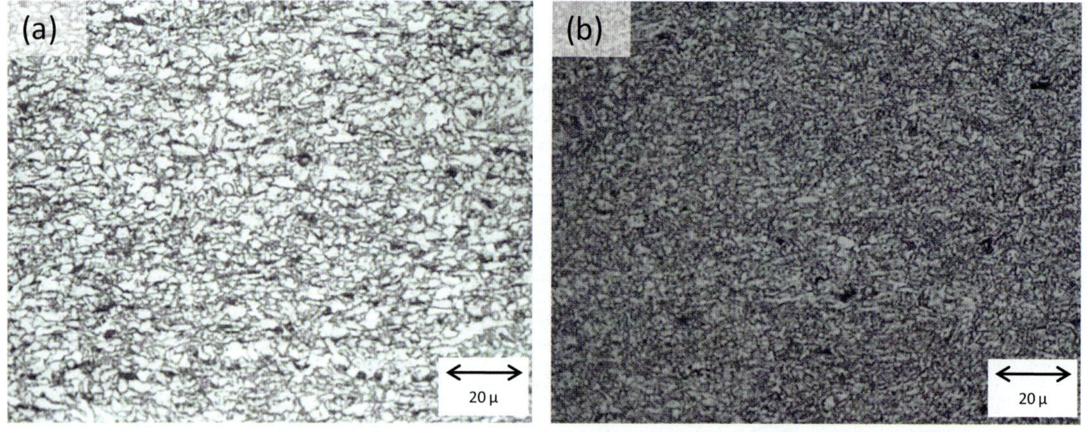

Fig. 5: Optical images taken at 500X; (a) X-70 steel and (b) X-80 steel

Validation of Mechanical Properties

Mechanical properties are calculated based on strengthening due to grain size, precipitates and phase's strength. The precipitation strengthening in ferrite is controlled by the volume fraction of the particles and the coarsening of precipitates. The volume fraction of particles can be estimated from the chemical composition and coarsening is calculated based on time-temperature history. The microstructure, nature of ferrite (polygonal or acicular), and second phase play a vital role for final mechanical properties. Multiple microscopic correlations were made before adjusting final equations for mechanical properties prediction.

Table 5 shows the measured and predicted mechanical properties for both X-70 and X-80 steels. The measured values are given in range as the model was validated against multiple coils properties and different locations in a coil. Tensile tests have been done in the transverse direction of rolling. The model has been validated and tuned to give reasonable results for mechanical properties.

Table 5: Measured and predicted mechanical properties for both grades

Mechanical Properties	X-70 Grade		X-80 Grade	
	Measured	Predicted	Measured	Predicted
YS (MPa)	533± 25	531	655 ± 25	650
UTS (MPa)	635 ± 20	639	756 ± 20	759
EL (%)	37 ± 3	36.9	32 ± 3	32.2

MODEL BASED INTERPRETATIONS

As already mentioned the complete model for microstructure and mechanical properties for X-70 and X-80 steels are made using HSMM interface. It is validated and tuned to give reasonable results for variability in rolling parameters and to some extent chemistry. Table 6 shows the model based interpretations for X-70 steel and Table 7 shows the same for X-80 steel.

The base simulation is considered same as mentioned previously. Five cases are considered against base simulation. The changing parameter is shown in bold letters. For all the other input parameters only two outputs are mentioned as these are most important for line pipe steels, i.e., yield strength and ferrite grain size. The aim of these case studies is to identify the suitable strategy for refinement of grain structure as required for low temperature applications.

Table 6: Model based interpretations for X-70 grade

Model based Result of hot rolling of a API X 70 steel slab from 220 mm thick in to a 10.5 mm strip (Coil)						
Parameters and Results	**Base Simulation**	**Case-1**	**Case-2**	**Case-3**	**Case-4**	**Case-5**
Reheating Temperature	1230°C	1230°C	1230°C	1230°C	1230°C	1230°C
Number of Roughing Passes	7	7	7	7	7	7
Temperature window for Roughing rolling	1205-1060°C	1205-1060°C	1205-1060°C	1205-1060°C	1205-1060°C	1205-1060°C
Thickness after roughing	40.5mm	40.5mm	40.5mm	40.5mm	**50 mm**	40.5mm
Time in between Roughing and Finishing mill	42 sec	42 sec	42 sec	42 sec	55 sec	80 sec
Temperature window for Finish rolling	925-835°C	925-835°C	925-835°C	925-835°C	915-835°C	895-835°C

Finishing exit speed	3 m/sec	3 m/sec	3 m/sec	3 m/sec	3 m/sec	**4 m/sec**
Finishing Temperature	830	**860**	830	830	830	830
Coiling Temperature	565	565	**595**	**535**	565	565
Ferrite Grain Size	5.4 μ	7.5 μ	6.1 μ	5.4 μ	3.5 μ	5.2 μ
Change in Yield Stress	0 basis	-9 MPa	+5 MPa	-17 MPa	+19 MPa	+20 MPa

Table 7: Model based interpretations for X-80 grade

Model based Result of hot rolling of a API X80 steel slab from 220 mm thick in to a 15 mm strip (Coil)						
Parameters and Results	**Base Simulation**	**Case-1**	**Case-2**	**Case-3**	**Case-4**	**Case-5**
Reheating Temperature	1240°C	1240°C	1240°C	1240°C	1240°C	1240°C
Number of Roughing Passes	7	7	7	7	7	7
Temperature window for Roughing rolling	1205-1055°C	1205-1055°C	1205-1055°C	1205-1055°C	1205-1055°C	1205-1055°C
Thickness after roughing	50.8 mm	50.8 mm	50.8 mm	50.8 mm	**60 mm**	50.8 mm
Time in between Roughing and Finishing mill	62sec	62sec	62sec	62sec	78sec	90 sec
Temperature window for Finish rolling	935-845°C	935-845°C	935-845°C	935-845°C	920-845°C	912-845°C
Finishing exit speed	1.85 m/sec	1.85 m/sec	1.85 m/sec	1.85 m/sec	1.85 m/sec	**2.5 m/sec**
Finishing Temperature	843	**873**	843	843	843	843
Coiling Temperature	571	571	**601**	**541**	571	571
Ferrite Grain Size	2.2 μ	2.4 μ	2.5 μ	2 μ	2.2 μ	2.1 μ
Change in Yield Stress	0 basis	-11 MPa	-24 MPa	+32 MPa	+2 MPa	+29 MPa

Case-1: If the finishing temperature is increased by 30 °C, it results in increase in ferrite grain size as the amount of retained strain in final austenite pancake grains reduced owing to high temperature deformation.

Case-2: If the coiling temperature is increased by 30 °C, it results in increase in grain size due to soft cooling. The strength of X-70 steel increases marginally due to more precipitation strengthening component of AlN precipitates [7]. While the strength of X-80 steel decreases due to grain coarsening.

Case-3: Decreasing the coiling temperature by 30 °C results in finer grains for X-80 steel and strength increases. While in X-70 steel no change in grain size could be observed and the strength decreased due to some of the interstitials (C and N) remain in solution and precipitation strengthening component decreases.

Case-4: Increasing the transfer bar thickness (by about 10 mm) after roughing has result in more deformation in the finishing stage and also increased low temperature deformation. This results in increased strength and refinement in grains for X-70 steel. While the X-80 steel does not show much change in the microstructure as its Tnr is very high (close to 1060 °C) [5]. Any deformation given below T_{nr} temperature gets stored within the material. Also, it is to be noted that, increasing transfer bar thickness will result in more plastic deformation in finishing stage, and hence more temperature gain in finishing passes. Finishing entry temperature is reduced to get the required finishing temperature by holding the transfer bar between roughing and finishing stages.

Case-5: If the finishing exit strip speed is increased (≈ 35%), some refinement in grain size can be observed for both the steels resulting in strengthening. This is due to increase in overall cooling rate; increasing speed reduces the cooling time to attain the same coiling temperature. To increase the finishing exit speed; the finishing stands entry speed also increases demanding further less finishing entry temperature to match the same finishing exit temperature.

Based on the observations, a suitable strategy for grain refinement in X-70 steel would be to increase the transfer bar thickness and for X-80 steel the lower coiling temperature.

One more case, not discussed here, is the effect of cooling power in the early stage of runout table. In all the previous cases, the cooling power of each header is kept constant. The increase or decrease in coiling temperature is attained through increasing or decreasing the number of headers ON in runout table.

The calculations were made for increasing the early cooling power by 30% keeping same FT and CT. For X-70 steel it results in final ferrite grain size of 2.9 μ and for X-80 steel it comes out to be 1.9 μ. This is a significant reduction in grain size. The effect of intensive early cooling for grain refinement in line pipe steel is reported elsewhere [8].

CONCLUSIONS

- A complete mathematical model using HSMM interface has been developed to simulate the hot rolling of steels equivalent to API X-70 and X-80 steel.
- The austenite recrystallisation during hot rolling has been validated using actual roll forces. The predicted and actual flow stresses during rolling are in close proximity.
- The phase transformation model is validated with the actual microstructure which is close to the predicted one.
- The effect of different rolling conditions on final microstructure is shown to get the directions for improvement in microstructure and properties of these steel.
- Intensive early cooling is desirable for refinement of grain structure in both the steels. Higher deformation below T_{nr} is desired for grain refinement of X-70 grade steel while lower coiling temperature (500-550°C) can help in improving the microstructure for X-80 grade steel.

REFERENCES

[1] Vladimir B. Ginzburg, Flat rolled steel processes- Advance Technologies, CRC Press Taylor and Francis Group, Boca Raton, London New York

[2] Richard A. Shulkosky, David L. Rosburg, Jerrid D. Chapman, INTEG process group inc." Validation of the Hot Strip Mill Model" AISI/DOE Technology Roadmap Program-final report-March 30, 2005 (www.osti.gov/bridge).

[3] M. Militzer, E. B. Hawbolt and T. R. Meadowcroft; Metallurgical and Materials Transactions A, Volume 31, Number 4 (2000), 1247-1259

[4] M. Militzer, M.G. Mecozzi, J. Sietsmac, S. vander Zwaag; Acta Materialia, Volume 54, Issue 15, September (2006), 3961–3972

[5] M. Militzer, R. Pandi and E. B. Hawbolt, "Microstructural model for hot strip rolling of high-strength low-alloy steels "Metallurgical and Materials Transactions A, Volume 27, Number 6 (1996), 1547-1556

[6] N. Saunders, U. K. Z. Guo, X. Li, A. P. Miodownik and J. -Ph. Schillé; JOM Journal of the Minerals, Metals and Materials Society, Volume 55, Number 12 (2003), 60-65

[7] Yonglin Kang, Hao Yu, Jie Fu, Kelu Wang, Zhongbing Wang; Materials Science and Engineering A351 (2003) 265-271

[8] Manuel Gomez, Pilar Valles, Sebastián F. Medina; Materials Science and Engineering A 528 (2011) 4761–4773

SECTION IV

Hot Deformation, Recrystallization and Grain Growth

Modelling Grain Size Distributions during TMCR of Nb-microalloyed Steels

M. Strangwood and C.L. Davis
School of Metallurgy and Materials, The University of Birmingham, Birmingham, UK, B15 2TT.

ABSTRACT

Models for recrystallisation and precipitation during hot deformation of HSLA steels, based on the original Dutta–Sellars approach, show good agreement with experimental data for compositions around 0.04 wt% Nb, strains around 0.3 and prior austenite grain sizes up to 600 μm. The experimentally determined recrystallisation start and finish temperatures have been compared with predictions made with the original and a modified Dutta-Sellars model for homogenised and segregated samples. These modifications include the use of Thermo-Calc to predict the solute Nb content and prior austenite grain size distributions, rather than an average prior austenite grain size, which have been shown to give consistently better predictions using literature data. The application of a Clyne-Kurz model has accounted for the amounts of solute-rich and solute-depleted material for steels with compositions solidifying through the peritectic transformation.

KEYWORDS: *Microalloyed Steels, Segregation, Recrystallisation, Hot Rolling*

INTRODUCTION

The combination of high strength and good toughness for high strength low alloy (HSLA) steels is achieved through the generation of fine ferrite grain sizes through application of controlled thermo-mechanical processing (TMP) schedules [1]. Steel grades amenable to this treatment will contain small amounts of some or all of Al, Ti, V, Mo and, especially, Nb in combination with C and N [2]. For a hot deformed and recrystallised (fully or partially) HSLA steel, the strength level is largely determined by the mode or average grain size (through the Hall-Petch relationship); toughness, however, is much more dependent on the population of larger grains in the grain size distribution. The grain size distributions are usually characterised as a log-normal distribution for full recrystallisation with the maximum grain size being 2 – 3 times that of the mean / mode [3]. This is a notable effect for unimodal grain size distributions, but bimodal grain size distributions the mixture of coarse and fine grains (banded or in patches) results in increased scatter in toughness values, although the strength levels are similar [4].

Niobium is the most potent microalloying addition for austenite grain size control through both solute drag and strain-induced precipitation (SIP) effects. During solidification, however, niobium shows a strong tendency to segregate into the liquid phase and its low diffusivity in austenite means that it will remain segregated through most of the TMP schedule resulting in spatial variations in the degree of retardation of recrystallisation.

In order to avoid low or scattered toughness values, then the grain size distribution during hot deformation needs to be controlled. In selecting TMP schedules then one on the more widely used predictive schemes is that of Dutta and Sellars (DS) [5]. The following equations are used to predict static recrystallisation and precipitation start time according to 5 % recrystallisation (R_s) and 5 % precipitation (P_s) based on classical nucleation theory.

$$R_s = 6.75x10^{-20} D_0^2 \varepsilon^{-4} x\, exp\left(\frac{300000}{RT}\right) x\, exp\left(\frac{2.75x10^5}{T\text{-}185}(Nb)\right)$$

$$P_s = A(Nb)^{-1} \varepsilon^{-1} Z^{-0.5} x\, exp\left(\frac{270000}{RT}\right) x\, exp\left(\frac{B}{T^3 (lnk_s)^2}\right)$$

where *[Nb]* is the amount of Nb in solution in austenite; ε is applied strain; T (K) is the deformation temperature; k_s is the supersaturation ratio; D_0 is the austenite grain size at the start of deformation; and Z is Zener-Holloman parameter.

This model was developed from experimental data based on deformation testing of steels with Nb contents between 0.03 - 0.045 wt %, starting prior austenite grain sizes of 100 – 450 μm and using a strain of 0.3 [5]. However, unimodal grain sizes are assumed along with a uniform Nb concentration. Earlier studies on a 0.46 wt % Nb steel that had been homogenised for 4 days at 1225 °C [6] showed that the DS equations using the parameters above overestimated the proportion of recrystallisation for a 10 s hold after 0.3 strain. The agreement, however, improved when grain size classes were used rather than just the average grain size.

The generality of this approach to dealing with grain size distributions and how chemical heterogeneities are dealt with is the subject of this paper.

MATERIALS AND EXPERIMENTAL PROCEDURE

The material used in this work was a 290 mm thick continuously cast slab of composition: 0.10 C, 1.51 Mn, 0.32 Si, 0.017 P, 0.005 S, 0.046 Nb, 0.002 Ti, 0.006 V, 0.045 Al and 0.008 N; all wt %. Specimens of size 10 x 15 x 20 mm (machined from the ¼ thickness position of the slab) from the slab were tested in the as-cast state and after being homogenised at 1225 °C for 4 days under vacuum and slowly cooled at a rate of 3 °C/min to remove the Nb segregation. The details of selection of homogenisation temperature and time have been discussed elsewhere [7]. The homogenised and segregated specimens were sealed inside silica tubing under a partial Ar atmosphere and reheated to 1150 or 1225 °C, soaked for 1 hour and then quenched (tube intact) into water to represent a commercial reheating temperature and time (to dissolve Nb-rich precipitates); quenching was used to preserve the Nb in solution prior to deformation trials. Thermo-Calc software (version Q) was used to determine the equilbrium Nb content in solution at 1150 / 1225 °C in the homogenised condition. The reheated specimens were subjected to plane strain compressive deformation in a Gleeble 3500 over a range of temperatures (975 °C to 1075 °C) and 0.3 strain being held at the chosen deformation temperatures for a range of times (4 to 10 s) at a strain rate of 10 s^{-1} after solutionising at 1150 or 1225 °C for 5 minutes.

Samples after reheating and Gleeble simulation were sectioned along their mid-thickness (normal to the compression axis) and mounted in Bakelite. The samples were metallographically polished to a 0.05 μm SiO_2 finish and etched in saturated aqueous picric acid solution at 60 °C to reveal the prior austenite grain structures. The heat-treated and deformed specimens were examined in a Leica DMRX microscope equipped with KS300 image analysis software. Quantitative determination of austenite grain size in terms of equivalent circle diameter (ECD) was carried out measuring about 1500 grains in terms of individual grain area and ECD (obtained directly from the KS300 software) to determine grain size distributions as an area fraction against size.

Results and Discussion

For homogenised material single hit Gleeble tests showed that the fully recrystallised grains showed a distribution that could not be modelled using equation of the type:

$$d_{rex} = D'\left(D_0\right)^{0.67} \varepsilon^{-1}$$

where drex is the average recrystallised grain size, D_0 the average starting austenite grain size and D' a constant with values quoted between 0.66 and 1.86. Inspection of the prior austenite grain size distributions for reheated (un-recrystallised) and fully recrystallised samples, Figures 1 and 2, shows that a good fit could be achieved by merely halving the original grain size on recrystallisation. This amounts to limiting recrystallisation to the formation of two recrystallised grains per deformed austenite grain on average and so may limit this approach to a certain range of starting austenite grains.

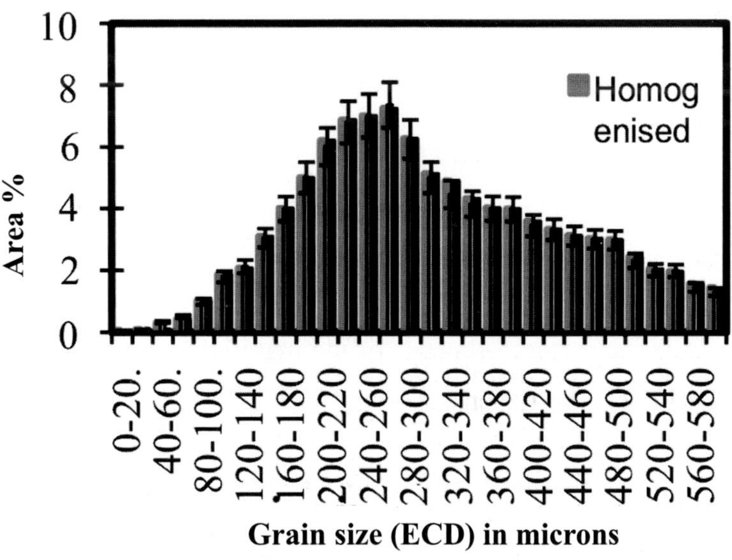

Fig. 1: Prior austenite grain size distribution for re-heated, homogenised sample

The good agreement between the predicted and experimental distributions in Figures 1 and 2 was obtained when the recrystallising grains were treated as separate classes, i.e. the Dutta-Sellars equations were applied to each 20 μm class, using the median size value as D_0 and then determining the fraction of that grain glass that had recrystallised under the thermomechanical simulation conditions. The fraction that had recrystallised was transferred to the class half the size of the original one, whilst the unrecrystallised fraction was left in the original class.

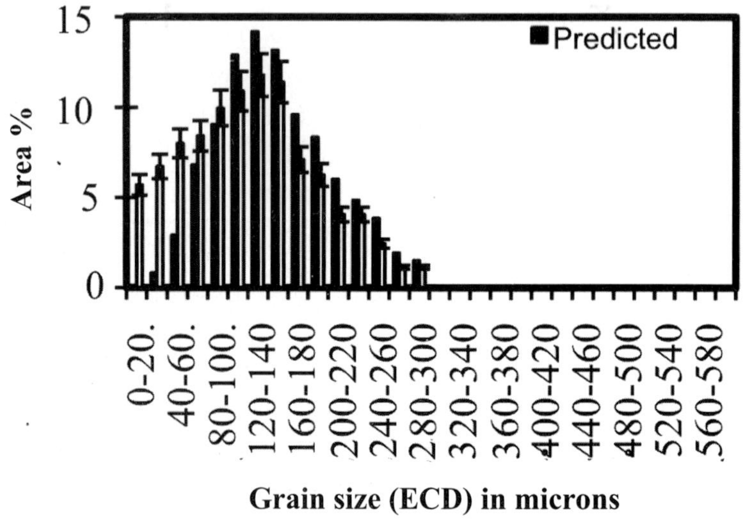

Fig. 2: Prior austenite grain size distributions (predicted and measured) for a fully recrystallised, homogenised sample.

This approach leads to good reproduction of grain size distributions but also a good fit with the recrystallised fractions, Table 1.

Table 1: Predicted (grain class approach) and measured volume fractions of recrystallised grains as a function of deformation temperature

Deformation T (°C)	Measured recrystallisation %	Predicted recrystallisation %
1075	100	93
1050	72	73
1025	42	44
1010	19	17
990	10	8
975	0	0

In the homogenised samples, all grains will have the bulk composition, particularly [Nb], which appears in the Dutta-Sellars equations. When dealing with segregated samples then the composition of the grains will vary depending on whether the grains are in solute-rich or solute-depleted regions of the material. The as-cast structure of the slab has been previously reported [7, 8]; the low [Ti] value in this steel meant that the amount of coarse TiN formed was low so that Nb(C,N) precipitated throughout the austenite grains on cooling rather than primarily on pre-existing TiN. This allowed energy dispersive X-ray spectroscopy

(EDS) to be used to determine [Nb], especially in the solute-rich areas. The composition profile for Nb was found to be a 'top-hat' with very consistent [Nb] in solute-rich and solute-depleted regions and a very steep change in composition between them. This indicated that, for this steel, the solute-rich and solute-depleted regions could be considered as behaving as material of different composition, uniform within each region (solute-rich and solute-depleted). The EDS analysis also revealed that around 25 % of the as-cast structure for this steel was solute-rich. In other HSLA steels with higher [Ti] then most Nb(C,N) is present as larger caps on the TiN particles so that EDS analysis only reveals detectable [Nb] at the TiN particles.

The average secondary dendrite arm spacing for this slab was 250 μm, which corresponds to a cooling rate of 0.182 K s^{-1}. Microsegregation in this system was modelled using a Clyne-Kurz [9] approach. This is based on mass balance and back diffusion of carbon during solidification of the steel to give a solidified fraction, f_s, as a function of temperature, T:

$$f_s = \left(\frac{1}{1-2\Omega k}\right)\left[1-\left(\frac{T_m - T}{T_m - T_L}\right)\right]^{\frac{(1-2\Omega k)}{(1-k)}}$$

where T_m is the melting temperature of pure iron, T_L is the alloy liquidus temperature, k is the equilibrium partition coefficient and Ω is the solidification factor given by

$$\Omega = a\left[1-\exp\left(-\frac{1}{a}\right)\right]-\frac{1}{2}\exp\left(\frac{1}{2a}\right)$$

where α is a function of the secondary dendrite arm spacing (SDAS), λ_s, local solidification time, t_f, and solute diffusivity in the solid phase, D_s:

$$a = \frac{D_s t_f}{(0.5\lambda_s)^2}$$

The assumptions made are that the first appearance of austenite via the peritectic transformation would form a low-diffusivity layer that would prevent further atomic migration between the dendrite and the interdendritic liquid. On this basis, the solute-depleted fraction would be the integrated fraction solidified by the start of the peritectic transformation; the dominant role of [C] on this transformation meant that the equilibrium value for the peritectic start temperature as predicted from Thermo-Calc was used to define the undercooling temperature range over which the dendrites solidified. Carbon diffusion in ferrite was also used for determination of D_s.

The fractions of solute-rich and solute-depleted determined this way deviated from the equilibrium values at the peritectic start temperature, but Thermo-Calc was used to establish the mass balance between delta-ferrite and liquid that would give these volume fractions (noting that the segregation profiles determined experimentally were top-hat functions). This method gives reasonable agreement between prediction and experiment, Table 2, and has been extended to steels with 0.04 – 0.1 wt % C and 0.019 – 0.094 wt % Nb, all of which involved the peritectic transformation during their solidification sequences.

The results shown in Table 2 show reasonable agreement with optical and scanning electron microscopy determination of the solute-rich and solute-depleted fractions, although differences of up to 6 % are observed. These differences may become significant when modelling a commercial multi-pass hot rolling schedule and cross-diffusion effects are being investigated using EDS mapping techniques in order to improve the fit between model and experiment.

Table 2: Predicted and measured solute-rich volume % for a range of C-Nb steels.

Alloy	Predicted	Measured
0.052 Nb – 0.06 C	7.6	8.3
0.046 Nb – 0.1 C	20.0	25.0
0.094 Nb – 0.04 C	22.0	28.0
0.019 Nb – 0.1 C	22.0	16.0

The modelling approach is also being expanded to deal with alloys solidifying without involving the peritectic transformation.

In applying the class-based Dutta-Sellars equations to segregated samples then the grain distributions need to be assigned a composition representative of solute-rich or solute-poor regions. For a unimodal grain size distribution, e.g. for re-heating at 1225 °C, then each grain size classification is assigned equally to solute-rich and solute-depleted. In this steel then 25 % of the grains in each grain size classification are assumed to have a [Nb] value representative of solute-rich material and 75 % a value representative of solute-depleted, Figure 3.

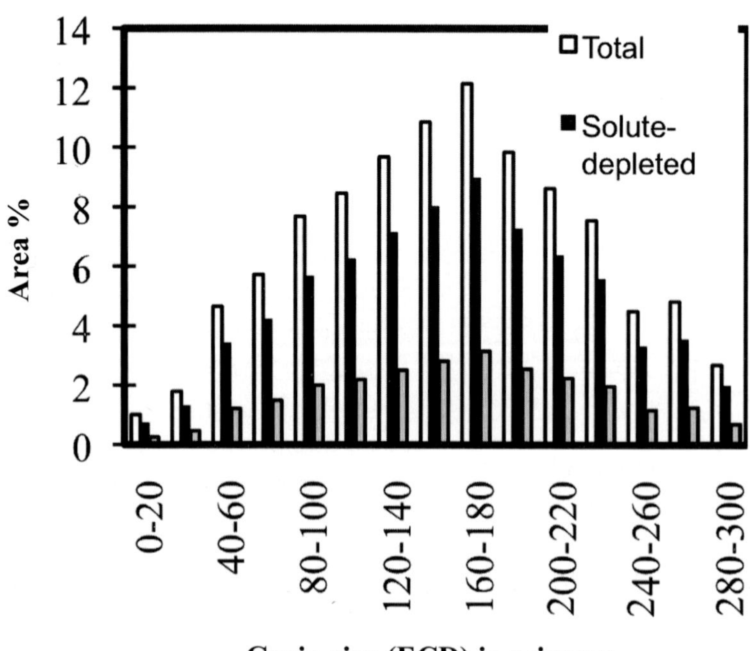

Fig. 3: Separation of grain size classes into solute-rich and solute-depleted for unimodal grain size distribution (re-heated to 1225 °C).

When reheating to 1150 °C, the situation is more complex as the resulting prior austenite grain size distribution is bimodal. EDS analysis of the coarse and fine grained regions in this distribution indicated that fine grains were observed in both solute-rich and solute-depleted material, but the coarse grained regions were exclusively solute-depleted. On this basis the coarse grains were assigned a [Nb] value representative of solute-depleted material, whilst the fine grained classes were split into solute-rich (33%) and solute-depleted (67%); the percentages taking into account the amount of solute-depleted material assigned to the coarse grain classes, Figure 4.

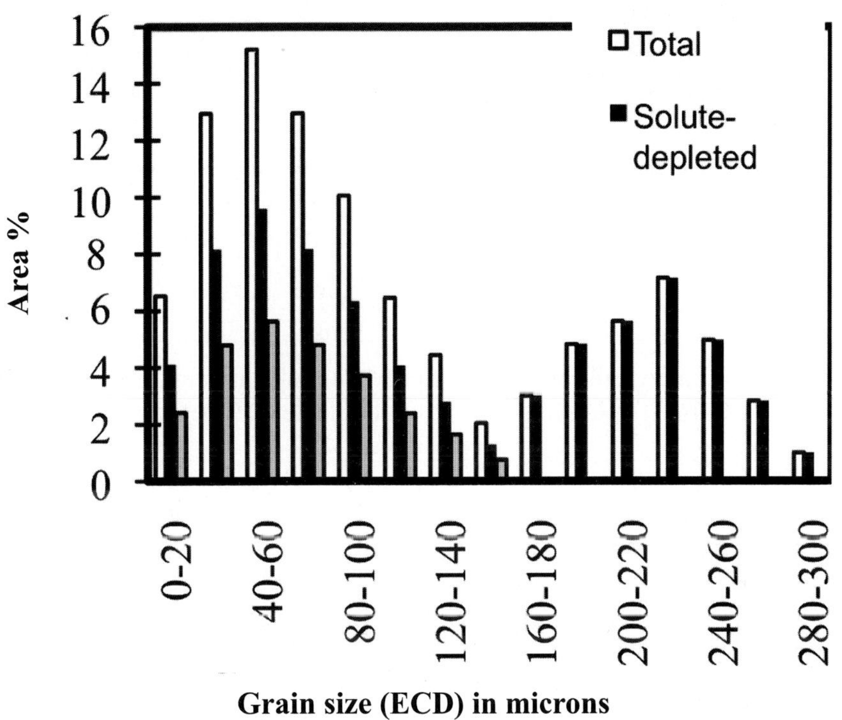

Fig. 4: Separation of grain size classes into solute-rich and solute-depleted for bimodal grain size distribution (re-heated to 1150 °C).

Taking segregation into account results in a number of effects. Firstly the partial recrystallisation range of temperatures is broadened on recrystallisation precipitation time temperature (RPTT) plots compared with an homogenised alloy. This can be predicted using this modification to the original Dutta-Sellars equations and so identifies processing windows at risk of developing bimodal grain structures, Figure 5.

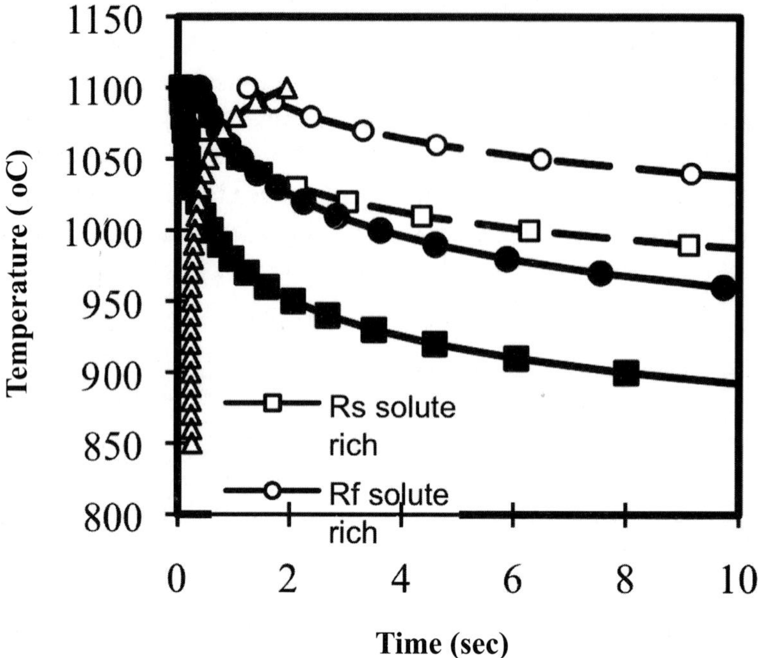

Fig. 5: Separation of recrystallisation start (square symbols) and finish (circular symbols) temperatures for solute–rich (solid symbols) and solute-depleted (open symbols) in a segregated steel sample.

Using the grain size class modification and the differing compositions allowed the grain size distributions to be determined as for the homogenised alloys. Although the fit is good it is not as good as in the homogenised case and this requires further work. Despite this the agreement between the recrystallised fraction predicted using differing compositions for the different grain classes was much better than when an average [Nb] value was used. The discrepancy for the use of an average [Nb] value was generally an underestimate by 30 – 50 %; this came down to an overestimate of 3 – 13 % when the different [Nb] values were used.

Conclusions and Further Work

The use of grain sizes classes instead of an average grain size gives better fit between a Dutta-Sellars-based prediction and experimentally determined recrystallisation fractions in an homogenised sample.

In a segregated steel sample, the use of different [Nb] in the modified Dutta-Sellars approach results in improved prediction of the amount of recrystallisation although the fit for grain size distributions is not as good as that seen for the homogenised case.

For steels solidifying via a peritectic transformation then a Clyne-Kurz approach gives reasonable prediction for the fraction of solute-rich material

The modelling approach is being improved by addressing the fit of grain size distributions; the effects of grain size on stored energy; variations in strain and composition; and solidification without the peritectic.

ACKNOWLEDGEMENTS

The authors wish to thank the support of the School of Metallurgy and Materials at the University of Birmingham and Tata Steel Europe and the results generated by Debalay Chakrabarti, Amrita Kundu and Dayue Zhang.

REFERENCES

[1] Sellars, C.M. The physical metallurgy of hot working. in Hot working and forming processes. 1979. University of Sheffield: The Metals Society, London.

[2] Deardo, A.J., Niobium in modern steels. International Materials Reviews, 2003. 48(6): p. 371-402.

[3] Fernandez, A., P.Uranga, B.Lopez and J.M.Rodriguez-Idabe, Static Recrystallization Behaviour of a Wide Range of Austenite Grain Sizes in Microalloyed Steels. ISIJ International, 2000. 40(9): p. 893 - 901.

[4] Wu, S.J. and C.L. Davis, Effect of duplex ferrite grain size distribution on local fracture stresses of Nb-microalloyed steels. Materials Science and Engineering: A, 2004. 387–389(0): p. 456-460.

[5] Dutta, B. and C.M. Sellars, Effect of composition and process variable on Nb(C,N) precipitation in niobium microalloyed austenite. Materials Science and Technology, 1987. 3: p. 197-206.

[6] Kundu, A., C. Davis, and M. Strangwood, Grain Size Distributions after Single Hit Deformation of a Segregated, Commercial Nb-Containing Steel: Prediction and Experiment. Metallurgical and Materials Transactions A, 2011. 42(9): p. 2794-2806.

[7] Kundu, A., Grain Structure Development During Casting, Reheating and Deformation of Nb-Microalloyed Steel, PhD thesis, School of Metallurgy and Materials 2011, University of Birmingham: Birmingham, UK.

[8] Davis, C.L. and M. Strangwood, Segregation behaviour in Nb microalloyed steels. Materials Science and Technology, 2009. 25(9): p. 1126-1133.

[9] Clyne, T.W. and W. Kurz, Solute Redistribution during Solidification with Rapid Solid-state Diffusion, Metallurgical Transactions A, 1981, 12, 965-971.

Investigation of Progress in Dynamic Recrystallization in Two Austenitic Stainless Steels Exhibiting Flow Softening

A. Sarkar and J.K. Chakravartty
Mechanical Metallurgy Division, Bhabha Atomic Research Centre, Mumbai-400085, India
Email: jayanta@barc.gov.in

ABSTRACT

The hot working behaviour of two austenitic stainless steels: SS304 and Ti-modified 15Cr-15Ni-2Mo (Alloy D9) were studied by hot compression test at temperatures of 900–1200 °C and at strain rates of 0.002–1 s^{-1}. The progress of dynamic recrystallization (DRX) was modelled by the Avarmi kinetics equation. The flow softening was directly related to the DRX volume fraction and the DRX time was determined by strain rate. For quantification of recrystallization rate, the time corresponding to the DRX fraction of 50% was used. Analysis of the sigmoid-shaped recrystallization curves revealed that the rate of DRX increases with increasing deformation temperature and strain rate. Avrami exponents obtained indicated heterogeneous grain or twin boundary nucleation during DRX of these steels. Results of the Avrami analysis were used to predict the DRX flow curves for different deformation conditions.

INTRODUCTION

Hot working is one of the important steps in fabrication of structural components used in various engineering applications. Dynamic recovery (DRV) and dynamic recrystallization (DRX) are the primary restoration mechanism during high temperature deformation of metallic materials. Micro mechanism of hot deformation in fcc metals and alloys is in general governed by their stacking fault energy (SFE). In some materials with high SFE such as aluminium, DRV can balance work hardening, and a plateau is achieved in the stress strain curves. However, in materials with low SFE such as austenitic stainless steels, the kinetics of DRV is low, and DRX can be initiated at a critical condition of strain energy accumulation. DRX is associated with softening in the stress-strain curves. DRX plays an important role in controlling microstructure and mechanical properties in hot working of many metallic materials [1–3]. Due to the great impact of DRX on the high temperature flow stress and its effect on the microstructure and properties of the material after processing, the evaluation of the rate and progress of DRX in terms of deformation conditions is of utmost importance. The rate of DRX depends on the chemical composition of the material, the mode of deformation, and the deformation conditions.

In the present work, the rate and the progress of DRX in two stainless steels namely SS304 and a Ti-modified 15Cr-15Ni-2Mo austenitic stainless steel (known as alloy D9) were studied during hot compression test.

EXPERIMENTAL PROCEDURES

Hot compression tests were conducted on solution-annealed (SA) SS304 and alloy D9 and specimens of 10-mm height and 5-mm diameter in a deformation dilatometer. The specimen was heated to the deformation temperature at a heating rate of 5°C/s and held there for 10 min. Single-hit hot compression tests were carried out at temperatures of 900-1200 °C with strain rates of 0.002–1 s^{-1} up to true strain of about 0.6.

RESULTS AND DISCUSSIONS

Stress-strain behavior

Flow curves obtained at different temperatures and at a strain rate of 0.01 s^{-1} for SS304 and alloy D9 are shown in Fig. 1 a and b respectively.

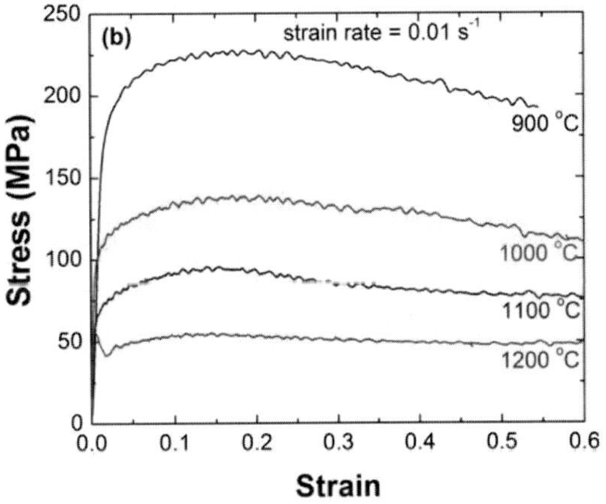

Fig. 1: Stress-strain curves for (a) SS304 and (b) alloy D9 deformed at a strain rate of 0.01 s^{-1} at different temperatures.

Most of the curves exhibit typical DRX behaviour with a single peak stress followed by a gradual fall towards a steady- state stress. However, the peak becomes less obvious when the strain rate increases or the deformation temperature decreases. The drop in flow stress with deformation temperature may be attributed to the enhancement in the rate of restoration processes and consequent decrease in the strain hardening rate. Since the formation of DRX nuclei becomes easier at higher deformation temperatures, the critical strain for initiation of DRX decreases. Moreover, the mobility of grain boundaries increases with increasing deformation temperature and hence the rate of DRX increases. Therefore, both the peak and steady state strains decrease with increasing the deformation temperature. The increase in the flow stress with strain rate can be ascribed to the decrease in the rate of restoration processes and increase in the strain hardening rate. The rate of DRV also decreases with increasing strain rates. Since the formation of DRX nuclei depends on the recovery substructures, the increase in the critical strain for initiation of DRX with increasing strain rate is expected. Moreover, the mobility of grain boundaries decreases with increasing strain rate, which in turn increases the peak and steady state strains. These observations are consistent with earlier observations on several alloy systems exhibiting flow softening during DRX [1-3].

Deformation microstructure

The hot-deformed samples were cut along the longitudinal direction using a precision cut-off machine. One half of the sample was taken to prepare specimens for electron backscattered diffraction (EBSD) investigation. The microstructures were examined in the maximum deformation zone of the specimens. EBSD scans were performed on all processed samples using a TSL-Orientation Imaging Microscopy (OIM) system attached to an FEI Quanta 200 scanning electron microscope (FEI, Eindhoven, Netherlands) operating at 30 kV. Samples for EBSD were polished up to 0.25-lm grit diamond paste using the standard metallographic polishing procedure. Finally samples were electro polished to ensure removal of any residual surface deformation. EBSD maps were collected from the processed samples using a step size of 0.5 μm using a hexagonal grid. Fig 2 shows the microstructure of the starting SS304 material.

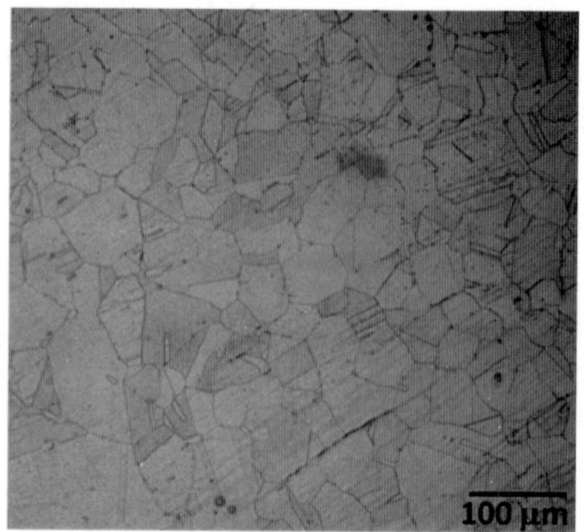

Fig. 2: Optical microstructure of the starting annealed SS304 sample.

Fig. 3 a and b show the EBSD maps of the SS304 samples deformed at 900°C and 1000 °C

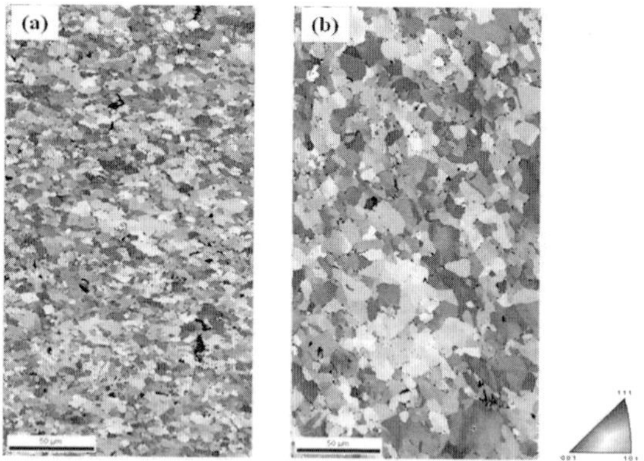

Fig. 3: EBSD images of deformed SS304 samples at strain rate of 0.01 s^{-1}(a) 900°C (b) 1000 °C.

Fig. 4 shows the EBSD maps of the starting alloy D9 material and the samples deformed at 1000 and 1100 °C at strain rate of 0.01 s^{-1}.

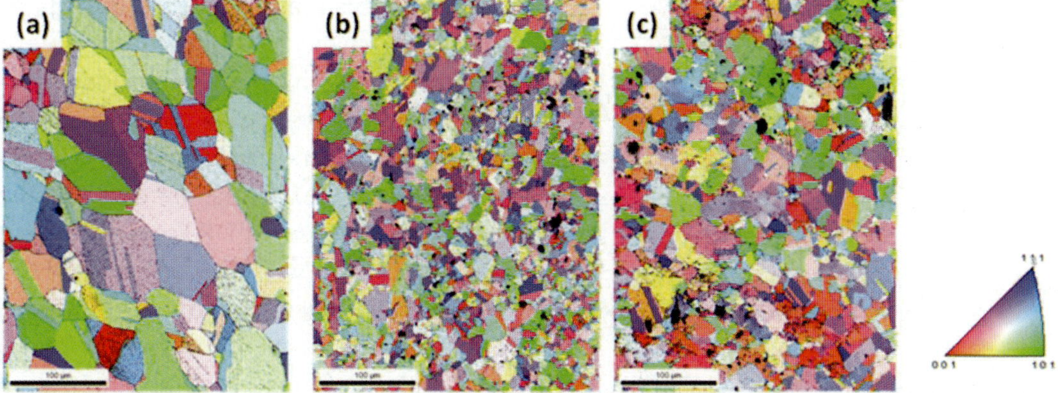

Fig. 4: EBSD images of (a) starting solution annealed alloy D9 and samples deformed at strain rate of 0.01 s^{-1} at (b) 1000°C (c) 1100 °C.

The starting solution annealed materials essentially comprised large grains (with an average diameter of about 35μm) along with large fraction of twin boundaries (Fig. 2 and 4a). It is seen from the Fig. 3a and 3b and 4 b and 4c that the initial structure is completely modified and the deformed microstructures consist of fine uniform grains. This is the typical characteristics of DRX. This type of microstructure modification and occurrence DRX during hot compression of steels has also been reported by Mandal et al. [4-7].

Avrami analysis of flow curves

Avrami relation [8-10] is a popular approach to study how solids transform from one phase (state of matter) to another at a constant temperature. It can specifically describe the kinetics of a phenomena which is governed by nucleation and growth processes. It is generally accepted that the progress in DRX also follows Avrami equation. However, the study of DRX in terms of Avrami equation is not straight forward. Various computational methodologies have been proposed to predict DRX and microstructure evolution such as Vertex models [11], Potts-type Monte Carlo [12], phase field models [13] and cellular automata [14]. Results obtained from these simulation studies provided a reasonable understanding of the DRX kinetics. However, it is always advantageous to investigate the evolution of DRX from experimental flow behaviour. Recently Jonas et al. [15] has developed a novel formalism to investigate the Avrami kinetics of DRX. They analyzed various flow curves of different steels using their method. Here in this study we apply the formalism of Jonas et al. to investigate the DRX kinetics of alloy D9.

In this approach, an experimental flow curve is considered to be the net result of the simultaneous operation of dynamic recovery and DRX. This has been illustrated by schematic curves in Fig 5.

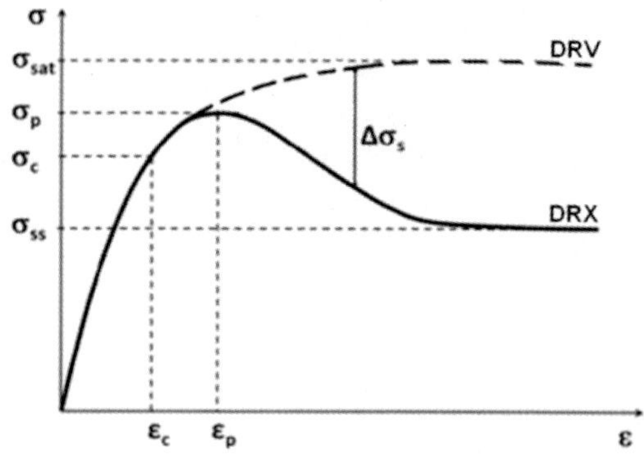

Fig. 5: Schematic representation of flow curves during dynamic recovery and dynamic recrystallization defining various stress and strain parameters involved in Avrami analysis.

It explains the difference between the features of flow curves obtained during operation of dynamic recovery and DRX. DRX curve is obtained experimentally by carrying out compression tests at different strain rates and temperatures. However, the recovery curve (σ_{DRV}) which is regarded as resulting from dynamic recovery alone is not available directly from the experiment. The recovery curve is assumed to represent the work hardening behavior of the unrecrystallized region. Now it is well known that during the deformation, DRX initiates after a certain critical stress (σ_c). The work hardening behavior of the recovery curve is same as that of the experimental curve prior to the critical strain (ε_c), where ε_c is strain corresponding to σ_c. The recovery curve finally reaches saturation (σ_{sat}) when there is a balance between the generation and annihilation rate of dislocation.

Identifying the ε_c and the corresponding σ_c is the first step toward DRX analysis. Flow curve analysis is now a well established technique for DRX analysis. The basis of the flow curve analysis method is that the DRX (as a restoration process) affects the flow curve shape by changing the rate of work hardening through the introduction of new strain free grains. To determine ε_c, we calculate the work hardening rate i.e. $\theta = d\sigma/d\varepsilon$ from the experimental data and plot against σ. The onset of DRX corresponds to a deviation in this work hardening curve. Fig 6a shows the typical plot of work hardening rate $\theta = d\sigma/d\varepsilon$ vs the flow stress σ for the D9 samples deformed at 1100°C at a strain rate of $0.002s^{-1}$. These work hardening rate curves consist of two parts. The initial rapid decrease of θ with σ corresponds to dynamic recovery. The change of slope of this curve at a certain σ is attributed to the initiation of dynamic recrystallization. Thus the change of slope in the work hardening curves is used to identify the critical stress σ_c and the corresponding critical strain ε_c. Poliak and Jonas [16] showed that the derivative of the work hardening rate ($-d\theta/d\sigma$) vs σ curve has a minimum at the inflection point of the work hardening curve. This is a better way to identify the σ_c. Inset in the Fig 6a show the plot of $-d\theta/d\sigma$ vs σ obtained from the experimental data.

In the approach of Jonas et al., the description of work hardening is based on the Estrin-Mecking [17] equation for evolution of dislocation density with strain:

$$\frac{d\rho}{d\varepsilon} = h - r\rho \tag{1}$$

Where, h is the athermal work hardening rate and r is the rate of dynamic recovery. Based on this equation Jonas et al. [15] derived the equation for flow stress for dynamic recovery as:

$$\sigma = (\sigma_{sat}^2 - (\sigma_{sat}^2 - \sigma_0^2)\exp(-r\varepsilon)\ ^{1/2} \tag{2}$$

where σ_0 is the yield stress and can be determined from the experimental flow curve in terms of a 2% offset in the total strain.

Using some simple algebraic substitution they also obtained a useful relation:

$$\sigma\frac{d\sigma}{d\varepsilon} = \sigma.\theta = 0.5r\sigma_{sat}^2 - 0.5r\sigma^2 \tag{3}$$

It can be seen from eqn (3) that r and σ_{sat} can be obtained respectively from the slope and intercept of $\sigma.\theta$ vs σ^2 curve.

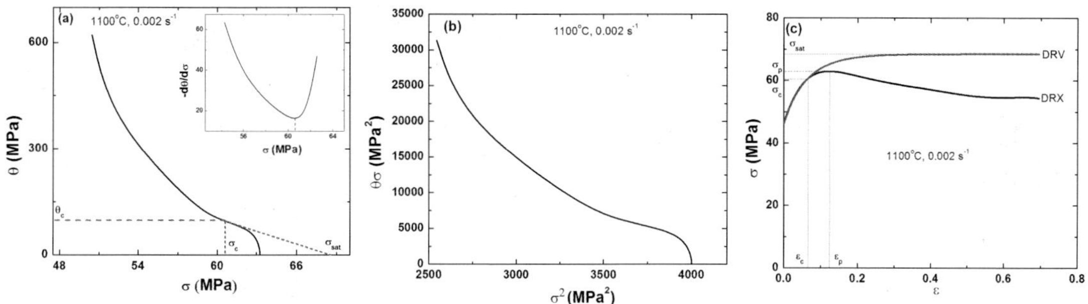

Fig. 6: (a) θ vs. σ plot, (b) $\theta.\sigma$ vs. σ^2 plot and (c) reconstructed dynamic recovery (DRV) curve plotted along with the experimental dynamic recrystallization (DRX) curve for alloy D9 deformed at 1100°C, 0.002 s^{-1}.

Fig 6b shows a typical plot $\sigma.\theta$ vs σ^2 for alloy D9 deformed at 1100°C at a strain rate 0.002 s^{-1}. Fig 6c shows the corresponding work hardening or recovery curve obtained using the r and σ_{sat} values along with the experimental DRX curve.

The recrystallized volume fraction is considered to be responsible for the difference between the σ_{DRV} and σ_{DRX} curves. The difference between these two curves ($\Delta\sigma_s$) is the net softening and is directly attributed to DRX. The maximum value of $\Delta\sigma_s$ is (σ_{sat} - σ_{ss}), where σ_{ss} is the steady state stress under DRX conditions.. The evolution fractional softening with strain is expressed as X= $\Delta\sigma_s/(\sigma_{sat} - \sigma_{ss})$. Once the recovery curve is derived for a particular deformation condition the evolution $\Delta\sigma_s$ with (ε-ε_c) can be obtained in a straightforward way.

The Avrami equation describing the SRX is expressed as:

$$X = 1 - exp(-kt^n) = 1 - exp\left(-0.693\left(\frac{t}{t_{50}}\right)^n\right) \tag{4}$$

Here X represents the recrystallized volume fraction, k is the Avrami constant and n is termed as the Avrami time exponent. t_{50} is the characteristic time corresponding to 50% recrystallized volume fraction and is specified as [15]:

$$t_{50} = A\varepsilon^{-p}\dot{\varepsilon}^q d_o^v \, exp\left(\frac{Q_{SRX}}{RT}\right) \tag{5}$$

Here A is a constant, d_0 is the grain size, p, q and v are the strain, strain rate and grain size exponents respectively and Q_{SRX} is the activation energy for SRX.

We apply the same formalism to investigate the evolution of DRX in SS304 and alloy D9. Avrami plots of X vs (ε-ε_c) are derived for different deformation conditions. Fig. 7 shows typical Avrami plots for SS304 deformed at 1200°C at a strain rate of 0.1 s^{-1}. Fig. 8 shows the Avrami plots for alloy D9 samples deformed at strain rate of 0.002 s^{-1} at temperatures 900°C, 1000°C and 1100°C.

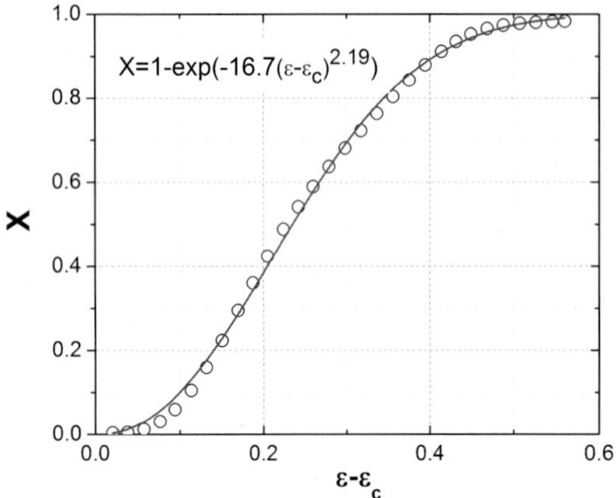

Fig. 7: Variation of fraction recrystallization w.r.t. (ε-ε_c) for SS304 deformed at 1200°C at a strain rate of 0.1 s^{-1}.

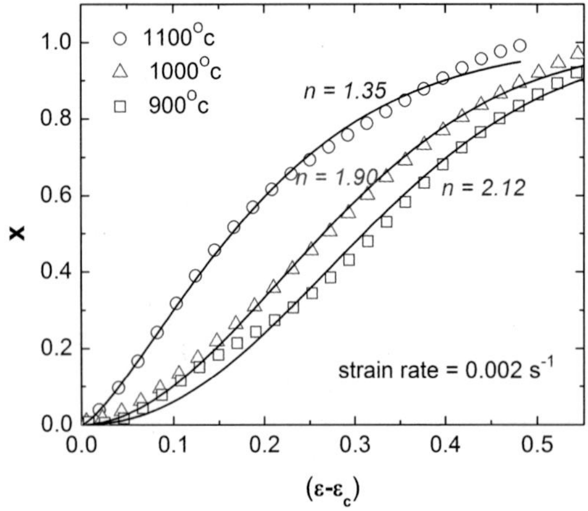

Fig.8: Variation of fraction recrystallization (X) w.r.t. (ε-ε_c) for alloy D9 at different deformation conditions.

The Avarmi time exponents (*n*) obtained from the slope of the curves are shown in the figures. It can be seen that the values of *n* lie between 1.35-2.12 and increases with decreasing temperature. The low value (~1) of *n* indicates heterogeneous site saturated nucleation of recrystallized grains. On the contrary *n* ~3, correlates with ideal Avrami recrystallization for homogeneous, constant rate nucleation [18]. The *n* values obtained in this study (1<*n*<3) signify that prior grain boundaries or twin edges are the nucleation cites for DRX in alloy D9. It is worth mentioning here that the detailed EBSD studies on the DRX behaviour of this steel by Mandal et al. [4-7] indicated heterogeneous nucleation at grain boundaries and formation of necklace structure. EBSD investigation also confirmed that twins play an important role during nucleation and subsequent expansion of the DRX process in alloy D9 [5].

In Fig. 9, the strain rate dependence of the characteristic time t_{50} for the flow curves determined at different temperatures is displayed for alloy D9. The value of the strain rate exponent *q* is ~ 1 and decreases with increasing temperature. The values of q are of the same order for that has been reported during the DRX of other fcc materials [15, 19] and are high compared to values reported in case of SRX [20].

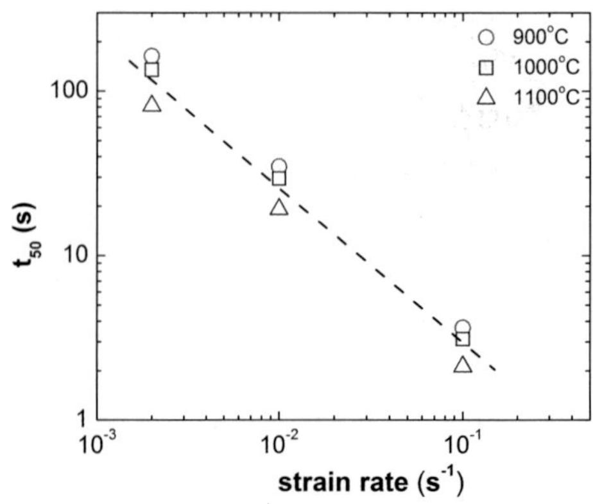

Fig. 9: Dependence of the characteristic time (t_{50}) on strain rate ($\dot{\varepsilon}$) for alloy D9.

The results of this analysis are also important as this will permit prediction of the DRX flow behavior at rates beyond that is achievable in laboratory test machines. This is significant from the industrial point of view for example to design the hot-forming process, predicting forming loads and power requirements at the relevant rates of deformation for alloy D9.

In order to generate the DRX flow curve at any desired rate of deformation it is required to establish the dependence of the parameters h and r on σ_p as has been suggested by Jonas et al. [15]. The various stress parameters associated with the DRX curve can be expressed in terms of σ_p using the experimental data. Figure 10 shows the variation of σ_{ss}, σ_{sat} and σ_c with σ_p for alloy D9.

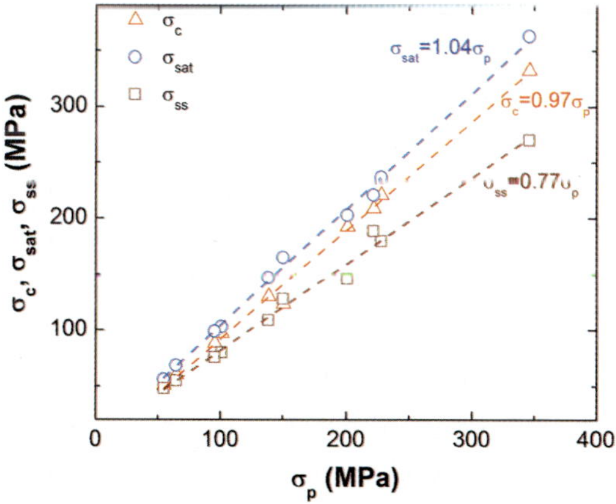

Fig. 10: Plots showing the variation of critical stress (σ_c) and steady state stress for dynamic recovery (σ_{sat}) and steady state stress for dynamic recrystallization (σ_{ss}) with peak stress (σ_p) for alloy D9.

A least square fit of the data gives the following relation between the stress parameters $\sigma_{ss} = 0.77\sigma_p$, $\sigma_{sat} = 1.04\,\sigma_p$, $\sigma_c = 0.97\,\sigma_p$. Similar relations have also been reported in other materials [15,21-24]. These relations will be useful for modeling purposes. Following Jonas et al. [15], the athermal work hardening parameter h can be determined using the relation:

$$h = r\frac{\sigma_{sat}^2}{(\alpha\mu b)^2} \qquad (6)$$

where α is a constant (taken as equal to 1) [15,25] and μ the shear modulus (81 GPa) [26] and b is the magnitude of Burgers vector (0.258 nm for alloy D9). Fig. 11 shows the dependence of r and h on σ_p for alloy D9.

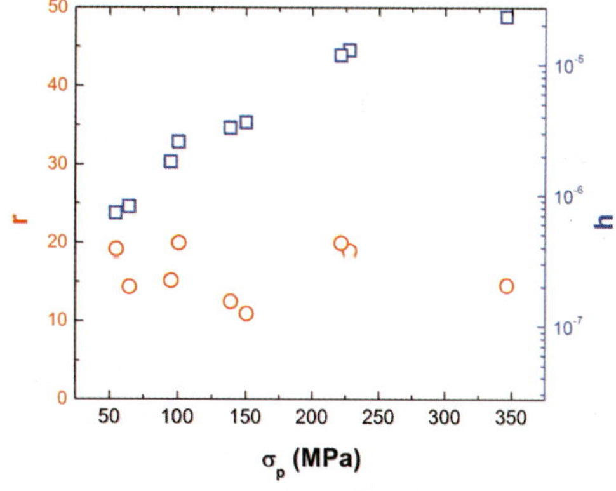

Fig. 11: Dependence of the dynamic recovery coefficient r and the athermal work-hardening parameter h on σ_p for alloy D9.

Now, σ_p can be related to T and $\dot{\varepsilon}$ by the well known relation:

$$Z = \dot{\varepsilon}\exp(Q/RT) = A'\left(\sinh(\alpha'\sigma_p)\right)^{n'} \qquad (7)$$

where $Z = \dot{\varepsilon}\exp(Q/RT)$ is the Zener-Holomon parameter and Q is an activation energy. A', α' and n' are constants derived from the nonlinear fit of the experimental data. Fig. 12 shows the fit of Eqn. (7) to the experimental data. The best fit gives the values: $A' = 4.0^{10}$, $\alpha' = 0.0078$, Q= 284 kJ/mol and $n' = 5.43$.

The constants thus evaluated can be utilized to determine σ_p for alloy D9 at desired T and $\dot{\varepsilon}$ using Eqn. (7). Once the dependences of *r*, *h* and the other parameters on σ_p is established and the constants in Eqn. (7) are determined, it is possible to construct the DRV curve and subsequently the DRX curve for the desired deformation conditions of industrial processing. The predicted and the experimental DRX curve for different deformation conditions are shown in Fig. 13.

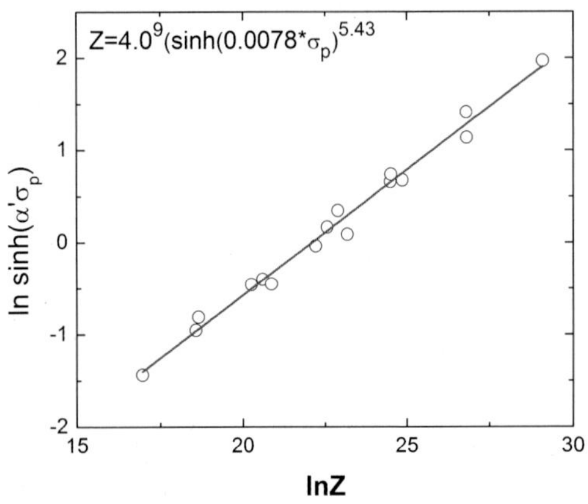

Fig. 12: Plot of $\ln(\sinh(\alpha'\sigma_p)$ vs. lnZ.

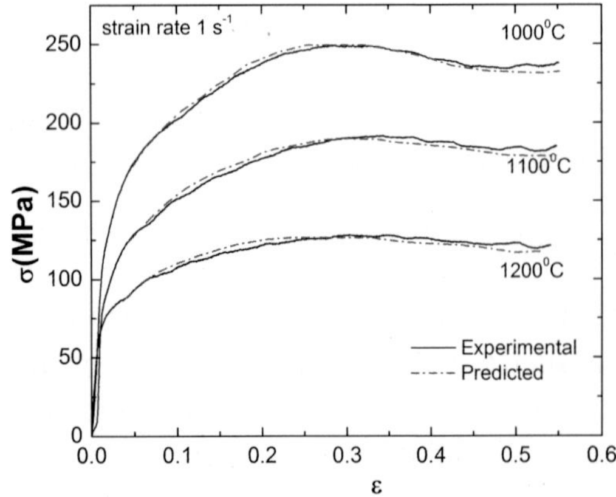

Fig. 13: Predicted and experimental stress-strain curves for strain rate 1 s^{-1} at different temperatures for alloy D9.

CONCLUSIONS

The DRX behavior of SS304 and Ti-modified austenitic stainless steel (alloy D9) are investigated under different deformation conditions through hot compression tests. The results of these tests are summarized as follows:

1. The stress–strain curves of SS304 and alloy D9 exhibited typical DRX behaviour with a single peak stress followed by a gradual fall towards a steady-state stress. At high temperature and low strain rate multiple oscillations were observed in the stress-strain curve.
2. The progress of DRX in the steels during hot deformation can be well predicted on the basis of the Avrami relation in conjunction with features of the flow curve and work hardening data.
3. The rate of DRX increases with deformation temperature. By increasing the deformation temperature, the recrystallization curve shifts to lower strains and recrystallization times.
4. The values of the Avrami exponent for different conditions ($1<n<3$) signify that prior grain boundaries or twin edges are the nucleation cites for DRX in these steels.
5. Flow curves for alloy D9 during hot working can be predicted well based on the analysis. These results will be useful to design the hot-forging/ extrusion/rolling processes for alloy D9.

REFERENCES

[1] T. Sakai, J. J. Jonas, Acta Metall., Vol. 32, 1984, p. 189.

[2] A.Sarkar, R. Kapoor, A. Verma, J.K. Chakravartty, A.K. Suri, J. of Nuclear Mater., Vol. 422, 2012, p. 1.

[3] J.K. Chakravartty, S. Banerjee, Y.V.R.K. Prasad, M.K. Asundi, J. of Nuclear Mater., Vol. 187, 1992, p. 260.

[4] S. Mandal, A.K. Bhaduri, V. S. Sarma, Metal. Trans. A, Vol. 43, 2012, p. 2056.

[5] S. Mandal, A.K. Bhaduri, V. S. Sarma, Metal. Trans. A, Vol. 43, 2012, p. 410.

[6] S. Mandal, A.K. Bhaduri, V. S. Sarma, Metal. Trans. A, Vol. 42, 2011, p. 1062.

[7] S. Mandal, P. V. Sivaprasad, V. S. Sarma, Mater. Manuf.Proc., Vol. 25, 2010, p. 54.

[8] M. Avrami, J Chem. Phys. Vol. 7, 1939, p. 1103.

[9] M. Avrami, J Chem. Phys., Vol. 8, 1940, p. 212.

[10] W. A. Johnson, R. F. Mehl, Trans. AIME, Vol. 135, 1939, p. 416.

[11] F.J. Humphreys, Scripta Metall. Mater., Vol.27, 1992, p. 1557.

[12] D. Raabe, Computational Materials Science: The Simulation of Materials Microstructures and Properties, Wiley-VCH Verlag GmbH, D-69469 Weinheim, 1998.

[13] L.Q. Chen, Y.Z. Wang, J. Miner., Met. Mater. Soc., Vol. 48, 1996, p. 13.

[14] N. Yazdipour, C.H.J. Davies, P.D. Hodgson, Compt. Mater. Sci., Vol. 44, 2008, p. 566.

[15] J. J. Jonas, X. Quelennec, L. Jiang, E. Martin, Acta Mater., Vol. 57, 2009, p. 2748.

[16] E. I. Poliak, J. J. Jonas, Acta Mater., Vol. 44, 1996, p. 127.

[17] Y. Estrin, H. Mecking, Acta Metall., Vol. 32, 1984, p. 57.

[18] J. W. Cahn, Acta Metall., Vol. 4, 1956, p. 449.

[19] C. Yue, L. Zhang, S. Liao, J. Pei, H. Gao, Y. Jia, X. Lian, Mater. Sci. Eng. A, Vol. 499, 2008, p. 177.

[20] M.R. Barnet, G. L. Kelly, P. D. Hodgson, Metal. Mater. Trans. A, Vol. 33, 2002, p. 1893.

[21] A. Dehghan-Manshadi and P. D. Hodgson, ISIJ International, Vol. 47, 2007, p. 1799.

[22] C. M. Sellars, Czech. J. Phys. B, Vol. 35, 1985, p. 239.

[23] N. D. Ryan, H. J. McQueen, Can. Metall. Q, Vol. 29, 1990, p. 147.

[24] A. Sarkar, J. K. Chakravartty, B. Paul, A. K. Suri, Phys. Status Solidi (a), Vol. 208, 2011, p. 814.

[25] X. Quelennec, N. Bozzolo, J. J. Jonas, R. Loge, ISIJ International, Vol. 51, 2011, p. 945.

[26] H.J. Frost, M.F. Ashby, Deformation-mechanism maps: The plasticity and creep of metals and ceramics, Pergamon Press, Oxford, 1982.

Comparison between Conventional Annealing and Rapid-Transformation Annealing of Low-Carbon Steel

A. Karmakar and D. Chakrabarti*

The Department of Metallurgical and Materials Engineering, Indian Institute of Technology (I.I.T.) Kharagpur, 721-302, Kharagpur, West Bengal, India.

*Email. debalay@metal.iitkgp.ernet.in

ABSTRACT

A comparative study between conventional annealing (CA) and rapid transformation annealing (RTA) of cold-rolled, low-carbon (microalloyed) steel having different starting microstructures showed that fine-ferrite grain sizes (4-6 μm) with hard-phases such as, martensite and bainite in RTA treated samples offered much higher strength (YS>800 MPa) and satisfactory elongation (e_t>15%) compared to the CA samples. RTA also resulted in better strain-hardening ability, absence of yield point phenomenon and comparable tensile toughness to CA.

KEYEYWORDS: *Cold-Rolling, Rapid Transformation Annealing (RTA), Conventional Annealing (CA), Low-Carbon Steels.*

INTRODUCTION

Search of high-strength, cold-rolled and annealed steels for automotive applications are driven by the potential reduction in car-weight, improvement in fuel-efficiency and crash-resistance. Conventional annealing (CA) of low-carbon (C ≤ 0.10) steels in batch-annealing furnace using slow heating rate (<10°C/s) and prolonged holding (> 1 hr.) at a temperature (600-700°C) below A_{c1} offers yield strength (YS) of 200-300 MPa [1]. Inter-critical annealing in continuous annealing furnace at ~730-830°C for short holding times (< 2 min), followed by water-quenching of cold-rolled, dual-phase grades increased YS to ~300-600 MPa [2, 3]. Ultra-fine grained (grain size < 3 μm) steels developed by severe plastic deformation offered a significant rise in YS to ~800 MPa, but hampered the strain-hardening ability and uniform elongation (e_u ≤ 5%) [4]. TRIP and TWIP steels on the other hand, show excellent combination of strength and ductility (YS of 400-600 MPa, UTS of 700-900 MPa, e_t ~ 25-40% [5], however, weldability of those high alloy grades remains an issue.

In order to improve the strength after conventional (recrystallisation) annealing different starting microstructures have been used before cold-rolling [6-8]. Due to the requirement of lesser rolling load and less chance of cracking, rolling of mixed ferrite-martensite structure is easier than that of fully hardened (i.e. martensitic) structure. Ferrite-martensite starting structure showed better tensile properties (YS of 370-470 MPa and uniform elongation, e_u of 16-22%) compared to ferrite-pearlite starting structure (YS of 270-330 MPa and e_u of 14-20% e_u) [7, 8]. Improvement of YS to ~ 470 MPa [8] was however, not to the level of strength obtained in ultra-fine grained, DP or TRIP steels [2-5].

In the present study, in search of higher strength, cold-rolled samples having different starting microstructures were subjected to rapid-transformation annealing (RTA) treatment. RTA is characterized by rapid heating (heating rate, R_H = 200-300°C) of cold-rolled steels above A_{c1}, followed by short-holding (5-60 s) and rapid cooling (cooling rate, R_C > 100°C) [9-11]. Much higher heating rate is used in RTA compared to the inter-critical annealing of DP or TRIP steels. Beneficial effect of RTA is well known [9-12], although comparative assessment between RTA and CA treatments has rarely been carried out in a systematic way and that makes the objective of the present study. Use of different starting microstructures also provided an additional opportunity for the improvement in mechanical properties and theoretical understanding of the RTA treatment.

EXPERIMENTAL DETAILS

Samples from a 6 mm thick hot-rolled strip containing 0.1 C, 0.33 Si, 1.42 Mn, 0.01 P, 0.003 S, 0.035 Al and 0.007 N (all wt%) has been reheated to 1150°C, soaked for 30 min and subjected to three different heat-treatments, namely furnace-cooling (FC), step-quenching (SQ) and intermediate-quenching (IQ). Details of those treatments are reported in [8]. Heat-treated samples were cold-rolled to 80% reduction and were annealed either inside a muffle furnace at 600°C for 1-10 hr. (CA treatment) or immersed inside a slat-bath at T_A = 850°C for t_A = 5-60 s (RTA treatment), before water-quenching. Selection of annealing temperature (T_A) was based on the dilatometric study carried out using BAHR®-Thermoanalyse (DIL 805A/D model) dilatometer. K-type thermocouple, which continuously monitored the sample temperature during annealing, showed a heating rate (R_H) of ~30°C/s for CA treatment and ~300°C/s for RTA treatment. Cross-section of the annealed strips were prepared following standard metallographic techniques and investigated by optical microscope, scanning electron microscope (SEM) and image analysis. Determination of average ferrite (α) grain size was based on the measurement of at least 500 grains in terms of equivalent circle diameter (ECD) size. Tensile specimens have been prepared following ASTM E-8M standard and tested in an Instron-8862 Universal Testing Machine (10 ton capacity) at room temperature (25°C) using a cross-head velocity of 1 mm / min.

RESULTS AND DISCUSSION

Microstructural Characterisation

FC, SQ and IQ heat-treatments on the as-received strip resulted in three different starting-microstructures namely, ferrite + pearlite, ferrite + blocky-martensite and ferrite + fibrous-martensite, respectively. Detailed discussion and characterization of the initial microstructure are reported earlier [8]. Microstructures of cold-rolled samples and cold-rolled and annealed samples are presented in Fig. 1 and Fig. 2, respectively. In case of cold-rolled ferrite + pearlite structure, increase in R_H from 30°C/s to 300°C/s resulted in ~30-40°C

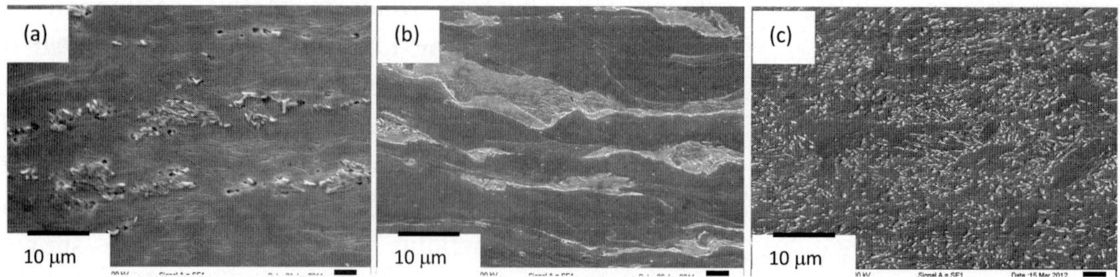

Fig. 1: Microstructures of (a) ferrite + pearlite, (b) ferrite + blocky-martensite and (c) ferrite + fibrous-martensite samples after 80% cold-rolling.

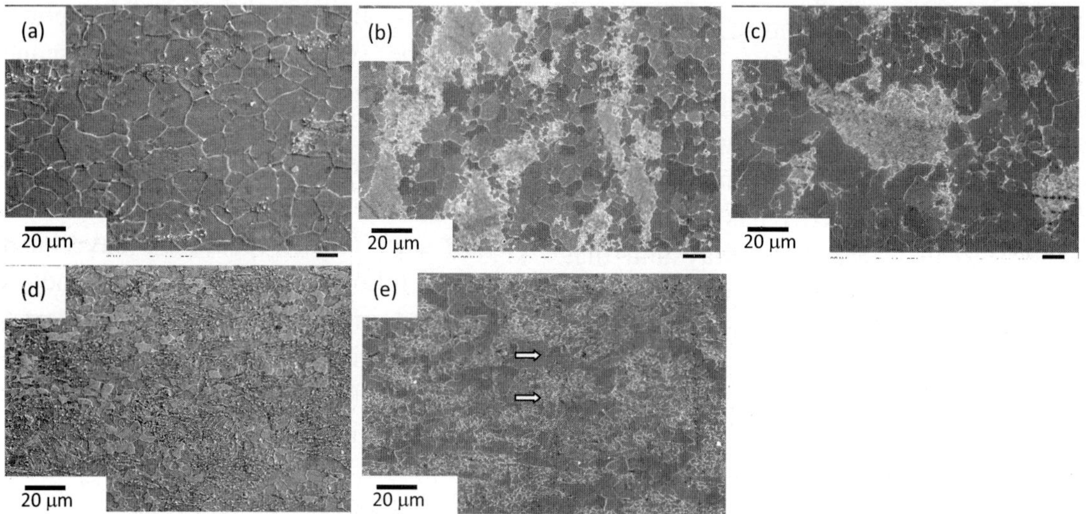

Fig. 2: Microstructures of cold-rolled and annealed samples for different starting structures: (a) ferrite + pearlite after conventional annealing (CA) at 600°C for 6 hrs. (b) ferrite + pearlite, (c) ferrite + blocky-martensite, (d) ferrite + fibrous-martensite after RTA at 850°C for 5 s, and (e) ferrite + fibrous-martensite after RTA at 850°C for 30 s. Coarse- and fine-grain regions in the bimodal grain structure in (e) are indicated by arrows.

rise in A_{c1} from ~750°C to ~780°C and A_{c3} from ~905°C to 930°C. Those values were much higher than the equilibrium transformation temperatures (A_{e1} ~ 680°C and A_{e3} ~ 860°C), predicted from the Thermo-Calc® software, which is expected [13-16]. Annealing temperature for RTA (T_A = 850°C), therefore, lies in the inter-critical temperature range.

Besides bending and thinning of cementite (θ) lamellae, dislocation pile-up against the lamellae in ferrite + pearlite structure can disintegrate them upon cold-rolling, Fig. 1a [17]. Cold-rolling of ferrite + blocky-martensite structure leads to heterogeneous plastic-flow with higher strain-accumulation inside the softer α-phase, Fig. 1b. Due to fibrous morphology with alternate layers of α and martensite, more homogeneous plastic deformation is expected in ferrite + fibrous-martensite sample, Fig. 1c. Being the supersaturated solution of C, annealing of deformed martensite results in the precipitation of finer θ-particles on the dislocation nodes, compared to that obtained after the fragmentation of pearlite in ferrite + pearlite structure [4, 6, 8, 17], Fig. 2a. Lower strain-accumulation in ferrite (α) and the presence of coarser θ-particle size can explain the coarser α-grain size obtained after CA in ferrite + pearlite samples, Table 1. More uniform strain-distribution and higher stability of the θ-particles resulted in finer α-grain sizes in ferrite + fibrous-martensite samples after CA treatment, than that in ferrite + blocky martensite, Table 1. Higher stability of θ-particles in ferrite + fibrous-martensite as reported earlier [8], compared to the other structures has been attributed to the partitioning of Mn during IQ heat-treatment, resulting in Mn enrichment within the θ-particles after cold-rolling and annealing [8].

Table 1: Comparison between CA and RTA samples in terms of average ferrite grain size ($D_α$) and tensile properties, for different starting microstructures.

CA	$D_α$ (μm)	YS (MPa)	UTS (MPa)	e_u (%)	e_t (%)
Ferrite+pearlite	14-17	320-356	368-396	8.1-11.3	26.3-28.7
Ferrite+blocky-martensite	7-9	382-436	429-463	18.8-21.0	25.5-34.4
Ferrite+fibrous-martensite	5.2-6.3	440-487	477-517	20.1-22.0	28.9-32.7
RTA					
Ferrite+pearlite	9-11.5	654-826	835-911	9.7-10.9	12.6-15.0
Ferrite+blocky-martensite	6-10	850-1080	1040-1269	8.0-10.6	15.7-19.7
Ferrite+fibrous-martensite	3.6-5.8	841-1210	995-1293	8.3-15.0	12.0-24.0

Microstructures obtained after the RTA was comprised of untransformed ferrite grains and a mixture of fine-ferrite, bainte and martensite (Fig. 2(b-e)), which formed during water-quenching of austenite (γ). With the increase in annealing time (t_A) from 5 s to 60 s, fraction of transformed phases increased continuously from 37% to 65% in ferrite + pearlite, from 28 to 78% in ferrite + blocky-martensite and from 28 to 65% in ferrite + fibrous-martensite structure. Presence of ~35%, ~5%, ~25% unrecrystallised α-grains in ferrite + pearlite, ferrite + blocky-martensite and ferrite + fibrous-martensite structures, respectively, after 5 s annealing and nearly complete recrystallisation after 60 s annealing indicated to the interaction between α-recrystallistaion and α→γ transformation, which is a characteristic feature of the RTA treatment [9-14]. α-Recrystallisation kinetics, γ formation kinetics and the interaction between the both depend on several factors, as discussed below.

Rapid heating is expected to slow-down the non-isothermal recrystallisation kinetics of deformed α compared to the slow-heating rate [14-16]. Higher the strain-accumulation in α, faster will be the recrystallisation kinetics and lesser will be the interaction, which holds true for Ferrite+blocky-martensite. Presence of fine-microalloy precipitates, such as, Nb(C,N) and V(C,N) and fine, θ-particles, which formed during reheating of deformed martensite, can also retard the α-recrystallisation and therefore, promote the interaction to take place [9-11].

Presence of unrecrystallised α-grains, on the other hand, may encourage the strain-induced α→γ transformation [9, 10]. Kinetics of γ formation will also depend on the available ferrite-carbide interface area, which is the nucleation site for γ [9, 15]. Rapid heating prevents the spheroidization of carbides in deformed ferrite + pearlite structure, and unspheroidized carbides transform to γ at a faster rate [15], which can explain the higher γ fraction observed in ferrite + pearlite at the initial stage of annealing. As Mn diffusion is much slower than C diffusion, higher Mn enrichment inside the θ-particles in $\alpha+M_f$ [8] could have delayed the dissolution of those particles during annealing, which resulted in slower rate of austenitisation in ferrite + fibrous-martensite, with respect to ferrite + blocky-martensite. Due to the higher heating rate and small dwell time used in RTA, θ-particles did not get time to distribute uniformly throughout the microstructure, resulting in coarse blocks of γ formed by the short-range C-diffusion, covering prior-pearlite or prior-martensite regions in ferrite + pearlite and ferrite + blocky-martensite samples, respectively, Fig. 2(b and c) [11, 15]. Fibrous martensite morphology in Ferrite + fibrous-martensite sample resulted in more uniform distribution of θ-particles and hence, γ islands during RTA, compared to the other starting structures, Fig. 2(d and e).

Etching the RTA treated samples in hot picric-acid solution showed the presence of fine γ-grain sizes (3-8 μm), which indicated to the increased γ-nucleation during rapid-heating [9-12]. Localised austenitization in ferrite + pearlite and ferrite + blocky-martensite samples also restricted the γ grain-growth as the γ grains impinged early. In ferrite + fibrous-martensite samples γ nucleated around the fine θ-particles (distributed quite homogeneously) and eventually consumed the particles during annealing, forming small γ-islands.

In general, γ transformation under the rapid-heating is known to be governed by para-equilibrium condition, which requires the diffusion of C without the diffusion of substitutional solutes (such as, Mn and Si) [9, 13, 15, 18]. Segregation of Mn and Si at the α / γ interface increased the hardenability of γ [9, 10, 13]. On the other hand, fine γ-grain size can reduce the hardenability. Fast-heating and short holding time could have been insufficient for the homogenization of coarse γ-blocks in terms of C, contributing to the formation of different phases (fine-ferrite, bainite and martensite) in ferrite + pearlite and ferrite + blocky-martensite structures after water-quenching [14]. Mixed (or bimodal) ferrite-grain structures has been observed in ferrite + fibrous-martensite sample, with coarse α-grains (6-10 μm) being present along with the ultra-fine α-grains (1-2 μm), Fig. 2e. Ultra-fine grain regions coincided with the regions richer in martenstic islands (i.e. prior γ-islands), which prevented the α-grains to grow [18]. Therefore, RTA treatment on cold-rolled samples of different starting microstructures can produce different microstructural morphologies.

The difference in average α-grain sizes after RTA for different starting microstructures, Table 1, can be explained in view of recrystallistaion-transformation interaction and the distribution of θ-particles and hence, γ-islands. The interaction was more severe in both ferrite + pearlite and ferrite + fibrous-martensite, than that in ferrite + blocky-martensite. However, more uniform distribution of fine γ-islands in ferrite + fibrous-martensite (than that in ferrite + pearlite) resulted in finest-α grain size in ferrite + fibrous-martensite by restricting the grain growth, Table 1. Beneficial effect of rapid heating has been explored by heating a 80% cold-rolled ferrite + pearlite sample to 850°C at a slow rate (R_H = 1°C/s) and water-quenching after isothermal holding for 30 s. The α-grain size obtained after slow heating (20 μm) was nearly two-times higher than that obtained after RTA, which can be attributed to the delayed α-recrystallisation and grain-growth under rapid heating. Higher transformed fraction (50%) in RTA sample compared to the slowly heated sample (40%) was an interesting observation and is under investigated. Such a difference could have resulted from the different conditions for austenite formation: para-equilibrium under fast heating and ortho-equilibrium under slow heating.

Evaluation of mechanical properties:

Typical engineering stress-strain curves of the CA and RTA samples are shown in Fig. 3. As the time and temperature used in RTA and CA were entirely in different range, Hollomon-Jaffe parameter (HJ-Parameter [8]) has been used to compare their strength, Fig. 4a, and ductility, Fig. 4b, using the same scale along the abscissa. With respect to CA treatment, RTA offered much higher YS and UTS (~400 MPa higher) due to fine α-grain size and presence of harder phases (bainite and martensite) in the microstructure [9-12]. Taking account of the high-strength of RTA samples, average e_u~10% and average e_t~20% can be considered as satisfactory, Table 1. The scatter in mechanical properties in Fig. 4(a and b) resulted from the use of different microstructures before cold-rolling. Comparing between different starting structures, the best tensile properties obtained in α+M_f samples (Table 1) can be attributed to finest α-grain size along with finest θ-particle size in case of CA samples

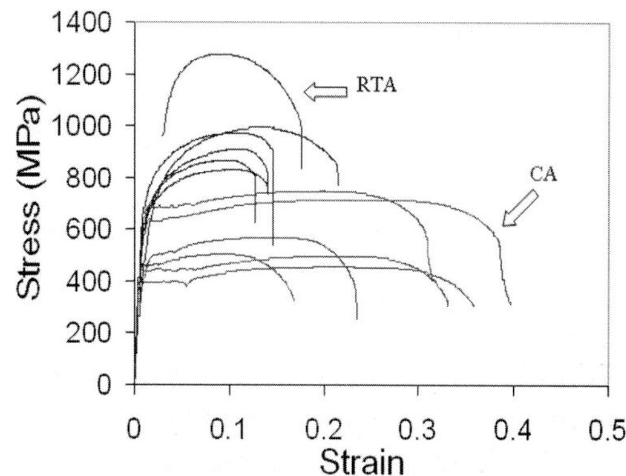

Fig. 3: Engineering stress-strain curves of the selected CA samples (annealed at 600°C for 6 hrs.) and RTA samples (annealed at 850°C for 15-30 s) for different starting microstructures.

and more uniform distribution of small martensitic islands in case of RTA samples.

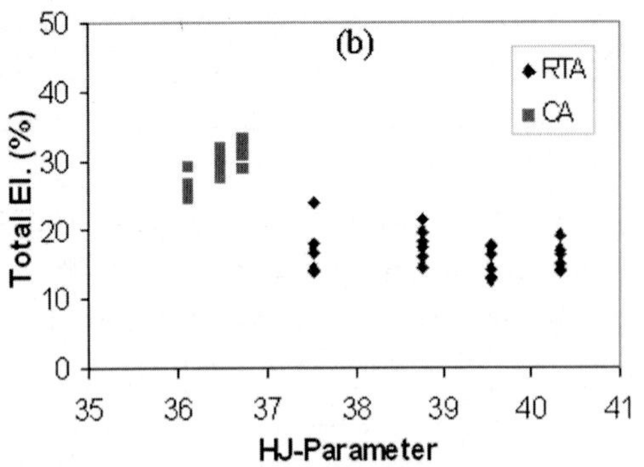

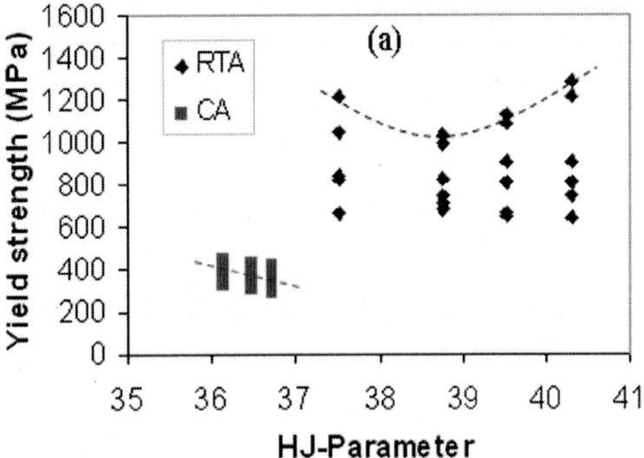

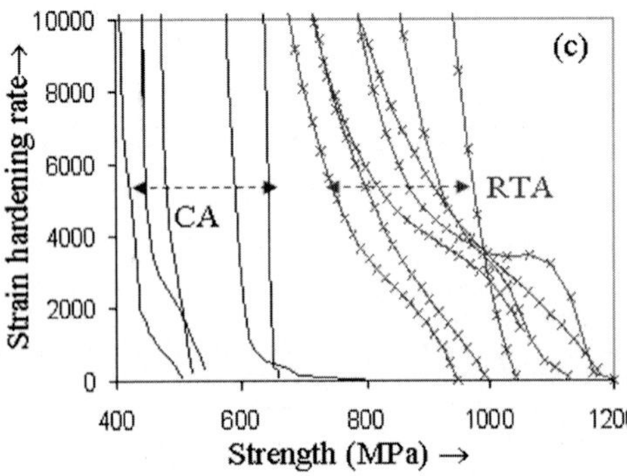

Fig. 4: Variation in (a) yield strength and (b) total elongation of RTA and CA samples with HJ-parameter. (c) Comparison between the RTA and CA samples in terms of strain-hardening rate.

Decrease in strength and increase in elongation of CA samples with the increase in HJ-parameter (Fig. 4(a, b)) can be related to the increase in recrystallistallised fraction and coarsening of the θ-particles. Change in the mechanical properties of RTA samples was controlled by a more complex interaction between α-recrystallistaion and austenite (γ) formation. Initial decrease in YS (and also UTS) with the increase in the annealing time to 15 s, as indicated by the dotted lines in Fig. 4a can be attributed to the softening of α-grains by recrystallisation. Subsequent rise in strength with further increase in annealing time till 60 s could be dictated by the increase in γ fraction, which contributed to the increased fraction of hard phases (bainite and martensite).

Fine α-grain size (≤ 5 μm) not only contributed to ~350 MPa as grain boundary strengthening but also can ensure decent ductility by activating more number of slip-systems during plastic deformation, compared to the coarser grains. Bimodal α-grain structures with ultra-fine grains (Fig. 4e) can also provide an ideal combination of strength and ductility [7]. Hard phases such as, bainite and martensite not only increased the strength, but also improved the strain-hardening ability of RTA samples, with respect to CA samples, Fig. 4c, by generating geometrically necessary dislocations in ferritic regions adjacent to the hard-phases [19]. As a result, RTA showed higher yield to tensile strength ratio (0.75 ± 0.08) than that in CA (0.87 ± 0.03), which makes RTA samples a better choice from the formability point of view. Formation of geometrically necessary dislocations in RTA samples also eliminated the yield point phenomenon, which can be in seen in the stress-strain curves of CA samples, Fig. 3. If the yield point elongations (4-7%) are not considered, then the amount of homogeneous plastic strain in CA and RTA samples become comparable, Fig. 3 and Table 1. Average tensile toughness, represented by the area under the stress-strain curves, for the RTA samples (160 ± 51 MPa) is similar to that for the CA samples (167 ± 65 MPa). Future studies will compare the CA and RTA treated samples in terms of impact toughness and fracture toughness.

In spite of several advantages of the RTA treatment over CA treatment with respect to the microstructure and tensile properties, the industrial applicability of RTA treatment and its effect on flatness and surface quality of cold-rolled sheets remains a challenging task. Although, recent interest on flash processing and rapid annealing provide a ray of hope [14], future research and industrial projects need to concentrate on this aspect.

ACKNOWLEDGEMENT

Council of Scientific and Industrial Research (CSIR), New Delhi for the financial support.

CONCLUSIONS

In summary, comparative assessment between conventional annealing (CA) and rapid transformation annealing (RTA) treatments on cold-rolled samples of different starting microstructures showed that RTA samples offered higher strength (YS>800 MPa) along with better strain-hardening ability, absence of yield point phenomenon and comparable tensile toughness, with respect to the CA samples (YS<500 MPa). Amongst different starting microstructures, ferrite + fibrous-martensite structure showed the best combination of strength and ductility after RTA.

REFERENCES

[1] P. Ghosh, B. Bhattacharya, R. K. Ray: Scripta Met. 56 (2007) 657-660.

[2] S. G. Chowdhury, E. V. Pereloma, D. B. Santos: Mat. Sci. and Eng. A 480 (2008) 540-548.

[3] R. O. Rocha, T. M. F. melo, E. V. Pereloma, D. B. Santos: Mat. Sci. And Engg. A 391 (2005) 296-304.

[4] R. Song, D. Ponge, D. Raabe, J. G. Speer, D. K. Matlock: Mat. Sci. and Eng. A 441 (2006) 1-17.

[5] A. K. Srivastava, D. Bhattacharjee, G. Jha, N. Gope, S. B. Singh: Mat. Sci. And Eng. A 445-446 (2007) 549-557.

[6] R. Ueji, N. Tsuji, Y. Minamino, Y. Koizumi: Acta Materialia 50 (2002) 4177-4189.

[7] H. Azizi-Alizamini, M. Militzer, W. J. Poole: Scripta Mater. 57 (2007) 1065-1068.

[8] S. M. Hasan, A. Haldar, D. Chakrabarti: Mat. Sci. Technol., Accepted paper, (2012), DOI: 10.1179/1743284711Y.0000000113.

[9] C. Lesch, P. Alvarez, W. Bleck, J. Gil Sevillano, Metall. Mater. Trans. 38A (2007) 1882–1890.

[10] P. Alvarez, C. Lesch, W. Bleck, H. Petitgand, J. Schöttler, J. Gil Sevillano, Mater. Sci. Forum 500–501 (2005) 771–778.

[11] V. Massardiera, A. Ngansopa, D. Fabrèguea, J. Merlin, Mater. Sci. Eng. A 527 (2010) 5654–5663.

[12] R. A. Grange, Metallurgical transactions. 2 (1971) 65-78.

[13] V. Andrade-Carozzo, P. J. Jacques: Mat. Sci. Forum 539-543 (2007) 4649-4654.

[14] T. Lolla, G. Cola, B.Narayanan, B. Alexandrov, S. S. Babu: Mat. Sci. and Technol. 27 (2011) 863-875.

[15] J. Huang, W. J. Poole, M. Militzer: Metall. Mater. Trans. A, 35 (2004) 3363-3375.

[16] F. G. Caballero, C. Capdevilla, C. Garcia De Andres: ISIJ Int. 43 (2003) 726-735.

[17] K. Hono, M. Ohnuma, M. Murayama, S. Nishida, A. Yoshie and T. Takahashi: Scripta Mater. 44 (2001) 977- 983.

[18] H. Azizi-Alizamini, M. Militzer, W. J. Poole: ISIJ Int. 51 (2011) 958-964.

Microstructure, Deformation and Fracture Behavior of Cr-Mo-V Steels

C. Gupta#, J.K. Chakravartty and S. Banerjee
Materials Group, Bhabha Atomic Research Centre, Trombay, Mumbai - 400086
#On Sabbatical as JSPS Post-Doctoral Fellow to 3D/4D Material Science Lab Toyohashi University of Technology, Japan

ABSTRACT

The paper reports the uniaxial deformation and fracture behavior of two Cr-Mo-V steels used for nuclear structural components. The two steels are – a new member of low alloy steel in the 3Cr-1Mo class and a plain tempered martensitic stainless steel containing 12%Cr. The microstructure investigations were carried out on the 3Cr-1Mo steel by establishing the continuous cooling transformation diagram using dilatometry complemented by optical microscopy studies. The nature of continuous cooling transformation diagram showed typical characteristics of bainitic transformation over wide range of cooling rates for the 3Cr-1Mo steel. The deformation behavior of the two steels were characterized using a series of tensile tests carried out in the temperature range 25 – 600°C and over a wide range of strain rate. The characteristic anomalies observed from the temperature and strain rate dependence of strength and ductility parameters obtained from the tensile tests provided indication to the presence of Dynamic strain aging (DSA) phenomenon in both steels. It was found that the embrittling features of DSA phenomenon was rather innocuously manifested in 3Cr-1Mo steel with bainitic structure as compared with the 12%Cr steel with tempered martensitic steel. The extent of reduction in fracture resistance of the tempered martensitic steel in the DSA regime was characterized in the J_{IC} format using high temperature fracture tests. The studies revealed that a reduction in fracture resistance in the DSA regime occurred primarily due to the presence of cracking instabilities during the fracture tests. The difference in the manifestations of DSA in the two steels was rationalized on the effectiveness of the interstitial impurities to interact with dislocations in the martensitic and bainitic microstructure.

KEYWORDS: *Cr-Mo-V Steels, Dynamic Strain Aging, Bainite, Martensitic Steels Deformation, Fracture Toughness*

INTRODUCTION

Ferritic steels are well known materials of construction of various structural components in nuclear reactors [1,2]. Reactor pressure vessels and reactor support structures in the PWR's and BWR's, heat transport piping and end fitting in PHWR's are some of the important examples of applications where ferritic steels have been employed as component materials. There exists a great concern over the influence of embrittlement phenomena on the ability of the material to discharge their functional duties during the lifetime of the critical components in use in reactors. This has grown a fertile area of research dealing with the influence of embrittlement due to thermal, irradiation and environmental sources on the components [1,2].

The presence of plastic instabilities is known to influence the performance of structural materials adversely [1,3]. These steels possess nano/ultrafine grained microstructures with excellent combinations of strength and ductility[3,4]. A related but distinctly different phenomenon is the Portevin Le Chatelier (PLC) effect, which also has been known to occur in structural steels. It was first reported by Le Chatelier in 1909 and subsequently by Portevin and Le Chatelier in 1924 [3]. Estrin and Kubin [3,4] have classified the various types of plastic instabilities into those that are caused when the work hardening has been suppressed and those that result from strain rate softening. The Lüders band formation during the occurrence of yield point phenomena belongs to the former while the PLC phenomena to the latter. The PLC effect has been regarded as a particularly insidious manifestation of the dynamic strain aging (DSA) phenomena [3]. The physical origin of DSA phenomena is due to the dynamic interaction of diffusing solute atoms with moving dislocations [4,5]. The manifestation of DSA in several alloy systems such as Fe, Ti, Al, Au, Zr alloys has been studied [4-10]. These studies convey that despite the early knowledge and investigations into the occurrence of DSA, the manifestations of PLC depend on the alloy system undergoing DSA. As a result there is a sustained need for characterizing the manifestation of DSA in the either newly developed alloys destined for structural applications or in those materials being used structural components, where it has not been studied as yet. Further, the influence of the microstructure on the manifestations of DSA in ferritic steels has not received serious attention. Hence the scope of investigations includes the study of relative manifestations of DSA in two different structural steels with nearly similar interstitial impurity content but with significant differences in alloy content and microstructure produced from $\gamma \rightarrow \alpha$ shear transformation in steels. Such a comparison is expected to elucidate the influence of microstructure on the on the extent of degradation of tensile and fracture properties due to presence of DSA in structural steels.

EXPERIMENTAL

The materials used for the investigations are a new member of 3Cr-1Mo class of low alloy steel and a plain 12Cr tempered martensitic steel. This 3Cr-1Mo type steel is a new generation nuclear reactor pressure vessel material with optimized composition to mitigate temper and irradiation embrittlement [1]. The AISI 403 steel is used for end fitting applications in pressurized heavy water reactor. The respective chemical compositions of the two steels are given in table 1. A deformation dilatometer Bahr DIL 805 was operated in the quenching mode for the dilatometry studies. Hollow tubular samples of outside diameter (O.D) 4.0 mm, 250μm thickness and 10 mm length were used for the tests. The sample was placed between two alumina rods, one of which was connected to a LVDT with less than 1μm resolution for measuring the change in length during the phase transformations. A thermocouple was spot welded to the O.D of the sample and it was inductively heated to the austenitic temperature. The heating rate of 100°C/min was employed for all tests for carrying the austenitising of the sample at 1100°C for 10 minutes. For the purpose dilatometry studies a series of samples were cooled over the cooling rate range 1075°C/min to 2.2°C/min. The sample change in length (ΔL) and temperature (T) were acquired online by a 16 bit DAQ during the entire heat-hold-cool cycle.

Table 1: Chemical composition of the steels in %wt

element	C	Cr	Mo	V	Mn	Ni
AISI 403	0.11	12.0	-	-	0.5	-
3Cr-1Mo type	0.15	3.0	0.7	0.28	0.4	0.7

The tensile tests were carried out using a floor model of instron make universal testing machine. Uni-axial tensile tests comprised of monotonic pull as well strain rate change tests over the temperature range 25 – 600°C. The tensile parameters of strength and ductility were derived from the digital data acquired during the tests. A three zone furnace was used to heat the sample, the temperature of which was monitored using a K type thermocouple attached to sample guage. The high temperature fracture tests were carried out in J_{IC} format using 1 inch thick fatigue pre-cracked compact tensile samples with a/W ratio of 0.6. The unloading compliance technique as well as DCPD technique was used for characterizing the fracture toughness as a function temperature. For the high temperature tests the crack length was monitored using both COD guage as well as using the direct current potential technique. The cracklength conversion from the measured potential drops was estimated from the Johnson's formula. A custom made furnace was made to accommodate the COD guage arms and DCPD current and potential leads inside the furnace. The electronics of the COD guage was air cooled.

Results and Discussion

The dilatometry studies on 3Cr-1Mo type steel yielded the temperature range of austenite transformation as a function of the prescribed cooling rates from the non-linear regions in the change in length with temperature plots. The transformation start and fininsh points plotted on the loci of the cooling curve in a semi-log temperature – time axis provided the the CCT diagram of the 3Cr-1Mo steel. The CCT diagram of the 3Cr-1Mo steel is shown in figure 1. It is clearly seen that a single bay in the transformation is seen to occur over a wide range of cooling rates. At slower cooling rates the austenite decomposition occurs at much higher tempertaure into two products with clear demarcated regions in the temperature – time space.

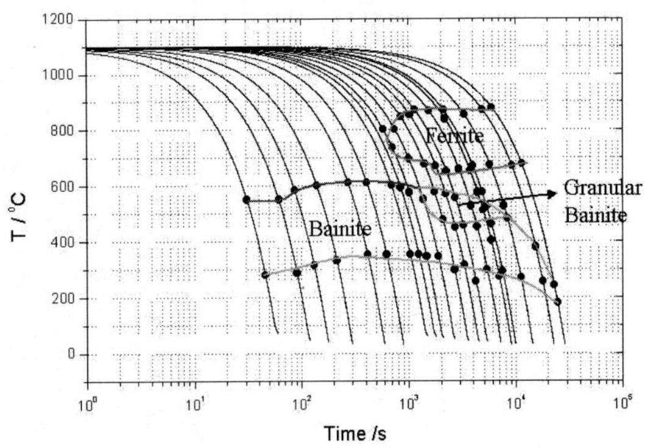

Fig. 1: CCT diagram of the 3Cr-1Mo type steel.

The optical microstructures of the samples exposed to selected cooling rates are shown in figure 2. At the faster cooling rates above 35°C/min the microstructure appears similar to the bainitic microstructure typically found in low alloy steels. Further the CCT diagram showing a single bay with transformation start temperatures to be insensitive to the cooling rates is typical of non-isothermal transformation characteristics of bainitic transformation seen in many steels [11]. At the slower cooling rates the optical microstructure shows that formation of pro-eutectoid ferrite (typical C- curve characteristics in the CCT diagram) followed by the formation of granular bainite. Within this region, the high resolution of the dilatometer detected multiple transitions below the bainite transformation start temperatures. Further investigations using interrupted quench studies need to be carried out establish the origin of these transitions. In this two phase region when the primary ferrite formation precedes the bainitic transformation, the transformation start temperatures for the granular bainite decreases with slower cooling rate.

Thus the overall characteristics of the CCT diagram and optical microscopy studies clearly establish that the dominant microstructure over a wide cooling rate is predominantly bainitic. This view is consistent with other microstructural studies on low alloy steels with similar alloy content for reactor pressure vessel applications[11]. The Bainite microstructure is a multi-

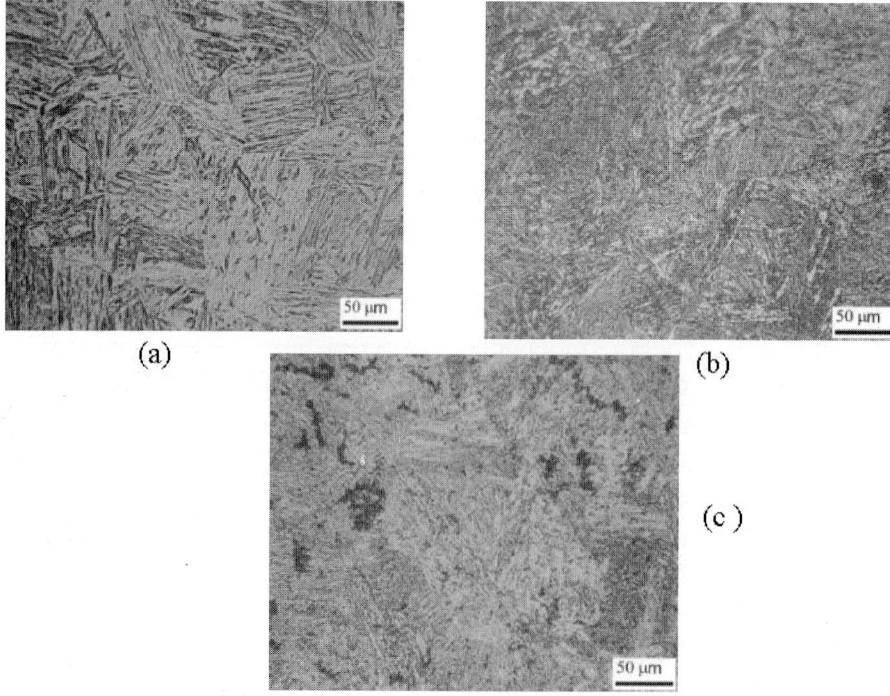

Fig. 2: Optical microscopy of the dilatometry sample cooled at (a) 1080°C/min (b) 108°C/min (c) 43°C/min

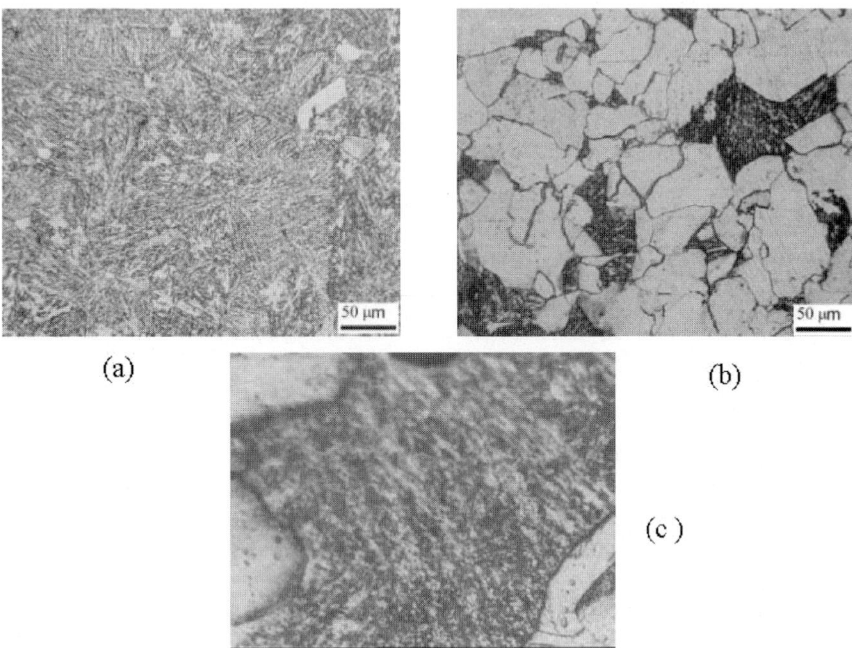

Fig. 3: Optical microscopy of the dilatometry sample cooled at (a) 12°C/min (b) 6.6°C/min (c) A magnified picture granular bainite (500x).

component system with unique morphology of the bainitic ferrite having typical spatial orientation relationships with carbides, and austenite. TEM studies are envisaged to reveal these important microstructural characteristics in the bainitic structure of the new steel.

The tensile results for the two steels carried out are summarized in the figure 4. The temperature variation of strength and ductility parameters for the respective steels has been plotted. It is seen that for the AISI 403 there is a sharp increase in the tensile strength values accompanied by a decrease in ductility and stasis in the temperature dependence of yield strength in the range 250 – 400°C. In comparison it can be seen that the 3Cr-1Mo type steel displays no such increases in the tensile strength while undergoing similar reductions in the ductility parameters as AISI 403.

It is to be noted however that, within the temperature ranges 250 – 400°C the uniform ductility for the 3Cr-1Mo type steel improves before sharply reducing at higher temperatures. The total ductility contrastingly shows significant upswing at temperatures beyond 400°C for both steels. The occurrence of serrated flow occurred in both steels a typical example is shown in figure 5 for AISI 403. It needs to be mentioned however that the in-homogenous flow was quite damped in case of 3Cr-1Mo steel, with no clear delineation of critical strain.

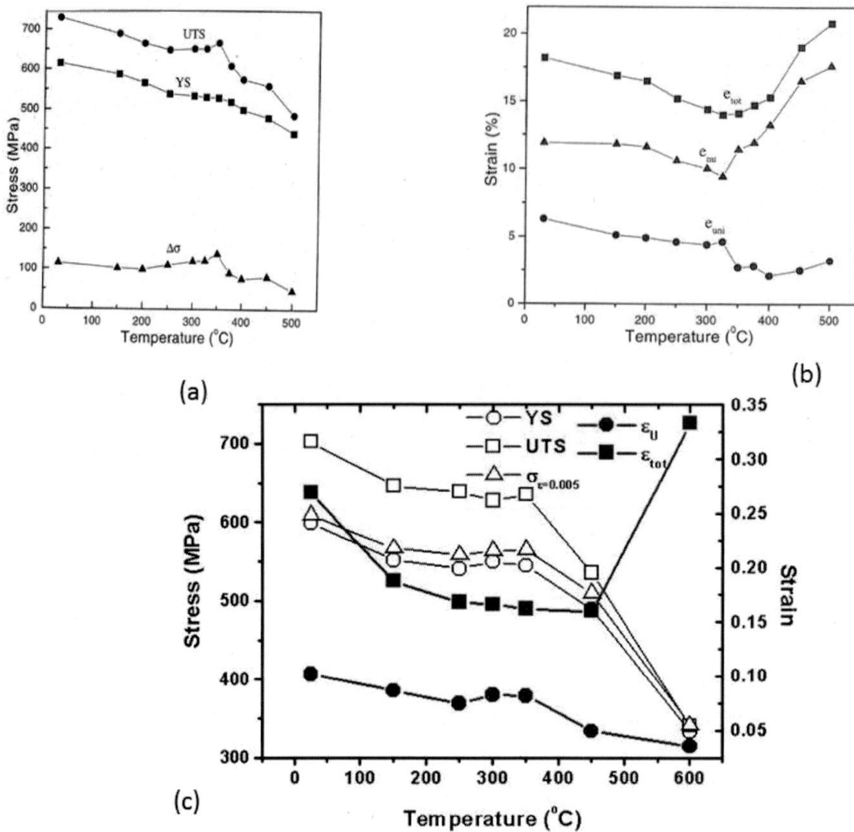

Fig. 4: Variation of (a) UTS and YS (b) elongation (total, uniform and non-uniform) with test temperature for AISI 403. (c) Variation of UTS, YS, total and uniform elongation with temperature for 3Cr-1Mo type steel.

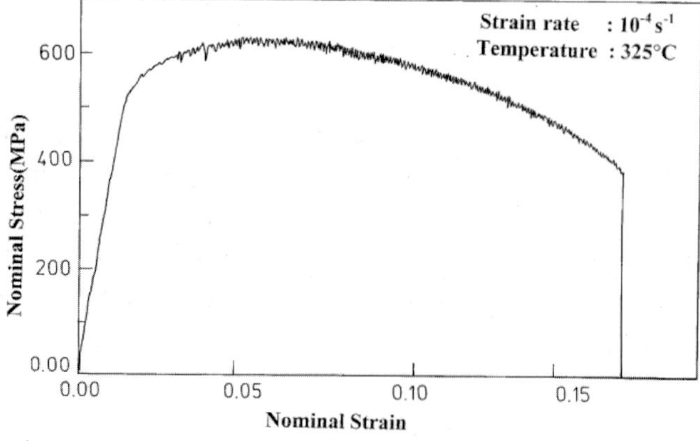

Fig. 5: Stress-strain plot of AISI 403 showing serrated flow.

The strain rate change tests carried out on both the steels enables characterization of the strain rate sensitivity as a function of temperature. The results are shown in figure 6. It can be clearly seen that both the steels show a reduction in the strain rate sensitivity towards negative values within the range 250 – 400° C.

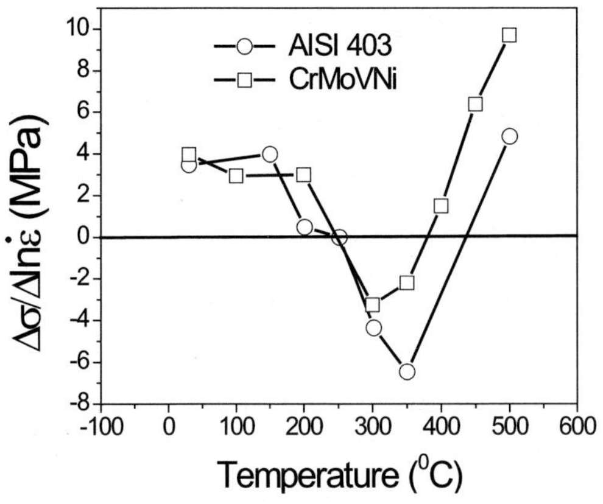

Fig. 6: Plot comparing the strain rate sensitivity variation with temperature for the two steels.

It is to be noted that the SRS values sharply decline to lower values for the AISI 403 steel than the 3Cr-1Mo type steel. The comparison of the tensile results between the two steels can summarized by

the temperature dependence of the ratio $\frac{\sigma_{UTS} - \sigma_{Y}}{\sigma^{o}_{UTS} - \sigma^{o}_{Y}}$, which gives an idea of the extent of work hardening with the respect to that at the ambient conditions. The ductility parameters has been compared in terms of the ratio $\frac{\varepsilon^{1}_{Uni}}{\varepsilon^{2}_{Uni}}$, which gives an idea of the relative magnitudes of uniform ductility obtained from tensile tests carried out on AISI 403 and Cr-Mo-V steels. The plot is shown in the figure 7. It can be seen from the figure that within the DSA temperature regime, while there is dramatic increase in $\frac{\sigma_{UTS} - \sigma_{Y}}{\sigma^{o}_{UTS} - \sigma^{o}_{Y}}$ ratio to levels even greater than that at room temperature for AISI 403, the ratio for 3Cr-1Mo type steel shows a gradual decrease with increase in temperature.

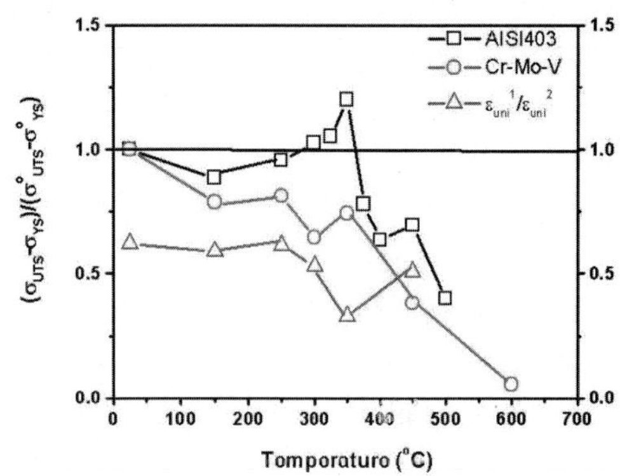

Fig. 7: Comparing the temperature variation of the ratios pertaining to strength and ductility for the two steels.

These trends imply that work hardening for AISI 403 with martensitic structure is significantly higher than that of 3Cr-1Mo type steel with baintic structure. A comparison of ductility ratio shows that it is less than unity at all temperatures, signifying that uniform elongation is smaller for the AISI 403 as compared with the 3Cr-1Mo type steel. It is to be noted that in the DSA temperature regime the ductility ratio decreases and reaches minimum at the temperature where the ratio $\frac{\sigma_{UTS} - \sigma_{Y}}{\sigma^{o}_{UTS} - \sigma^{o}_{Y}}$ for AISI 403 is the maximum. Thus the maximum difference in the ductility between the two steels occurs at the temperature where the work hardening for AISI 403 is the highest. As the uniform ductility for 3Cr-1Mo type steel remains relatively constant in the temperature range 250 – 350°C (Fig. 4(c)) it clearly implies that there has been greater embrittlement of AISI 403 as compared with 3Cr-1Mo type steel. These results bring out the fact that the AISI 403 steel have been more susceptible to extreme manifestations of DSA as compared with 3Cr-1Mo type steel. This has resulted in the greater deterioration of ductility in the temperature ranges where DSA was observed to occur for AISI 403 as compared with that of 3Cr-1Mo type steel.

The tensile results also show that the uniform elongation and total elongation curves diverge sharply beyond the DSA temperature range (figure 4). At the temperatures, where the SRS was found to be positive the contribution of non-uniform elongation to the total ductility of the steels is observed to be higher. This has been reported earlier in steels [12] and could be rationalized in the following way. It is well known that the attainment of tensile strength is controlled by both work hardening rate and strain rate hardening. However, only the latter remains active beyond UTS. As SRS controls the extent of strain rate hardening. The contribution of non-uniform elongation to the total elongation is directly related to its magnitude and sign of SRS. When it is negative, (i.e. serrated flow behavior) localization of deformation in the form propagative bands (eg. Lüders bands) and increase in work hardening promote development of local zones of damage, leading to decrease in both the ductility parameters. Beyond the DSA temperatures, a positive SRS would increase the strain rate hardening over that of work hardening rate. This is further exacerbated by processes such as dynamic recovery that promotes work softening [12]. While, these factors results in reducing the uniform elongation by inducing premature necking, they would be beneficial for non-uniform elongation, leading to the observed divergence between total and uniform elongation parameters as observed in figure 4.

In order to explore the extent of embrittlement of AISI 403 high temperature fracture toughness tests were carried out in the temperature range where serrations were observed in the steel. The strain rate - temperature domain of the occurrence of serrations in AISI 403 are shown in the figure 8, superimposed in which are the test

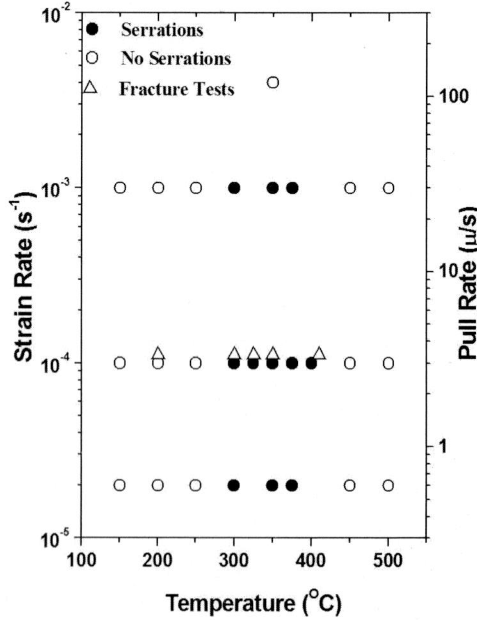

Fig. 8: Plot showing the temperature – strain rate combinations where serrated flow (marked as filled circles) has been found in the tensile tests in AISI 403. The corresponding pull rates have also been shown.

conditions where the fracture toughness tests were carried out. The calculation of the effective crack tip strain rate was carried out from formula derived considering fully plastic solutions at crack tip [13] and was estimated as $3.8x10^{-3}$ s^{-1} for the applied pull rate shown in figure 8.

The load displacement curve and the corresponding DCPD signal is shown in figure 9 for two pull rates applied during the fracture toughness tests. It is clearly seen that the crack growth stage of the test (which occurs after the maximum load) shows perturbations in load and DCPD signal for the lower pull rate. This pull rate was within the DSA temperature – strain regime of AISI 403 where serrated flow was observed. On increasing the pull rates by a factor of 10, such perturbations in the load and DCPD signal was not observed. The estimation of the elastic plastic fracture toughness parameters from the fracture test from the unloading compliance method and

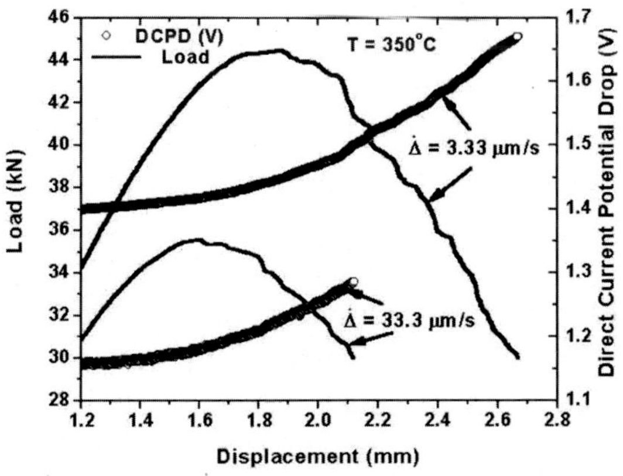

Fig. 9: An enlarged portion of the load displacement plot obtained from the DCPD tests at 350°C and pull rate of 3.33 miron/s and 33.3 micron/s. The corresponding DCPD signal is also shown for each test.

DCPD technique enabled the temperature variation of J_{IC} and dJ/da to characterized for AISI 403. Figure 10 shows the plot of J_{IC} and dJ/da with temperature. It is clearly seen that both the fracture toughness parameters decrease quite significantly in the temperature range 250 – 400°C.

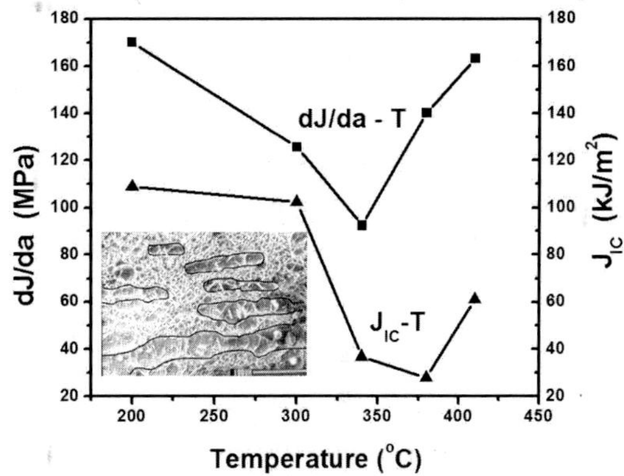

Fig. 10: The evolution of fracture toughness in J_{IC} format and crack growth resistance in the dJ/da format over the temperature range 200°C to 425°C carried out using pull rate of 3.33 micron/s. Inset shows the SEM fractograph picture of the sample displaying the lowest fracture toughness value.

The inset in the figure 10 shows the SEM picture of the fracture surface of the sample tested at the temperature corresponding to the lowest fracture toughness value. The SEM fractograph clearly shows the arrangement of under developed dimples interspersed between well developed dimpled areas. The width of these relatively flat areas in the SEM photograph is about 40μm.

It is therefore seen from the results obtained from the fracture toughness tests as well as fractographic examination that effect of DSA on fracture behaviour in AISI 403 results in embrittlement by way of reduction in the fracture parameters in terms of both initiation and propagation toughness. The fractographic observation clearly show evidences of flat featureless areas corroborating the 'jerky' load-displacement record obtained in the crack growth tests. The feature-less areas containing few ductile dimples are signatures of the occurrence of the crack jumps, because as pointed out by Mohan and Marschall [14], they result in rapid separation of the crack faces leaving less time for large number of voids to nucleate during the ductile crack growth.

The presence of these ductile cracking instabilities can also be associated with excessive machine compliance [14]. However in the multi pull rate crack growth tests the 'jerky' load-displacement record appeared only in the lower pull rate that gave an average crack tip strain rate within the DSA favorable strain rates. This rules out the influence on machine compliance. The second reason for these cracking has been attributed in ealrier studies [15,16] , primarily carried out on low carbon steels for piping applications, to DSA although the exact mechanism leading to periodic crack jumps has remained unclear.

A possible rationalization of the tensile test results, the appearance of cracking instability and its implication on the fracture parameters is now presented. Earlier investigations carried out by Mukherjee and Sellars [17] on Fe-Cr-C steels with tempered martensitic structure on a range of Cr contents (0.87% - 11.7%) attributed the dynamic strain aging to occur as a consequence of formation of Cr-C complexes, which enabled interstitial locking of dislocations upto 400°C. Thus the trends seen in AISI 403 are consistent with earlier investigations where the role of Cr has been unequivocally established. In the case of low alloy steels, to which the 3Cr-1Mo type steel belongs, studies have shown that they are susceptible to DSA due to interaction of dislocations with C and N [15-17]. The chemical nature of carbides in the tempered bainitic structure shows that they are Cr rich carbides[32] . As a result the elements such as Mn and Ni are free to form X-N or X-C complexes (X = Mn/Ni), thereby curtailing their migration to the dislocations. The formation of complexes of C and N with Mn are known to reduce the mobility of these interstitial atoms (i.e. C and N) [18]. Thus, as a consequence of the formation of Cr rich carbides, elements such as Ni/Mn could form complexes with C/N atoms, which may lead to a reduction in the effectiveness of dislocations pinning by these solutes.

The macroscopic effect of solutes pinning dislocations is to reduce the strain rate sensitivity. Louchet and Brechet [19,20] have studied theoretically the possible influences of strain rate sensitivity on toughness. They have argued that the strain rate sensitivity influences the action of crack tip relaxation process such as shielding and blunting [19]. Conditions that result in small values of strain rate sensitivity promote flow localization and retard the crack tip relaxation process, which can act as a driving force for crack extension by increasing the rate of stored elastic energy. Further, they have shown that in-homogenous deformation, such as the release of PLC bands at the crack tip lowers the stresses during their scanning time at the tip [20]. Considering the inferences from these studies [19,20] it may be proposed that in the fracture toughness tests carried out on AISI 403, with the release of PLC bands during DSA, the extent of plastification at the crack tip may be affected. With repeated release of these bands perturbations in the plastic zone size ahead of the tip may appear, which will impair the relaxation of local crack tip stresses. As a result an increase of the elastic strain energy rate would occur and provide the driving force for the crack extension in the form of jumps. The relationship between the plastic zone size and ability for crack tip shielding has been shown analytically by Chen et. al. where formation of reduced plastic zone size resulted in lower ability to relax the crack tip stresses [21]. As a result, the embrittling manifestations of DSA on the mechanical properties of AISI 403 steel are significantly increased.

SUMMARY

The results in this work compares the manifestations of DSA in two steels with tempered martensitic and bainitic structures and attempts to show that phase transformation products also exert a non-trivial influence on the anomalies in tensile properties caused by the operation of the phenomena. Earlier studies where the austenite transformation products were ferrite, pearlite, martensite or a combination of them, chemistry control was found to be the controlling factor to reduce the influence of DSA on the tensile properties of the material. The results being reported in this work on the new generation low alloy steel show that chemistry control along with the presence of bainitic phase can be used to limit the full extent of manifestations of DSA on tensile properties. Further, as the potential for DSA in the material is pushed at higher temperatures, the ascendant influence of recovery processes (eg dynamic recovery) would suppress the embrittling features of the phenomena.

ACKNOWLEDGEMENTS

CG gratefully acknowledges Japan Society for Promotion of Science (JSPS) for the JSPS fellowship award and grant-in-aid for scientific research (No. 22.00384).

REFERENCES

[1] C. Gupta, Deformation and Fracture of Advanced Structural Steels: Effect of Dynamic Strain Aging, Ratcheting and Creep. Lambert Academic Publishers, Germany, 2012 pp. 20-35.

[2] K. L. Murty and C. S. Seok, JOM, Vol. 6 2001 p. 23.

[3] Y. Estrin and L. P. Kubin in: Continuum Models for Materials with Microstructure ed. H. B. Mühlhaus, John Wiley and Sons, Germany 1995 pp. 385-450.

[4] Y. Estrin and L. P. Kubin, Mat. Sci. Eng. Vol. A137 1991 p. 125.

[5] K. L. Murty J. Metals Vol. 16 1985 p.34.

[6] A. Gysler, G. Lutjering, and V. Gerold Acta. Metall. Vol. 22 1974 p.901.

[7] S. Banerjee and U. Naik, Acta. Mater. Vol 44, 1996 p.3997.

[8] P. Rodriguez, Bull. Mater. Sci. Vol 6 1984 p.653.

[9] E. Pink and A. Grinberg Mater. Sci. Eng Vol 51 1981 p.1.

[10] A. Van Den Beukel, Phys. Stat. Sol. (a) Vol 30, p.197.

[11] H. K. D. Bhadesia Bainite in Steels 2nd ed. IOM London 2001.

[12] W.P. Longo and R. E. Reed Hill, Scripta. Mat. Vol 6 1972 p.833.

[13] J. H. Yoon B. S. Lee, Y. J. oh and J. H. Hong, Int. Jour. Press. Vess. Pip Vol. 76 1999 p.150.

[14] R. Mohan and C. Marschall Acta. Mater. Vol 46 1998 p. 1933.

[15] B. Mukherjee, Nucl. Eng. Des. Vol. 111 1989 p. 63.

[16] J. W. Kim and I. S. Kim, Nucl. Eng. Des. Vol 172 p.49.

[17] T. Mukherjee and C. M. Sellars, Met. Trans Vol 3 1972 p. 953.

[18] C. C. Li and W. C. Leslie Met. TransA Vol. 9 1978 p.1765.

[19] Y. Brechet and F. Louchet, Acta. Metall. Mater. Vol 41 1993 p. 783.

[20] Y. Brechet and F. Louchet, Acta. Metall. Mater. Vol 41 1993 p. 793.

[21] J. Chen and S. Kitaoka, Int. Jour. Fract Vol 100 1999 p.307.

Microstructural Evolution and Mechanical Properties of Type 304L Stainless Steel Processed in Semi-Solid State

Dipti Samantaray[1], Vinod Kumar[2], A.K. Bhaduri[1] and Pradip Dutta[3]
[1]Materials Development and Technology Group, IGCAR, Kalpakkam-603102, India
[2]Steel Products Group, R&D Centre for Iron & Steel, SAIL, Ranchi-834002, India
[3]Department of Mechanical Engineering, Indian Institute of Science, Bangalore-560012, India
Email: deepsaroj@igcar.gov.in.

ABSTRACT

The paper discusses the effect of semi-solid processing on the microstructural evolution and mechanical property of 304L stainless steel. For the study, steel specimens were partially melted and cooled to room temperatures in different cooling medium. The effect of temperature, time and cooling medium on microstructural evolution was studied by using optical microscopy. It was found that melting begins with the nucleation of liquid phase at the triple junctions and grain boundaries followed by propagation of the liquid phase along grain boundaries. The mechanical behavior of the semi-solid processed material was compared with that of the conventionally processed material with regard to their tensile properties and hardness. The semi-solid processed material shows better ductility and reduced YS and UTS than the conventionally processed counterpart. The correlation of the tensile properties and evolved liquid content shows that the UTS of the material decreases, while the YS increases with liquid fraction. The study also shows that there is a significant effect of cooling medium on the microstructural evolution and hence on the mechanical properties.

KEYWORDS: *304L Stainless Steel, Semi-Solid Processing, Microstructure, Tensile Properties, Hardness.*

INTRODUCTION

AISI type 304L austenitic stainless steel (304L SS) is widely used in power, chemical, petrochemical [1] and nuclear industries [2]. Conventionally, stainless steels are fabricated by casting and forging processes. However, continuous efforts are being made to find alternatives manufacturing routes to accomplish many objectives such as reducing energy consumption, enhancing workability, producing near-net shape products, and minimizing intermediate process steps [3]. Semi-solid processing is one such promising alternative to conventional manufacturing techniques [4], which not only fulfills the above objectives but also combines the benefits of both conventional casting and forging processes. This technology has already been adopted in an industrial scale for manufacturing of various non-ferrous alloys [4], such as aluminum [5] and magnesium alloys [6, 7]. The suitability of this technology has also been investigated for producing some components in laboratory scale from the ferrous alloys, such as X210Cr [3, 8] and bearing steels [9]. Therefore, adoption of this technology for other grades of steels has drawn worldwide attention. The major constraints in semi-solid processing of steels arise from high melting point of these alloys, narrow semi-solid temperature range, and phase transformation during melting and solidification [10]. In addition to the processing-related issues, assessing the suitability for semi-solid processing of steels also requires evaluation of post-processing mechanical properties. Therefore, the present work focuses on the microstructural evolution in 304L SS after cooling from the semi-solid state and on the resultant mechanical properties. A comparative study of the microstructure and mechanical properties of semi-solid and conventionally processed 304L SS is presented and discussed.

EXPERIMENTAL

For the present study, commercially available solution annealed 304L SS plates, with chemical composition Fe-18.52Cr-7.85Ni-0.016C-1.59Mn-0.29Si-0.003S-0.016P, have been used.

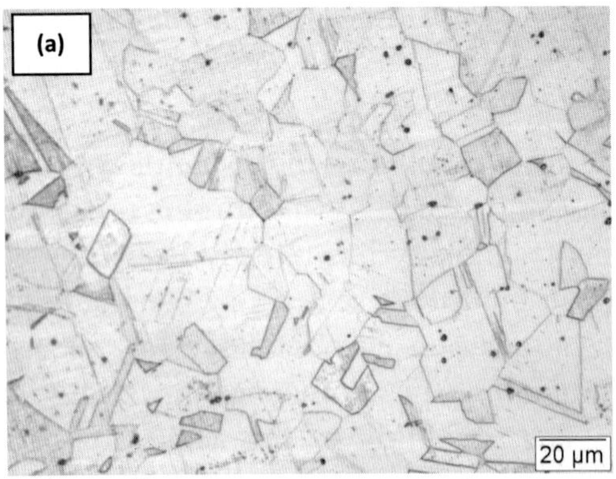

Fig. 1: Microstructure of solution annealed SS 304LTable 1: Parameters used for heat treatment of SS 304L

The microstructure of the steel in as-received condition is shown in Fig.1. Small bars of size 15mm × 15mm × 120mm were cut from the plates and used to carry out the experiments in the semi-solid state. The bars were partially melted in a muffle furnace and then allowed to cool to room temperature inside the furnace. To investigate the effect of higher heating and cooling rates on the microstructural features of the material, a few cylindrical specimens of size of 10 mm diameter and 15 mm height were heated in a Gleeble thermo-mechanical simulator to the semi-solid temperature, held at the targeted temperature for 30s, and then quenched in water. The experimental details are given in Table.1 After cooling the bars and cylinders to room temperature, a small sample of each specimen was taken for microstructural investigation. The diamond polished metallographic samples were etched in 10% oxalic acid solution at 2V, and microstructures were observed using an optical microscope.

Table 1: Parameters used for heat treatment of SS 304L

Sp. No.	Heating Rate (°C/sec)	Temper-ature (°C)	Soaking Time (mins)	Cooling Condition
S1	300	1400	5	Furnace Cooled
S2	300	1400	15	Furnace Cooled
S3	300	1400	25	Furnace Cooled
S4	300	1410	5	Furnace Cooled
S5	300	1420	5	Furnace Cooled
G1	5	1400	½	Water quenched
G2	5	1410	½	Water quenched
G3	5	1420	½	Water quenched

The mechanical properties of the material processed in semi-solid state were evaluated by carrying out tensile tests at strain rates 0.01 s^{-1} at room temperature using a 100kN tensile testing machine. Standard (as per ASTM standard E8M) specimens fabricated from furnace-cooled bars were used for the tensile tests. Vicker's hardness of the furnace-cooled bars as well as the water-quenched specimens was measured in hardness testing machine under a load of 10 kg.

RESULTS AND DISCUSSIONS

Microstructural Evolution

Microstructural evolution of any material depends on its chemical composition [11] and various parameters defined by the processing route such as heating rate, temperature of processing [12], cooling rate [13], mechanical loading [12], and so on. In any investigation on the effect of the processing route or the parameters involved in the process on the microstructure, it is always necessary to understand the effect of composition, as it defines the various phases that form within the processing temperature regime. In semi-solid processing, the material is heated to a temperature between the solidus and liquidus temperatures and then cooled to room temperature. Hence, it is of paramount interest to know the possible phases that can form between room temperature and the liquidus temperature. It is well known that the 304L SS retains its austenite phase at room temperature and undergoes phase changes at high temperatures, depending on the Cr-equivalent (Cr_{eq}) to Ni-equivalent (Ni_{eq}) ratio of the steel [14]. In order to calculate the Cr_{eq} and Ni_{eq} for different grades of 304 SS, the following equations are used [14]:

$$Ni_{eq} = Ni + 30 \times C + 30 \times N + 0.5 \times Mn \quad (1)$$

$$Cr_{eq} = Cr + Mo + 1.5 \times Si + 0.5 \times Nb \quad (2)$$

where Ni, C, N, Mn, Cr, Mo, Si and Nb are the wt% of the elements present in the steel. Using the above equations, it was found that the Cr_{eq}/Ni_{eq} ratio for the present grade of 304L SS is 2.02. This indicates that the steel under investigation follows a primary ferritic mode of solidification [14]. Upon heating, the austenite (γ) completely transforms to delta-ferrite (δ) and further heating causes melting. The transformation follows the reverse order during cooling. At the solidus temperature, the liquid phase is nucleated at triple junctions and grain boundaries depending on the solid-liquid interface energy (γ_{SL}) and the grain boundary energy (γ_{GB}) [15]. With further increase in temperature, the liquid phase propagates along the grain boundaries and forms a continuous network along the grain edges. The growth of liquid phase along the grain edges causes a continuous decrease in size of the grains which gradually approach a critical size. Once the grains achieve the critical size, rapid melting occurs and the grains disappear in the molten metal pool [16]. However, in this study, the material has not been allowed to undergo complete melting. The material has only been allowed to melt partially, after which it was cooled to room temperature. The microstructures of samples that were furnace-cooled after partial melting (Fig.2) indicate the solidified fraction of the liquid around the grains.

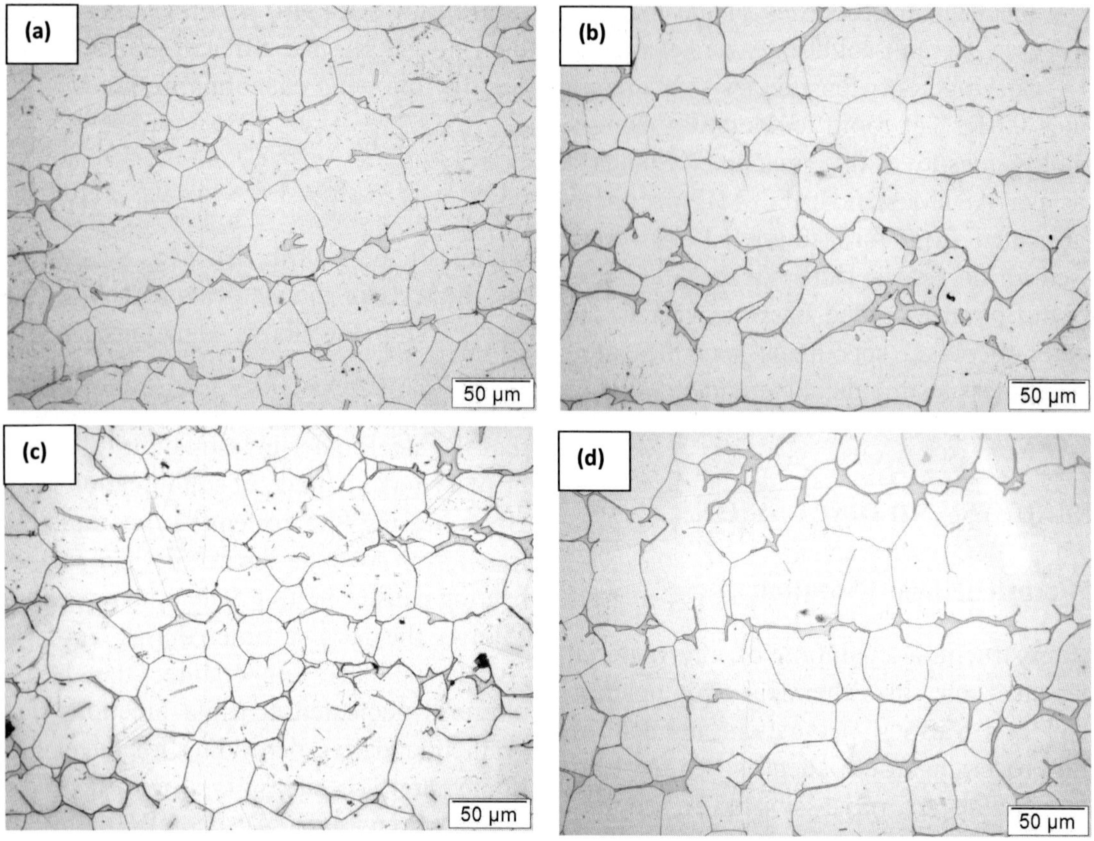

Fig. 2: Microstructural evolution in 304L due to partially melting at (a) 1400°C/5 mins (b)1410°C/ 5min (c) 1420°C /5min (d) 1400°C/ 25 min followed by furnace cooling

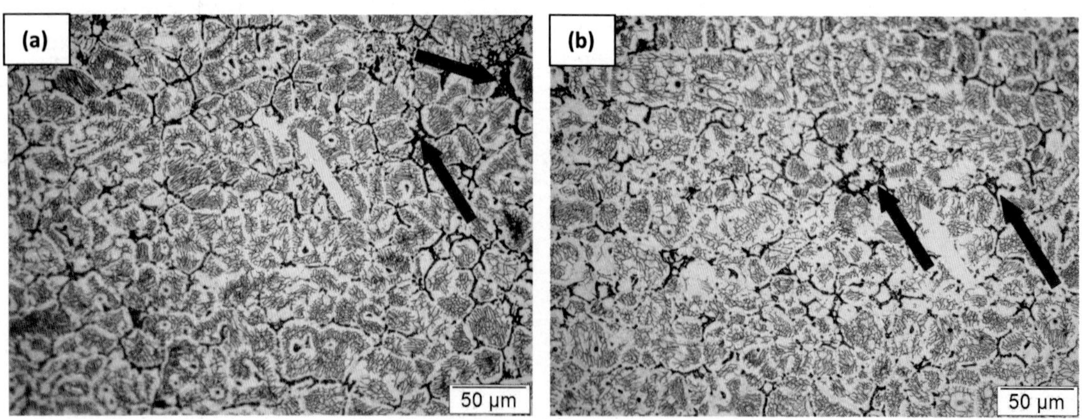

Fig. 3: Microstructural evolution in 304L due to partially melting at (a) 1410°C (b)1420°C followed by water quenching

It could be observed that these samples contain bigger grains that vary from ~25 to 60μm, whereas the average grain size of the sample was 17μm before processing. This shows that grain growth occurs in the material before initiation of melting. The nucleation of the liquid phase at the triple junctions and grain boundaries can be seen in the Fig.2(a), while increase in the liquid phase and its propagation along the grain boundaries with temperature and time of holding can be observed from Figs.2(b)-2(d). From these figures, it is observed that with increase in temperature from 1400 to 1420°C, there is a significant increase in the number of liquid phase entrapments inside the solid grains. However, no such increase in sites of interior grain melting was obtained with increase in soaking time. Only growth of the nucleated liquid phase by melting of the interface boundary was observed with prolonged soaking time.

On comparing the microstructures shown in Fig.2, it can also be noticed that the grain edges become smoother with increase in liquid fraction. In many metals, the appearances of facets have been witnessed either on melting or on solidification [16]. The appearance or disappearance of the facets during melting and solidification depends on the thermal environment at the solid-liquid interface [17]. Though the facet disappearance is easily noticeable in the microstructure of semi-solid processed 304L SS, it is difficult to conclude that the disappearance has occurred during melting as the micrographs have been taken after solidification of the molten fraction.

The microstructure of the water-quenched samples after partial melting is shown in Fig.3, in which the liquefied regions around the grains are shown by red arrows. A visual comparison of these micrographs with those of the furnace-cooled samples indicates that the liquid fraction is relatively low in the water-quenched samples. This could be attributed to the smaller holding time of the samples at the melting temperature. The other noticeable features in these micrographs are the small network of delta ferrite inside the grains (shown by yellow arrows). The occurrence of the fine delta-ferrite networks could be ascribed to the insufficient time for the $\delta \rightarrow \gamma$ transformation due to high cooling rate (~1500°C/s) obtained on water quenching by spray nozzles in the Gleeble thermo-mechanical simulator. These delta-ferrite networks are not visible in the microstructures of the furnace-cooled samples, as in this case, the material does get sufficient time for $\delta \rightarrow \gamma$ transformation.

Mechanical Properties

Tension test is widely used to provide the important information on the strength of a material for the design of the product or for making structures using the material [18]. Therefore, to study the effect of the process parameters as well as the microstructure on the strength, it is required to evaluate the tensile properties of the material. However, it is always not possible to measure the tensile properties of a small volume of material produced in a laboratory scale. Therefore, hardness testing provides an alternative means of assessing their mechanical properties [18]. There are several relationships available to correlate the hardness with the yield strength and ultimate tensile strength of materials [18, 19]. In this investigation, both tensile properties and hardness have been evaluated for the samples that were furnace-cooled after partial melting, while only the hardness has been determined for the samples that were water-quenched after partial melting, due to size limitation of the samples used. The yield strength (YS), tensile strength (UTS) and the ductility (percent uniform elongation) obtained from the tensile tests performed on the as-received material (conventionally processed) and on the semi-solid processed material (with furnace- cooling) are given in Table 2. From the table, it is observed that the material processed in the semi-solid range shows better ductility than the conventionally processed (by hot rolling followed by solution annealing) material, i.e. However, these materials show reduced YS and UTS compared to the conventionally processed one. Similar comparison of the tensile properties of a cast product of 304L SS reported in literature [20] shows that the semi-solid processed material shows better UTS and ductility with a comparable YS value. The correlation between the fraction of liquid and the tensile properties are shown in Fig.4. From this figure an overall decrease in UTS is observed over the entire domain of study, while the YS is found to have improved with increase in liquid content. On the other hand, an initial decrease in the ductility is observed up to a liquid fraction of 14% beyond which the ductility increases with increase in liquid fraction. Here, it is to be noted that the liquid fraction in the specimens have been calculated using image analysis technique. For calculation of the liquid fraction in a particular specimen, an average of 15 micrographs of each specimen has been used. The micrographs have been taken from various locations of the specimen at different magnifications. The errors involved in the calculation could be minimized by carrying out thermal analysis of the material in the solidus–liquidus temperature domain. Table 2 also shows hardness values of these materials. The hardness values of the furnace-cooled

Table 2: Mechanical properties of the conventional and semi-solid processed specimens

	Liquid Fraction	YS (MPa)	UTS(MPa)	Uniform Elongation (%)	VHN (10kg)
As-Recieved		275	715	58	186
S1	10±2%	205	682	75	144
S2	13±3%	220	664	70	134
S3	16±4%	235	671	68.5	131
S4	14±2%	246	652	67	157
S5	21±3%	254	651	70	143
G1		-	-	-	162
G2		-	-	-	162
G3		-	-	-	165
CF-3[20]		248	531	60	140

material are found to be lower than those of the as-received material; while these are found to be similar to the hardness value of the cast steel. The variation of hardness with the liquid fraction shows a similar trend as that of the YS. These results agree with the earlier findings on the relationship between the hardness and YS of austenitic steels [19].

On comparison of the hardness values of furnace-cooled specimens with the specimens that were water quenched after partial melting, it is noticed that the water-quenched specimens are harder than the furnace-cooled ones. As the hardness of the material reflects the trend in YS, higher YS is expected in water-quenched specimens. Higher strength of the specimens could be correlated to the presence of the fine delta-ferrite networks in the microstructure. At room temperature, a higher shear stress is required to cause slip in a bcc material than in a fcc material [18]. As delta-ferrite has a bcc structure [21], the presence of which in the fcc austenite matrix increases the average shear stress required to cause plastic deformation in the material. In addition, the presence of finely distributed second phase in the matrix is known to contribute to the strengthening of the material in many ways [22], most significantly by pining the dislocations and reducing their mobility [23]. Thus, the higher resistance to plastic deformation can be attributed to the fine network of the delta-ferrite in the austenite matrix in the water-quenched specimen. The mechanical properties of these materials have not been correlated to the liquid fraction as the evolved liquid fraction in these specimens could not be calculated using the image analysis technique owing to the presence of fine delta-ferrite networks in the microstructure. In this case also, the hardness of the semi-solid processed material is found to be less than that of the as-received material; however, its harness is higher than that of the cast 304L SS.

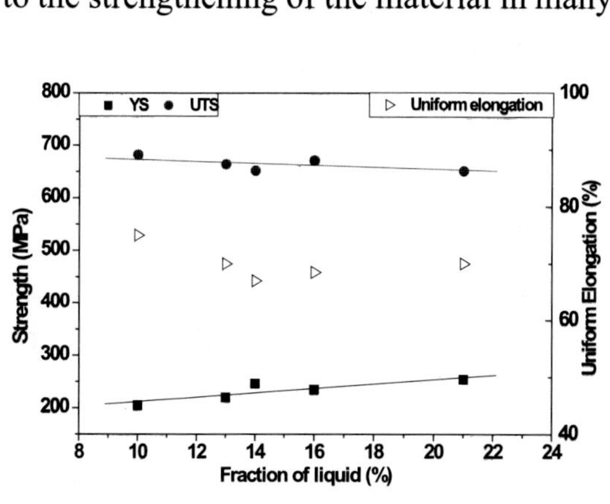

CONCLUSIONS

The effect of semi-solid processing on the microstructural evolution and mechanical properties of a type 304L SS has been studied. Towards this end, the partially 304L SS were cooled to room temperature using different cooling medium. Tensile properties and hardness of the specimens were evaluated to compare the mechanical behaviour of the semi-solid processed material with the conventionally processed material. Also, the effect of evolved liquid content during semi-solid processing on the mechanical properties of this material has been studied. The following conclusions can be drawn from the above investigation.

1. The solidification mode of type 304L SS is primary ferrite. With increase in temperature, severe grain growth occurs in the specimen, and melting

begins with the nucleation of liquid phase at the triple junctions and grain boundaries followed by propagation of the liquid phase along the grain boundaries.

2. An increase in melting temperature causes the nucleation of liquid phase at the grain interiors along with the triple junctions and grain boundaries, whereas increase in soaking time causes growth of the nucleated phase.
3. The semi-solid processed material shows better ductility and reduced YS and UTS than their conventionally processed counterpart that was solution annealed after hot rolling.
4. An overall decrease of UTS in observed over the entire domain of study; the YS is found to improve with the evolved liquid content.
5. Hardness of water quenched specimens is superior to that of furnace cooled specimens. The superior resistance to plastic deformation has been attributed to the presence of fine delta-ferrite networks in the water quenched specimens.

REFERENCES

[1] Woei-Shyan Lee, Jen-I Cheng, Chi Feng Lin, Material Science and Engineering A, Vol.381, 2004, p. 206.

[2] A. Renault, J. Malaplate, C. Pokor, P. Gavoille, Journal of Nuclear Materials, Vol. 421, 2012, p. 124.

[3] D.I. Uhlenhaut, J. Kradolfer, W. Puttgen, J. F. Loffler, P.J. Uggowitzer, Acta Materialia, Vol. 5, 2006 p. 2727.

[4] H.V. Atkinson, Progress in Material Science, Vol. 50, 2005, p. 341.

[5] C. Chayong, H. V. Atkinson, P. Kapranos, Material Science and Engineering A, Vol. 390, 2005, p. 3.

[6] Y. Chino, M. Kobata, H.Iwasaki, M. Mabuchi, Acta Materialia, Vol. 51, 2003, p.3309.

[7] Y. Birol, Journal of Alloys and Compound, Vol. 455, 2008, p. 178.

[8] B. Masek, D. Aisman, M.Behulova, H. Jirkova, Trans. Nonferrous Met. Soc. China, Vol. 20, 2010, p. 1037.

[9] W. Puttgen, B. Hallstedt, W. Bleck, J. F. Loffler, P.J. Uggowitzer, Acta Materialia, Vol. 55, 2007, p.6553.

[10] P. Cezard, R. Bigot, V. Favier, M. Robelet, Advanced Methods in Metal Forming, pp309-320.

[11] P. Kapranos, Thixoforming of Steel, Ed. H.V. Atkinson, A. Rassli, Shaker Verlag, Aachen, 2012, pp. 13-35.

[12] D. Samantaray, S. Mandal, V. Kumar, S.K. Albert, A.K. Bhaduri, T. Jayakumar, Materials Science and Engineering A, Vol. 552, 2012, p. 236.

[13] J. W. Elmer, S.M. Allen, T. W. Eagar, Metallurgical Transactions A, Vol. 20,1989, p.2117.

[14] J.C. Ma, Y.S. Yang, W.H. Tong , Y. Fang, Y. Yu, Z.Q. Hu, Materials Science and Engineering A, Vol. 444, 2007, p. 64.

[15] G.A. Chadwick, Liquid in Solids, Solidification, ASM, USA, 1971, pp. 99-120.

[16] M.E. Glicksman, Direct observation of solidification, Solidification, ASM, USA, 1971, pp.155-200.

[17] D.P Woodruff, Philosophical Magazine, Vol. 18, 1968, p123.

[18] G.E. Dieter, Mechanical Metallurgy, McGraw-Hill Book Company, London, 1988, pp. 275-336

[19] J. T. Busby, M. C. Hash, G. S. Was, Journal of Nuclear Materials, Vol. 336, 2005, p.267.

[20] J.M. Svoboda, High Alloy steels, ASM Handbook, Vol-15, CASTING, ASM International, USA,1998, pp.721-739.

[21] P. Marshall, Austenitic Stainless Steels: Microstructure and Mechanical Properties, Elsevier Applied Science Publishers, London and New Tork, 1984, pp.1-20.

[22] Donald Peckers, The Strengthening of Metals, Reinhold Publishing Corporation, New York-Amsterdam-London, 1967, pp.163-199.

[23] G.E. Dieter, Mechanical Metallurgy, McGraw-Hill Book Company, London, 1988, pp. 103-240.

Thermo-Mechanical Simulation of Hot Extrusion Process for Zr-2.5Nb-0.5Cu

S.K. Jha[1], Kumar Vaibhaw[1], V. Raizada[1], N. Saibaba[1], D. Srivastava[2] and G.K. Dey[2]
[1]Nuclear Fuel complex, Hyderabad, India
[2]Materials Science Division, Bhabha Atomic Research Centre, Mumbai, India
Email: skjha4652@yahoo.com

ABSTRACT

Garter spring is one of the important components made of Zr-2.5wt%Nb-0.5wt%Cu which separates pressure tube from calandria tube in pressurized heavy water reactors (PHWRs). Garter spring which supports pressure tube both in initial condition and later after creep sag of the pressure tube should have adequate strength. Addition of small amount of Copper to Zr-2.5%Nb, which used as pressure tube material in PHWR, improves its strength by precipitation of Zr2Cu on ageing the ternary martensite. Addition of copper also improves the corrosion properties of the garter spring. Although physical metallurgy and characterization of Zr-2.5 %Nb has been studied significantly, however the studies on Zr-2.5Nb-0.5%Cu are rather limited. The properties of Zr-2.5Nb-0.5Cu alloy are very sensitive to the thermo mechanical fabrication steps including hot and cold working due to their anisotropic deformation behavior and presence of the intermetallic phase.The fabrication of garter spring involves alarge number of thermo-mechanical treatments, and the finalmicrostructure and texture developed as a result of these fabricationsteps determines the mechanical properties which is essential for production of quality product. Therefore, it is important to understand the influence of important variables of fabrication steps on evolution of microstructure and crystallographictexture.

The present paper is aimed to establish hot deformation behavior in extrusion process at different strain rates (1 to 10 s^{-1}) and temperature (700-925°C) by employing Gleeble-3500 thermo mechanical simulator. FEM based HyperXtrude program was used in conjunction with the above determined flow properties to optimize the extrusion parameters such as extrusion temperature, ram speed etc.

KEYWORDS: *Thermo-Mechanical Simulation, Hot Extrusion, Zirconium Alloy*

INTRODUCTION

Zr-2.5Nb-0.5Cu alloy garter spring is an important structural component of pressurized heavy water reactors (PHWRs). It separates pressure tube from calandria tube.It is a helical extension or compression springs whose ends are connected in order to form each spring into a circle which exerts radial forces around pressure tube. Garter spring needs adequate strength to support pressure tube and it should prevent contact between pressure tube and calandria tube in the event of diametral creep of pressure tube. Copper is added to Zr-2.5%Nb alloy used as pressure tube material in PHWRs toimprove its strength by precipitation. The corrosion properties are also improved due to addition of copper. Although physical metallurgy and characterization of Zr-2.5%Nb has been studied significantly, however the study on Zr-2.5Nb-0.5%Cu is rather limited. Both metastable and equilibrium phase transformation behavior of Zr-2.5%Nb is known to be affected by presence of alpha and beta phase stabilizing elements. The properties of Zr-2.5Nb-0.5Cu alloy are very sensitive to the thermo mechanical fabrication steps.Starting with hot extrusionfollowed by multiple working with intermediate annealing.In extrusion, correct design and control requires, among other things, the determination of the deformation mechanics involved in the process. Without the knowledge of the influence of variables such as friction process conditions, material properties and work piece geometry on theprocess mechanics, it would not be possible to design the dies and the equipment adequately, or to predict and prevent the occurrence of defects.

The use of FEM in the extrusion process provides an excellent means of obtaining information on how the material flows, how the desired geometry can be achieved, and expected mechanical properties in the extrudate. Such objectives essentially consist of, a) establishing the kinematic relationships (velocities, strain-rates) between the deformed and unreformed part, i. e. predicting metal flow, b) establishing the limits of formability, i. e. establishing, whether it is possible to form the extrudate without surface and\or internal defects and c) predicting the forces and stresses necessary to execute the process so that tooling and equipment can be designed or selected accordingly. Present study is aimed to simulate the hot extrusion of Zr-2.5Nb-0.5Cu alloy. Flow behavior of this alloy as a function of strain rate and temperature has been obtained by using Gleeble-3500 and these flow stress data has been utilized to simulate hot extrusion.

SIMULATION OF HOT EXTRUSION

The importance of modeling and simulation in the metal forming industry has increased heavily during the last decades. Process simulation using the finite element method (FEM) is now accepted as an important tool for product and process development. However, the introduction of computer simulation in extrusion technology has not been as fast as in other parts of the manufacturing industry. This is mainly due to the very large deformations which make thesesimulations technically challenging and computer intensive. The simulation of hot extrusion processes is clearly one of the most difficult problems in process modeling [1].Nevertheless, the potential of using numerical methods for analysis of extrusion is large. The whole process, including the important initial non-steady state of extrusion, can be analyzed. The evolution of, for example, stress, strain rates and temperatures in the material during the process can be studied in detail. As the capacity of computing hardware and software is increasing, itis no doubt that the FEM will be a tool of great use for development and optimization of extrusion. In the Aluminum industry, on the other hand, extensive activities have been devoted to modeling and simulation, and in the recent years great progress has been made in this area. Simulation of Magnesium extrusion has also been reported [2] FEM modeling of deformation process are based on mainly Lagrangian, Eulerian and more recently

Arbitrary Lagrangian - Eulerian (ALE) techniques.

In Lagrangian FE codes, the mesh moves with the material and deforms with the material flow. The quadrature points also move with the material which means that the constitutive equations are evaluated at the same material points through the whole analysis. This approach is very useful for extrusion analyses, since the thermo-mechanical history during the process can be studieddirectly and the free surface of the extrudate can be followed.Lagrangian methods are efficient and suitable for handling non-linear problems in which small strains prevail, where boundary conditions do not change with the course of deformation, and where mesh distortion is not a critical factor in the analysis. As these conditions are usually essential to any metal forming analysis this can cause severe problems. In particular, the issue of mesh distortion and subsequent remeshing is still a challenge and requires further understanding. This approach requires higher computing machines due to complexities associated with remeshing etc. In hyper extrude the constitutive models, boundary conditions, solution algorithm of flow analysis and variational formulation has been discussed in the published literature [3-7].

The Eulerian technique avoids the difficulties associated with remeshing as it uses a mesh that is fixed in space for all time, through which the material flows, as in many fluid mechanics problems. Although many researchers have used the Lagrangian formulation there has been less effort concerning the development of a consistent Eulerian formulation in solid mechanics applications. In the Eulerian approach, it is necessary to properly determine the state of the material associated properties, i.e. stress and strain, at the beginning of each time-step, and although the method has generally been used for the study of flow problems and steady state metal forming processes, it has not been used in problems where domain boundaries of interfaces move substantially. However, the Eulerian route assumes that the die is entirely rigid and this may not be the case.

HyperXtrude10.0 software is based on arbitrary Lagrangian-Eulerian formulation. This software has been effectively utilized to model mainly extrusion process for Aluminium alloys. This is the first attempt to simulate extrusion of Zr-2.5%Nb-0.5 Cu alloy. In Hyperextrude Models the fundamental equations for modeling incompressible viscous flow are based on the conservation of mass, momentum, and energy. These equations are written in terms ofprimitive variables (velocity components, pressure, and temperature) with reference to an Eulerian reference frame, i.e., a space-fixed system of coordinates through which the fluid flows.

HyperXtrude software makes use of finite element model based on following empirical ZenerHolomanconstitutive relation.

$$\sigma = \frac{1}{\alpha}\sinh^{-1}\left\{\left(\frac{Z}{A}\right)^{1/n}\right\}$$

$$Z = \dot{\varepsilon}e^{(Q/RT)}$$

Where,

Q = Activation Energy = 402,200 (J/mol)

R = Universal Gas Constant = 8.314 $(J/mol\text{-}K)$

n = Stress Exponent = 4.3

A = Reciprocal Strain Factor = 2.61×10^{15}

α = Reciprocal Stress Factor = 3.69×10^{-09} $(1/Pa)$

T = Temperature (K) }- Computed

$\dot{\varepsilon}$ = Strain Rate (1/s) }- Computed

In Hyperextrude Software the constitutive models, boundary conditions, solution algorithm of flow analysis has been discussed in the published literature[3-10].

EXPERIMENTAL

Hot Compression Experiments OnGleebleThermo Mechanical Simulator

The Gleeble machines typically have a high speed heating system, a servo hydraulic system and a computer control data acquisition system. Cylindrical samples from the beta quenched Zr-Nb-Cu alloy billets were machined in the dimension of 10mmØ x 15 mm L for compression between the two anvils. Hot deformation behavior has been determined at different strain rate (1 to 10 sec-1) and temperature (700-925°C) by employing Gleeble-3500 machine. To keep the deformation uniform during compression, temperature within the volume of the specimen was made uniform by employing resistance type heating and allowing soaking up to 60 secs at the test temperatures. Barreling of specimensafter deformation was minimized with the applications of graphite foil based lubricants at the interface between the anvils. Flow stress was determinedon the basis of this study. The test matrix is given in Table 1.

Table 1: Gleeble Hot Compression Test Data

Gleeble Hot Compression Test Data (MPa)						
Strain	Strain rate,s^{-1}	Temperature (°C)				
		700	750	815	850	925
0.1	1	270.77	166.29	83.08	69.62	49.05
	5	293.58	186.91	110.12	89.77	70.54
	10	293.4	229.77	130.09	105.12	80.7
0.2	1	239.64	152.38	79.85	67.43	48.43
	5	272.72	173.18	102.69	88.65	69.86
	10	282.35	209.91	120.6	101.54	79.8
0.3	1	211.99	138.4	75.96	65.68	48.16
	5	245.7	162.25	97.47	86.304	68.48
	10	265.69	191.89	112.38	97.59	78.11
0.4	1	195.94	126.38	71.65	63.34	48.03
	5	225.27	152.47	92.77	84.58	67.91
	10	247.05	177.27	107.33	95.09	76.88
0.5	1	180.42	120.64	67.08	63.43	47.356
	5	210.22	142.51	90.105	81.76	66.262
	10	229.28	164.65	103.47	92.692	75.348

Modeling and simulation of Zr-2.5Nb-0.5Cu Hot Extrusion

Computation was made for velocity, temperature, stress, strain and pressure in the flow field and based on the stresses, forces on the tool were also determined. While deciding the meshing both tetrahedron and hexahedron elements were tried. The tetrahedrons however showed a stiff behavior and the extrusion force was heavily overestimated by these elements whereas hexahedral elements indicated reasonably good results hence hexahedral elements were finally used in the extrusion simulation.

During hot processing, the temperature has a significant influence on structural state and mechanical properties of billet. Furthermore plastic strains during hot extrusion processes are much larger than the elastic ones, those leads to the necessity of considering a thermoplastic process. Moreover, flow stress depends on effective strain, effective strain rate and temperature. Other mechanical properties are temperature dependent. The initial temperature was taken as input before submitting for each simulation run. Cooling of the surface in contact with the tools, as well as of the free surfaces takes place during the process. Convection heat exchange between the work piece and the tools and between the work piece and the surrounding area was considered. Furthermore, the contact condition of surfaces between the work piece and tools was also considered. FEM based HyperXtrude program was used in conjunction with the determined flow properties to optimize the extrusion parameters. Direct flow data table was used for simulation purpose and its validity was checked using extrusion force directly obtained from Gleeble hot compression data.

Hot Extrusion of Zr-2.5Nb-0.5Cu Billet

β-quenched billets of 142mm OD X 550mm length were mechanically jacketed with seamless copper tubes. Jacketed billet was preheated in a resistance heating furnace and extruded in a 3780T Horizontal Extrusion Press (HEP) maintaining extrusion ratio of 44:1 and ram speed in the range of 20 to 50 mm/sec. Billets were preheated in a resistance heating furnace and temperature were varied in the range of 700 to 850°C. Copper jacketing of billet prevents oxidation of Zr-2.5Nb-0.5Cu billet and also provides the additional lubricant with oil based graphite. Copper cladding prevents direct contact of material with the extrusion tooling surfaces which otherwise would stick because of galling nature of Zirconium alloys. Copper jacketing remained on therod surface after extrusion was removed by dissolution into HF& HNO_3 solution.

RESULT AND DISCUSSION

Flow Properties of Zr-Nb-Cu

Hot compression testing was carried out at four different temperatures using gleeble-3500 viz.700, 750, 815, 850 and 925°C and constant true strain rates of 1.0, 5.0 and 10 S^{-1}. Cylindrical hot compression samples of size 10mm diameter and 15 mm length were used for compression testing. True strain value of 0.5 was used for conducting the test at all temperature and strain rates. Fig. 1a, b & c indicate the plot of true stress and true strain at varying temperatures and constant strain rate of 1,5 and $10S^{-1}$ respectively. Whereas Fig. 2a, b & c indicate the plot of true stress and true strain at varying and constant temperature of 815, 850 and 925°C respectively.

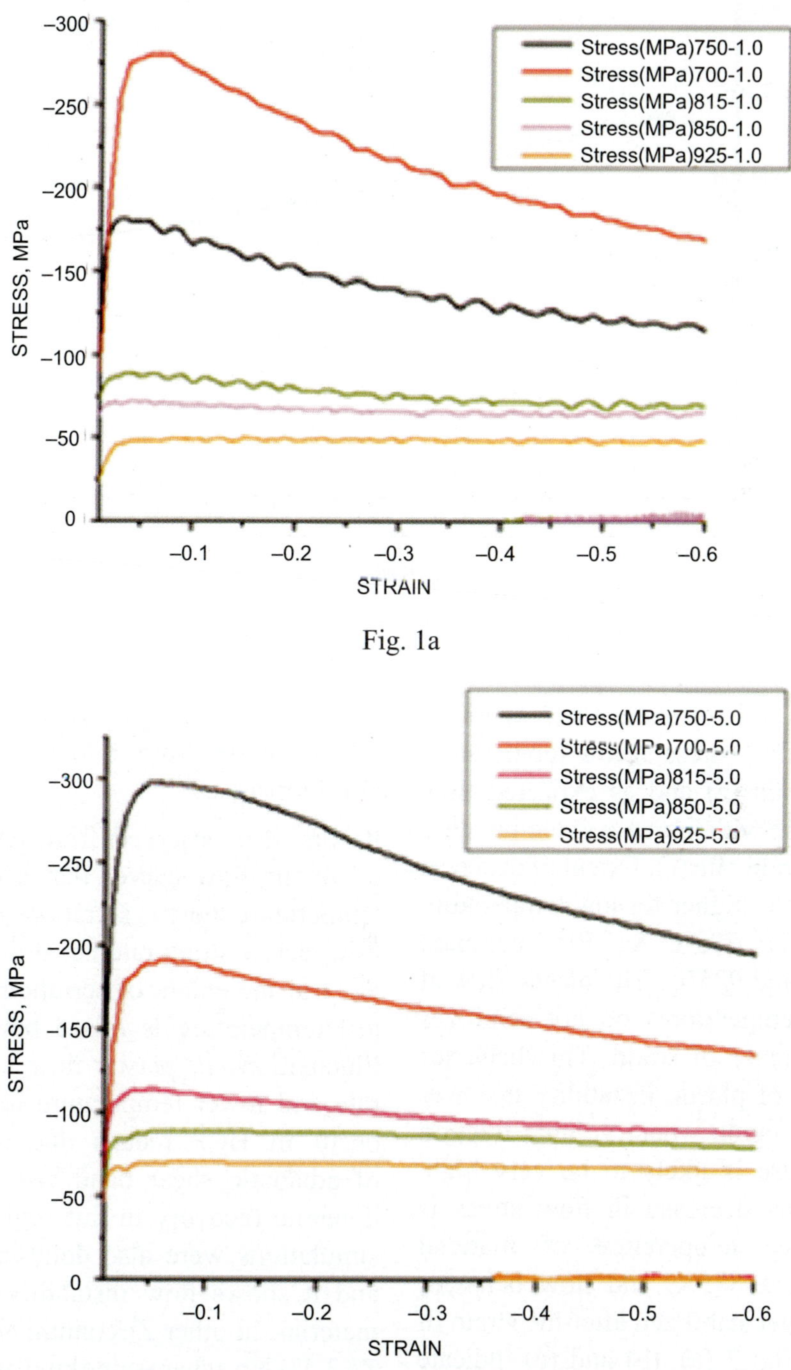

Fig. 1a

Fig. 1b

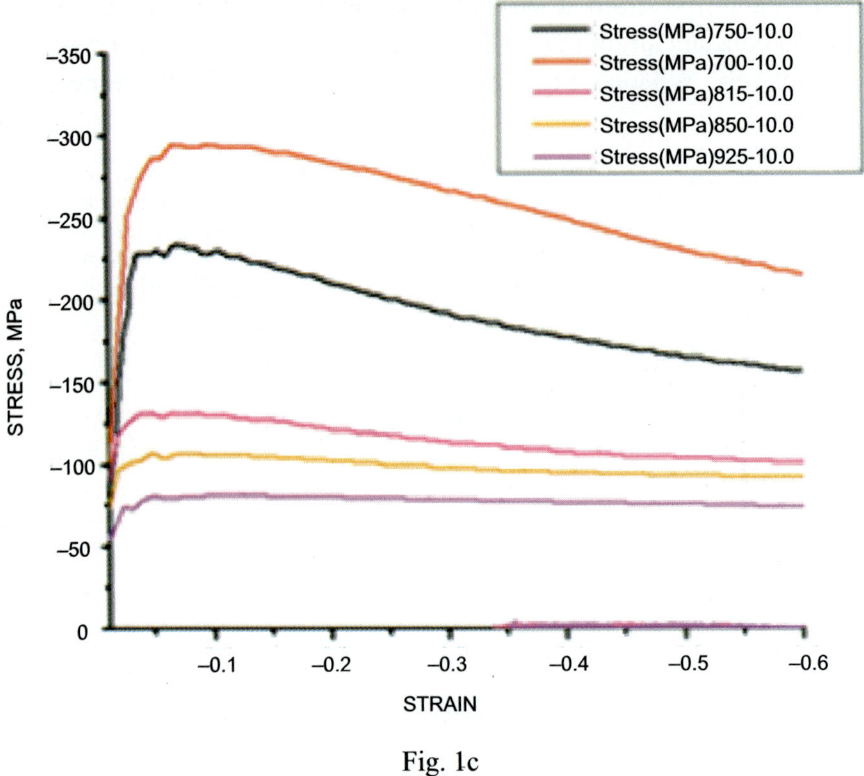

Fig. 1c

Fig. 1(a),(b) and (c) show there is lot of variation of flow stress at low temperature with true strain (ε) and as expected flow stress decrease with increase in temperature for a given strain rate ($\dot{\varepsilon}$). Extent of decrease in flow stress is higher for low temperature flow curves i.e. 700°C & 750°C compare to 815, 850 and 925°C. The plastic flow at lower two temperatures do not stabilizes with the increase of strain. This behavior is indicative of plastic instability in lower temperature region where volume fraction of alpha phase is likely to be very high. However, this decrease in flow stress is reduced once temperature of material increases from 815°C and flow behavior indicates to get stabilized after the strain of 0.3 to 0.4. Fig. 2 (a), (b) and (c) indicate that at constant temperature as the strain rate increases from 1 to10S^{-1}, flow stress also increases.

It was also observed from characteristics of all the flow curves that irrespective of temperature there is serrations in the plastic flow at low strain rate i.e. strain rate of 1.0 S^{-1}. But the extent of serrations reduces as test temperature is raised beyond 815°C. Fluctuations in plastic flow at low strain rate and lower temperature region mainly occur in HCP metals due to formation of adiabatic shear band also presence of dynamic recovery in this region.Extrusion simulations were also done in this region and it shows flow instability in extruded material. In other Zirconium alloys such as Zr-2.5% Nb, where same kind of studies have

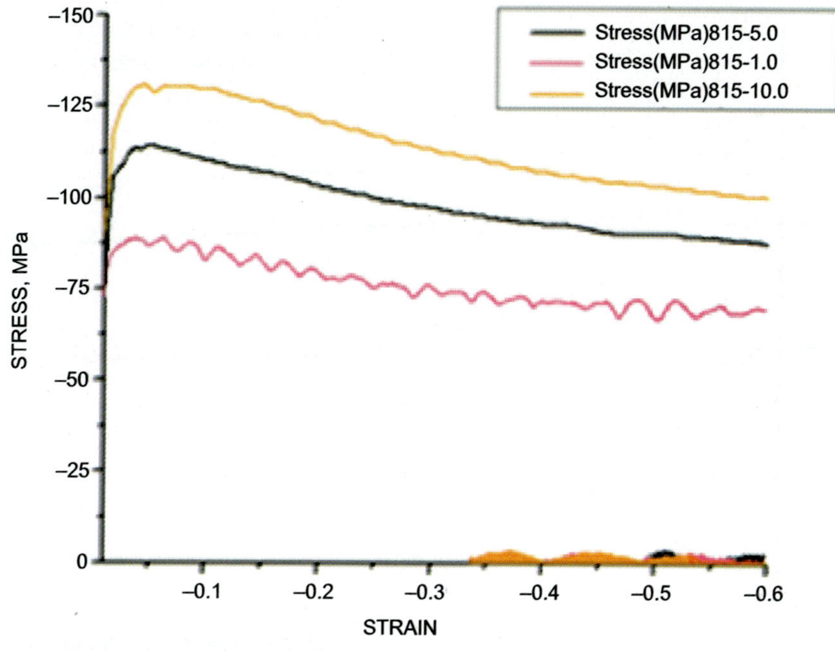
Stress(MPa)815-5.0
Stress(MPa)815-1.0
Stress(MPa)815-10.0
–150
–125
–100
–75
–50
–25
0
STRESS, MPa
–0.1
–0.2
–0.3
–0.4
–0.5
–0.6
STRAIN

Fig. 2a

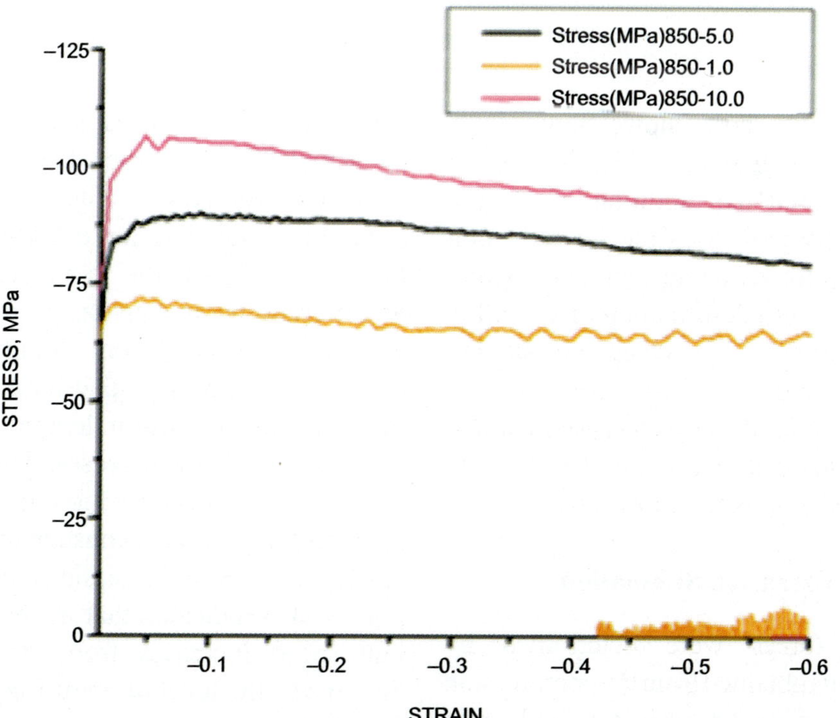
Stress(MPa)850-5.0
Stress(MPa)850-1.0
Stress(MPa)850-10.0
–125
–100
–75
–50
–25
0
STRESS, MPa
–0.1
–0.2
–0.3
–0.4
–0.5
–0.6
STRAIN

Fig. 2b

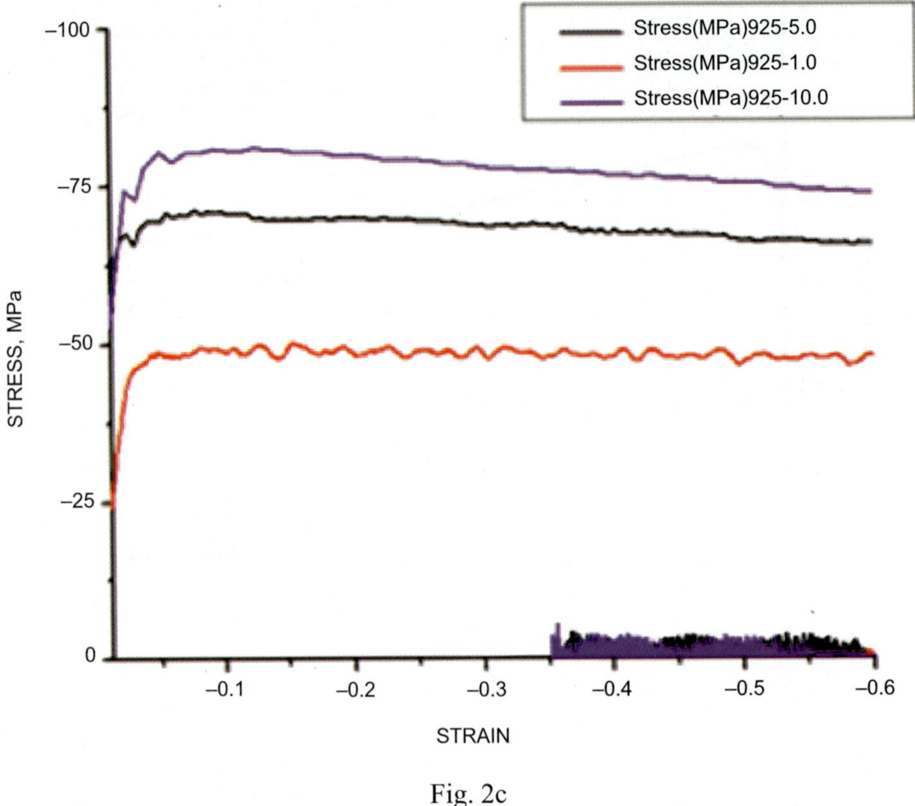

Fig. 2c

been made elsewhere shows flow curves having less fluctuation. So, in this case it can be assumed that addition of Copper in Zr-2.5%Nb is responsible for reduction in tendency for dynamic recrystallization. High temperature deformation of this alloy in higher strain rate provides very smooth flow curve and it is considered as flow curve suitable for dynamic recrystallization. Based on this data entire simulation study was planned at higher temperature.

Results of Extrusion Simulation

Extrusion forces were determined by inserting data obtained from hot compression testing for different temperature and strain rate. Since, for Zr-2.5%Nb-0.5% Cu no constitutive equation is available hence, the actual flow properties data obtained by hot compression tests in Gleeble-3500 were used for simulation for extrusion model. Fig. 3(a) shows the computed results at different parameters as a function of extrusion ram speed from 20 mm/sec to 50 mm/sec. It can be seen that for a given ram speed as the extrusion length decreases, the extrusion load increases. This trend is contrary to the observation made in hot compression testing at constant temperature of 815o C and with increasing strain(fig 2a). Fig 3a also indicates that as the extrusion ram speed increases from 20mm/sec to 50mm/sec, the level of extrusion load rises. This behavior is quite similar to what was observed in fig1a to 1c where at constant temperature as the flow stress increases with increase of strain rate from 1 to $10S^{-1}$.

It was analyzed that during extrusion of billet at a constant ram speed cumulative heat loss from the billet to the tooling over extrusion time duration is dominant and increasing the flow stress as compared to softening effect of deformation temperature rise.

The peak extrusion loads at 815°C for different ramspeed is shown in fig 3b. It can be seen that with increasing initial speed the peak load requirement for extrusion increases almost linearly, this suggest that for a given extrusion temperature there will be a maximum ram speed.

Simulation results for the temperature variation across the radius billet during hot extrusion havebeen shown in fig 4 (a)-(e). This temperature distribution across the radius has been plottedfor after 2 secs and 20 secs of extrusion at 815°C, i.e at the start and almost end of extrusion respectively. This can be interpreted in terms of temperature variation between the leading and the trailing end of the extrudate.Graphs 4b-e clearly indicates that for a given extrusion temperature as the extrusion speed increases from 20 mm/sec to 50 mm/sec the temperature variation between leading end and trailing end nearer to surface over the length reduces from about 15°C to 5°C. This is due to higher deformation heat rise at higher extrusion ram speed compensating the billet heat loss to the tooling to the greater extent.

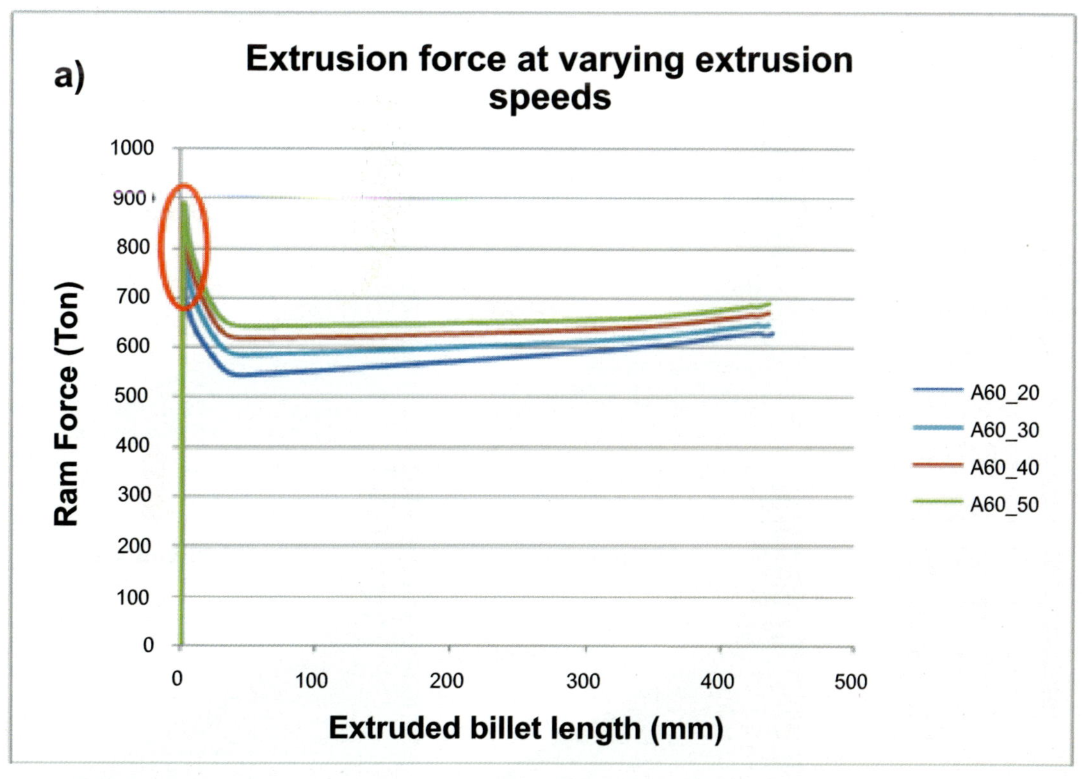

b)
Peak Ram Force at varying extrusion speed
Ram Force (Ton)
900
880
860
840
820
800
780
760
15
35
55
Extrusion Speed (mm/sec)
Peak Ram Force

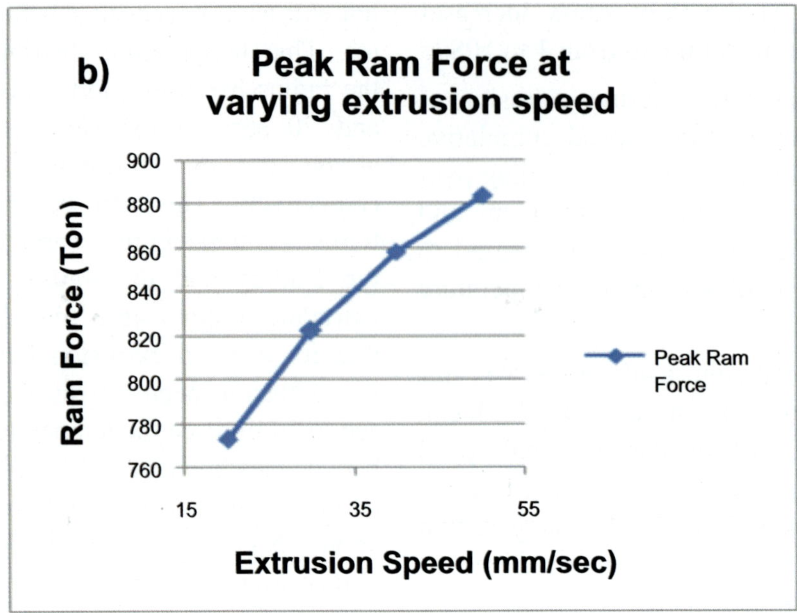

Fig. 3a & b

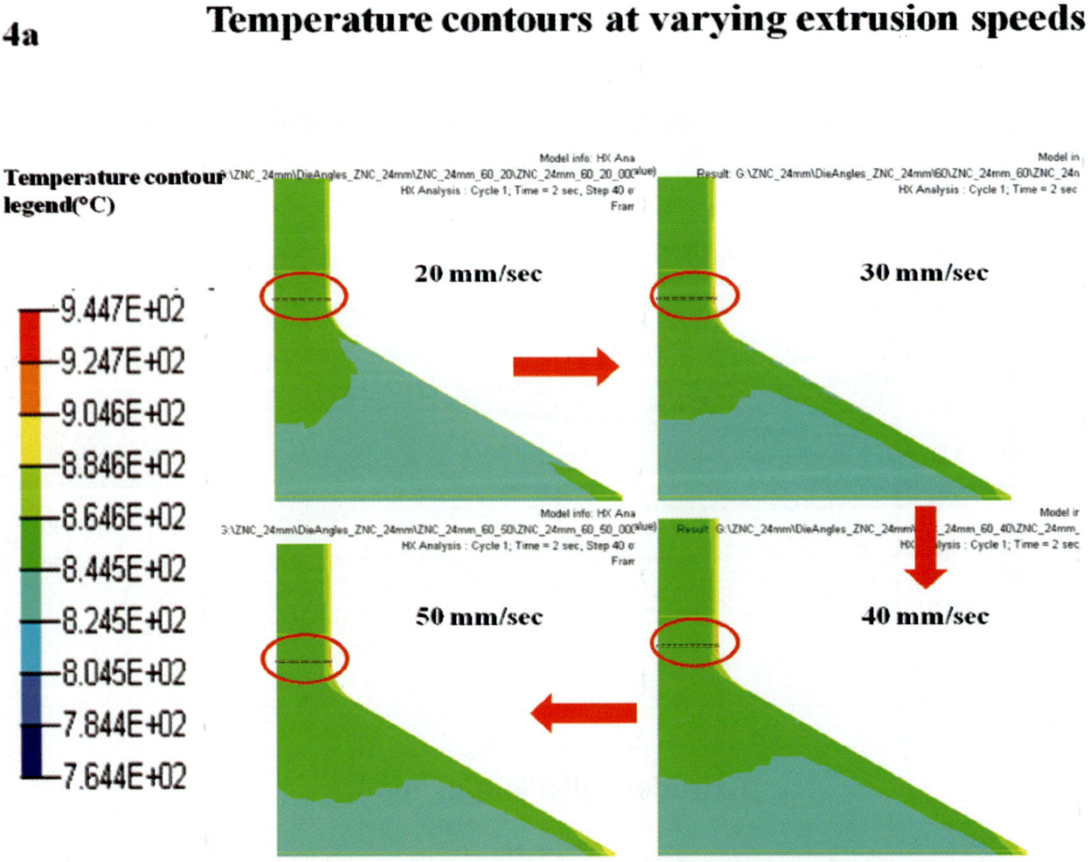

4a
Temperature contours at varying extrusion speeds
Temperature contour legend(°C)
9.447E+02
9.247E+02
9.046E+02
8.846E+02
8.646E+02
8.445E+02
8.245E+02
8.045E+02
7.844E+02
7.644E+02
20 mm/sec
30 mm/sec
40 mm/sec
50 mm/sec

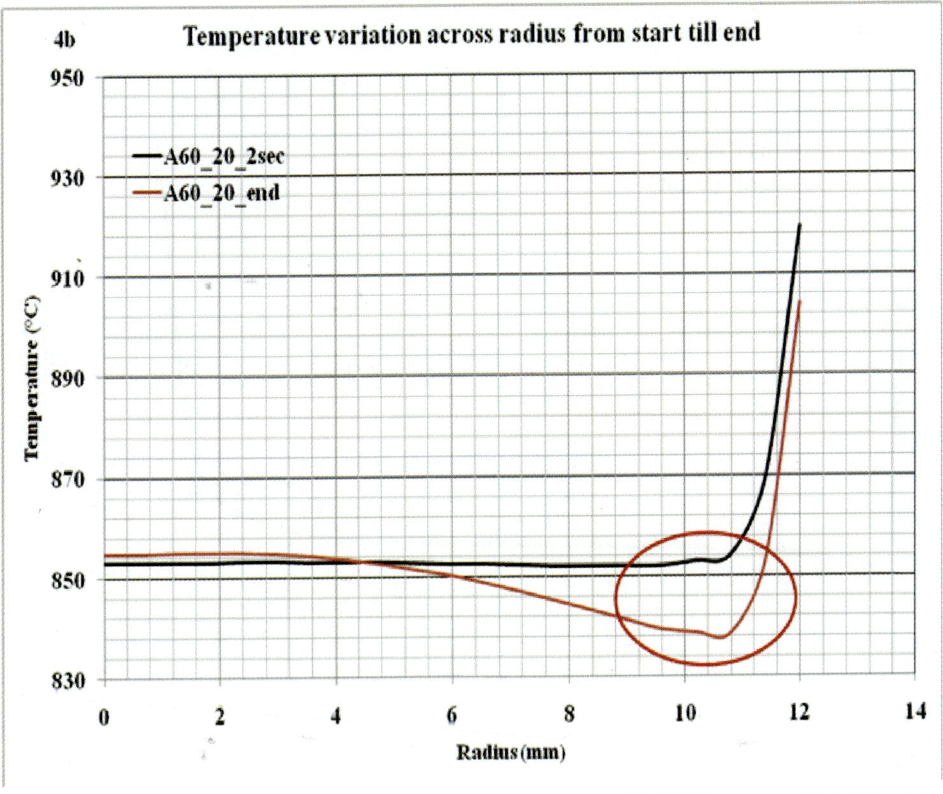
4b
Temperature variation across radius from start till end
A60_20_2sec
A60_20_end
Temperature (°C)
950
930
910
890
870
850
830
0
2
4
6
8
10
12
14
Radius (mm)

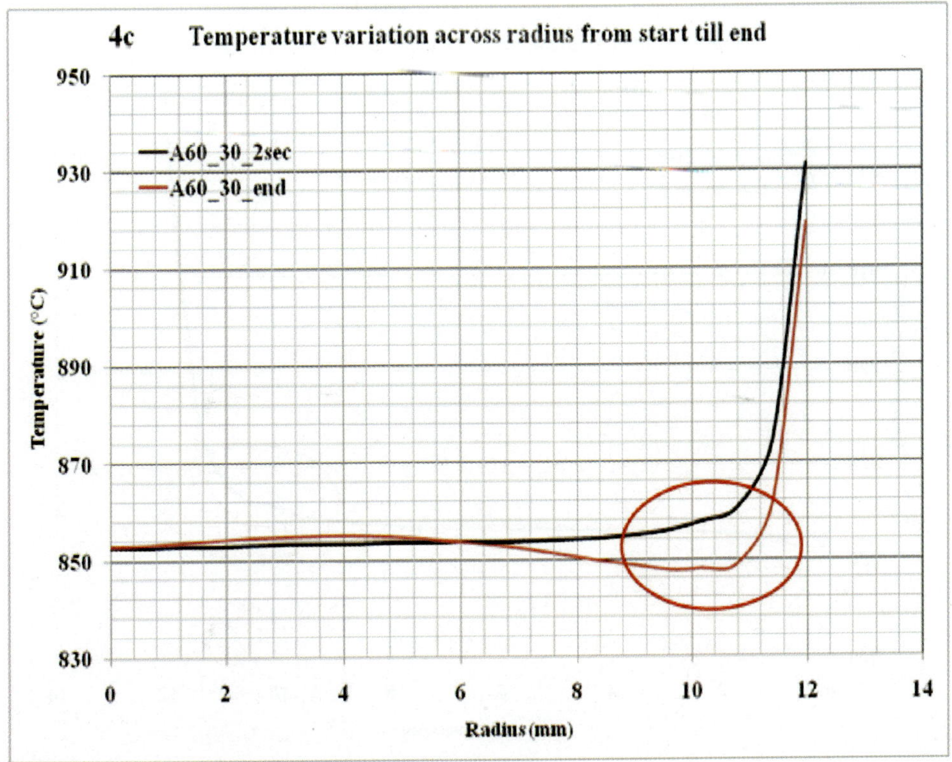
4c
Temperature variation across radius from start till end
A60_30_2sec
A60_30_end
Temperature (°C)
950
930
910
890
870
850
830
0
2
4
6
8
10
12
14
Radius (mm)

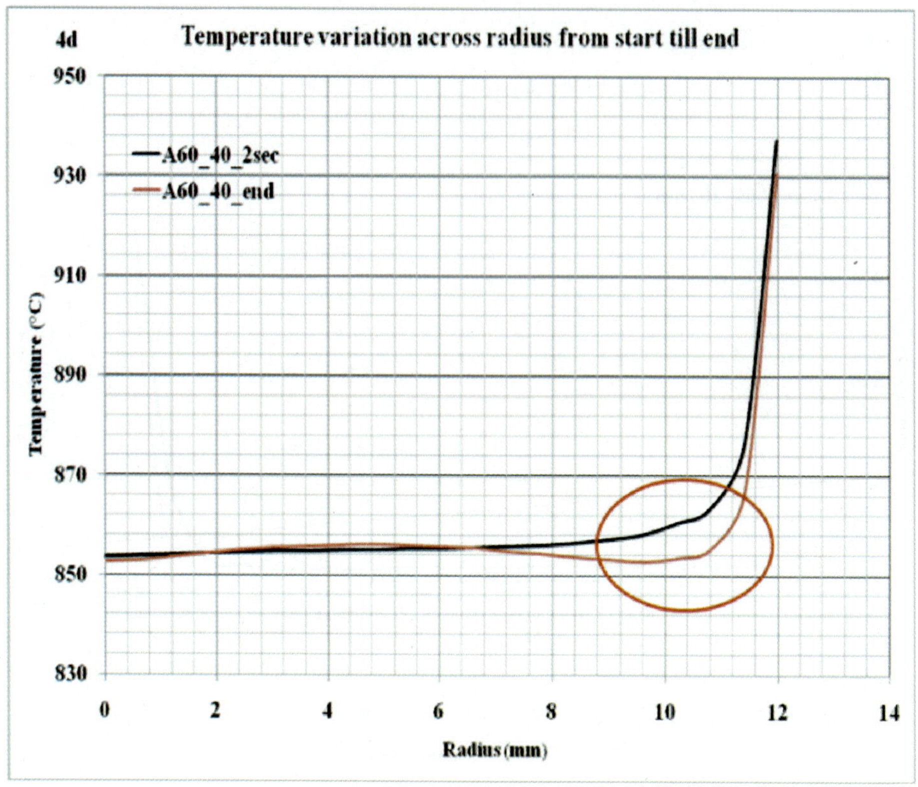
4d
Temperature variation across radius from start till end
A60_40_2sec
A60_40_end
Temperature (°C)
Radius (mm)
950
930
910
890
870
850
830
0
2
4
6
8
10
12
14

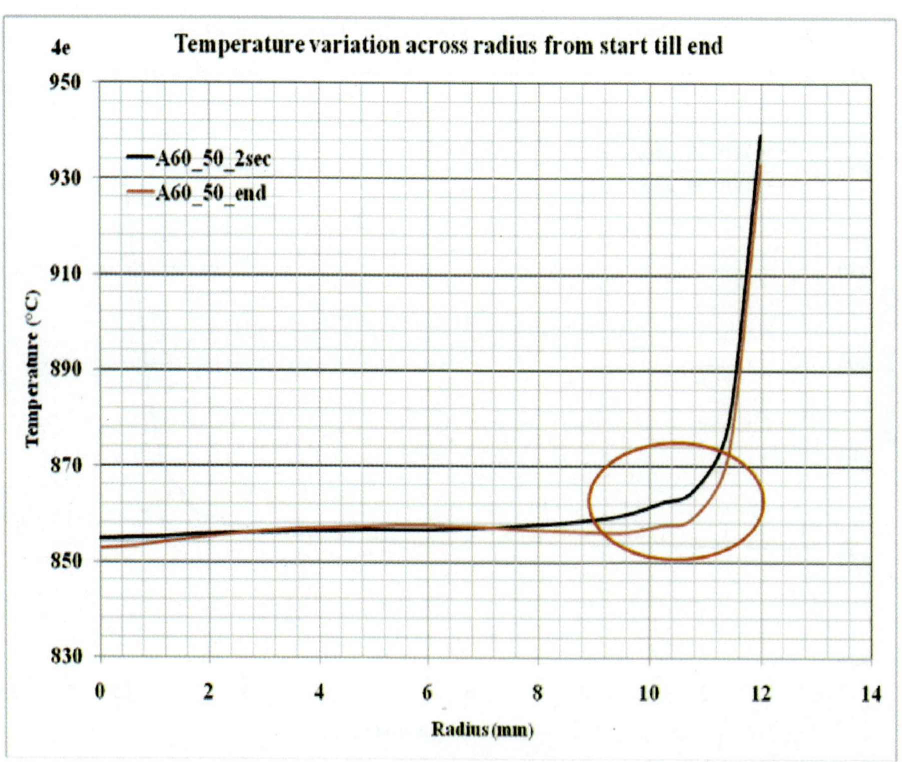
4e
Temperature variation across radius from start till end
A60_50_2sec
A60_50_end
Temperature (°C)
Radius (mm)
950
930
910
890
870
850
830
0
2
4
6
8
10
12
14

CONCLUSIONS

a) In case of Zr-2.5Nb-0.5Cu alloy deformation at lower temperature results in flow instabilities which was seen in flow behavior studies by Gleeble as well as load curve obtained in simulation of extrusion.

b) Strain rate less than equal to 1.0 S-1 resulted in fluctuations of flow stress which is not good for desirablemicrostructure of product and this requires further studies of microstructural studies at different stages of testing.

c) Optimizationof extrusion temperature and ram speed was done with this work for extrusion of Zr-2.5Nb-0.5Cu alloy.

REFERENCES

[1] A. Paine, M. Aloe, and J. Walters. Simulation of hot extrusion processes. Aluminium Extrusion, 7(1):39–44, 2002.

[2] Sofia Hansson, Doctoral thesis, Modeling of the Stainless Steel Tube Extrusion Process, Lulea University of Technology, Department of Applied Physics and Mechanical Engineering, Division of Material Mechanics.

[3] Mahender P. Reddy Computational Mechanics Company, Austin, TX.78757 and Olivier P. Hardy, Harold L. Thomas, Pramod R. Bangarpet Altair Computing, Inc., World Head Quarters, Troy MI, 48084, Optimal Design Of Extrusion Dies Using Hp-Adaptive Finite Element Model, Aiaa-98-4790.

[4] Monica Relli, Francesco Agostoni, Mahender P. Reddy, HakanCarlsson, Ola Bergqvist, Numerical Simulations of Flow Near Bearings During Aluminum Extrusion, 5th World Congress "Aluminium 2000". Rome, Italy. 18-22 March 2003.

[5] Mahender P. Reddy, Edward G. Schaub, Computational Mechanics Company, Inc., Austin, TX Louis G. Reifschneider, Industrial Technology Department, Illinois State University, Normal, IL Harold L. Thomas, Altair Computing, Inc., Irvine, CA, Design And Optimization Of Three Dimensional Extrusion Dies Using Adaptive Finite Element Method, Presented At Antec'99 New York.

[6] Mahender P. Reddy, Ravisankar Mayavaram, David Durocher, HakanCarlsson, Ola Bergquist ,Altair Engineering, USA, Altair Engineering AB, Sweden, Analysis and Design Optimization of Aluminum Extrusion Dies, 8th International Extrusion Technology Seminar – Orlando, FL – May 2004.

[7] Mahender P. Reddy, Osama N. Hamzeh, Computational Mechanics Company, Inc., Austin, TX Charlotte Svensson, HakanCarlsson, Altair Engineering AB, Lund, Sweden, Optimization Of Bearing Surfaces In Aluminum Extrusion Die Design, Presented At Opticon '99.

[8] PHLEXprocess 1.0 An hp-adaptive Finite Element Program for Materials

Processing, ComputationalMechanics Company, Inc. Austin, TX. (1998).

[9] Reddy, J. N., and Reddy, M.P., "Three Dimensional Finite Element Analysis of Non-Newtonian Fluid Flows," in Fluid Dynamics Review, W. G. Habashi (ed.), John Wiley & Sons, pp.437-485, (1995).

[10] Ainsworth, M., and Oden, J. T., "A posteriori error estimation in finite element analysis," Comput. Methods Appl. Mech. Engg., Vol 142,pp. 1-88,(1997).

Optimization of Batch Annealing Parameters for Low Carbon Steel

M.B.N. Raju, Dharmendra Kumar, Ravi Prakash, N. Rajesh and Geert Paulussen
Tata Steel Limited, Jamshedpur-831001, India
Email: mbn.raju@tatasteel.com

ABSTRACT

Market demand for automotive steels has been ever increasing. It is imperative for steel makers to produce the same in higher volumes in order to sustain the desired market share. However, low carbon automotive steel grades require higher annealing time and temperatures when produced through batch annealing. This is because adequate time and temperature is required for the steel to evolve out with the necessary microstructure and adequate mechanical properties, suitable for end use. These steels therefore have comparatively lower productivity/ throughput. Production of these steels is therefore challenging, even more, when product mix is varied and compliance to customer requirements has to be met within time.

With increasing volumes of low carbon steel production and increase in base capacity of batch annealing furnaces, demand for hydrogen consumption also increases. This gives rise to further bottleneck in the process, as generation of hydrogen in a specific unit is limited and depends on capacity of the unit. Such a circumstance calls for optimization of process parameters during batch annealing without impacting desired end properties and customer quality requirements. This was the motivation for the work and the paper would try to deal with the same.

In this work, the authors studied the effect of hydrogen consumption during batch annealing vis a vis end properties and surface characteristics in selected grades of steel. During the process, the authors have tried to optimize the consumption of hydrogen required during batch annealing for low carbon steel, including activity on benchmarking.

KEYWORDS: *Low Carbon Steel, Automotive, Batch Annealing, Hydrogen Consumption*

INTRODUCTION

In CRM, Tata Steel, Jamshedpur, with the increase of EDD/EIF volume and overall productivity, Hydrogen becomes a concern and intensive pressure to optimize the hydrogen consumption without affecting the customer's quality requirements.

Recently we installed two new bases in FY-12 which enhanced to add 3,500 t in CRCA production. And it requires extra volume of Hydrogen (Approx 14000 Nm3).

Nitrogen purging is needed at the start and end of the batch annealing process to prevent mixing of air and hydrogen. Hydrogen is a highly flammable gas, and a mixture of air and hydrogen is highly explosive. To prevent the possibility of an explosion, the nitrogen purging practice is part of the safety measure of the annealing bay control.

Because hydrogen is light, has a low viscosity and actively removes carbon from the strip surface, the first priority in the annealing process is to establish a high hydrogen content in the protective atmosphere. As a first step the nitrogen from the safety purge needs to be flushed out. With the introduction of hydrogen in the gas atmosphere the recirculation fan will start pumping a higher volume at lower power consumption. As a direct effect, heat transport from the inner cover to the coils improves and the heat-up rate of the coils increases. The hydrogen permeates between the coil wraps, which results in a radial conductivity of heat in the coil. Hydrogen atmosphere not only conducts heat more easily to the coils, it also improves heat transport from the coil surfaces to the coil core.

For direct anneal coils, the strip surface is contaminated with traces of rolling emulsion. The moisture will generally evaporate at 100°C. To remove as much of the oil as possible, the annealing cycle is generally designed to have steady heat input between 300°C and 500°C. Although oil is not present on ECL cleaned material the same purging practice is often adopted. The reason for this is that cleaned material often has bound moisture (silicate) on the strip surface which is only released at elevated temperatures, typically around 250°C. At this point in the process a low amount of hydrogen purging is sufficient to flush out oil vapour or water vapour and keep the hydrogen concentration below the inner cover at a high level.

EXPERIMENTAL

Nitrogen is used as a cheap flushing agent for evaporating water until a reducing atmosphere is needed (i.e. from 400°C gas temperature onwards). All material processed in the annealing bay is direct anneal material.

The hydrogen feed is provided by two shutoff valves of unequal size. Opening the large valve will result in a purge rate of approximately 16 m^3/h. Opening the small valve results in a purge rate of 8m^3/h. Opening both results in a purge rate of approximately 24 m^3/h.

The programmed purging practice is as follows for all cycles:

- At a gas temperature of 275°C (rising): start nitrogen flow (~30 m3/h) with a minimum of 1 hour duration. This purge pattern may be delayed until the nitrogen safety purge has finished. The trigger temperature is reached almost immediately when the furnace is fired up.
- At a gas temperature of 400°C (rising): start high hydrogen flow (24 m3/h) with a minimum of 5 hours total duration of high flow.
- At a gas temperature of 500°C (rising) start low hydrogen flow (8 m3/h) with a minimum of 6 hours duration of low flow.
- Start hydrogen supply (no set flow) directly following the previous segment
- At 5°C below soaking set point (rising) start low flow for 0 hours. Effectively no purge is done because the set duration is 0 hours. The programmed pattern prevents the gas programme from progressing into the cooling stage until soaking is actually reached.
- At a gas temperature of 50°C below set point (dropping): start low hydrogen flow (8 m^3/h) for 6 hours.

- Start hydrogen supply (no set flow) directly following the previous segment, until the nitrogen safety purge in preparation of breaking the inner cover seal.

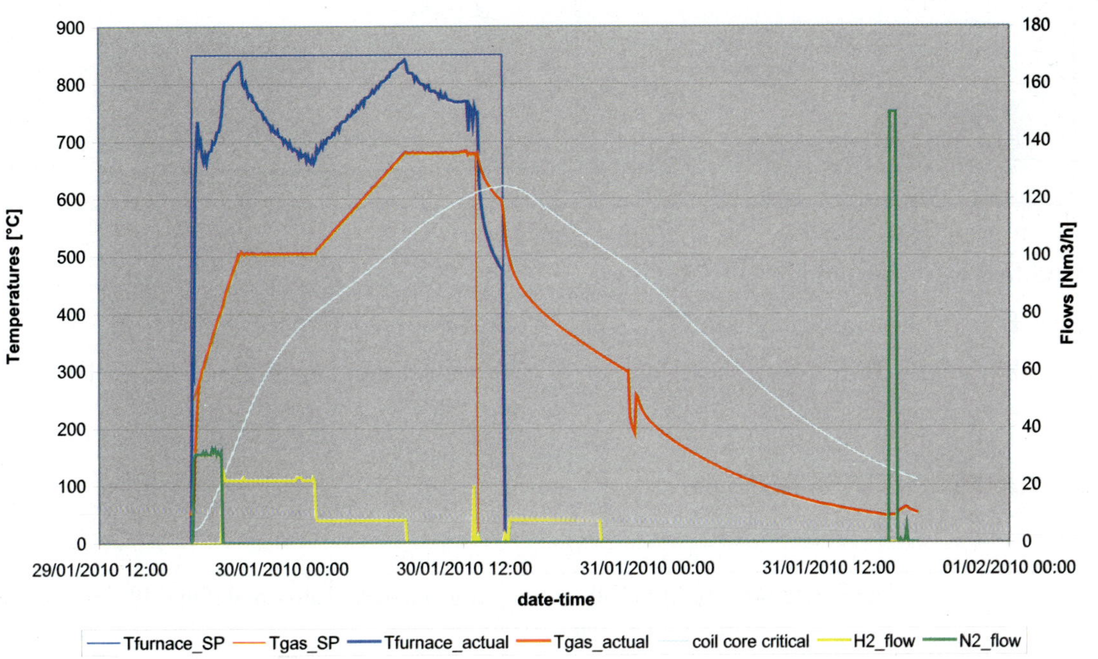

H_2 purge pattern in EAL/EIF/HIF grade before modification

Hence, first of all we have selected EDD/ EIF grades which are ECL cleaned coils to optimize Hydrogen consumption which was similar to BAF direct coils as follows:

i. For first 8 hours @ 30 Nm3/hr,

ii. For next 12-15 hrs @ 15 Nm3/hr

iii. For next 10 hrs @ 10 Nm3/hr,

iv. For next 10 hrs @5 Nm3/hr,

v. For next 8 hrs @0 Nm3/hr,

vi. At the end of cooling 2 hrs @10 Nm^3/hr.

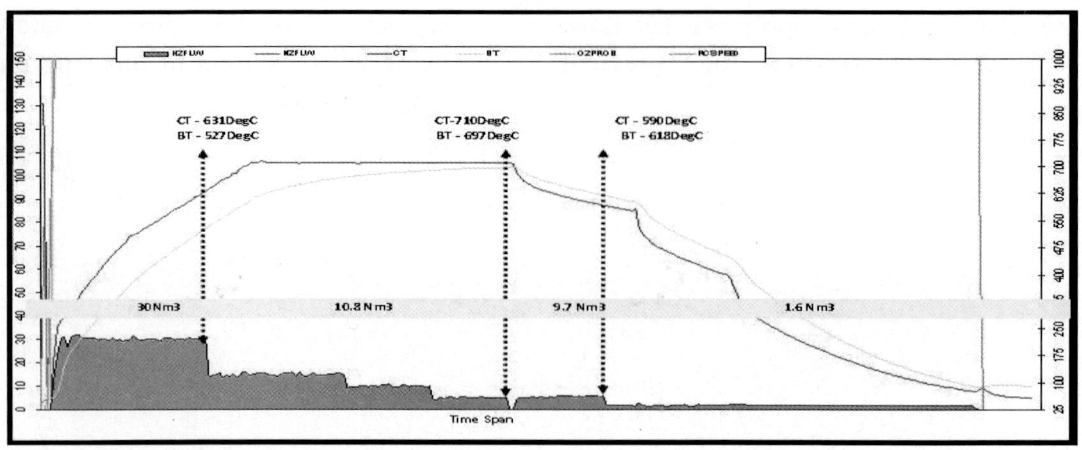

Modified H_2 Purge pattern in EDD Grade

I. For first 8 hours @ 20 Nm3/hr,

II. For next 8-10 hrs @ 10 Nm3/hr

III. For next 20 hrs @ 5 Nm3/hr,

IV. For next 10 hrs @ 0 Nm3/hr,

V. At the end of cooling 2 hrs @ 10 Nm3/hr.

RESULTS AND DISCUSSION

	Past Period (Apr '10 - Jun '10)				Current Period after trial (Sept '10 - Nov '10)			
Grades	Total Production	No. of Stacks	H_2 Consumption (Total in Nm^3)	H_2 Consumption (Nm^3 / t)	Total Production	No. of Stacks	H_2 Consumption (Total in Nm^3)	H_2 Consumption (Nm^3 / t)
EIF	18500	185	68265	**3.69**	21900	219	78402	**3.58**
HIF	4300	43	17415	**4.05**	6400	64	24640	**3.85**
EDD	24700	247	100035	**4.05**	26700	267	103062	**3.86**
Total	47500	475	185715	**3.93**	55000	550	206104	**3.76**

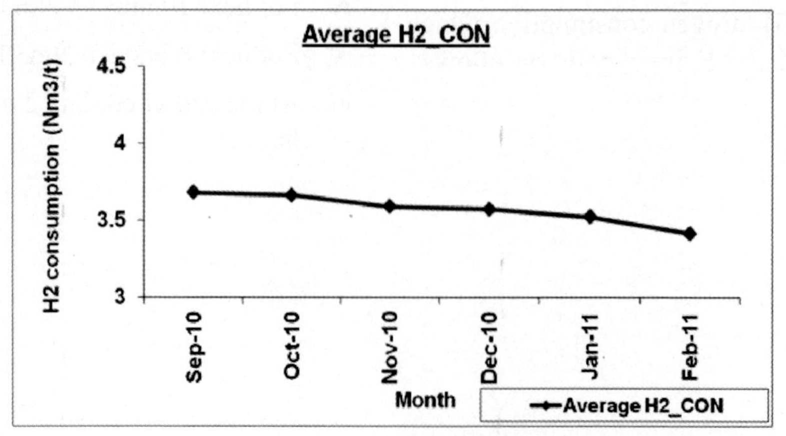

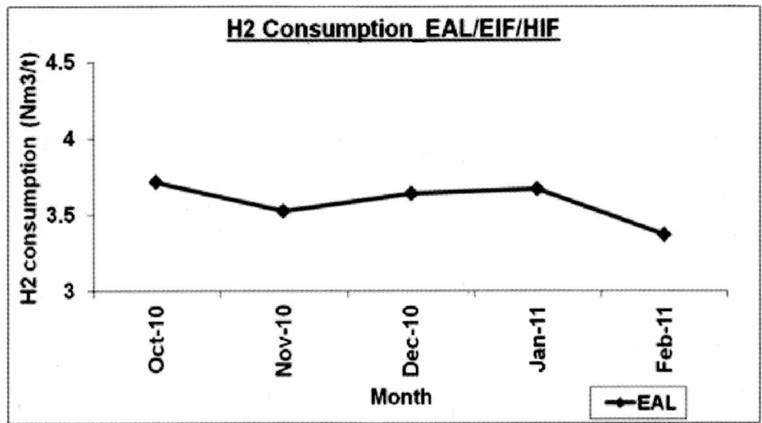

Based on the experimental studies, the following standard operating procedure has been formulated:

- Start with a fast purge of the nitrogen atmosphere at the start of the heating cycle (i.e. apply 150 m^3 of hydrogen at a high flow rate) and keep the remainder of the cycle at a low purge rate (i.e. 8 m^3/h hydrogen) until the end of the cycle.
- If surface cleanliness results remain good: shorten the low purge length during cooling until 1-2% of methane can be detected in the protective gas, or the outside coil surface starts being contaminated with loose soot. This loose soot is not a problem because it easily wipes off and will not result in build-up in the skin pass mill (SPM). The worst case charge to determine this limit with is with a charge of small width, and large coil diameters and a thin gauge to maximise the exposed surface.
- If surface cleanliness results remain good: shorten the low purge length during heating until the outside coil surface starts being contaminated with soot from cracking oil vapour. An alternative is to filter protective gas over a white paper filter. As long as there is a visible yellow discoloration of the filter, oil is evaporating and low purging should continue. Loose soot will only be present on the outside surfaces of the coil, so should not be a problem for the skin pass mill (SPM). The worst case charge to determine this limit is with a charge of large width, and large coil diameters and a thin gauge to maximize the exposed surface and thus the amount of evaporating oil.

CONCLUSIONS

The benefits of reduced hydrogen consumption are two-fold. Obviously hydrogen has a cost per unit, so a reduction in consumption will result in lowered costs. The major gain however, will be the fact that at significant savings in consumption, erection of a second hydrogen plant to fulfil the current need will no longer be necessary.

ACKNOWLEDGEMENT

Authors are grateful to the management of TATA STEEL for the permission to publish this paper. Contribution from colleagues for necessary help in preparing the paper is acknowledged with thanks.

SECTION V

Emerging Trends in Steel Products

Evaluation of DQ&P Processing Route for the Development of Ultra-high Strength Tough Ductile Steels

Mahesh C. Somani[1], David A. Porter[1], L. Pentti Karjalainen[1] and Devesh K. Misra[2]

[1]University of Oulu, Department of Mechanical Engineering, Centre for Advanced Steels Research, Post Box 4200, 90014 Oulu, Finland

[2]University of Louisiana at Lafayette, Dept. of Chemical Engineering & Centre for Structural and Functional Materials, P.O. Box 44130, Lafayette, LA 70504-4130, USA

Email: mahesh.somani@oulu.fi

ABSTRACT

A novel concept of direct quenching and partitioning (DQ&P) processing route has been explored with the specific aim of producing ultra-high strength structural steels with yield strengths ≈1100 MPa combined with good uniform and total elongations and impact toughness. The approach used was to design suitable compositions based on high silicon and/or aluminium content, establish the DQ&P parameters with the aid of physical simulation on a Gleeble simulator and finally, design the DQ&P processing route for trials on a laboratory rolling mill. Dilatation experiments were made on a Gleeble simulator to determine the appropriate cooling rates and quench stop temperatures for obtaining martensite fractions in the range 60-90%. Two types of dilatation tests were conducted starting with either strained or unstrained austenite prior to quenching to roughly simulate industrial rolling with low and high finish rolling temperatures, respectively. Laboratory rolled samples were direct quenched in water to desired quench stop temperatures followed by partitioning in a furnace held at the quench stop temperature. Some samples were cooled very slowly thus simulating the cooling of the coiled strips in industrial rolling. Metallographic studies showed that the desired martensite-austenite microstructures were achieved. Besides high strengths, the ductility (including uniform elongation) and impact toughness were significantly improved. Straining of austenite prior to DQ&P led to extensive refinement of martensite packets/blocks and laths and hence resulted in improved mechanical properties, particularly impact toughness. No adverse effect of prolonged partitioning simulating coiling has been noticed suggesting new possibilities for developing tough ductile structural steels both for strip and plate products.

KEYWORDS: *Direct Quenching, Quenching and Partitioning, Thermomechanical Processing, Martensite, Retained Austenite, Finish Rolling Temperature*

INTRODUCTION

There has been an ever-increasing demand for developing ultra-high strength steels with excellent combinations of mechanical properties such as high strength, ductility, toughness and formability. This, however, requires optimal design of new, inexpensive compositions and/or thermomechanical processing (TMP) routes to meet the new challenges. Conventionally, quenching and tempering is employed to realize high-strength structural steels with good impact toughness. Generally, the ductility of these steels in terms of their elongation or reduction of area to fracture in uniaxial tensile testing is acceptable, but their uniform elongation, i.e., the work hardening capacity is relatively low. Such a deficiency limits the wider application of these steels, because strain localization during fabrication or as a result of overloading can be detrimental to the integrity of the structure. In recent years, a novel concept of quenching and partitioning (Q&P) has been proposed as a potential processing route for improving the balance of elongation to fracture and tensile strength for advanced high-strength steels [1-3]. In essence, the Q&P processing of a cold-rolled sheet involves austenitization at a desired temperature, followed by quenching to a temperature between the martensite start (M_s) and finish (M_f) temperatures and holding at a suitable temperature for an appropriate time to allow the partitioning of carbon from martensite to austenite, which can thereby be partly or fully stabilized down to room temperature. The formation of iron carbides and the decomposition of austenite are intentionally suppressed by the use of Si, Al or P alloying [3-6] unlike in the case of tempering.

In a recent study, the potential use of thermomechanical rolling combined with DQ&P to improve the work hardening, uniform elongation and impact toughness of a high-strength hot-rolled structural steel has been presented [7]. The specific aim was to establish a direct quenching and partitioning (DQ&P) route for the production of a steel with a yield strength ≥ 1100 MPa combined with good ductility and impact toughness. A martensitic matrix has the potential to provide the required high strength, while a small fraction of finely divided retained austenite stabilized between the martensitic laths through carbon partitioning was expected to provide the desired uniform elongation and improved work hardening via TRIP effect. Dilatometer measurements and electron microscopy studies were combined with published information in a fresh attempt to understand the possible microstructural mechanisms operating during Q&P and thereby get closer to the thorough understanding needed to design optimized compositions and/or DQ&P processing routes [8]. This paper presents a brief account of the salient results encompassing the dilatation behaviour, hot rolling aspects, various microstructural mechanisms and mechanical properties.

EXPERIMENTAL PROCEDURES

Two experimental compositions were designed for the DQ&P experiments, one based on a high Si content and one on a high Al content, in wt.% respectively Fe-0.2C-1.5Si-2.0Mn-0.6Cr and Fe-0.2C-0.5Si-1Al-2.0Mn-0.5Cr-0.2Mo, hereinafter referred as High-Si and High-Al steels. The chosen compositions were cast as 70 kg slabs. Pieces of 200 x 80 x 60 mm from these castings were homogenized and hot rolled to a 14 mm thick plate from which cylindrical specimens Ø6x9 mm or Ø5x7.5 mm were cut with the axis of the cylinders transverse to the rolling direction in the rolling plane. These were used for thermomechanical treatment and dilatation measurements on a Gleeble 1500 simulator. Preliminary CCT measurements suggested that cooling at 30°C/s would be appropriate to avoid transformation prior to reaching the M_s temperature. Two austenite states were investigated prior to quenching, recrystallized and strained, in order to roughly simulate industrial rolling with high and low finish rolling temperatures.

In the case of High-Si steel, for undeformed austenite, samples were reheated at 20°C/s to 1150°C, held for 2 min prior to cooling at 30°C/s to a quench-stop temperature (T_Q) below the M_s temperature giving initial martensite fractions in the range 60-85%. The samples were then partitioned for different times (Pt) in the range 10 - 1000 s at T_P ($\geq T_Q$), followed by cooling in air to room temperature (RT) at ~10-20°C/s down to 100°C. In the case of deformed austenite, samples were cooled at 2°C/s to 850°C, held 10 s, and then compressed with three hits each having a strain of ~0.2 at a strain rate of 1 s^{-1}. The time between hits was 25 s. The specimens were then held 25 s prior to cooling at 30°C/s to a quenching temperature below M_s giving initial martensite fractions of 75-85%, Fig. 1.

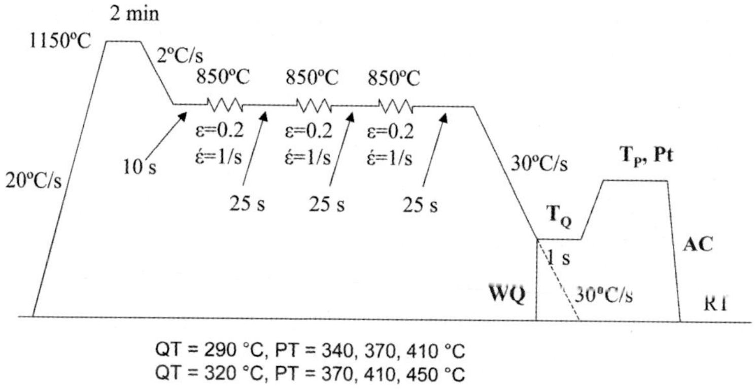

Fig. 1: Thermomechanical simulation schedule

In case of High-Al steel, the difference was essentially in reheating conditions of the specimens (10°C/s to 1000°C, held for 2 min) prior to subjecting to Q&P schedules for both unstrained and strained specimens. As compared to a more detailed evaluation of Q&P processing in the case of High-Si steel, the study was restricted to roughly determine appropriate rolling conditions for High-Al steel and hence, specimens were partitioned for 10 - 1000 s only at the quench stop temperature, i.e., $T_P = T_Q$ giving initial martensite fractions of about 70-90%.

Vickers hardness measurements were made on all dilatation specimens using a 5 kg load. This was followed by microstructural characterization using a light optical microscope (LOM). The volume fraction and mean carbon content of the retained austenite in the dilatation specimens was determined using X-ray diffraction (XRD; $18.5° < 2\theta < 42°$, Mo_{Ka} radiation, 40 kV, 40 mA). Selected specimens were subjected to more detailed metallographic examination using transmission electron microscopy (TEM) and scanning electron microscopy combined with electron backscatter diffraction (SEM-EBSD).

Based on the results of dilatation experiments, rolling trials were made on both the steels in a two-high, reversing laboratory rolling mill with blocks of 110 x 80 x 60 mm cut from cast ingots. The temperature of the samples during rolling and quenching was monitored by thermocouples placed in holes drilled in the edges of the samples to the mid-width at mid-length. The samples were heated at 1200°C for 2 h in a furnace prior to two-stage rolling. In the first stage, hot rolling was carried out in the recrystallization regime in 4 passes to a thickness of 26 mm with about 0.2 strain/pass with the temperature of the fourth pass at about 1030-1050°C. The second stage comprised waiting for the temperature to drop to ≈900°C and then rolling to a thickness of 11.2 mm with 4 passes of about 0.21 strain/pass and a finish rolling temperature (FRT) at 800-820°C. Immediately after rolling, the samples were quenched in a tank of water close to desired T_Q and then subjected to partitioning treatment in a furnace at the same temperature for 10 min (i.e., $T_P = T_Q$), Fig. 2. Some rolled samples were cooled very slowly over 29-30 hours by switching off the furnace in order to simulate the cooling of the coiled strips in actual industrial rolling.

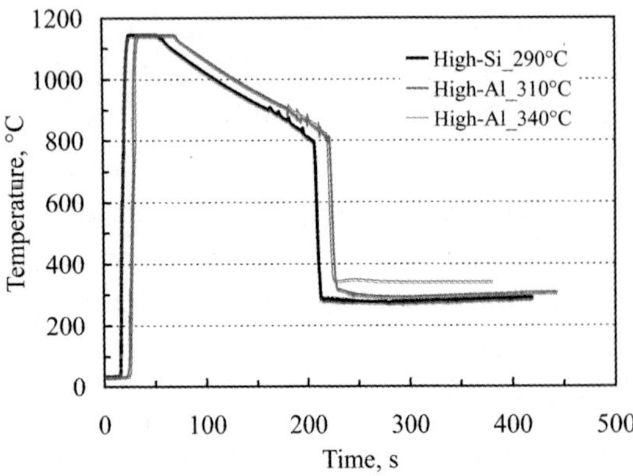

Fig. 2: Examples of recorded rolling and cooling cycles during DQ&P processing

In order to compare the mechanical properties of the laboratory rolled DQ&P samples with those obtainable on a composition giving similar strength properties, a steel containing 0.14C-1.13Mn-0.2Si-0.71Cr-0.15Mo-0.03Ti-0.0017B was also laboratory rolled in the same way as described above and directly quenched to room temperature in water.

RESULTS AND DISCUSSION

Retained Austenite and Hardness

The dilatation curves of specimens cooled at 30°C/s to RT showed that there was no transformation on cooling prior to reaching the M_s temperature at about 395 and 405°C respectively, for High-Si and High-Al steels, respectively. A simple analysis of the dilatation curves for High-Si steel indicates that quenching at 340, 320 and 290°C should produce initial martensite fractions of 60, 75 and 84%, respectively. Likewise, for High-Al steel, quenching at 340, 310 and 290°C gave initial martensite fractions of 70, 86 and 91%, respectively. Subsequent partitioning of High-Si Q&P specimens at the quenching or higher temperatures up to 450°C resulted in various measured fractions of austenite, but the variations in the retained austenite fractions were greater than the expected accuracy of the XRD measurements (about 2%), see Fig.3 for instance. Likewise, no systematic variations were observed for High-Al steel held for different partitioning times at the quenching temperature. A detailed explanation for the variations could be decomposition of austenite into bainite and/or isothermal martensite or due to carbide precipitation despite the high Si content in the steel [1-6]. Further work carried out to understand the microstructural mechanisms operating during partitioning treatment is published elsewhere [8] and will be briefly presented in subsequent section.

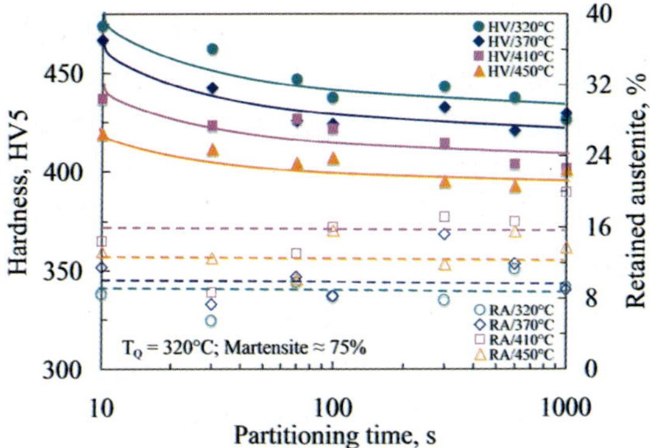

Fig. 3: Hardness (HV5) and retained austenite (RA) vs. partitioning time in unstrained Gleeble specimens quenched to 320°C

Table I shows that as expected mean retained austenite contents increase as T_Q and T_P increase, however prior strain has only little effect. It is, therefore, expected that the amounts of retained austenite will generally be low in rolled DQ&P samples, as partitioning temperatures would normally be ≤ T_Q and for long durations, such as in case of coiled strips. The mean carbon content of the retained austenite as estimated from lattice parameter measurements does not show any systematic dependence on holding time for both steels, Table I. The average carbon content of the Gleeble tested Q&P specimens (0.7-1.2%,irrespective of the steel type and/or austenite state) is much lower than would be expected from simple calculations suggesting that not all carbon is partitioning into austenite. Moreover, the dilatation curves of Q&P specimens showed that there is a gradual expansion during partitioning and required systematic study owing to widely differing opinions in literature, cf. [2,9,10,11].

Table I: Average retained austenite and corresponding carbon contents of Q&P specimens partitioned 10-1000 s.

Steel	T_Q, °C	RA /CC, %	Unstrained						3x0.2 strained			
		T_P, °C	290	320	340	370	410	450	340	370	410	450
High-Si	290	RA	6.7		8.9	11.1	11.3		8.8	11.1	12.3	
		CC	0.68		-	1.2	1.1		0.92	1.16	0.86	
	320	RA		8.4		10.5	15.1	13.2		11.2	14.6	16.4
		CC		0.57		0.8	0.8	0.9		0.92	0.86	0.89
	340	RA			8	10.1						
		CC			0.87	0.8						

Steel	T_Q, °C	RA /CC, %	Unstrained			3x0.2 strained		
		T_P, °C	290	310	340	290	310	340
High-Al	290	RA	7			7.1		
		CC	1.2			1.2		
	310	RA		6.6			8.8	
		CC		0.66			0.81	
	340	RA			7.4			8.8
		CC			1.0			1.1

RA = Retained austenite, CC = Carbon content

An example of hardness vs. partitioning time for unstrained austenite in the case of High-Si Q&P specimens is given in Fig. 3. As expected, holding at the partitioning temperatures, particularly at high temperatures leads to softening, i.e., tempering of the martensite, with the rate of softening decreasing with holding time (note the logarithmic scale). As expected, lower quenching temperatures produce higher hardness due to the higher initial martensite fractions. It is interesting to note that prior straining of the austenite had only little effect on final hardness [7].

Microstructural Mechanisms

An example of gradual expansion during partitioning is depicted in Fig. 4, which shows how the specimen diameter varies with time up to at least 1000 s after reaching T_Q = 320°C. After 1 s at T_Q, heating to T_P occurs at 30°C/s, which explains most of the expansion immediately after 1 s. Apart from this, on a logarithmic scale, the diameter increases slowly (region I) and

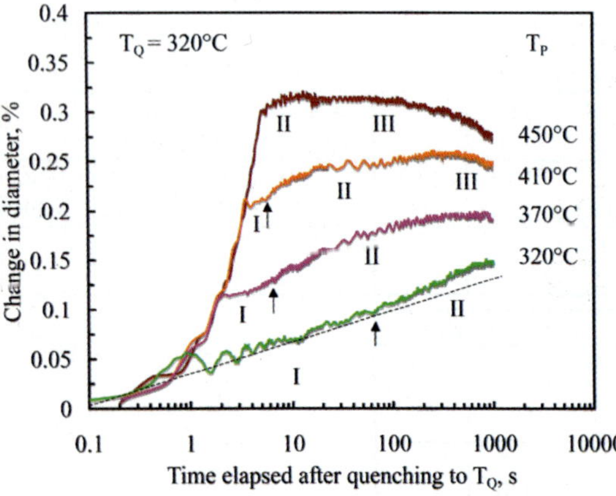

Fig. 4: Change in specimen diameter with time at different partitioning temperatures (T_P) following quenching at T_Q = 320°C

then more rapidly (region II). In some cases, a third region can be identified (region III), in which the steel contracts.

Carbon partitioning begins immediately below Ms during cooling to T_Q and continues thereafter. It can be shown that the linear dilatation caused by complete partitioning for the present steel will only be 0.027-0.033% [12]. Therefore, other microstructural changes are also affecting the dilatation apart from thermal expansion on heating from T_Q to T_P and carbon partitioning into austenite.

One of these changes is probably the formation and migration of ledges that allow the martensitic laths to grow into the austenite isothermally, thus producing the irregular laths with ledge-like protrusions and constitutes together with carbon partitioning, the expansion observed in region I of Fig. 4 [8,9]. With time, some austenite pools transform to lower bainite, i.e., ferrite laths containing carbides aligned in the same direction, marked as region II in Fig. 4. The time required for this to initiate decreases as T_P is increased indicating that the chosen values of T_P are largely below the lower bainite nose of the TTT diagrams. The austenite pools into which lower bainite grows will tend to be the alloy-rich regions of the original dendritic structure, and they will see developing carbon concentration profiles due to the carbon leaving the untransformed regions. In effect, there will be a range of "C-curves" for bainite that will vary with positions and time. It is also expected that the C curves will be shifted to earlier times than would be predicted on the basis of the local composition, due to the accelerating effect of the martensite on the nucleation of bainite [13].

The concentration seen in region III in Fig. 4, especially at the highest partitioning temperatures suggests that martensite is contracting due to tempering. No evidence for carbide precipitation in martensite was found in TEM study indicating that carbon in the martensite laths is perhaps clustering, but this needs further investigation.

During the final cooling at about 20°C/s, some bainite formation is possible in the case of high T_p, followed by the formation of untempered, in some cases, twinned, high carbon martensite starting at temperatures close to 200°C. The corresponding carbon contents (0.7-0.9%) computed from standard equations given in literature are comparable with the estimated carbon contents of the average retained austenite of the Q&P samples partitioned for 10-1000 s, Table I [8].

Microstructures of Dilatation Specimens

It has earlier been presented in the case of High-Si steel that 3x0.2 straining in austenite prior to Q&P processing resulted in finer packers and blocks of fine martensite laths [7]. Similar effects of prior straining on the microstructural refinement were noticed in the case of High-Al steel too. A typical example of the optical microstructures of the unstrained specimens following the Q&P process is presented in Fig. 5a, which shows mainly coarse packets and blocks of martensite laths. Specimens that were compressed at 850°C and subjected to a similar Q&P process typically showed finer packets and blocks of fine martensite laths, shortened and randomized in different directions, Fig. 5b. Such a microstructure should enhance the impact and fracture toughness due to the need for more tortuous crack path propagation [14]. Since the martensite/austenite interface is effectively constrained, LOM does not show much change in microstructure following holding at high temperatures and/or long times except for slight truncation of the martensite laths.

Select Q&P specimens from High-Si steel without prior straining were examined in a FEG-SEM. Fig. 5c presents an example of a phase contrast image taken on an unstrained specimen quenched to 290°C and partitioned at 410°C for 100 s. Elongated pools of austenite can be seen between the martensitic ferrite laths. TEM investigations also showed the presence of fine interlath austenite. A typical example is shown in Fig. 5d for a specimen quenched at 320°C and partitioned at 410°C for 600 s. The dark field image at the same location further reveals the interlath austenite clearly, Fig. 5e. Further microstructural characterization of strained specimens is yet to be completed.

Property Evaluation of Rolled DQ&P Samples

Microstructural features in respect of martensite packets and block sizes were carried out on select samples from both High-Si and High-Al steels following laboratory rolling and DQ&P processing. The structural features were quite similar to those seen in optical microstructures of Gleeble simulated specimens, suggesting possibilities of controlling fairly accurately the deformation conditions in rolling as well as direct quenching at high rates (average cooling rates about 30-35°C/s

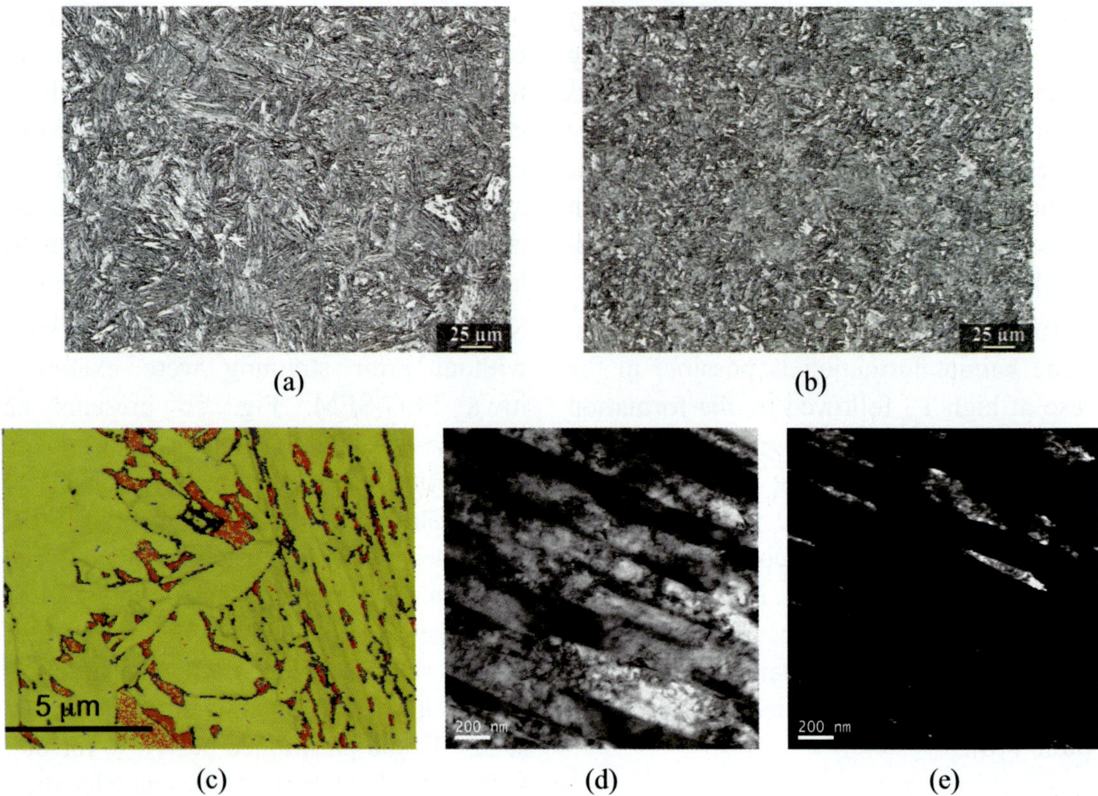

Fig. 5: Typical microstructures of Gleeble simulated DQ&P specimens. Light optical microscopy showing microstructures of unstrained (a) and strained (b) specimens in High-Al steel. Interlath austenite revealed in High-Si steel by FESEM-EBSD (c) and TEM (d and e).

up to about 400°C) close to partitioning temperatures. Typical austenite contents as measured by XRD were in the range 6-9% and 4.5-7% for High-Si and High-Al steels, respectively, irrespective of partitioning temperature. As mentioned earlier, some of the rolled samples were subjected to long time partitioning, while being cooled extremely slowly over 29-30 hours by switching off the furnace thus simulating the cooling of the coiled strips in industrial rolling. Hardness measurements made on rolled samples showed closely similar levels of hardness as compared to those observed in Gleeble specimens, particularly for High-Si steel (about 440-450 HV). However, some High-Al steel samples showed hardness values as low as 390 HV, as a consequence of very fine strain induced ferrite noticed in some sections of the rolled plates suggesting drop in local temperatures during controlled rolling and the need for slightly increasing the FRT in the case of High-Al steel.

Examples of typical engineering stress-strain curves recorded at room temperature on controlled rolled, direct quenched and partitioned samples are given in Fig. 6. For comparison, typical tensile curves from the lab-rolled 0.14C DQ martensitic steel of similar yield strength have also been plotted. The advantages of DQ&P

processing over simple DQ in respect of total ductility and uniform elongation are evident, even though the tensile strength of High-Al steel is somewhat lower and may require further improvement. However, there was no significant influence of coiling simulation (long time slow cooling over 29 hours in furnace) on tensile properties of either steels, as will be shown later. XRD measurements near the fractured ends of some tensile specimens of High-Si steel revealed that the retained austenite content had dropped during straining from the original 7.5% to about 3.5% showing that a large fraction of the original retained austenite transformed during straining.

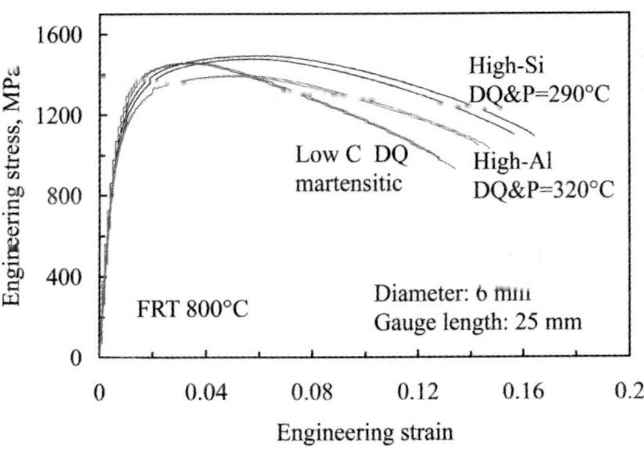

Fig. 6: Typical engineering stress-strain curves; specimens transverse to rolling direction

Fig. 7 presents Charpy impact energy as a function of test temperature for both High-Si and High-Al steels lab-rolled down to low finish rolling temperature, followed by direct quenching and partitioning at 320°C. In the case of High-Al steel, the DQ&P sample was subjected to coiling simulation (CS; cooled slowly in furnace allowing non-isothermal partitioning to occur over long time). For comparison, the impact toughness data obtained on specimens from a High-Si steel sample rolled at a high finish rolling temperature (recrystallized unstrained austenite) followed by direct quenching and partitioning at 320°C are also presented in the same graph. As can be discerned from Fig. 7, both the steels have shown comparable impact toughness properties at sub-zero temperatures. However, the upper shelf energy of High-Al steel is significantly lower than that of High-Si steel.

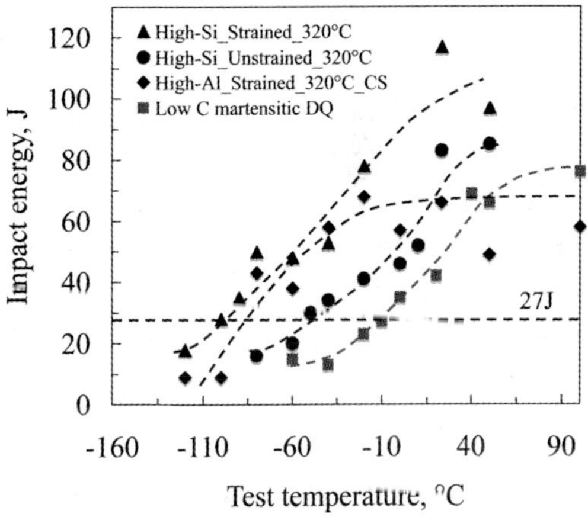

Fig. 7: Typical Charpy V impact energy as a function of test temperature (10 x 10 mm specimens; transverse to rolling direction)

The advantage of rolling at low temperatures in the no-recrystallization temperature range on impact toughness is clearly evident: the 27J transition temperature (T27J) is lowered by about 55°C in the case of High-Si steel. This is consistent with the observed refinement of the martensite packet and block sizes, and the shortening and randomization of the lath structure caused by straining of the austenite prior to quenching. It is also possible that retained austenite has a beneficial effect on the transition temperature. Fig. 7 also shows that DQ&P processing improves the transition temperature compared to that of the lower carbon fully martensitic DQ steel. Results presented in Figs. 6 and 7 show that good combinations of strength, ductility and impact toughness are possible with laboratory rolling, despite the fact that the total hot rolling reduction is much lower than that available in an industrial process. Optimization of the hot rolling and cooling conditions should allow even better combinations of properties to be produced on an industrial scale. The Q&P concept has recently been considered in literature in context of hot-strip mill processing, whereby a desired fraction of martensite forms during cooling to the coiling temperature, followed by non-isothermal partitioning of carbon from supersaturated martensite to remaining austenite, thereby stabilizing it to room temperature [15].

Table II presents a summary of mechanical properties obtained on rolled DQ&P samples from both the steels. Also included are the tensile and impact property data of the lower carbon, fully martensitic DQ steel. The 0.2% yield strength ($R_{p0.2}$) is marginally lower than the target level (1100 MPa) but for example temper rolling should be able to raise the yield strength to the required level (e.g. 1.0% proof strength ($R_{p1.0}$) values are of the order of 1300 MPa). Referring to the High-Si DQ&P samples, both yield and tensile strengths obtained on recrystallized DQ&P specimens (FRT =1000°C) are marginally lower than those finish rolled at the lower FRT of 800°C. While the total elongation (A) and reduction of area to fracture (Z) vary in a narrow range, the total uniform elongation (A_{gt}) and the plastic uniform elongation (A_g) are somewhat better at the lower quenching temperature of 290°C (5.5% and 4.7%, respectively). Even the sample quenched to 290°C, followed by long-time non-isothermal partitioning during cooling in furnace (coiling simulation) too did not show any significant change in properties, except marginally lower uniform elongation (A_{gt} = 4.6%, A_g = 3.9%). The elongation to fracture with a proportional gauge length (A) is the standard value (EN 10002-1) used for hot rolled materials in Europe. The temperatures

Table II: Tensile* and Charpy impact properties of laboratory rolled DQ&P plates.

Steel	FRT (°C)	T_Q / T_P (°C)	$R_{p0.2}$ (MPa)	$R_{p1.0}$ (MPa)	R_m (MPa)	A (%)	A_{gt} (%)	A_g (%)	Z (%)	T27J (°C)	T50% (°C)
High-Si DQP	800	290	1054	1339	1489	13.5	5.5	4.7	53	-99	-28
		290 CS	1029	1355	1505	12.8	4.6	3.9	54	-108	0
		320	1047	1323	1391	12.9	3.2	2.5	55	-100	-6
	1000	320	953		1393	13.8	4.4	3.7	56	-44	15
High-Al DQP	800	340	1074	1310	1344	12.3	2.9	2.2	53	-55	-7
		360	1076	1283	1306	11.7	2.1	1.5	56	-41	15
		270 CS	1180	1413	1495	9.9	3.9	3.2	46	-87	0
		320 CS	1020	1271	1393	11.1	4.4	3.7	48	-90	-6
	890	310CS	1116	1343	1393	9.3	3.0	2.3	52	-67	-4
Low C DQ	800	DQ	1108		1452	11.5	3.5	2.8	57	-12	25

* Average of 3 tests CS = Coiling simulation

corresponding to 27J impact energy (T27J) and 50% shear fracture (T50%) are quite low and distinctly improved over those of the low C martensitic DQ steel (see also Fig. 7). For the plates with FRT 800°C, the reasons for the lower tensile strength and uniform elongation values obtained with T_Q/T_P=320°C compared to 290°C need further investigation.

Referring to High-Al steel, the DQ&P samples quenched at 340 and 360°C showed relatively lower tensile strength as compared to that obtained on DQ martensitic steel. Even the total and plastic uniform elongations were at the same level as the DQ steel or even lower. However, the properties were better at lower quenching temperatures of 320 and 270°C. Despite long-time non-isothermal partitioning (coiling simulation) at these temperatures, the tensile properties including the total uniform elongation (A_{gt}) and plastic uniform elongation (A_g) were reasonably improved. In particular, the temperatures corresponding to 27J impact energy (T27J) and 50% shear fracture (T50%) are quite low and comparable to those obtained on High-Si DQ&P steel and distinctly improved over those of the martensitic DQ steel. Tiny fractions of strain-induced ferrite noticed in some locations of a few High-Al plates may have influenced the mechanical properties and calls for adjusting the chemical composition or modifying the thermomechanical processing conditions prior to DQ&P processing. An attempt to raise the finish rolling temperature to 890°C, however, was counterproductive and marginally reduced the uniform elongation (A_{gt} = 3.0%, A_g = 2.3%) as well as raised the T27J temperature (-67°C), possibly as a consequence of deformation in partial recrystallization regime. This needs further investigation.

SUMMARY AND CONCLUDING REMARKS

Preliminary results in respect of microstructures and mechanical properties of two experimental High-Si and High-Al steels processed through direct quenching and partitioning processing schedules have been presented. On the basis of Gleeble simulations, laboratory rolling experiments were designed to achieve either recrystallized or non-recrystallized austenite prior to direct quenching and partitioning in a furnace at the quenching stop temperature. Metallographic and XRD studies revealed clearly the presence of extended pools of retained austenite between martensite laths. Also, the advantage of straining in austenite in respect of the refinement of martensite packets/blocks and lath structure was clearly evident and resulted in significant improvement in mechanical properties, particularly the impact toughness. Dilatometric measurements and electron microscopy have shown that besides carbon partitioning, isothermal martensite and bainite form at or close to the partitioning temperature. At high partitioning temperature, martensite tempering competes with the partitioning process. Bainite as well as high carbon martensite can form on cooling from the partitioning temperature. Promising results in respect of microstructures and mechanical properties as compared to those of a lower carbon, fully martensitic DQ steel, indicates that there are possibilities for developing tough ductile structural steels through the DQ&P route.

ACKNOWLEDGEMENTS

The authors would like to express their gratitude to the FIMECC Oy (Finnish Metals and Engineering Competence Cluster Ltd.), Tekes (Finnish Funding Agency for Technology and Innovation) and Rautaruukki Oyj for funding this research activity under the auspices of Light and Efficient Solutions Program.

REFERENCES

[1] J.G. Speer, D.V. Edmonds, F.C. Rizzo and D.K. Matlock, Current Opinion in Solid State Mater. Sci., Vol. 8 , 2004, p. 219.

[2] H.Y. Li, X.W. Lu, W.J. Li and X.J. Jin, Metall. Mater. Trans. A, Vol. 41A, 2010, p. 1284.

[3] E. De Moor. S. Lacroix, A.J. Clarke, J. Penning and J.G. Speer, Metall. Mater. Trans. A, Vol. 39A, 2008, p. 2586.

[4] J.G. Speer, F.C.R. Assunção, D.K. Matlock and D.V. Edmonds, Mater. Research, Vol. 8, No. 4 , 2005, p. 417.

[5] A.J. Clarke, J.G. Speer, M.K. Miller, R.E. Hackenberg, D.V. Edmonds, D.K. Matlock, F.C. Rizzo, K.D. Clarke and E. De Moor, Acta Mater., Vol. 56, 2008, p. 16.

[6] D.V. Edmonds, K. He, F.C. Rizzo, B.C. De Cooman, D.K. Matlock and J.G. Speer, Mater. Sci. Eng. A, Vol. 438-440, 2006, p. 25.

[7] M.C. Somani, L.P. Karjalainen, D.A. Porter and D.K. Misra, Mater. Sci. Forum, Vol. 706-709, 2012, p. 2824 .

[8] M.C. Somani, D.A. Porter, L.P. Karjalainen and D.K. Misra, Proc. Int. Symp. on Recent Developments in Steel Processing, MS&T'12 Conf. & Exhibition, Pittsburgh, PA, 2012 (in press).

[9] D. Kim, J.G. Speer and B.C. De Cooman, Metall. Mater. Trans. A, Vol. 42A, 2011, p. 1575.

[10] J. Speer, D.K. Matlock, B.C. De Cooman and J.G. Schroth, Acta Materialia, Vol. 51, 2003, p. 2611.

[11] S.M.C. van Bohemen, M.J. Santofimia and J. Sietsma, Scripta Mater., Vol. 58, 2008, p. 488.

[12] M.J. Santofimia, L. Zhao and J. Sietsma, Mater. Sci. Forum, Vol. 706-709, 2012, p. 2290 .

[13] V. Smanio and T. Sourmail, Solid State Phenomena, Vol. 172-174, 2011, p. 821.

[14] K. Takahashi and M. Hashimoto, Int. Conf. on Thermomechanical Processing of Steels and Other Materials (THERMEC'97), Wollongong, Australia, The Minerals, Metals & Materials Society, 1997, pp. 395 – 401.

[15] G.A. Thomas, J.G. Speer and D.A. Matlock, Proc. Int. Conf. on New Developments in Advanced High Strength Sheet Steels, Orlando, Florida, AIST, 2008, pp. 227 – 236.

Structure-Property Correlation in TRIP Aided Steels

Shiv Brat Singh[1], Tanmay Bhattacharyya[1, 2], Ravi Ranjan[1], Sandip Bhattacharyya[2]
and Debashish Bhattacharjee[2]
[1]Department of Metallurgical and Materials Engineering, IIT Kharagpur
[2]R&D and SS, Tata Steel Limited
Email: sbs22@metal.iitkgp.ernet.in

ABSTRACT

The need for light-weight auto bodies with improved crash resistance and other safety parameters has encouraged the development of a new family of multi-phase steels having higher strength and better formability. High strength TRIP aided steel is the ideal material for such applications, but developing a grade that has good formability, coatability and weldability is a major challenge. The present work has been undertaken to address the issues of coatability and weldability as well as to target a high strength - ductility balance. The work includes Thermo Calc calculations to study the phase relationships, thermo-mechanical simulation, prediction of microstructure through artificial neural network and laboratory experiments in hot-dip galvanising simulator. Routine characterisation was done for the assessment of microstructure and properties.

Three experimental steels were prepared in a vacuum induction furnace. The as-cast ingots were forged and then hot and cold rolled. Laboratory salt bath heat treatment of the cold rolled samples was carried out following the standard two-step heat treatment cycle consisting of intercritical annealing and isothermal bainitic transformation to obtain the desired microstructure and target mechanical properties. As a part of weldability assessment, the heat treated samples were spot welded.

A difficulty with TRIP steels is their poor wettability during galvanising due to the formation of oxides of silicon on the surface. Two of these three test grades had aluminium as the replacement for silicon to improve wettability. Dew point during thermal processing plays a critical role and affects the wettability. The two-step heat treatment described above followed by hot dip galvanising was simulated with varying dew points in hot dip galvanising (Rhesca) simulator to assess the coatability of the samples.

Very good strength-elongation balance was obtained for all the three steels and the samples with lower silicon and higher dew point showed better wettability and coatability. Weldability studies revealed that the introduction of post weld tempering cycles improves the weld nugget geometry, breaking load and failure mode under tensile shear loading. Tensile studies at high strain rate revealed a satisfactory performance of the steel even at 100 s^{-1} suggesting the fitness to be used at the crash prone zones of automotives.

KEYWORDS: *TRIP (Transformation Induced Plasticity), Wettability, Weldability, Crash Resistance Crumpling*

INTRODUCTION

The need for addressing stringent environment and safety norms has compelled the automakers to design light-weight auto bodies with enhanced crash resistance. A new family of multi-phase steels having higher strength and better formability, named Transformation induced plasticity (TRIP) aided steel is gaining popularity to satisfy the demand as it has been adjudged as the most promising solution for the production of cars with low body mass because of the combination of high strength and large uniform elongation that they offer. The TRIP effect arises from deformation induced transformation of retained austenite to martensite [1]. It is accompanied by an invariant plane strain shape deformation as well as a volume expansion. This results in a higher strain hardening rate that delays the onset of necking, eventually resulting into higher uniform and total elongation [2]. An important feature of steels of this genre is enhanced ductility at a very high strength level [3]. TRIP effect enhances the mechanical properties by two mechanisms [4, 5]: (i) composite strengthening via the formation of hard martensite particles dispersed in the ferrite matrix and (ii) formation of dislocations around newly formed martensite regions as a result of the volumetric expansion during the austenite to martensite transformation.

This steel is ideal for passive safety structural applications like bumper reinforcement, door impact beam etc., due to its high strain hardening rate and dynamic energy absorption capacity [6]. The microstructure of conventional TRIP aided steels comprises ferritic matrix (~55-65%) along with bainite (~25-35%) and metastable retained austenite (~5-20%) [7]. The TRIP effect depends on the amount of retained austenite and its stability to deformation induced transformation.

Conventionally, TRIP aided steels contain about 1.5 wt% silicon which enhances the volume fraction and stability of the retained austenite by suppressing cementite formation during the isothermal bainitic transformation [8]. The solubility of Si in cementite is very small and therefore formation of a cementite nucleus in the presence of Si requires the diffusion controlled ejection of Si at the transformation front, causing a build up in the concentration of Si around the cementite nucleus. This locally increases the activity of carbon resulting in lowering of the flux of carbon and hence inhibits the development of cementite embryos [9]. Inhibition of carbide formation allows austenite to inherit a large amount of carbon, which lowers the M_s (martensite start) temperature and thus a considerable amount of metastable austenite is retained at room temperature [10].

Conventional TRIP aided steel with high silicon (1.5 wt%) suffers from poor wettability during galvanising due to deleterious effect of silicon. It is reported that silicon in concentration higher than 0.5 wt% is detrimental to coatability [11]. High Si leads to poor weldability as well [12]. To avoid these problems silicon can be partially replaced by aluminium [13, 14, 15]. Aluminum, like silicon, is insoluble in cementite and helps in retarding cementite formation. Al facilitates coatability by forming an inhibition layer on the steel surface, which in turn prevents the formation of Si/Mn oxides and bare spots. However, aluminum is a weak solid solution strengthening element; complete replacement of silicon by aluminium is not suggested as these Al-containing steels cannot reach the strength level possible with Si alloyed steel [1]. Moreover, Chen et al. [16] have shown that Si level should be more than 0.8 wt% to obtain reasonable amount of retained austenite. Of late, addition of microalloying elements for enhancing the strength level has become very popular [17]. Niobium is considered to be very effective amongst these, especially for the steels where silicon is replaced either partially or fully by aluminum [17]. The effect of Nb addition is rather complex. It helps refine austenite formed during the intercritical annealing [18] which can lead to formation of some pro-eutectoid ferrite during cooling to the isothermal bainite transformation temperature and consequent carbon enrichment of the untransformed austenite. The finer grain size increases the stability of austenite by reduction of M_s (martensite start) temperature [18]. Niobium in solid solution retards recrystallisation of austenite, and the kinetics of austenite to ferrite transformation [17]. It also inhibits recrystallisation by strain induced precipitation. Furthermore, it increases the relative amount of retained austenite by approximately 25%, which can significantly improve the mechanical properties of the steel [19, 20]. Niobium accelerates the bainite reaction, particularly at the temperature of galvannealing [21]. As a result of increased bainite formation at a faster rate, it is possible to enrich the remaining austenite with carbon and stabilize it. It is thus possible to retain austenite in the final microstructure, even at the relatively low times of holding during galvannealing in a continuous line [21]. Therefore, microalloying, especially with niobium, attracts research studies in the area of development of new generation TRIP aided steels.

P added TRIP-assisted steel [1] imparts solid solution strengthening and increases the stability of retained austenite. Oxides of phosphorous segregates at α-Fe grain boundary resulting in reduction of boundary-flux and the extent of oxide formation on surface [22]. It also decreases the inter-diffusion of Fe and Zn, favours G1 formation and hinders Γ. It is more amenable to wetting by molten zinc. It is reported that P is not harmful from the point of view of embrittlement and as well as spot weldability up to 0.1 wt% for such steel compositions [23].

Apart from alloy engineering, process optimization is also considered to be an important means for improvement of coatability. It is reported that if the hot dipping is done after a combination of intercritical annealing and isothermal bainitic transformation in a furnace atmosphere with a high dew point, the wettability improves significantly for conventional 1.5 wt% Si TRIP-aided steel. However, still some bare spots are unavoidable for CMnSi TRIP steel. On the contrary, the Si free CMnAl TRIP steel has a much better wettability when annealed at a low dew point. Therefore, dew point plays a crucial role for the surface quality of the steel and the wetting phenomenon in zinc bath during hot dip galvanizing [24].

Integration of automotive structures made up of high strength/ advanced high strength steels are predominantly carried out through resistance spot welding (RSW) process due to consistency in quality as well as high productivity. But there is a large gap area to specify optimal welding conditions for TRIP-aided steels having relatively richer alloying additions, such as high Si, always deteriorate weldability [25, 26]. Better coatability during galvanizing of TRIP-aided steel necessitates major alloy engineering. It has invited renewed challenge for welding of TRIP-aided steel [27].

The weldability of 'low carbon low alloyed' TRIP-aided steel faces severe complexities. After solidification, welds become very hard and can show a brittle behavior. The hardness of the heat affected zone (HAZ) attains to the tune of 500 HV and more [28], and cold cracking phenomenon is very prone to occur. During resistance spot welding, especially, the interface between the plates can act like a notch and promotes fracture of the weld. It becomes very severe during coach peeling/ shear tensile fracture which usually produces partially interfacial fracture (inferior in terms of ductility) as opposed to full button peel off/ plug type (superior) [29].

The high strain rate properties of steel indicates the performance of the same under crash condition. It is already established that TRIP-aided steels are used in the crumpling zone of a car because of their high strain rate performance and ability of large dynamic energy absorption. Therefore, dynamic testing is of special interest for the steel grades under discussion. Yield stress and ultimate tensile strength of TRIP aided steel increase continuously as it occurs for ferritic steels, but showing a smaller slope. At very high strain rate the elongation values of these steels are as high as under quasistatic testing conditions or even higher. This behavior is probably correlated to adiabatic heating. The used test set up and specimen size can lead to a local temperature increase up to 120 K. TRIP steel shows very good energy absorption quality even if at 200 s^{-1} strain rate [30]. Both temperature and strain rate affect the retained austenite transformation. At high strain rates, the uniform elongation decreases, whereas the total elongation and energy absorption increase. With raising test temperature, the tensile strength is reduced and the mechanical properties generally deteriorate, especially at 110 °C. However, excellent mechanical properties are obtained at 50 °C and 75 °C [31].

High strain rate properties being the indicator of performance under crash condition, a detailed study on tensile behaviour under dynamic loading has been presented for a simulated performance under crash condition with an overview on the study of wettability (coating performance during hot dip galvanising) and weldability.

SELECTION OF MATERIALS AND PROCESSING PARAMETERS

Three steels were selected for this work. The compositions are shown in Table 1.

Table 1: Composition (wt%).

Grade	C	Mn	S	P	Si	Al	Nb
Grade 1 CMnSiAlP	0.22	1.43	0.007	0.065	0.62	1.18	0.003
Grade 2 CMnSiAl Nb	0.21	1.58	0.008	0.017	0.66	1.30	0.032
Grade 3 CMnSiNb	0.20	1.36	0.015	0.015	1.69	0.009	0.035

As reported in previous works [32, 33] the details about transformation phenomena are collated in Table 2.

Table 2: Transformation temperatures, phase fraction calculation in the intercritical region and predicted values of retained austenite for the selected process parameters

Grade	Transformation Temperature (°C)	
	Thermo-Calc	Gleeble
1: CMnSiAlP	Ae_1:731 Ae_3:1093	Ac_1:742 Ac_3:1125
2: CMnSiAlNb	Ae_1:735 Ae_3:1100	Ac_1:739 Ac_3:1135
3: CMnSiNb	Ae_1:709 Ae_3:867	Ac_1:746 Ac_3:985

Based on the philosophy elucidated in previous works [32- 34], the process parameters were set with the help of simulated information, Artificial neural network (ANN) work on prediction of microstructure, existing facilities for laboratory heat making, forging, hot rolling, cold rolling as well as two stage annealing followed by hot dip galvanizing in hot dip process simulator (Rhesca). The process flow including assessment plans is depicted in Fig.1.

In the present work, samples were selected from the two stage heat treatment cycle as elaborated in the tables. For quasistatic tensile test, the samples were collected from HDPS (Hot Dip Process Simulator, Rhesca) trial

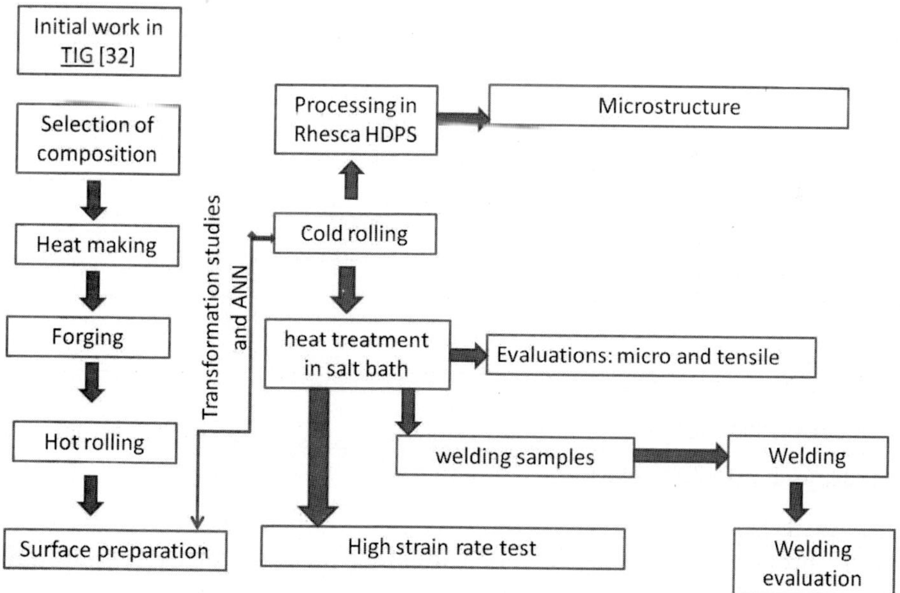

Fig. 1: Process flow

Table 3: Results of mechanical test of the samples from HDPS (Rhesca) trial (average of three samples from every individual cycle, IBT: Isothermal bainitic transformation 450°C/120s, IA: Intercritical annealing: mentioned in the Table)

Grade	Yield strength (YS) (MPa)	Ultimate tensile strength (UTS) (MPa)	Uniform elongation (%)	Total elongation (%)	n- value	UTS x Total elongation (MPa %)
Grade 1 CMnSiAlP IA: 800°C/60s	485	795	31	36	0.34	28620
Grade 2 CMnSiAlNb IA: 800°C/60s	511	849	23	29	0.27	24621
Grade 3 CMnSiNb IA: 790°C/60s	521	959	19	23	0.26	22057

The quasistatic tensile test results revealed a good combination of strength and elongation and a logically good uniform elongation and n- value. The microstructural evaluation (Table 4) revealed presence of reasonable amount of retained austenite which could lead to have TRIP properties. For further work, under the purview of the paper, samples processed through the above cycles (grade specific) had been chosen.

Table 4: Comparison of the amount of experimentally measured retained austenite with that predicted using ANN

Sample	%RA in microstructure (based on optical vis-a-vis XRD examination)	%RA from ANN
Grade 1 CMnSiAlP For IA: 800°C/60s	9	12
Grade 2 CMnSiAlNb For IA: 800°C/60s	13	12
Grade 3 CMnSiNb For IA: 790°C/60s	8	6

(IBT for all cases 450°C/120s)

As reported previously [33], a very good match was observed between the retained austenite predicted through ANN and actually achieved experimentally [Table 4]. Also, The volume fraction of retained austenite measured through optical metallography had a reasonably good match with the value obtained through XRD technique [33].

PERFORMANCE APPRAISAL OF WETTABILITY AND WELDABILITY STUDY

Wettability (coating performance during hot dip galvanizing in Rhesca HDPS):

An assessment [34] of the coating quality revealed that the steels with silicon level lower than that of conventional (1.5% silicon) TRIP-aided steel showed better wettability, coating thickness and adherence property. At higher dew point (+ 5° C) during, the coatablity was much improved than that at the lower dew point (- 40° C). The rationale for the same is as follows: for the steel grades where silicon is partially replaced by aluminium, some silicon and manganese would not be oxidised internally and hence some enrichment of these elements on the surface will occur during the annealing at a low dew point. Therefore better coating occurs at higher dew point.

Weldability

Weldability studies were carried out with samples heat treated in salt bath furnaces. Literature review [30-31] and a number of trials suggested that introduction of a post welding heating cycle could improve the weldability in terms of nugget diameter. Accordingly, a post welding heating was introduced in the final trial. The welded samples were subjected to tensile load in Instron 8801with a cross head speed of 1mm/minute. The test results revealed significant improvement in terms of nugget formation and geometry, failure mode (exhibiting achievement of expected plug type mode of failure) and load of weld at failure [35].

No major disparity was apparent for the three different grades for the weld zone and the HAZ. However, superior properties were observed for grade 1 and grade 2 to that of grade 3.

ASSESSMENT OF STRUCTURE PROPERTY CORRELATION UNDER HIGH STRAIN RATE TESTING

For high strain rate testing, another set of samples, cold rolled samples, cut in to required size (Fig. 2), for all three grades were heat treated as per above cycles in two adjacent salt bath furnaces for intercritical annealing and isothermal bainitic transformation (allowing shortest possible manual-shifting time from IA to IBT bath) followed by oil quenching to the room temperature. For high strain rate testing, the surfaces of the heat treated samples were made free from loose scales by grinding of emery papers and cleaned with acetone.

Characterisation

All tests for the high strain rate work were accomplished on a servo hydraulic Zwick/ Roell Rel 1852 high-speed tensile testing [locate at IEHK, RWTH University, Aachen] machine including a digital measuring and controlling unit. The Vmax and Fmax for the machine are 10 m/s and **12 kN** (to be reconfirmed) respectively. The tests had been carried out at strain rates 1 and 100 s^{-1} (average). Each test had been done for three times for Grade 1 and 2, and twice for Grade 3 due to paucity of material. German specification for dynamic tensile test **[36]**, was followed for this work. The test pieces were made in accordance with the dimensions specified in

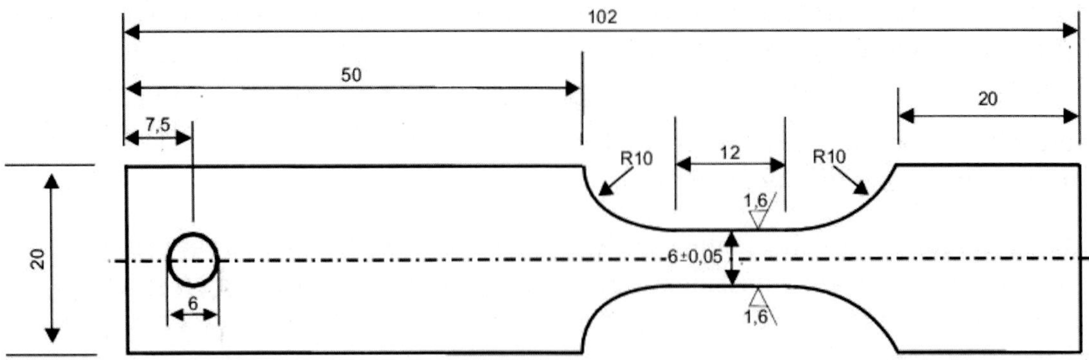

Fig. 2: Specified sample geometry for high strain rate testing. All dimensions are in mm. The thickness of all the samples varied between 1.2 to 1.4 mm.

The strain rate was determined based on the available condition of the equipment. However, published experimental data and internal circulations of leading steel manufacturers [37] are in agreement with the maximum strain rate (100 s^{-1}) of the range selected here, to have resemblance with the crashing condition of passenger car components. For quasistatic test with a strain rate of 3.33 x 10^{-4} s^{-1}, data has been captured from Table 3.

The force measurement during the test was done using a piezoelectric load cell instrumented with a Weg Zimmer (opto-elektronisch) extensometer. The storage of test data was done with a 4-channel transient recorder (sampling rate 1 MHz, 10 bit analogue/ digital converter) and later for further calculation on a personal computer. To evaluate the results, the recorded data were first converted into technical stress strain diagrams allowing the determination of the mechanical properties by use of special software developed by IEHK, Aachen. At high strain rates the force signal was superimposed by oscillation, caused by the inertia of the test equipment, making the direct determination of the characteristics values impossible. Therefore a cubic spline was utilized to approximate the unfiltered stress-strain signal, allowing the gathering of the strength properties. Earlier works showed good conformity of results obtained in this way, using the load cell instrumented with strain gauges plus spline approximation, compared to the results achieved using the strain gauges fixed directly on the specimen [38].

Results and discussion

The details of the test results for higher strain rate from dynamic tensile testing (1 and 100 s^{-1}) vis-à-vis the quasistatic properties are tabulated below (Table 5):

Table 5: High strain rate test results vis-à-vis quasistatic test results

Grade	Strain rate (s^{-1})	YS (MPa)	UTS (MPa)	YS/ UTS	UTS/ YS	Ag (%)	A_{Tot} (%)	n	Energy at 10% elongation (kJ/kg)	$Energy_{Tot}$ (kJ/kg)	UTS x A MPa*%
1	QS	485	795	0.61	1.64	31	36	0.34	5.706	26.045	28,620
	1	493	764	0.65	1.55	24.3	35	0.298	8.689	32.017	26,740
	100	562	815	0.69	1.45	22.3	33	0.312	8.795	31.616	26,895
2	QS	511	849	0.60	1.66	23	29	0.27	6.313	21.087	24,621
	1	498	814	0.61	1.63	18.3	27.4	0.234	9.685	26.822	22,304
	100	580	858	0.68	1.48	18.5	28.4	0.255	9.352	28.485	24,367.
3	QS	521	959	0.54	1.84	19	23	0.26	6.989	18.955	22,057
	1	451	914	0.49	2.03	16.5	26.7	0.196	11.085	29.556	24,404
	100	539	971	0.56	1.80	15.2	27.8	0.21	10.62	31.089	26,994

(Grade 1: CMnSiAlP , Grade 2: CMnSiAlNb, Grade 3:CMnSiNb)
EnergyTot : area under the stress strain curve

The yield strength for Grade 1 consistently increased from quasistatic to 100 s^{-1} whereas UTS initially decreased at 1 s^{-1} and then increased at 100 s^{-1} crossing the quasitatic value (Fig. 3). But for grades 2 and 3, YS and UTS decreased at the intermediate level (i.e. at 1 s^{-1}) and then increased significantly at 100 s^{-1}, crossing the quasistatic value (Fig. 4 and 5).

Uniform elongation for all grades decreased from quasistatic through 100 s^{-1} (Fig. 6 to Fig. 7). Total elongation for Grade 1 decreased at 1 and 100 s^{-1} from the QS value consistently (Fig. 6). On the other hand, though it initially decreased at 1 s^{-1}, recovered at 100 s^{-1} and attained almost at the same level of the QS value for Grade 2 (Fig. 7). For Grade 3, the value increased from QS to 100 s^{-1}through 1s^{-1} consistently (Fig.8).

YS
UTS
Strength/ MPa
Strain Rate/ s^{-1}
Grade 1 CMnSiAlP, QS: Quasistatic

Fig. 3: Variation of strength with respect to strain rate_ Grade1

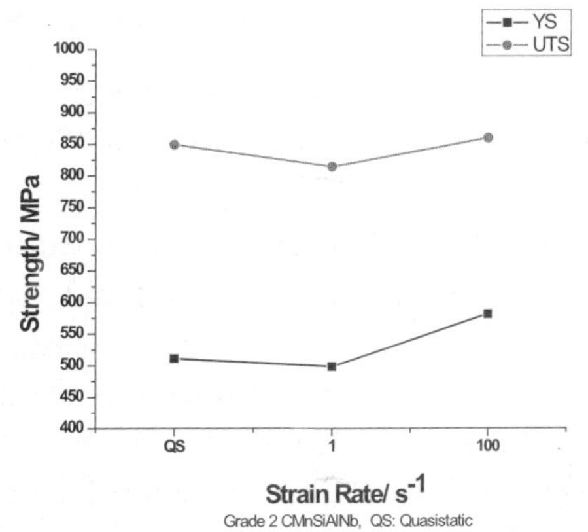

Fig. 4: Variation of strength with respect to strain rate_ Grade2

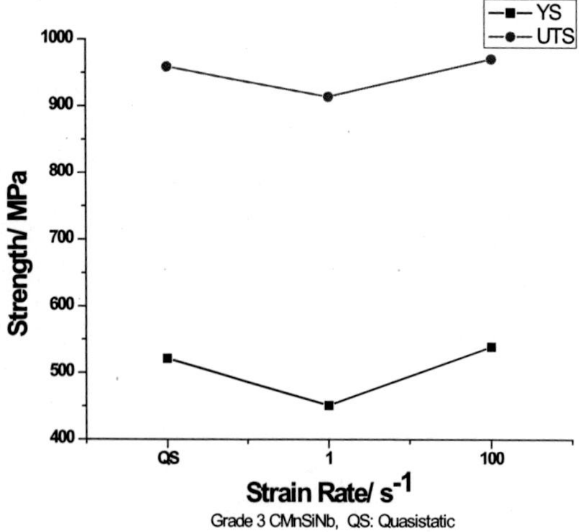

Fig. 5: Variation of strength with respect to strain rate_ Grade3

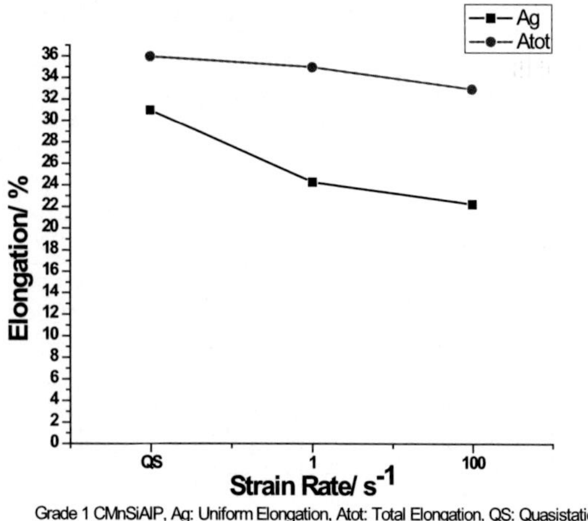

Fig. 6: Variation of elongation (total and uniform) with respect to strain rate_ Grade1

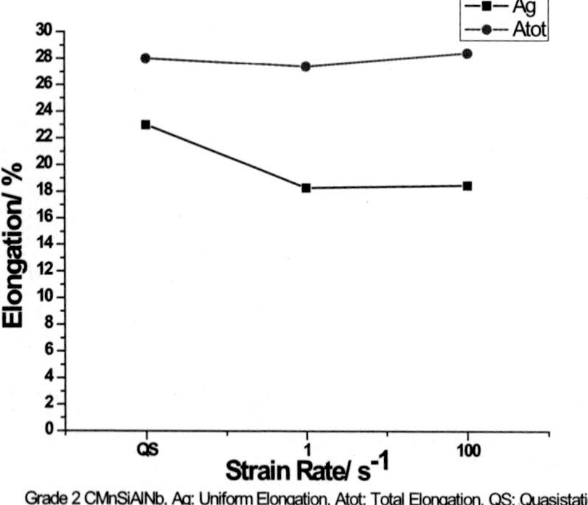

Fig. 7: Variation of elongation (total and uniform) with respect to strain rate_ Grade2

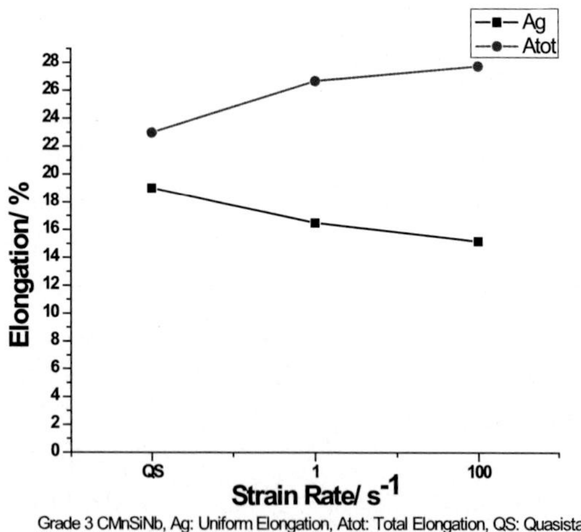

Fig. 8: Variation of elongation (total and uniform) with respect to strain rate_ Grade3

Energy absorbed at high strain conditions reflects the ability to withstand crash conditions. Usually, energy absorbed during tensile test is expressed as the area under the stress strain diagram, derived by integration. The multiplied value of the UTS and %total elongation is also considered as an indicator of the same.

This is also reported that the change of energy absorption with changing strain rate is predominantly guided by the elongation values. It is also estimated by several reported works that crash events involves a strain around 10% and the parts damaged in a crash will never be deformed until fracture. Therefore a rationale has been in place to compare the energy absorption of different steels by at a fixed elongation 10%. This is also envisaged that the adiabatic effect and the consequence of statistical scatter of extension is diminished [39].

For Grade 1, as observed in Fig. 9, the total energy (area under the engineering stress strain diagram) increases at strain rates 1 s^{-1} as well as 100 s^{-1}. Whereas, it has shown decrements at higher strain rates for UTS*% total elongation. Energy at 10% elongation increase from quasistatic to 1% s^{-1} and 100 s^{-1} and it is more or less unaltered for 1% s^{-1} and 100 s^{-1} strain rate.

For two other grades, same trend has been observed for energy at 10% elongation. The values were found in ascending order from Grade 1 to Grade 3 (Table 5 and Fig.9 to Fig. 11). For Grade2, the total energy is in continuous ascending order at 1s^{-1} and

100 s^{-1} though UTS*% total elongation decreased at 1s^{-1} and then crossed the QS value marginally (Fig. 10). For Grade 3, for total energy as well as UTS*% total elongation, noteworthy ascent has been observed (Fig. 11) as the initial energy at QS is lower than two other grades. This phenomenon has been collated in comparative manner in Fig. 11.

Initial n-value at QS is highest for Grade 1 and lowest for Grade 3. At strain rate 1s^{-1}, n-value decreased for all three grades and then it increased at 100 s^{-1} though it was lower than the initial (i.e. QS) values.

In general, it is evident from the assessment that reasonably high energy absorption for all strain rates is accomplished by TRIP steel. The absorption (UTS x Total Elongation) drops to a lower level during dynamic testing at the intermediate strain rate (here 1 s^{-1}) with an increase when strain rate rises (100 s^{-1}) except the Grade3 where an altogether steep rise has been observed. However, this development has likeliness with the trend of variation of strength and elongation.

For Grade 1 and Grade 2, the ratio of UTS and YS decresad continuously and on the contrary, the ratio of YS and UTS increased. Whereas for Grade 3 the ratio of UTS and YS increased at 1s^{-1} and then decreasd. Just opposite trend has been observed for the YS and UTS ratio. High UTS and YS ratio extends aassurance for significant energy absorption and dissipation occurs during inelastic deformation. It also helps to avoid premature failure due to strain concentrations. Better material flow during forming is resulted from this also. However at high strain rate, it is lowered to some extent which has direct relation with the lower uniform elongation and n-value.

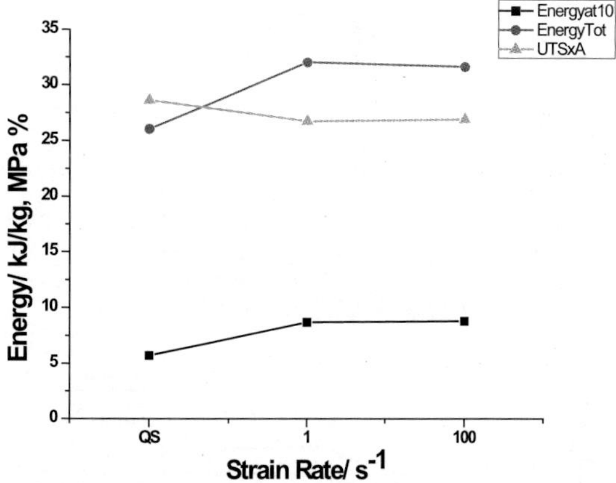

Fig. 9: Variation of energy with respect to strain rate_ Grade1

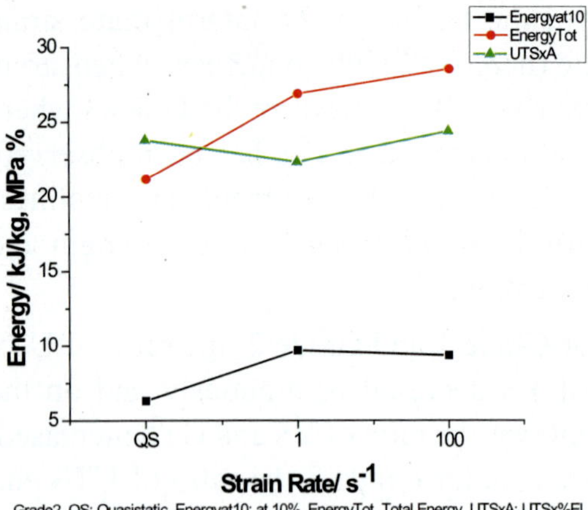

Fig. 10: Variation of energy with respect to strain rate_ Grade2

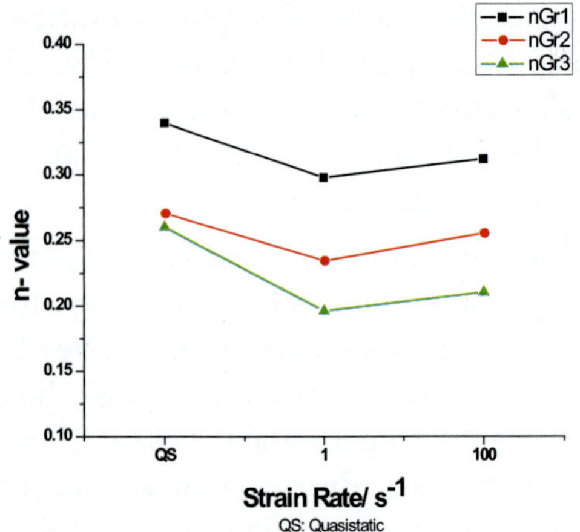

Fig. 13: Comparative diagram on variation of n-value with respect to strain rate

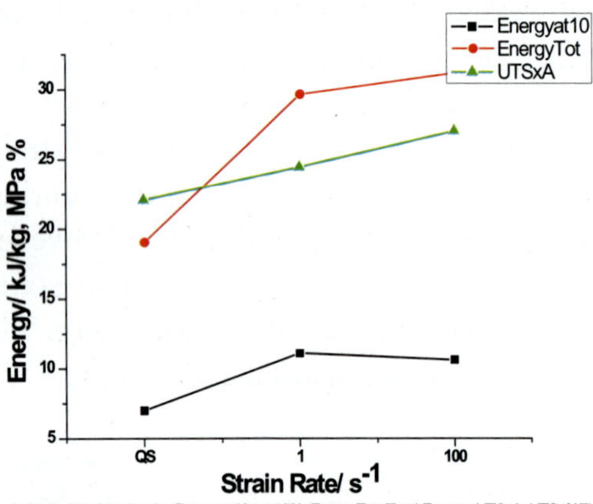

Fig. 11: Variation of energy with respect to strain rate_ Grade3

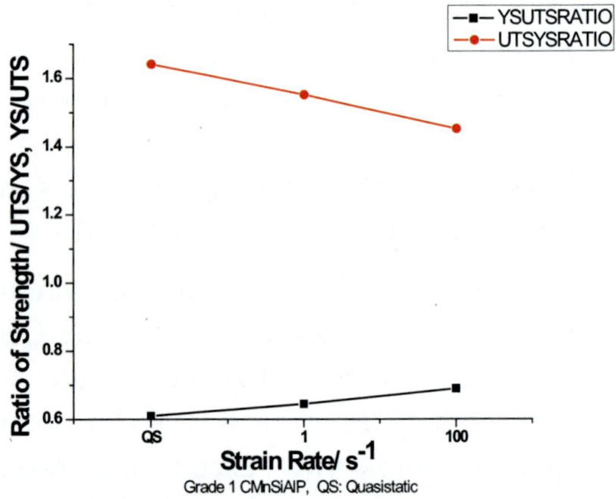

Fig.14: Variation of strength ratio with respect to strain rate_ Grade1

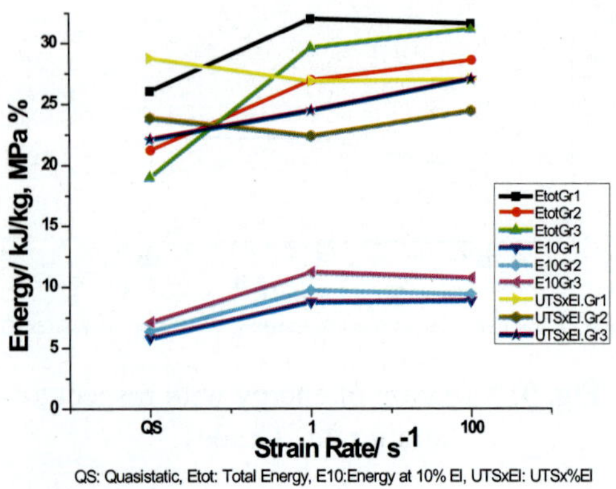

Fig. 12: Comparative diagram on energy variation with respect to strain rate

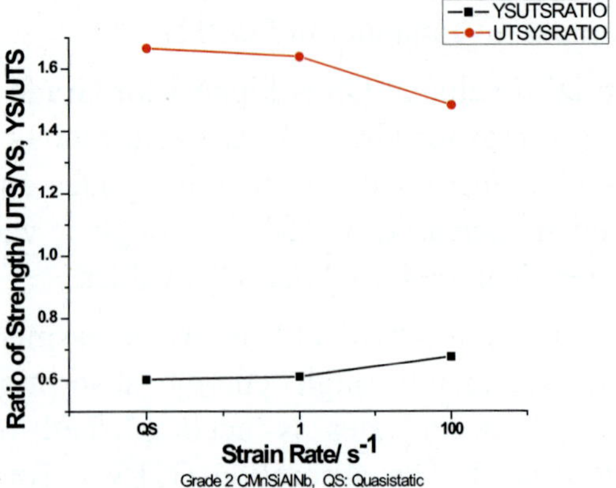

Fig. 15: Variation of strength ratio with respect to strain rate_ Grade2

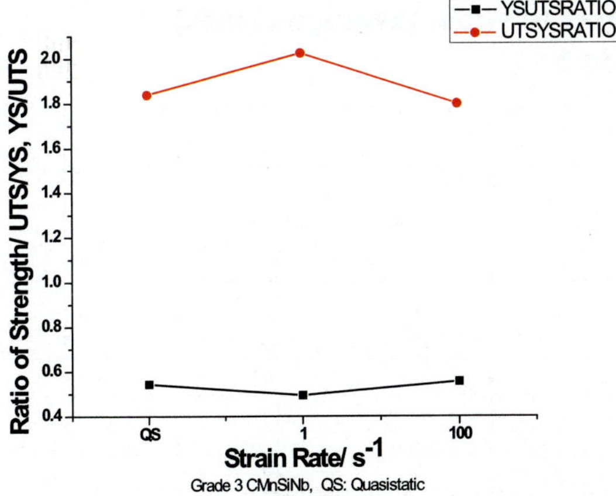

Fig. 16: Variation of strength ratio with respect to strain rate_ Grade3

The basic difference between the high strain rate test and the quasi-static test is that inertia and wave propagation effects are more distinct at higher strain rates. These dynamic affects on test results at higher strain rates. When the strain rate is increased through the medium strain rate regime, the measurement of load is the first to be affected by stress wave propagation. The ratio between the strength at the higher strain rate (here 100 s^{-1}) against the lower strain rate (here 1 s^{-1}) at different strains are described as dynamic factor. This factor has got its highest value at yield point (Fig. 17). Again this dynamic factor is highest in Grade 3 CMnSiNb and lowest in Grade 1 CMnSiAlP. It might have a direct correlation with the maximum change of tensile strength vis-à-vis the energy absorption in Grade 3.

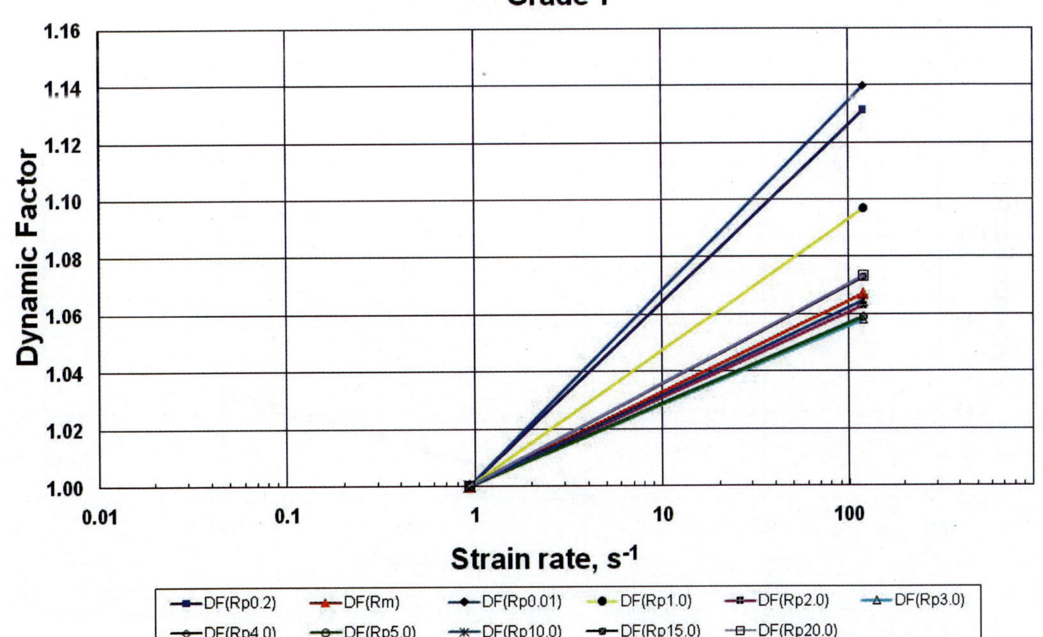

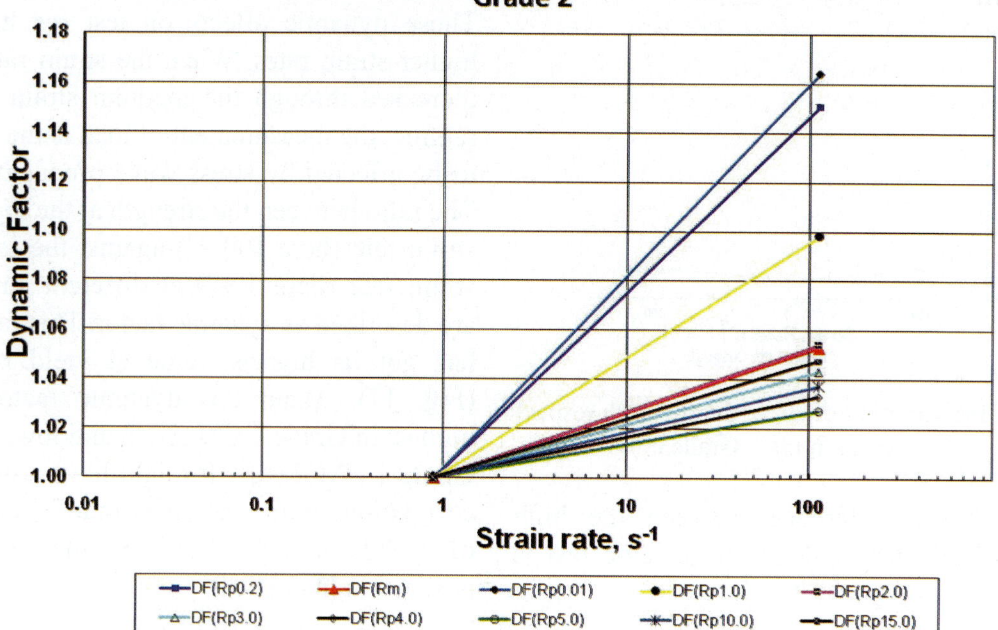

Dynamic Factor (DF) with respect to strain (average values)
Grade 3

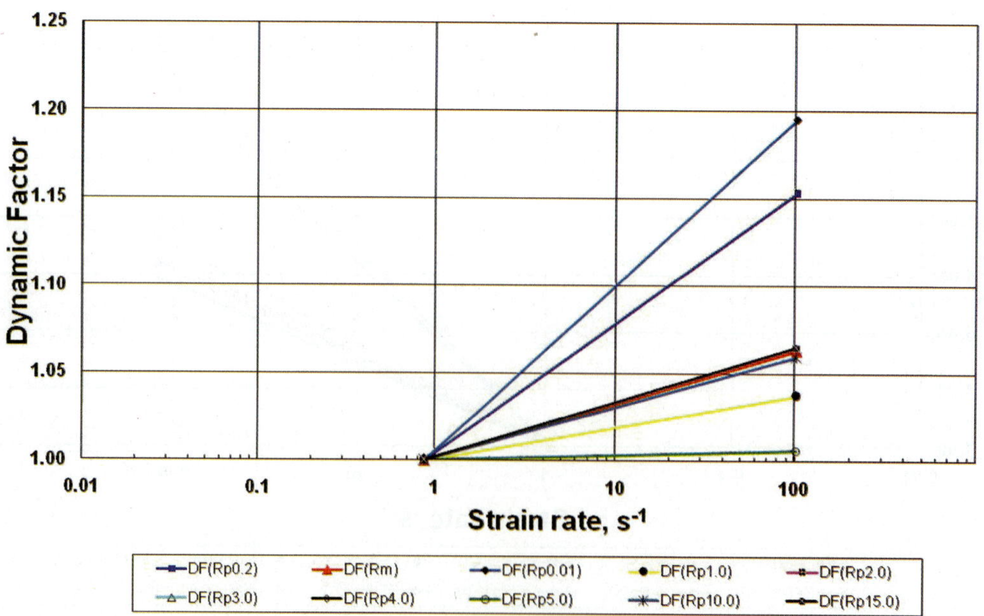

Fig. 17: Variation of Dynamic Factor against strain rate: a. Grade1, b. Grade 2, c. Grade 3

Yield plateau is remarkably higher for dynamic tensile test (high strain rate testing) as observed in Fig. 18 to Fig. 20. This is maximum in 100 s^{-1}. During high speed deformation, it is envisaged that a huge accumulation of dislocations around hard phase for a very short time raises obstacle towards diffusion of dislocation as well as strengthening of ferrite substrate. The pronounced yield phenomenon is resultant of this effect.

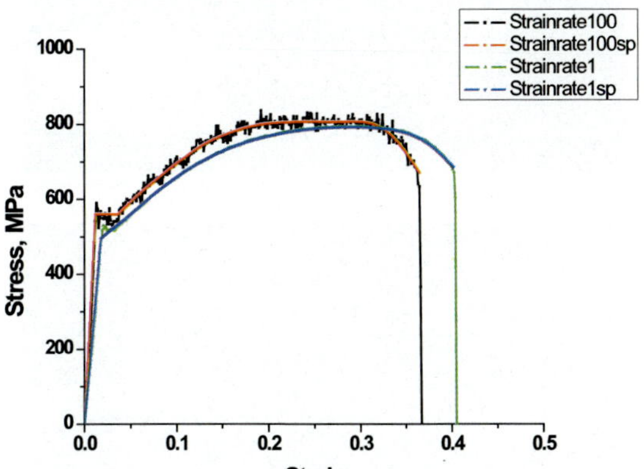

Fig. 18: Engineering stress strain diagram for strain rates 1 and 100 s^{-1} with and without spline (sp) for Grade 1_Specimen 1

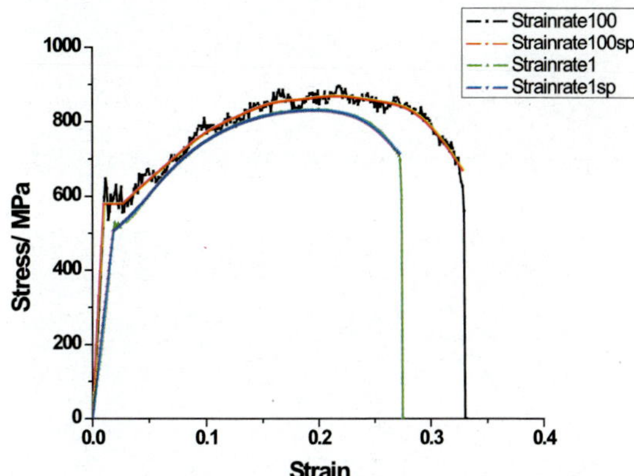

Fig. 19: Engineering stress strain diagram for strain rates 1 and 100 s^{-1} with and without spline (sp) for Grade 2_Specimen 1

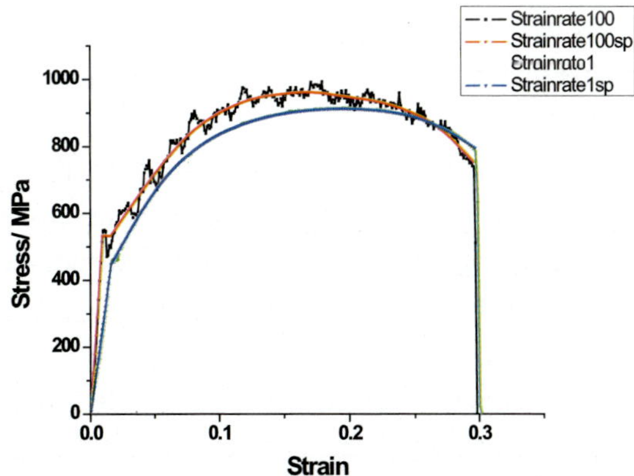

Fig. 20: Engineering stress strain diagram for strain rates 1 and 100 s^{-1} with and without spline (sp) for Grade 3_Specimen 1

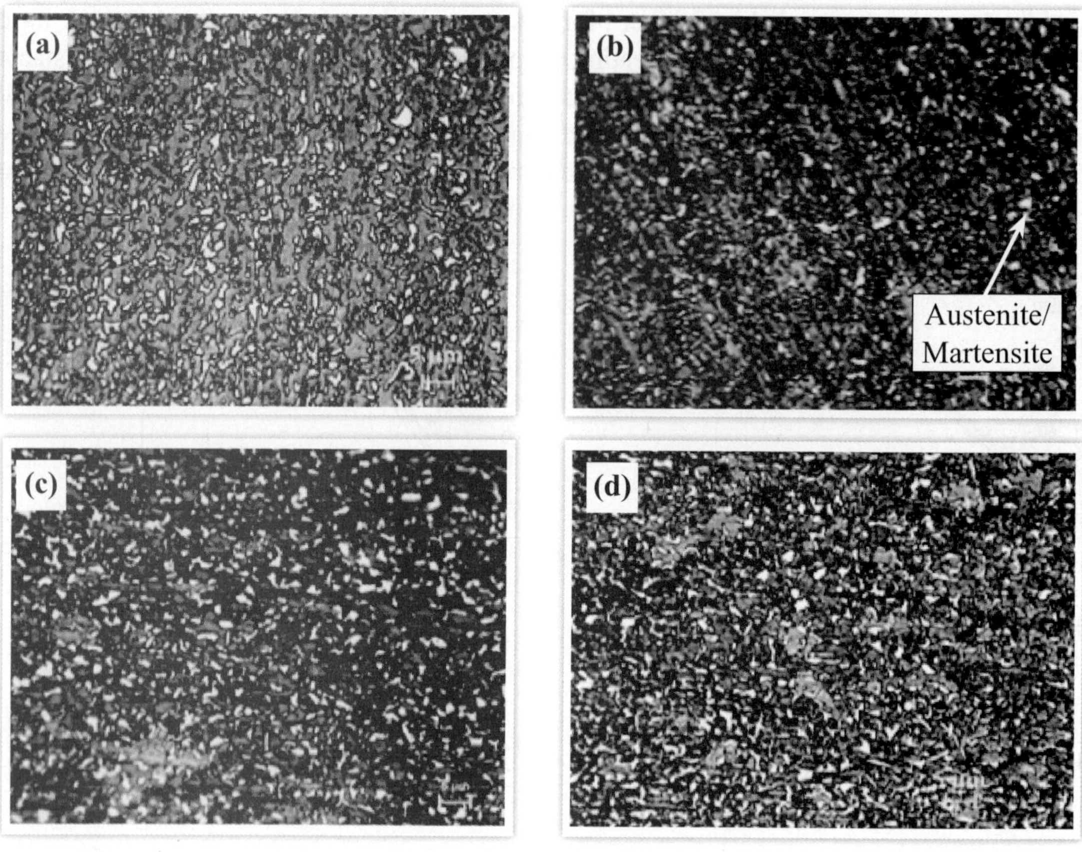

Fig. 21: a to d show microstructure for Grade 2 after tensile test at 100 s^{-1} From fracture face towards the grip. Microstructure: Ferrite + Bainite + Austenite/ Martensite (white phase)

Microstrcuture of a representative sample showed presence of retained austenite even after high strain rate testing (Fig. 21a to 21 d, Table 6]. This carries the signature of transformation behaviour of austenite to martensite during deformation which is having a direct correlation with the TRIP property.

Table 6: % Retained austenite in microstructure for Grade 2 after tensile test at 100 s^{-1}

Distance from fracture face (mm)	Fig.	% Reatined austenite from XRD
3	21 a	NT
5	21 b	4
8	21 c	11
11	21 d	14

NT : not traceable properly

CONCLUSIONS

Three low carbon TRIP-aided steels have been developed through cold rolling, two step heat treatment and galvanising route. The material under study has shown very good performance under quasistatic as well as high strain rate tensile test resulting in high ultimate tensile strength with high total and uniform elongation. The material also has a high strain hardening exponent which has direct correlation with the availability of retained austenite in the microstructure after heat treatment.

The volume fraction of retained austenite measured through optical metallography had a reasonably good match with the value obtained through XRD technique. These values corroborated the amount of retained austenite predicted through artificial neural network model.

Changing the dew points, the coatability have been improved and it has been observed that the samples with lower silicon and treated at higher dew point showed better coatability during hot dip galvansing coating.

Trials for resistance spot welding have revealed that the steels are weldable. Introduction of post weld tempering cycles improves the weld nugget geometry, breaking load and failure mode under tensile shear loading

It is observed form the results that in general at high strain rates, the uniform elongation had shown decrement, whereas the total elongation and energy absorption were increased (except some deviations). The observations has corroborated various other works [30].

It is envisaged that during the high strain rate testing, temperature increases sharply which remains within the system. So this deformation mechanism is an adiabatic one which may play a significant role in restricting the extent of transformation of austenite to martensite [31], and thus, the reason behind the decrease in tensile strength during this adiabatic deformation can be explained due to the decrease of flow level and less transformation to martensite. The decrease in strength due to adiabatic deformation could also be due to the decrease in the flow level in accordance with the temperature dependence of yield point as well as due to decrease in strength decreases as exhibited in the temperature dependence of the strengthening graph [40].

For 1 s^{-1}, the impact of adiabatic heating being high, a trend of decrease in strength is very common. Whereas, for 100 s^{-1},as the testing time is extremely low, that might not have much impact over the transformation of austenite to martensite resulted in higher martensitic transformation except Grade 3. Microstructure of Grade 3 has revealed that it had relatively lower austenite and there could have been higher martensite as it had been evident from calculation of Ms temperature with the help of x-ray diffraction as well as calculation through Thermo-Calc.

Energy absorption is a function of tensile strength [41]. The tensile strength having been linearly proportional to the volume fraction of martensite in the structure [42, 43], presence of martensite from higher transformation of austenite or due to preexisting higher martensite, the energy level becomes higher. This has been reflected for higher energy levels for Grade

2 at $100s^{-1}$, and for Grade 3, both at 1 and $100\ s^{-1}$.

It is also established from this work that the behaviour at higher strain rate of the three developed steels, especially in terms of energy, have made them qualified to be applied to manufacture crash resistant components over the conventional high strength steels [44].

ACKNOWLEDGEMENT

Authors are personally indebted to Prof. Wolfgang Bleck, Head, Department of Ferrous Metallurgy, IEHK, RWTH Aachen University for their support to carry out the work. Authors thankfully acknowledge Dr P K Banerjee, Chief (Actg.), R&D and Scientific Services, Tata Steel Ltd., Prof. S. Das, Head, Department of Metallurgical and Materials Engineering, IIT, Kharagpur, Authors are thankful to the Department of Science andTechnology and Ministry of Steel, Government of India, International Bureau of the Federal Ministry of Education and Research, Germany for their partial financial support for the work. Authors also thank Prof. Subhabrata Datta of School of Materials Science and Engineering, Bengal Engineering and Science University, Shibpur, Howrah, Dipl. Ing. Christoph Keul, RWTH Aachen University, Dr. Monideepa Mukherjee of Tata Steel Ltd. for their hearty support.

REFERENCES

[1] B. C. De Cooman, Current Opinion in Solid State and Material Science 8 (2004) 285-303.

[2] V.F. Zackay, E.R. Parker, D. Fahr and R. Busch, Transactions of the American Society for Metals 60 (1967) 252-259.

[3] B. Mintz and J.C. Wright, J.Iron and Steel Institute 208 (1970) 401-405.

[4] E.M. Bellhouse and J.R. McDermid, Metallurgical and Materials Transaction A, Volume 41A (June 2010) 1460-1473.

[5] P. Jacques, Q. Furne´mont, A. Mertens, and F. Delannay: Philos. Mag. A, Vol. 81 (2001) 1789–1812.

[6] J. H. Jung et al., 2008 International Conference on New Development in Advanced High Strength Steels, 25-37.

[7] T. Heller and A. Nuss, Proceedings of the International Symposium on Transformation and Deformation Mechanisms in Advanced High Strength Steels, eds. M. Militzer, W. J. Poole and E. Essadiqi, CIM, Montreal (2003) 7-20.

[8] O. Matsumura, Y. Sakuma and H. Takechi, Transactions ISIJ 21 (1987) 570-579.

[9] W. S. Owen, Transactions of the American Society for Metals, 46 (1954) 812-829.

[10] E. Girault, P. Jacques, Ph. Harlet, K. Mols, J. Van Humbeeck, A. Aernoudt, and F. Delannay, Materials Characterization 40 (1998) 111-118.

[11] B. Mintz, International Materials Review 46 (2001) 169-197.

[12] ASM Handbook, Vol. 6, Welding, Brazing and Soldering, ISBN 0-87170-383-3 (Dec. 1993) 405-407.

[13] J. Maki, J. Mahieu, B. C. De Cooman, and S. Claesseur, Materials Science and Technology, Vol. 19 (Jan. 2003) 125-131.

[14] W. Bleck, A. Frehn, and J. Ohlert, Proceedings of the International Symposium on Niobium, Orlando (2001) 727-752.

[15] J. Gao and M. Ichikawa, Proceedings of the International Conference on Advanced High Strength Sheet Steels for Automotive Applications, AIST, Winter Park, Colorado (2004) 107-116

[16] Chen et al., SE AISI Q (April 1991) 44-49.

[17] D.Krizan, B. C De Cooman and J. Antonissen, Proceedings of the International Conference on Advanced High Strength Sheet Steels for Automotive Applications, AIST, Winter Park, Colorado (2004) 205-216.

[18] K. Hulka, W. Bleck and K. Papamantellos, Proceedings of the 41st Mechanical Working and Steel Processing Conference, Iron and Steel Society, Baltimore (1999) 67-88

[19] X. Liu, P. Karjalainen, and J. Pertula, Thermo-mechanical Processing in Theory, Modelling and Practice" ed. W. Hutchinson et al., (1996), Stockholm, Sweden, 240-248

[20] M. Onink, Th. M. Hoogendoorn and J. Colijn, Material Science Forum, 284-286 (1998), 185-192.

[21] Debanshu Bhattacharya, Proceedings of the International Conference on Microalloyed Steels: Emerging Technologies and Applications (2007), Ed. Debashish Bhattacharjee et al., 50-60.

[22] H. K .D. H. Bhadeshia and R. W. K. Honeycombe, Steels Microstructure and Properties, Third Edition, 2006, Elsevier Ltd.

[23] Liesbeth BARBÉ, Kim VERBEKEN and Emiel WETTINCK, ISIJ International, Vol. 46 (2006), No. 8, 1251–1257.

[24] J. Mahieu, S. Claessens, B.C. De Cooman, Metallurgical and Materials Transaction A Volume 32A, November 2001, 2905-2908.

[25] ASM Handbook, Vol. 6, Welding, Brazing and Soldering, ISBN 0-87170-383-3 (Dec. 1993) 405-407

[26] T. Herai et al, Resistance spot welding of high strength steel sheets, IIW Doc. III, 664-80

[27] M. I. Khan, M. L. Kuntz, E. Biro and Y. Zhou, Materials Transactions, Vol. 49, No. 7 (2008) pp. 1629-1637.

[28] Cretteur Laurent et al., Steel Research, ISSN 0177-4832, 2002, vol. 73, pp 314-319.

[29] Yun Peng et al., Materials Science Forum Vols. 638-642 (2010) pp 3591-3596.

[30] Rong TIAN, Lin LI, BC De Cooman, Xi-chen WEI and Peng SUN, Journal of Iron and Steel Research, International,Volume 13, Issue 3, May 2006, Pages 51-56.

[31] Wolfgang Bleck and Ingo Sacheal, Steel Research 71(2000) No. 5—32.

[32] Tanmay Bhattacharyya, Shiv Brat Singh, Sourav Das, Arunansu Haldar, Debashish Bhattacharjee, Materials Science and Engineering A 528 (2011) 2394–2400.

[33] Tanmay Bhattacharyya, Shiv Brat Singh, Swati Sikdar Dey, Sandip Bhattacharyya, Wolfgang Bleck, Debashish Bhattacharjee, Materials Science and Engineering A, MSEA D12 02974 (under publication).

[34] Indian Patent Application No. 633/ KOL/10 of 11 .06. 2010.

[35] Indian Patent Application No.1356/ KOL/2011 of 21. 10. 2011.

[36] Standard for high strain rate test, "Stahl Eisen Prüfblatt SEP 1230".

[37] Internal Document, Corus

[38] D. Stark-Seken, W. Bleck, W. Dahl, Materialprufung, 37 (1995) 165-168.

[39] I. D. Choi, D. M. Bruce, S. J. Kim, C. G. Lee, S. H. Park, D. K. Matlock, J. G. Speer, ISIJ International, 42 (2002) 1483-1489.

[40] W. Bleck, Text Book on Material Testing for RWTH Students, Aachen University.

[41] *R.G. Davies*, Metall. Trans. A, 10 (*1979*) 1549–1555.

[42] *R.G. Davies*, Metall. Trans. A, 9 (*1978*) 671-679.

[43] J. F. Butler, J. H. Bucher, Seminar on Vanadium Cold Passing and Dual Phase Steel, 18-20, Oct. 1978, Berlin.

[44] R. G. Davies, C. L. Magee, Proc. Symp. On Structure and Properties of Dual Phase Steel, February 19-21, 1979, New Orleans, The Metal. Soc. AIME.

Copper Precipitation in IF Steels

O.N.Mohanty[1], R.Rana[2] and S.B.Singh[3]

[1]RSB Group, Bhubaneswar, India, omkarmohanty@gmail.com; [2]RD&T, Tata Stel, Ijmuiden, The Netherlands, radhakanta.rana@tatasteel.com ,[3]Metallurgical Engg. Department, IIT Kharagpur, Email: sbs22@metal.iitkgp.ernet.in

ABSTRACT

The thermo electric power (TEP) measurement has been successfully used in a IF-Cu steel for studying the copper precipitation during isothermal ageing. Prior to the same, a theoretical treatment of the influence of copper precipitate on TEP was done and the predicted values of TEP agreed well with the experimental data. Primarily, samples that had gone through the continuous annealed (CA) and the full annealed (FA) conditions following cold-rolling had sufficient copper in solution in the matrix. These were subjected to TEP during ageing at various temperatures . Extensive copper precipitation associated with a marked increase in hardness was observed in these CA and FA materials, aged between 500ºC and 600ºC. The TEM study of the precipitates did show a mix of spherical and rod-like precipitates at 600ºC, where as at 700ºC, the precipitates were predominantly rod-like and are less copious. The evolution of TEP during isothermal ageing was investigated in conjunction with the hardness change and appropriately reconciled.

KEYWORDS: *IF-Cu Steels, Spherical and Rod-Like Copper Precipitation, Thermo Electric Power (TEP)*

INTRODUCTION

A major current interest in copper for the steel-makers stems primarily from the fact that a good amount of the steel scraps being used are sourced from electrical industry and as such would contain copper; eliminating the same would involve cost and time. Consequently, finding a way to use them beneficially would be attractive. Strength-increase exceeding 100 H_v, through precipitation of copper in the iron-base was known since long [1]. The copper containing HSLA grades of plates were initiated by the US Navy in 1984 (e.g. HSLA-80 with an Y.S. of ~ 550 MPa, subsequently HSLA-100 with Y.S. of 690 MPa and then with even higher strength plates) for highstrength and toughness in combination with enhanced weldability (compared to the earlier HY-grades) and consequently, resulting in appreciable cost -reduction [2].

The mechanism of strengthening by copper (which is softer than the matrix iron) is fairly complex in nature. and has been the subject of a number of studies. A range of models of copper precipitate strengthening effects (e.g. chemical, misfit strain, elastic modulus difference and frictional force and so on) have been used to explain the strength increase. An excellent analysis and comparison of the existing theories with rigorous experimental results has been provided recently by Takahashi et. al. [3]. One of the difficulties of tracking the nature of copper precipitation in the early stages has been its extremely fine size (<3 nm) and poor diffraction contrast (due to coherent bcc clusters); hence the poor amenability of the early precipitation to TEM examination. The techniques used to follow this stage have included small angle neutron / X-Ray scattering (SANS / SAXS), field ion microscopy (FIM) and in more recent times, three dimensional atom probe (3DAP) and the related technique of atom probe tomography (APT) and so on [3]. The other complexity with the copper precipitation is the multiple steps associated with copper precipitation, viz. bcc clusters à 9R →3R →fcc ε Cu.

The interstial-free (IF) steels, with extremely low-carbon (~ 0.002 C) content and possessing very high planar-anisotropy values (with r-bar > 2.0), meet the most stringent drawability requirement for auto-body applications [4-6]. However, their low strength levels (Y.S. ~220 MPa) has limited their use since weight reduction (and hence lower thickness with higher strength) has become essential. Copper additions into these steels has been attempted to enhance the strength through ageing treatments following easier forming of the softer initial condition [7-11]. Further, studies on possible hot-shortness induced by copper [12] and its minimization [13] have contributed to the understanding in this field. The affordable, copper- containing IF steel grades with minimal or no nickel (usually added to prevent hot-shortness due to copper) are thus potential candidates for many critical applications. They would find broader utility if the strength could be enhanced through precipitation hardening based on copper.

Techniques, such as electrical resistivity have been used effectively to track precipitation of copper during ageing of HSLA steels and elucidation of precipitation kinetics has been done using such data [14-16]. Now, Thermo Electric Power (TEP) technique has also emerged as a strong tool for characterization of phenomena such as solid solution, precipitation, cold-working and annealing etc. [17-23]. An extensive investigation was

conducted on copper precipitation in a binary Fe-1.4 wt. pct. Cu alloy, combining a number of techniques including the TEP measurements [24]. It was therefore considered appropriate to subject copper precipitation in a newly introduced IF grades [8] to TEP measurements; this work was recently published [25].

The present paper would show some results of a the above TEP study and examine them in the light of the TEP data from a similar investigation in a binary Fe-Cu alloy [24]. Additionally, an analysis of the hardening behaviour in the present IF-Cu steel in conjunction with the age-hardening data from the earlier HSLA-100 steel would be made.

MATERIAL &EXPERIMENTAL TECHNIQUES

Materials and treatments

The material chosen was a laboratory melted, copper-bearing, Ti-stabilized IF-Cu steel; details of alloy making etc. are given elsewhere [25]. The chemical composition of the steel is given in Table I, along with that of a binary Fe-Cu alloy. The IF steel was hot-rolled in a lab. rolling mill, finish rolled at 900°C (above A_{e3} of the steel) and air-cooled to room temperature. The hot rolled strips were then cold rolled in a laboratory mill into 0.4 mm thick sheets giving ~80% reduction in thickness. The samples were then heat treated inside salt bath simulating continuous annealing (CA) process, for 1 minute at 820ºC, then quenched in water. The quenched samples were further held in salt bath at various aging temperatures in the range of 500ºC to 700ºC, times going up to 24 hours and were eventually quenched in water. The CA-simulation thus done, was above the solvus temperature of 780ºC [7], within the single-phase ferrite region, with the intention of dissolving the entire copper in the ferrite. The Full Annealing (FA) simulation was done in the single-phase austenite region and would thus take all copper into solid solution.

Table I: Chemical Composition (wt. %) of IF-1.2%Cu steel & Fe-1.4% Cu alloy

Elements	**Material**	
	IF-1.2% Cu Steel	**Fe-1.4% CuAlloy**
C	0.002	0.0004
Mn	0.080	—
S	0.007	0.0001
P	0.012	—
Si	0.023	—
Al	0.011	—
Cu	1.20	1.40
Cr	0.024	—
Ni	0.013	—
Nb	<0.005	—
Ti	0.043	—
N	0.0014	0.0001
Fe	balance	balance

Thermo Electric Power (TEP) Measurements

The principle of TEP is based on the Seebeck effect.[26]. If a temperature difference (ΔT) is applied between two junctions of the steel sample with two pure iron blocks, generally called cold and hot junctions, a voltage (ΔV) is generated across the two junctions because of Seebeck effect. The Seebeck coefficient, S, represents the difference between the absolute TEP of the steel (S*) and the absolute TEP of pure iron (S_0*). Thus, the relative TEP of the investigated steel with respect to pure iron was measured in this study and is defined as follows[22]

$$\Delta S = S^* - S_0^* = \Delta V / \Delta T \qquad (1)$$

The value of S is sensitive to the defects present in the matrix such as alloying elements in solid solution, dislocations and precipitates. etc. [18,22,23,27] . It can be considered as a sum of various contributions and can be expressed as:

$$\Delta S = \Delta S_{ss} + \Delta S_d + \Delta S_{pp} \qquad (2)$$

The terms on the right representing the contributions from solid solution, defects (e.g. dislocations) and precipitation respectively. The elements in solid solution generally have a strong influence (either positive or negative) on TEP and are described by the Nirdheim – Gorter Law [28] .The following equation was used to quantitatively estimate the influence of Cu in solid solution in a binary Fe-Cu alloy :

$$1/\Delta S_{ss} = -(0.1086/Cu_{wt.\%}) - 0.04276 \; (\mu V/K)^{-1} \qquad (3)$$

The specific TEP contribution from Copper, as seen from above, is negative. It is can therefore be expected that the dissolution of copper precipitates in the iron matrix will lead to a decrease in the TEP value and the precipitation of copper will result in an increase in TEP. One could use the eqn. (3) above to determine the solubility limit of copper after complete precipitation at different temperatures, assuming that the final TEP value at each temperature depends only on the copper content in solid solution and is thus equal to ΔS_{ss}. It may be noted that the solubility limits thus determined were confirmed by Atom Probe Tomography (APT) [29,30].The influence of dislocations on TEP is negative[27] as they act as additional scattering centres for electrons and phonons. Consequently, cold working of steel results in a decrease and recovery leads to an increase in TEP; all these processes have been followed effectively by TEP.

When the precipitates are fine coherent or semi-coherent, they have a strong influence on the TEP as they cause a distortion of the lattice due to the coherency strain[23]. However, the coarse, incoherent precipitates have little effect on the TEP. Thus, the TEP would be generally insensitive during coarsening of precipitates. However, indirectly, when the solute content changes in the matrix during coarsening, the TEP values would get affected.

Supporting Measurements

In order to study the nature of precipitates in the IF-1.2 Cu steel, TEM was conducted on thin films and extraction replicas. Additionally, the replicas were subjected to field emission SEM. The in-situ chemical analysis of these samples were conducted using EDAX both in the TE & SEM. Further, optical microscopy as well as hardness studies on samples were conducted. The details of the sample preparation methods are given in the recent journal paper[25].

RESULTS AND DISCUSSION

Evolution of TEP as a function of time during aging at 550ºC, after annealing under various conditions and water-quench, for the IF-1.2 Cu steel as well as binary Fe-1.4Cu binary alloy are given in Fig. 1. The batch-annealed sample shows almost a flat behaviour owing to the fact that most of the Copper are already precipitated out (during the lower temperature annealing treatment) and as such, very little change in the solute Cu-content takes place. On the other hand, both in the continuously annealed and fully annealed condition of IF steel, as well as in the binary Fe-Cu alloy, there is a steady increase pointing to the precipitation phenomenon in these conditions. Further, the fully annealed (FA)and the continuously annealed (CA) IF steel reflect almost an identical behaviour. For the Fe-1.4 Cu binary alloy, at time t=0, using the eqn. (3), one estimates a TEP value of - 8.3μV/K ,which is in excellent agreement with the experimental value of -8.6 μV/K (S at t=0). For the IF-1.2 Cu steel, one obtains a TEP value of - 7.5 μV/K from the eqn.(3), while the experimental value is - 8.0 μV/K. (at t=0) for the CA / FA conditions; this is a good agreement considering the presence of other alloying elements as well. This would further suggest that in the IF steel, most of the copper is present in solid solution in the initial CA / FA conditions. It may be stated that the presence of other prominent alloying elements such as Mn & Ti (in low concentrations in solution in ferrite) have only a limited influence on the TEP value; the effect of Ti was already known to be low in literature[18]. The Fig.2 (a) & (b) present the data of Fig.1 for the IF-Cu in the CA / FA conditions and the binary Fe-Cu alloy, in a format amenable to possible kinetic analysis for copper precipitation. One finds from the Fig. 2 (b) that the kinetics of Cu-precipitation is identical in the two materials (steel & binary alloy), for ageing at 550ºC.

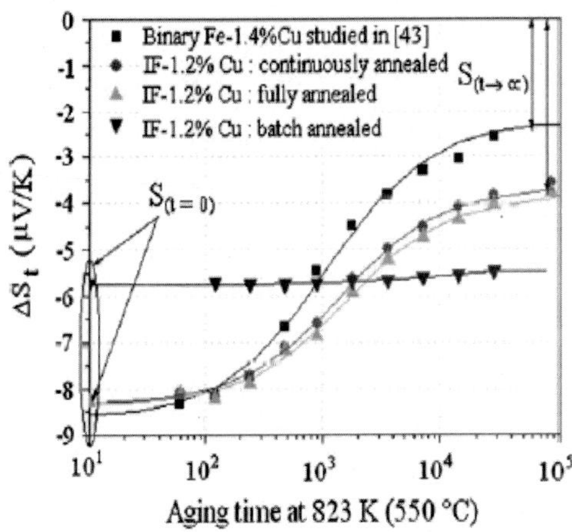

Fig. 1: TEP Values for the IF-Cu s and for Fe-Cu binary alloy, during Ageing at 550ºC.

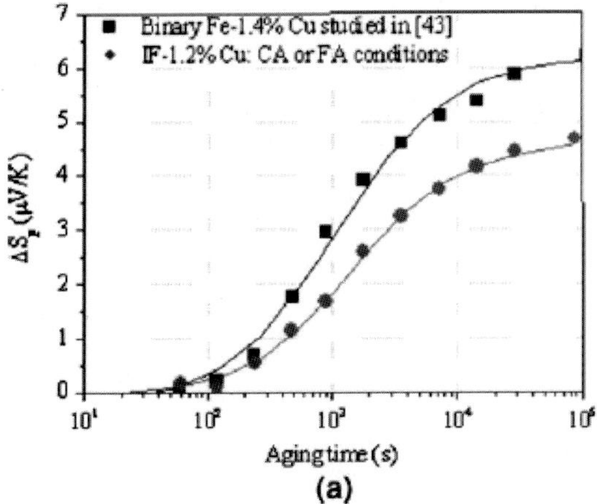

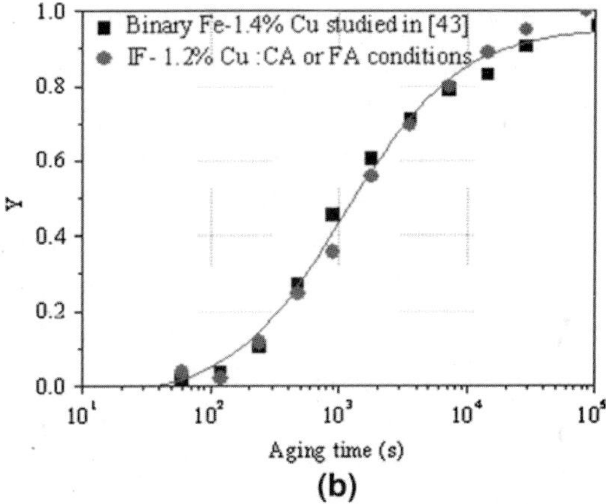

Fig.2: (a) Relative TEP Values for the IF-Cu steel and for Fe-Cu binary alloy during Ageing at 550ºC (derived from Fig.12) and (b) corresponding fraction (Y) transformed

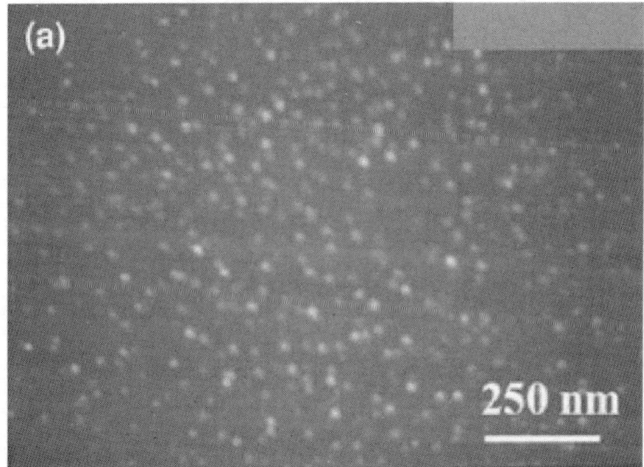

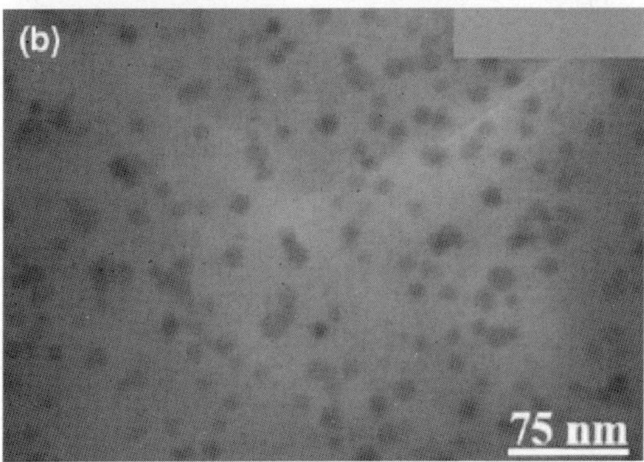

Fig. 3: TEM observations on thin foils of IF-1.2 pct Cu steel, CA and aged 24 h at 500ºC; (a) dark field ; (b) bright field

The TEM observations of continuously annealed (CA) IF-Cu steel aged for 24 hrs. at 500°C are shown in Fig.3 (a) and (b); the EDAX confirmed the precipitates as copper; there were other elements also present indicating that copper could have nucleated on other known precipitates (e.g. TiC, TiS or Ti4C2S2 etc.).in the Ti-bearing IF-Cu steels. From their habit plane (used in the beam orientation for the dark-field), the copper precipitates are believed to be f.c.c. □-Cu [25].

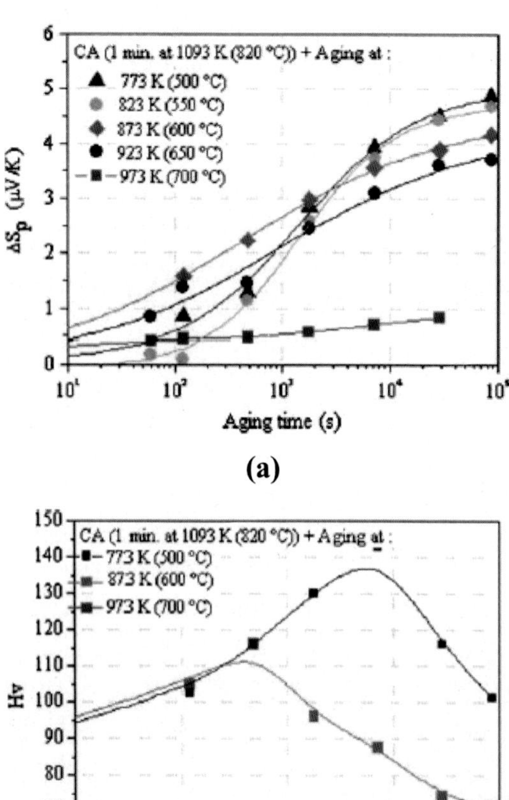

Fig. 4: Isothermal ageing studies on IF-Cu steel, CA at various temperatures with time; (a) TEP (b) Hardness

Isothermal ageing studies were done at various temperatures and ageing times. The evolution of TEP and hardness are shown in Fig.4 (a) and (b) respectively. It may be observed that while the hardness data (Fig. 4b) reflect the classical age-hardening characteristics (e.g. peak & over-ageing ; higher peak for lower temperature of ageing etc.); the TEP plots, also shown earlier in Fig. 2,do show a continuous increase with ageing time (in the temperature-zones presented here), ostensibly into the domain of precipitate coarsening as well, indicating continued precipitation of copper from the matrix. Further, it is to be noted that at the lower temperatures of ageing (e.g. 500 & 550°C) the TEP plots display a typical sigmoidal shape (nucleation-growth process). The rise in TEP appears accelerated, only in the initial phase, for higher ageing temperatures up to around 600°C.

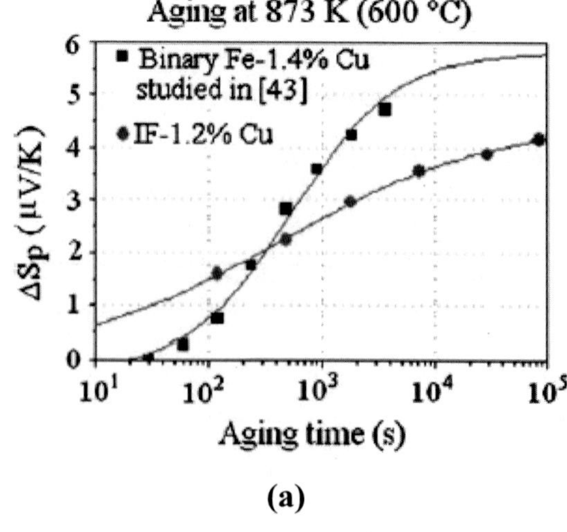

(a)

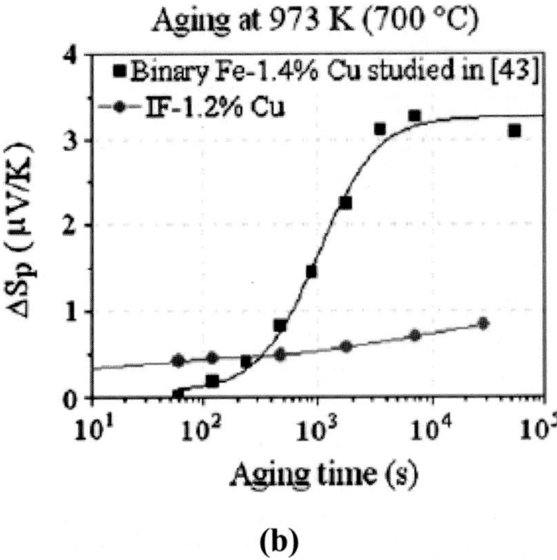

(b)

Fig. 5: Comparison of TEP values between IF-Cu steel (CA) & the Fe-Cu binary alloy, for ageing at high temperatures, (a) 600°C and (b) 700°C

In Fig. 5, a comparison is given between the IF-Cu steel and the Fe-Cu binary alloy for the higher temperatures (600 & 700°C) of ageing. It may be noted that unlike the situation at 550°C- ageing where the two materials displayed a similar kinetics for copper precipitation, tracked through TEP [compare Fig.2 (b)], at 600°C the TEP data for IF-Cu steel shows a more sluggish rise with time compared to the binary Fe-Cu alloy. When the aging temperature increases from 500°C to 600°C, the time associated with the peak hardness decreases from 2 hours to about 8 minutes, owing to more rapid diffusion of copper in the matrix (at 600°C). This confirms the earlier observation from TEP variation that, among the temperatures investigated, the precipitation kinetics is the fastest at 600°C. However, the value of TEP corresponding to the time of peak hardness at 600°C (i.e. $\Delta Sp \approx 2$ μV/K) is lower than that corresponding to the time at peak hardness at 500°C (i.e. $\Delta Sp \approx 4$ μV/K). This also indicates that the amount of copper precipitation responsible for the peak hardness could be lower at 600°C. In addition, the size of the copper precipitates is expected to be coarser at 600°C. Correspondingly, the hardness-increase is also lower at 600°C. For the case of ageing at 700°C, there is only a limited increase in TEP for the IF-Cu steel; the hardness also shows a marginal drop with ageing, suggesting early onset of the coarsening phenomenon. On the other hand, the rise in TEP for the Fe-1.4%Cu binary alloy at 700°C is appreciably higher, and the typical sigmoidal character is still observed. One needs to do further work to elucidate this difference in the precipitation behaviour between the IF-Cu steel and the Fe-Cu binary alloy at very high temperatures (e.g. 700°C).Copper precipitation in the IF-Cu steel at 600 & 700°C was studied with TEM and the results are presented in Fig.6 (a) &(b) respectively; while Fig.6 (c) shows the chemical analysis of a rod-like precipitate through EDAX. The TEM features point to two distinct morphologies of precipitates, spherical & rod-like. This observation is consistent with the information in literature [31,32,33] that beyond around 30 nm and at higher temperature of ageing, the precipitates tend to be rod-like. Thus, while the 600°C ageing shows both spherical and rod-like precipitates, the latter type are only seen at 700°C. Further, the amount of precipitates is much less, compared to those for the 600°C ageing.

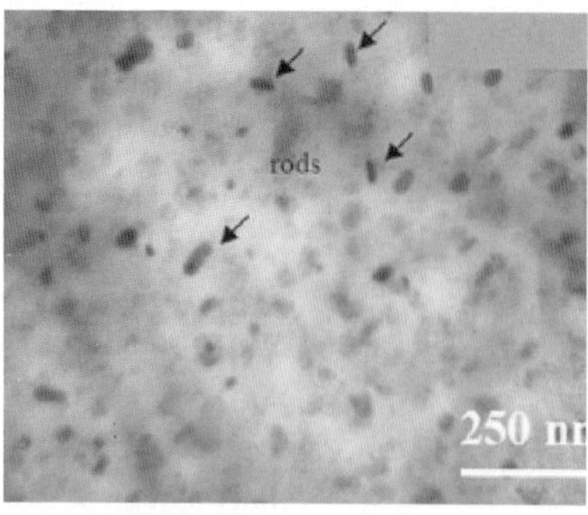

(a)

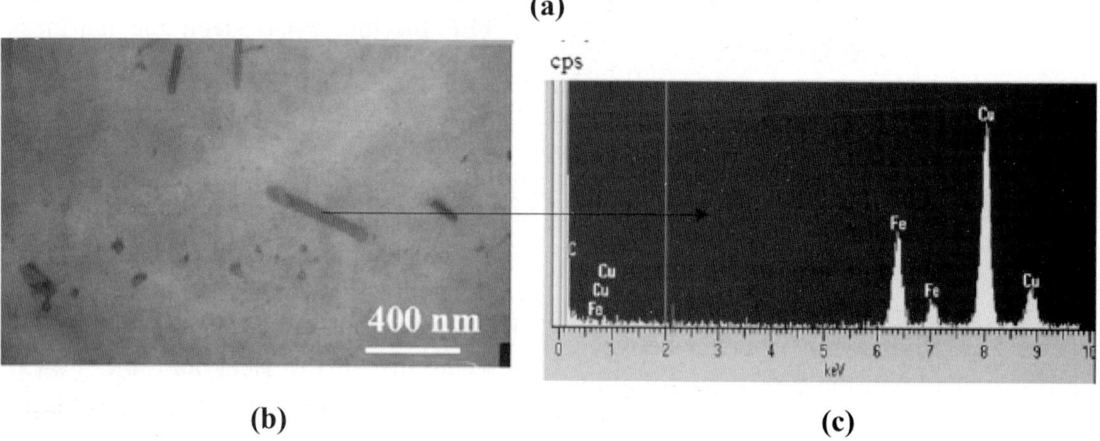

(b) (c)

Fig. 6: TEM pictures of IF-1.2 % Cu foils, (a) aged for 24 hrs./ 600°C, (b) aged for 8 hrs. at 700°C and (c) EDAX for the rod-like ppt. at 700°C

COMPARISON OF COPPER PRECIPITATION IN HSLA & IF STEELS

Strength Behaviour

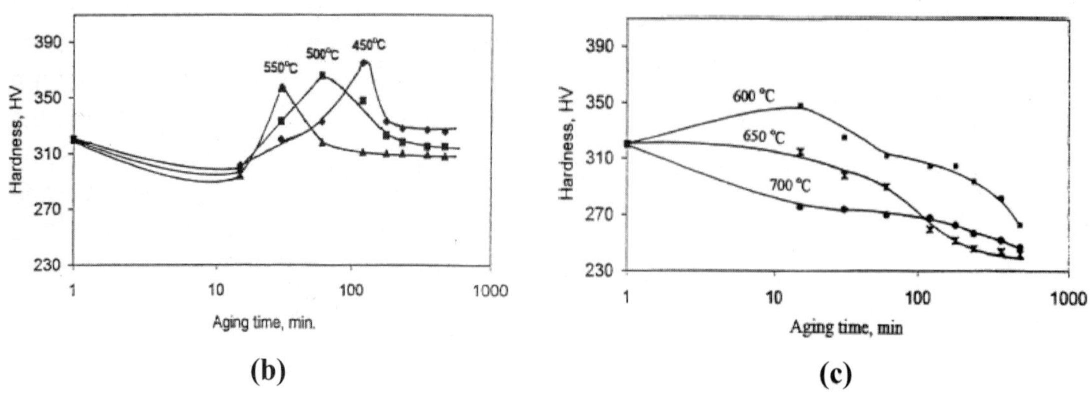

(b) (c)

Fig.7(a) &(b): Hardness in HSLA-100 as a function of ageing (a) &(b)[15]

In both HSLA-100 (containing 1.98 %C) and IF steels, the presence of copper induces classical age-hardening effect. Interestingly, the time to attain the peak hardness in the low temperature (< 600°C) range is approximately same [compare with Fig.4 (b)]. Again, the higher temperature (> 600°C) behaviour in both cases are similar: at 600°C there is a small rise and then the over-ageing sets in; at still higher temperatures, there is a continuous fall in hardness. On the other hand, both the electrical resistivity in HSLA[15]and TEP (in IF-Cu steel) show some changes indicative of continued copper precipitation in the over-aged condition. The coarser size of precipitates at the higher temperatures of ageing is believed to be the cause of this behaviour. Additionally, as would be expected, the absolute value of hardness in the HSLA is much higher (≈ 300 H_v) compared to that in the IF_Cu steel (≈ 100 H_v); however, the enhancement in hardness for the highest peak is ~ 30 %. (the Cu-content in the chosen IF-Cu steel is ~ 1.4 % while in the selected HSLA it is ~ 2 %). From the hardness data therefore, the over all broad kinetics of copper precipitation responsible for the strengthening is similar. The estimation of activation energy from the resistivity data of the HSLA and its analysis have been provided elsewhere [15] and it is only expected that in finer aspects, the precipitation behaviour in HSLA would be would differ from that in the IF-Cu case as the diffusion behaviour of copper in the two matrices could be different.

Impact toughness

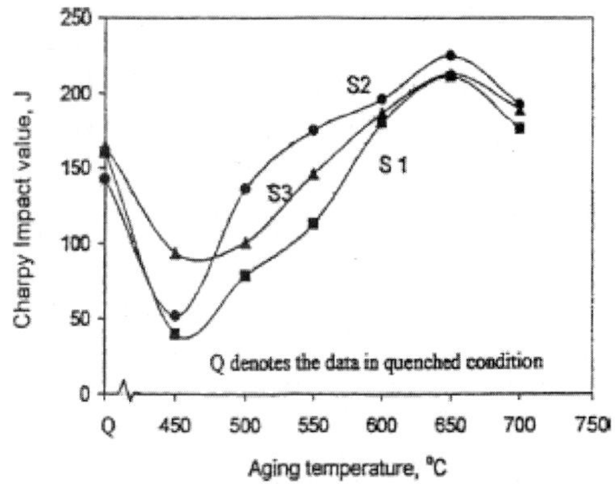

Fig.8: Impact Values of three HSLA materials after ageing for 1 hr. at each temp. [15]

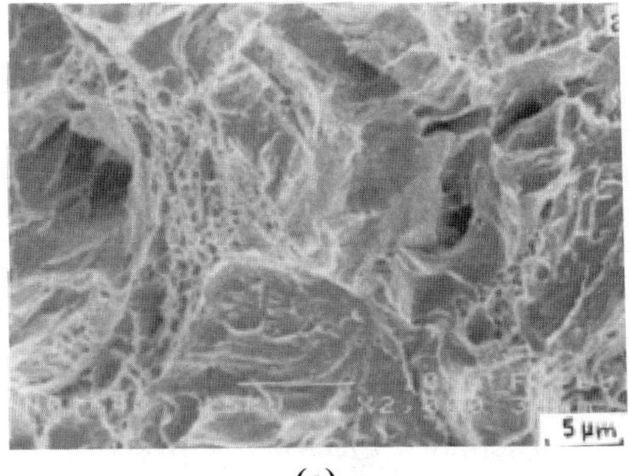

(a)

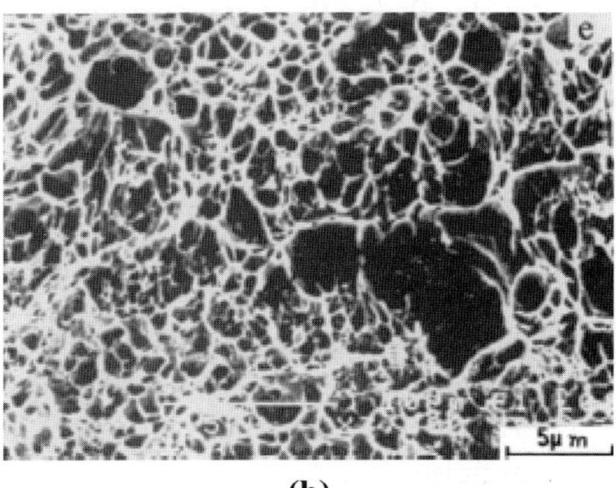

(b)

Fig. 9: Structure of impact fractured surfaces of HSLA sample, aged at 450°C (a) and at 650°C (b) [15]

The impact strength of the HSLA as a function of the ageing temperature is shown in Fig.8 and some of the corresponding fractographs are shown in Fig. 9(a) and (b). It is to be noted that the condition of peak hardening is associated with a sharp fall in the impact value, also corroborated by the cleavage fracture. On the other hand, in the over-aged condition, one observes appreciable increase in the impact energy of absorption and a concurrent shift to a dimple mode of fracture. The harmful effects of coherent copper precipitates and the improvements with appearance of semi-coherent/incoherent precipitates on the impact values have also been dealt with in the literature [34,35,36]Although no impact data for the peak-hardened IF-Cu steel is presented here, it is expected that, one needs to optimise the strength and toughness by judiciously selecting the ageing conditions.

SUMMARY & CONCLUSION

- Thermo Electric Power (TEP) techniques has been applied effectively in elucidating the copper precipitation behaviour in the industrially potent IF-Cu steel. Even the early stages of the ageing could be investigated.
- Apart from qualitatively characterizing the process of precipitation, TEP can yield data that could be subjected to quantitative (such as kinetic) analysis; one needs however, similar data from a reference materials for such an exercise.
- Supporting techniques (e.g. hardness, optical/TEM etc.) have been used in order to interpret the TEP results. Copper precipitation show spherical as well as rod-like morphologies; at high temperature (~ 700ºC) of ageing, it is predominantly rod-like.
- The generally apprehended hot-shortness issue of using copper in steel (in the absence of Ni), can be avoided and exploiting strengthening from copper precipitation is still possible by optimizing the amount of copper.
- The lowering of impact toughness at peak hardness values (as shown in the HSLA steel) need to be examined more carefully for the IF-Cu steel as well. Perhaps by sacrificing some strength and resorting to over-aged conditions the brittle condition, if encountered could be avoided. More study is needed in this direction.

REFERENCES

[1] E. Hornbogen, R.C. Glenn, Trans. TMS-AIME, Vol. 218 ,1960, p. 1064–1070.

[2] E. J Czyryca: Advances in Low Carbon High Strength Ferrous Alloys, eds. O. N. Mohanty, B.B.Rath, M.A.Imam,C.S. Sivaramakrishnan. Key Engineering Materials,Trans. Tech. Pub., Switzerland, Vol. 84–85, 1993, p.491-520

[3] J. Takahashi, K. Kawakami, Y. Kobayashi Matl. Sc. & Engg. A, vol. 535,2012, p.144-152

[4] S. Hoile: Mater. Sci. Technol., vol. 16, 2000, p. 1079–93.

[5] K. Tsunoyama: Phys. Stat. Sol. (a), vol. 167,1998, p. 427–33.

[6] H. Takechi: ISIJ Int., vol. 34, 1994, p. 1–8.

[7] R. Rana, W. Bleck, S. B. Singh, and O. N. Mohanty, Steel Res.Int., vol. 78, 2007, p. 612–21.

[8] R. Rana, W. Bleck, S.B. Singh, and O.N. Mohanty, Mater. Lett., vol. 61,2007, p. 2919–22.

[9] R. Rana, W. Bleck, S. B. Singh, and O. N. Mohanty, Mater. Sci.Technol., vol. 23, 2007, p. 561–66.

[10] R. Rana, S. B. Singh, W. Bleck, and O. N. Mohanty, Metall. Mater. Trans. A, vol. 41,2010, p.1483–92.

[11] R. Rana, W. Bleck, S.B. Singh, and O.N. Mohanty, Steel Res. Int., 2007, vol. 78, p. 622–30.

[12] D. T. Llewellyn: Ironmaking Steelmaking,vol. 22, 1995,p. 25–34.

[13] K. Shibata, S.J. Seo, M. Kaga, H. Uchino, A. Sasanuma, K.Asakura, and C. Nagasaki: Mater. Trans., vol. 43, 2002, p. 292–300.

[14] A.N.Bhagat,S.Ranganathan, O. N. Mohanty, Mat. Sc. & Tech., vol. 19, 2003, p.343 - 346

[15] A. N. Bhagat, S.K.Pabi, S. Ranganathan, O. N. Mohanty, ISIJ Intl., vol.44, 2004, p.115-122

[16] A. N. Bhagat, S. K. Pabi, S. Ranganathan, O. N. Mohanty, Matl. Sc. & Tech., vol.23, 2007, p.1-7

[17] A. Brahmi and R. Borrelly: Acta Mater., vol. 45, 1997, p. 1889–97.

[18] S. Carabajar, J. Merlin, V. Massardier, and S. Chabanet, Mater.Sci. Eng. A, vol. 281,2000, p.132–142.

[19] M. Houze´ , X. Kleber, F. Fouquet, and M. Delnondedieu, Scripta Mater., vol. 51, 2004, p 1171–1176.

[20] N. Lavaire, V. Massardier, and J. Merlin: Scripta Mater., vol. 50, 2004, p. 131–135.

[21] V. Massardier, N. Lavaire, M. Soler, and J. Merlin: Scripta Mater., vol. 50, 2004, p. 1435–1439.

[22] V. Massardier, V. Gue´ taz, J. Merlin, and M. Soler: Mater. Sci.Eng. A, vol. 355, 2003,p.299–310.

[23] J.M. Pelletier, G. Vigier, J. Merlin, P. Merle, F. Fonquet, and R.Borrelly, Acta Metall., vol. 32, 1984, p. 1069–1078.

[24] M. Perez, F. Perrard, V. Massardier, X. Kleber, A. Deschamps, H.De Monestrol, P. Pareige, and G. Covarel: Phil. Mag., vol. 85, 2005, pp. 2197–2210.

[25] R. Rana, V.Massardier, S.B.Singh, O.N.Mohanty, Metall. Mater. Trans. A,2012, DOI: 10.1007/s11661-012-1386-z

[26] F.J. Blatt, P.A. Schroeder, C.L. Foiles, and D. Creig: Thermoelectric Power of Metals, Plenum Press, New York, 1976, p. 264.

[27] N. Lavaire, J. Merlin, and V. Sardoy: Scripta Mater., vol. 44, 2001, p. 553–60.

[27] L. Nordheim and C.J. Gorter: Physica, vol. 2, 1935, p. 383–90.

[28] D. Blavette, A. Bostel, J. M. Sarrau, B. Deconihout, and A.Menand: Nature,vol. 363, 1993, p. 432–35.

[29] P. Pareige, P. Auger, P. Bas, and D. Blavette: Scripta Metall.Mater., vol. 33, 1995, p. 1033–36.

[30] G.R. Speich and R.A. Oriani: Trans. Metall. Soc. AIME, vol. 233, 1965, p. 623–31.

[31] P.J. Othen, M.L. Jenkins, G.D.W. Smith, and W.J. Phythian: Phil.Mag. Lett., vol. 64, 1991, p.383–91

[32] M. Charleux, F. Livet, F. Bley, F. Louchet, and Y. Brechet: Phil. Mag. A, vol. 73,1996, p.883–97.

[33] M. R. Krishnadev: INCRA Report No. 244, New York, 1974.

[34] F. Leysen., J. Neutjens., K. Mols., S. Vandeputte and Y. Houbaert: Mater. Sci. Forum, Switzerland, vol.263, 1998, p.284–286

[35] L. Skoufari-Themistou, B. Mintz and D. N. Crowther: Mater. Sci. Technol., vol.15, 1999, p.1069.

Ratcheting Behaviour of Tensile Pre-strained Interstitial Free Steel

Partha Sarathi De[1], P.C. Chakraborti[1], M. Shome[2], B. Bhattacharya[2] and D. Bhattacharjee[3]
[1]Metallurgical and Material Engineering Department, Jadavpur University, Kolkata 700032, India
[2]Research and Development Division, TATA Steel Ltd., Jamshedpur 831001, India
[3]Research, Development and Technology, IJmuiden Technology Centre, TATA Steel Ltd., The Netherlands
Email: pravashchandrachakraborti@hotmail.com

ABSTRACT

Effect of tensile pre-strain in the range of 2.5 to 15 pct on ratcheting behaviour of a titanium stabilised interstitial free steel has been investigated at two different stress amplitudes of 187.46 and 158.62 MPa keeping mean stress constant at 28.84 MPa. Ratcheting strain evolution direction is found to depend on the amount of pre-strain and stress amplitude. It is observed that the number of cycles required for 16.28 pct ratcheting strain accumulation exponentially increases with the amount of pre-strain for both the stress combinations. Further marginal variation in post-ratcheting tensile properties as a function of tensile pre-strain is observed.

KEYWORDS: *Interstitial Free Steel, Ratcheting, Stress Amplitude, Bauschinger Effect*

INTRODUCTION

Asymmetric stress cycling in the elastic-plastic domain results in the evolution of ratcheting strain in the direction of mean stress which is accumulated in the components over a period of stress cycles. Such progressive accumulation of strain may continue until fracture [1]. It is, therefore essential to understand the ratcheting behaviour of materials for structural integrity purpose. Over the last three decades numerous studies have been carried out to understand the ratcheting behaviour of different materials. All these studies can be broadly classified into two: (i) experimental studies to investigate the effect of stress parameters and test conditions on the uniaxial and multiaxial ratcheting response of different materials, and (ii) constitutive modeling to qualify the ratcheting behaviour of different materials under uniaxial and multiaxial loading conditions through modifications of the original Armstrong-Frederick [2] nonlinear kinematic hardening model.

However, although it is known that most of the components receive different degrees of deformation in the fabrication stage the effect of such prior deformation on the ratcheting response has rarely been looked into. Studies on the effect of prior deformation on subsequent cyclic behaviour of different materials, e.g. low carbon steel [3], pure iron and spheroidized medium carbon steel [4], AISI 430 Stainless Steel [5], SA333 Gr.6 C-Mn steel [6] are mostly limited to symmetrical stress or strain cycling condition. All these studies revealed that accumulation of plastic strain (cyclic creep) occurs in the direction opposite to the prior deformation.

Pioneering work of Feltner and Laird [7,8], and that of Christ et.al. [9] provide valuable information on the effect of different amounts of prior deformation on completely reversed strain-control fatigue behaviour of copper. Feltner and Laired [7,8] observed that cyclic stress-strain curve in the region of high plastic strain amplitude is not influenced by the low to intermediate amounts of prior deformation. Dislocation structure formed during intermediate level prior deformation is converted to persistent slip bands (PSBs). With higher deformation the prior dislocation structure is found to be retained with more domination [9]. Jia and Fernandes [10] have also reported that with increasing pre-strain complete modification of the equiaxed and disorganised cell structure in copper formed during prestraining is resisted during subsequent fatigue cycling.

Interstitial Free (IF) steels find extensive application in automotive industries for their very good cold-forming characteristics. Components made of this variety of steel are often subjected to cyclic loading during service. Hence an understanding of the cyclic deformation behaviour of such a highly ductile material seems relevant. Recently interests among researchers, steel producers and car manufacturers are noticed in areas of fatigue performance and fatigue deformation mechanisms of IF steel. But the response of this variety of steel under asymmetric cyclic loading condition has received very little attention so far except the one that describes the anisotropy effect on the ratcheting behavior of IFHS steel [11] and the other by the present authors [12]. Since, components made of interstitial free steels receive various levels of cold deformation and are also subjected to cyclic stressing during service the effect of such prior deformation on the ratcheting response is worth investigating. Present investigation has been carried out with an objective to understand the effect of different levels of tensile pre-strain on the ratcheting response of a titanium stabilized interstitial free (Ti-IF) steel at two different stress amplitudes keeping the mean stress constant.

EXPERIMENTAL PROCEDURE

Cold rolled and batch annealed 2.3 mm thick Titanium stabilised interstitial free (Ti-IF) has been used in the present investigation. The chemistry of the steel in weight percent is: C-0.002, Mn-0.04, Si-0.004, S-0.007, P-0.015, Al-0.0359, Nb-0.001, Ti-0.063, N-0.024 and balance Fe.

Specimens for tensile and ratcheting tests with rectangular cross-section were fabricated from the as-received steel sheets keeping the long axes of the specimens parallel to rolling direction of the steel sheets. Apart from tensile test of the steel in the as-received condition until complete fracture a number of tensile specimens were pre-strained to 2.5 to 15 pct with a strain rate of 10^{-3} sec^{-1} and then slowly unloaded. The pre-strained specimens were subsequently strained to fracture under tensile loading keeping nominal strain rate fixed at 10^{-3} sec^{-1} as before.

Engineering stress-control ratcheting tests have been done on specimens with and without any prestrain at two different stress amplitudes of 187.46 and 158.62 MPa keeping mean stress constant at 28.84 MPa. Tests have been carried out using a triangular load waveform at a stress rate of 250 MPa

sec^{-1}. During cyclic stressing strain was measured with the help of an extensometer of ± 20 pct capacity. Tests were limited to cycles corresponding to +18 pct strain limit in the positive part of the asymmetric loading cycles. This, in fact, results in the evolution of 16.28-pct cumulative ratcheting strain. Following ratcheting tests the specimens were unloaded and then fractured under tensile loading at a strain rate of 10^{-3} sec^{-1}.

RESULTS AND DISCUSSION

Microstructure and Tensile Properties

Optical micrograph shown in Fig.1 reveals that microstructure of the steel consists of equiaxed ferrite grains of varying size distribution. The average ferrite grain size has been measured both by planimetric and duplex method using an automated image analyzer, Leica Materials Workstation. The measured average grain size is found as ASTM 8. This corresponds to an average grain diameter of 22 μm. Hardness of the steel measured with 1 kgf laod in Leica VMHT microhardness tester is 96.2 VPN. The tensile properties of the investigated steel are: Yield strength: 144.2 MPa, Tensile strength: 276.16 MPa, Uniform strain 26.98% and Total strain: 54.44%.

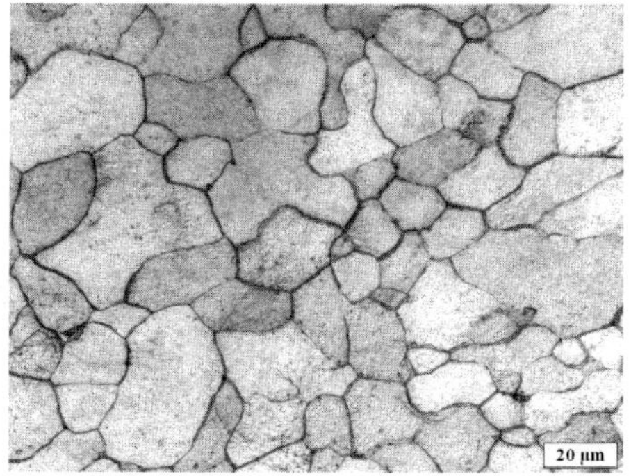

Fig. 1: Optical microstructure of the investigated Ti-IF steel.

Engineering stress-strain curves of the investigated Ti-IF steel with and without pre-strain are shown in Fig. 2. It is observed that although the steel used is almost free of interstitial contents, yield point appears on reloading the specimens with higher amounts of pre-strain. This is attributed to the development of coarse dislocation cell structure during pre-straining stage. The appearance of yield point has been shown in the inset of Fig. 2.

The variation of strength and ductility parameters of the steel as a function of the degree of pre-strain is shown in Fig. 3 and Fig. 4 respectively. It is observed that with increase in the amount pre-strain both yield strength and tensile strength increase with simultaneous reduction in ductility. Increase in yield strength with the amount of pre-strain is found to occur in two different stages; initially there is a rapid increase of yield strength up to 7.5 pct pre-strain, followed by more gradual increase with higher amounts of pre-strain.

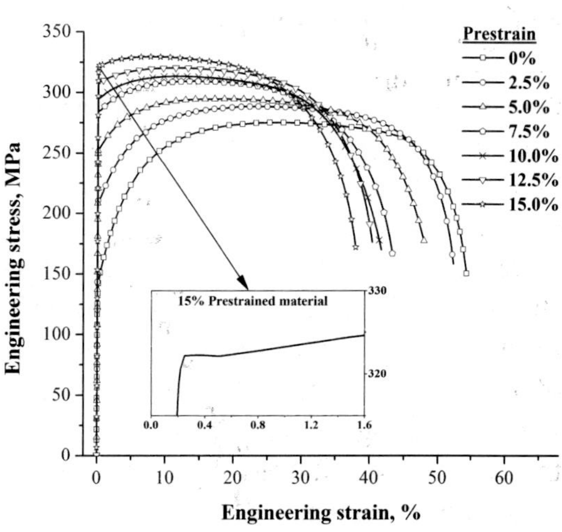

Fig. 2: Engineering stress strain curve of different pre-strained specimen.

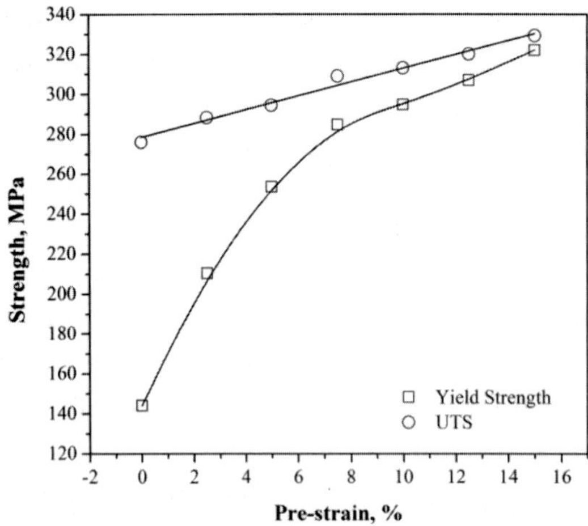

Fig. 3: Variation of strength with amount of pre-strain.

The evolution of dislocation structure at small plastic strain studied for OFHC copper [10], low carbon steel [13] and interstitial free steel [14,15] reveals that irregular structure with uniform dislocation density, more precisely the dislocation tangles, forms from straight-line dislocations at small plastic strain produced either by rolling or tensile deformation. With higher amounts of deformation non-uniform dislocation cell structure develops. Antoine et al. have [14] reported that with about 5 pct deformation almost uniform entangled dislocation structure is transformed to low energy non-uniform dislocation cell structure in case of interstitial free steel. Dislocation density increases uniformly with increasing pre-strain up to 7-10 pct. As a result resistance to start plastic deformation on reloading of pre-strained material increases sharply with increasing pre-strain. However, once the dislocation cell structure of low energy configuration appears dislocation density increases non-uniformly. The large amounts of dislocations are collected on the cell walls while less at the cell interiors. After about 7-10 pct. strain non-uniform increase of dislocation density leaves enough spaces, where mobile dislocation can move freely. So pre-straining in excess of some critical amount (7-10%) the number of barriers against dislocation movement increases slowly than in the early stage of deformation. This causes a relatively slow rise of yield strength above some critical amount of pre-strain. In the present investigation the transition in the rising behaviour of yield strength with the amount of pre-strain is found to take place at 7.5 pct. It is, therefore, logical to consider that the effect of pre-straining on subsequent two stage yielding behaviour as observed in the present investigation is due to the formation of stable dislocation cell structure during pre-straining by about 7.5 pct and above.

The variation of tensile strength with the amount of pre-strain, however, does not show two-stage behaviour; instead tensile strength increases with pre-strain following a perfectly linear relationship (Fig.3). The extrapolated tensile strength (= 280.5 MPa) at zero pre-strain is found to closely match with the tensile strength data of the steel without any pre-strain (= 276.2 MPa). On the other hand, while a there happens a linear decrease in total strain with the amount of pre-strain, uniform strain decreases with increase of pre-strain in two stages (Fig.4). Up to 7.5 pct pre-strain the uniform strain decreases very rapidly. The rate of decrease of uniform strain is comparatively slow with further increase in the amount of pre-strain.

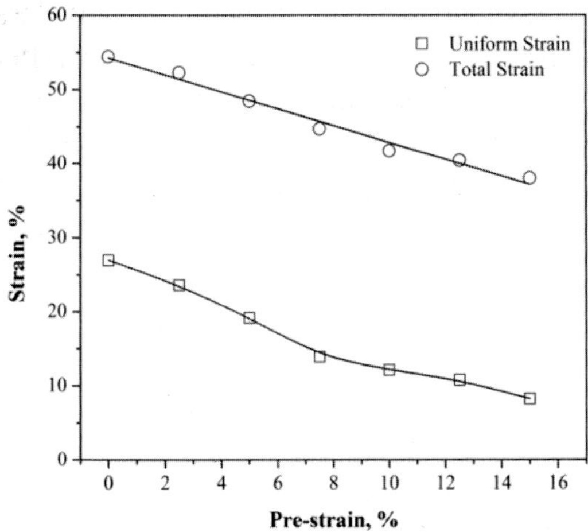

Fig. 4: Variation of ductility with amount of pre-strain.

Effect of Tensile Pre-Strain on Ratcheting

The cycle-by-cycle evolution of ratcheting strain in the specimens with and without any pre-strain has been found from the maximum and minimum strain data recorded during stress cycling with the help of an extensometer attached to the specimen. From the recorded strain values ratcheting strain in each case has been calculated using the following formula:

$$\varepsilon_r = \frac{1}{2}\left[\ln(1+e_{max}) + \ln(1+e_{min})\right] \quad (1)$$

where, ε_r is the true ratcheting strain, and e_{max} and e_{min} are extensometer readings for maximum strain and minimum strains at peak maximum and minimum stresses respectively.

The cumulative ratcheting strain accumulation curves for different pre-strained specimens including the one with zero pre-strain for the stress combination of 28.84±187.46 MPa are shown in Fig.5. The initial stage of the ratcheting strain accumulation pattern has been shown in the inset of Fig.5. It is known that during asymmetric stress cycling in the elastic-plastic domain ratcheting deformation occurs in the direction of mean stress. It means that in presence of tensile mean stress elongation of the specimen would occur, and in presence of compressive mean stress ratcheting deformation would take place in the compressive direction resulting in shortening of the specimen. In the present investigation it is found that initially ratcheting strain evolution occurs in the compressive direction even in presence of tensile mean stress when the amount of pre-strain is in excess of 2.5 pct. In specimens with zero and 2.5 pct. tensile pre-strain the ratcheting deformation occurred in the tensile direction from the very beginning of stress cycling as normally expected.

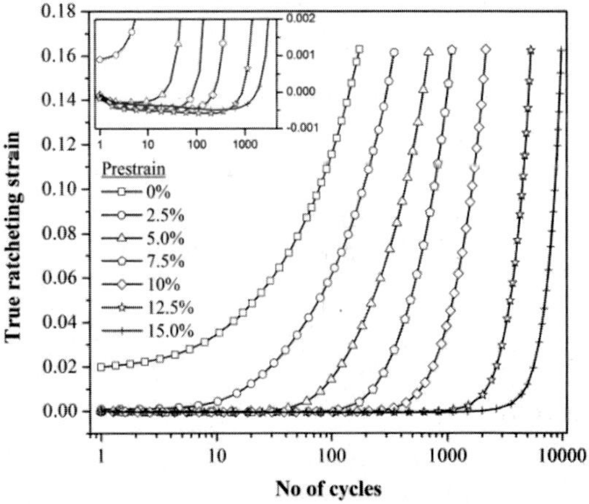

Fig. 5: Evolution of ratcheting strain with number of cycles.

A representative cumulative ratcheting strain accumulation curve in 12.5% tensile pre-strained specimen (Y.S = 307.20 MPa) for the stress combination of 28.84±187.46 MPa has been shown in Fig. 6. The cyclic stress-strain hysteresis loops corresponding to certain points of interest of this ratcheting curve are also shown in the inset of Fig. 6. It is observed that with progression of stress cycling the hysteresis loop is widened. This shows that cyclic softening occurs during asymmetric stress cycling with a maximum stress of 216.30 MPa, which is 90 MPa less than the corresponding yield strength under 12.5-pct pre-strained condition. Moreover, the evolution of compressive ratcheting strain during the initial cycles is quite obvious here.

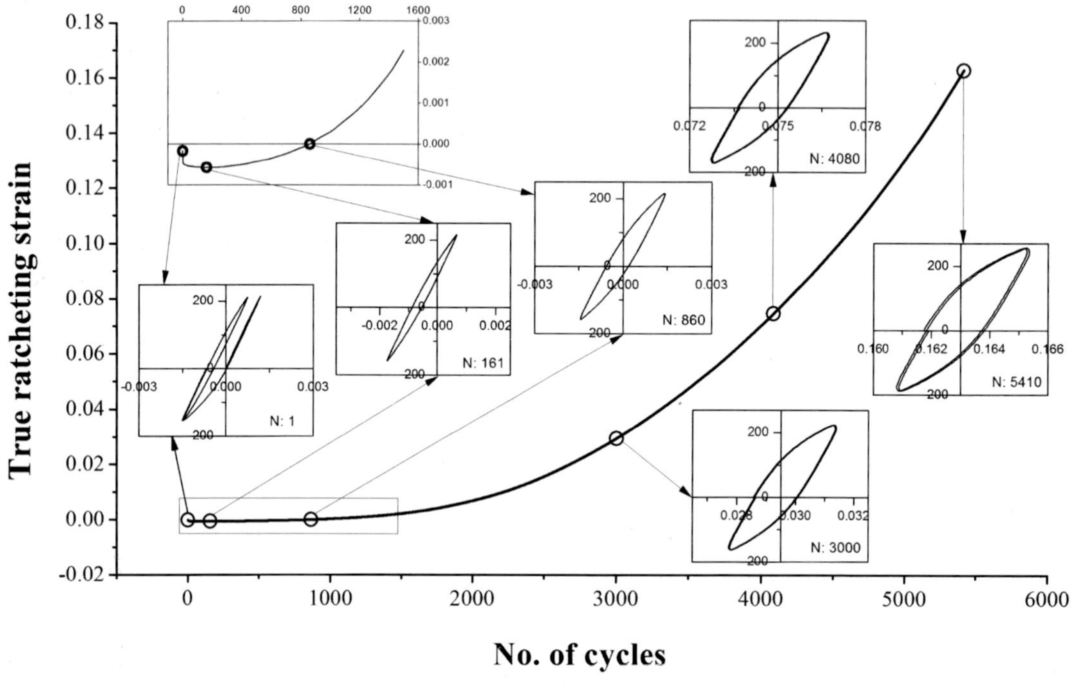

Fig. 6: Evolution ratcheting strain at 28.84±187.46 MPa stress combination in 12.5% prestrained specimen.

Cyclic stress-strain hysteresis loops developed during cyclic stressing of 2.5 and 5-pct pre-strained specimens at 28.84±187.46 MPa for the first 10 cycles have been presented in Figs.7(a-b). It is observed that the hysteresis loops shift in the compressive direction with substantial yielding in the reverse loading path when the amount of pre-strain is more than 2.5 pct. This happens due to Bauschinger effect arising out of back stress developed as a direct consequence of the dislocation pile-up during tensile pre-straining.

Early investigations revealed that back stress developed during prestraining promotes the evolution of compressive plastic strain during symmetrical stress cycling [3,4,16]. Christ et al [9] observed the presence of tensile mean stress during completely reversed plastic strain controlled fatigue cycling of pre-strained copper. The existence of tensile mean stress in case of completely reversed strain cycling is due to the development of compressive back stress during pre-straining.

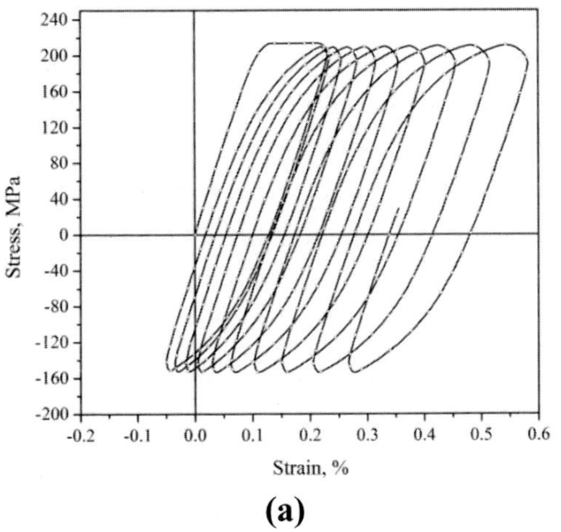

(a)

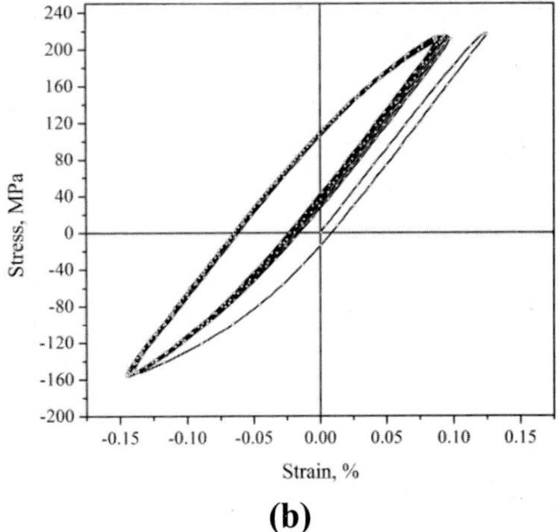

(b)

Fig. 7(a-b): Hhysteresis loops of first 10 cycles at 28.84±187.46 MPa stress combination: (a) 2.5% and (b) 5% pre-strained condition

In the present investigation it is observed that the Bauschinger effect in the first cycle is much more explicit, which is drastically reduced on subsequent cycling and finally gets exhausted. During symmetrical stress cycling Chai and Laird [3], and Ni and Wang [4] also observed the occurrence of ratcheting strain evolution in the opposite direction of the prior deformation in case of low carbon steel, pure iron and spheroidised medium carbon steel.

In case of prestraining by an amount of 2.5% the evolution of ratcheting strain is found to occur in the direction of imposed mean stress, i.e., in the tensile direction, from the very beginning of cyclic stressing (Fig.7b) as that also observed for specimen with zero pre-strain. This is understood on consideration of the amount of pre-stress (= 204.20 MPa) corresponding to 2.5-pct tensile pre-strain and the maximum stress (= 216.30 MPa) of the imposed stress cycle, which is higher than the prestress amount. Hence, it is quite logical to expect that the imposed stress cycle (28.84±187.46 MPa) would overcome the back stress developed during pre-deformation stage. It is indeed observed that in case of specimen with 2.5% pre-strain relatively large amount of plastic deformation occurs in the very first quarter cycle due to comparatively low yield strength (= 210.50 MPa). On reverse loading, although the compressive yielding occurs at an early stage due to Bauschinger effect this is not of sufficient magnitude to shift the hysteresis loop in the compressive direction to cause the evolution of compressive ratcheting strain in the beginning of stress cycling.

The present results clearly revels that the ratcheting strain evolution direction in pre-deformed materials is the result of the predominance between imposed mean stress and the back stress developed during pre-deformation stage. As long as back stress overpowers the imposed tensile mean stress ratcheting strain evolution occurs in the compressive direction. However, on continued cyclic stressing the effect of back stress appears to get exhausted, presumably because of unlocking of the dislocations from different barriers along with associated change in the dislocation substructure, whereas the imposed mean stress remains always constant. The imposed mean stress gradually becomes effective with softening of the pre-deformed structure through generation of (mobile) dislocations, which piled up against different kinds of barriers resulting in the development of back stress. As and when the conflict between the imposed tensile mean stress and back stress is neutralized the cumulative ratcheting strain accumulation rate ($d\varepsilon_r/dN$) becomes zero, and at this condition the cumulative compressive ratcheting strain becomes maximum. Following this the imposed mean stress becomes dominant over back stress facilitating the evolution of ratcheting strain in the direction of mean stress, i.e., in the tensile direction.

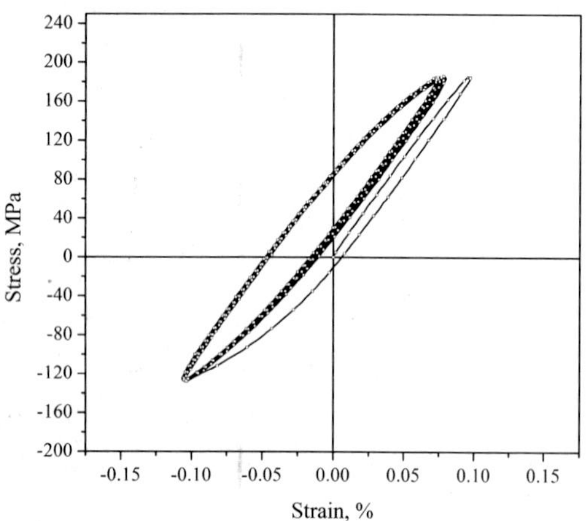

Fig. 8: Hysteresis loops of first 10 cycles for 2.5% pre-strained specimen (28.84±158.62 MPa)

In contrast to the ratcheting strain evolution direction in 2.5-pct pre-strained specimen under the stress combination of 28.84±187.46 MPa it is observed that in case of stress cycling at 28.84±158.62 MPa evolution of compressive ratcheting strain occurs from the very beginning (Fig.8). It is observed that at a stress amplitude of 158.62 MPa the Bauschinger effect is quite predominant as compared to higher stress amplitude (187.46 MPa). It should be carefully noted that while the amount of pre-stress corresponding to 2.5 pct pre-strain is 204.20 MPa the imposed maximum stress under which specimens with 2.5 pct. prestraining were subjected to stress cycling were 216.30 MPa and 187.46 MPa respectively. The imposed maximum stress (187.46 MPa) for 28.84±158.62 MPa stress combination being at a lower level than the yield strength under 2.5% pre-strained condition (210.50 MPa) the back stress developed during prestraining cannot be surpassed at the very beginning of stress cycling at 28.84±158.62 MPa. Hence compressive ratcheting strain evolution occurs right from the very first cycle because of the strong dominance of the Bauschinger effect in contrast to the condition when stress cycling is done at 28.84±187.46 MPa, the corresponding maximum stress being 216.30 MPa. However, on continued cycling the effect of back stress for the stress combination of 28.84±158.62 MPa is also exhausted and finally ratcheting strain evolution occurs in the direction of mean stress, i.e., in the tensile direction.

It is, therefore, concluded that for a given pre-strain condition the peak maximum stress during forward loading would control the ratcheting strain evolution direction at the very start of asymmetric cyclic stressing.

Net Plastic Strain Amplitude Response

The evolution of net plastic strain amplitude during asymmetric stress cycling of specimens with and without pre-strain at 28.84±187.46 MPa is shown in Fig. 9. It is observed that the net plastic strain amplitude curves in case of pre-strained specimens lie at a lower level as compared to the zero pre-strain condition. As the amount of pre-strain increases net plastic strain amplitude evolution curves shift downwards. However, for all pre-strain conditions little hardening occurs during the initial stage of stress cycling. Chai and Laird [3] have also reported this phenomenon in low carbon steel. Entrapment of the mobile dislocation formed during pre-straining is considered to be the possible reason for such hardening.

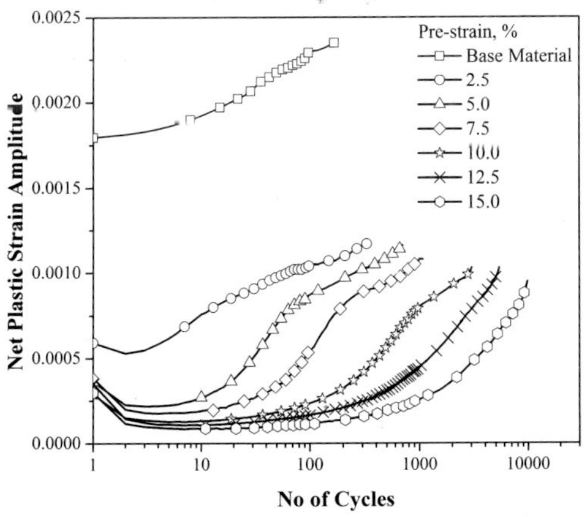

Fig. 9: Effect of pre-strain on the evolution of net plastic strain amplitude.

Following the initial hardening behaviour softening occurs in all cases on continued stress cycling. It is observed that the rate of softening becomes low as the degree of prestraining is increased. The continuous softening behaviour after the initial small hardening is an indication of reorganization of the dislocation substructure formed during pre-straining. It is interesting to note that the strain amplitude response curves for different amounts pre-strain seem to reach to the same level after the evolution of 16.28 pct true ratcheting strain. This indicates that with a cumulative tensile ratcheting strain of 16.28 pct the dislocation substructure becomes almost similar irrespective of the degree of pre-strain. In absence of TEM results post ratcheting tensile tests have been done to arrive at definite conclusions.

3.5 Relation between tensile pre-strain and number of cycles for 16.28 pct ratcheting strain evolution

In the present investigation it is observed that ratcheting strain evolution rate is hindered as the amount of pre-strain increases. Fig. 10 showing the effect of pre-strain on the number of cycles required for evolution of 16.28 pct ratcheting strain (N_{16}) reveals that N_{16} increases with the increase in the amount of pre-strain following a perfect exponential relationship for both the stress amplitudes.

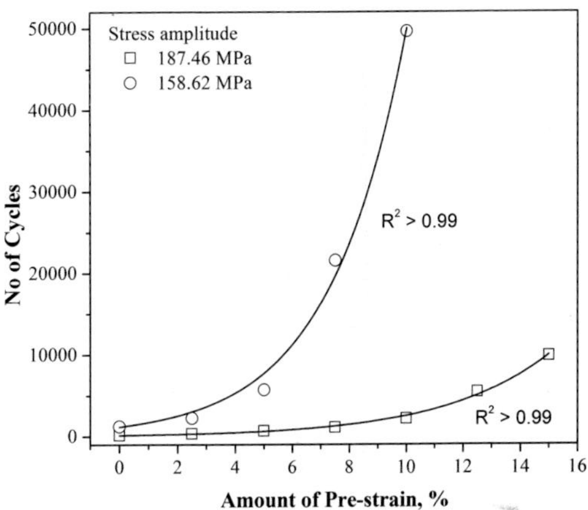

Fig. 10: Effect of pre-strain on the number of cycles required for evolution of 16.28 pct tensile ratcheting strain.

Since with increasing pre-strain the dislocation structure become complex and more stable modification of the pre-deformed structure during subsequent stress cycling becomes difficult when the amount of pre-strain is large. As a result ratcheting strain evolution which occurs through breakdown of dislocation cell structure and formation of new dislocation cells [3] is delayed when the amount of tensile pre-strain is high. Christ et al. [9] reported that with increasing pre-strain it is difficult to destroy the pre-deformed dislocation structure during subsequent fatigue loading. This means that the softening of the pre-deformed structure is retarded with increasing pre-strain. Consequently ratcheting strain accumulation rate becomes less with increasing pre-strain. Fig. 10 also reveals that stimulating the ratcheitng strain accumulation at lower stress amplitude through modification of the pre-strained structure is much more difficult than that at higher stress amplitude.

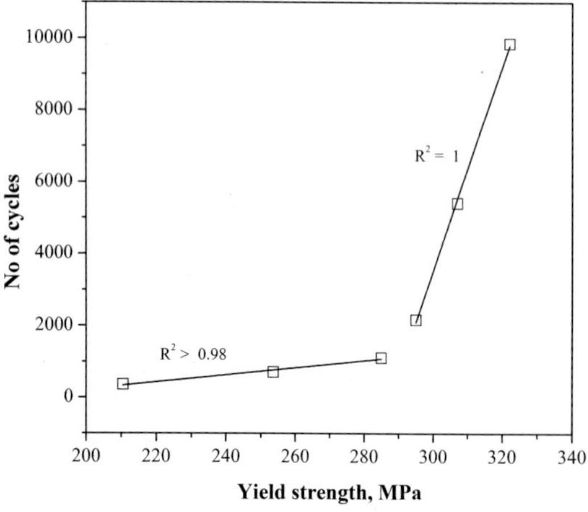

Fig. 11: Variation of N_{16} with corresponding yield strength under pre-strained conditions.

On plotting the N_{16} values for the stress combination of 28.84±187.46 MPa against the yield strength under pre-strained conditions reveals that N_{16} increases with yield strength in a perfectly bilinear manner (Fig. 11). It is observed that initially N_{16} increases slowly with increase of yield strength and then increases sharply with increase of yield strength. This transition from one stage to another occurs at around 7.5 pct pre-strain. As mentioned earlier, up to 7-10 % pre-strain the dislocation structure is non-uniform and less stable dislocation cell structure is generally observed. So modifying this dislocation structure during asymmetric stress cycling would be much easier resulting in faster accumulation of ratcheting strain. After 7-10% pre-strain stable dislocation cell structure is formed. Hence to modify this pre-existing stable dislocation cell structure during asymmetric stress cycling would require higher number of cycles at any stress combination.

Post-Ratcheting Tensile Properties:

Figure 12 shows that post-ratcheting (28.84±187 MPa) yield strength and tensile strength are increased very slowly maintaining a steady linear relationship with pre-strain. The extrapolated value to zero pre-strain matches well with the post ratcheting yield strength and tensile strength data of specimen without any pre-strain. On the other hand, total and uniform strain is found to decrease linearly with the amount of pre-strain. The same trend in post-ratcheting tensile properties is also maintained for the stress combination of 28.84±158.62 MPa.

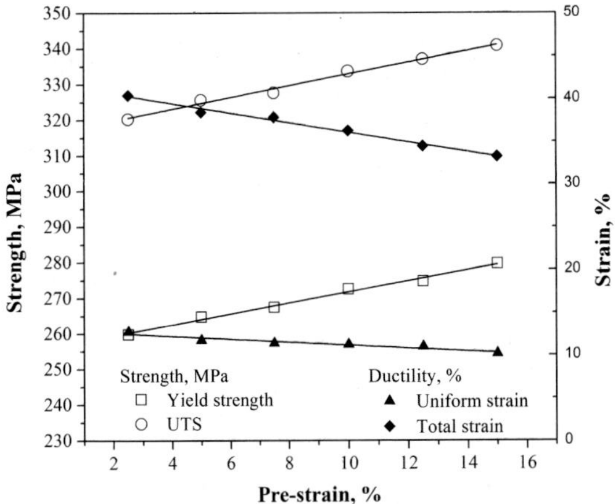

Fig. 12: Variation of post ratcheting tensile properties with pre-strain (28.84±187.46 MPa)

However, actual post-ratcheting tensile test data reveals that post-ratcheting yield and tensile strength vary within ± 10 MPa about a mean value of 270 and 330 MPa respectively. Along with such low variation in yield and tensile strength very marginal change in post-ratcheting tensile ductility with the amount of prestrain indicate that ratchcting deformation modifies the pre-deformed dislocation substructures for all pre-strain levels to the same dislocation configuration. However, detail TEM study is necessary to firmly establish the dislocation configurations, both after pre-straining and after pre-straining followed by ratcheting.

CONCLUSIONS

Ratcheting strain evolution direction in pre-strained specimens depends upon the amount of pre-strain and the imposed maximum stress.

An exponential relationship between N_{16} and amount of pre-strain is maintained. Bilinear variation of in N_{16} with yield strength is observed. Very sharp increase in N_{16} is noted on crossing the yield strength corresponding to 7.5 pct. pre-straining.

Marginal variation in post-ratcheting tensile properties as a function of tensile pre-strain is due to the effect of stress cycling in evolving the dislocation substructure to the same configuration irrespective of the amount of pre-straining.

ACKNOWLEDGEMENTS

One of the authors (PSD) acknowledges CSIR, New Delhi for the award of Senior Research Fellowship to carry out the present investigation at Jadavpur University, Kolkata, India.

REFERENCES

[1] X. Yang, International Journal of Fatigue, Vol. 27, No. 9, 2005, p. 1124.

[2] N. Ohno, Journal of the Society of Materials Science, Japan, Vol. 46, No. 3Appendix, 1997, p. 1.

[3] H.-F. Chai and C. Laird, Materials Science and Engineering, Vol. 93, No. 0, 1987, p. 159.

[4] H. Ni and Z. Wang, Materials Science and Engineering: A, Vol. 314, No. 1–2, 2001, p. 12.

[5] Y.-C. Chiou, Y.-M. Jen and W.-K. Weng, Engineering Failure Analysis, Vol. 18, No. 2, 2011, p. 766.

[6] S. K. Paul, S. Sivaprasad, S. Dhar and S. Tarafder, Materials Science and Engineering: A, Vol. 528, No. 24, 2011, p. 7341.

[7] C. E. Feltner and C. Laird, Acta Metallurgica, Vol. 15, No. 10, 1967, p. 1621.

[8] C. E. Feltner and C. Laird, Acta Metallurgica, Vol. 15, No. 10, 1967, p. 1633.

[9] H. J. Christ, G. Hoffmann and O. Öttinger, Materials Science and Engineering: A, Vol. 201, No. 1–2, 1995, p. 1.

[10] W. P. Jia and J. V. Fernandes, Materials Science and Engineering: A, Vol. 348, No. 1-2, 2003, p. 133.

[11] S. K. Paul, Materials Science and Engineering: A, Vol. 538, No. 0, 2012, p. 349.

[12] P. S. De, P. C. Chakraborti, B. Bhattacharya, M. Shome and D. Bhattacharjee, ***accepted*** in Metallurgical and Materials Transactions A, 2012,

[13] J. Pal'a, O. Stupakov, J. Bydžovský, I. Tomáš and V. Novák, Journal of Magnetism and Magnetic Materials, Vol. 310, No. 1, 2007, p. 57.

[14] P. Antoine, S. Vandeputte and J. B. Vogt, ISIJ International, Vol. 45, No. 3, 2005, p. 399.

[15] Q. Chen and B. Duggan, Metallurgical and Materials Transactions A, Vol. 35, No. 11, 2004, p. 3423.

[16] Z. Yang and Z. Wang, Materials Science and Engineering: A, Vol. 210, No. 1–2, 1996, p. 83.

API X 70 Grade HR Coils for ERW Pipes

Santosh Kumar[1], S.K. Shukla[1], S.K. De[1], Atul Saxena[1], B.K. Jha[1],
B. Mishra[2], Alok Verma[2] and S. Mallik[2]
[1]R&D Centre for Iron & Steel, Steel Authority of India Limited, Ranchi,
[2]Bokaro Steel Limited, Steel Authority of India Limited, Bokaro Steel City
Email: santosh-kumar@sail-rdcis.com

ABSTRACT

For supply of API 5L Grade X70 (PSL 2) 1610 mm wide hot rolled (HR) coils of 10.4, 11.2 and 12.8 mm thickness, trial rolling was undertaken at Bokaro Steel Limited (BSL). Two heats were made and cast into 200 x 1630 mm section slabs with designed alloy chemistry. Rolling parameters of exit roughing (R5) temperature, finish rolling temperature (FRT), and coiling temperature (CT) were kept in the range of 1080 – 1100 °C, 860 – 890 °C, and 600 – 630 °C respectively. Coiling of high strength 10.4, 11.2 and 12.8 mm thick coils of 1610 mm width at such low temperature was carried out for the first time in SAIL.

Microstructure of hot rolled coils comprised ferrite and pearlite. Average grain size for ferrite varied between 6.3 – 9.2 micron. Microstructure of Centreline segregated zone of cast slab showed presence of primary carbides of microalloying elements. Optical micrograph clearly shows formation of coarse primary carbides at inter dendritic region. SEM and EDAX examinations confirmed presence of Nb and V in these coarse carbides. Genesis of these coarse primary carbides rich in Nb & V is attributed to segregation and subsequent slow cooling. To ensure complete dissolution of coarse primary carbides of Nb and V, a suitable time-temperature combine was adopted during reheating. Elimination of primary carbides in the rolled stock was testified through microstructural evaluation. High yield strength (~ 540 MPa) could be achieved accordingly. Excellent YS / UTS ratio (< 0.85) has been achieved. Charpy impact properties of HR coils has been > 100 J even at – 40 °C . Lowering of FRT and CT led to a concurrent increase in strength and toughness. At ambient temperature, 100% shear area was observed after drop weight tear test (DWTT) whereas at 0 °C, shear area was above 95%.

After pipe forming, tensile properties of the parent material and that of fusion-line were found satisfactory. Toughness of parent material were found to be adequate (> 200 J at 0 °C and > 170 J at – 30 °C). There had, however, been a few cases of inferior impact resistance of weld-seam. Metallurgical investigation established the genesis of inferior impact toughness of fusion line to be linked with incidence of penetrators. Originating crack on the detrimental flow lines propagate easily in presence of non-fusion / penetrators resulting from improper selection of welding parameters viz. heat input, pressure and speed of welding etc. During process of ERW pipe making, these inclusions must be ejected. Modified alloy design of API X 70 ensures liquid state of silicates at the end of ERW process which ejects out easily to provide penetrators free seam leading to adequate toughness of the weld seam. With modified alloy design, more than 4000 T of API X 70 grade HR coils have been supplied for successful production of pipe, meeting all the metallurgical requirements.

The process technology for production of API X 70 grade HR coil has been established. API X 70 is now being regularly made against the commercial order at BSL.

KEYWORDS: *Linepipe Steel, Microallying, Impact Toughness, Microstructure, Precipitates, Electrical Resistance Welding (ERW), Penetrators*

INTRODUCTION

High volume transportation of crude, oil and gases necessitate use of larger diameter linepipes and higher-pressure booster pumps. In addition to meeting the operational engineering requirements like weldability, recent generation high strength linepipe steels having guaranteed impact toughness, provide a superior option for pipe-laying projects owing to substantial material and cost savings when compared with linepipe steels of yester years1. In line with recent developments, Bokaro Steel Limited (BSL) initiated development program on high strength line pipe steels which included development of API X 70 grade HR coils for linepipes.

Alloy Design

Alloy chemistry was designed not only to achieve the aimed properties of API X – 70 HR coils as shown in Table 1 which ensured meeting the end properties required in a API X – 70 grade linepipe, but also it ensured that the operating window of steel processing for API X – 70 grade is suitable for BSL conditions of steel making, casting and the hot rolling processes. Table 2 summarises the approach of alloy design. In view of strong tendency of Mn to segregate, Mn content was kept range bound between 1.3 – 1.5%, while content of C, S, P and N was adjusted in the alloy in accordance with the steel making practices of BSL. Addition of small amount of Ti (0.01 – 0.02%) was proposed in view of earlier experience of lamination originating from the pre-existing subsurface defects / cracks in slab. Addition of Ti was intended to tie up N in steel and reduces AlN / Nb(CN) precipitation along austenite grain boundary. TiN precipitates as large sized precipitates within the grains and thereby reducing chances of cracking caused by nitride formation at grain boundaries of prior austenite leading to reduction in hot ductility of cast structure. Addition of V (0.04-0.05%) was proposed for V provides precipitation hardening without increasing the mill load.

Table 1: Aim Properties of API X – 70

	Aim	Worst	API X – 70 Spec.
YS, MPa	540	500	483 – 621
UTS, MPa	640	580	565 – 758
% Elong.	> 30	30	-
CIE (J), 0oC	> 100	100	-

Table 2: Alloy Design Considerations 2-8 of API X – 70 grade

Elements	***Effect on Steel characteristics***
C	Lower C improves hot ductility impact resistance & weldability
Mn	Higher tendency of segregation of Mn is important
Si	Si provides ferrite strengthening
Al	Al required for killing. AlN responsible for high temperature ductility trough
S	Lower S improves impact toughness and HIC-resistance
P	Control of P important for impact resistance
V	V useful for precipitation hardening without overloading of mill
Nb	Nb Important for grain refinement
Ti	Ti combines with N and enhances hot ductility
N	A lower N content is desirable for improved hot ductility

EXPERIMENTAL

Heat making & Casting

BOF-LF-CC Route was followed for making of API X – 70 grade steel using combined blowing technology. Two heats were made and cast into 200 x 1630 mm section slabs. A typical chemical analysis is shown in Table 3. Typical casting parameters comprised controlled and steady casting speed of 1.2 M/min., super heat < 20 C and soft secondary cooling practice. The cast slab section was 200 x 1630 mm. these slabs were stack cooled for 60 hours. Slab surfaces and cast structure were inspected for assessing slab quality.

Hot Rolling

Central portion of slab and corresponding earlier processed rolled plates were examined metallographically for primary carbides. SEM-EDAX analysis was used to confirm the primary carbides. In view of the observations, temperature of soaking was selected on the basis of the temperature of dissolution of primary Nb(CN) precipitates. Thickness of slab was reduced from 200 mm to ~38 mm in 5 passes with individual reduction > 24%. Rolling parameters of exit roughing (R5) temperature, finish rolling temperature (FRT), and coiling temperature (CT) were kept in the range of 1080 – 1100 C, 860 – 890 C, and 600 – 630 C respectively. Coiling of high strength 10.4, 11.2 and 12.8 mm thick coils of 1610 mm width at such low temperature was carried out for the first time in SAIL.

Characterisation of HR coil

Samples were drawn from HR coils. Tensile, impact, ductile brittle transition temperature (DBTT), drop weight tear test (DWTT) and

Table 3: Chemistry (Ladle Analysis in wt.%)

	C	Mn	P	S	Si	Al	Nb	V	Ti
Heat 1	0.084	1.39	0.014	0.002	0.322	0.056	0.045	0.049	0.016
Heat 2	0.088	1.31	0.019	0.003	0.264	0.026	0.041	0.05	0.015

bend properties were evaluated. Optical and SEM examination were carried out on longitudinal section. Standard procedures of polishing were adopted. 2% natal was used as etchant. Samples for TEM were prepared by jet polishing of 3mm disc cut from thin foil made from the hot band in 95 : 5 acetic & perchloric acid at 10 – 15 C temperature. TEM study was carried out in a 120kV transmission electron microscope (HITACHI 7600 Model).

Performance Trial

After edge preparation though milling and online ultrasonic testing over 100% surface of the material, ERW pipe of 18" dia ERW line pipes was formed. Fig. 1 depicts the processing steps of pipe making and various testings. Post high frequency induction resistance welding welding, seam was induction heated for normalizing treatment of seam and HAZ. When the seam temperature dropped below ~ 200 °C, samples were drawn for quality assessment as shown in Fig. 1.

Quality Assessment of Formed Pipe

Besides ultrasonic testing of the HR Coil before pipe rolling and eddy current test, formed pipes were subjected to drop weight tear testing of parent material of the pipe, impact testing of parent and fusion line portion, tensile testing of parent and weldment seam, 0oC, 90oC and reverse bend tests (flattening tests).

Investigation of ERW weld seam

Weld seams having inferior impact toughness was examined through both optical and scanning electron microscope. Presence of penetrators and their qualitative chemical analysis was ascertained to arrive at the genesis of the inferior impact toughness of weld-seam.

RESULTS AND DISCUSSION

Primary Carbide

In the mid-thickness area of rolled stocks, coarse broken carbides of Nb were

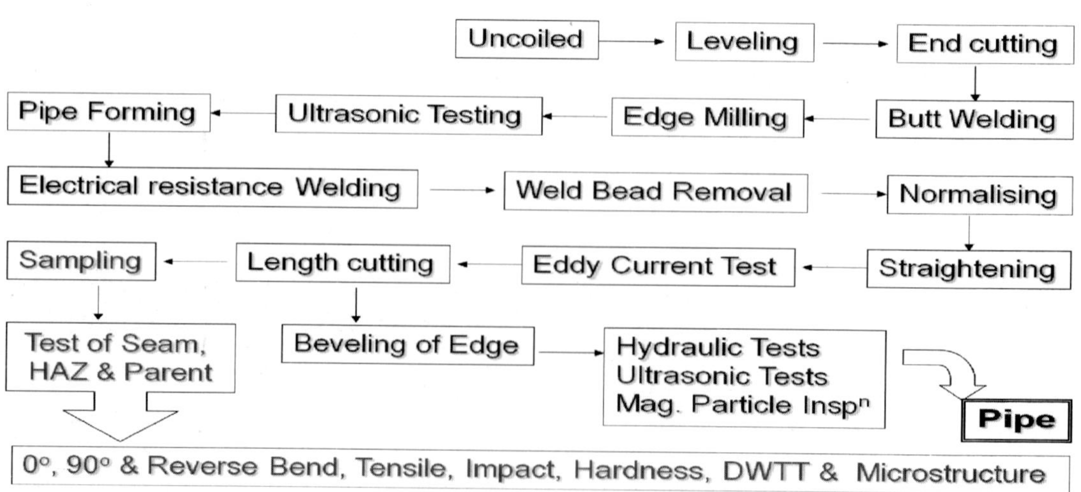

Fig. 1: Schematic Depiction of the processing steps of Pipe Making and various testings of pipe

observed (Fig.2) in a Nb micro-alloyed steel. As these carbides donot contribute to strengthening process, these are drain on expensive ferro-Nb added to API X70 grade steels. To confirm their genesis to these coarse carbides, microstructure of Centreline segregated zone of cast slab was examined, which showed presence

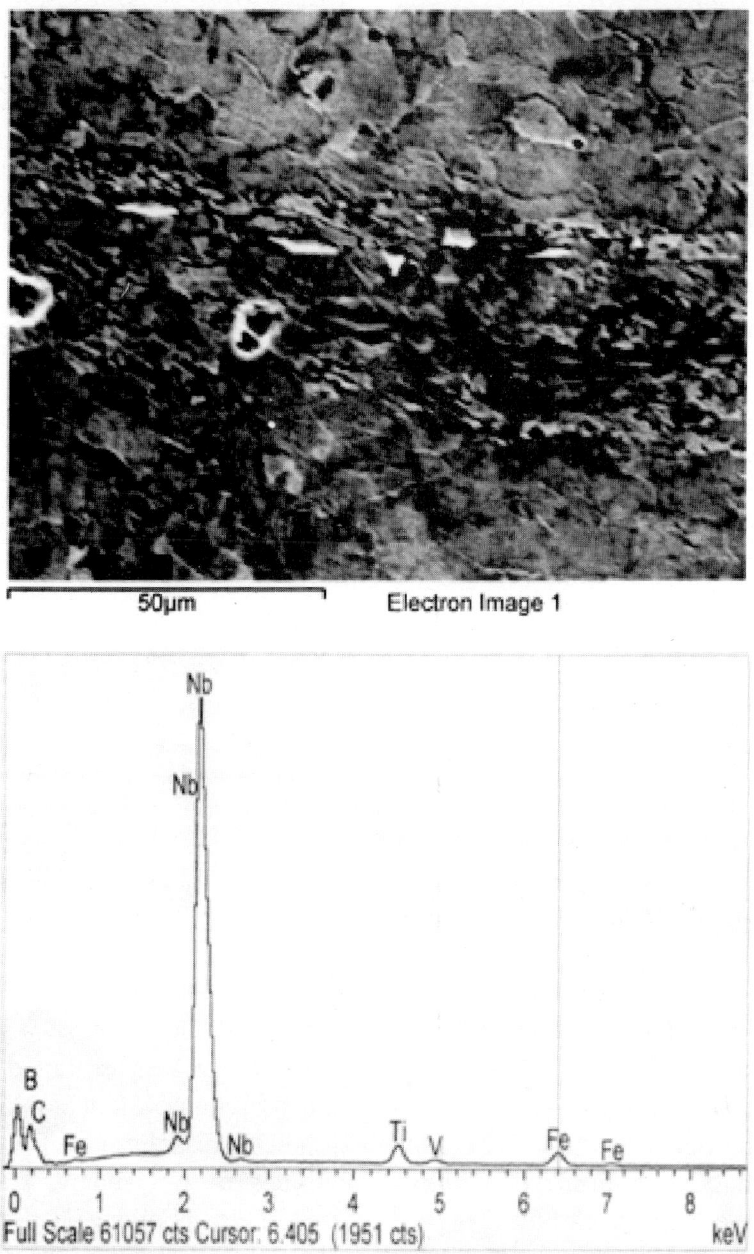

Fig. 2: (a) Scanning Electron Micrograph depicting white coarse Nb rich phase (b) EDAX of the the white coarse phase confirming presence of Nb carbide

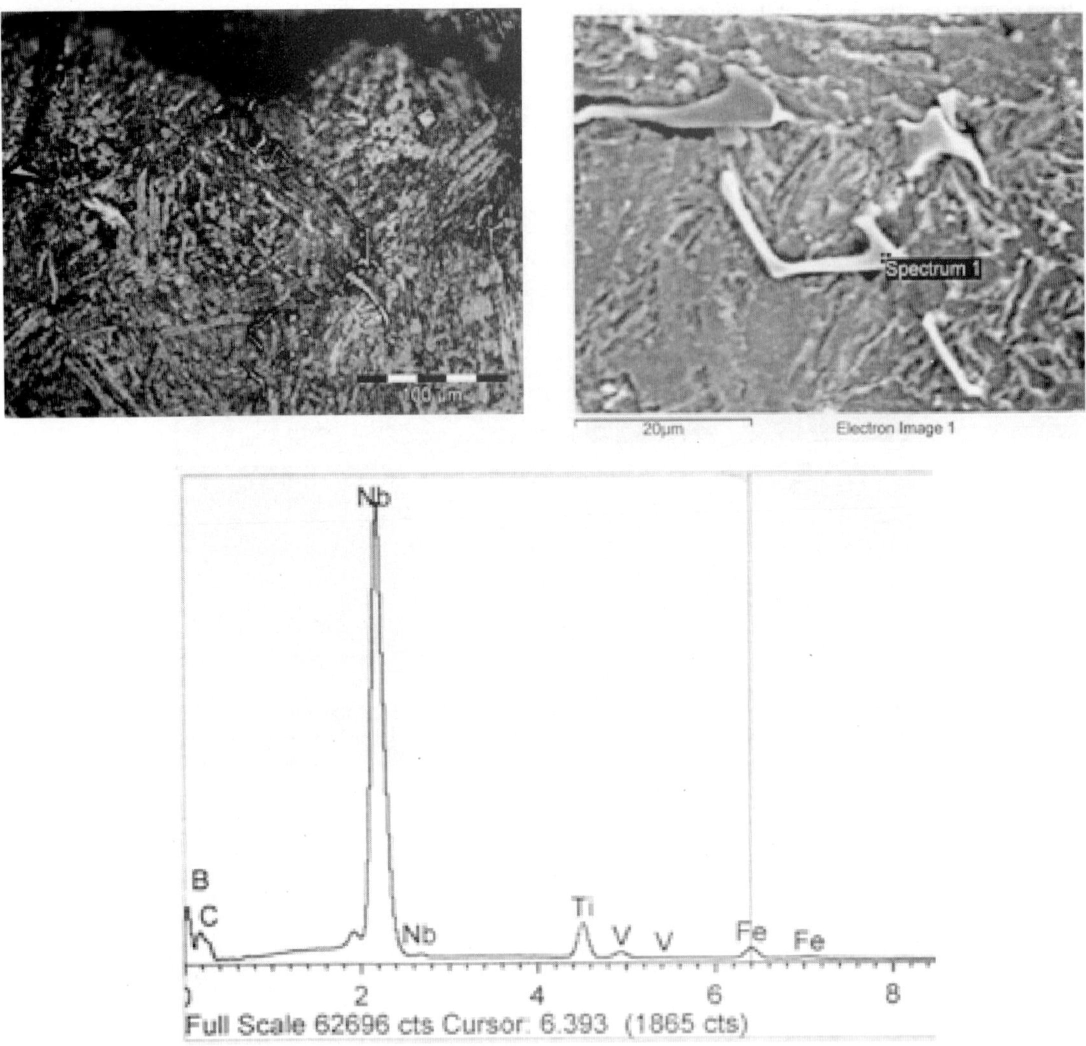

Fig. 3: Primary carbides in central segregated zone of slab (a) Optical Micrograph, (b) Scanning Electron Micrograph and (c) EDAX Analysis on point 1 marked in Scanning Electron Micrograph

of primary carbides of microalloying elements. Optical micrograph (Fig. 3a) clearly shows formation of coarse primary carbides at inter dendritic region. SEM and EDAX examinations confirmed presence of Nb and V in these coarse carbides (Figs. 3 b&c).

Genesis of these coarse primary carbides rich in Nb & V is attributed to segregation and subsequent slow cooling. To ensure complete dissolution of coarse primary carbides of Nb and V, a suitable time-temperature combine is therefore vital to microalloyed highstrength steels.

Characterisation of HR Coil & Performance after Pipe making

Microstructure of hot rolled coils comprised ferrite and pearlite. Average grain size for ferrite varied between 6.3 – 9.2 micron. A typical microstructures and ferrite grain size distribution are depicted in Fig.4. Elimination of primary carbides in the rolled stock using proper soaking conditions was testified through microstructural observation across thickness. TEM observation (Fig. 5) shows presence of nano-sized Nb-rich precepitates. Epitaxial growth of Nb rich area on the TiN coarse precipitate are also observed in Fig. 5.

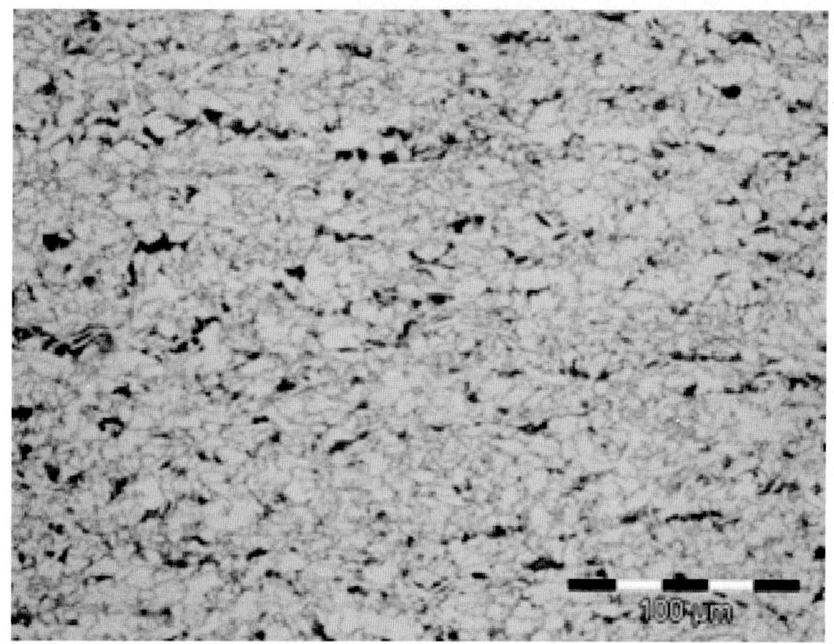

Fig. 4: Typical Microstructure of hot rolled HR coils

Table 4: Typical Mechanical Properties of Parent Material

Coil No.	t, mm	YS, MPa	UTS, MPa	%El (50 GL)	YS / UTS	2t Bend	CVN at -40C, J			Hardness, VHN
Ht. No. 1	11.2	541.25	634.94	38	0.852	ok	153	166	210	221, 216, 219
Ht. No. 1	12.8	529.52	624.99	39	0.847	ok	193	173	180	232, 218, 216
Ht. No. 2	10.4	540.08	633.25	38	0.85	ok	102	126	153	228, 215, 212
Ht. No. 2	11.2	519.36	621.15	36	0.836	ok	140	153	166	221, 233, 228

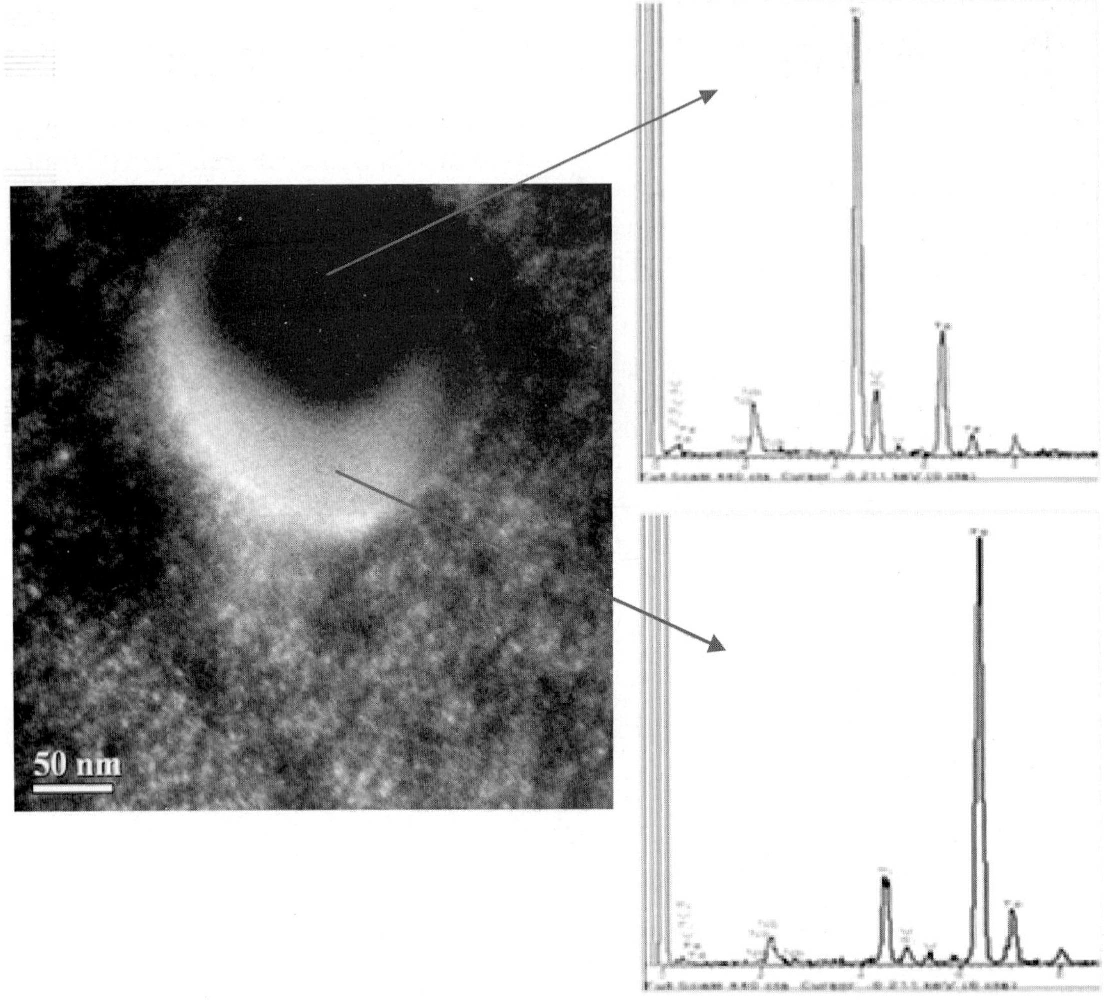

Fig. 5: Evidence of presence of nano-sized Nb-rich precipitates

Table 3 summarises mechanical properties of some of the HR coils. Yield strength varied between 519 – 542 MPa whereas tensile strength is observed to vary between 621 – 635 MPa with excellent YS / UTS ratio (< 0.85). Percent elongation of more than 36% is indicative of excellent ductility needed for pipe forming. Accordingly bend results have been satisfactory at 2T (Fig. 6).

Charpy impact properties of HR coils are varied between 102 – 210 J and even individual observations were > 100 J at – 40°C indicating that ductile brittle transition temperature is adequately low (~ – 60°C) (Fig. 7). Lowering of FRT and CT led to a concurrent increase in strength and toughness (Fig. 8a & b). Drop weight tear test (DWTT) is an important test for linepipe steels. As per quality requirements of the final pipe, the test pieces are drawn after pipe forming and the parent material flattened prior to DWT test. DWTT at 0 oC should show more than 85% shear area which is indicative of ductile mode

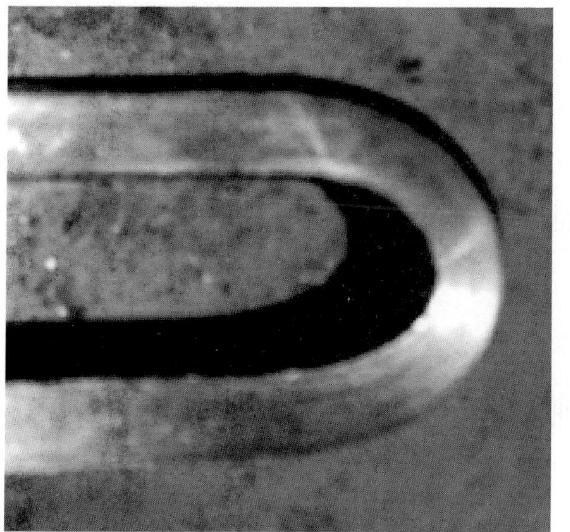

Fig. 6: 2T Bend Tested Specimen

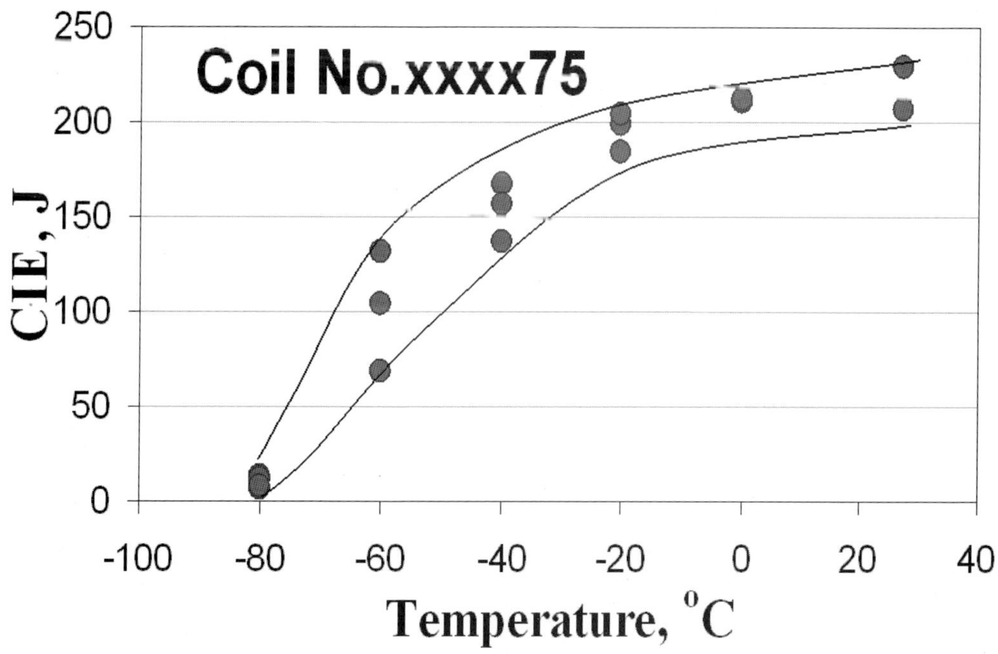

Fig. 7: A typical Plot of Ductile Brittle Transition Temperature

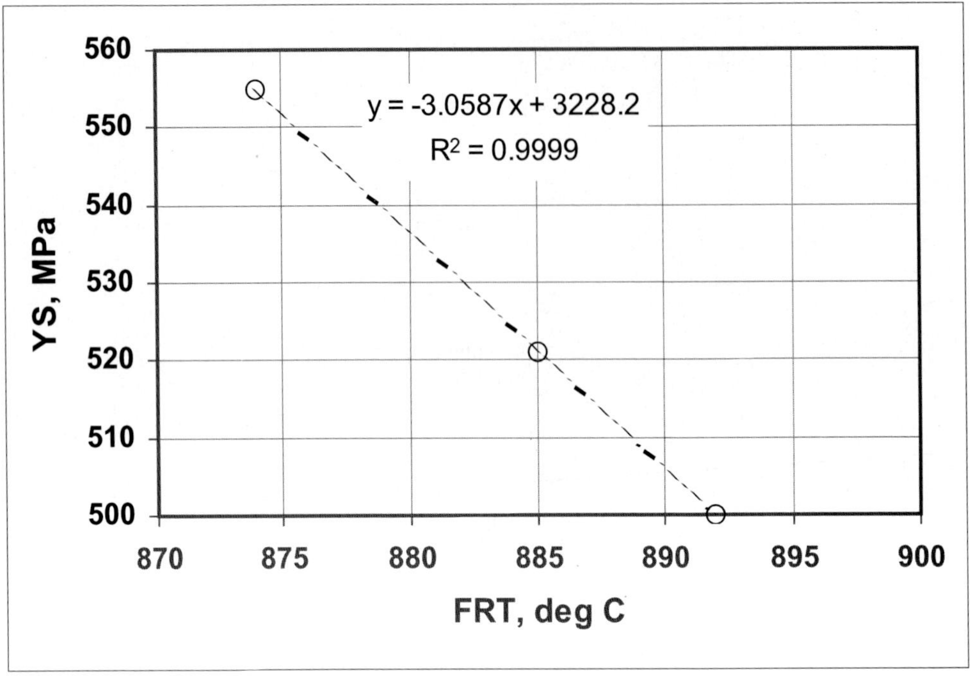

(a) Influence of FRT on YS

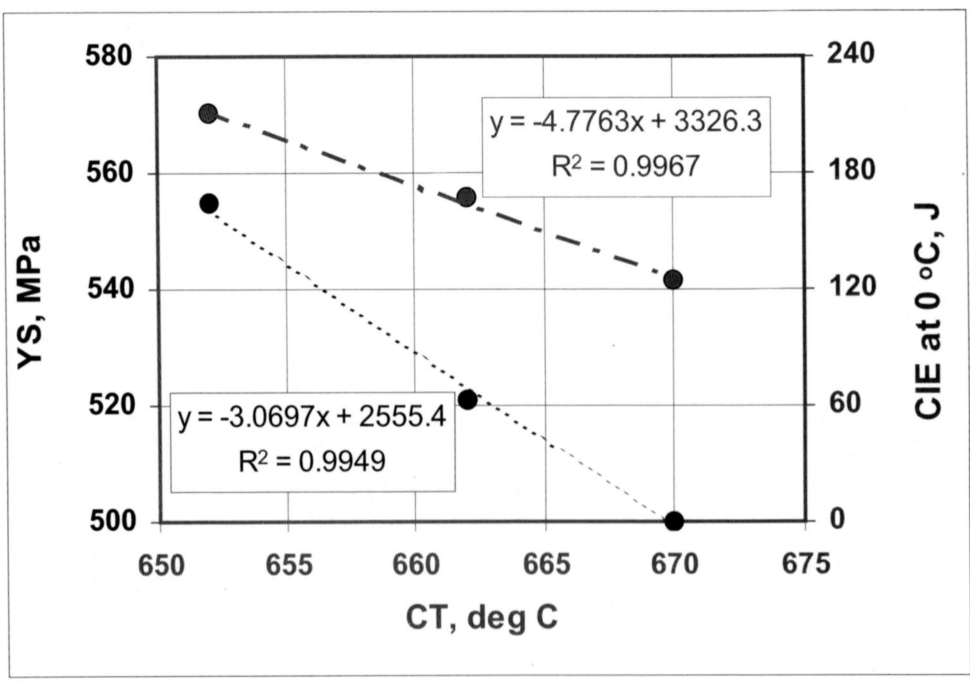

(b) Influence of CT on YS and CIE

Fig. 8: Plots depict influence of FRT and CT on YS and CIE values

(a) Ambient Temperature

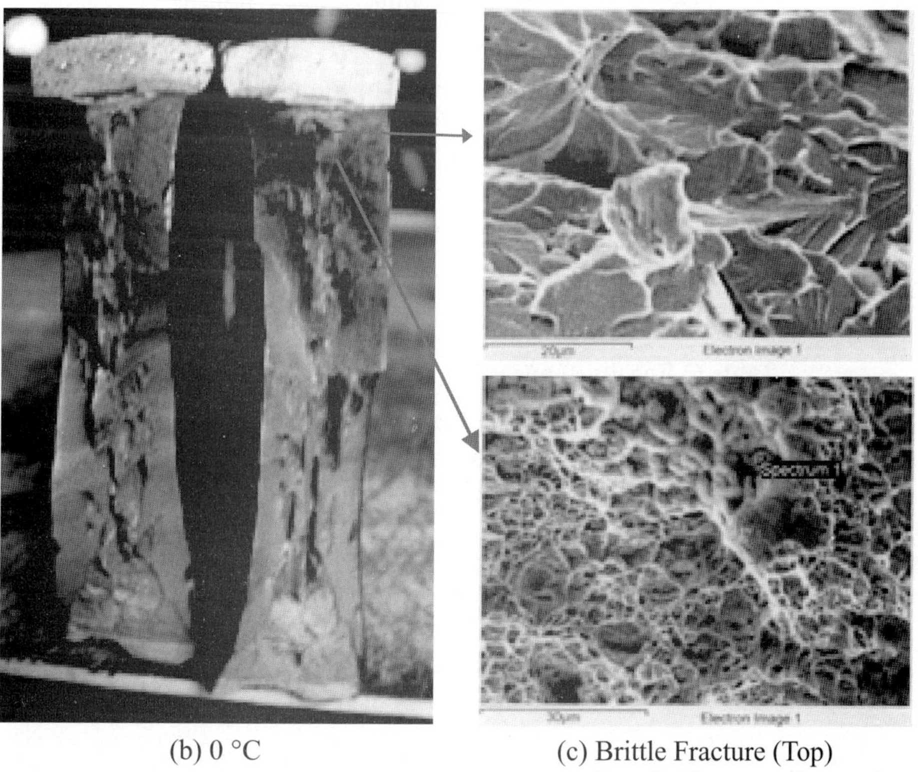

(b) 0 °C

(c) Brittle Fracture (Top)
Ductile Fracture (Bottom)

Fig. 9: Fractured surfaces of Drop Weight Tear tested Specimen of failure. Even in cases of 12.8 mm thick HR coils, 100% shear area is observed at ambient temperature (Fig. 9a) after DWTT whereas at 0 C, shear area is above 95% (Fig. 9b).

Table 5 shows tensile properties of formed pipe- parent material and seam whereas Table 6 depicts two cases : first case shows typical impact toughness at fusion and HAZ while another relates to inferior results observed in some pipes. Tensile properties of the parent material and that of fusion-line were found satisfactory. Toughness of parent material were found to be adequate with Charpy impact energy values of > 200 J and > 170 J at 0 °C and – 30 °C respectively. Although, toughness of weld seam and HAZ were found to be largely satisfactory as shown in case 1 of Table 6, there had been some cases of inferior impact resistance of weld-seam which were also found associated with inferior bend properties (Case 2 in Table 6).

Weld seams having inferior impact toughness was examined through both optical and scanning electron microscope. Presence of penetrators and their qualitative chemical analysis was ascertained to arrive at the genesis of the same.

Optical microscopy revealed presence of about 40 micron wide decarburized band parallel to seam (as evident from absence of any pearlite). Metallurgical investigation established the genesis of inferior impact toughness of the fusion line to be linked with incidence of penetrators (Figs. 10 & 11). Eccentric seam normalizing of weld seam which allows non-homogeneity and high

Table 5: Tensile Properties of API X – 70 linepipe

	YS-P(MPa)	UTS-P(MPa)	El.-P(%)	YS/UTS-P	UTS-W(MPa)
Pipe Sample 1	545	653	36	0.85	657
Pipe Sample 2	557	651	34	0.84	660
Pipe Sample 3	516	659	35	0.82	659
P stands for parent, and W stands for weldment					

Table 6: Results of impact and bend test of formed Pipes

Coil No.	Charpy Impact (J) at 0 °C	Charpy Impact (J) at - 30 °C	Bend
N****26 (Case 1) (Typical Cases)	W : 95, 109, 80 H : 306, 280, 247	W : 77, 123, 107 H : 280, 173, 296	90° : OK RBT : OK
N****22 (Case 2) (Failed Case)		W : 5.32, 5.32, 39.9 H : 6.65, 46.55, 7.98	90°: Partial OK (3 OK, 2 Fail) RBT: Fail
W : Weldment, H : HAZ (Heat affected zone), RBT : Reverse Bend Test			

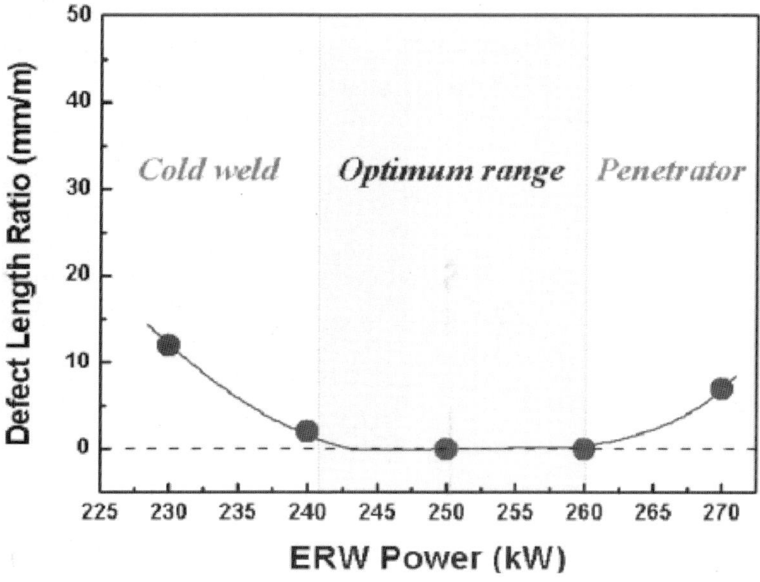

(a) Effect of ERW power (as per literature)

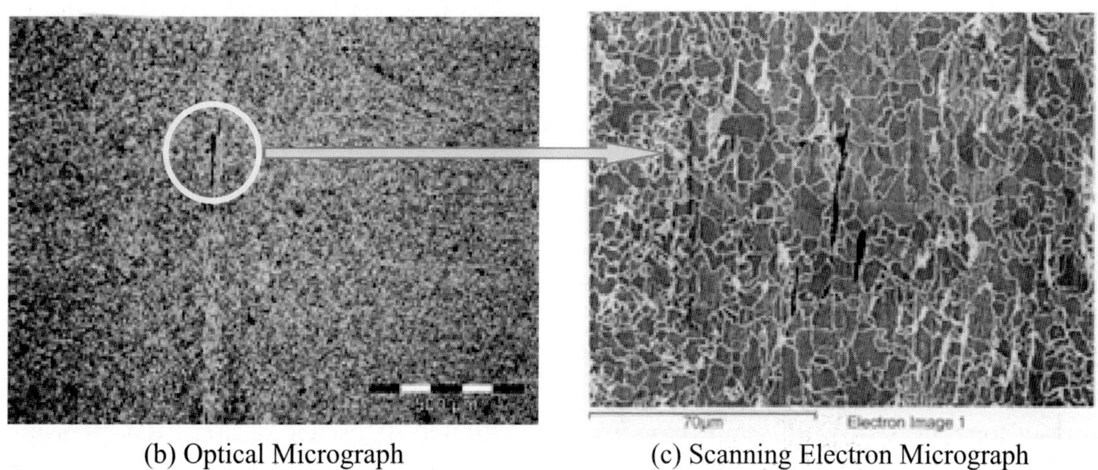

(b) Optical Micrograph (c) Scanning Electron Micrograph

Fig. 10: Penetrator at fusion line of ERW pipe

propensity of flow lines in structure near seam are detrimental. Originating crack on these flow lines propagate easily in presence of non-fusion / penetrators resulting from improper selection of welding parameters viz. heat input, pressure and speed of welding etc. At the time of ERW process, Mn and Si having higher affinity for oxygen, oxidises in preference to iron in absence of protective atmosphere of weld seam during ERW process. Thus forms Mn silicates at weldment. During process of ERW pipe making, this oxide phase must be ejected from fusion line through proper selection of ERW parameters like heat input, welding speed etc. to ensure a good impact

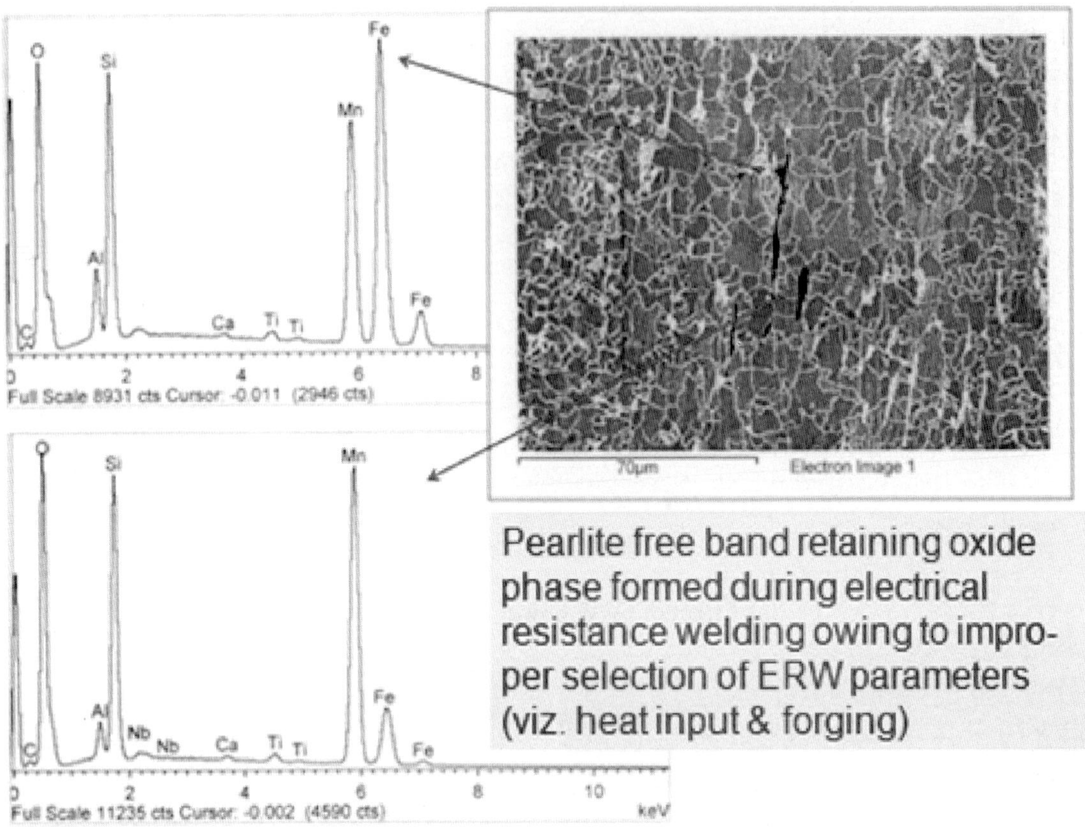

Fig. 11: SEM & EDAX analysis of penetrators found at fusion line of ERW pipe

toughness of weldment in ERW pipes. Wherever, the suitable ERW parameters has been there, fusion line has been free from such silicate stringers leading to acceptable impact toughness of weld seam (case 1 of Table 6). Also it is observed that absence of the penetrators, concurrent acceptability in various flattening tests results.

Fine tuning of alloy design for ERW pipe

Innovative solution to improve impact toughness of weld seam in ERW pipes through modified alloy design was achieved. Liquefying oxide phase at the time of steel solidification through lowering of Si content has been the innovation that facilitated its escape during ERW pressing operation which ensure a perfect seam with good impact toughness and bend-characteristics. Based on theoretical calculation using thermodynamic software package 'FactSage', it has been observed that inclusion composition forming in earlier steel corresponds to vertical down arrow in Fig. 12 whereas lowering of Si in steel alters the inclusion chemistry which enables it to be liquid at solidus temperature of steel during ERW process (as shown by broad arrow). Adoption of this modification led to total elimination of failure at weld seam and more than 4000 T of API X 70 grade HR Coils could be rolled into ERW pipes.

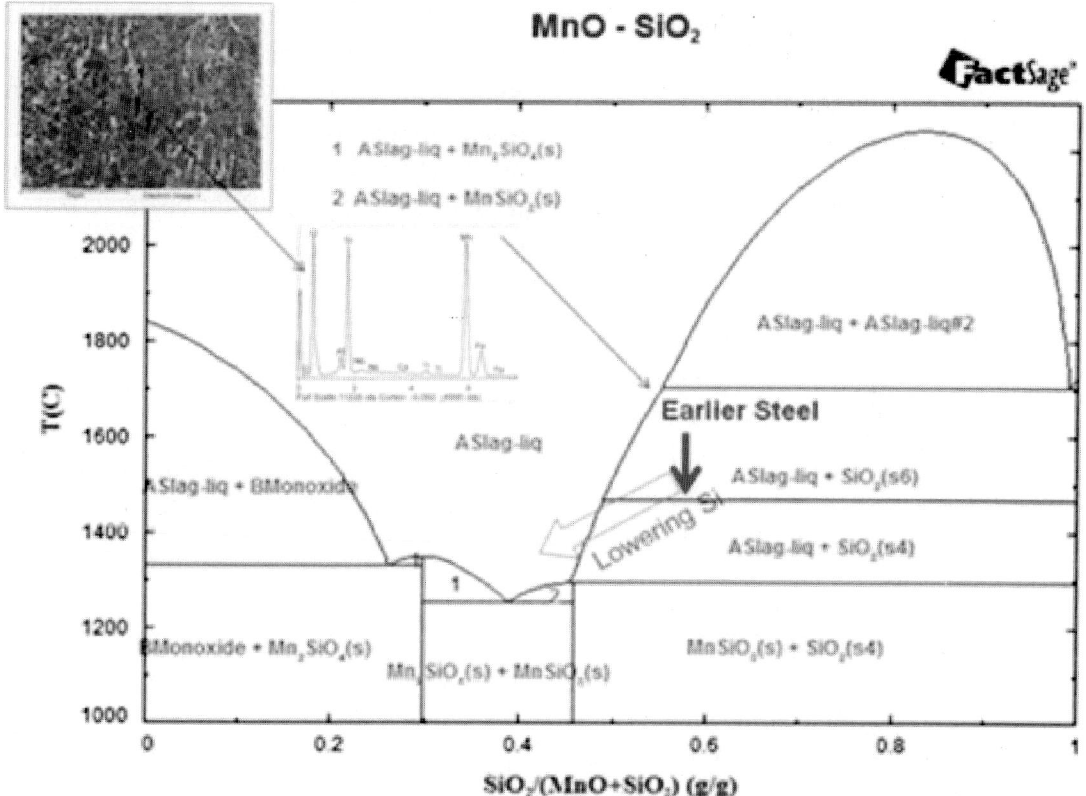

Fig. 12: Pseudo-binary diagram with indicative position of penetrator (oxide phase)-composition before and after innovation (depicted by broad arrow)

In view of the occasional incidence of failure in achieving toughness properties because of improper ERW parameters, alloy design of API X 70 grade HR coil was fine tuned to ensure that the oxide phases forming during the welding process are liquid in state at the time of metal squeezing taking place at the end of ERW process. Based on the modified alloy design, more than 4000 T of API X 70 grade HR coils have been commercially supplied and the same have been rolled into ERW pipes successfully. All metallurgical parameters of finally pipe were met and customer feedback had been satisfactory.

CONCLUSIONS

1. Trends of variation in yield strength and Charpy impact toughness underlined significance of lower FRT (< 880 °C) and CT (< 680 °C).
2. Suitably tailored alloy design and selected process parameters ensured

excellent microstructural features both at optical (fine ferrite size) and TEM level (nano precipitates) in HR coils of thicker gauge.

3. Properties achieved have been well above the lower limits of the specification of API X – 70 for HR coils up to 14.4 mm thickness. HR Coils of BSL meet all quality requirements of API X 70 specification after pipe making. Developing microstructure of hot band underlines suitability of alloy design and proper selection of process parameters.
4. Eccentric seam normalizing of weld seam which allows non-homogeneity and high propensity of flow lines in structure near seam are detrimental. Originating crack on these flow lines propagate easily in presence of non-fusion / penetrators resulting from improper selection of welding parameters viz. heat input, pressure and speed of welding etc. Under suitable operating window of ERW process, meeting all properties of seam is possible.
5. Processing technology for production of API X-70 grade HR coils at Bokaro Steel Limited has been established. More than 4000 T of API X 70 grade HR coils have been commercially supplied to roll ERW pipes successfully.

ACKNOWLEDGEMENTS

Authors are grateful to the management of SAIL for providing opportunity to work for development of API X 70 grade HR coil. We acknowledge help received from BSL collectives of SMS, HSM and RCL who helped in implementing the requisite suggestions. Help and support received from M/s Maharastra Seamless Pvt. Ltd, Nagothane, Mumbai and M/s Jindal India Pvt. Ltd., Kolkata towards pipe making are acknowledged with pleasure. Last but not the least, our sincerest thanks are due to RDCIS laboratory staff for characterization of material.

REFERENCES

[1] Hiroo Akasaki, "Progress in Pipe and Tube Technology and Its Future Prospect – Application and Manufacturing, Nippon Steel Technical Report No. 90 July 2004, pp 75 – 88

[2] F.B. Pickering: Microalloying 75, Union Carbide Corp., New York, NY, (1977), 9.

[3] L. Meyer, "Grundlagen der Werkstoffentwicklung durch Verknüpfung thermischer und mechanischer Vorgänge beim Warmumformen", Stahl und Eisen, 101 (7) (1981), 483 – 491.

[4] A.J. DeArdo, "The Physical Metallurgy of Thermomechanical Processing of Microalloyed Steels, "Thermec '97, eds. T. Chandra and T. Sakai, (1) (12) (TMS publications, 1997), 13-30.

[5] T. Gladman, "Precipitation hardening in metals", Materials Science and Technology, 15 (1999), 30 – 36.

[6] L. J. Cuddy: Thermomechanical Processing of Microalloyed Austenite, TMS, Warrendale, PA, (1982), 129.

[7] K. Fujiwara, S. Okaguchi and H. Ohtani: ISIJ Int., 35 (1995), No. 8, 1006.

[8] D. Russwum & P. Wille, Microalloying 95 Conf. Proc., Iron and Steel Society, Pittsburgh, USA, June 1995, p 37

Influence of Ti Microalloying on the Microstructures and Mechanical Properties of Hot Rolled C-Mn LPG Steel

S.N. Prasad, B.K. Jha, Santosh Kumar and A.S. Mathur
Research and Development Centre for Iron and Steel, Steel Authority of India Limited
Ranchi-834002, India
Email: snprasad@sail-rdcis.com

ABSTRACT

In order to improve the cold forming property, fixing of nitrogen by microalloying with Ti (0.01-0.02%) was considered as practical solution for commercial production of LPG (Liquified Petroleum Gas) cylinder manufacturing grade steel. A number of Ti bearing LPG heats with reduced Mn content (lower than the amount prescribed for C-Mn variety of LPG steel) were processed through BOF-LF-CC route. Ti microalloying resulted in significant improvement in press forming characteristic due to considerable decrease in strain ageing index (from more than 15 to 10% max.) of the steel even with high nitrogen levels. About 10 % decrease in forming stress improved the die/punch life during drawing of half cylinders. Successful demonstration trials were carried out at the works of leading LPG cylinder manufacturers. Reduction in incidence of failure during forming of cylinder halves has resulted in increased yield (1.4 cylinder/ton of steel).

KEYWORDS: *Hot Rolled Steel, Cold Forming, Mechanical Properties, Strain Ageing Index, Microalloying*

INTRODUCTION

Hot rolled C-Mn steel is used for manufacturing liquefied petroleum gas (LPG) cylinders for domestic cooking purpose. Formability of such steel is considered as one of the important parameters to be satisfied for trouble free drawing during cylinder manufacturing. Formability can be seriously impaired by the free nitrogen available in the ferrite phase [1]. Presence of free nitrogen leads to manifestation of higher strain aging index (SAI), which deteriorates cold forming capacity of the material and also mechanical property of the formed component. Many researchers [2-6] in the past have linked formation of crack running vertically downwards during press forming of sheet with this phenomenon.

Conventionally, Al is used for tying up of free nitrogen available in the steel. The influence of aluminum content on precipitation ratio of AlN (AlN/total N) in hot rolled steel is well documented [7]. It has been shown by the researchers [7] that addition of aluminum more than 0.05% is considered very effective in fixing nitrogen as AlN. However, fixing of nitrogen by Al requires an optimized temperature management in hot strip mill (low slab reheating and high coiling temperature). Lowering of reheating temperature in the shop has obviously certain limitations particularly in controlling mill load within the permissible limit of the finishing strands.

Similarly, coiling temperature as high as 710-730°C is necessary for precipitation of AlN [7]. Such high coiling temperature leads to unfavorable pickling behavior and limits uniformity in mechanical properties along the strip length. Additionally, the strength level may be lowered and it can also produce mixed grain structure, which is undesirable for forming.

In order to over come these quality related problems, addition of Ti has been considered as an affective means for fixing nitrogen as TiN [8]. It is envisaged that the formability of HR coils increases efficiently not only by enhancing of fixing of nitrogen but also through effective modification in the shape of inclusions. The following equation [8] is employed for assessing the amount of Ti that would remain in solution after combining with N, S and C.

Ti in solution = Total Ti-3.42 N-1.5 S-4C

The above equation suggests that about 3 to 4 times the total amount of nitrogen present in the steel, Ti is required to be added for fixing free nitrogen and any amount of excess Ti will then combine with sulphur and carbon.

In the light of the above, the influence of Ti microalloying on microstructure, mechanical properties and forming characteristics of C-Mn steel has been studied and discussed in this paper.

EXPERIMENTAL PROCEDURE

The Ti bearing LPG steels (Table 1) were made in LD converter, refined in ladle furnace (LF) and continuously cast to slabs. The slabs were reheated at 1200 °C and finish rolled in hot strip mill 860-880°C and coiled at 590°C. However, the conventional C-Mn LPG steel is made through the same LD-LF-CC route and generally finished rolled at 880 °C and coiled at relatively higher temperature (690 °C).

Table 1: Chemistry (weight %) of LPG steels

Type of LPG steels	C	Mn	P	S	Si	Al	Ti	N (ppm)
Conventional C-Mn LPG steel	0.05	0.60	0.02	0.013	0.026	0.039	-	-
1st Ti-LPG heat with low nitrogen	0.05	0.35	0.02	0.012	0.02	0.043	0.013	30
Ti-LPG heats with higher nitrogen	0.06-0.08	0.37-0.38	0.02 max	0.02 max	0.026-0.037	0.033-0.068	0.013-0.018	70-92

Specimens cut from the rolled samples were mechanically polished and etched with 2% nital. The optical microstructures of both the steels were recorded with a NEOPHOT light optical microscope. The ferrite grain size was measured using linear intercept method. Transmission electron microscope JEOL Model 400 EX was used for transmission electron microscopy (TEM) examination. Specimens for TEM study were cut by slicing 0.1 mm thick disk in an isomet slow speed saw. The slices were mechanically thinned to 0.05 mm thickness and electrolytically polished at 30

V using a solution of 95% acetic acid and 5% perchloric acid maintained at -10 °C.

The tensile properties were evaluated using an Instron machine (model 1273) at a constant cross head speed (2 mm/min). Standard specimens of 50 mm gauge length were used for tensile tests at room temperature. The strain ageing index (SAI) for both the steels were determined by pre-straining the tensile specimens with 7% deformation followed by heating of these specimens at 100°C in boiling water for an hour and finally testing them again till breakage. Strain ageing index was estimated by reporting the ratio (expressed in percentage) of increase in flow stress immediately on reloading, as a result of ageing i.e. heating the sample in boiling water for an hour, to the original flow stress value after 7% pre deformation. Further, for the formability studies, blanks cut from the hot rolled coils were grid marked with 5 mm diameter interlocking type circular grids. The accumulated strain distribution was measured using a flexible calibrated transparent scale after drawing of blanks in half cylinders.

Further, cold forming properties of Ti microalloyed hot rolled (HR) sheet were evaluated at the premises of different LPG cylinder manufacturers in the country. During this exercise, the parameters that were monitored are forming load and percentage rejection at the different stages of manufacturing and these were correlated with the overall yield (cylinders/ton of steel).

RESULTS AND DISCUSSION

Mechanical Properties

The typical tensile properties and hardness values of the hot bands of conventional C-Mn LPG steel are YS: 300 MPa, UTS: 400 MPa, EL.: 31%, bend: close and hardness: 60 HR_B. The strain-aging index is 17%. In order to reduce the strain-aging index (SAI) and improve cold formability of the coil, a Ti microalloyed (0.014%) LD heat of LPG steel was continuously cast, finish rolled and coiled at 880 and 590°C respectively in the hot strip mill. The chemistry of the steel is given in Table 1. The Mn content of the steel was reduced to 0.35% from 0.55%, which is usually aimed in conventional C-Mn grade LPG steel (Table 1), to offset the cost of addition of Ti in the new grade. The typical properties of a coil from the above Ti bearing heat are compared with that of conventional LPG in Table 2. The cold formability improved markedly as reflected by higher elongations, higher n, lower YS/UTS and lower SAI. The lower yield and tensile strength than that of conventional LPG steel is due to nitrogen fix up [1] and lower Mn content.

Table 2: Comparison of mechanical properties of Ti bearing LPG steel with that of conventional LPG steel

Steel Type	YS (MPa)	UTS (MPa)	%El. (50mm GL)	n	YS/UTS	SAI (%)
Conventional LPG	306	398	30.4	0.206	0.77	17.0
Ti-LPG with low nitrogen	260	370	39.0	0.226	0.70	10.0
Ti LPG with higher nitrogen	284-333	353-392	40-44	-	-	2-10

Additional benefit of longer die life during forming of half cylinders could be achieved, owing to lower amount of loose scale generation in the die cavity during forming. Adherent thinner scale of 2-6 micron thickness was observed for Ti bearing steel where as it was 8-10 micron for conventional C-Mn steel. This is attributed to lower coiling temperature employed while processing Ti bearing steel. The extent of deterioration in tensile property due to strain aging phenomenon was quantified in terms of increase in UTS and decrease in elongation value of both conventional as well as Ti bearing LPG steels. In case of Ti bearing LPG steel, the increase in UTS and decrease in % El were nominal compared to that of conventional LPG steel (Table 3). The total % El after aging was substantially higher (35.2 %) for Ti bearing steel than that for the un-aged conventional LPG grade steel (30.4 %).

The mechanical properties of Ti-LPG steel are in the range of YS: 260-310 MPa, UTS: 360-380 MPa, %El.: 41-48 (50 mm GL) and SAI 6-10%. Thus fixing of free nitrogen with Ti has resulted in reduced SAI from more than 15% in conventional LPG steel to 6-10 % in the Ti bearing LPG steel. However, lower carbon (0.05 %) and manganese (0.35 %) in Ti bearing LPG steel also reduce the yield strength.

Table 3: Measurements on deterioration in tensile properties as a result of strain aging

Steel Type	Increase in UTS (MPa) (% increase)	Decrease in El. (50 mm GL) (% decrease)
Ti-LPG with low nitrogen	370 to 380 (2.6)	39 to 35.2 (9.7)
Conventional LPG	398 to 435 (9.3)	30.4 to 22.4 (26.3)

In view of encouraging results of first trial heat, additional 4 heats with reduced Mn and higher nitrogen content (Table 1) were made. They were finish rolled and coiled at 860 and 590°C respectively. The mechanical properties as shown in Table 2 were in the range of YS: 284-333 MPa, UTS: 353-392 MPa, % El.: 40-44 and the SAI: 2-10% only in spite of higher nitrogen content of the steel. The yield strength has increased due to favorable Ti/N ratio (1.9) for grain refinement [9].

Microstructure

As shown in Fig. 1a, the microstructure of conventional C-Mn LPG steel consists of equiaxed ferrite and pearlite. The grain size of ferrite is 12.5 μm. The microstructure of low nitrogen Ti bearing LPG steel is shown in Fig. 1(b). The ferrite grain size is 12.0 μm, which is comparable to that of conventional LPG steel (Fig. 1a). The microstructure clearly shows that no appreciable grain refinement could be achieved in the present case with N: 30 ppm and Ti: 0.014%. The size of TiN particles plays an important role in the grain refinement of prior austenite grains. Madina etal [9] reported that maximum grain refinement takes place when fine TiN is precipitated and the ratio of Ti to N is about 2.

Fig. 1: Typical micrographs of LPG steels showing (a) ferrite pearlite in conventional steel (b) microstructure of Ti bearing steel which is comparable to that of conventional steel (c) finer grain size with 1.9 Ti/N ratio in higher nitrogen steel.

In the additional 4 heats with reduced Mn and higher nitrogen content (≥ 70 ppm) (Table 1), Ti/N ratio is maintained close to 2. This results in formation of fine TiN precipitates (<50 nm) which led to finer (8 μm) ferrite grains [9], Fig. 1 (c). However, increase in Ti/N ratio to 4.7 with low nitrogen content of 30 ppm, resulted in formation of coarser ferrite grains (~ 12.0 μm) shown in Fig. 1 b. Transmission electron microscopy was carried out to ascertain the size and type of precipitate present. Figure 2(a-c) shows typical electron micrographs (both bright & dark fields) and its diffraction pattern taken from steel sample of low nitrogen content with Ti/N ratio of 4.7.

The precipitates are TiN as analyzed in the selected area diffraction pattern, Fig. 2 (c) which is schematically represented in Fig. 2(d). The precipitate size is of the order of 100 nm, which is not expected to contribute significantly for grain refinement [8]. These results are very important because maintaining nitrogen level consistently less than 30 ppm in LPG steel is not practicable at Bokaro Steel Plant. The microalloying with Ti contributes to lower SAI resulting in better formability even when the nitrogen level is as high as 70-92 ppm and also results in higher yield strength due to grain refinement.

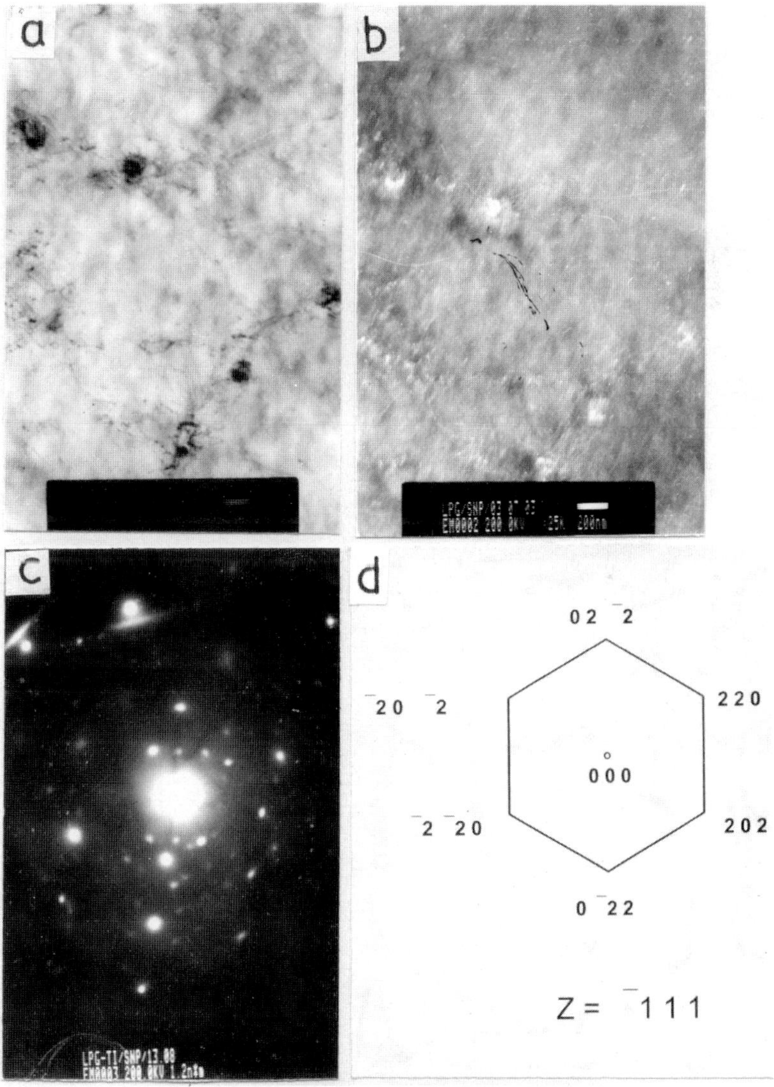

Fig. 2: TEM micrographs of Ti bearing steel with low nitrogen content and Ti/N ratio of 4.7 showing (a) and (b) TiN particles dark and bright fields respectively (c) corresponding SAD (d) its schematic

Formability

Grid marked blanks of low nitrogen Ti bearing and C-Mn LPG steels were formed to half cylinders. The drawing pressure for Ti-LPG was 10 % lower than that for conventional one. Fig. 3 shows the typical formed components with grid marks on them. The circular grids deformed into ellipses. Measurements of the major and minor axes of these ellipses were used for computing the values of major and minor strains respectively. Fig. 4 shows the major and minor strain values for conventional and low nitrogen Ti-LPG steels. The critical strain required for the onset of necking was much higher for both the conventional and Ti bearing LPG steels than the actual measured strain associated with press forming. However, for a particular minor strain value the major strain value in case of Ti bearing steel is lower than that for conventional one indicating its better formability.

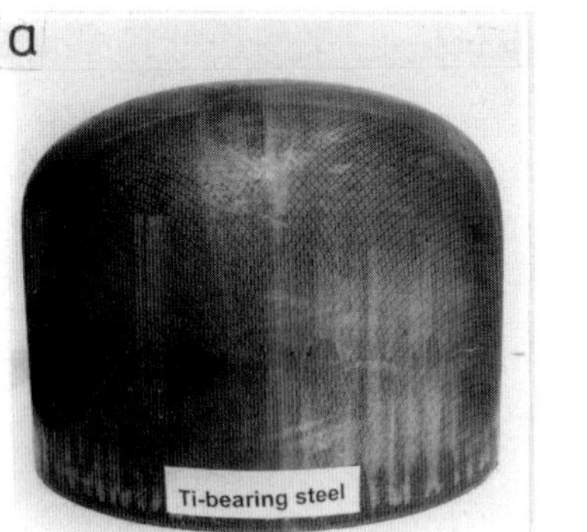

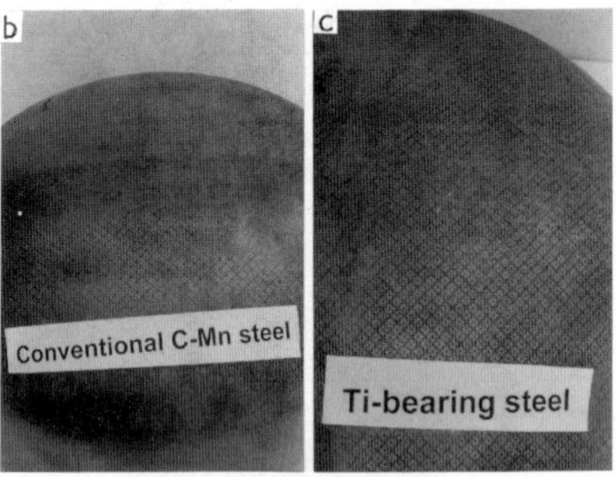

Fig. 3: (a) Complete half cylinder of Ti bearing steel (b) and (c) parts of the half cylinders of C-Mn conventional and Ti bearing steels respectively showing clearly grid marking on them after forming

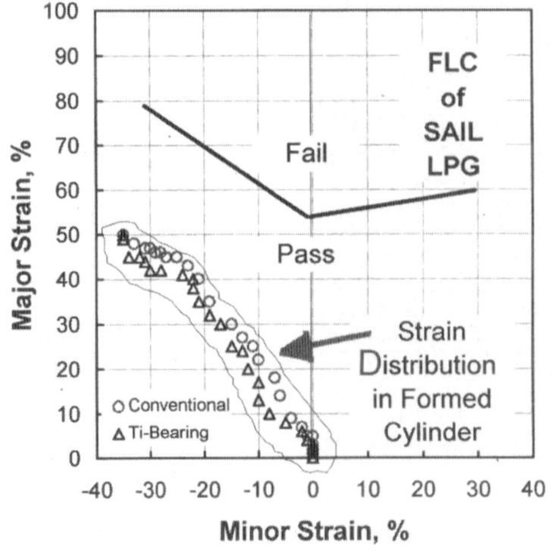

Fig. 4: Strain distribution in the formed cylinder

Thickness and Hardness Variation

The patterns of thinning/thickening and hardness variation experienced in the formed half cylinders of C-Mn conventional as well as Ti bearing LPG steels have been shown in Fig. 5 (a and b). The input thickness was 2.85 and 2.84 mm for conventional and Ti bearing LPG steels respectively. The maximum thinning to the extent of 0.15 mm was near the circular base portions of the half cylinders of both the steels. The maximum hardness values were also comparable at the same locations which were 342 and 350 LD for Ti bearing LPG and conventional steels respectively.

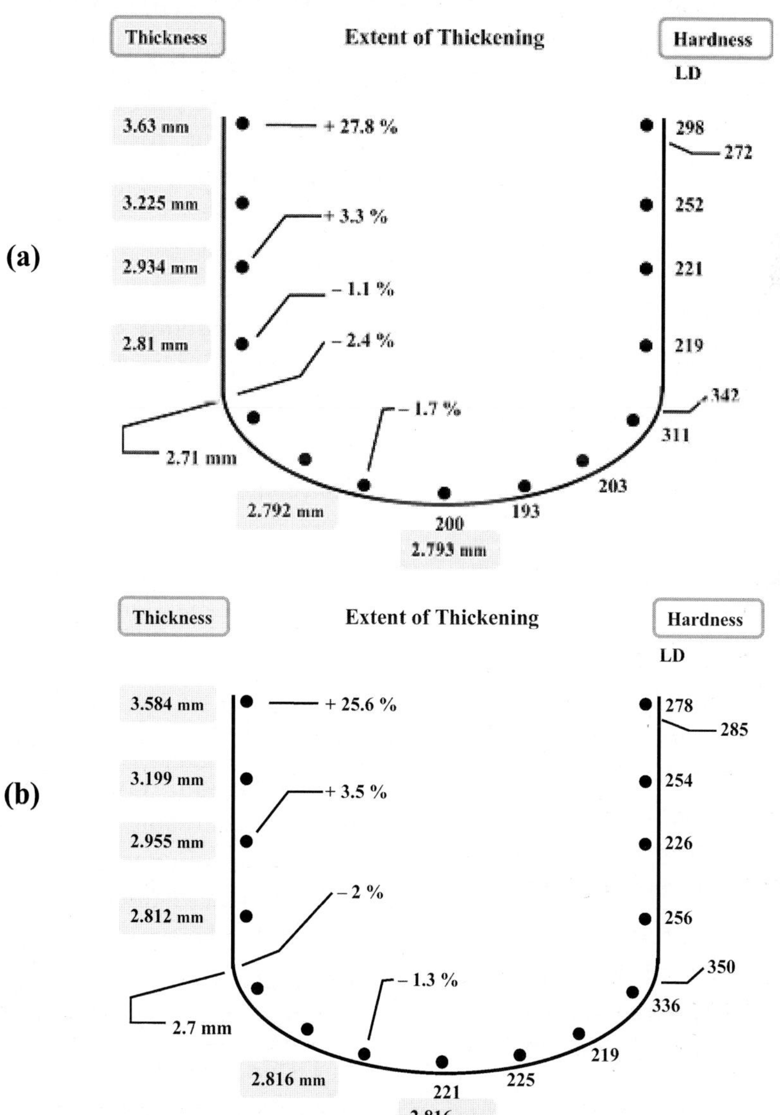

Fig. 5: Pattern of thinning/thickening and hardness variation in formed half cylinders (a) Ti bearing steel (b) C-Mn conventional steel.

Field Trials

Subsequently, thousands of tons of Ti bearing HR coils of LPG steel were supplied to various customers all over the country. The analysis of the detailed feed back reports from a dozen of customers in the country indicates that most of the customers found improvement in the formability as there was lower forming stress resulting in improved die life and also higher yield (1.4 cylinders/ ton of steel) due to lower rejections in the Ti bearing LPG steel.

CONCLUSIONS

Microalloying of Ti (0.01-0.02%) in C-Mn LPG steel is beneficial in binding free nitrogen in steel resulting in reduced SAI (from more than 15 to 10 % max.) which improves formability. About 10 % decrease in forming stress in the case of Ti bearing LPG steel results in increase in the die/ punch life during drawing of half cylinders.

The microstructure of conventional as well as Ti bearing LPG steels consists of equiaxed ferrite and pearlite. The grain size of ferrite in Ti bearing HR sheet is also comparable with that of conventional LPG steel (about 12 μm). However, steels with reduced Mn and high nitrogen content (≥ 70 ppm) with Ti/N ratio close to 2 results in formation of finer (8 μm) ferrite grains.

The customer feedback reports show that there is improvement in yield (1.4 cylinders/ ton of steel) due to improved formability leading to reduction in % rejection of half cylinders.

There is no extra cost involved in the production of Ti bearing LPG steel as Mn is reduced to 0.35% from 0.55% in conventional LPG steel.

The microalloying with Ti contributes to lower SAI (≤ 10 %) resulting in better formability even when the nitrogen level is as high as 70-92 ppm and also results in higher yield strength due to grain refinement.

ACKNOWLEDGEMENTS

The authors express gratitude to the management of RDCIS, SAIL, Ranchi and Bokaro Steel Plant, for their support and encouragement.

REFERENCES

[1] W. Muschenborn et al, Conference Proceedings, Microalloying 95, ISS, 1995, pp. 35-41.

[2] A. R Perrin, M. Wolosiuk, and A. McLean, Mechanical Working and Steel Processing, XXIII, ISS, 1986, 3.

[3] W.C. Leslie et al, Trans. American Society of Metals, Vol. 46, 1954, p. 1470.

[4] R. Koyama et al, Sumitomo Metals, Vol. 48, 1996, p. 192-194.

[5] Qian Yonglan, Proceedings, HSLA Steels Processing, Properties and Application, Edited by G. Ticher and Z. Shouhua, Minerals, Metals and Materials Society, 1992, pp. 209-214.

[6] S. Gunduz, Industrial lubrication and tribology 57/4, 2005, pp. 145-149.

[7] K. Ushioda et al, Development in the Annealing of Sheet Steels, Edited by R. Pradhan and I. Gupta, The Minerals,

Metals and Materials Society, 1992, pp 261-286.

[8] I. Gupta and D. Bhattacharya, Metallurgy of Vacuum Degassed Steel Products, Proceeding, R. Pradhan, Editor, TMS, Warrendale PA, 1990, pp. 43-72.

[9] S. F. Medina et al, ISIJ International Vol. 39, 1999, p. 930.

Efficient Mobility Solution Using New Concept

Dhananjay Kumar
KLT Automotive & Tubular Products Ltd., Pune, India
Email: dhananjay.kumar@kltgroup.net

ABSTRACT

Physical mobility is a necessity for both goods & services and in the era of virtualization, it is still necessary and considered as an important parameter of development. Total energy requirement of planet Earth is nearly 16 terra unit/day and atleast 20% of the same is consumed in facilitating mobility only. Railroad, water ways, road transport & aircraft are the common mode of mobility and their energy efficiency are also in decreasing order. For any mobility means, it has to carry itself and with growing requirements of comfort along with safety, it bounds to get more inefficient while the cost of energy along with concern of environment demands better efficiency. So, the material scientist along with system designer has task defined to come out with higher performance material with efficient design which is not enough and so out of the box thinking is require to come out with new concept/means/devices for efficient mobility solution

Rail road car is the most efficient mode of mobility for both goods and services where its self weight is 25-30% against the payload. So, there is a challenge to reduce the self weight to less than 15% which will improve not only the payload capacity by the same percentage but also it will improve the energy efficiency when it is running empty.

Similarly, road transport vehicle for both passenger and commercial segment and in the category of light, medium and heavy and in inter and intra-city category can be made more efficient through targeting a weight saving of 20% which will have improvement in fuel efficiency by 4 to 5%, besides improvement in comfort and safety.

At any point of time, Earth receive nearly 12 terra unit of energy/hr and so mere 2 hr. is sufficient to meet our all energy requirements. So, there is need to utilise solar radiation in captive power generation in mobility system and here 3D solar has major role to play. Our conventional IC engine which is powering the major chunk of mobility solution is considered to be less than 40% efficient and atleast 5% power is further wasted in the transmission. Braking losses takes away almost half of the energy in city transport while highway it is 25-30% and further 5 to 10% losses comes from thrust & drag. So, our actual energy utilization is not more than 5 to 10% irrespective of the research performed in last 130 years.

So, there is a need to dwell on all these aspect to come out with a revolutionary mobility solution which has been extensively discussed in this paper based on author 27 yrs of research experience in diverse engineering field.

KEYWORDS: *Electric Vehicle, Motor Generator, Solar PV, Vector Control PMBM,UHSS, CFRP, Power Electronics Control, Battery Management*

INTRODUCTION

The world, and with it the sphere of individual mobility, is in a state of environmental, economic and social upheaval. Global developments such as climate change and dwindling resources are encouraging many people to act with greater social responsibility and aspire to a more aware and sustainable lifestyle.

The fuel use projection results from projections of vehicle stocks, travel, intensity, and fuel types for each mode, from 2000 through 2050 is shown in figure 1. Year 2000 fuel use breakout by mode shows that LDVs accounted for nearly half of transport energy use, and all road vehicles accounted for over three-fourths. The breakout by mode and fuel type shows that LDVs represented about 80% of total gasoline use and trucks about 75% of diesel fuel use. In the Reference Case projection, global transportation fuel use increases by a factor of nearly 2.5 between 2000 and 2050. Use of gasoline, diesel fuel, and jet fuel grows substantially, while other fuels retain a tiny share. Alternative fuels and vehicles do not penetrate significantly in

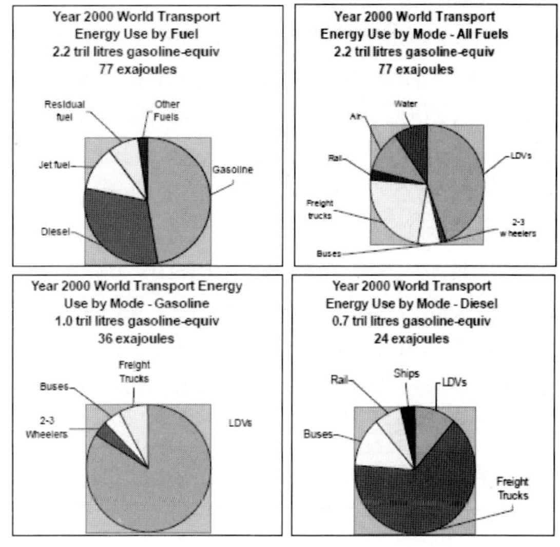

Fig. 1: Global Transportation Fuel use

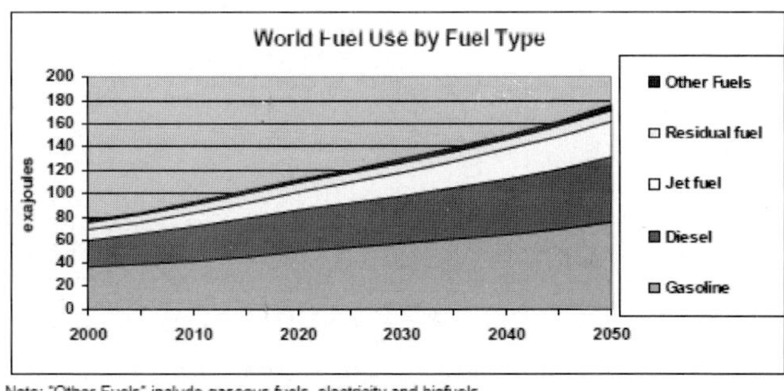

Note: "Other Fuels" include gaseous fuels, electricity and biofuels

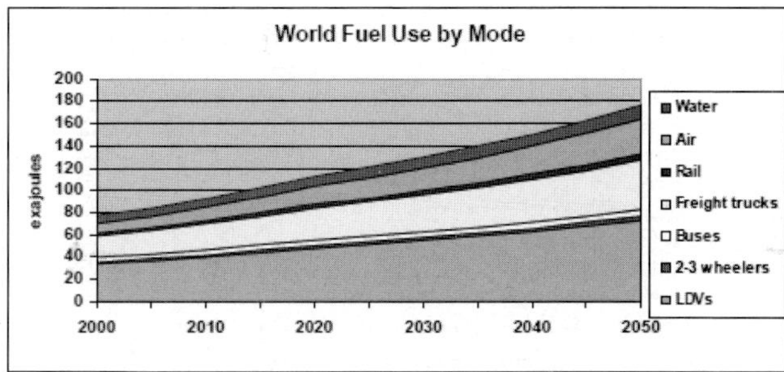

Fig. 2: World Fuel Use pattern

the reference case. As shown in the figure 2 below, total fuel use by mode will grow significantly for all modes but buses, with the biggest growth occurring for light-duty vehicles, freight trucks, and air travel [1].

LDVs provide the biggest overall increase, accounting for nearly 40%of total increase of 99 exajoules. Though air has the highest growth rate and more than triples its fuel use between 2000 and 2050, its overall increase (22.6 exajoules) is significantly less than LDVs (38.5 exajoules), and somewhat less than freight trucks (26.8 exajoules). In terms of this increased energy use over the 50 years (year 2050 energy use minus year 2000 energy use), the figure 3 pie chart shows where most of the additional energy is being used.

Presently transport (well-to-wheels) CO2 accounts for about 24% of all energy-related CO2 emissions. This includes all transport modes except domestic and international shipping, which when added will increase the transport share by a couple of percentage points. (It also does not include energy used during pipeline transport, not included in this study). Compared to the IEA 2002 World Energy Outlook reference case projection across all energy-use sectors, our transport CO2 projection of well-to-wheels CO2-equivalent emissions is a fairly constant share through 2030, about 28% (Figure 4 & 5) [1].

As a rough guide, the table 1 indicates the approximate rate of change in car ownership with change in income (as an elasticity). Until the regional average income reaches \$5k per year, ownership growth is very slow, but takes off at this point. It then has

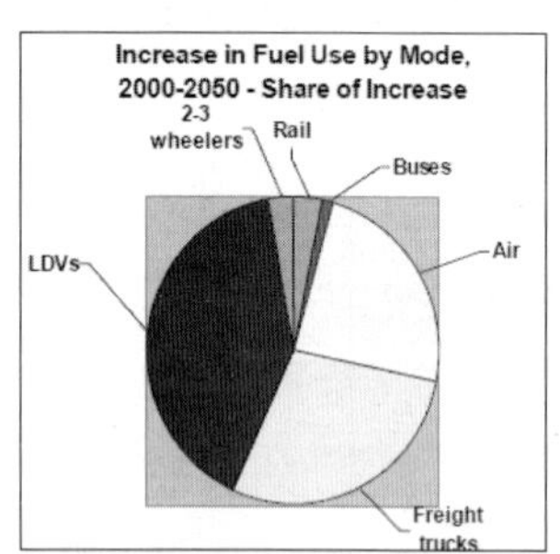

Fig. 3: Share of Increase of Fuel by Mode.

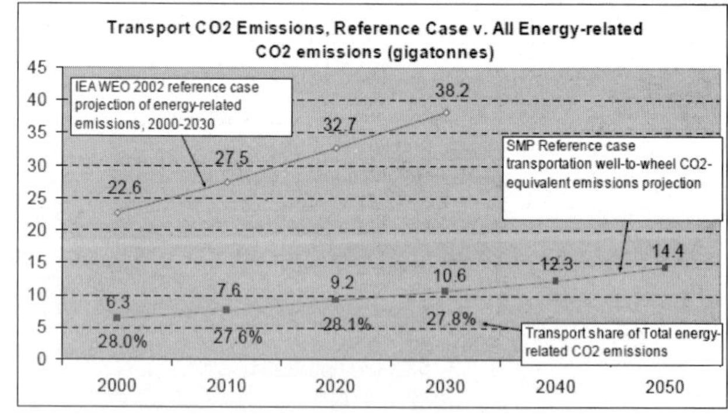

Fig. 4: Share of Mobility in total CO2 emission.

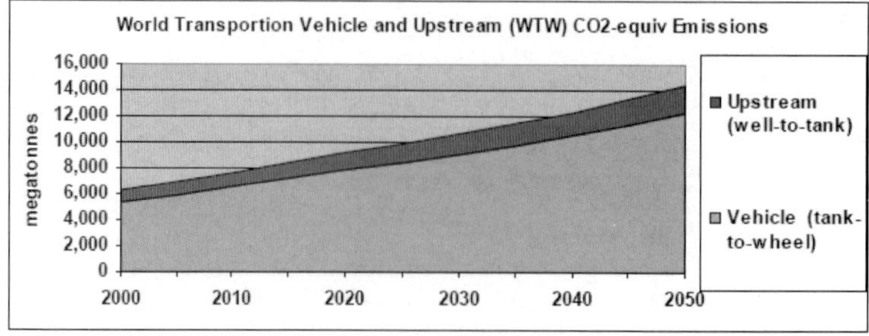

Fig. 5: Share of Emission of Upstream in Mobility

Table 1: LDV ownership growth rates at different income & car ownership level.

	Ownership growth elasticity relative to income growth (pct basis)	Maximum ownership level for this growth rate
until $5k income	0.30	no maximum
>$5k	1.30	300 cars per cap
>$5k	0.60	500 cars per cap
>$5k	0.25	600 cars per cap
>$5k	0.10	700 cars per cap

Table 2: LDV Car ownership rate (per 1000)

	2000	2005	2010	2015	2020	2025	2030	2035	2040	2045	2050
OECD North America	618	626	636	644	651	659	667	674	682	690	698
OECD Europe	390	417	451	468	483	497	511	525	540	555	570
OECD Pacific	438	459	477	494	510	527	545	562	581	599	618
FSU	100	108	133	168	212	258	310	340	373	408	447
Eastern Europe	201	218	235	304	345	391	440	466	493	522	551
China	13	17	26	37	50	66	86	111	142	181	231
Other Asia	21	25	29	33	37	46	56	68	82	98	117
India	10	14	17	21	26	30	40	51	66	83	105
Middle East	42	45	47	51	57	63	68	74	79	85	91
Latin America	78	87	101	118	136	157	181	209	240	276	317
Africa	20	23	27	31	34	38	42	46	50	54	58

a very high elasticity of 1.3% change in car ownership for each 1% increase in average income until the regions reaches 300 cars per capita. It then slows, reflecting an inflection in the logistic curve and the beginning of asymptotic (saturation) behavior [1].

Table 2 and figure 6 show how these growth rates translate into ownership estimates. The colour coding in the table matches that in the table 1, in order to indicate the situation different regions are in at different times.

EFFICIENT MOBILITY SOLUTION

No doubt Railway is the most efficient mobility solution (Table 3) [5] for both freight and passengers and with High speed railway, metro and mono rail and many other variants it is getting efficient and environment friendly. Though the cost of infrastructure is high and installation takes longer but once rails are layed then it becomes easier to maintain. Further from space point of few, it can be layed on ground, under-ground and also on raised column. Normally, it uses electric drive train and so depending upon the source of power generation it can have fairly low carbon foot print and if the source is green then it can have minimal environmental impact. There is further scope of improving the performance by using UHSS and there is scope to reduce the wagon weight by more than 30% and also loading volume by nearly 40% (Table 4). Using Rare Earth & sensor less PMBM (permanent magnet brushless motors) efficiency of the drive train can be further improved upto 30 to 35%. Breaking losses which is more common in Urban transport and also in passenger long distance train can be improved by using efficient Motor-Generator which will power back the grid and so no need to store the same, unlike electric drive train in automotive. Also there exist a great scope using micro turbine to regenerate the thrust and drag which is the major source of losses and considering high surface area, 3D PV solar cell can be an efficient mode of power

Table 3: Comparative efficiency of different mode of freight.

	Truck	Train	Aircraft	Ship
Truck	10.52		0.22	2.73
Train		1.22		
Aircraft			196.65	
Ship (sea)				3.20
Modal transfer			0.12	0.12
Total tonnes CO_2-eq	10.52	1.22	196.99	6.05

Environmental cost of transporting 100 t of bulk heavy goods from Amsterdam to Madrid (Source: www.ecotransit.org)

Table 4: Improved Wagon design using UHSS.

Specification	E71 reference	Solution 1 (same volume as E71)	Ermewa Group solution (43% increase in volume)
Weight of wagon empty	27 t	18 t (-33%)	22.5 t (-17%)
Weight of load (max)	53 t	53 t	67.5 t (+27%)
Total weight	80 t	71 t (-11%)	90 t (+12.5%)
Load volume (max)	70 m^3	70 m^3	100 m^3 (+43%)
Material costs	Reference	-34%	-5%

Comparison of the upscaled Ermewa Group FreightRail wagon

generation which again will power back the grid.

CHANGING PREMISES OF MOBILITY

Mobility is an important parameter of development and mode is primarily driven by individual need and taste. Tough terrain mobility solution (Figure 7) is different from a family transport system (Figure 8) and Multi utility vehicle used in farm sector as mobility solution of passenger and cargo as again different from fun loving young passenger who look for fancy gadget capability along with speed. Similarly, grown up passenger will look for utility in driving the entire family on week end and at the same time it meets the regular requirement of week days of not only buying utilities but also attending different business function and at the same time looking efficient and easy to maintain.

Fig. 7: Tough terrain mobility solution

Flexibility is another requirement where the same family mode of transport which add thrill, passion, comfort and safety can turn to a utility vehicle which can as well carry feedstock, taking produce to market place and many other application which require flexible mobility. Such extreme requirements invigorate the automotive designer to have lateral thinking and innovation which is a complex process as it is shown in figure 9 [4]. Provocative Lateral thinking give rise to 100s of ideas which once put through the cross functional requirements and constraints, it reduces to 10s of workable concept and out of the same depending upon the priority few are selected and undergoes the grinding of prototyping and testing and after successful proving it

Customer Segment A:
Love to Drive

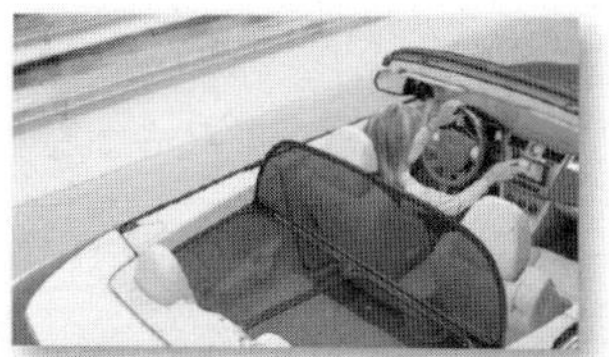

Customer Segment C:
Love the thrills

Opportunity Focus B: Dirty Cargo Capability

Some Possible Solutions (I)

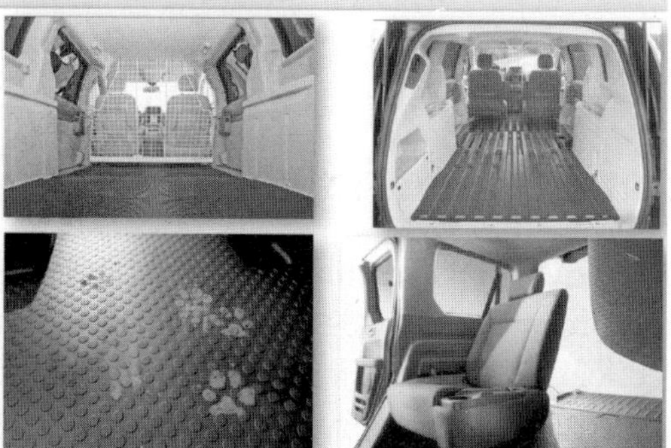

Customer Segment B:
Love Innovation and Creativity

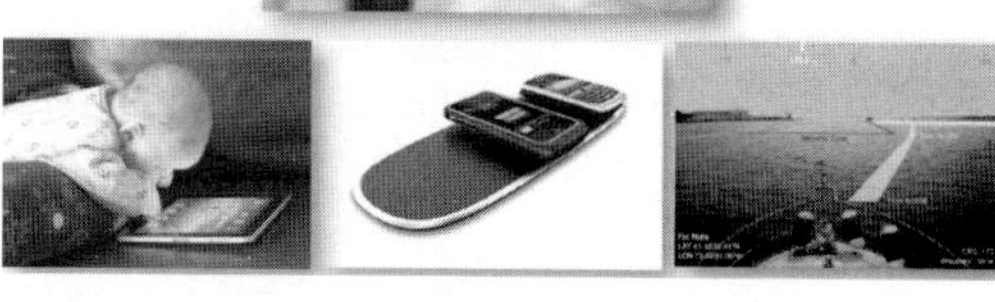

Opportunity Focus A: Need for Pass Through

Some Room for Pass Through

Some Possible Solutions (II)

Fig. 8: Innovation in Customer centric mobility solution

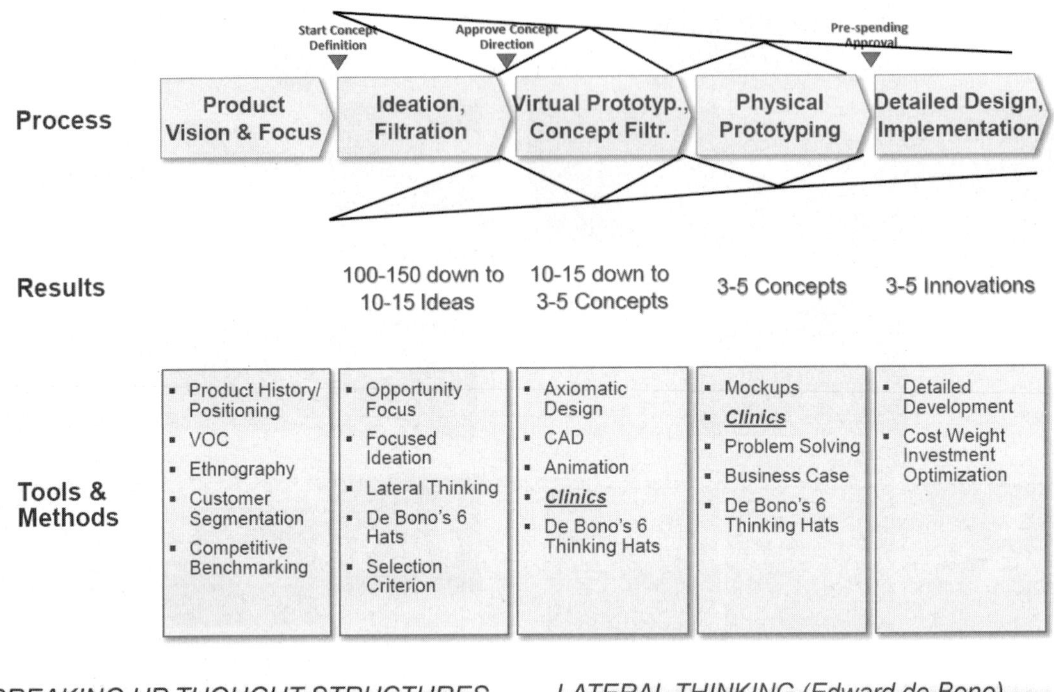

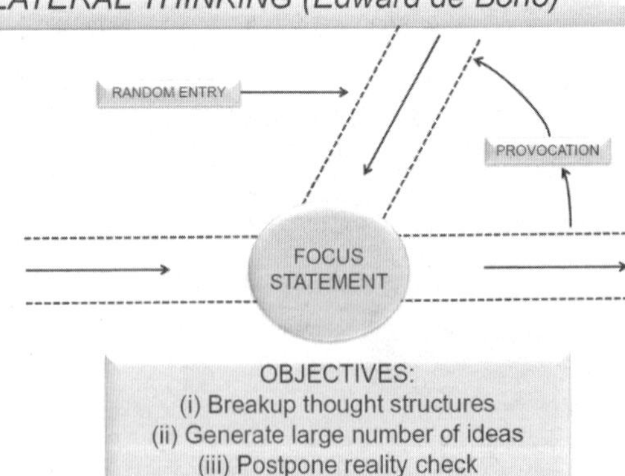

Fig. 9: Dynamics of Innovation Process

leads to innovative product which again checked on financial feasibility to come out with final product which ultimately wait for customer reaction to make the same successful.

GREEN MOBILITY SOLUTION

The advancing greenhouse effect and finite resources have made it a key priority to reduce CO2 emissions significantly and quickly, to drive forward the switch from fossil fuels to renewable energy sources, and to increase the efficiency of all energy consumption. The governments of various countries have recognised this too and are responding with ambitious fleet legislation and the promotion of zero-emission vehicles. With growing urbanization, half of the world's population already lives in cities and the numbers are rising daily. Cities are drawing in urban inhabitants like a magnet, resulting in more and more people living together in increasingly confined spaces. Along with the cities, the challenges are growing too: altered infrastructures, lack of space and environmental pollution demand new solutions for mobility. Mobility for the future requires a new balance between global requirements and individual needs. What are needed are new solutions for personal, and above all sustainable, mobility in urban areas. The Automotive industry has recognised these needs and is developing a vehicle that opens up new possibilities in this area and can adapt to evolving customer requirements.

The most efficient IC engine only able to convert 1/3rd of fuel energy into mechanical energy and the remaining 2/3rd goes wasted as exhaust gases. Thomas Seebec has first detected EMF generation against temperature difference in some of the semi-conductor junction in 1821 and this concept has been exploited to generate electricity between the temperature difference in exhaust gases and coolant which has been stored in battery and subsequently used in powering the engine during acceleration/ idling which has ultimately improved the engine efficiency to the extent of 5%. The concept can be shown as Figure 10 where with heat flow across the semiconductor causes electron flow and EMF generation which can be stored in the battery and the same will be used for faster heating the air during start of combustion, idling of the engine while stopping at signal, acceleration where momentary extra power is required and also running the compressor while stopping and in all it can have considerable improvement in fuel efficiency to the extent of 5% [3].

The Automotive industries see electric mobility as another possible way of meeting future demand for personal mobility. Here, one major advantage lies in zero local

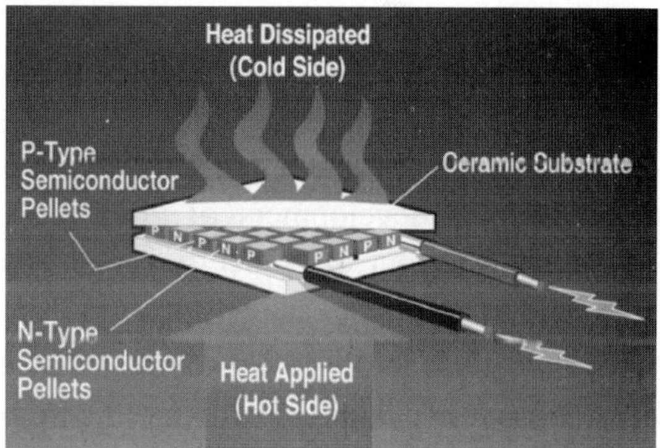

Fig. 10: Basic Concept of ThermoElectric generation

Fig. 11: Typical Chassis & drive train of a conventional vehicle

emissions. Since electro-mobility involves electric current rather than fuel being converted into propulsion, no climate-harming gases are created during the journey. If the energy to drive the vehicle is obtained from a renewable source, e.g. from wind, solar or water power, e-mobility is climate-neutral and conserves natural resources. But electro-mobility is not just emission-free, it also offers a unique and exciting driving experience. On the one hand electric vehicles move almost silently, and on the other the entire torque of the electric motor is available from a standing start. That makes the car extremely agile and provides a high fun factor.

Electro-mobility is just at the beginning of its development, and consequently there is further innovation work yet to be done in certain areas. The greatest challenge is clearly the development of the energy storage device. Due to its specific properties with regard to energy density and weight, this is currently the limiting factor of e-mobility. Up to now a battery has only been able to store a limited amount of energy as the energy density of the cell complex is comparatively low, though the high efficiency of an electric motor – up to 96% – partly compensates for this. Admittedly, the range of an electric vehicle today is still not comparable with that of a combustion engine, but significantly expanded research efforts over the next few years will lead to further leaps in technology and enable greater ranges.

The weight of the energy storage component is the second factor that restricts range. Due to its low energy density, a battery to drive a car is approximately the size of a large suitcase, as well as being very heavy. To avoid having to increase the size of the battery further and thus making it even heavier, the development engineers are endeavoring to make the best possible use of the battery capacity available and to extend the range through such measures as systematic lightweight construction and intelligent strategies for battery charging and usage. Work is also being targeted at significantly reducing recharging times; at the moment, several hours of recharging time, and thus idle time, are still needed to fully charge an EV.

A modern and efficient mobility solution using high speed and long range 4KW hub motor with big torque 60Ah lithium battery with 5-6 hours charging time max. speed can be 60km/h and range of 110kms and climbing ability of 20%, 60V/600W onboard charger, Aluminum framework with light weight, Front wheel fork+hydraulic helix spring shock absorber, rear alloy rocker+hydraulic helix, 5-point

Fig. 12: Electric motorcycle for efficient mobility solution.

seatbelt, Front and rear disc brake is shown in figure 12.

ELECTRIC HEARTBEAT – THE MOTOR

The electric motor with its associated power electronics is the heartbeat of the electric drive system. It converts electric energy into propulsion. In simplified terms, the latest-generation electric motor consists of a tubular stator fixed to the casing and a rotating cylinder inside the stator (the rotor). The rotor is connected to the gear assembly and, through this, to the driven wheels. Inside the stator are coils in which a magnetic field is generated through current flow. On the rotor, meanwhile, are one or several magnets with fixed polarity. Switching on the current causes the south pole of the magnetic field generated in the stator to attract the north pole of the rotor magnet.

However, before the north pole of the rotor reaches the south pole of the stator, the south pole is switched onto the next phase. As a result, the rotor also continues to turn and "chases" the alternating magnetic fields produced by the stator. Through its rotational movement the rotor transfers the mechanical energy required for propulsion. The fundamental ingredient in a functioning electric motor when it comes to providing optimum performance is the correct rotation of the magnetic field around the stator. This important task is carried out by a special power electronics control unit, which ensures the rotating field is switched at the desired speed and with the necessary field strength. In this way it ensures that the rotor spins at the required speed and delivers the desired torque.

THE BATTERY – THE ELECTRIC VEHICLE'S "FUEL TANK"

An extremely strong electric current is required to drive the motor of an electric vehicle. Currents of up to 400 amperes are activated for each phase, which equates to roughly 25 times that of a domestic power outlet. At up to 400 volts, voltages are also almost double those of the conventional power supply of everyday devices. A large number of battery cells connected in series and divided into smaller modular units are used to store this energy and make it available as required. The paramount consideration in the development and design of the energy storage device and the vehicle's high-voltage system is to guarantee the safety of occupants. Numerous safety measures are therefore in place for the energy storage and high-voltage system to ensure optimum protection of passengers. Comprehensive validation tests are required to demonstrate that all legal and internal crash safety requirements are met.

CHALLENGES FOR THE FUTURE

The energy density of the energy storage system in the electric vehicle has not yet reached levels comparable with that of a full tank of conventional fuel. A battery with 22 kWh contains energy equivalent to approximately 2.5 litres of premium unleaded petrol – and the distances possible in electric cars are currently much less as a result. Having said that, an electric motor has an efficiency level of up to 96%, much higher than a combustion engine, which is capable of 40% at best. So an electric vehicle with only this small amount of energy on board will take much further than a similarly powered vehicle fitted with a combustion engine. This extraordinarily high level of efficiency means that the distances possible in an electric vehicle are already sufficient for 90% of customers, making the vehicle fully suited for everyday use.

However, the central question remains: how can range be further extended? The developers are pursuing two directions here. The key is to minimise vehicle weight through the application of lightweight design principles wherever possible and the intelligent use of materials. This is also applied to the drive-train wherever possible. Increasing the battery capacity is another conceivable option, though a larger battery would add weight to the vehicle, in turn reducing the range. As a result, the engineers are looking at ways of exploiting the available battery capacity as effectively as possible. There are two ways of achieving this: on the one hand, the batteries are discharged as far as possible. On the other, intelligent charge control techniques, need-based battery temperature control and efficient operating strategies help to minimise energy consumption while on the move. That includes measures for the efficient use of heating / climate control, lighting and infotainment systems. However, the longer-term future lies in the further development of energy storage devices to increase energy density.

RANGE EXTENDER – SMALL ENGINE, LONG RANGE

One particular approach to increasing the distance the vehicle can cover comes in the form of a range extender. Here, a combustion engine produces electricity via a generator in order to charge the battery during a journey or maintain it at a constant charge. This could add a considerable distance to the vehicle's range. With a full-size electric motor already fitted, this combustion engine only needs to be relatively small. With output of 20 to 30 kW, the range extender would supply enough energy to fulfil the customer's driving requirements. The moderate output of the combustion engine, moreover, means relatively low levels of fuel consumption. Although the range extender is a thoroughly conceivable short-term answer to the issue of increasing vehicle range, the development engineers are focusing firmly on the further development of battery technology. With energy storage system technology for vehicles gaining in momentum, we can look forward to further significant advances in development.

The electrification of a vehicle requires new concepts for the vehicle architecture and body design in order to fully exploit the potential of e-mobility. So how does a functional and effective body construction for an electric vehicle shape up? With the revolutionary "Life-Drive" [2] concept the engineers have completely rethought the vehicle architecture to adapt it to the demands and realities of future mobility. They created a body concept which is squarely geared to the vehicle's future purpose and area of usage and which offers innovative use of materials.

Similarly to vehicles built around a frame (Figure 11), the Life-Drive concept consists of two horizontally separated, independent modules. The Drive module – the chassis – forms the solid foundation of the vehicle and integrates the battery, drive system and structural and basic crash functions into a single construction. Its partner, the Life module, consists primarily of a high-strength and extremely lightweight passenger cell made from carbon fibre-reinforced plastic (CFRP). With this innovative concept, it lends a totally new dimension to the areas of lightweight design, vehicle architecture and crash safety.

LIGHTWEIGHT DESIGN MAKES A WEIGHTY DIFFERENCE

For a vehicle powered by an electric drive system, lightweight design is crucially important because, alongside battery capacity, weight is the key limiting factor when it comes to the vehicle's potential range. The lighter a vehicle, the longer the distance it will be able to travel – simply because the electric drive system will have less mass to move. However, the drive-train of an electric car is almost 100 kg heavier than that of a vehicle with a combustion engine, full tank included. That is primarily due to the weight of the battery. To compensate for this, innovative light weight material is to be considered and CFRP plays a pivotal role in weight saving. The large-scale use of this material, as in the Life Module, is unique in volume car production to date. While it is at least as strong as steel, CFRP is approximately 50% lighter. Aluminium, on the other hand, would save "only" 30% in weight terms over steel. That makes CFRP the lightest material that can be used in body construction without compromising safety. A further benefit is that it is extremely corrosion-resistant and does not rust, giving it a far longer lifespan than metal. Furthermore, it retains its integrity under all climatic conditions.

CONCLUSION

Efficient mobility solution is one of the most happening things today and it is the time for gasoline based IC engine to be on the receiving end. Future is of green mobility where without having any emission imprint efficient mobility solution will be the challenge.

ACKNOWLEDGEMENT

I take this opportunity to thank the CMD of KLT, who is allowed me pursue some of my interest which may have long term commercial impact as part of diversification program.

REFERENCES

[1] IEA/SMP Model Documentation and Reference Case Projection L. Fulton, IEA / G. Eads, CRA July 2004.

[2] BMW: Mobility of the Future, by Horatiu Boeriu, July 2nd 2010.

[3] Yang J., 24th Int. Conf. on Thermoelectrics, ICT2005, pp. 170 - 174.

[4] AMS Conf. Dec. 2011, New Delhi, India.

[5] PR-LF-FRR-BN-02/2012, Flat Carbon Europe, Arcelor Mittal.

SECTION VI

Process Simulation

Improved Simulation of Continuous Casting to Predict Transverse Corner Cracking in Microalloyed Steels

K.M. Banks[+], A.S. Tuling[+] and B. Mintz*

[+]Industrial Metals and Minerals, University of Pretoria, SA, *City University, London, UK.

ABSTRACT

This paper reviews recent work concerning improvements to the hot ductility test to simulate straightening during continuous casting of steel slabs and billets. Various thermal paths were tested for their suitability in assessing the likelihood of transverse cracking in a variety of mostly microalloyed steels. For non-Ti steels with reasonably high Mn contents, the simple procedure of reheating to dissolve all the microalloying additions, followed by cooling to the test temperature is adequate. However, tests that include *in-situ* melting, followed by primary rapid cooling in the mould and foot-rolls and slow secondary cooling are generally recommended. Whilst the inclusion of thermal oscillations to simulate slab-roll contact improves accuracy, they are only recommended for steels with high transformation-start temperatures. The following systematic studies on as-cast hot ductility were performed using the improved cycle method: i) The influence of Ti in low C-Nb steel slabs - here, the ductility decreased markedly with the addition of Ti using simple tests. However, when the improved thermal cycle was applied, Ti additions were found to be beneficial up to a Ti:N ratio of at least 7.5:1. ii) The influence of V and N content in Nb-Ti peritectic steel slabs – the ductility improved when V was added to Nb-Ti steels and was attributed to a delay in precipitation and a decrease in fine particle fraction. To avoid transverse cracking in these steels containing high N, hypo-stoichiometric additions of Ti are required. iii) The influence of high N content in a range of medium C microalloyed billets - it was shown that Ti additions to V- or Al-containing grades, although beneficial at low N levels, are detrimental to hot ductility probably due to a higher precipitate volume fraction. Ti-high B-high N steels produced the poorest hot ductility. Low total B content and near stoichiometric Ti:N ratios significantly improved ductility by reducing the potential for harmful BN and AlN precipitation. Grades containing moderate amounts of V resulted in good ductility due to the suppression of precipitation. Thermal modelling and industrial observation indicated that corner temperatures of microalloyed steel slabs should be below the Ae_3 during unbending to prevent corner cracking. Conversely, high unbending temperatures corresponding to the upper portion of the ductility curve should be aimed for in billet casting.

KEYWORDS: *Transverse Cracking, Hot Ductility, Microalloyed Steels*

INTRODUCTION

The simple hot tensile test has long used to estimate the likelihood of transverse cracking occurring during the continuous casting of microalloyed steels. Although an over-simplification of the industrial continuous casting operation, this test successfully ranked C-Mn-Al, C-Mn-Nb and C-Mn-V steels in order of their susceptibility to transverse cracking. More sophisticated tests for direct application to commercial billet and slab casting operations are reviewed here[1-5].

The original hot tensile test consisted of heating to sufficiently high temperatures (1250-1300°C) to dissolve all carbonitride and AlN precipitates and generate a coarse grain size to approximate the as-cast microstructure. The tensile specimen was then cooled at a constant, *average* cooling rate experienced by the strand from the melt to the test temperature, this being taken as typically 60°C/min for a 220mm thick strand. For these conditions a reduction-in-area, RA, above 35-40% was necessary to prevent cracking [6,7].

A 2-D temperature model [8] showed that the surface of the strand cools rapidly in the mould after pouring and reaches a temperature trough at the foot-rolls. As the strand moves down the caster, the surface temperatures initially rise again due to the hotter interior but eventually fall slowly during secondary cooling. When the hot strand is in contact with the colder guide-rolls, the surface temperature oscillates significantly for a short period.

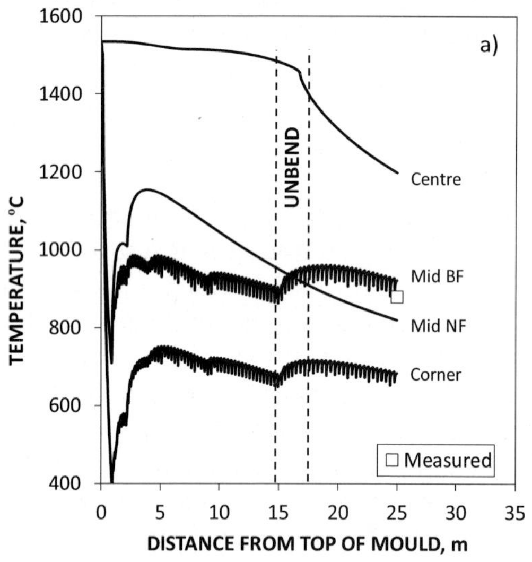

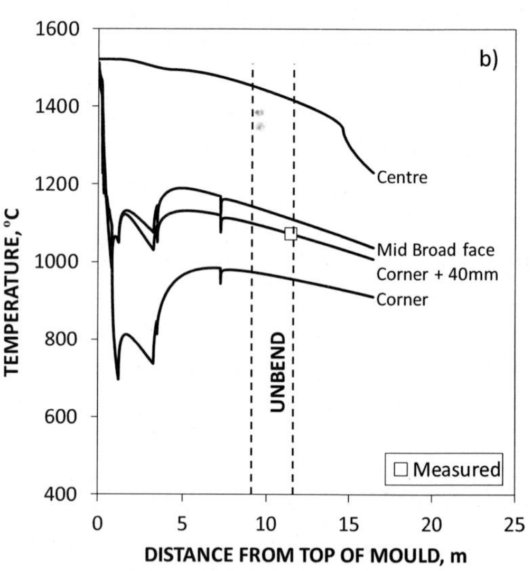

Fig. 1: 2-D continuous casting temperature model for a) steel C6, 240 x 900mm^2 strand: speed = 1m/min, radius = 10m, b) steel M5, 140mm^2 billet: speed = 2.3m/min, radius = 7m. BF = broad face, NF = narrow face. [8]

Cooling at the corners region is the fastest and, due to geometry, cooling rates in billets are higher than in slabs. Depending on the severity of the cooling pattern, the temperature in the corner region can fall into the austenite-ferrite transformation zone at the end of primary cooling, thereby refining the grain size and promoting precipitation of carbides and nitrides. A small amount of transformation is detrimental to ductility since the strain concentrates in the thin ferrite films, whilst a large, soft ferrite fraction is beneficial since much more strain can be absorbed during unbending.

The thermal path selected for the tensile test approximates a specific position in the strand. The most relevant thermal path to simulate corner cracking is open to question and accounts for the wide range of cooling rates used (10-1200°C/min) [7]. Although cracks originate at oscillation marks on the surface, the location to which they propagate is likely to be below the surface, where the chill, fine-grained zone merges into the coarse, columnar region.

EXPERIMENTAL

The steel compositions studied, together with their calculated Ae_3 temperatures [9], are shown in Table I and include low C, peritectic C and medium C microalloyed steels. Additional results on medium C-Ti-B steels, not published previously, are also briefly discussed. "Peritectic" here refers to the C content but may actually be non-peritectic due to other additions. Various cooling profiles were applied using a Gleeble 1500D, fig.2, to determine hot ductility behaviour. A typical unbending strain rate of $3x10^{-3}s^{-1}$ was used in all tests.

Average cooling rate – cycles A,B,C: In fig.2a, cycle *A* represents the older test method, where the specimen was reheated to 1300°C and cooled at a constant cooling rate of 300°C/min. to the test temperature. In cycle *B, s*pecimens were *in-situ* melted and directly cooled at 300°C/min. to the test temperature.

Table I: Composition of steels tested

Steel	Code	Type	C %	Mn %	Si %	Nb %	Ti %	V %	N ppm	B ppm	Ti:N	Ae_3 °C
E1*	AC2-34	ELC	0.02	0.2	-	-	-	-	48	-	-	844
L1	K42	LC-Nb	0.04	1.6	-	0.059	-	-	32	-	-	810
L2	K24	LC-Nb	0.05	1.7	-	0.033	-	-	82	-	-	802
L3	K40	LC-Nb	0.04	1.6	-	0.055	-	-	66	-	-	809
L4	K66	LC-Nb-Ti	0.06	1.7	-	0.044	0.013	-	72	-	1.8	792
L5	K14	LC-Nb-Ti	0.03	1.6	-	0.031	0.013	-	42	-	3.1	810
L6	K15	LC-Nb-Ti	0.04	1.6	-	0.031	0.012	-	77	-	1.6	805
L7	K57	LC-Nb-Ti	0.04	1.6	-	0.046	0.036	-	48	-	7.5	795
C1	RC1-28	C-Nb	0.13	1.2	-	0.022	-	-	95	-	-	780
C2	RC1-83Nb	C-Nb	0.17	1.5	0.3	0.021	-	-	74	-	-	769
C3	RC1-83	C-Nb-Ti	0.16	1.4	0.3	0.028	0.028	-	60	-	4.7	766
C4	RC1-83	C-Nb-Ti	0.17	1.4	0.3	0.021	0.016	-	74	-	2.2	766
C5	RC1-83	C-Nb-Ti	0.16	1.4	0.3	0.023	0.023	-	60	-	3.8	766
C6	RC1-85	C-Nb-Ti	0.13	1.5	0.3	0.034	0.025	-	62	-	4.0	770
C7	RC1-14	C-Nb-Ti-V	0.13	1.5	0.3	0.039	0.034	0.030	80	-	4.3	765
C8	RC3-02	C-V	0.14	1.5	0.3	-	-	0.050	105	-	-	777
M1	SAE5160	MC-V	0.56	0.8	0.3	-	-	0.028	42	-	-	721
M2**	080A42	MC-V-Ti	0.38	0.8	0.3	-	0.019	0.025	90	-	2.1	745
M3	15B36	MC-Ti-B	0.34	1.4	0.3	-	0.028	-	119	31	2.4	730
M4	51B40	MC-Ti-B	0.39	0.8	0.3	-	0.032	-	116	36	2.8	748
M5	51B40	MC-Ti-B	0.42	0.8	0.3	-	0.035	-	140	30	2.5	742
M6	51B40	MC-Ti-B	0.39	0.8	0.3	-	0.045	-	145	19	3.1	745

All steels contained 0.03-0.05%Al, 0.01-0.02%P, S < 0.005%, except where indicated.

* S=0.019%. **Average of eight casts tested.

In cycle *C*, specimens were melted and cooled at a relatively fast secondary cool rate of 24°C/min directly to the test temperature.

Primary & secondary cooling–cycles D,E: Cycle *D* approximated the thermal path at the sub-surface corner region of slabs. This quasi-industrial cycle, fig.2b, included *in-situ* melting, initial cooling at 600°C/min to the end of the foot-rolls, a temperature rise due to heating of the surface from the molten core and subsequent secondary cooling at 12°C/min to the test temperature. The corner temperatures at the end of initial cooling in these tests was chosen to fall in the austenite region, although these may be lower depending on cooling severity. Cycle *D* for simulating 140mm^2 billet casting included faster speeds and faster secondary cooling rates (2.2m/min, 24°C/min), fig.2c.

Cycle *E* was similar to cycle *D* for slabs but included a number of thermal oscillations to simulate contact of the strand surface with guide-rolls on the broad faces.

RESULTS AND DISCUSSION

Influence of thermal cycle

Fig.3 shows the effect of thermal cycle on the RA value of various steels at 800°C. Generally, melt cycle *B* showed inferior ductility in all steels when compared to reheated cycle *A*. The poor ductility found in melted and directly cooled tests was observed previously by Mintz *et al* [7] and indicated that a more accurate simulation of continuous casting was required. This is especially true for Ti-microalloyed steels, where good ductility experienced under industrial (melted) conditions must be replicated [10-12].

In fig.3a, the RA values for steel E1 after cycles *B-D* are similar, but significantly lower than that found after cycle *E*. The improvement in hot ductility of steel E1 when cycle *E* was used is a result of its narrow trough [13]; thermal oscillations to 750°C, coupled with deformation at 800°C - temperatures well below the Ae_3 of 844°C - producing sufficient ferrite prior to deformation, resulting in excellent ductility.

For this reason, hard cooling is often applied industrially to low C - low Mn grades. At test temperatures above Ae_3, *i.e.* in the single-phase austenite, ductility was expected to be similar after all cycles.

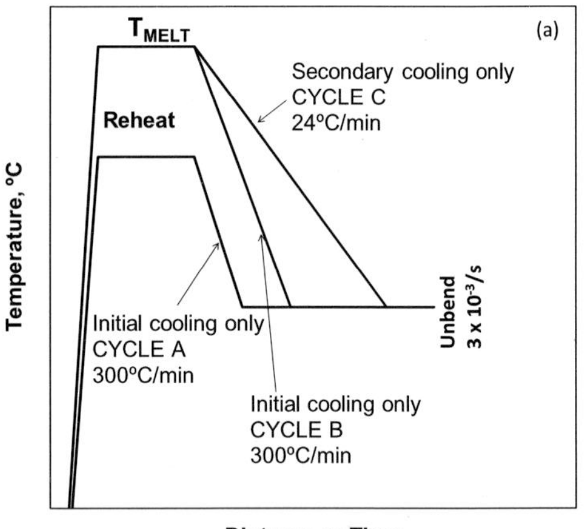

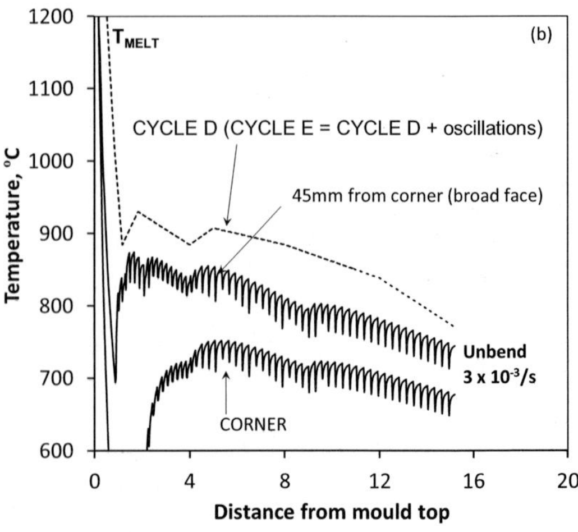

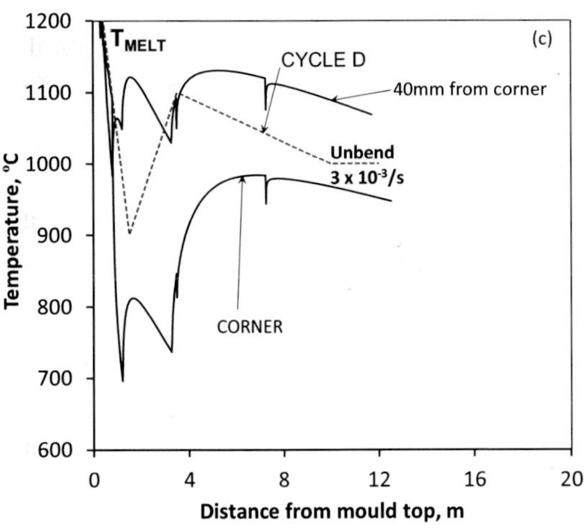

Fig. 2: Gleeble thermal cycles to simulate continuous casting: a) Average cooling cycles *A,B,C*, b) Cycles *D, E* for slabs - 600°C/min primary and 12°C/min secondary cooling rate and c) Cycle *D* for 140mm² billets - 600°C/min primary and 24°C/min secondary cooling rate.

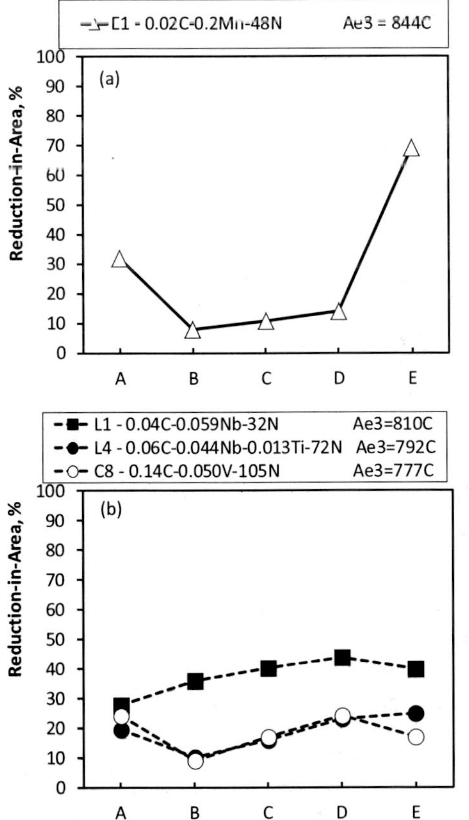

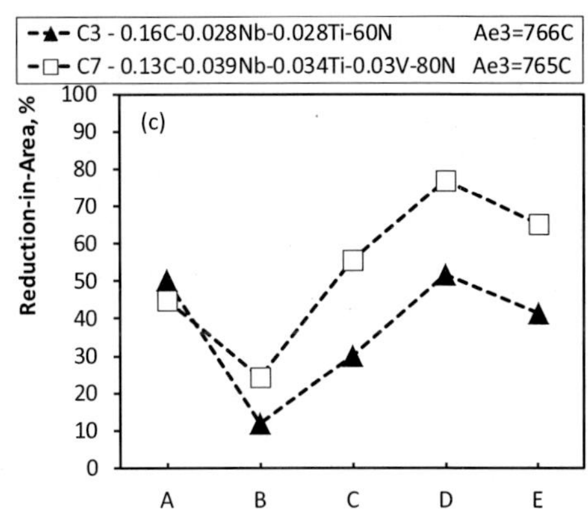

Fig. 3: Influence of thermal cycle on RA at 800°C. a) low C-low Mn, b) low C-Nb-, V- and low Ti-Nb steels, c) peritectic Nb-Ti steels. The experimental error on %RA was ±10%.

The inclusion of thermal oscillations to simulate slab-roll contact improves accuracy but is only recommended for steels with a high Ae_3. The effect of slab-roll contact was confirmed in fig.4, where SEM showed smooth ferrite growth planes on the austenite grain boundaries, which have only been observed when the ferrite films are thin. They were present at the surface of steel E1 after cycle *D* but absent after cycle *E* when large amounts of ferrite had formed. Cycle *C* had little effect on ductility suggesting little influence of cooling rate on this grade.

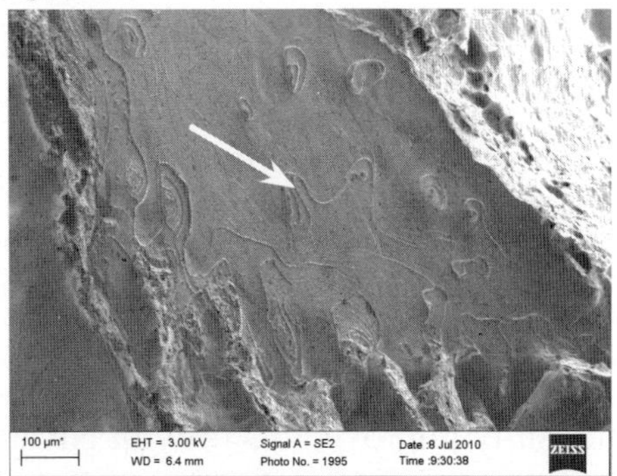

Fig. 4: Fracture surface showing smooth ferrite growth planes normally only observed when thin films are present. Cycle *D* in low C- low Mn steel El. Tested at 800°C.

Fig.3b showed little variation in RA for any cycle applied to steels L1, L4 and C8 given an average error of about 10% typically found in RA tests. It appears that when the test temperature is marginally lower than the Ae_3, then all thermal cycles produce similar ductility, implying that when the unbending temperature is close to the Ae_3, then any of these ductility tests are fair predictors of poor ductility.

Fig.3c showed that, in peritectic C-Nb-Ti steels C3 and C7, cycle *A* produced relatively high RA values at 800°C, whilst cycles *B* and *C* deteriorated ductility only to recover dramatically in cycles *D* and *E*. The good ductility after cycle *A* was attributed to the limited dissolution of Ti and N at typical reheat temperatures. Stable TiN restricted the austenite grain size and limited fresh precipitation of fine TiN, which resulted in good ductility [10]. However, this cycle does not occur in practice. The excellent ductility found after quasi-industrial cycles *D* and *E* was due to coarsening of fresh precipitates during slow secondary cooling compared to the poor ductility in cycle *B* and average ductility in cycle *C* which indicated a strong dependence of these steels on secondary cooling rate.

i) Low C–Nb-Ti slabs: Hot ductility curves after cycles *B* and *D* are shown in fig.5 for low C-Nb and low C-Nb-Ti steels.

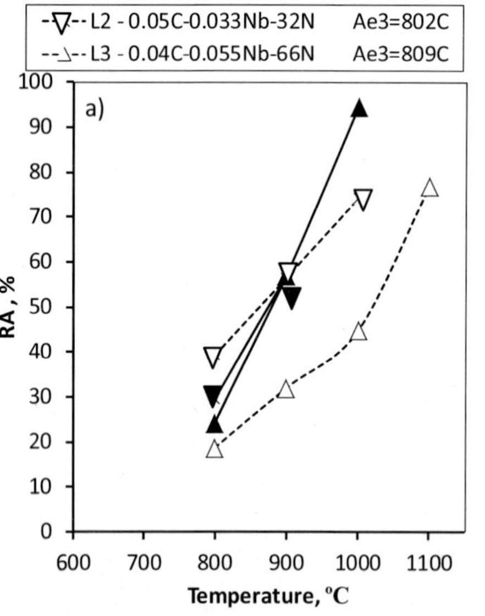

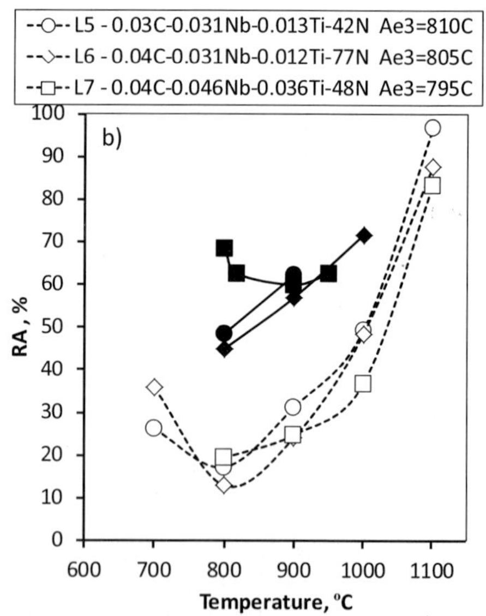

Fig. 5: Hot ductility curves for a) low C-Mn-Nb and b) low C-Mn-Nb-Ti steels. Dashed and open symbols: Cycle *B* , Solid and closed symbols: Cycle *D*, *cf.* fig.2b.

Fig.5a shows that, although cycle *D* significantly improved ductility in Nb steel L3, the RA was about 20% at 800°C and hence, crack susceptible. There was almost no change to RA in Nb steel L2, probably due to the low N content. With a minimum RA of 30% after cycle *D,* transverse cracking is thus not expected to occur in this grade.

Fig.5b showed that cycle *D* significantly improved ductility compared to cycle *B*. For cycle *B*, Ti content resulted in a slight loss in ductility, whilst in cycle *D*, Ti level improved RA, especially in the lower austenite region – the RA between 800 and 900°C was above 40% - sufficient to avoid cracking. This cycle explains the benefit of Ti additions to hot ductility of steels under industrial conditions and stresses the importance of accurate simulation of the surface and corner temperatures. The positive effect of Ti was more substantial at a higher Ti:N ratio of 7.5:1, even in high Nb steels L3 and L7 - The RA increased from 24% to 60% at 800°C.

The variation in Nb level in fig.5b did not appear to affect ductility possibly because in melting, some Nb may precipitate as coarse, harmless eutectic [14].

ii) Peritectic microalloyed slabs: Fig.6 shows hot ductility curves for peritectic Nb-microalloyed steels with and without Ti. Although the low Ti:N ratio improved hot ductility in steel C4, it was not as effective as the hyper-stoichiometric Ti:N ratio in steels C5 and C6, which contained sufficient Ti to combine with all the N. However, the ductility at 750°C was still low in all the steels. The ductility trough in C5 and C6 occurred below the Ae_3 (766°C) but started to recover again at 700°C.

The N level in these steels is important and small changes (0.001%) can have a large influence on hot ductility[4]. The slow secondary cooling rate of 12°C/min allowed the Ti-Nb rich precipitates to coarsen sufficiently so that they had less influence on the hot ductility in hyper-stoichiometric steels C5 and C6 [3]. Generally, when the average precipitate size is larger than 20nm, as was found in these steels [6], only a small deterioration in hot ductility occurred. Higher Ti:N ratios improve the hot ductility and have been found to apply to steels with up to 0.007%N. N contents above 0.008% lead to unacceptably large precipitate volume fractions of TiN which, even if coarse, can be detrimental to ductility. A high volume fraction of fine precipitates increases the matrix strength, leading to enhanced grain boundary sliding [6]. Thus, if high N levels in peritectic steels are unavoidable, then Ti contents should be limited to about 0.015%.

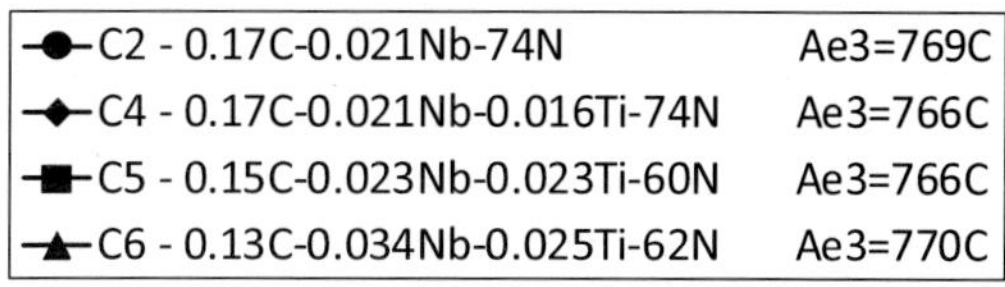

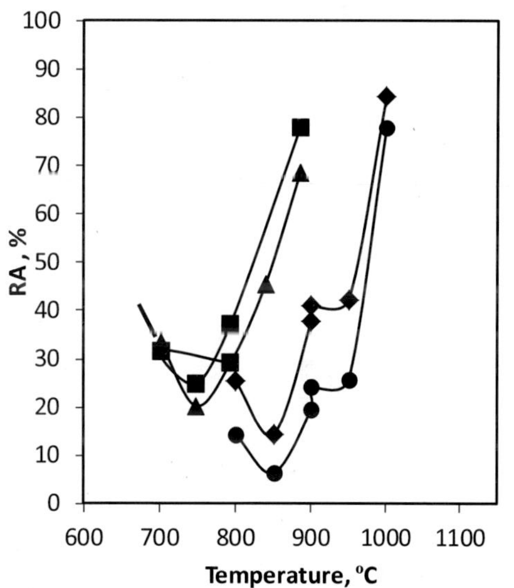

Fig. 6: Hot ductility of peritectic C-Nb and peritectic C-Nb-Ti steels - Cycle *D*, fig.2b.

Nb-Ti-V microalloyed steel C7 gave substantially better ductility than C-V steel C8 which, in turn, was superior to C-Nb microalloyed grade C1, fig.7. Steel C7 was also superior to the Nb-Ti steels in fig.6 at 800-850°C but RA decreased to a minimum of 28% below the Ae_3 of 765°C.

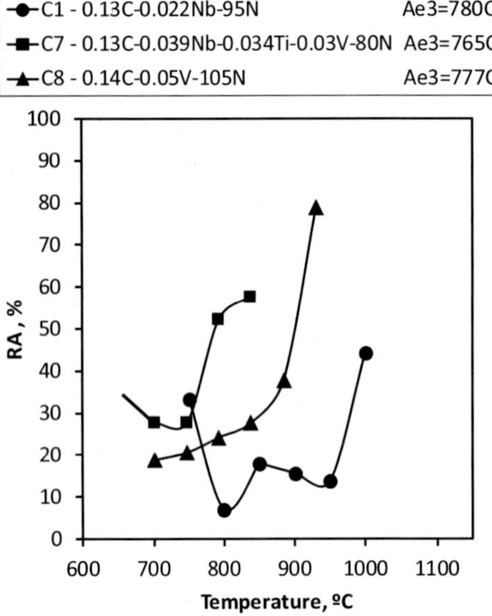

Fig. 7: Hot ductility of high N peritectic microalloyed steels. *cf.* Cycle *D*, fig.2b.

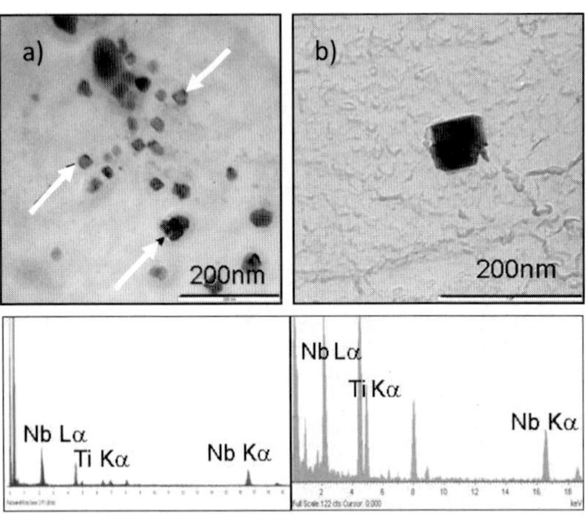

Fig. 8: TEM micrographs showing a)Nb-Ti microalloyed steel C6 with copious, fine NbTi(C,N) precipitation and b) a coarse NbTi(C,N) particle in Nb-V-Ti microalloyed steel C7. Tested at 800°C.

Fig.8a shows fine carbonitride precipitation in Nb-Ti steel C6 tested at 800°C, which explains the poor ductility – 30% - found at this temperature. Fig. 8b shows a typical coarse, isolated carbonitride in Nb-V-Ti steel C7 at test temperature of 800°C, which is consistent with the relatively high RA of 50%, fig.7.

Previous work [15] showed that, using a constant cooling rate from the melt (cycle *B* type test), Nb-V combinations result in better ductility than either Nb-only or V-only steels. Vanadium, when added either by itself or with Nb, forms a coarser, less detrimental precipitate and retards precipitation during unbending [16,17].

The risk of industrial transverse corner cracking in steel C6 is illustrated in fig.9 by comparing the predicted temperature profiles of two slab casters via thermal modelling.

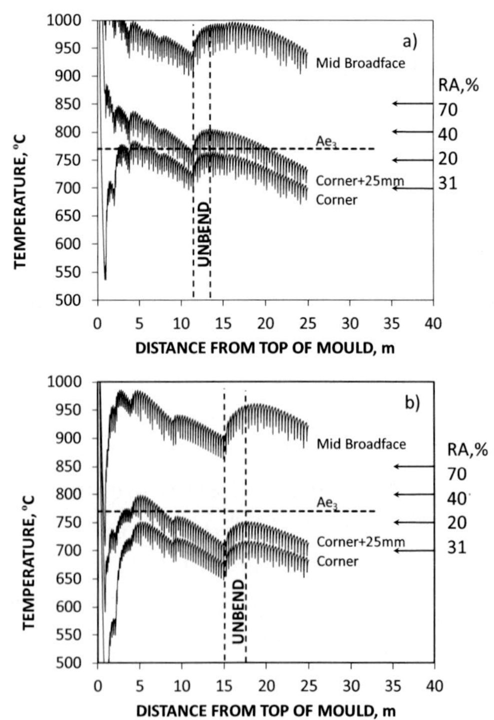

Fig. 9: Thermal profiles for steel C6 slabs (240mm x 900mm, speed = 1m/min) for a) caster I - softer cool, early unbending and frequent cracking and b) caster II - harder cool, late unbending with no observed cracking.

The relatively soft cool at Caster I (7.5m radius) resulted in corner and off-corner temperatures during unbending to approach the Ae_3 and produced frequent transverse cracks. The harder cooling at caster II (10m radius) forced the corner region temperatures below Ae_3 during unbending and produced crack-free slabs, presumably due to the formation of sufficient ferrite at this stage of casting. This was confirmed by the recovery in ductility below 750°C to above 30%, fig.6, the RA value below which transverse cracking in slabs is thought to occur. Similar behaviour was found for all the peritectic steels studied, which all had Ae_3 temperatures in the 765-780°C range. Thus, the interaction between steel chemistry and cooling patterns needs to be carefully characterized to avoid transverse cracking.

iii) Medium C billets: Cycle *D* was used to study a cracking problem in medium C microalloyed billets containing high N (≥0.008%). Fig.1b showed that the corner region in 140mm² square billets is in the 900-1000°C range during the unbending operation for a withdrawal speed of 2,2mm/min, significantly higher than those found in 240mm thick slabs. Hot ductility curves for 0.35%C-V steels with and without Ti are shown in fig.10. The loss in hot ductility generally occurred at higher temperatures than those found for the low or peritectic C steels and well above their Ae_3 temperatures. The high temperature ductility troughs in the medium C steels is thought to be due to grain boundary sliding [18] in coarse-grained austenite and smaller critical strains at higher C levels [19], rather than ductility loss due to strain-induced ferrite films found in low C and peritectic micro-alloyed grades.

V-only steel M1 resulted in good ductility at above 900°C, due to the suppression of precipitation, fig.11. However, only when the temperature in the corner region drops to below 900°C, fig.1b, in steel M1 does the RA fall to dangerously low levels. The addition of Ti to medium C-V-N steel M2, although beneficial at low N levels, impairs hot ductility due to a higher precipitate volume fraction at higher N levels. Evidence to support this is the billet rejection histogram in fig.12 reported by Alvarez de Toledo *et al* [20].

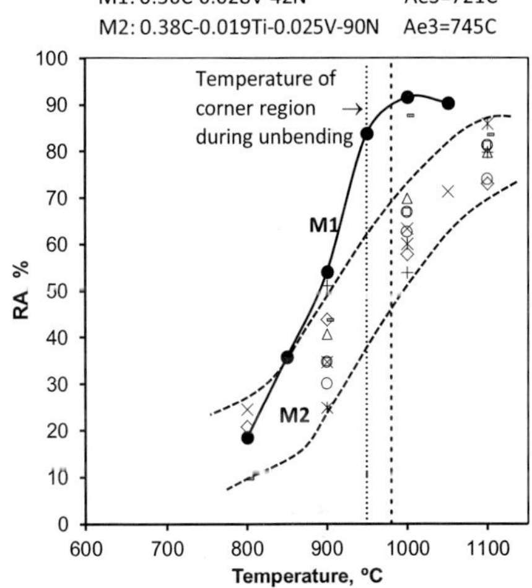

Fig. 10: Hot ductility curves for medium C-V-steel M1 and C-V-Ti steel M2 (eight steels). Cycle *D,* fig.2c.

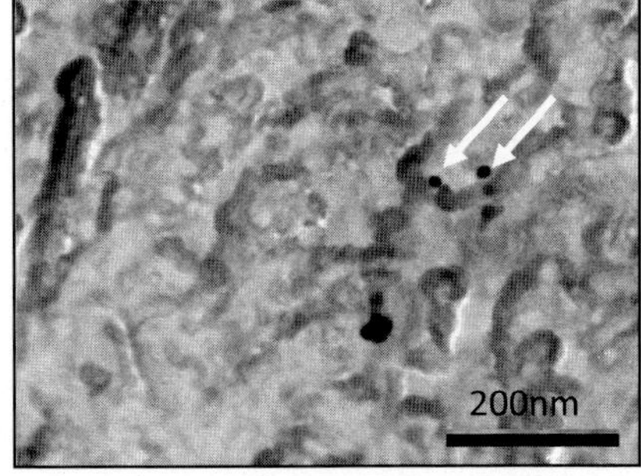

Fig. 11: TEM micrograph showing sparse, relatively coarse precipitation in V-only grade M1. Tested at 900°C.

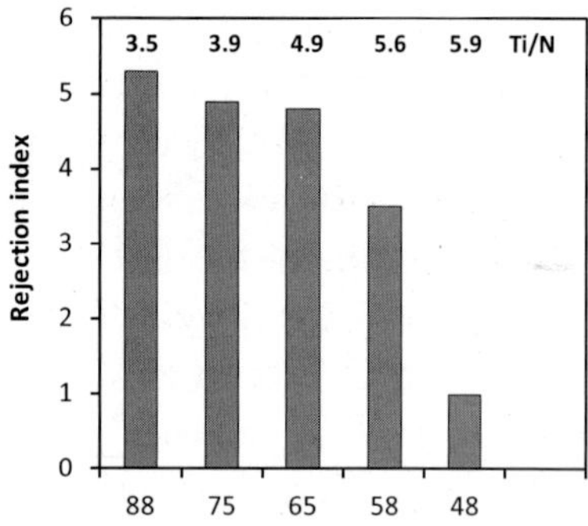

Fig. 12: Influence of total N content on rejection rate in medium C-Ti steels [20].

Ti-high B steels M3-M5 had the poorest hot ductility in medium C steels containing high N and consequently the greatest risk for transverse cracking since RA fell to below 20% at the temperature expected in the corner region during unbending, fig.13. In steel M6, the lower total B content and near stoichiometric Ti:N ratio increased the solute B content to significantly improve ductility in the unbending temperature region and is expected to reduce transverse cracking in this steel type, especially when high N contents are unavoidable.

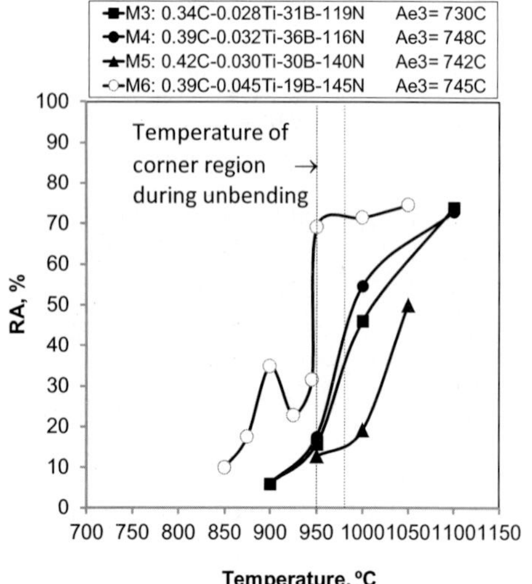

Fig. 13: Hot ductility curves for medium C-Ti-B steels M3-M6. Cycle *D*, fig.2c.

Boron is beneficial to hot ductility provided the cooling rate is restricted to below 100°C/min [7]. Solute B improves hot ductility because of the segregation of B atoms to grain boundaries, where vacancies are occupied and the formation and propagation of microcracks is prevented [21]. Fast cooling rates allow fine BN precipitation to occur in the matrix. The small interparticle spacing would allow cracks to link up easily, promoting cracking due to microvoid coalescence [22]. The Ti content was too low to precipitate all N as TiN in steels M3-M5 and precipitation of BN and AlN was expected. A high rejection rate was experienced in these grades in the form of transverse cracks in billets and seams in rolled bars.

The fracture surfaces for steels M5 and M6 are shown in fig.14 after testing at 950°C. Steel M5, with low RA, fractured predominantly along the weak grain boundaries probably due to a low solute B content. Steel M6 had a high RA and experienced mostly high temperature ductile rupture. Thus, the higher solute B in steel M6 was expected to strengthen grain boundaries and improve ductility.

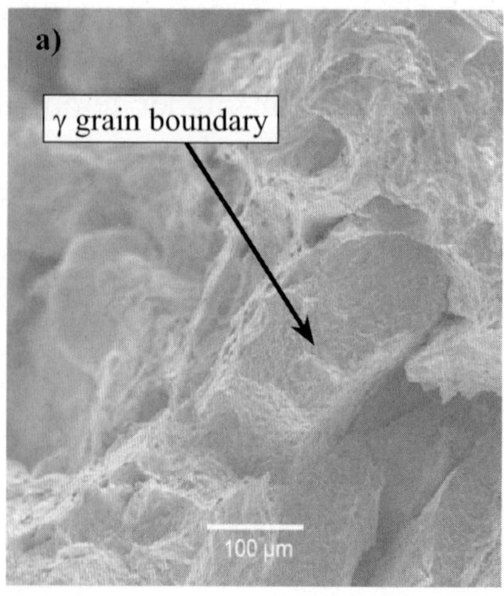

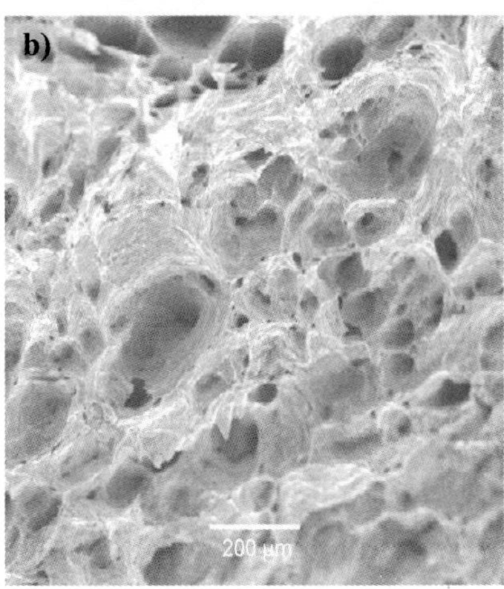

Fig. 14: SEM micrographs of fracture surfaces of Ti-B steels tested at 950°C: a) M5- fracture along austenite grain boundaries and b) M6 - high temperature ductile rupture.

CONCLUSIONS

1. Simple reheating, followed by a constant cooling rate, can be used to assess transverse corner cracking in C-Mn steels containing combinations of Al, V and Nb. However, for improved simulation of transverse crack susceptibility of Ti-microalloyed steels, melted specimens, combined with rapid primary cooling and slow secondary cooling, should be used.
2. In low C-Nb-Ti steels, the positive effect of Ti additions on reduction-in-area was shown to hold up to a Ti:N ratio of at least 7.5:1.
3. In peritectic microalloyed steel slabs,
   a) containing Nb-(Ti), the Ti:N ratio should be stoichiometric and in high N steels, the Ti content should be limited to about 0.015% for acceptable ductility.
   b) containing V-Nb-Ti, the ductility is higher at 800°C than in V-N, Nb-N and Nb-Ti steels due to a delay in sparse, fine precipitation.
4. In medium C Ti-B billets with high N, good ductility can be achieved by limiting the total B content to about 20ppm and ensuring a near stoichiometric Ti:N ratio.
5. When the unbending temperature of the corner region is near the Ae_3, the risk of transverse cracking increases. For slab casting, the critical hot ductility for the avoidance of transverse corner cracking is estimated to be about 30%.

ACKNOWLEDGEMENTS

Thanks go to ArcelorMittal (SA) for permission to publish this work.

REFERENCES

[1] K.M.Banks, A.S.Tuling and B.Mintz, Mat. Sci. and Tech, Vol. 28, 2012, pp.536-542.

[2] K.M.Banks, A.S. Tuling, C. Klinkenberg, B.Mintz, Mat. Sci. and Tech, Vol. 27,2011, pp537-545.

[3] K.M. Banks, A.S. Tuling and B. Mintz, Mat. Sci. and Tech. Vol. 27, 2011, pp1309-1314.

[4] K.M. Banks, A.S. Tuling and B. Mintz, J. of South African Mining and Metallurgy, Vol.111, 2011, pp. 711-716.

[5] K.M.Banks, A.S. Tuling and B.Mintz, Mat. Sci and Tech., 2012. In print.

[6] B. Mintz, S.Yue and J.J. Jonas, Int. Mat. Reviews, Vol.36, 1991, pp. 187-217.

[7] B. Mintz and D.N. Crowther, Int. Mat. Reviews, Vol.55, 2010, pp168-196.

[8] K.M. Banks. 2013. To be published.

[9] K.W. Andrews. J. Iron Steel Inst., 1965,Vol.203, 721.

[10] B.Mintz and D.N.Crowther, Titanium Technology in Microalloyed Steel ed. T.N.Baker. Inst. of Materials, London (1997) pp98-114.

[11] R. Abushosha, O. Comineli and B.Mintz. Mat. Sci. and Tech., Vol. 15, 1999, pp278-286.

[12] O. Comineli, R. Abushosha and B.Mintz, Mat.Sci. and Tech, Vol.15, 1999, 1058-1067

[13] B.Mintz and A.Cowley, Mat.Sci. and Tech. Vol.22, 2006, 279-292.

[14] R.Abushosha, R.Vipond and B.Mintz, Mat.Sci. and Tech., Vol.7, 1991, pp1101-1107.

[15] K. Banks, A. Koursaris, F.Verdoorn and A.Tuling, Mat.Sci.and Tech., Vol.17, 2001, pp1596-1604.

[16] B. Mintz and R. Abushosha. Ironmaking and Steelmaking, 1993, 20, 445-452.

[17] M.G. Akben, B. Bacroix and J.J. Jonas, Acta Metall., Vol.31, 1983, pp161-174.

[18] B.Mintz, Mat. Sci. and Tech, Vol.24. No.1, 2008, pp112-120.

[19] T.Matsumiya, M.Ito, H.Kajioka, S. Yamaguchi and K. Nakamura. Trans. ISIJ, Vol.26, 1986, p540.

[20] G. Alvarez de Toledo, A. Arteega and J.J. Laraudogeitia, Mater. Sci. Forum, 2005, pp500-501

[21] N.E. Hannerz, Trans. ISIJ, Vol. 25, 1985, pp149-158.

[22] K. Cho, D.J. Mun, J.Y. Kim, J.K. Park, J.S Lee and Y.M Koo, Met. And Mat. Trans. A, Vol. 14A, June, 2010, pp1421-1428.

Thermodynamic and Kinetic Modeling Tools and their Application to the Life Cycle of Stainless Steel

Åke Jansson[1], Anders Engström[1], Johan Bratberg[1], Qing Chen[1], Paul Mason[2] and Pingfang Shi[1]
[1]Thermo-Calc Software AB, Norra Stationsg, 93, Plan 5, SE-113 64, Stockholm, Sweden
[2]Thermo-Calc Software Inc., 4160 Washington Road, McMurray PA 15317, USA
Email: ake@thermocalc.se

ABSTRACT

Computational thermodynamics has evolved over the past four decades to become a powerful, flexible tool for the study of a variety of scientific and technological areas of industrial importance. For example, it was rather early on successfully applied in alloy development for stainless steels, and has since then continuously developed towards use in more and more applications within the lifecycle of stainless steel.

Computational kinetics is today also successfully applied within several areas of the lifecycle of stainless steel, e.g. simulation of; micro-segregation during solidification, inter-diffusion during post-weld heat treatment, and evolution of microstructure during service in high temperature applications.

This presentation aims to describe some recent advances in the application of computational thermodynamics and kinetics to complex situations within the lifecycle of stainless steel, ranging from alloy design to scrap re-cycling.

KEYWORDS: *Computational Thermodynamics, Computational, Kinetics, Modeling, Stainless Steel, Alloy Design, Scrap Recycling*

INTRODUCTION

The world of metals and materials research and development is currently experiencing a strong trend towards modelling and simulation. Since the early 1990's there has been an exponential type of growth in the number of citations and applications of computational thermodynamics, and more specifically CALPHAD (Computer cALculation of PHAse Diagrams) [1], to problems of scientific and industrial importance. This already on-going trend has gained added momentum in the last few years with the publication by the National Academies report on Integrated Computational Materials Engineering (ICME) in 2008 [2], and then the announcement by President Obama of the Materials Genome Initiative (MGI) in June 2011 [3].

Computational thermodynamics including the CALPHAD technique are core components of ICME since it fundamentally links the phases that form, and hence microstructure, to the chemical composition of an alloy, and also the temperature variation that a material may be subjected to. It is also essentially the driving force for many of the phase transformation reactions that take place during materials processing.

THE MATERIALS LIFE CYCLE

A nice way of summarizing and describing the stages and processes a material will go through during its existence is the so called materials life cycle. Fig. 1 shows one example of the materials life cycle can be shown. It is interesting to note that thermodynamic and/or kinetic software tools can be applied to increase the understanding of all of the 11 segments in Fig. 1.

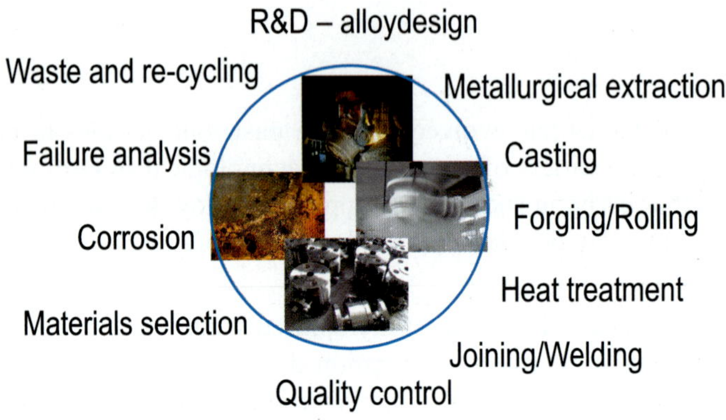

Fig. 1: The Materials Life Cycle

EXAMPLES WITHIN THE LIFE CYCLE OF STAINLESS STEEL

The perhaps most obvious of all the segments of the materials life cycle when it comes to applying thermodynamic calculation is in the research stage, what might be called Alloy Design.

In the present paper we will however not give any example in this segment, but instead move on to Metallurgical Extraction, where there are several examples to be found. These include Controlling and designing slag systems, Interaction between slag and liquid metal and Influence of an external pressure. Fig. 2 shows an example of steel/slag interaction in a steel in a property diagram, calculated with Thermo-Calc [4]. Fig. 3 shows the pressure effect on solubility of N in a stainless steel calculated using Thermo-Calc.

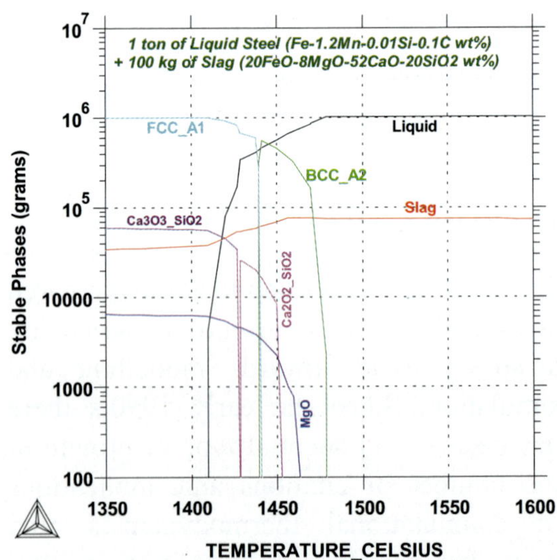

Fig. 2: Amount of stable phases as a function of temperature in a steel/slag system.

Casting is a process which most, but not all, metallic materials go through. Even materials which are not cast with few exceptions go through a transition from liquid to solid state. The solidification process is very suitable for study by both thermodynamic and kinetic modelling.

The solidification simulations shown in Fig. 4[5] are calculated using Thermo-Calc and DICTRA [4], and compares equilibrium calculation with Scheil-Gulliver simulation and so called partial equilibrium calculation which is a Scheil simulation which allows equilibration for fast diffusing elements such as carbon. Finally a DICTRA simulation which takes into account the full diffusional transformation from liquid to is included.

Using some modern databases such as TCFE7 [6] or TCNI5 [7] which contain volume data, it is also easy to calculate not only volume fractions but also density and thermal expansivity as a function of temperature. Other usage in the area of casting includes the possibility of generating data for exporting to dedicated casting codes. Such data may include solidus and liquidus temperature, fraction of solid, specific heat and more.

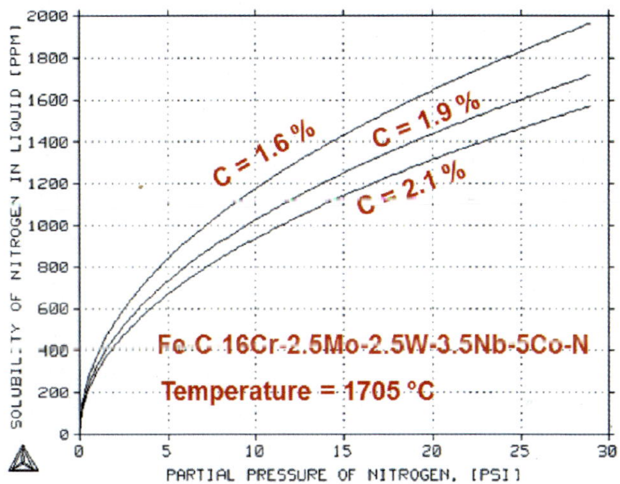

Fig. 3: The influence of Nitrogen (N_2) pressure on the liquid solubility of N in a ferritic stainless steel.

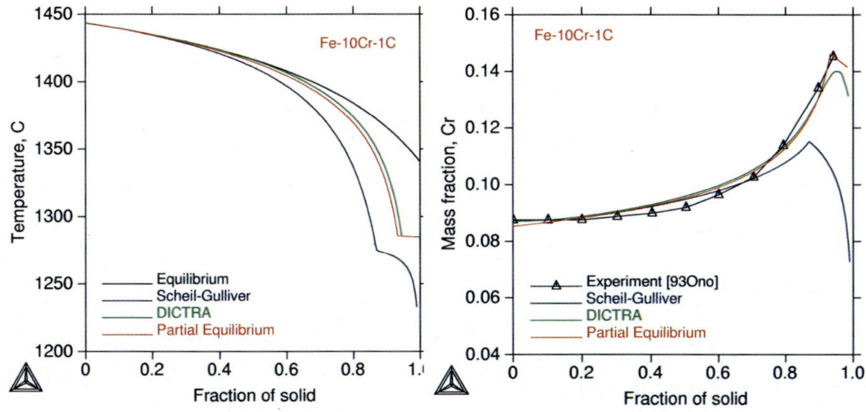

Fig. 4: Solidification curve and calculated micro segregation for Fe – 10Cr – 1C (wt-%)

Forging and rolling are probably the two most important metal forming processes. How a property diagram calculated with Thermo-Calccan give information on rolling is shown in Fig. 5 [8].

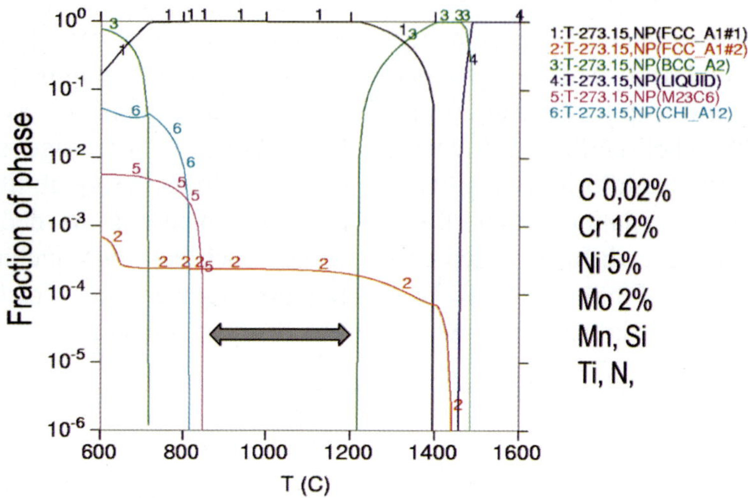

Fig. 5: Safe rolling temperatures for this supermartensitic stainless steel is indicated by the arrow in the diagram. In this safe zone, only austenite and a small amount of cubic carbonitride are stable. Below in temperature M23C6 becomes a stable phase, and above ferrite. Note log scale on y-axis.

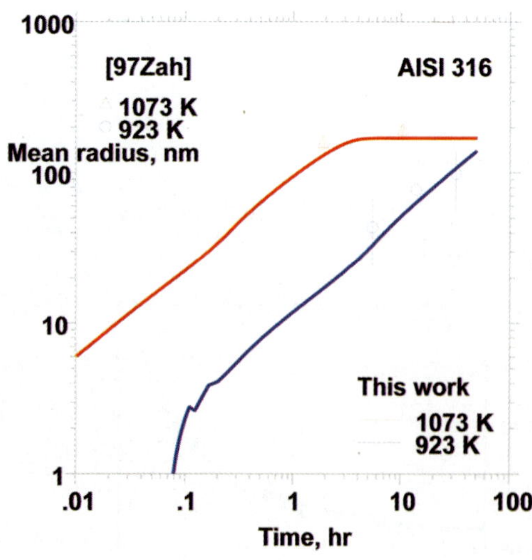

Fig. 6: Growth of average particles of M23C6-carbide in stainless steel with 18 Cr – 12 Ni – 2 Mo – 1.5 Mn – 0.08 C- bal. Fe (wt-%) simulated at two different temperatures.

Heat treatment, for example normalizing, aging, tempering, surface hardening, etc., is usually a thermochemical diffusion process, and thus suitable for handling with kinetic or sometimes thermodynamic modeling.

Fig. 6 shows two different calculations on the precipitation of M23C6 carbide in stainless steel made with TC-PRISMA [9], showing excellent agreement with experimental data.

Surface hardening is simulated in Fig. 7, where a DICTRA calculation [4] of carburization of a stainless steel is shown compared to and agreeing with experimental data [10]. Carburization to such high concentrations of carbon requires the formation of various carbides, in this case M7C3 and M23C6. This is handled by the DICTRA software.

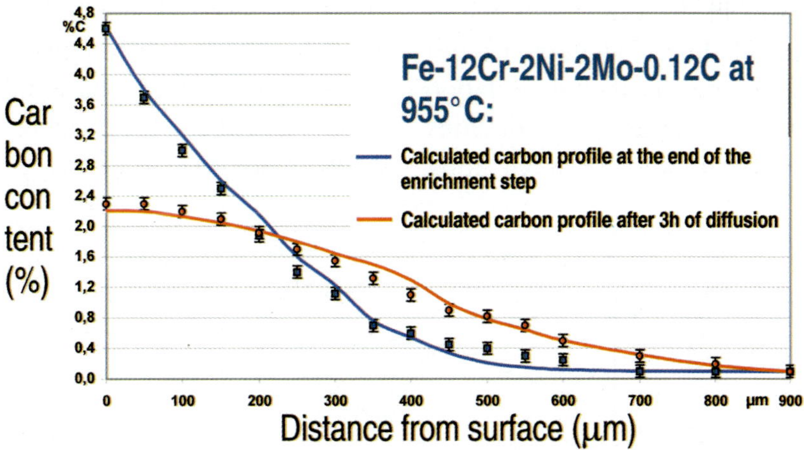

Fig. 7: Simulation of carbon diffusion with formation of carbides in a martensitic high alloyed steel, as performed by French company Aubert & Duval [10].

Other usage related to heat treatment includes calculations for gas phase reactions and oxide scale formation. The latter is shown in Fig. 8 for a Mn-Cr steel. This calculation is thermodynamic using Thermo-Calc which will give composition and conditions of existence of the oxidic (and other) phases.

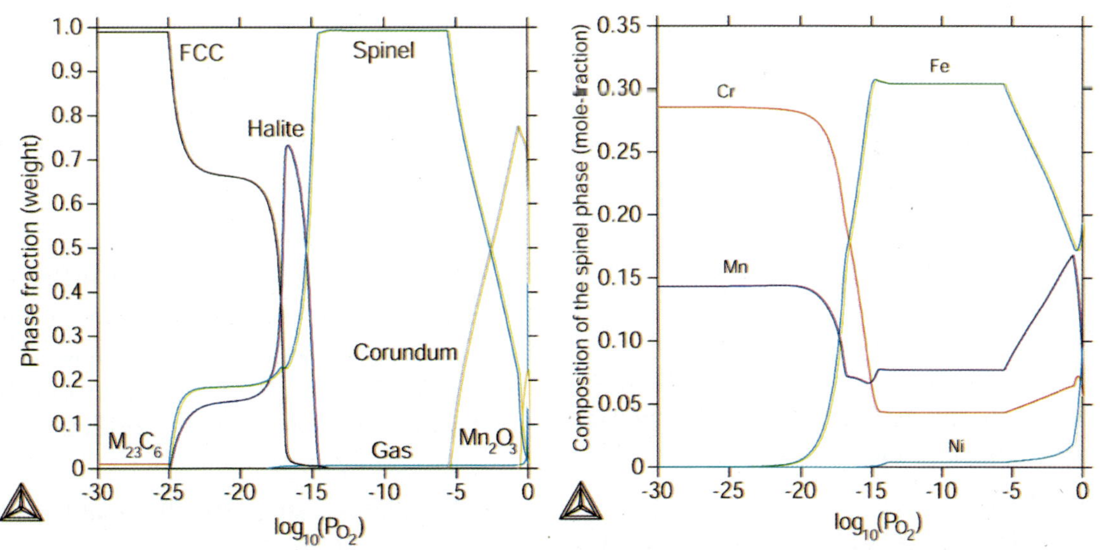

Fig. 8: Oxide formation at 900 °C on a steel with 17.8% Mn, 9.5% Cr, 1.0% Ni and 0.27% C. The right hand diagram shows that below the outer scale (rich in corundum) a spinel phase is stable and should form, and that a halite is stable next to the metallic substrate. The left hand diagram shows that the spinel is Fe-Mn rich close to the corundum (high oxygen potential) and Cr-Mn rich next to the metal (low oxygen potential). This is verified experimentally in the work by Douglas et al [11].

Another segment in the Life Cycle wheel is Joining and Welding. Calculation types pertaining to this category include Relation between flux composition and liquidus temperature, Inter-mixing between alloy and filler material and Influence of alloy additions to welded material micro-structure (or rather phase make-up). An example of the second calculation type is the phase diagram in Fig. 9.

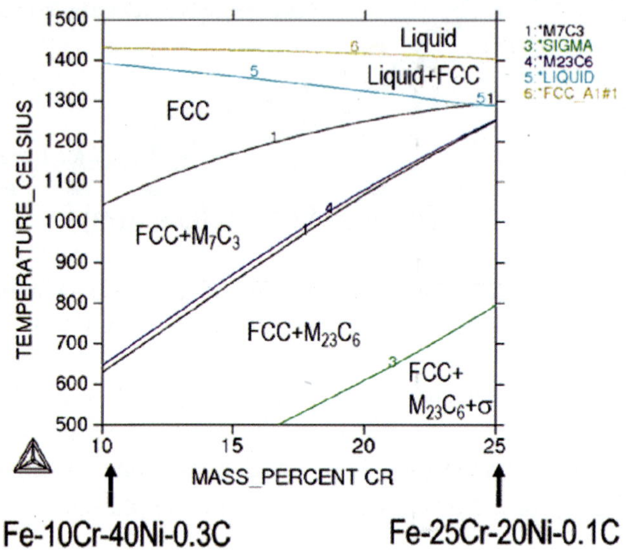

Fig. 9. Calculated phase diagram where the left vertical axis represents the filler material, rich in Ni, and the right vertical axis represents the substrate, a high-alloyed austenitic stainless steel. Moving right to left from axis to axis should cover the weld from pure substrate to pure filler. Different carbides are predicted to form in different regions. Note that the x-axis could as well be given in Mass-percent Ni, but if so it would be scaled from 40 to 20.

BENEFITS OF SIMULATION

The six bullets below are of course not limited to the field of stainless steel, but apply to the greater world of materials development, metallurgy and materials science as well.

- The return on investment has been estimated to \$3-\$9 for every \$1 spent on simulation. (REF61)
- Simulations should give the possibility of accelerated development and implementation of new materials into plant and products.
- Better understanding of the material through simulation will lead to for instance stronger patents and the control of tail effects of compositional variation that can be difficult to determine experimentally.
- Simulation supports the innovative process. Calculations can be much more far ranging and speculative than is possible using incremental development methods. This may lead to break-through ideas.
- Relevant experiments are not always possible to perform, but a simulation

is. One such, current and very important, example is determining the life span of canister material for nuclear waste storage.

- It is important to note that simulation does not replace the need for experiments and testing, but complements it by better targeting of experiments.

REFERENCES

[1] N. Saunders and A.P. Miodownik, "Calphad (Calculations of Phase Diagrams): A Comprehensive guide," Pergamon Materials Series, vol. 1, ed. R. W. Cahn (Oxford, OX: Elsevier Science Ltd, 1998).

[2] National Research Council. *Integrated Computational Materials Engineering: A Transformational Discipline for Improved Competitiveness and National Security.* Washington, DC: The National Academies Press, 2008.

[3] http://www.whitehouse.gov/sites/default/files/microsites/ostp/materials_genome_initiative-final.pdf

[4] Andersson J.O., Helander T., Höglund L., Shi P.F., and Sundman B. Calphad, 26, 2002, pp. 273-312.

[5] Q.Chen and B.Sundman, Met. Transactions A, Vol. 33, 2002, p. 551.

[6] Thermo-Calc Software TCFE7 Steels/Fe alloys database version 7, http://www. thermocalc.com (Accessed 9 Oct. 2012)

[7] Thermo-Calc Software TCNI5 Ni alloys database version 5, http://www.thermocalc.com (Accessed 9 Oct. 2012)

[8] Andre Luis Coste e Silva, UFF, Rio de Janeiro, Brasil, Private communication.

[9] TC-PRISMA User's Guide, http://www.thermocalc.com (Accessed 9 Oct. 2012)

[10] T. Turpin, J.Dulcy and M. Gantois, Met. Transactions A, Vol. 36, 2005, pp. 2751-2760.

[11] D.L. Douglas, F. Gesmundo and C. De Asmundis, Oxidation of Metals, Vol. 25, 1986, pp. 235-268

[12] IDC White paper on Modeling and Simulation: The Return on Investment in Materials Science, July 2004; sponsored by Accelrys Inc.

Introduction of Materials Modelling into Processing Simulation – Towards True Virtual Design and Simulation

Zhanli Guo, Nigel Saunders, A. Peter Miodownik and Jean-Philippe Schillé
Sente Software Ltd., Surrey Technology Centre, Guildford GU2 7YG, U.K.
Email: zhanliguo@gmail.com

ABSTRACT

Computer-aided engineering (CAE) tools are increasingly being used in industry to speed up the design and simulation in metal processing and manufacturing. However such design and simulation at present is not yet completely "virtual", in that the essential material property data required for CAE simulation can only be obtained through experimental measurements. This paper describes the development of material models that can calculate most of the material data required in CAE simulation, including TTT/CCT diagrams, physical and thermophysical properties, and strength and flow stress curves. The calculated material data have been used by various CAE tools, and two case studies are presented for the simulation of hot stamping and quench distortion.

KEYWORDS: *CAE Simulation, Material Data, TTT/CCT Diagrams, Quench Distortion, JMatPro.*

INTRODUCTION

Introduction of materials modelling into CAE simulation has become popular in recent years, whereas the fundamental technical challenge is the development of material models to calculate the physical, thermophysical and mechanical properties that are essential for design, manufacture and simulation. To overcome the problem of lacking material models and provide reliable and cost effective material data for processing simulation, computer-based models have been developed and now most of the material data required for CAE simulation can be readily calculated.[1,2]

This paper reviews our recent development on material modelling and its application in processing simulation. The first part introduces the main physical phenomena involved in the numerical simulation of metal processing, how they interact and how the simulation of each phenomenon is carried out. The material models used to calculate the property data for the above simulations were introduced in the second part. To make them easy to use, these models have been implemented in computer software JMatPro,[1] where the material data calculated can now be organised in formats that can be directly read by many simulation packages. The third and final part features two case studies to demonstrate how JMatPro's data have been used in various CAE packages for the simulation of hot forming and heat treatments.

NUMERICAL SIMULATION

The physical phenomena involved in metal processing as well as their interactions and interdependencies are shown in Fig. 1. An accurate coupling of these phenomena is essential to achieve reliable simulation.

The mathematical treatment of heat transfer (temperature field calculation) and deformation mechanics (stress-strain field computation) are usually dealt with in CAE simulation packages. Nonetheless, their governing equations are described below. CAE packages do not have the capacity of calculating transformation kinetics. Instead they rely on external user inputs typically in the form of TTT or CCT diagrams, the modelling of which is given in Session 3.

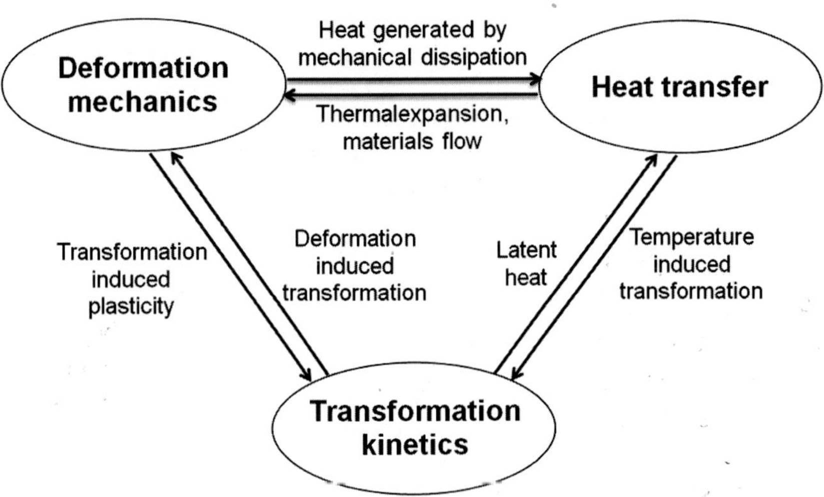

Fig. 1: Interactions between the three main physical phenomena in materials processing.

Heat transfer

The temperature field is governed by the following equation:[3]

$$\rho c\dot{T} = \frac{\partial}{\partial x_i}(k\frac{\partial T}{\partial x_i}) + \sigma_{ij}\dot{\varepsilon}_{ij}^{p} + \sum L_{IJ}\dot{\xi}_{IJ} + \dot{Q} \quad (1)$$

where ρ, c, k and L_{IJ} are density, specific heat, thermal conductivity and latent heat from phase I to phase J, produced by the progressive J^{th} constituent with volume fraction ζ_{IJ}, respectively. Q is the heat generated by external heat sources. $\dot{\xi}_{IJ}$ is the time differentiation of ζ_{IJ} and is termed the transformation rate. The 1st term on the right hand side is from Fourier's law of heat conduction. The 2nd, 3rd, and 4th terms represent the heat due to plastic deformation, the heat absorbed or released during phase transformation, and other heat sources.

Deformation mechanics

The (tensor of) total strain generated during processing can be decomposed into various individual strain contributions as follows, given in the form of strain rate for ready adaptation in CAE analysis:[4]

$$\dot{\varepsilon} = \dot{\varepsilon}^{e} + \dot{\varepsilon}^{p} + \dot{\varepsilon}^{\theta} + \dot{\varepsilon}^{tr} + \dot{\varepsilon}^{tp} \quad (2)$$

where ε^{e}, ε^{p}, ε^{θ}, ε^{tr}, and ε^{tp} are the elastic, plastic, thermal, phase transformation, and transformation plasticity strain contributions, respectively. The calculation of ε^{tr} and ε^{tp} is described below.

Transformations from austenite to ferrite, pearlite, bainite and martensite give rise to additional strain. This additional strain is one of main factors causing local deformation of the steel part and is proportional to the relative difference in the volume fraction of each phase transformed as follows:

$$\dot{\varepsilon}_{ij}^{tr} = \sum \beta_{IJ} \dot{\xi}_{IJ} \delta_{ij} \tag{3}$$

where β_{IJ} is transformation expansion coefficient and it assumes that all of the phase I transforms to phase J instantaneously, and δ_{ij} represents the Kronecker delta.

Transformation plasticity is a deformation that appears in a transforming material under an applied stress even for the stresses lower than the yield stress of the phases. [5] It will be presented in the directions of the deviatoric stress components. The transformation plasticity strain in rate term during transformation from phase I to phase J is expressed as:[4,5]

$$\dot{\varepsilon}_{ij}^{tp} = 3K_{IJ}(1-\xi_{IJ})\dot{\xi}_{IJ}\dot{s}_{ij} \tag{4}$$

where K_{IJ} and s_{ij} represent the transformation plasticity coefficient and the deviatoric stress, respectively.

MODELLING OF MATERIALS PROPERTIES

Simulation of heat transfer requires the knowledge of thermal conductivity and heat capacity, and deformation simulation requires elastic modulus, Poisson's ratio, thermal expansion coefficient, strength and flow stress curves. All of these properties are required as a function of temperature and/or strain rate. Phase transformation simulation requires the knowledge of transformation kinetics, as well as heat evolution and volume change during transformations. This session briefly introduces the material models developed to generate the above property data. An alloy used in later case studies is AISI 4140 (composition in wt%: Fe-0.40C-0.20Si-0.85Mn-0.95Cr-0.20Mo), so the example calculations shown in this session are for this alloy.

TTT/CCT diagrams

Significant work has been undertaken to develop material models that can calculate TTT and CCT diagrams for steels, and now such calculation can be performed for a wide range of steels, including medium to high alloy types, tool steels and stainless steels. [6] The present model for the transformation from austenite to ferrite, pearlite and bainite is based on a previous model proposed by Kirkaldy et al.,[7] and it takes on the following form, which calculates the time (τ) to transform x fraction of austenite at a temperature T,

$$\tau(x,T) = \frac{1}{a\, D\Delta T^{q}} \int_{0}^{x} \frac{dx}{x^{2(1-x)/3}(1-x)^{2x/3}} \tag{5}$$

where $a=\beta 2^{(G-1)/2}$, β is an empirical coefficient, G is the ASTM grain size, D is an effective diffusion coefficient, ΔT is the undercooling, and q is an exponent dependent on the diffusion mechanism. Once the TTT diagram is calculated, the CCT diagram can be obtained using well-established additivity rules.[7]

An accurate description of the martensitic transformation is of great importance because of the large volume

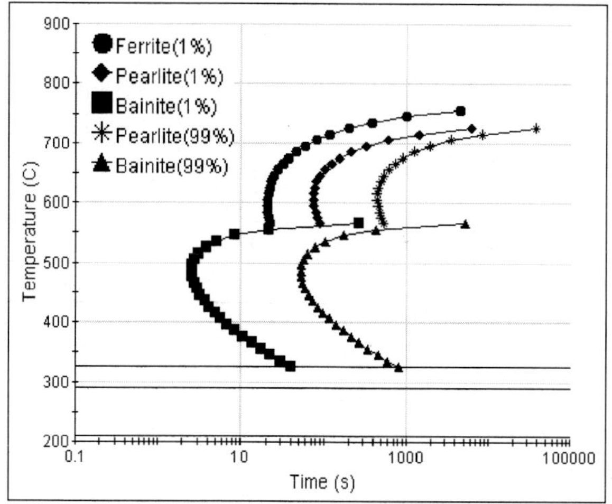

Fig. 2: Calculated TTT diagram of steel 4140.

change associated with this transformation. The amount of martensite, f_M, as a function of the undercooling ΔT below martensite start temperature *Ms* is calculated using the following equation:

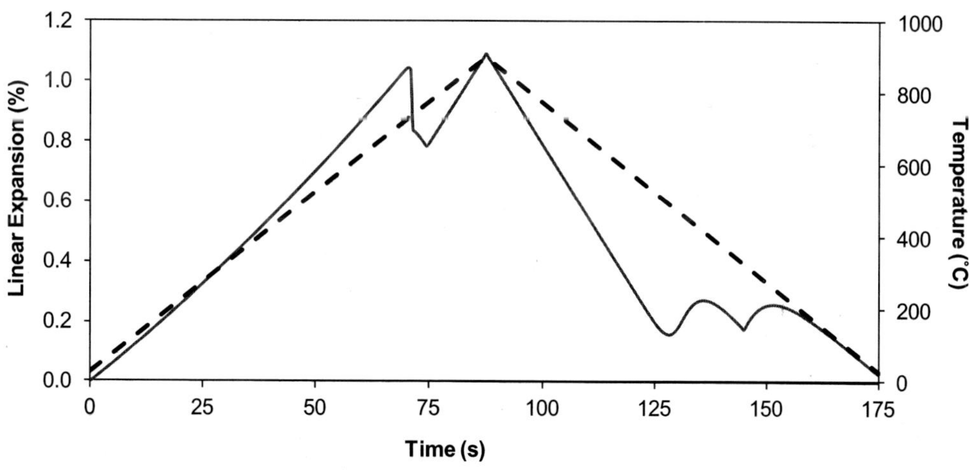

Fig. 3: Linear expansion of alloy 4140 during a heating and cooling cycle, both at 10°C/s.

$$f_M = 1 - \exp(-c \cdot \sqrt{Ms} \cdot \Delta T) \qquad (6)$$

The calculated TTT diagram of AISI 4140 steel is shown in Fig. 2, based on austenite grain size ASTM 9 and austenisation temperature 900°C.

Physical/thermophysical properties

JMatPro's ability to calculate physical and thermophysical properties has been well documented in previous work for various metallic systems [8]. The calculation basically consists of three steps. Firstly, the microstructure is calculated based on thermodynamics and phase transformation kinetics, which gives the amount and constituent of each phase in the alloy. Secondly, for each individual phase, its property is calculated based on its constituent known from the first step. Once the property of each individual phase is defined, finally, the property of the alloy can be calculated via appropriate mixture models.[9,10] The properties that can be modelled include density, molar volume, thermal expansion coefficient, thermal conductivity, Young's modulus, Poisson's ratio, specific heat, and latent heat. Naturally such material properties can be provided for each phase when needed.

Heat treatments consist of heating and cooling and it is important to know the materials properties in such cycles. The linear expansion of alloy 4140 during a thermal cycle is shown in Fig. 3, where both heating rate and cooling rate are set as 10°C/s.

Mechanical properties

The calculation of strength/hardness of an alloy during cooling has been described previously.[11,12] JMatPro is capable of providing the strength of either the overall microstructure or that of each phase. Fig. 4 is the calculated $\sigma_{0.2}$ proof stress as a function of temperature for each phase in AISI 4140 during cooling at 10°C/s.

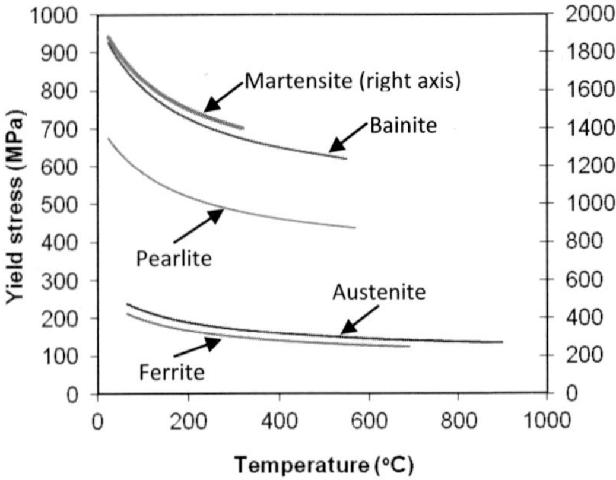

Fig. 4: Proof stress of various phases during cooling at 10°C/s.

The approach for the calculation of flow stress models has been described in Ref.13. The flow stress curves at a fixed strain rate

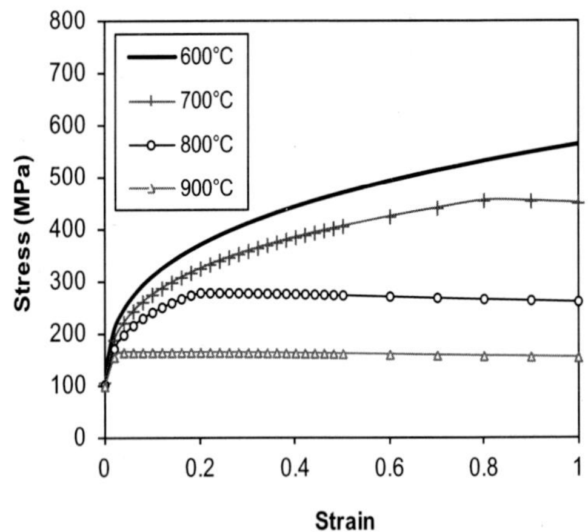

Fig. 5: Flow stress curves of austenite at various temperatures, strain rate 1.0 s^{-1}.

of 1.0 s^{-1} of the austenite phase at various temperatures is shown in Fig. 5, and it can be seen the model calculates three types of curves: work-hardening only (e.g. 600°C), a mixture of work-hardening and flw softening (e.g. 700°C and 800°C), and flow softwening only (e.g. 900°C).

CASE STUDIES – LINKING JMATPRO WITH CAE PACKAGES FOR PROCESS SIMULATION

The material data calculated using JMatPro have been used as inputs to a number of CAE packages for the simulation of casting,[14,15,16] heat treatments,[17,18] and forming.[19] Two case studies are provided here to highlight this application.

Prediction of quench distortion

The standard Navy C-ring test is used in this work to evaluate quench distortion of an AISI 4140 steel. The C-ring was first heated to 900°C in a furnace and held for 1 hour, followed by oil quenching. The calculated material data have been used as inputs to DEFORM™-HT [20] to carry out the heat treatment simulation, and and details of this study can be found in Ref.18.

Fig. 6 shows the simulated martensite distribution on the C-ring surface at different stages of quenching, as well as the dimensional displacement. The specimen was initially expanded when heated to become fully austenitised. After 4 seconds into quenching, martensite starts to form in the region close to the C-ring gap. After 10 seconds, practically all the thermal expansion disappears, but martensite formation is still in progress, which is responsible for the further distortion. As can be seen from the microstructure after 25 seconds of cooling that the martensite formation in the thicker portion of the C-ring is associated with the final opening of the ring gap and the outside diameter of the ring. A comparison between the experimental and predicted values for the gap opening and the outside diameter was given in Table 1. It can be seen that the relative difference for gap opening and outside diameter increase are both smaller thn 7%.

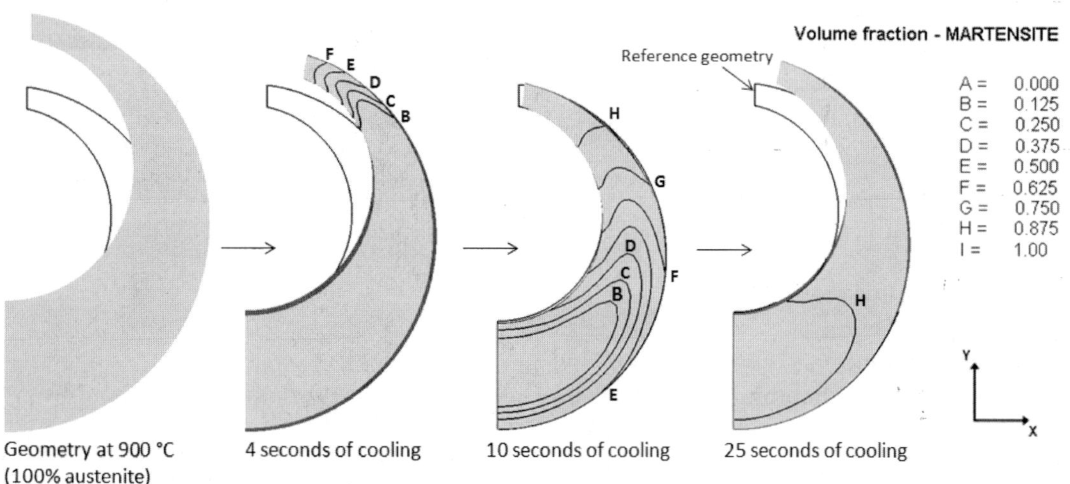

Fig. 6: Martensite formation on the C-ring during the cooling simulation.

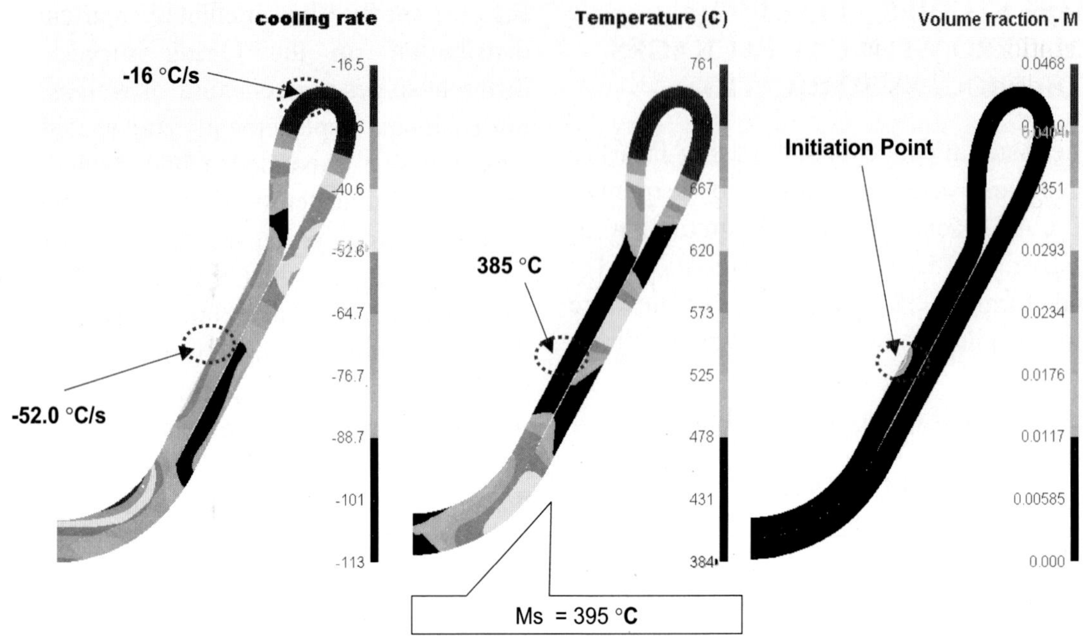

Fig. 7: Temperature field, cooling rate distribution and onset of martensite formation (about 1.6 seconds into quenching).

Table 1: Comparison between experimental (Exp.) and simulated (Sim.) results.

Dimension change	Exp.	Sim.	Dif.
Gap opening (mm)	0.43	0.402	6.5%
Outside diameter (mm)	0.09	0.084	6.7%

Simulation of hot stamping

The hot stamping operation investigated in this work is a V-bending process of an alloy 22MnB5: Fe-0.24C-0.17Cr-1.14Mn-0.27Si-0.036Ti-0.003B in wt%. The CAE package used here is DEFORM™-HT, and details of this study can be found in Ref.19.

The temperature field, cooling rate at each position and volume fraction of martensite formed in the part are shown in Fig. 7, respectively. It corresponds to 1.6 seconds into quenching, i.e. the onset of martensite formation. Fig. 8 compares the experimental springback effect with simulation. The die angle is 60°. In experiment, the final angle changed from 60° to 61.5° due to springback, whereas that change in simulation is from 60° to 62°.

SUMMARY

Reliable simulation of metal processing requires accurate material data, the lack of which has been a common problem, even though there are many simulation packages available. The success in the development of computer software JMatPro for material property modelling has provided a viable solution to this problem and now most of the data required by processing simulation can be calculated. The calculated property data have been used as direct inputs to CAE simulation packages to perform forming and heat treatment simulation software. This has significantly reduced the amount of experimental work required before any simulation attempts, and is a big step forward towards true virtual design and simulation.

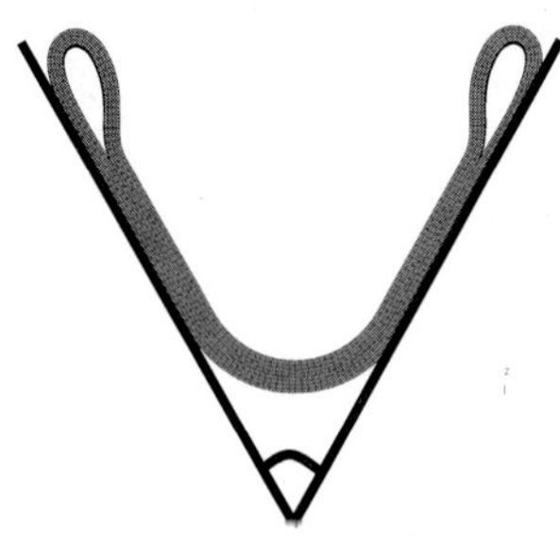

60°→62° from simulation

60°→61.5° from measurement

Fig. 8: Comparison of experimental and simulated final shape.

REFERENCES

[1] http://www.sentesoftware.co.uk, Sente Software Ltd., U.K. 2012.

[2] Z. Guo, N. Saunders, A.P. Miodownik, J.P. Schillé: Mater. Sci. Eng. A, 2009, 499: 7-13.

[3] K. Arimoto et al.: SFTC Paper #347, Scientific Forming Technologies Co.

[4] K. Arimoto et al.: SFTC Paper #370, Scientific Forming Technologies Co.

[5] S. Denis, E. Gautier, A. Simon, G. Beck: Mater. Sci. Technol., 1 (1985) 805-814.

[6] N. Saunders, Z. Guo, X. Li, A.P. Miodownik, J.P. Schillé: The calculation of TTT and CCT diagrams for general steels, Internal report, Sente Software Ltd., U.K., 2004.

[7] J.S. Kirkaldy, D.Venugopolan, in: Phase transformations in ferrous alloys, eds. A.R. Marder and J.I. Goldstein, AIME, (Warrendale, PA: AIME, 1984) 125.

[8] Z. Guo, N. Saunders, A.P. Miodownik, J.P. Schillé: Mater. Sci. Eng. A, 413-414 (2005) 465-469.

[9] Z. Fan, P. Tsakiropoulos, A.P. Miodownik, *J. Mat. Sci.*, 29 (1994) 141.

[10] A.P. Miodownik, N. Saunders and J.P. Schillé, unpublished research.

[11] Z. Guo, N. Saunders, A.P. Miodownik, J.P. Schillé, in: The 2nd International Conference on Heat Treatment and Surface Engineering in Automotive Applications, 20-22 June, 2005, Riva del Garda, Italy.

[12] Z. Guo, N. Saunders, A.P. Miodownik, J.P. Schillé: International Journal of Microstructure and Materials Properties, 4 (2009) 187-195.

[13] Z. Guo, N. Saunders, J.P. Schillé, A.P. Miodownik, in: MRS International Materials Research Conference, 9-12 June, 2008, Chongqing, China.

[14] K.D. Carlson, S. Ou, C. Bechermann, Metall. Mater. Trans. B, 2005, 36B: 843-856.

[15] X. Zhang et al., Israel Journal of Chemistry, 2008, 47 (3-4): 363-368

[16] Z. Guo, N. Saunders, E. Hepp and J.P. Schillé, Modelling of material properties - A viable solution to the lack of material data in casting simulation, in: The 5th Decennial International Conference on Solidification Processing, 23-25 July, 2007, Sheffield, U.K.

[17] J. Fu et al., Advanced Materials Research, 2011, 317-319: 19.

[18] A.D. da Silva et al., Distortion in Quenching an AISI 4140 steel C-Ring – Predictions and Experiments. Accepted for publication in Materials & Design, 2012.

[19] Z. Guo, G. Kang, N. Saunders, Modelling of materials properties used for simulation of hot stamping, in: The 13th International Conference on Metal Forming, 19-22 September 2010, Toyohashi, Japan

[20] DEFORM™-HT version 10.1, Scientific Forming Technology Corporation, Columbus, Ohio, USA.

Property Enhancement in Metastable 301LN Austenitic Stainless Steel through Strain-Induced Martensitic Transformation and its Reversion (SIMTR) for Metro Coach Manufacture

S. Srikanth[1], P. Saravanan[1], Vinod Kumar[1], D. Saravanan[2], L. Sivakumar[2], S. Sisodia[2], K. Ravi[1] and B.K. Jha[1]

[1]R&D Centre for Iron and Steel, Steel Authority of India Limited, Ranchi-834 002, India

[2]Salem Steel Plant, Steel Authority of India Limited, Salem-636 013, India

Email: srikanth@sail-rdcis.com

ABSTRACT

Metastable austenitic stainless steel type 301LN is widely used for fabrication of structural components of Metro Coaches. The steel exhibits both high strength and enhanced plasticity due to strain hardening as well as formation of strain-induced martensite (α') during cold deformation (TRIP effect). The current market requirement, as projected by many of the ongoing Metro Rail Projects in India, calls for manufacture of this steel with ultimate tensile strength (UTS) in excess of 1000 MPa and yield (YS/UTS) ratio of less than 0.8, as this would facilitate substantial reductions in tare weight and crash-worthiness of metro coaches. The typical property requirements in high temper (HT) as per one typical Metro Coach specification are: Yield Strength (YS) ~ 751-921 MPa, Ultimate Tensile Strength (UTS) ~ 1001-1151 MPa, Elongation ~ 22% min, Hardness ~ 36 HRC max and YS/UTS ratio <0.8. Previous plant experience has shown that the maximum attainable UTS in this grade through cold rolling is only to the tune of 970 MPa and any excess deformation severely impairs both ductility (< 22%) and hardness (> 36 HRC) beyond acceptable limits. In the present work, an innovative thermomechanical processing (TMP) methodology has been evolved for the attainment of this seemingly unlikely combination of properties through experimental cold rolling and short annealing simulations in Gleeble 3500 C thermomechanical simulator. The novel process methodology entails imparting heavy cold reductions (CR) of 45-50% in Sendzimir Mill followed by brief/ short reversion annealing treatments (80-160 s) by means of single furnace operation at 750 °C at standard line operating speeds for 300 series in Annealing Pickling Line-1 (AP Line-1) of Salem Steel Plant (SSP) in India. The improvement in properties (strength-ductility combination) has been attributed to grain refinement through formation of submicron grained austenitic (γ) microstructure by controlled reversion of strain-induced martensite (α') during the short annealing treatment. The process is distinct from conventional long annealing treatments (300-360 s), which are employed to soften the steel after cold rolling by means of recovery and recrystallization processes.

KEYWORDS: *Austenitic Stainless Steel, Cold Rolling, Strain-Induced Martensite, Reversion Annealing, Grain Refinement*

INTRODUCTION

Austenitic stainless steels (ASS) are traditionally used in various applications requiring good corrosion resistance and formability. However, their biggest demerit is that the yield strength is quite low, e.g., 230-260 MPa for EN 1.4301 (AISI 304), and 350-380 MPa for EN 1.4318 (AISI 301LN) steel. This limits their use for structural applications. The tensile strength of these steels is in the range of 600 to 800 MPa and total elongation varies between 45 to 60%. With respect to the ratio of strength and elongation, however, stainless steels are located well above the ranges of carbon steels[1,2].

In recent years, there has been a growing interest in developing the "third generation" of advanced high strength steels, including stainless steels, for light weight constructions. These steels possess nano/ultrafine grained microstructures with excellent combinations of strength and ductility[3,4]. Since austenitic stainless steels (ASSs) do not undergo any transformation during hot deformation, the grain refinement may be achievable only through the occurrence of recrystallization in the microstructure[5]. According to literature, the level of grain refining achieved by these processes is very limited (about 5 μm)[6]. Fig.1 depicts the classical "Banana Curve" showing strength-ductility combinations of various advanced high strength steels (AHSS) including typical property range for conventional annealed as well as temper-rolled metastable austenitic stainless steels.

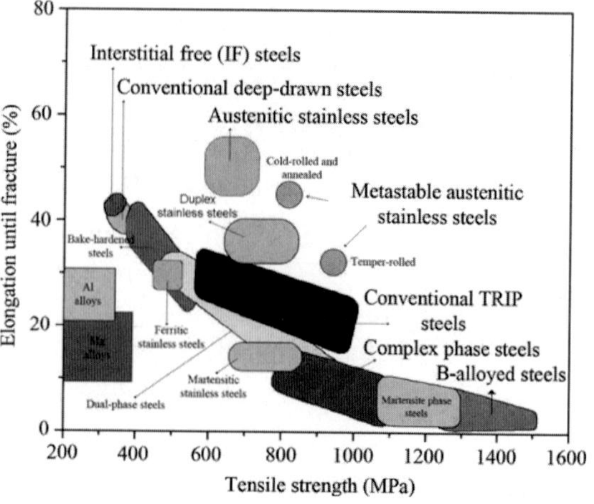

Fig.1: "Banana Curve" showing strength-ductility combinations of various AHSS including austenitic stainless steels[7]

Metastable austenitic stainless steels (ASS) are important grades of the austenitic family in which γ-austenite can be transformed to α′-martensite during cold deformation due to their relatively lower Ni equivalents. Hence, metastable grades exhibit higher tensile strength and better formability than the stable austenitic grades. Metastable ASSs are employed in various structural applications, such as railway and automotive structural components, due to the necessity of weight reduction and crash safety of automobiles and coaches. The martensite formation, resulting from plastic deformation of metastable austenite, is of great interest for producing high strength and ductility combinations as well as TRIP effect in austenitic stainless steels (delay in onset of necking). Such steels exhibit high strength and enhanced plasticity due to both strain hardening (increase in dislocation density, twins, stacking faults, or the Suzuki locking) and formation of strain-induced martensite (α′) during cold deformation (TRIP effect). Substantial strengthening can be obtained by plastic deformation below M_d temperature (M_d is the highest temperature at which the martensitic transformation can be induced by plastic deformation) to produce bcc α' and hcp ε martensite. The transformation of fcc austenite γ to α' and ε martensite depends on the alloy composition, stacking fault energy, degree of deformation, plastic strain, strain rate, grain size, temperature, etc.[8,9].

A novel processing route of controlled reversion annealing of the heavily cold-deformed martensite in metastable Cr-Ni austenitic stainless steels has been employed resulting in highly refined austenite grain size by a number of researchers[2,10-13]. In the first stage, the deformation of austenite in the vicinity of room temperature (RT) leads to strain-induced shear transformation of austenite to martensite, and then upon annealing, this heavily deformed martensite transforms back to fine-grained austenite either through a martensitic shear or diffusional reversion mechanism. An optimized annealing schedule leads to the formation of ultrafine (nano- or submicron) grained structure leading to dramatic improvements in strength-ductility combinations[2,10-13].

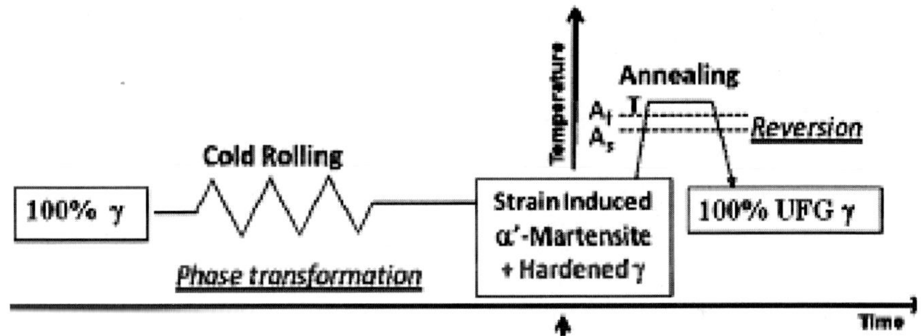

Fig. 2: TMP to obtain UFG austenitic structure through reversion of deformation-induced martensite in metastable ASS

Fig.2 illustrates the thermo-mechanical processing (TMP) used to obtain ultrafine grained (UFG) or nano-grained (NG) austenitic structure through controlled reversion of deformation-induced martensite in metastable ASS.

Amongst the wide variety of austenitic stainless steels (ASS) produced at Salem Steel Plant (SSP) in India, the metastable austenitic stainless steel type 301LN is widely used for manufacture of rail car structural components and automotive trim applications. Currently, there exists a substantial demand for this grade for fabrication of Metro Coaches suited for urban rail transportation. The steel has relatively low yield strength of 350-380 MPa and tensile strength of 600-800 MPa with total elongation of 45-60% in annealed condition. However, the current market requirement, as projected by many of the ongoing Metro Rail Projects in India, calls for the manufacture of this steel with UTS in excess of 1000 MPa and yield (YS to UTS) ratio of less than 0.8 as this would facilitate substantial reductions in tare weight and crash-worthiness of the metro coaches. The cold rolled coils (CRC) of the steel are manufactured in a wide variety of thicknesses and temper designations (namely, high temper (HT), medium temper (MT), standard temper (ST), deadlight temper (DLT) and light temper (LT)) through cold rolling and concurrent annealing cycles. Table 1 gives the nominal chemical composition of 301LN metastable austenitic stainless steel (in wt%).

Table 1: Chemical composition of 301LN metastable austenitic stainless steel (in wt%)

C	Mn	Si	S	P	Cr	Ni	N
0.03 max	1.50-2.00	1.00 max	0.030 max	0.045 max	16.0-18.0	6.0-8.0	0.07-0.20

The cold deformation step is employed to increase the strength/ hardness by virtue of strain hardening (resulting from increased dislocation density, twins, stacking faults, or the Suzuki locking) with concomitant formation of strain-induced martensite (α′) from metastable austenite ($M_{d30/50}$ = 35°C; M_s = –115°C). The steel, thus, exhibits both high strength and enhanced plasticity due to strain hardening as well as formation of strain-induced martensite (α′) during cold deformation (TRIP effect). Normally, annealing cycles are subsequently employed to attain properties in all the tempers except high temper (HT), where the steel is produced in as-cold rolled condition.

The salient combination of properties required in HT temper for coil thickness of 4/ 4.5 mm as per prevalent specification for Metro Coach manufacture are: YS ~ 751-921 MPa, UTS ~ 1001-1151 MPa, Elongation ~ 22% min, Hardness ~ 36 HRC max and YS/UTS ratio <0.8. Practical experience has revealed that the maximum attainable UTS in this grade through cold rolling is only to the tune of 970 MPa and excess deformation severely impairs both ductility (< 22%) as well as hardness (> 36 HRC) beyond stipulated limits for HT temper. Accordingly, the present work was taken up to develop a suitable processing methodology for manufacture of 301LN coils with UTS > 1000 MPa, YS/UTS ratio < 0.8 and maximum hardness of 36 HRC at SSP.

EXPERIMENTAL

Base period study

The operational capability of Sendzimir Mill and Annealing Pickling Line-1 (AP Line-1) were studied to explore the feasibility of employing severe plastic deformation and short duration reversion annealing treatments for effecting grain refinement and property enhancement in 301LN ASS. Cold reductions to the tune of 45-50% are achievable in Sendzimir Mill; the limiting factors for higher reductions being the mill capacity and resulting strip flatness and hardness. For these reasons, thinner gauges of coils facilitate even higher reductions. The viability of short annealing treatments in AP Line-1 was also similarly examined. Fig.3 shows the schematic layout of annealing section of AP Line-1 at SSP with its discrete three-furnace arrangement and strip cooling/ quenching section. It was observed that: (1) the lowest achievable annealing temperature in AP Line-1 was around 730-750 °C, (2) the line speeds could be varied between 6 to 45 mpm (metre per minute) for coil thicknesses ranging from 4.0 to 0.3 mm, and (3) the three furnace arrangement of AP Line-1 proffers the flexibility to operate one, two or all the three furnaces, each furnace being 16 m in length. By restricting the annealing treatment to single furnace operation, short annealing cycles could be effected involving time scales of 80-160 s.

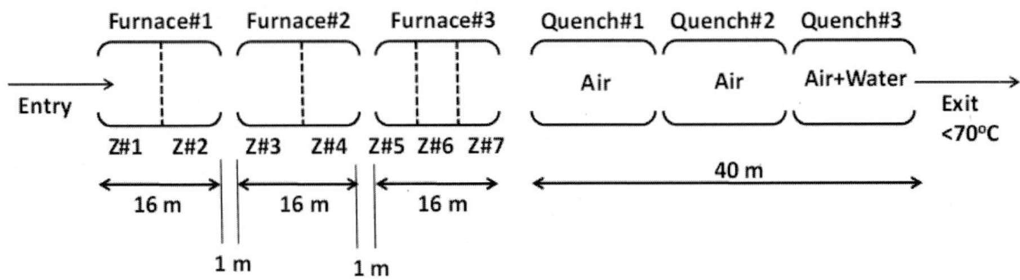

Fig. 3: Schematic layout of annealing section of AP Line-1 at Salem Steel Plant

Material

Commercial 301LN austenitic stainless steel for experimentation was provided by Salem Steel Plant (SSP). Hot rolled strip samples of size 300 mm (length) x 275 mm (width) x 8.5 mm (thickness) were sourced from an solution-annealed and pickled hot rolled coil (HRC) with No.1 finish. The chemical composition of the steel is given in Table 2.

Table 2: Chemical composition of annealed and pickled 301LN HRC sample (in wt%)

C	Si	Mn	P	S	Cr	Mo	Ni	N
0.007	0.36	1.63	0.024	0.009	16.66	0.13	6.80	0.136

Experimental cold rolling

The strips were cold rolled in Hillé, UK laboratory rolling mill by imparting overall cold reductions of 45 and 50% through multipass, accumulative cold rolling in as many as 60-80 passes. Roll gap was progressively set in the mill to achieve effective thickness reductions between 3-8% in each pass. The overall thickness reduction was accomplished using progressively decreasing 10 to 12 roll gap settings. The cold rolled strips were adequately lubricated with cold rolling oil and intermittently cooled under running water to control the adiabatic heating in between passes. During rolling, the roll gap, rolling load and output thickness of the strip were continuously monitored for controlled rolling.

Strip annealing simulation

Specimens of size 190 x 25 mm were sectioned transverse to the rolling direction from cold rolled strips, subjected to 45 and 50% overall cold reductions. The reversion annealing experiments were conducted in Gleeble 3500C thermomechanical simulator by imparting relatively short duration annealing treatments of 80-160 s. Two annealing temperatures viz. 750 and 800 °C were chosen for the strip annealing simulation and a single furnace length of 16 m was assumed. The annealing times were so chosen as to simulate line speeds of 6-12 mpm to comply with the requirement for 4 mm coils concomitant with single furnace operation in AP Line-1. From a practical perspective and to accord the simulation a more pragmatic approach, it was further assumed that the peak annealing temperature is attained just at 16-metre furnace exit.

The uniform temperature zone in the middle of the rectangular strip specimens was estimated to be about 25 mm in length in contrast to that reported by previous researchers[2,12,13], given the fact that thicker specimens exhibit a wider uniformly heated zone at the centre. Post annealing, the strip samples were cooled in an air-mist blow, down to about 50-100 °C at a cooling rate of 200 °C/s. Fig.4 shows a schematic illustrating the annealing cycle employed in Gleeble simulations.

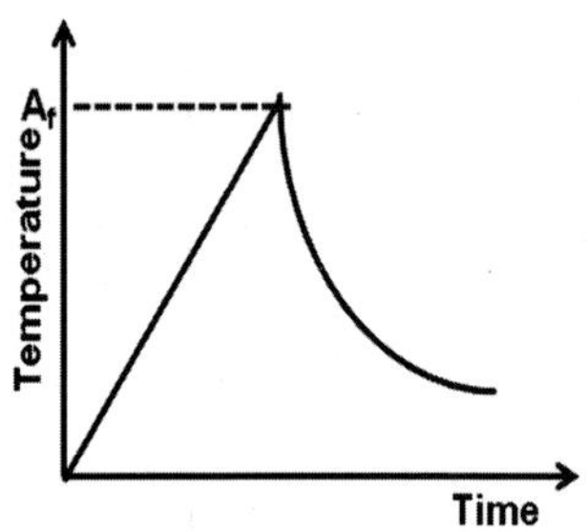

Fig. 4: Schematic illustrating annealing cycle employed in Gleeble simulation

Metallurgical characterization

The mechanical properties of as-received, cold rolled and reversion annealed specimens were essentially determined through tensile testing in 10-ton static tensile testing machine, INSTRON 1195. All tensile specimens were machined and prepared to dimensions according to ASTM A370-07a. While standard sized tensile specimens of 50 mm gauge length were prepared from as-received and cold rolled stainless steel samples, subsized specimens with 25 mm gauge length were machined from steel strips reversion annealed in Gleeble simulator. The subsize specimens were used owing to the limitation of narrow uniformly-annealed zone at the middle in Gleeble tested specimens. All tensile specimens were extracted transverse to the rolling direction except for as-received material, in which case the tensile specimens were machined in both rolling as well as transverse directions, to examine the variation in mechanical properties. Fig.5 depicts a schematic illustrating the location of tensile and metallographic samples in Gleeble test specimen.

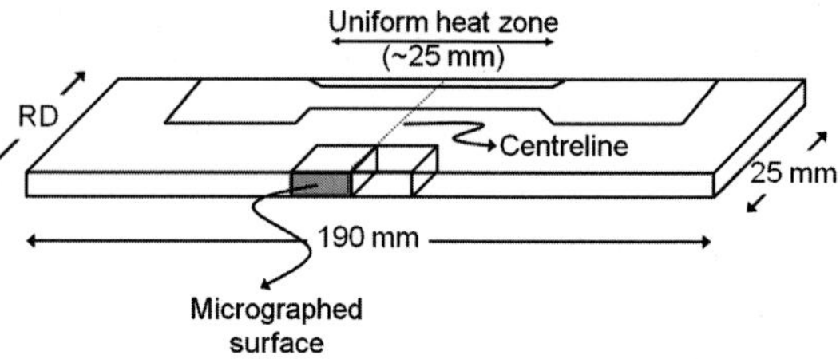

Fig. 5: Schematic illustrating the location of tensile and metallographic samples in Gleeble test specimen

The microstructures of as-received, cold rolled and reversion annealed steels were examined both under Olympus optical microscope and JEOL JSM-840A scanning electron microscope (SEM) for qualitatively resolving and imaging the annealed, cold worked and reverted submicron austenite grain structures. In each case, metallographic specimens were sectioned and prepared carefully using conventional grinding-polishing procedures and electrolytically etched in 10% oxalic acid solution at 6 V for 30 s for microscopic observation. The volume fraction of strain-induced martensite (α′) at each stage was determined using a Helmut Fischer MP3 ferritescope. The ferritescope readings were converted to α′-martensite volume fractions by multiplying with a correction factor of 1.71, as suggested by previous researchers[14]. For further corroboration of results, Rockwell hardness measurements were also effected on as-received, cold rolled and reversion annealed specimens in INSTRON Wolpert hardness tester.

RESULTS AND DISCUSSION

Initial microstructure and properties of as-received material

Table 2 gives the chemical composition of as-received annealed and pickled 301LN hot rolled coil (HRC) sample (in wt%) supplied with No.1 finish. The austenite stability index in terms of $M_{d30/50}$ temperature as proposed by Angel works out to be 35 °C, which is sufficiently above room temperature, and the M_s temperature calculated according to Eichelmann and Hull equation is -115 °C, which is well below the room temperature. The steel fulfills the basic pre-condition for austenitic metastability and grain refinement through "Strain-induced Martensitic Transformation and Reversion to Austenite" (SIMTR) technique, as proposed by Tomimura et al[11].

Fig.6 shows the microstructures of as-received hot rolled, solution-annealed and pickled stainless steel as observed under optical microscope and SEM. The grain size of the annealed microstructure was considerably coarse and varied over a wide range of 10-40 μm. Table 3 gives the corresponding tensile properties, hardness and α′-martensite content of the as-received material. The steel did not exhibit any marked variation of tensile properties in longitudinal and transverse directions. The yield strength of annealed stainless steel is quite low at 331-352 MPa, from the standpoint of engineering requirement of 751-921 MPa for a Metro Coach. The tensile strength is in the range of 812-831 MPa, also well below the requirement for HT temper viz. 1001-1151 MPa. However, the overall elongation to fracture and the strain hardening characteristics of the steel are quite superior, well corroborated by elongation and strain hardening exponent (n) values of 62-64% and 0.62-0.64, respectively.

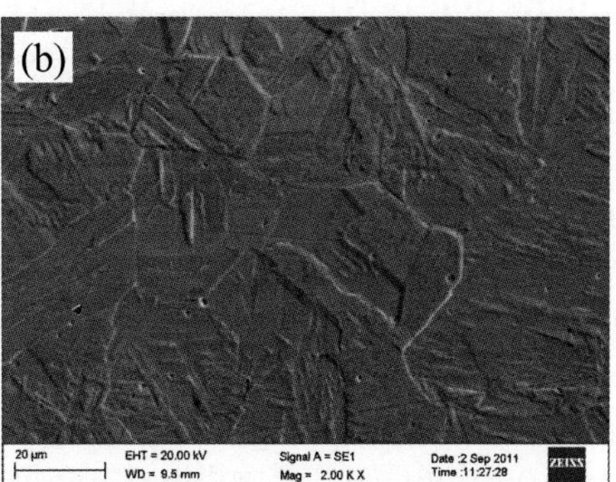

Fig. 6: Microstructures of as-received hot rolled, solution-annealed and pickled 301LN stainless steel: (a) Optical image (b) Secondary electron image

Table 3: Tensile properties and hardness of as-received solution-annealed and pickled 301LN HRC samples

Direction	YS (MPa)	UTS (MPa)	YS/UTS ratio	% El	Strain hardening exponent (n)	Hardness	α′-Martensite content (%)
Longitudinal	332.06	811.77	0.41	63.79	0.62	96.2 HRB	1.6
Longitudinal	330.69	819.33	0.40	61.74	0.64		
Transverse	349.03	831.10	0.42	61.80	0.64		
Transverse	351.78	831.00	0.42	63.05	0.64		

The steel was found to exhibit an increase in strain hardening exponent during the course of tensile straining signifying the transformation-induced plasticity (TRIP) effect associated with strain-induced martensite (α′) formation. This property renders the steel capable of absorbing higher amounts of plastic energy before the onset of necking, a property of vital importance for crash worthiness of Metro Coach structural members. Incidentally, the annealed steel exhibited hardness of 96 HRB and low α′-martensite content of 1.6%.

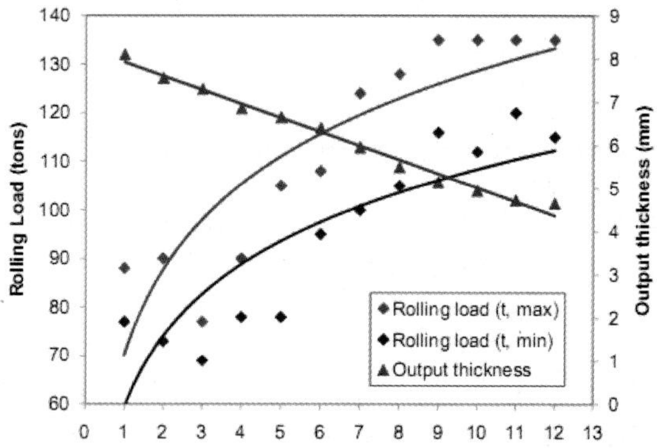

Fig. 7: Variation in rolling load with passes (Overall cold reduction: 45%)

Findings from experimental cold rolling

Fig.7 graphically depicts the variation in rolling load with passes for an overall cold reduction (CR) of 45%. It is evident that there is an abrupt increase in rolling load after about 22% cold reduction, which presumably signifies the onset of formation of strain-induced martensite (SIM) in the steel. The formation of SIM saturates at around 39% cold reduction and at a saturating strain (ε_s) of about 0.39 with present chemistry and grain size of steel and for present set of rolling conditions (temperature, strain and strain rate).

Since the cold deformation was continued beyond the saturating strain (ε_s) by about 5 and 10% for 45 and 50% reductions respectively, it is safe to presume that the martensite was effectively conditioned and heavily dislocated through crushing and fragmentation during cold rolling. The formation of heavily-dislocated martensite is one of the essential precursors for accomplishing effective and successful reversions of SIM to ultrafine grained austenite for enhancement in mechanical properties in 301LN austenitic stainless steel[3,12,15].

Metastable austenitic stainless steels such as 301LN contain thermodynamically metastable austenite at room temperature, which can be transformed to α′-martensite (strain-induced martensite, SIM) by deformation below the M_{d30} temperature. The volume fraction of SIM increases with increasing strain. The martensite formation becomes saturated at a specific strain called "saturating strain" (ε_s). The martensite crushes subsequently during deformation, increasing the concentration of lattice defects (dislocations, dislocation cells and stacking fault bundles) inside the SIM[3,12,15].

Microstructure and properties after cold rolling

Fig. 8 depicts the microstructural evolution in 301LN austenitic stainless steel with 45 and 50% cold rolling of the solution-annealed hot rolled steel. The light optical micrographs and the secondary electron images of the cold reduced steels revealed pronounced fibering/ texturing in the microstructure, resulting from heavy cold rolling of the annealed stainless steel.

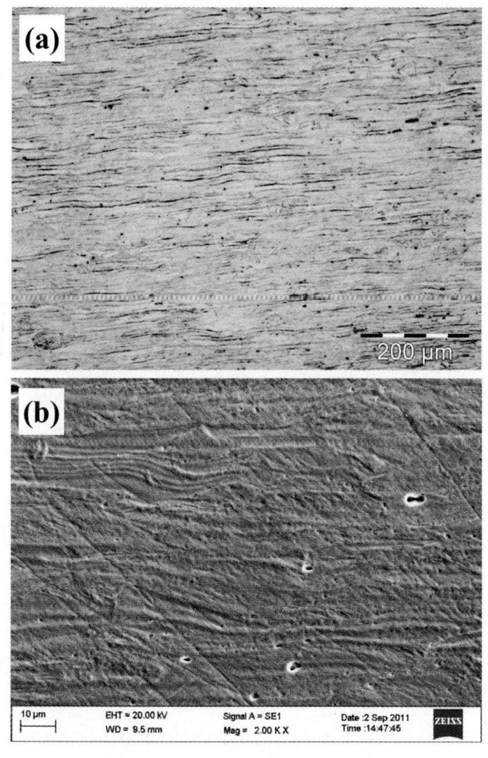

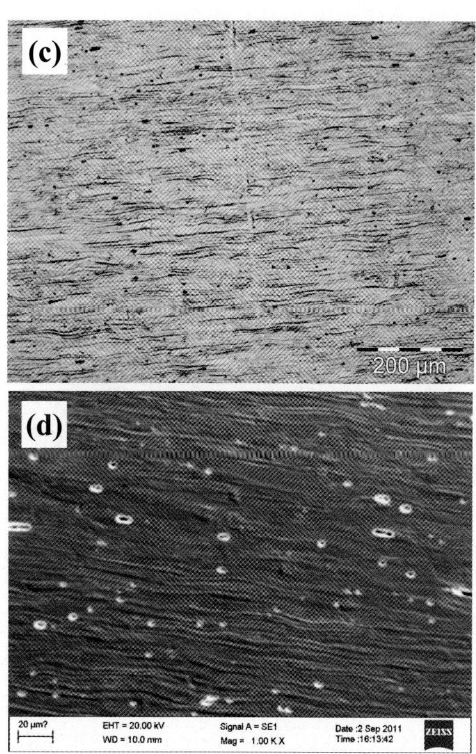

Fig. 8: Microstructures of cold rolled 301LN stainless steel: (a) Optical image after 45% cold reduction (b) Secondary electron image after 45% cold reduction (c) Optical image after 50% cold reduction (d) Secondary electron image after 50% cold reduction

Table 4 gives the tensile properties of 301LN coil samples after 45 and 50% cold reductions performed in Hillé experimental cold rolling mill. The steel exhibited an yield strength (YS) of 1260-1324 MPa and an ultimate tensile strength (UTS) of 1476-1488 MPa after 45% cold reduction. In comparison, the 50% cold reduced steel revealed YS of 1180-1201 MPa and UTS of 1548-1557 MPa.

Table 4: Tensile properties of 301LN coil samples after cold rolling in Hillé experimental cold rolling mill

% Cold redn	Direction	YS (MPa)	UTS (MPa)	YS/UTS ratio	%El	Strain hardening exponent (n)	Hardness (HRC)	α′-Martensite content (%)
45	Transverse	1259.60	1488.17	0.85	6.30	0.25	46.4	98.4
	Transverse	1324.35	1476.40	0.90	4.89	0.22		
50	Transverse	1180.14	1556.84	0.76	5.65	0.30	46.5	~100
	Transverse	1200.74	1548.01	0.78	3.66	0.43		

It appears from the results that with a 5% increase in cold reduction, the yield strength is marginally lowered while the tensile strength is somewhat improved. Understandably, both steels revealed very inferior elongations of 3.7 to 6.3%, very low strain hardening exponents ranging between 0.22-0.43 and high hardness values of ~46 HRC due to excessive strain hardening and strain-induced martensitic transformation in the steel as a result of very heavy cold reductions. The overall elongations of the cold rolled steels are well below the stipulated requirement of 22% minimum in HT temper for Metro Coach application. Likewise, the hardness values are also well beyond the acceptable limit of 36 HRC maximum as stipulated by specification for HT temper. The cold rolled steels exhibited near complete transformation of austenite to α′-martensite content of 98 and ~100%, respectively, after 45 and 50% cold reductions. One of the other precursors for grain refinement by Strain-Induced Martensitic Transformation and its Reversion to austenite (SIMTR) is the near complete SIM transformation of the austenite during cold rolling[11,16].

The volume fraction of martensite and ε_s markedly influence the achievement of nano/ ultrafine structure in Strain-Induced Martensitic Transformation and its Reversion to austenite (SIMTR). These two parameters also depend on other process variables including deformation conditions (strain, strain rate, strain path, rolling temperature), austenite stability (composition and initial grain size) and primary dislocation density[17-21].

It is imperative that the metastable austenite be almost completely transformed to α'-martensite during cold deformation for effective reversion of martensite to austenite with marked grain refinement. Therefore, processing parameters in cold rolling must be carefully chosen and controlled in order to obtain more than 90% α'-martensite from cold deformation[22].

As mentioned, the formation and the amount of strain-induced martensite depends on the austenite stability (chemical composition and initial austenite grain size) and the cold deformation conditions (the deformation temperature, strain and strain rate). The strain-induced martensitic transformation is suppressed with increasing deformation temperature. This has been attributed to an increase in the stacking fault energy (SFE) and decrease in the available chemical driving force for the transformation. It has also been mentioned that the SIM transformation is suppressed with increasing strain rate. This latter effect was primarily attributed to a decrease in the chemical driving force of the transformation due to adiabatic heating[1]. The martensite content is increased when the amount of deformation is high or the deformation temperature is low. Moreover, by increasing the ASTM austenite grain size number, the amount of strain-induced martensite is reduced[22]. Thus, a coarse grain size would be preferable from view point of strain-induced martensitic transformation and its reversion (SIMTR).

Microstructure and properties after short reversion annealing treatments

Table 5 summarizes the properties achieved in 301LN austenitic stainless steel samples through experimental cold rolling and short reversion annealing simulations in Gleeble 3500 C thermo-mechanical simulator.

Table 5: Properties achieved in 301LN ASS strips through experimental cold rolling and short annealing simulations in Gleeble 3500 C thermo-mechanical simulator (based on single furnace length of 16 m in AP Line-1)

Steel	Peak temp (°C)	Heat rate (°C/s)	Anneal time (s)	Simulated line speed (mpm)	YS (MPa)	UTS (MPa)	% El	Strain hardening exponent (n)	YS/ UTS ratio	Hardness (HRC)	α′-Martensite content (%)
301LN (45% CR)	800	5	160	6	681.99	999.64	46.32	0.3036	0.68	34.5	3.7
		5.83	137	7	679.24	1037.90	41.14	0.2885	0.65	33.8	2.8
		6.67	120	8	746.74	1058.50	38.73	0.2604	0.71	31.0	2.7
		7.5	107	9	661.68	1017.30	39.14	0.2763	0.65	28.4	2.0
		8.33	96	10	679.34	1030.05	41.44	0.3113	0.66	31.9	2.1
		9.17	87	11	617.15	1012.39	46.47	0.3140	0.61	30.5	2.4
		10	80	12	644.22	1017.30	45.57	0.3430	0.63	32.4	2.4
	750	4.69	160	6	809.13	1125.21	36.81	0.2274	0.72	36.0	8.9
		5.47	137	7	779.40	1122.26	36.98	0.2285	0.69	30.7	8.1
		6.25	120	8	736.73	1115.40	39.14	0.2146	0.66	35.5	7.0
		7.03	107	9	785.29	1088.91	34.75	0.2145	0.72	36.5	6.5
		7.81	96	10	783.13	1099.70	36.69	0.2192	0.71	33.8	5.6
		8.59	87	11	636.57	1075.18	39.00	0.2266	0.59	36.6	5.9
		9.38	80	12	748.11	1068.31	NA	NA	0.70	32.8	6.0

301LN (50% CR)	800	5	160	6	798.44	1056.54	37.57	0.2048	0.76	32.4	3.5
		5.83	137	7	712.40	1033.97	43.10	0.2978	0.69	29.9	2.3
		6.67	120	8	696.90	1034.96	42.56	0.3059	0.67	29.3	2.4
		7.5	107	9	772.64	1023.18	41.10	0.2194	0.76	34.2	2.7
		8.33	96	10	646.48	1029.07	42.36	0.3019	0.63	29.5	2.4
		9.17	87	11	685.03	1013.37	38.92	0.2793	0.68	36.6	6.9
		10	80	12	644.52	1035.94	41.14	0.2652	0.62	33.0	2.4
	750	4.69	160	6	778.42	1056.54	32.40	0.2111	0.74	37.4	9.0
		5.47	137	7	770.18	1090.87	32.95	NA	0.71	37.9	8.4
		6.25	120	8	695.73	1096.76	32.67	0.1845	0.63	34.1	4.3
		7.03	107	9	681.60	1063.40	34.94	0.2153	0.64	36.7	4.7
		7.81	96	10	842.38	1100.68	35.08	0.2223	0.77	30.7	7.1
		8.59	87	11	768.61	1050.65	36.22	0.2377	0.73	36.8	6.2
		9.38	80	12	708.58	1041.82	35.39	0.2180	0.68	30.0	2.2

The following observations can be made from the table:

1. The 301LN ASS exhibits YS of 617-777 MPa, UTS of 1000-1197 MPa, overall elongation of 39-47%, YS/UTS ratio of 0.61-0.71 and hardness of 28-35 HRC for starting cold reduction of 45%, annealing temperature of 800 °C and simulated line speeds of 6-12 mpm assuming the furnace length of 16 m in Gleeble simulation.
2. The steel shows YS of 637-809 MPa, UTS of 1068-1125 MPa, overall elongation of 35-39%, YS/UTS ratio of 0.59-0.72 and hardness of 31-37 HRC for cold reduction of 45%, annealing temperature of 750 °C and simulated line speeds of 6-12 mpm..
3. The steel reveals YS of 645-798 MPa, UTS of 1013-1057 MPa, overall elongation of 38-43%, YS/UTS ratio of 0.62-0.76 and hardness of 29-37 HRC for cold reduction of 50%, annealing temperature of 800 °C and simulated line speeds of 6-12 mpm.
4. The steel displays YS of 682-842 MPa, UTS of 1042-1101 MPa, overall elongation of 32-36%, YS/UTS ratio of 0.63-0.77 and hardness of 30-38 HRC for cold reduction of 50%, annealing temperature of 750 °C and simulated line speeds of 6-12 mpm.

Fig. 9 shows light optical micrographs of some of the Gleeble tested 301LN specimens including the micrograph of the as-received, solution-annealed and pickled 301LN stainless steel for comparison. The micrographs of the Gleeble tested 301LN specimens reveal composite microstructures comprising very fine, partially recrystallized austenite grains generated through partial reversion of strain-induced martensite (α') formed during cold rolling. The occurrence of fibering in the microstructures is presumably the remnant textural feature of prior cold rolling and is indicative of partial reversion of strain-induced martensite (α′) that is attributable to the short annealing treatment conferred to the steel at relatively lower annealing temperature. The presence of untransformed strain-induced martensite (α′) was subsequently corroborated through ferritescope measurements. Table 5 also gives the measured α′-martensite content in the reversion-annealed steels, which was found to vary in the range of 2-9%.

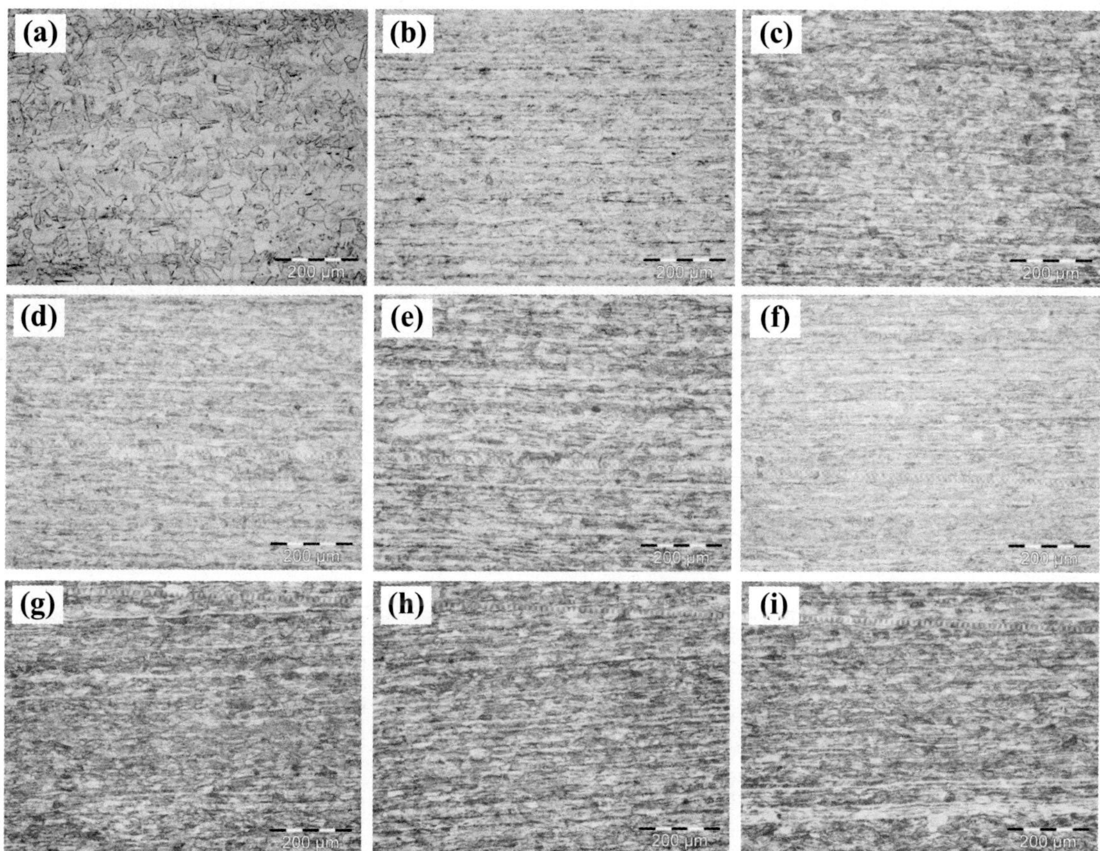

Fig. 9: Light optical micrographs of Gleeble tested 301LN ASS specimens showing refinement in reverted austenite grain size with thermo mechanical treatment: (a) Solution annealed (b) 45% CR/ 800 °C/ 6 mpm (c) 45% CR/ 750 °C/ 6 mpm (d) 45% CR/ 800 °C/ 9 mpm (e) 45% CR/ 750 °C/ 9 mpm (f) 50% CR/ 800 °C/ 7 mpm (g) 50% CR/ 750 °C/ 7 mpm (h) 50% CR/ 800 °C/ 11 mpm (i) 50% CR/ 750 °C/ 11 mpm

Fig.10 depicts the secondary electron images (SEIs) of some of the Gleeble tested 301LN specimens. The electron micrographs clearly reveal the nucleation of sub-micron sized austenite (γ) grains formed through controlled reversion of strain-induced martensite (α′) by means of thermo-mechanical processing (TMP) comprising 45 and 50% cold reductions followed by short reversion annealing treatment at 750 and 800°C at simulated line speeds of 6-12 mpm (annealing time: 160-80 s) assuming single furnace operation (Furnace length: 16 m) in AP Line-1 of SSP. The SEI of as-received, solution-annealed and pickled 301LN stainless steel is also included for comparison.

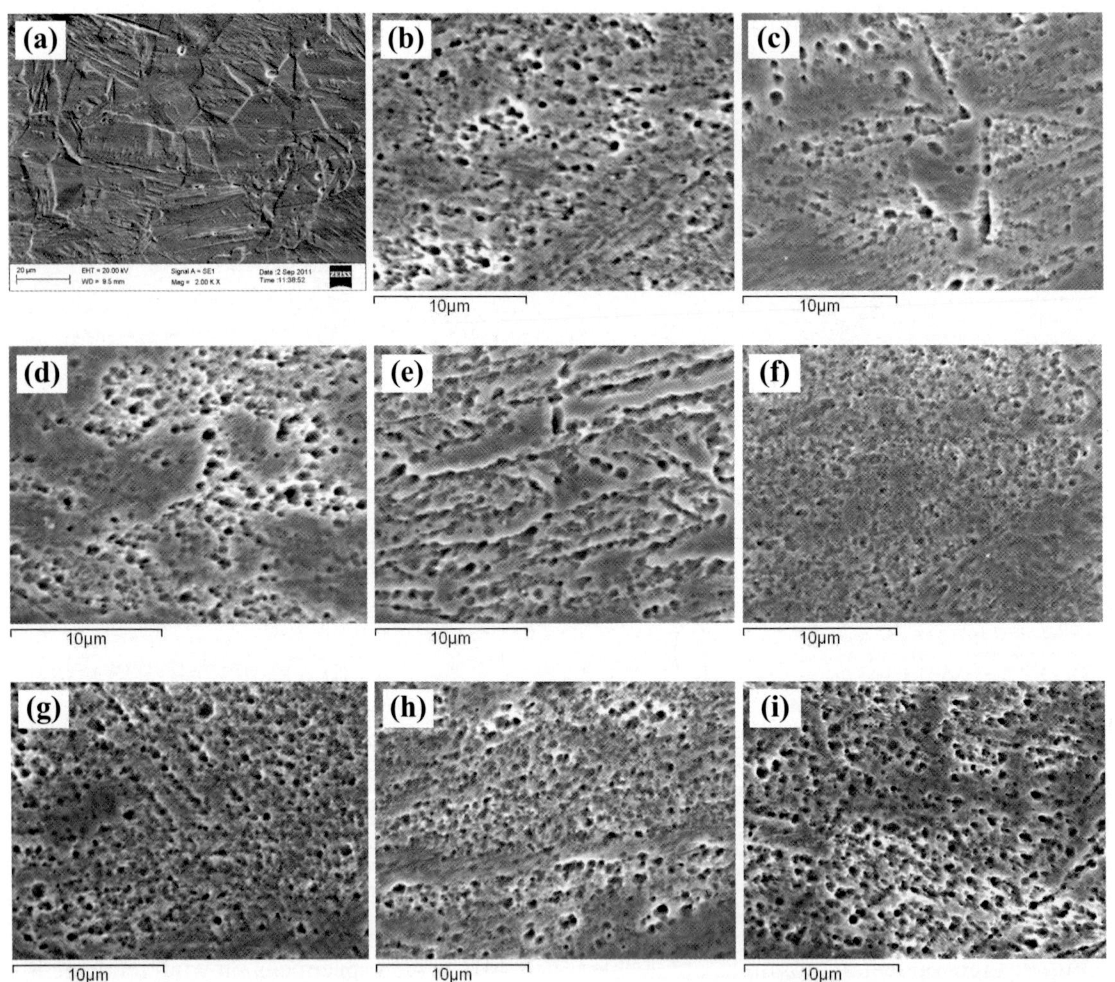

Fig. 10: Secondary electron images of Gleeble tested 301LN ASS specimens showing sub-micron sized austenite (γ) grains obtained through controlled reversion of strain-induced martensite (α′) by means of thermo-mechanical processing (TMP): (a) Solution annealed (b) 45% CR/ 800 °C/ 6 mpm (c) 45% CR/ 750 °C/ 6 mpm (d) 45% CR/ 800 °C/ 9 mpm (e) 45% CR/ 750 °C/ 9 mpm (f) 50% CR/ 800 °C/ 7 mpm (g) 50% CR/ 750 °C/ 7 mpm (h) 50% CR/ 800 °C/ 11 mpm (i) 50% CR/ 750 °C/ 11 mpm

Fig.11 graphically depicts the variation in hardness of 301LN samples after experimental cold rolling and reversion annealing treatments in Gleeble 3500C. Most of the measured hardness values lie well within the stipulated limit of 36 HRC for Metro Coach specification in HT temper, underlining the reproducibility of the thermomechanical processing (TMP) methodology. Fig.12 graphically presents the strain-induced martensite (α') contents in as-received No.1 finish, laboratory cold rolled and reversion annealed 301LN samples determined through Ferritescope measurements. The graph illustrates the near complete transformation of austenite (γ) to strain-induced α'-martensite with 45% cold reduction and the rapid kinetics of near complete reversion back to austenite (γ) at 750 and 800 °C at simulated line speeds of 6 and 9 mpm (Annealing time: 160 and 107 s) assuming single furnace operation (Furnace length: 16 m) in AP Line-1 (Annealed value refers to the α'-martensite content in as-received, solution-annealed and pickled 301LN HRC sample with No.1 finish).

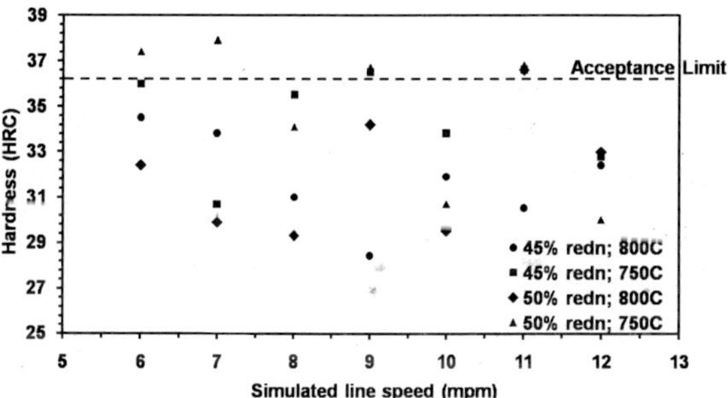

Fig. 11: Variation in hardness of 301LN ASS samples after experimental cold rolling and reversion annealing treatments in Gleeble 3500C

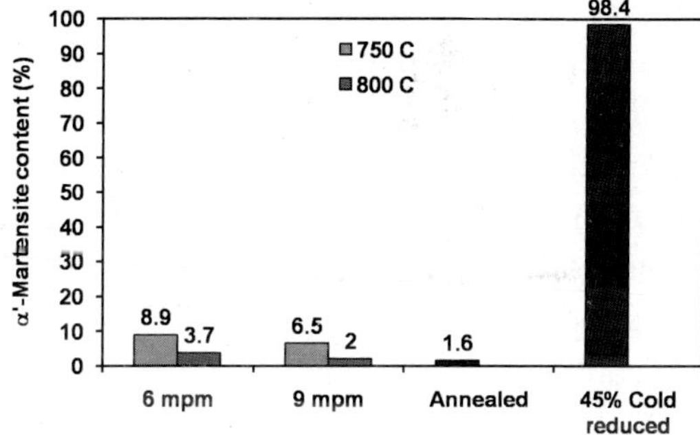

Fig. 12: Strain-induced martensite (α') content in as-received No.1 finish, laboratory cold rolled and reversion annealed 301LN ASS samples determined using Ferritescope

The results from the annealing simulations clearly show that the seemingly unlikely combination of properties stipulated by Metro Coach specification in HT temper viz. Yield Strength: 751-921 MPa, Ultimate Tensile Strength: 1001-1151 MPa, Elongation: 22% min, Hardness: 36 HRC max and Yield (YS/UTS) Ratio: <0.8 can be achieved in 301LN grade austenitic stainless steel by imparting heavy cold reductions to the tune of 45-50% in Sendzimir Mill followed by brief/ short annealing treatments at 750 °C using single furnace (F/c #3) operation (16 m furnace length) at normal line speeds followed as per standard operating practice (SOP) for 300 series in Annealing Pickling Line-1 at SSP.

The thermomechanical process (TMP) methodology involving heavy cold rolling and short annealing treatment serves to revert the deformation/ strain-induced martensite (α'), formed during heavy cold rolling/ deformation, to strain-free, sub-micron sized austenite grains, which contribute to improvement in mechanical properties through grain refinement/ grain boundary strengthening mechanism. The strength, ductility and toughness of such ultra fine-grained austenitic microstructure is undoubtedly superior to that achieved through deformation-induced martensitic strengthening of metastable austenite accomplished through mere cold rolling.

CONCLUSIONS

Combination of properties viz. YS: 751-921 MPa, UTS: 1001-1151 MPa, Elongation: 22% min, Hardness: 36 HRC max and Yield ratio: <0.8 can be attained in 301LN grade coils by imparting heavy cold reductions to the tune of 45-50% in Sendzimir Mill followed by brief / short annealing treatments (80-160 s) by means of single furnace operation at 750 °C at standard line operating speeds stipulated for 300 series in Annealing Pickling Line-1 (AP Line-1) at SSP.

Short annealing treatment would serve to revert strain-induced martensite (α'), formed during heavy cold rolling, to strain-free, sub-micron sized austenite grains, which contribute to improvement in mechanical properties (strength-ductility combination) through grain refinement/ grain boundary strengthening mechanism.

Strength, ductility and toughness of ultra fine-grained austenitic microstructure are superior to that achieved through deformation-induced martensitic strengthening of metastable austenite.

The acceptable combination of properties in HT temper, especially those stipulated by Metro Coach specification, necessitate the achievement of composite microstructure comprising ultrafine grained austenite and some amount of untransformed strain-induced martensite (SIM less than about 10 vol%) through partial reversion.

The process is distinct from conventional long annealing treatments (300-360 s), which are conventionally employed to soften the steel after cold rolling by means of recovery and recrystallization processes. Thus, conventional annealing softens the ASS whereas reversion annealing serves to strengthen steel through grain refinement.

ACKNOWLEDGEMENT

The authors wish to acknowledge the whole-hearted support and encouragement of managements of R&D Centre for Iron and Steel and Salem Steel Plant of Steel Authority of India Limited for pursuing and publishing this work.

REFERENCES

[1] L.P. Karjalainen, T. Taulavuori, M. Sellman and A. Kyröläinen: Steel Res. Int., 79 (2008), No. 6, pp. 404-412.

[2] M.C. Somani, P. Juntunen, L.P. Karjalainen, R.D.K. Misra and A. Kyröläinen: Metall. Mater. Trans. A, Vol. 40a, March 2009, pp. 729-744.

[3] M. Eskandari, A. Najafizadeh, A. Kermanpur and M. Karimi: Mater. Des. 30 (2009), pp. 3869-3872.

[4] A. Rezaee, A. Najafizadeh, A. Kermanpur and M. Moallemi: Mater. Des., 32 (2011), pp. 4437-4442.

[5] T. Maki: Curr. Opin. Solid State Mater. Sci., 2 (1997), pp. 290-295.

[6] A.D. Manshadi, M.R. Barnett and P.D. Hodgson: Mater. Sci. Eng. A, 485 (2008), pp. 664-762.

[7] S. Rajasekhara, L. P. Karjalainen, A. Kyröläinen and P. J. Ferreira: Advanced Steels, Eds. Y. Weng, H. Dong and Y. Gan, Metallurgical Industry Press, Springer, p.372.

[8] M. Moallemi, A. Najafizadeh, A. Kermanpur and A. Rezaee: Mater. Sci. Eng. A, 530 (2011), pp. 378-381.

[9] A.A. Lebedev and V.V. Kosarchuk: Int. J. of Plast., Vol. 16 (2000), No. 12, pp. 749-767.

[10] K. Tomimura, S. Takaki, and Y. Tokunaga: ISIJ Int., Vol. 31 (1991), No. 12, pp. 1431-1437.

[11] K. Tomimura, S. Takaki, S. Tanimoto, and Y. Tokunaga: ISIJ Int., Vol. 31 (1991), No. 7, pp. 721-727.

[12] R.D.K. Misra, S. Nayak, S.A. Mali, J.S. Shah, M.C. Somani and L.P. Karjalainen: Metall. Mater. Trans. A, Vol. 41a, Jan 2010, pp. 3-12.

[13] R.D.K. Misra, S. Nayak, S.A. Mali, J.S. Shah, M.C. Somani and L.P. Karjalainen: Metall. Mater. Trans. A, Vol. 40a, Oct 2009, pp. 2498-2509.

[14] J. Talonen, P. Nenonen, G. Pape, and H. Hänninen: Metall. Mater. Trans. A, Vol. 36a, Feb 2005, p.423.

[15] A. Rezaee, A. Kermanpur, A. Najafizadeh and M. Moallemi: Mater. Sci. Eng. A, 528 (2011), pp. 5025-5029.

[16] B. Ravi Kumar, Sailaja Sharma, B. Mahato: Mater. Sci. Eng. A, 528 (2011), pp. 2209-2216.

[17] J. Talonen and H. Hanninen: Acta. Mater., 55 (2007), pp. 6108-6118.

[18] K. Nohara, Y. Ono and N. Ohashi: ISIJ Int, 63 (1977), pp. 212-222.

[19] V. Shrinivas, S.K. Varma and L. Murr: Metall. Trans. A, 26 (1995), pp. 661-671.

[20] T. Angel: Iron Steel Inst, 177 (1954), pp. 165-174.

[21] M. Shimojo, T. Inamura, T.H. Myeong and K. Takashima: Metall. Trans. A, 32 (2001), pp. 261-265.

[22] H. Mirzadeh and A. Najafizadeh: Mater. Charac., 59 (2008), pp.1650-1654.

Evolution of Dual Phase Microstructure in Compact Strip Process by Simulation of Hot Strip Mill Parameters with Alloy Design

Mrigandra Singhai, Srimanta Sam, Neel kant, R. Krishna, Vurity Saibabu, Anil Gargi, Ashok K. Vishwakarma and Pradip Patra
R&D and Product Development, JSW ISPAT Steel Limited, Dolvi, Maharastra
Email: pradip.patra@jswispat.in

ABSTRACT

Development of Dual phase steel from a thin slab caster and with shorter run out tables has always been a challenge. Instead of opting of alloying approach to the problem, JSW ISPAT Steel Ltd (JISL) opted the thermo mechanical approach by using a leaner chemistry. Chemistry and process design was carefully selected to suit the plant constraints. As against the conventional line, JISL has a shorter ROT line and engineering modifications were required to achieve the desired process parameters. Indigenous design of ultra fast cooling headers was successfully done and mathematical modeling was done of ROT line to predict the temperature and desired processing parameters. A two steps quenching was used along with lower FT and CT. First to reduce the temperature below Ar1 to nucleate ferrite. Mill speed was so chosen that around 80% ferrite transformation is complete before the second fast cooling. Second ultra fast cooling step ensures avoidance of Bainite nose and forms Martensite. The product meets requirement of EN 10346 and IS 1079 for hot rolled DP steel. The microstructure consists of 80% ferrite and rest is Martensite + some Bainite with YS < 420 & UTS > 590 MPa. Subsequent trials are planned for further fine tuning of microstructure with the help of Gleeble Simulator at JSW Vijayanagar.

KEYWORDS: *Dual Phase Steel, Compact Strip Process, Thin Slab Caster, Microstructure Evolution, ROT Cooling*

INTRODUCTION

In order to meet requirement of weight reduction, safety and good formability, Dual Phase steel is a preferred material for new generation automotive vehicles [1]. The steel with low YS/ UTS ratio, high strength, good formability is projected as one of the most important raw material for in coming years for auto industry. In India, currently this grade has found its main application in wheel rims and this grade is developed at JISL focusing this particular application.

Batou steel (China) has commercially development Dual phase steel in the CSP line with C-Mn steel with two step cooling and the product is used for wheel disc of car beam and heavy duty trucks. Hot rolled dual phase steel can be manufactured by alloying (Mo & Cr) so as to significantly shift the nose of CCT curve to right and then subsequent adopting a cooling strategy over a long ROT to achieve desired volume fraction of different phases[2]. JISL decided to adopt a thermo mechanical approach with leaner chemistry instead of addition of costly Molybdenum. Additionally, JISL had a disadvantage of a shorter ROT line which made development of this grade further challenging. Cooling strategy was carefully designed based on chosen chemistry and arrangements were made by indigenous developed ultra fast cooling headers to give desired cooling profile in ROT. Mathematical modeling of ROT cooling was done and several cooling trials were done in mill in actual operation to arrive at operational parameters to achieve the cooling profile. These experimental trials with regular CG/CR grades (in DP cooling strategy) were analyzed for micro structural changes and mechanical property change which gave very useful insight into micro structural evolution and these learning were further extended in other products which demands specific mechanical properties which are difficult to achieve with regular laminar cooling[3].

EXPERIMENTAL TRIAL

Ultra fast cooling header was indigenous designedanddevelopedtoachievethecooling path required for achieving the desired microstructure. Chemical composition and mill parameters were designed based on data from previous trials. Instead of opting for a rich chemistry with Molybdenum, JISL opted for thermo mechanical approach and a leaner chemistry [2]. Cooling headers were specifically designed and developed. Mathematical modeling of ROT cooling was done to identify processing window for each mill parameter. Following chemistry was selected for the pilot trial:

Table 1: Chemical composition of DP steel

Elements	C	Mn	Si	Al	N (ppm)	Cr
% Composition max	0.16	2.0	0.8	0.04	70	1.0

One full heat of approx. 200 tons was processed and all the critical process parameters were closely monitored from steel making stage to final product. Slabs were cast in thin slab caster which was further routed through a tunnel furnace to the six stands tandem mill. Slab was reduced from 62 mm to the final thickness of 4- 4.5mm and subsequently cooled in ROT with two steps quenching process as schematically shown below:

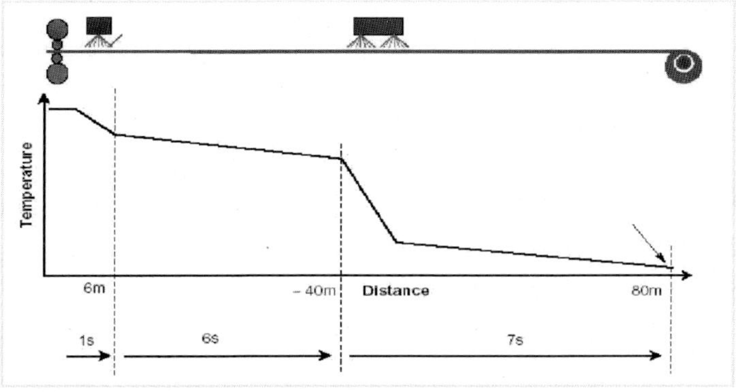

Fig. 1: Schematic cooling strategy for DP steel

Subsequently samples were analyzed for mechanical properties and sent to an outside reputed lab also, for verification of test results.

Micro structural characterization was done with the help of Le Pera etchant to resolve phase boundaries. Subsequently, these microstrural images were processed using image analyzer to obtain volume fraction of different phases [4].

RESULTS

Mechanical property results of a typical sample is tabulated below. The product meets requirement of EN 10346 and IS 1079 for hot rolled DP steel as well as it shows very good uniform elongation required for wheel rim application.

Table 2: Mechanical properties of DP01 trial.

YS (offset 0.2%) (MPa)	UTS (MPa)	% Elongation	YS/UTS	n value
388	610	24.2%	0.64	0.18

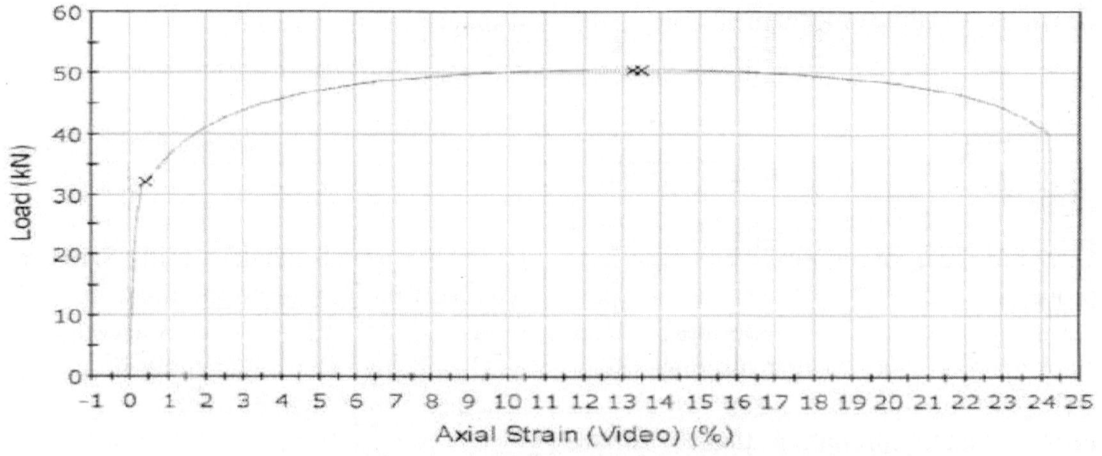

Fig. 2: Stress strain graph of DP steel sample

MICROSTRUCTUTAL ANALYSIS

Micro structural analysis was done using Le Pera reagent and volume fraction of phases were calculated using image analyzer.

Typical microstructure is as below at 200X & 500X. Average volume fraction of different phases is as below:

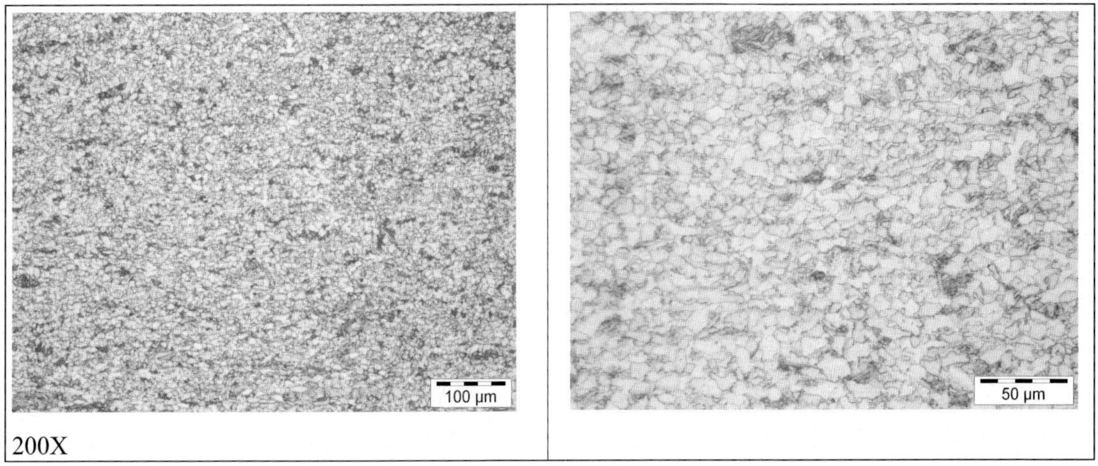

Fig. 3: Microstructure of DP steel samples

Table 3: Volume fraction of phases in different samples

Phase	Ferrite	Bainite	Martensite
Volume fraction %	77	10	13

DISCUSSION

Desired microstructure was achieved by finishing at lower temperatures, thus providing adequate sites for ferrite nucleation. Constraint of a shorter ROT was overcome by two step cooling. Initial fast cooling after finishing brings temperature sufficiently down forcing ferrite nucleation to accelerate and subsequent delay in further cooling over ROT allows time for ferrite stabilization. Mill speed and temperatures are so chosen to ensure around 80% ferrite transformation to complete before second ultra fast cooling which brings the sheet temperature below Martensite start temperature (Ms) after it. The processing has resulted in approx 80 % ferrite with remaining Martensite + some Bainite which gives the desired mechanical properties.

CUSTOMER FEEDBACK

Samples were sent to major wheel rim manufacturers and upon receiving the feedback and order, entire 200 tons was supplied. Customers are satisfied with mechanical properties.

CONCLUDING REMARKS

Plant trial of DP steel is successful at JISL and customer feedback is encouraging. Challenges lies in achieving these properties on a consistent basis. Temperature measurement and control in intermediate zone and at CT is a challenge due to the presence of thick layer of water and steam. The mill was not designed to make DP steel through and level 2 programming needs to be upgraded to achieve these cooling settings which were done manually during the trial. Furthermore, Gleeble simulation is planned at JSW R&D at Bellary to further optimize the process parameters and to develop a full series of DP steel to meet future market requirements. Customer requirement in Indian market also needs to mature to encourage steel manufacturers to venture into development of advanced steel grades for mutual benefits.

ACKNOWLEDGEMENT

Authors are grateful to the management of JSW ISPAT Steel Lld for the permission to present the paper in the International Conference on "Thermo-mechanical Simulation and Processing of Steels (SimPro'12)."

REFERENCES

[1] Hofmann Harald, Mattissen Dorothea, Schaumann T.W., Steel Research Int. Volume 80 (2009) Issue 1. p.22-28

[2] Lopez-Baltazar A., Salinas-Rodriguez A.,Nava-Vazquez E. Material Science Forum Vol. 560 (2007) p.79-84.

[3] Tulio M.F. Melo, Fernando S. Costa: International Conference on New Developments in Advanced High-Strength Sheet Steels, AISTECH, Issue June(2008), p.115-124.

[4] Rui-feng DONG, De-gang LI, Li-gang SUN, Guo-dong, Proceedings of the International Symposium on Thin Slab Casting and Rolling, (2009). p.396-399.

Heat Affected Zone (HAZ) Thermal Simulation Studies on High Strength Structural Steel

B. Basu and K.K. Sarkar
Naval Materials Research Laboratory, Ambernath
Email: kksnmrl@gmail.com

ABSTRACT

The objective of this paper is to study the property variation across the width of a typical HAZ formed during welding of a high strength steel used for naval structural application. The different regions of HAZ, corresponding to various peak temperatures, attained during a welding process, were reproduced through thermal cycle simulation, using a Gleeble machine. Thermal cycle simulation was done on samples having the size of typical charpy specimens over a range of peak temperature of 1000°C -1300°C at a given $t_{8/5}$ of ~14 sec. To simulate multipass welding, thermal simulation of second pass was also done to determine its effect on its preceding first pass. The toughness of the simulated HAZ region was determined using charpy impact testing at -50°C and correlated with the microstructure of the simulated HAZ regions. For single cycle thermal simulation, a substantial lowering of toughness was observed in intercritical HAZ (760°C) compared to a small drop in toughness at grain coarsened HAZ (1300°C). For double cycle thermal simulation, the FGHAZ (T_P in between 900°C-1000°C) a substantial improvement in toughness has been observed. But the intra critical coarse grained HAZ (CGHAZ) had absorbed lesser energy in contrast to the subcritical HAZ (SCHAZ) region which absorbed comparatively higher impact energy.

KEYWORDS: *Welding, Heat Affected Zone, Microstructure, Ferrite, Pearlite, Bainite, Multipass, Thermal Simulation*

INTRODUCTION

The quality of a typical weld is a complex interaction of many variables with a concurrent microstructural changes in the weld metal. During welding, the base metal adjacent to the fusion zone experiences thermal and stress excursions. These occur due to a steep and wide range temperature gradient as a result of severe thermal cycles due to very rapid heating and cooling in the fusion zone. Alteration of the original microstructure occurs and a gradient of microstructures form, from the fusion boundary to the unaltered base metal called Heat Affected Zone (HAZ).

The HAZ produced in ferritic steels consist of several regions, typical of which are Coarse Grained HAZ (CGHAZ), Fine Grained HAZ (FGHAZ), Intercritical HAZ (ICHAZ) and Subcritical HAZ (SCHAZ) as shown in **Fig 1 (1-6)**. These various regions correspond to a definite peak temperature (T_P) experienced at a given distance from the fusion line during welding. Each of these regions are characterised by a unique microstructure and thereby different mechanical properties. These regions are generally too narrow to study the behaviour of a specific HAZ region without the interference from the adjacent regions having subtle different microstructure /

mechanical properties. To characterise each HAZ region, from the point of views of microstructure / mechanical properties, an accurate thermal simulation is more important.

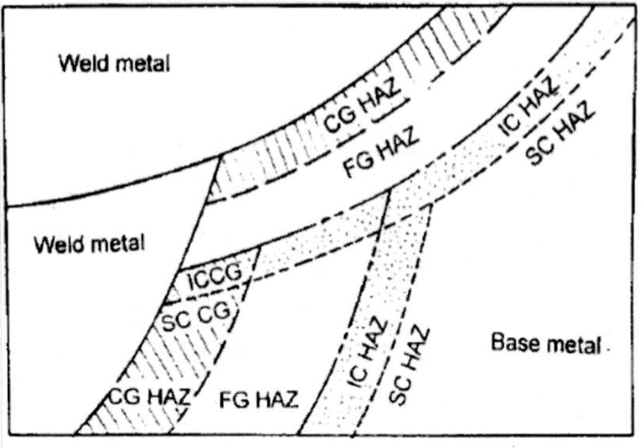

Fig. 1: Schematic illustration of HAZ regions in a ferritic steel weld (1)

During thermal cycle simulation of a multipass welding, the thermal cycle of subsequent passes impose a series of heat treatments on the underlying passes that may successively alter the microstructures in the HAZ, as shown in **Fig. 2(7)**. The nature of the alterations is complicated by the differences among the structures that have been reheated as well as by the range of temperature to which they are heated by the localised heat sources of the weld **(7)**. The microstructure and the resulting mechanical properties in the HAZ get, in most of the cases, unfavourably altered after undergoing reheat treatments from the subsequent thermal passes.

WELD THERMAL SIMULATION

In any typical HAZ of a structural steel the total distance over which microstructural alterations occurs is generally small and few mm wide. This has often possessed difficulties in microstructural and toughness evaluation of a real weld HAZ **(8)**, are mainly due to the following reasons

(i) width of HAZ is small, compared with that of the weld and

(ii) heterogeneity within HAZ microstructure.

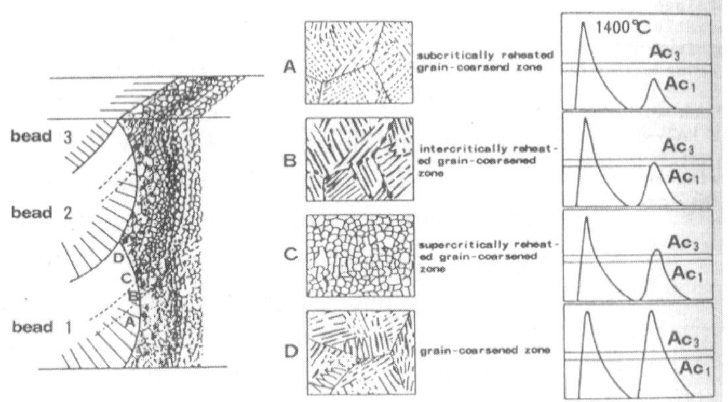

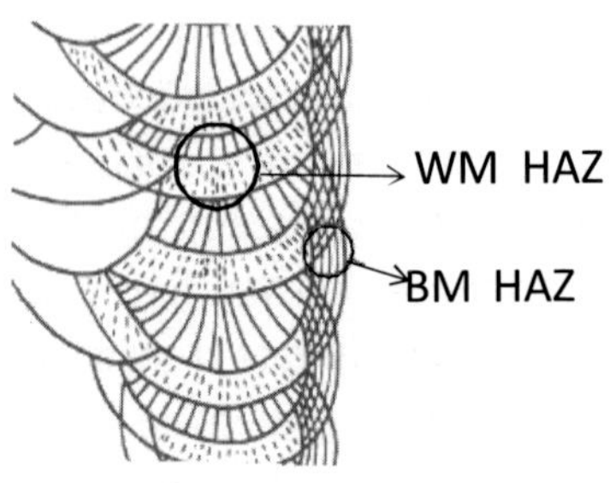

Fig. 2: Schematic illustration on effect of multipass welding conditions on microstructure of HAZ (7)

As a result it becomes practically difficult to locate charpy notch at any desired microstructural region within a HAZ. But different microstructural regions in a typical HAZ of a welded joint can be simulated using a Thermal Cycle Simulator (TCS). It is a relatively inexpensive and quick means of generating specific HAZ microstructures uniformly throughout the specimen thickness and also over a certain width of the specimen. The inherent problems associated with multipass weld, e.g. microstructural inhomogenity, uncertainty in crack tip locations etc may be taken care when using thermally simulated specimens.

EXPERIMENTAL

In this investigation, a high strength low alloy steel, DMR 249A was used for carrying out the thermal simulation study using a Gleeble 3800 Thermo Mechanical Simulator. The chemical composition and mechanical properties of DMR249A steel is as shown in **Table 1** and **2**. Samples of size 60 x11 x 11 mm were machined from base plate. The specimens were then subjected to different weld thermal cycles, as tabulated in **Table 3**. Single thermal cycles were simulated with different peak temperatures (T_p) The cooling rate represented by $t_{8/5}$, was kept constant at 14 sec, $t_{8/5}$ which represents the time required to cool from 800°C to 500°C, **Fig. 3**. $t_{8/5}$ of ~14 sec corresponds to a Heat Input of 1.5 kJ/mm typically used for a MMAW process for a plate thickness of 12 mm.

Table 1: Chemical Analysis of Base steel

C	Mn	Si	S	P	Ni	Cu
0.08-0.1	1.2-1.5	0.15-0.3	0.01max	0.015max	0.6-0.9	0.02-0.03

Table 2: Specified mechanical properties of low alloy Base steel

YS (MPa)	UTS (MPa)	El (%)	Toughness (J)
>390	>510	>22	>78J@-600C

Subsequently during double cycle simulation 1st peak temperature T_{P1} was kept constant and T_{P2} varied from 650°C to 1000°C, while $t_{8/5}$ was kept constant at ~14 sec, **Table 3**. **Fig 3** shows a typical double thermal cycle simulation.

Table 3: Thermal cycle parameters

Sr. No.	Peak Temp. TP(°C)		T8/5 (sec)
	T_{P1}(°C) (1st pass)	T_{P2}(°C) (2nd pass)	
1	1300	-	14
2	1000	-	14
3	900	-	14
4	800	-	14
5	650	-	14
6	1300	1000	14
7	1300	900	14
8	1300	700	14

After Thermal simulation the specimens were machined to (10x10x55) charpy impact specimens. Impact testing was carried out at -50°C as per ASTM E23 standard. Microstructural observations were carried out on surfaces very close to the surfaces fractured during impact testing.

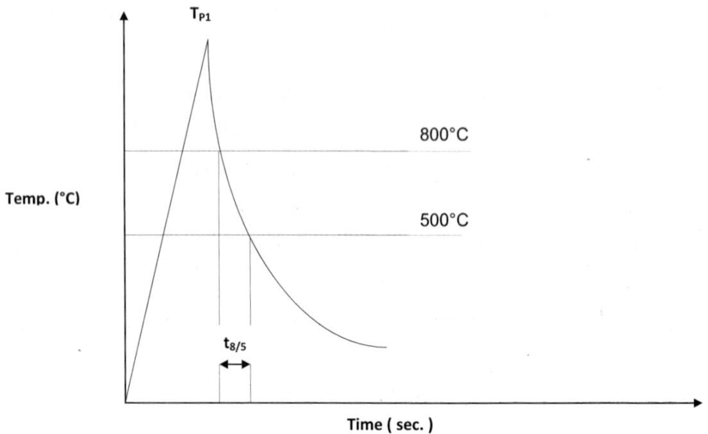

Fig. 3a: Schematic illustration of thermal cycle (single) used for weld simulation

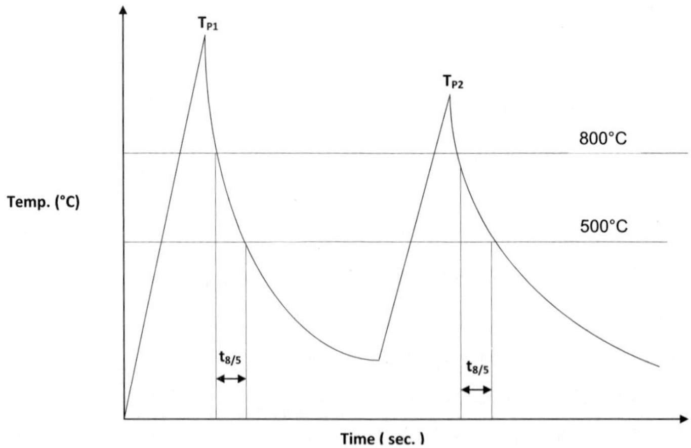

Fig. 3b: Schematic illustration of thermal cycle (double) used for weld simulation

RESULTS AND DISCUSSION

Impact toughness of single-pass simulations i.e. single cycle, as well as two pass simulations with $t_{8/5} =$ 14 sec are given in **Table 4** and graphically represented in **Fig.**4., A marginal drop in toughness value was observed in case of low temperature (650°C) subcritical treatment but a Substantial drop in toughness was observed at the grain coarsening temperature of 1300°C when compared with that of unaffected base metal. In comparison, in the region of FGHAZ ($\mathbf{T_p}$ in between 900°C -1000°C) a substantial improvement in toughness has been observed.

Table 4: Impact Toughness of thermally simulated specimens

Sr. No.	Peak Temp. TP(0C)		Impact Toughness at -500C (J)
	T_{P1}(°C) (1st pass)	T_{P2}(°C) (2nd pass)	
1	1300	-	8
2	1000	-	72
3	900	-	93
4	800	-	50
5	650	-	40
6	1300	1000	39
7	1300	900	35
8	1300	700	7

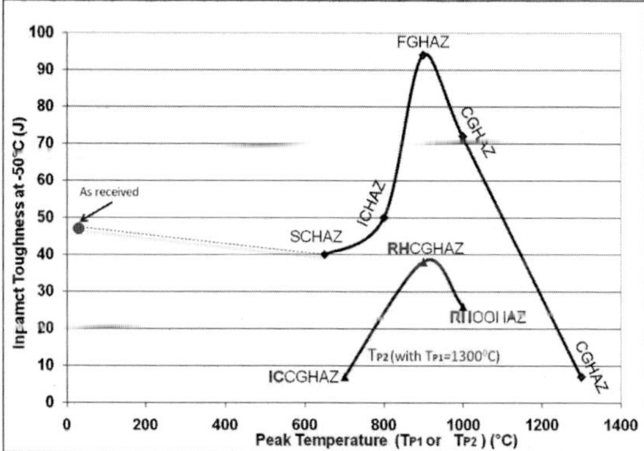

Fig. 4: Graphical representation of variation of impact toughness of thermally simulated specimens with peak temperature

The microstructure of base metal is as shown in **Fig.5**. The microstructure of sample heated to different reheating temperatures is shown in **Fig.6-10.** At intercritical temperature microstructure showed the formation of pearlite nodules inside the prior austenite grains. The sample simulated to grain coarsening temperature of 1300°C, an upper bainitic microstructure is obtained, **Fig. 6**. This microstructure has

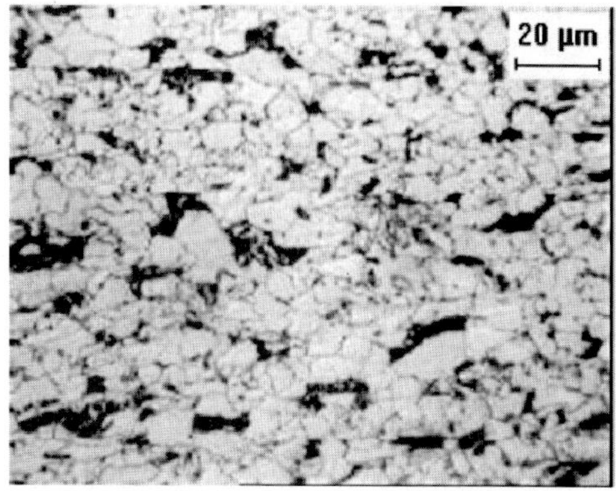

Fig. 5: Microstructure of DMR249A base metal showing ferrite – pearlitic microstructure

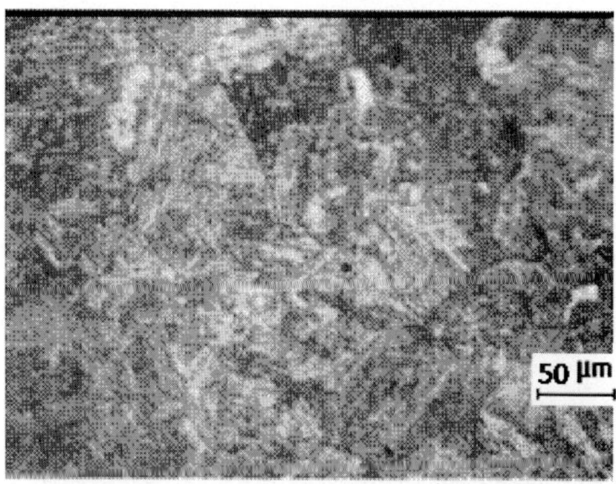

Fig. 6: Microstructure of DMR249A thermally simulated (T_{P1} = 1300°C) steel showing coarse grain prior Austenite grains

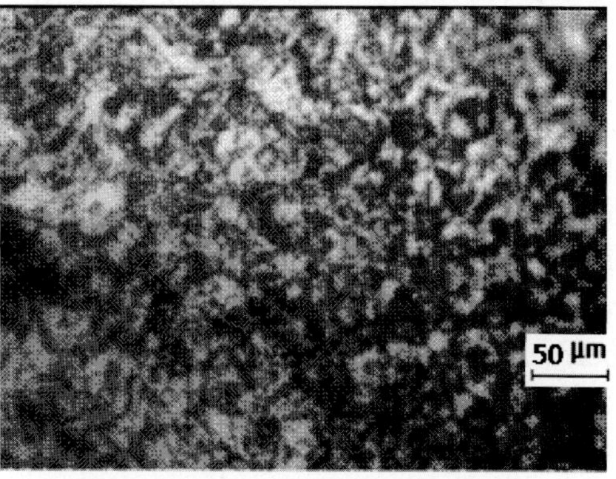

Fig. 7: Microstructure of DMR249A thermally simulated (T_{P1} = 1000°C) steel showing coarse grains HAZ

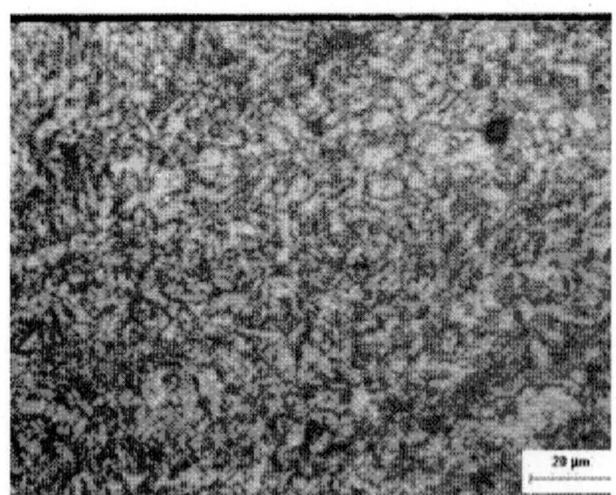

Fig. 8: Microstructure of DMR249A thermally simulated (T_{P1} = 900°C) steel showing presence of fine grains HAZ

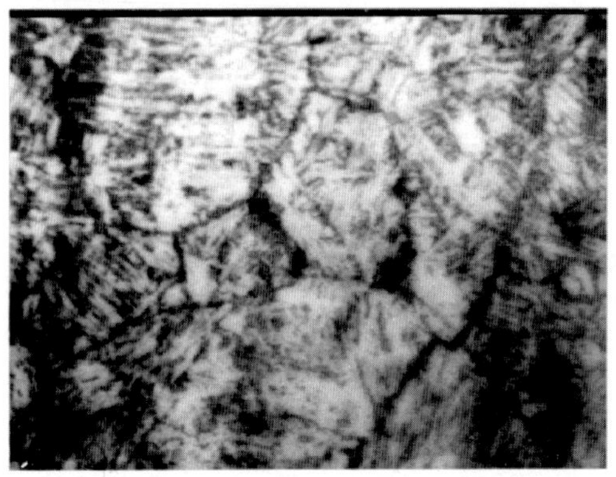

Fig. 9: Microstructure of DMR249A thermally simulated (TP1= 1300°C, TP2= 1000°C) steel to super critical reheated CGHAZ showing thick precipitation along prior Austenite grain boundary.

low impact toughness when compared to original base metal toughness. During double pass simulation, once the upper bainitic microstructure is formed from grain coarsened austenite, the subsequent thermal cycles (T_{P2}=900-1000°C) could partially restore its toughness. But further deterioration in impact toughness was observed in case of intercritically reheated (ICCGHAZ) samples. A network of Martensite-Austenite (MA) islands along

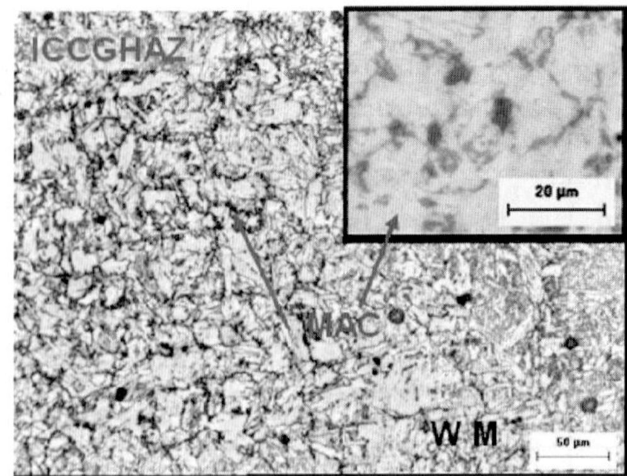

Fig. 10: Microstructure of DMR249A thermally simulated (TP1= 1300°C, TP2= 700°C) steel Intercritical reheated CGHAZ showing presence of MAC phases.

prior austenitic grain boundaries was observed to have formed, **Fig.10.** This type of dual phase microstructure was presumed to be the reason for deterioration in impact toughness **(9)**.

CONCLUSIONS

From the present investigation on the effects of single or multiple thermal cycles on microstructure and properties of a high strength structural steel following observations could be made:

(i) Toughness in entire HAZ deteriorated (except fine grained region) during single pass simulation

(ii) During second pass simulation, restoration of toughness deteriorated during first cycle occurred to a certain extent.

(iii) In ICCGHAZ, toughness deteriorated further during second pass simulation, due to formation of MA constituents along prior austenite grains boundaries.

REFERENCES

[1] C.D Lundin and G. Zhou, Acta Metallurgica Sinica (English Letters) Vol.13 No. 1 pp 223-232 February 2000,

[2] Shiga, C. : Proc. Int. Conf. on 'The Metallurgy, Welding and Qualification of Microalloyed (HSLA) steel weldments', Houston, Texas, USA, p327, Nov, 1990.

[3] Scoonover, T.M. : Proc. Int. Conf. on 'The Metallurgy, Welding and Qualification of Microalloyed (HSLA) steel weldments', Houston, Texas, USA, p276, Nov, 1990.

[4] Devillers. L., Kaplan, D. and Testard, P. : Welding in the world, 31(4), p256-67,1993.

[5] Akselsen, O.M., Solberg, J.K. and Grong, O. : Scand. J. Metall., 17, p194-200, 1988

[6] S. Moeinifar, A.H. Kokabi, H.R Madaah Hosseini, Materials and Design, 31 (2010) 2948-2955.

[7] Koo, J.Y. et. al. : Met. Trans. A, vol. 11A, p 852, 1980.

[8] Pisaraski, H.G. : Proc. Int. Conf. on the Metallurgy, Welding and Qualification of microalloyed (HSLA) steel weldments, Houston, Texas, USA,1990.

[9] B Basu, et. al., Proc. International Welding Conference IWC2001, held at New Delhi 2001.

Improvement in Mechanical Properties through Physical Simulation of Hot Rolling Process

H.A. Lanjewar[1], Archana R.[1], D. Satish Kumar[1], S. Manjini[1]
M. Singhai[2] and P.K. Patra[2]
[1]Research & Development and Scientific Services, JSW Steel Ltd, Bellary (KA), India.
[2]Product Development and Innovation Center, JSW Ispat Ltd, Raigad (MH), India.
Email: lanjewar.harishchandra@jsw.in

ABSTRACT

During hot rolling the combined effect of process variables such as chemical composition, finishing temperature in combination with reductions in each stand, ROT cooling rate and coiling temperature governs the final mechanical properties of the rolled products. Thermo-mechanical simulation is an important tool for understanding of DRX behavior and subsequent evaluation of microstructures under varying process parameters to achieve the desired properties. Present paper describes the rolling simulation approach using Gleeble for achieving the desired properties in an Nb –V micro-alloyed steel. For micro-alloyed steel, multi-hit hot compression tests followed by controlled quenching were carried out to simulate rolling and subsequent cooling. Effect of post-deformation cooling rate and coiling temperatures on grain refining and strengthening were studied for optimizing the process parameters.

KEYWORDS: *Physical Simulation, Thin Slab Direct Rolling, Microalloyed Steel.*

INTRODUCTION

In an Integrated steel plant, hot rolling process plays vital role in finishing the cast steel to the near net shape. Hot strip rolling is either through TSDR (Thin Slab Direct Rolling) or thick slab CCR (Cold Charge Rolling). Process parameters during rolling determine the final mechanical properties of the material and many a times it takes several test trials to reach at the satisfactory results. However, physical simulation proves to be useful tool to simulate the thermal and mechanical deformation cycles to study the effect varying process variables on the material microstructure and properties. In the present study, a microalloyed steel grade was investigated for improvement in the mechanical properties through physical simulation on Gleeble- 3800.

A Nb-V micro-alloyed steel widely used automotive grade for wheel rim and discs application for passenger cars, commercial and heavy duty vehicles, was studied for improvement in the tensile strength. The application entails high strength, good weldability and good formability. This steel is being produced through Compact strip casting (CSP) and TSDR process route. The Nb and V present in the steel strengthen the steel through fine precipitation of $M_x(C,N)$-type as well as solid solution strengthening [1, 2]. Nb and V during solidification precipitate on the TiN cuboid particles which formes at high temperature and acts as nucleation site [2]. In absence of Ti, AlN which has a higher solution temperature than Nb or V carbonitride, precipitates during continuous cooling and causes the precipitation of starlike particles of (NbV)(CN). Even though closed packed hexagonal structure poses difficulty for AlN to precipitate in austenite, the MnS particles formed before AlN acts as nucleation sites for AlN precipitation [2, 3]. The Nb(CN) or V(CN) coarse eutectic particle formed in austenite during continuous cooling does not contribute to the strengthening of the structure by reducing the microalloy content available for subsequent precipitation and solid solution strengthening [3, 4].

There has been studies investigating the effect of cooling rate and coiling temperatures on the mechanical properties of the steel [5], but not much effort has been put in studying the effect of accelerated cooling regime in conjunction with coiling temperature on the microstructures developed in microalloyed steel through TSDR route. The objective of the present study is to examine the effect of the post-deformation accelerated cooling rates and coiling temperature on the contribution of different micro-constituents in structure and microalloying elements to the strength of steel.

EXPERIMENTAL

The chemical composition of the investigated thin slab is given in Table 1. The Nb-V microalloyed steel slab was produced at JSW Ispat Ltd. through Compact Strip Process (CSP). The thermo-mechanical simulation studies are conducted using Gleeble 3800 system. The rolling hot deformation studies are performed by deforming the samples in axisymmetric hot compression tests. Cylinder samples of 15 mm in length and 10 mm diameter are machined from the steel slab. The steel samples are allowed to homogenize at the deformation temperature of 920 ^{0}C for 60 sec. The hot deformation is carried out by applying 0.4 engineering strain at the strain rate of 50s^{-1}. Post deformation incubation time of 2 sec is allowed before quenching the sample at different cooling rates (75, 100, 125, 150 ^{o}C/sec) to simulate the run out table cooling conditions and different coiling temperatures are applied (475, 500, 525, 550, 575, 600, 625 ^{o}C). Incremental effect of accelerated cooling rates combined with higher/lower coiling temperatures is investigated to study their effect on evolution of microstructure and tensile strength of the present microalloyed grade. With the intention of simulating the coiling conditions the samples are cooled at 0.5 ^{o}C/sec cooling rate till 200 ^{o}C and then air cooled to room temperature. The test conditions with two variable process parameters are as listed in Table 2.

Bulk hardness measurement is conducted at the center of the midplane perpendicular to the deformation axis. For microstructural characterization optical microscopy and SEM imaging is performed on JEOL 8230. X-ray diffraction measurements carried out to evaluate the dislocation/subgrain structure from ferrite peak broadening.

Table 1: Chemical Composition

%C	%Si	%Mn	%Nb	%V
0.060	0.16	1.46	0.027	0.051

RESULTS AND DISCUSSION

Mechanical Properties

Bulk hardeness measurement results are as shown in Table 2. Fig. 1 shows the bulk hardness measured at different coiling temperatures with incremental cooling rates. Sample A is taken as the reference condition and coiling temperatures are varied at different cooling rates under the investigation. The hardness increase is observed in sample D, I and J from the reference value of 183 Hv to almost 230 Hv. There are many empirical relations reported in the literature which relates the material strength with the bulk hardness of the material [6 - 8]. JMatPro – the materials property simulation package is used to calculate the strength from the measured bulk hardness, which clearly indicates the increase in yield and ultimate tensile strength of the samples D, I and J. The UTS calculated comes to about 747, 722 and 728 MPa respectively for samples D, I and J.

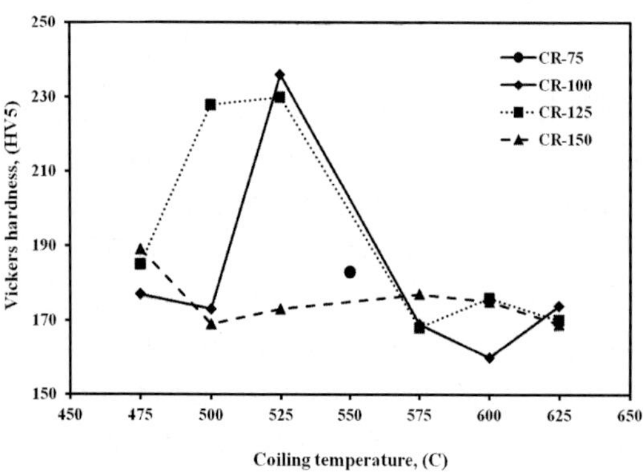

Fig. 1: Bulk hardness depending on coiling temperatures at different cooling rates.

The increase in the strength of the material can be due to one of many materials strengthening phenomena which are enlisted as below [9];

i) Grain boundary strengthening (High and low angle),
ii) Solid solution strengthening,
iii) Two phase aggregates,
iv) Fine particle strengthening,
v) Strengthening due to point defects,
vi) Strengthening due to phase transformation,
vii) Strengthening due to metal working,
viii) Strengthening due to preferred orientation.

X-Ray Diffraction

The strengthening of the material introduces lattice imperfections in the material which in turn causes the X-ray diffraction peak broadening [10, 11]. X-ray peak broadening is caused due to: (i) Instrumental broadening, (ii) Crystallite size and (iii) Strain [12]. In order to determine the contribution of the instrumental broadening to the diffraction peak broadening, sample of the same material is annealed, sufficiently above austenization temperature for 5 hrs and diffraction pattern is collected. The A_3 temperature calculated for the steel using JMatPro comes to about 845 ^{0}C and hence sample is annealed at 50^{0}C above A_3 temperature, i.e. at 900 ^{0}C. Fig. 2 shows the

Table 2: Test conditions for the study.

Sr. No.	Steel Sample	Cooling Rate, ^0C/sec	Coiling Temperature, ^0C	Vickers Hardness, Hv5
1	A	75	550	183
2	B	100	475	177
3	C		500	173
4	D		525	236
5	E		575	169
6	F		600	160
7	G		625	174
8	H	125	475	185
9	I		500	228
10	J		525	230
11	K		575	168
12	L		600	176
13	M		625	170
14	N	150	475	189
15	O		500	169
16	P		525	173
17	Q		575	177
18	R		600	175
19	S		625	169

FWHM (Full Width at Half Maximum) for different coiling temperatures. Increase in strength of the material coincides with concurrent peak broadening of the diffraction peaks, whereas in case of samples with cooling rate of 150^{0}C/s, no such dependence is observed. The peak broadening observed can be because of: (i) dislocations, (ii) stacking faults, (iii) twinning, (iv) microstresses, (v) grain boundaries, (vi) sub-grain boundaries, (vii) coherency strains, (viii) chemical heterogeneities and (ix) crystallite smallness [10].

Microstructures

JMatPro is used to calculate the Ar_3 and Ar_1 temperature under equilibrium conditions during continuous cooling, as shown in Fig. 3. The Ar_3 and Ar_1 temperature comes out to be 845^{0}C and 660^{0}C respectively. However, under rapid cooling conditions phase transformation temperatures decrease, whereas the time required for transformation decrease i.e. the rate of transformation is faster under accelerated cooling conditions [13].

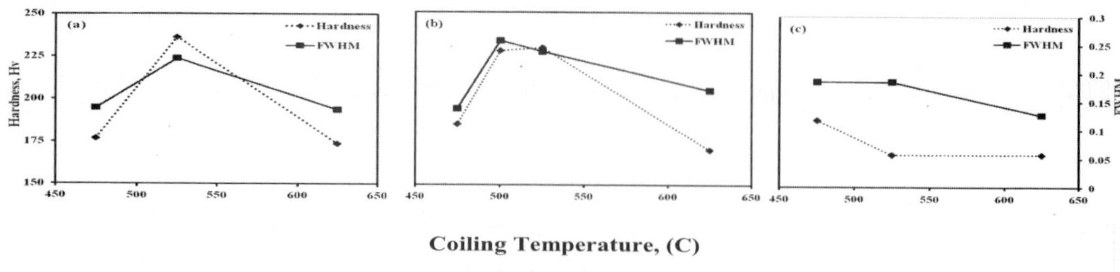

Fig. 2: FWHM at different coiling temperatures for cooling rate of (a) 100 ^{0}C/s, (b) 125 ^{0}C/s and (c) 150 ^{0}C/s.

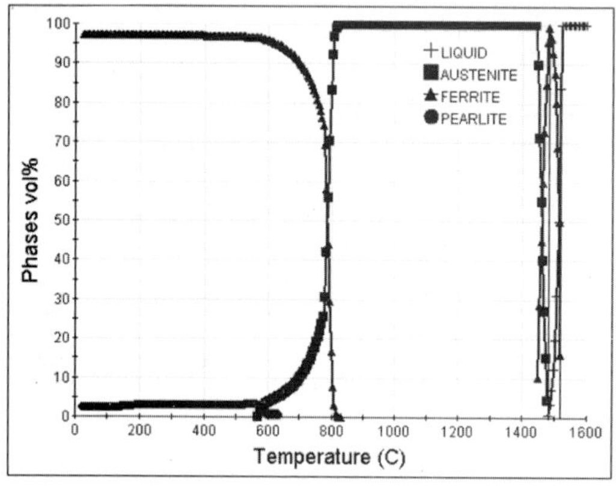

Fig. 3: Continuous cooling phase transformation

Presence of solute Nb too lowers the Ar_3 temperature as it segregates along the γ/α interphase boundary and thereby reduces ferrite growth rate through solute drag effect [13]. The net effect is lowering of Ar_3 temperature and the transformation temperature range which results in increased undercooling ($\Delta T = Ac_3 - Ar_3$) and nucleation rate coupled with intra-granular nucleation. At higher cooling rates and low transformation times pro-eutectoid ferrite formed during transformation decreases. The γ-α phase transformation during rapid cooling leads to shear stresses which are relaxed by the formation of GNDs (Geometrically Necessary Dislocation) [14], which again may cause the peak broadening. Ferritic microstructures formed after decomposition of austenite under accelerated cooling and lower transformation temperatures assume non-equiaxed morphologies which are characterized by high dislocation densities, presence of crystal defects, substructure and supersaturated carbon [15].

According to the classification proposed by ISIJ ferrite classification system [16], different ferrite morphologies identified are; Polygonal Ferrite (PF), Granular Ferrite (GF), Quasi-polygonal or Massive Ferrite (QF), Bainitic or Accicular Ferrite, Lamellar Pearlite (P), Degenerated Pearlite (DP) and Martensite-Austenite (M/A) Constituents. Fig. 4 shows the SEM images for various coiling temperatures at different cooling rates. The microstructures at lowest transformation temperatures show the presence of PF, GF, QF and DP showcasing a very heterogeneous structure with very irregular grain boundaries, though overall structure is fine. As already discussed, microstructures formed at lower coiling temperature are characterized by intra-granular nucleation of carbides as shown in Fig. 4 (a), irregular and jagged ferrite grain boundaries due to the solute drag effect,

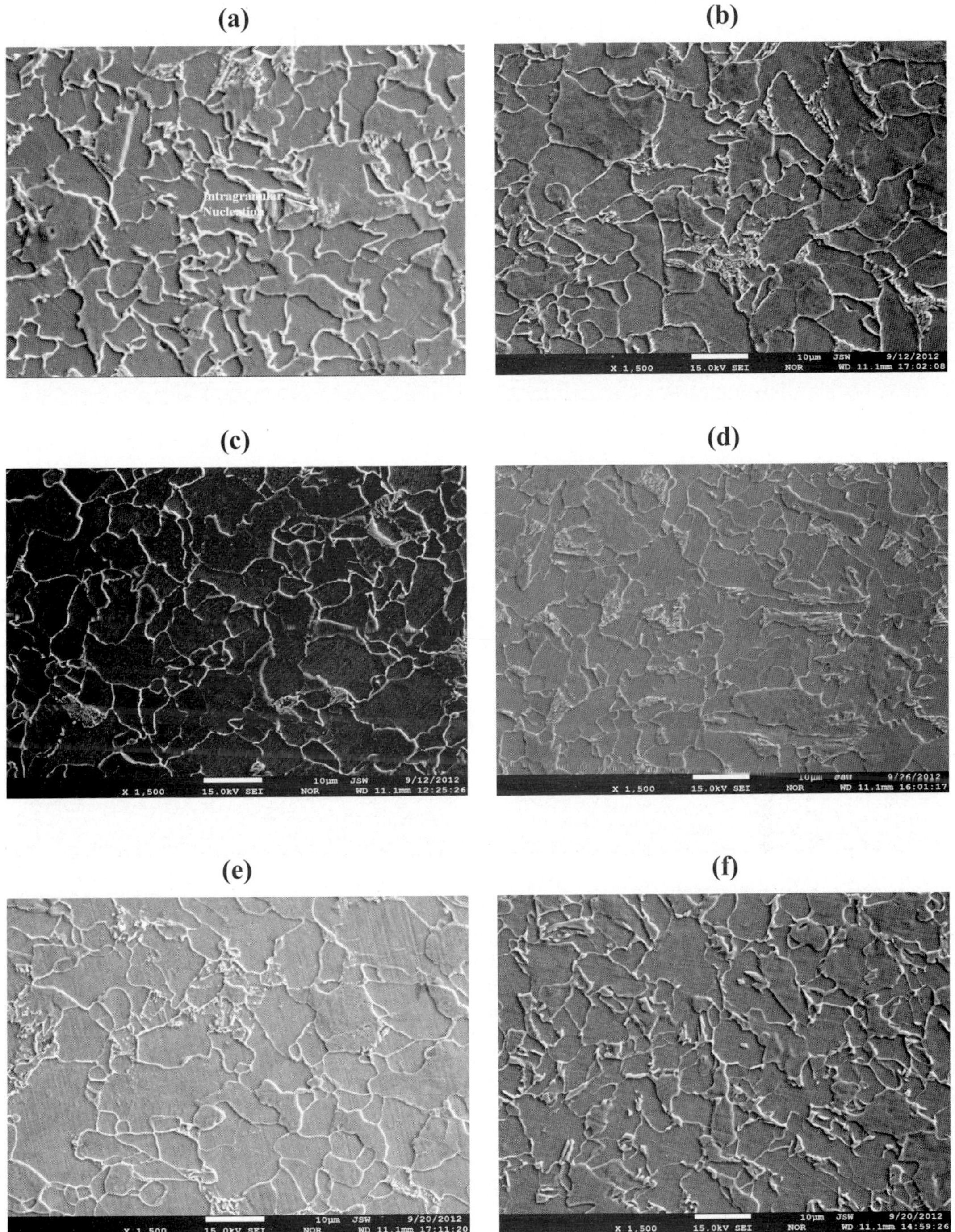
(a)
(b)
(c)
(d)
(e)
(f)
Intragranular Nucleation
X 1,500 15.0kV SEI NOR 10µm JSW WD 11.1mm 9/12/2012 17:02:08
X 1,500 15.0kV SEI NOR 10µm JSW WD 11.1mm 9/12/2012 12:25:26
X 1,500 15.0kV SEI NOR WD 11.1mm 16:01:17
X 1,500 15.0kV SEI NOR 10µm JSW WD 11.1mm 9/20/2012 17:11:20
X 1,500 15.0kV SEI NOR 10µm JSW WD 11.1mm 9/20/2012 14:59:26

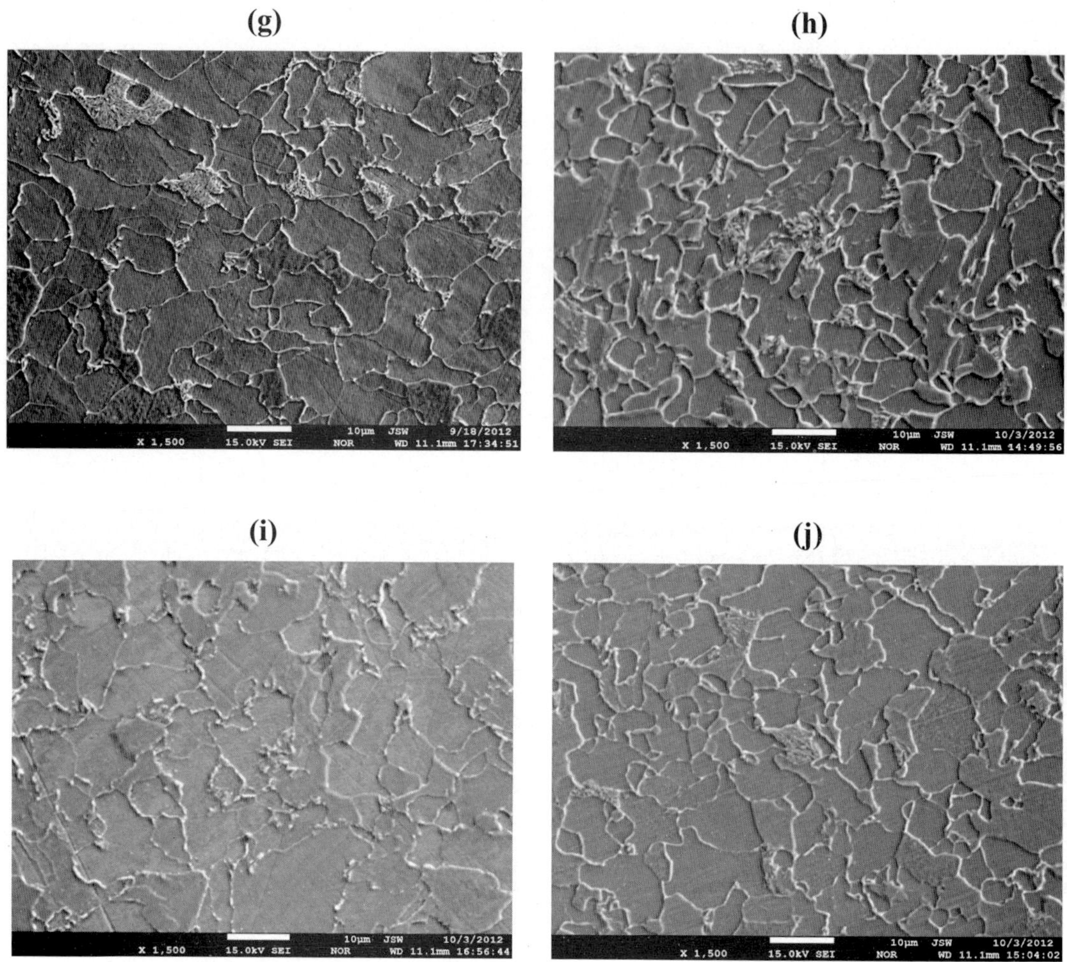

Fig. 4: SEM images for different coiling temperatures at different cooling; (a) 475 ^{0}C, 100 ^{0}C/s (Coiling Temperature), (b) 525 ^{0}C, 100 ^{0}C/s, (c) 625 ^{0}C, 100 ^{0}C/s, (d) 475 ^{0}C, 125 ^{0}C/s, (e) 500 ^{0}C, 125 ^{0}C/s, (f) 525 ^{0}C, 125 ^{0}C/s, (g) 625 ^{0}C, 125 ^{0}C/s, (h) 475 ^{0}C, 150 ^{0}C/s, (i) 525 ^{0}C, 150 ^{0}C/s and (j) 625 ^{0}C, 150 ^{0}C/s.

an attribute of the interstitial and Nb atoms. With increasing coiling temperatures the carbides form a more lamellar structure as is evident in Fig. 4 (c), (g) and (j). With increase in cooling rate, the transformed ferrite has more elongated plate-like shape and devoid of any polygonal ferrite. At 150 ^{0}C and lowest coiling temperature, the microstructure consists of GF, QF, DP and elongated ferrite crystals exhibiting very heterogeneous structure. Samples D, I and J shows a comparatively homogeneous structure consisting of GF, QF and DP in combination with fine ferrite grains (of about 2-3 µm), with unit size distribution narrower compared to lower transformation temperature structures. The grain boundaries in above three samples are well defined as compared to those observed at highest cooling rate and lowest coiling temperatures. Microstructures observed at the coiling temperature of 625 ^{0}C are characterized by the presence of PF, QF and P in conjunction with well delineated grain boundaries and comparatively coarser grain size. Hence in addition to the type of ferrite constituents present, their morphology and distribution seems to have an effect on the bulk hardness of the material. But the observed increase in hardness values cannot be entirely described by the microstructural changes occurring as a result of accelerated cooling and varying coiling temperatures.

The Nb added in the steel produces strengthening effects in the form of fine precipitates and through solute drag effect by pinning the crystal defects and dislocations in the matrix [4]. Even after the precipitation of Nb in the form of carbonitrides, sizeable amount of Nb remains in the solution. This microcontent of Nb is reported to have increased the strength of steel by about 55 MPa, while increase in strength due to the decreasing coiling temperature in the presence of same solute Nb content have also been reported [17]. With the increase in post-deformation cooling rates, increase in tensile strength has been reported in case of medium carbon steel, though the cooling rates used are too low than those used in the present study [18].

CONCLUSIONS

Physical simulations are carried out in Gleeble 3800, to study the effect of accelerated post-deformation cooling rates and coiling temperatures on the strength of Nb-V microalloyed steel. The conclusions can be drawn as follows:

1. With increase in cooling rates and decrease in coiling temperatures, strength of the Nb-V microalloyed steel is found to be increasing. However no such effect is observed at the cooling rates of 150 ^{0}C/s and strength actually decreased at the coiling temperature of 475 ^{0}C compared to the maximum strength obtained.
2. The different combinations of cooling rates and coiling temperatures result in differing microstructures consisting of PF, GF, QF, P and DP. The increase in strength is thought to be due to the presence of microstructure consisting of GF, QF and DP having a bimodal grain size distribution with well outlined grain boundaries.
3. Microcontent of Nb increases the strength of the matrix by effectively pinning the crystal defects and dislocations. Hence, increase in the strength of the samples can be attributed to the increase in crystal defects, dislocation density, micro-constituents present, their morphologies and effect of Nb microcontent through solute drag effect.
4. However, further investigation is needed to objectively determine the mechanism effective at different cooling rates and coiling temperature conditions in order to quantify their individual impact on the mechanical properties.

ACKNOWLEDGEMENTS

H. A. Lanjewar is grateful to Mr. Pranav Kumar Tripathi and Mr. Umashankar Keelu for their assistance during the experimental work and material characterization.

REFERENCES

[1] S. Shanmugam, N. K. Ramisetti, R. D. K. Misra, T. Mannering, D. Panda, S. Jansto, Materials Science and Engineering A, 460-461, 2007, p. 335-343.

[2] Y. Li, D. N. Crowther, P. S. Mitchell, T. N. Baker, ISIJ International, Vol.42, No.6, 2002, p. 636-644.

[3] J. S. Park, M. Ajmal, R. Priestner, ISIJ International, Vol.40, No.4, 2000, p. 380-385.

[4] Ruizhen Wang, C. I. Garcia, M. Hua, K. Cho, Hongtao Zhang, A. J. DeArdo, ISIJ International, Vol. 46, No. 9, 2006, p. 1345-1353.

[5] Sha Qing-yun, Li Gui-yan, Qiao Li-feng, Yan Ping-yuan, Proceedings of Sino-Swedish Structural Materials Symposium, 2007.

[6] E. J. Pavlina, C. J. Van Tyne, Journal of Materials Engineering and Performance, 17, No. 6, 2008, p. 888-893.

[7] P. Zhang, S. X. Li, Z. F. Zhang, Materials Science and Engineering A, 529, 2011, p. 62-73.

[8] W. D. Callister Jr., Materials Science and Engineering – An Introduction, 7th Ed., John Wiley, New York, 2007, p. 159-160.

[9] G. E. Dieter, Mechanical Metallurgy, McGraw-Hill Book Co., 1988, p. 184-240.

[10] T. Ungár, Scripta Materialia, 51, 2004, p. 777-781.

[11] T. Ungár, Materials Science and Engineering A, 309-310, 2001, p. 14-22.

[12] C. Suryanarayana, M. Grant Norton, X-Ray Diffraction: A Practical Approach, Plenum Press, New York, 1998. p. 207-222.

[13] P. A. Manohar, T. Chandra, C. R. Killmore, ISIJ International, Vol. 36, No. 12, 1996, p. 1486-1493.

[14] S. Zaefferer, J. Ohlert, W. Bleck, Acta Materialia, 52, 2004, p. 2765-2778.

[15] George. Krauss, Steven W. Thompson, ISIJ International, Vol. 35, No. 8, 1995, p. 937-945.

[16] T. Araki, I. Kozasu, H. Tankechi, K. Shibata, M. Enomoto, H. Tamehiro. Atlas for Bainitic Microstructures, Vol. 1, ISIJ, Tokyo, 1992.

[17] Qingbo Yu, Zhaodong Wang, Xianghua Liu, Guodong Wang, Materials Science and Engineering A, 379, 2004, p. 384-390.

[18] G. R. Ebrahimi, H. Arabshahi, Journal of Advanced Research in Mechanical Engineering. Vol. 1, Issue 4, 2010, p. 182-187.

SECTION VII

Mathematical Modeling

Modeling of Steelmaking Processes: Efforts at IIT-Kanpur

Dipak Mazumdar
INAE Chair Professor & Ministry of Steel Chair Professor
Materials Science and Engineering
Indian Institute of Technology, Kanpur-208016, India
Email: dipak@iitk.ac.in

ABSTRACT

Visual opacity, high operating temperature and relatively large size of steel processing reactors often preclude direct experimental observation and therefore, analysis, design, control and optimization of steelmaking rely overwhelmingly on modeling. The present author and in his associates have been actively engaged in steelmaking process modeling (both physical and mathematical modeling) for over three decades. Over the years, oxygen steelmaking, ladle and tundish metallurgy and continuous casting processes have been investigated through physical and mathematical modeling. In the present work, some unpublished work of the author and his associates are presented and discussed to primarily high light the application of mathematical modeling in refining and casting of steel. The specific examples chosen include,

1. Static control model and improvement in tap to tap time of an energy optimizing steelmaking converter,
2. Post combustion in an electric arc furnace,
3. Inclusion removal in slab casting tundish systems and
4. Grade intermixing and minimization of intermixed slab/bloom volume.

Shop floor data obtained from some steel industries as well as laboratory scale physical model results are presented to substantiate the role of modeling further.

KEYWORDS: *Steelmaking, Modeling, Process Analysis, Prediction, Design, Experiment, Industrial Trials*

INTRODUCTION

Steelmaking, from BOF to the caster, entails a variety of rate processes such as, melting, dissolution, chemical reaction, mixing, solidification and so on. While these help control composition, temperature and cleanliness of the melt, their rates determine shop floor efficiency and to a large extent, the performance of a steel mill. The efficiency of the associated rate processes, as it is well known, is intricately related to fluid flow and turbulence phenomena. As there is very little scope to map the prevalent flow in such systems experimentally, modeling of flow is therefore a pre-requisite to the study of steelmaking processes. As a consequence of such, efficiency of heat and mass transfer process in steelmaking has been generally investigated through physical and mathematical modeling. Popularity of modeling among researchers in steelmaking is readily evident from many review articles

[1-3] written on the subject of physical and mathematical modeling of ladle and tundish metallurgy as well as continuous casting of steel.

As pointed out already, transient, three dimensional, multiphase (i.e., involving solid slag, liquid slag, argon gas and bulk liquid steel), turbulent flow simulation is a pre-requisite to the modeling of steelmaking. These necessarily involve multi-physics and are therefore, generally complex, time and effort intensive. Furthermore, physically based and sound theoretical frame work is often not available to model steelmaking processes rigorously. Considerable idealizations have therefore been customarily applied to mathematically model a variety of steelmaking operations.

Table 1: Component of Mathematical Models in Different Problems Considered in this Work

Problem	Flow model	Associated models	Computa-tional platform
EOF static control model	Not applicable	Material and enthalpy balance	Developed in house
Post combustion in an EAF	Steady, homogeneous, non isothermal , reactive and turbulent	Thermal energy transport, species conservation and chemical reaction	Fluent®
Grade intermixing	Not applicable	Dimensional analysis and regression	developed in house
Inclusion removal	Steady, homogeneous and turbulent	Population Balance	Fluent® + procedure developed in-house

In the following sections, mathematical modeling of steelmaking, with reference to a set of four typical problems, is discussed. Specific examples chosen include:

1. Static control model and improvement in tap to tap time in an energy optimizing steelmaking converter,
2. Post combustion in an electric arc furnace,
3. Inclusion removal in slab casting tundish systems and
4. Grade intermixing and minimization of intermixed slab /bloom volume.

While model equations etc. have been deliberately avoided in the text for the sake of simplicity, essential components of process models, referenced to the four above mentioned problems, are summarized briefly in Table 1 to provide the readers some flavor of the subject.

STATIC CONTROL MODEL AND IMPROVEMENT IN TAP TO TAP TO TAP TIME IN AN EOF

Energy optimimizing furnace (EOF) steelmaking is a variant of the oxygen steelmaking process and employs both submerged tuyeres and supersonic lances to supply oxygen. With the atmospheric injectors in the upper part of the furnace, the carbon monoxide evolved during refining is burnt to carbon dioxide above the bath surface through post combustion. The heat thus liberated is used to preheat scrap and hence is the name. It has generally been acknowledged that bath mixing in EOF is sluggish, carbon boil at time is vigorous and as a result, considerable amount of Fe is lost through the slag door. To reduce excessive carbon boil and to minimize yield losses, an operator is often forced to blow oxygen at less than desired rate, prolonging tap to tap time. This is undesirable and affects plant performance adversely. A trace of a typical oxygen blow profile from a 60T capacity EOF is shown in Fig.1. There, reduction in oxygen flow rate beyond first 8 minutes is readily apparent. This has been a matter of grave concern for the plant.

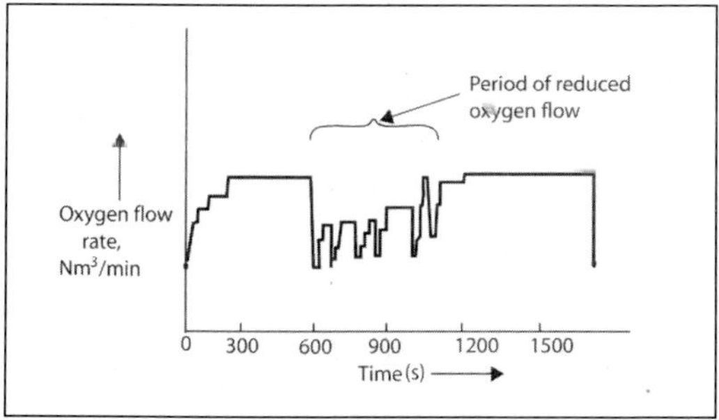

Fig. 1: The period of reduced oxygen flow rate to minimize excessive carbon boil

To address the problem and evolve remedial measures, a thermo-chemical model was developed in which, material and enthalpy balance equations were formulated. Schematics of the mass and energy flow into and out of an EOF is shown in Fig. 2. Based on such as well as rigorous water modeling trials, charge proportioning together with an angularly oriented pair of bottom tuyeres were advocated for plant scale operation. Extensive industrial trials were carried out to assess the adequacy of the model results. In Fig.3, as typical examples, predicted and measured DRI amount as well as oxygen volume are shown. These indicate that the thermo-chemical model developed in this work captures the characteristics of the EOF reasonably accurately.

(a) Mass input and output

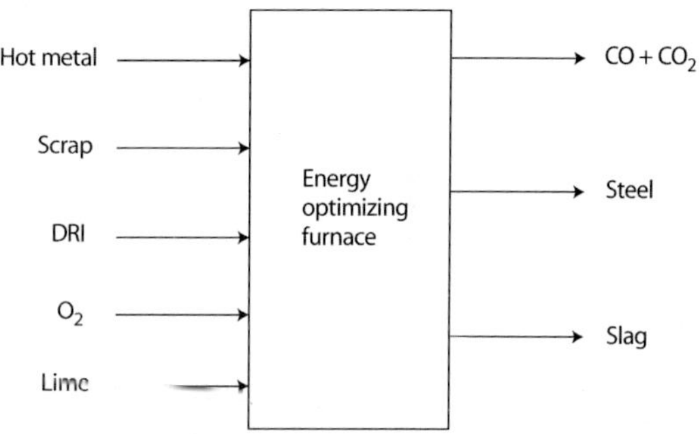

(b) Enthalpy input and output

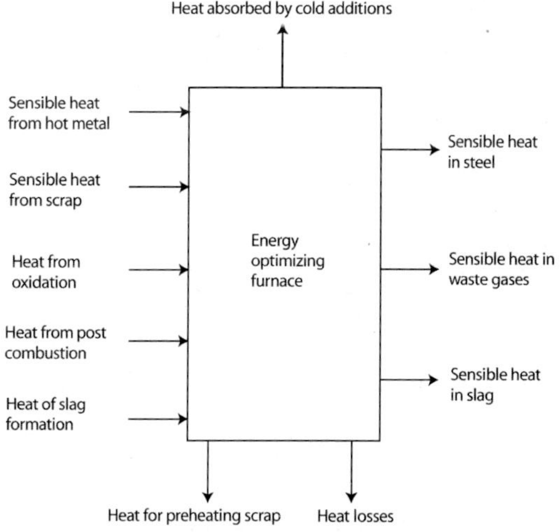

Fig. 2: (a) Material and (b) enthalpy input to and output from an energy optimizing steelmaking furnace

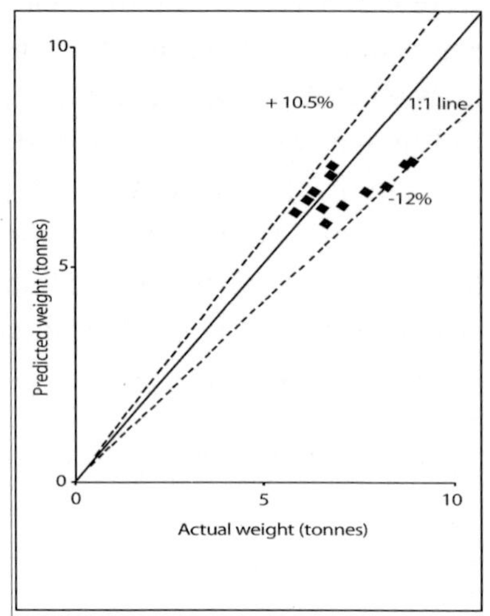

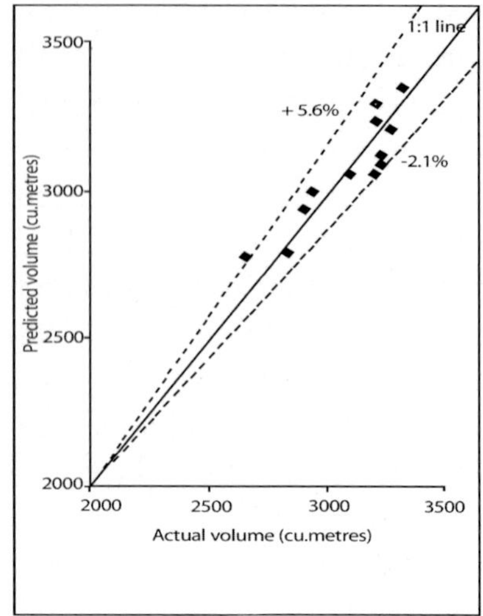

Fig. 3: Comparison between thermo-chemical model prediction and actual furnace operating conditions in terms of DRI amount and oxygen flow volumes[4]

In Fig. 4, industrial trial results before and after the implementation of modeling results (i.e., straight vs. inclined tuyere) are shown. There, as an example, improvement in tap to tap time is shown. As seen, blowing time is shorter since the same oxygen flow rate could be maintained throughout without any noticeable excessive carbon boil. This remarkable development has paved way for three additional, 60 tonnes heat on a daily basis offering huge economic benefits.

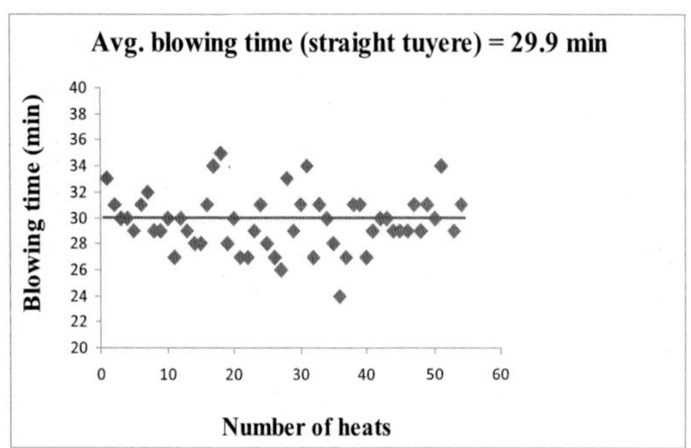

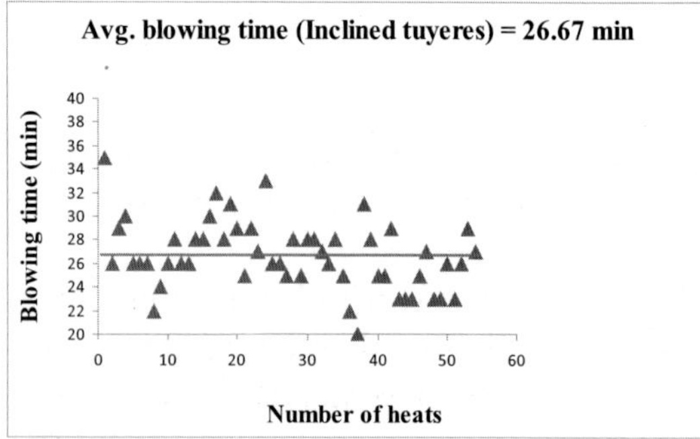

Fig. 4: Improvement in tap to tap time following the implementation of modeling results on shop floor

POST-COMBUSTION IN AN ELECTRIC ARC FURNACE

In primary steelmaking a process, carbon monoxide (CO) is generated when oxygen reacts with dissolved carbon. The reaction is highly exothermic and releases chemical energy. The CO gas can be further oxidized to CO_2 if mixed with Oxygen, before the former leaves the furnace. Carbon di-oxide thus produced, releases about three times the energy released due to the oxidation of carbon to carbon mono-oxide. The potential use of this energy and the attendant increase in productivity is the economic driving force behind post combustion technology in steelmaking. An adequate post combustion technology must meet the following two objectives, namely:

I. Efficient and optimized injection of oxygen and

II. Significant energy transport to melt or scrap

It is in such context, mathematical modeling can provide useful insight and guidelines. A substantially idealized post combustion scenario embodies the evolution of CO from the melt and its concurrent oxidation to CO_2 within the upper region of the vessel. Considerable amount of agitation in reality is present in the melt phase and this influences the rate of CO evolution, since the latter is known to be melt phase mass transport controlled. Rigorous modeling of post combustion therefore necessitate simultaneous prediction of fluid flow, heat and mass transfer within the entire vessel (comprising the melt and the upper gas phases) such that CO evolution, CO_2 generation and heat energy release are predicted with certainty. Alternatively, taking a simplistic stand and treating the melt phase as a stationary medium, post combustion phenomena can be mathematically modeled with an empirically determined rate of CO evolution, prescribed *a priori*. Such an approach simplifies the problem considerably and necessitates that flow, heat and mass transfer equations are solved for the gas phase alone, within the upper region of the vessel. The idealized post combustion model comprises of the following components:

I. A three dimensional, homogeneous turbulent flow model,

II. The standard coefficient two equation k-ε turbulence model,

III. Species conservation equations with chemical reaction and

IV. A thermal energy transport equation.

In addition many auxiliary equations / expressions are needed to represent the post combustion model in a closed form (viz., reaction rate, rate of heat generation due to chemical reaction etc.). A schematic of the EAF (calculation domain) is shown in Fig.5 while the principal operating conditions considered in numerical simulation are summarized in Table 2.

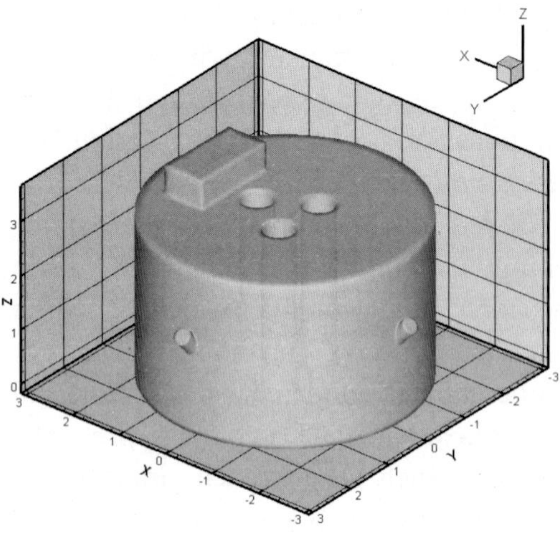

Fig. 5: A schematic of the EAF and the 3-D calculation domain

Table 2: Physical dimensions of the EAF and operating conditions

Furnace capacity	165 T
Internal diameter	5.0m
Internal height	3.0m
Electrode diameter	0.61m × 3
Vent size	1.7 × 0.5m
Oxygen injectors diameter	× 4 0.3m
Injection angle	0 or 45°
CO inflows Flow rate temperature	 2.5 kg/s 1773 K
Oxygen rate (kg/h per injector)	0.325

As a typical example of model capability, predicted gas temperature on the x-z plane is shown in Fig.6. There, large scale variation of gas temperature together with high relatively higher gas temperature in the vicinity of the melt surface is readily apparent.

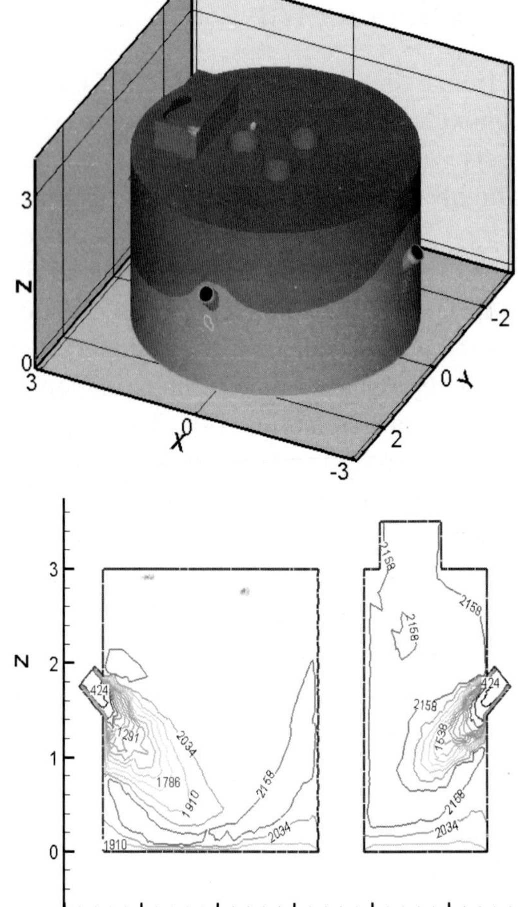

Fig. 6: Predicted gas temperature in the furnace as well as on the x-z plane in Kelvin

On the basis of the mathematical model, a parametric study was carried out to optimize nozzle position and number in the 165 tonne EAF. Results thus obtained were evaluated in terms of CO usage efficiency and oxygen consumption efficiency. These are summarized in Table 3. Evidently, the number and position of gas injectors have profound influence on the efficiency of post combustion in electric arc furnace steelmaking.

Table 3: The influence of position and number of injectors on the efficiecny of post combustion reaction in eaf

	CO (kg/s)	Oxygen injection (kg/s)	CO usage efficiency (%)	Oxygen consumption efficiency (%)
Case 1	2.51	**0.325 × 4 (45° angle)**	81.91	90.53
Case 2	2.51	**0.325 × 4 (parallel)**	71.9	78.54
Case 3	2.51	**0.26 × 5**	**87.3**	96.78
Case 4	2.51	**0.2167 × 6**	82	90.51

FLUID FLOW AND INCLUSION REMOVAL IN STEELMAKING TUNDISH SYSTEMS

Process performance of steelmaking tundish (viz., inclusion removal efficiency, slag entrainment into mold via drainage vortex, intermixed slab volume during grade change, temperature drop etc.) depends considerably on liquid steel flow characteristics which in turn depend on the tundish geometry as well as the flow modifiers. Consequently, flow conditions in any given tundish can be favorably altered by incorporating suitable flow modifiers and/or changing the geometry of the tundish, leading to improved process performance. It is, however, difficult to arrive at the optimal designs (of either tundish or flow modifiers) in the industry through direct plant scale trials, since the latter are cumbersome, and often, prohibitively expensive. Physical and mathematical modeling has therefore been customarily applied to investigate the process dynamics of steelmaking tundish system.

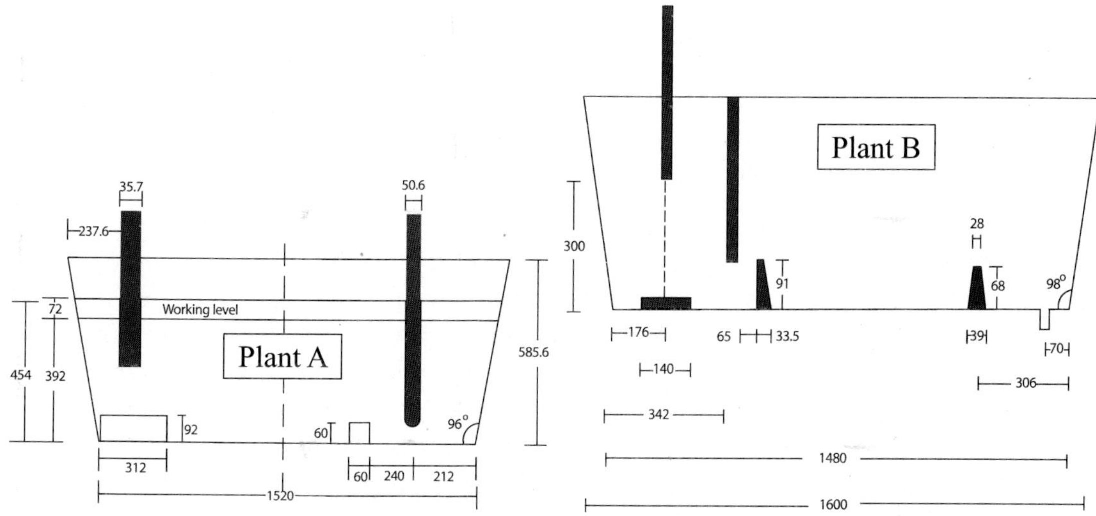

Fig. 7: Designs of slab casting tundish used in steelmaking industry (dimensions are scaled)

In steel industry, different types of products e.g., billet, bloom and slab are routinely cast through continuous casting. Since the section size is different, the number of strands, dimensions of the tundish and furniture therein vary greatly from one tundish system to another. Even for similar end products, say a slab, tundish operating practice and designs have been found to vary considerably. Interestingly, as seen from Fig.7, slab caster tundish systems deployed in the industry embody dissimilar tundish furniture. Thus, tundish in plant A while is operated with a pouring box and a near strand dam, tundish in plant B is operated with a pouring pad, a set of dams and weirs and a near strand dam. In this section, a generic slab casting tundish design has been mathematically investigated through CFD (computational fluid dynamics) and its metallurgical performance evaluated in terms of inclusion flotation efficiency.

Thus original designs of the two tundish systems shown above were modified by incorporating a basal wedge, an embedded pouring box and a naturally opening slotted, near strand dam. Their exact dimensions and placement were derived carefully by conducting a very large number of water model trials. The details of the modifications including the design and dimensions of the wedge, embedded pouring box and the near strand dam with a pair of holes are shown in Fig. 8.

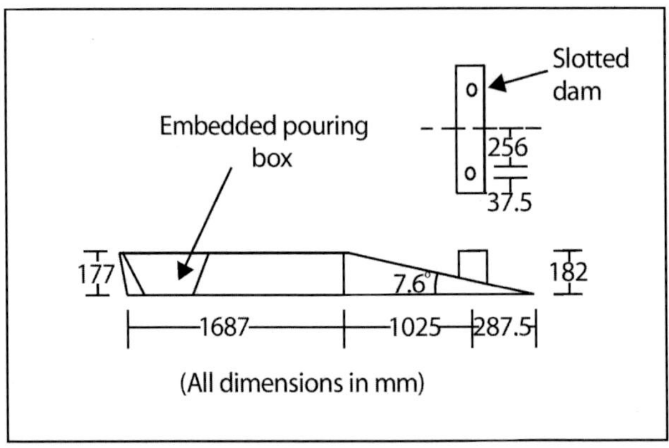

Fig. 8: A basal wedge, an embedded pouring box and a naturally opening slotted dam considered to evolve a generic slab casting tundish

Fluid flow and inclusion population distribution in the four distinct tundish configurations (two original and two modified designs) were mathematically modeled via Fluent™ (for flow) and a software developed in-house (for particle population balance). To assess the reliability of the calculation procedure developed, the experimental work on inclusion removal kinetics reported earlier by Taniguchi and coworkers [5] has been modeled first. In Table 4, present predictions are compared directly with their measurements. Very reasonable agreement between prediction and measurements is readily evident demonstrating essentially the adequacy of the present approach.

Table 4: Comparison between present prediction and experimentally measured [5] inclusion removal efficiency

Removal ratio due to	Previous work	Present work
Floatation	44	46.83
Wall adhesion	25	30.99
Total	69	77.82

To investigate this aspect further, alumina inclusion removal efficiency in the original and modified design tundish system B were predicted numerically as a function of bath depth and throughput rate. Results thus obtained are illustrated in Fig.9. There it is readily evident that modified and original tundish systems ensure practically similar inclusion removal efficiency over a wide range of operating conditions. This is to be expected since the volume reduction due to the wedge is less than 10% of the effective tundish volume. It is instructive to note here that with similar inclusion removal efficiency, modified design tundish systems led to substantial improvement in yields (due to reduction of tundish skull. Direct validation of computational results is presented in Table 5, where inclusion volume fraction, size and their types as obtained from solidified slabs are shown for operations, before and after modifications. More details of the study are available in ref. [6].

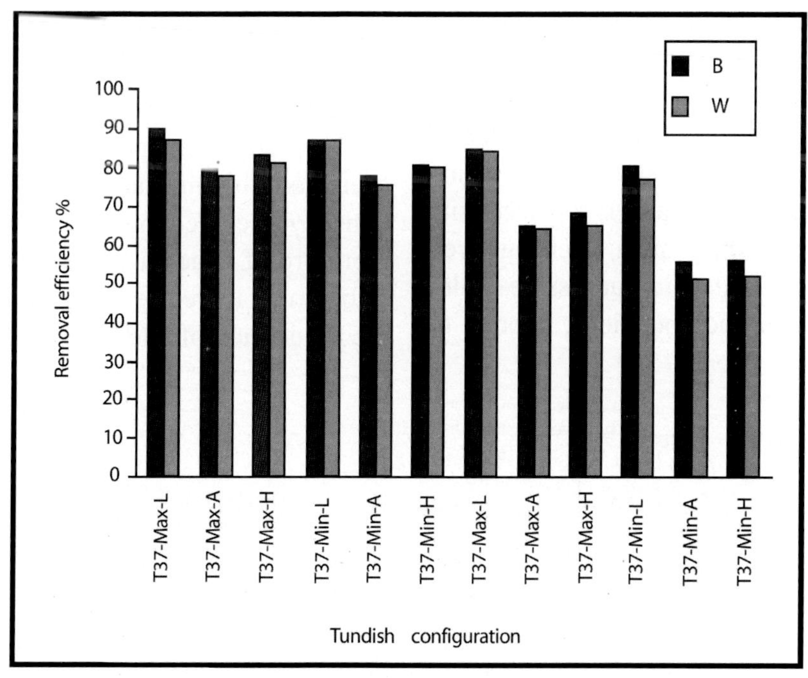

Fig. 9: Predicted inclusion removal efficiency in the original (B) and modified design tundish systems (W)

Table 5: inclusion diastribution characteristics in solidified slab samples collected from original and modified design tundish operations

Sample number	Condition /location	Inclusion size (μm)	Inclusion volume fractions (%)	Inclusion type
1	Steady/slab	43.10	0.42	Fe-Si-Al-O
2	Steady/slab	51.72	3.31	Fe-O-Si-Al
3	Steady/slab	44.82	2.82	Fe-Al-O, Al-O-Fe-S-Mn
4	Steady/Slab	45.17	2.80	Si-O-Al-Na-Fe
5	Steady/slab	31.03	1.18	Al-S-O-Fe-Mg, Al-O-Fe-Ca-Ti
6	Steady/slab	58.44	1.54	Fe-Si-O-Al

1-3 Original design tundish; 4-6 Modified design tundish

GRADE INTERMIXING AND MINIMIZATION OF INTERMIXED SLAB/BLOOM VOLUME

Slabs and blooms having mixed or transitory compositions are generally produced in continuous casting during ladle changeover and grade transition operation. During continuous casting, as steel in a teeming ladle is exhausted, the later is replaced with a new ladle. During such a changeover, if the melt contained in successive ladles are of different compositions, bloom or slab, having an intermediate composition between the two grades results, as is reflected from Fig.10. The transition bloom/ slab thus formed is often downgraded (i.e., rejected and recycled into BOF as scrap) and therefore, it is important to identify and restrict the extent of the transition bloom/ slab to the extent possible. Reduction in transition bloom/slab volume leads to considerable economic benefits and is therefore of concern to steelmakers. Every effort must therefore be made to minimize the production of intermixed bloom/slab.

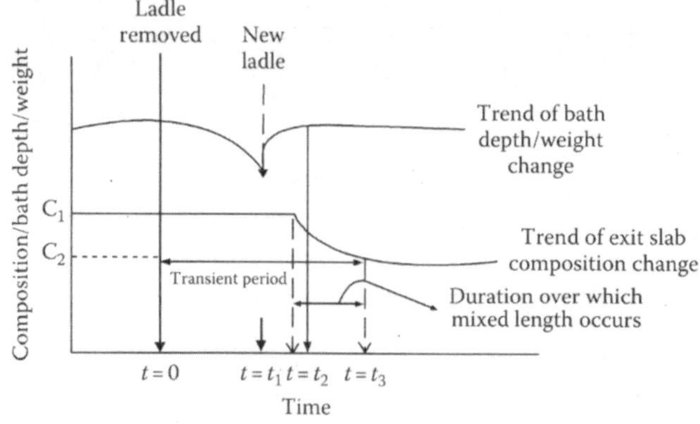

Fig. 10: Changes in melt composition, tundish depth and tundish weight during ladle change over operation

Grade intermixing in a 17 tonne, 4 strand bloom casting has been investigated in a ~0.3 scale, geometrically and dynamically similar, isothermal water model tundish system. Grade intermixing phenomena was quantified in terms of a mixing time (corresponding to a degree of 95% homogeneity) and studied by simulating mixing of water, having different dissolved tracer (electrolyte) concentration, and measuring the same near the strands as a function of time. Towards this, electrical conductivity measurement technique was employed. A typical tracer response curve obtained during grade intermixing is shown in Fig. 11.

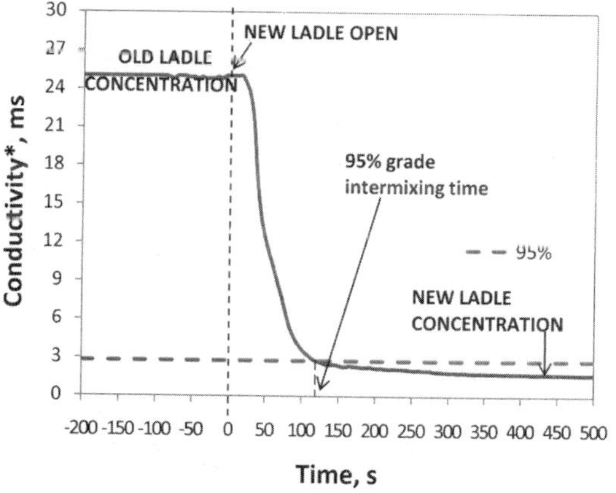

Fig. 11: A typical tracer response curve during grade intermixing experiment in a water model tundish

Influence of key operating variables such as, inflow rate from ladle, net outflow rate from tundish, residual liquid volume left over from the previous grade etc. on the intermixing phenomena was studied experimentally. It was observed that any increase in residual volume as well as inflow rate tends to prolong intermixing time. On the other hand, the outflow rate was found to have quite opposite effect. Furthermore, while strand to strand variation of intermixing time was only marginal, the tundish interior design (presence of flow modifiers, pouring box etc.) influence intermixing time significantly. Based on a large number of experimental data, dimensional analysis and regression were carried out to deduce an explicit correlation for the given tundish system. In SI unit, the same is expressed as:

$$\tau_{int.mix,95\%} = 6.38 \times V_{res}^{0.86} Q_{in}^{0.48} Q_{out}^{-1.32}$$

in which, Q_{in} is the input flow rate, $Q_{out,T}$ is the net outflow rate, V_{res} is the residual liquid volume and $\tau_{int.mix}$ is the intermixing time corresponding to a degree of 95% homogeneity. In terms of pertinent industrial variables, i.e., casting speed (U) and residual liquid depth (L), the above correlation for a given mold size (0.25mx0.25m) has been cast in the following form:

$$\tau_{int.mix}(s) = 3.59 \times L_{res}^{0.99}(mm) U_{cast}^{-0.84}(m/min)$$

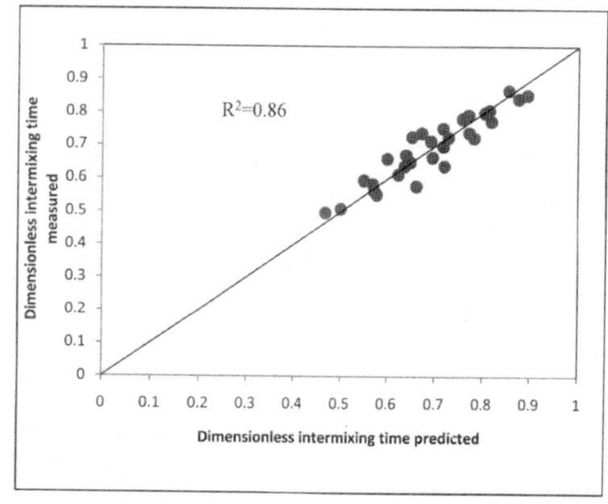

Fig. 12: Adequacy of the macroscopic grade intermixing model developed for a bloom casting tundish system

The adequacy of the proposed correlation is shown in Fig. 12. Based on such and taking into account the geometry of the caster and its relevant dimensions (distance between tundish base and SEN tip, mold length, linear distance of the gas cutter from the mold exit etc.), a bloom length optimization model has been developed such that intermixed region could be controlled and precisely restricted in a single bloom length. Industrial scale trials are currently being carried out to assess the adequacy of the model study program. More details of the study can be found in ref. [7].

CONCLUSIONS

The role of modeling in steelmaking has been highlighted through a set of examples. Four different industrially relevant problems, one each from the area of oxygen steelmaking, EAF steelmaking, tundish metallurgy and continuous casting were discussed and their modeling elucidated. Physical and mathematical model results as well as industrial scale measurements cited in this work confirm that physical and mathematical modeling are effective tools and form a useful basis for steelmaking process analysis and design.

ACKNOWLEDGEMENT

The work reported in this work has been generously supported by many Indian steel industries. In particular, the author acknowledges physical and financial support received from JSPL Raigarh, JSW Torangallu, RINL Vishakhapatnam and Hospet Steel, Karnataka.

REFERENCES

[1] D.Mazumdar and R.I.L.Guthrie, ISIJ International, Vol.35 (1),1995, pp. 1-20.

[2] D.Mazumdar and R.I.L.Guthrie, ISIJ International, Vol. 39(6), 1999, pp. 525-548.

[3] B.G. Thomas and L. Zhang, Iron Steel Institute Japan (ISIJ) International, 41:10, 1181-1193, 2001.

[4] A.Muthuchamy, A thermo-chemical model for charge proportioning in an energy optimizing furnace, M.Tech. Thesis, Dept. of MSE, IIT Kanpur, 2009.

[5] L. Zhang, S.Taniguchi and K.Cai, Materials and Metallurgical Transactions, Vol. 31B, 2000, pp. 253-266.

[6] K.Rajasekar, Modeling of inclusion removal kinetics in steelmaking tundish, M.Tech. Thesis, Dept. of MSE, IIT Kanpur, 2009

[7] K.Muralikrishna, Physical modeling of grade intermixing phenomena in a four strand bloom casting tundish, M.Tech. Thesis, Dept. of MSE, IIT Kanpur, 2012

Numerical Modelling of Complex Manufacturing Process Chains in the Metal Forming Industry

Gabriel Mc Bain, Hendrik Schafstall and Stefan Zimmer
Simufact Engineering GmbH, Tempowerkring 3, 21079, Hamburg, Germany
Email: office@simufact.de

ABSTRACT

Simufact Engineering GmbH is a global operating software and service company for process simulation in the manufacturing industry. For more than 15 years, the Hamburg (Germany) based company develop and support simulation solutions for the design and optimization of manufacturing processes in metal forming. Simufact is one of the leading companies worldwide in this area. Based on well established technologies, Simufact develops and globally markets industry solutions for the forming, welding and heat treatment. Simufact customers come from all areas of manufacturing technology: forging, cold massive forming, rolling, sheet metal forming, mechanical joining, heat treatment, and welding.

The global competition and new trends like "blue efficiency", "e-mobility" and "biological engineering" require a new approach in the product design and manufacturing. Companies must change from a geometry oriented manufacturing to a property oriented manufacturing. Lightweight design, cost-saving materials, high strength steel and multi-material components create new challenges for the manufacturing industry and for the simulation software.

Simufact identified the new demands for simulation software which lead to process chain simulation, high precise material data and heat treatment simulations. Using the Simufact product family our customers were the first who can simulate a complete process chain from the casted ingot to welding or mechanical joining in a user-friendly way. This requires highly accurate results and shortest computational times

Future product developments as well as the development of their related manufacturing processes will be driven by quality requirements, resource-effectiveness and targeted component properties. Already now, many manufacturing process steps are optimised, but this optimisation is done individually for each manufacturing step, typically without considering the interaction of individual steps with subsequent process steps. This limits the potential which can be taken advantage of by optimisation.

A typical, but complex manufacturing process chain in the metal forming industry begins with a casing process, is followed by open die forging processes in one or more heats. Then machining is carried out and finally the roughly machined part is subject to heat treatment. The overall material yield, energy input and final part properties are the determining factors for effectiveness of the manufacturing chain.

By targeted adjustments of entire process chains new significant potentials in respect to product properties and quality, energy consumption and manufacturing effort can be activated. This is in particular the case for process chains with high energy requirement, e.g. incremental forming processes like open die forging, shell forging and ring rolling. The prediction of the final part properties requires the knowledge of the property determining micro-structural mechanisms and for their precise modelling at an interdisciplinary level. Assuming homogeneous material properties and neglecting the impact of previous steps is not admissible for the task of optimising process chains. Also, the material scientist and the manufacturing specialist must work together.

A modern tool to exemplify and visualize these interactions is numerical process simulation. By means of process simulation feasibility studies as sensibility studies can be efficiently carried out and herewith a deeper process understanding be gained, which is the basis for further process optimisation.

Newly developed interfaces in Simufact.forming provide the possibility to import simulation results from casting simulations to be used as initial conditions for open die forging simulations. With the new material data infrastructure in Simufact.materials material properties based on the local concentration of alloying elements can be considered and mixed phases and the resulting material properties computed. Depending on the physical effects and the materials modelled the required metallurgical models are applied. Their management is carried out by a new integrative material data management tool which cross-links the material history across the entire manufacturing process chain. The control of the kinematically complex processes cogging and ring rolling is provided by Simufact.forming by already available control modules.

We will present our recent developments to support process chain developments and optimisations to meet future challenges of incremental forming processes by the help of numerical simulations.

KEYWORDS: *Steel, Numerical Modelling, Finite-Element-Method, Material Database, Material Models*

INTRODUCION

Figure 1 shows a typical process chain for the manufacturing of a metal component. Starting form a casted billet it is forged in several heats into a pre shape. The pre shape is machined and heat treated. A final machining makes the part ready for the use under operating loads.

Fig. 1: Process chain from the cast Ingot to the final heat treated product

During the individual process steps the microstructure is repeatedly impacted and modified by different mechanisms. This is during the casting process of the initial microstructure, which is not homogeneous. During the subsequent process steps plastic deformation, temperature changes and gradients of the chemical composition lead to the mechanisms shown in Figure 2.

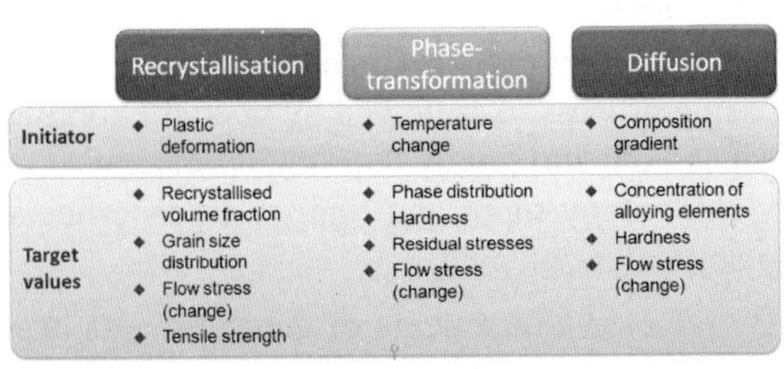

Fig. 2: Most important physical effects for the microstructure during forming and heat treatment

These mechanisms are considerably affected by the initial microstructure and the local composition at the beginning of the open die forging process and require to be considered in the simulation.

An example for this shows Terhaar et. al. (2010) presenting cogging simulations in which the initial microstructure defined by the solidification structure of the ingot leads to anisotropic plastic yield due to the crystal orientations and to locally different resistance to damage due to segregation. Combining both, the occurrence of damages in the core of ingots made of alloy 718 can be explained.

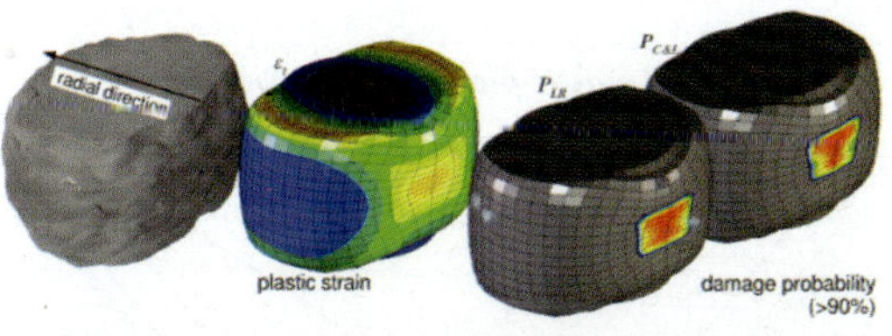

Fig. 3: Macroscopic blowholes from casting simulation as initial geometry for cogging simulation in Simufact.forming based on a hexahedral mesh.

For the consideration of the casted condition as the initial condition for the simulation of the subsequent metal forming and heat treatment steps an interface was developed with an industrial partner. This interface transfers all relevant result values. The results are imported and mapped to the mesh in Simufact.forming.

Figure 4 illustrates this on the example of the local contents of the alloying elements Carbon and Chromium. These local contents are traced during the entire process chain simulation and modified according to the physical properties (deformation, temperature and diffusion). This has in return an effect on the local material properties. More physical models have been integrated during the project, in order to calculate the properties of the product during the manufacturing process as well as of the finished product. This was carried out in close cooperation with nameable research institutes. The simulation provides the initial parameters which linked and interacting with the integrated material models are carried along the entire process chain simulation.

Die microscopic blowholes have been initially computed with MagmaSoft based on the Nijama criterion and are converted to the relative density for the subsequent simulation.

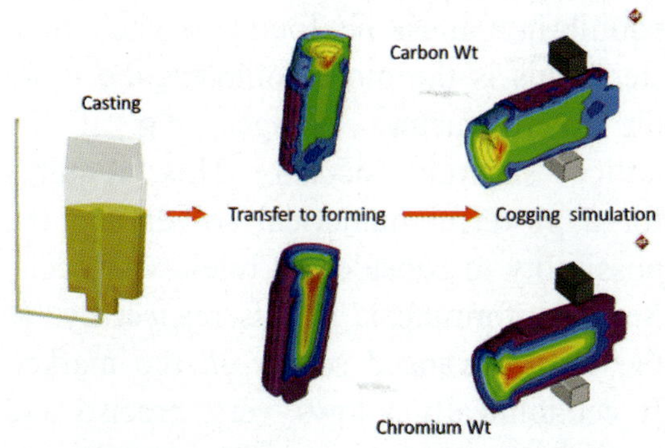

Fig. 4: Content of alloying elements

Fig. 5 shows the simulated temperature distribution (right) and the plastic strain during the open die forging in the first heat. It shows clearly the areas with different temperatures, which in turn superimpose segregations and lead to locally different conditions, which are also partially visible at the end of the workpiece.

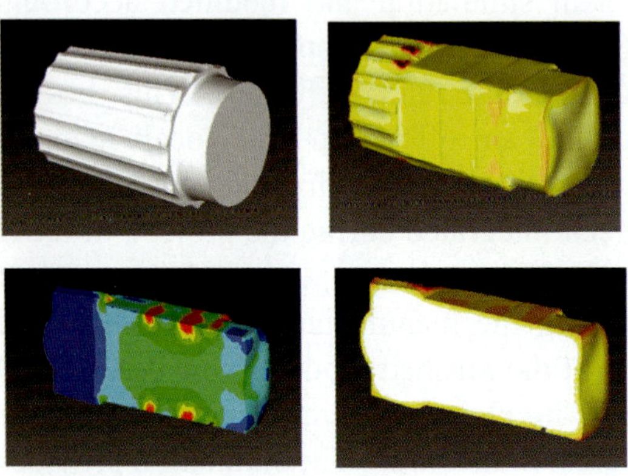

Fig. 5: Temperature distribution (right) and plastic Deformation (below, left) at the end of stage 1 for the first heat

SOFTWARE ENGINEERING REALISATION

The different material models and the incremental external loads result in ever changing conditions, so that a new equilibrium must be found in each time step. This is the more complex, the more the microstructure changes, e.g. if the lattice shearing occurs. This requires for a powerful numerical solver and the possibility to consider all relevant aspects. Simufact.forming is in this respect by far the most advanced solver on the market. It economically allows very precise and realistic computations. The simulations are carried out with hexahedral-elements, which provide a very precise computation of the stress and straining state. This results in precise predictions of the final component properties. Figure 6 shows the interdependencies and the different submodels implemented in Simufact. forming.

This is a core functionality and a basic necessity, because all models are meritless if no precise calculations are possible. Consequently, a real elasto-plastic material model is used to accurately compute the residual stresses. They are computed during all process steps until the final part, where they can be used as auxiliary stresses to bear the loads.

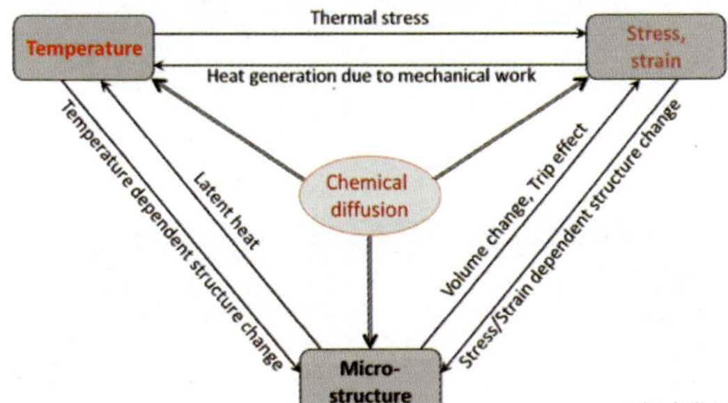

Fig. 6: Interdependencies and submodels for the material model

A new flexible data structure for the handling of material data required for these computations was created and implemented in Simufact.forming. This material data base considers the chemical composition and provides an infrastructure for phase dependent material properties. Based on the phase composition at the individual integration points microstructure dependent material properties are computed and used for the calculation of the stress state.

Next to the internal material data infrastructure an open and modular external material management tool was developed which centrally files and visualises all required material parameters (Fig. 10). New models can easily be added and coupled with the simulation. It stores all material properties required for the forming simulation as well as the parameters for the different microstructure and phase transformation models. The full integration to Simufact.forming allows the computation of the forming process at the same time with phase transformations and the microstructure development (see also Figure 6). These programs have been integrated in a modular software package called Simufact.premap (prediction of material properties). This concept allows to meet the future requirements of simulations in this field. The precision of the simulation results is determined to a large extent by the precision of the material parameters. For this task interfaces to different material data bases are provided to combine material data for the individual models. An innovative source of material data is e.g. JMatPro of SenteSoft, which computes the all relevant material properties based on the chemical composition for the entire temperature range and for all phases.

AUTOMATIC PROCESS CONTROL

The control of the open die forging process is another essential component of the simulation system for process chain modelling. The forming of the casted ingot requires a large number of operations in several heats. As soon as the formability is lost due to temperature loss the workpiece must be reheated in a furnace. The course of the forging steps is determines the soundness of the forging and the microstructure in which the defining parameters are mainly the static and dynamic recrystallisation due to plastic strain. But also at the edges of the workpiece or in the contact zone to the manipulator grips the faster cooling rate causes phase transformations.

The number of individual forming blows can easily reach one thousand. In between

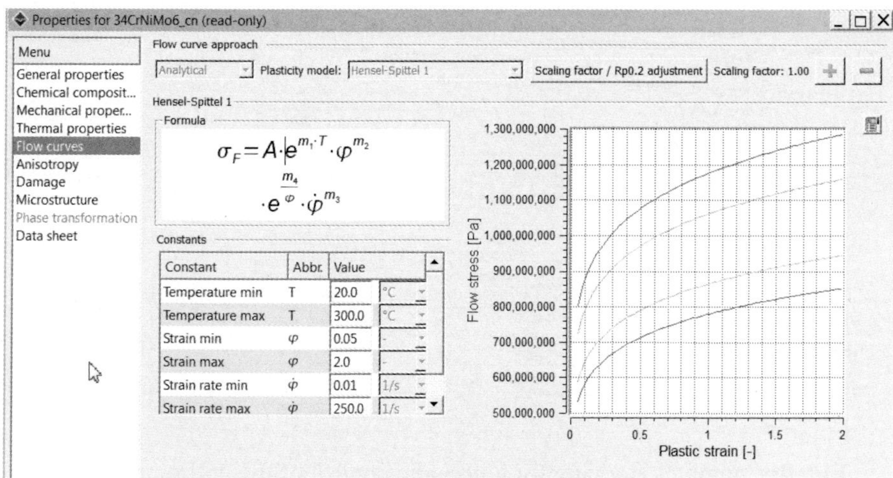

Fig. 7: Open material database Simufact.materials

the blows and after the pass pause times have to be considered as well as the transport and heating times in between the heats. The time history and subsequently the temperature field should be modelled as precisely as possible since they are determining the force requirement as well as the final product properties. Moreover, depending on the strategy the workpiece is subject to different reaction movements, which are to be considered.

Hence, in the last years an adaptive control was implemented in Simufact.forming providing an automatic cogging simulation for several heats. For this purpose a pass plan is roughly to be given (Fig. 8), the handling with one or multiple manipulators, the number of blows for each pass, etc. are automatically determined. This is possible due to the continuous evaluation of the simulated process on the basis of the geometry, forces, strains and temperature among others. These characteristics are used as criteria for the further course. Thereby the Elasticity of the manipulator can be considered. The kinematic control is a modular concept and can be individually extended, even to the point of self-optimizing systems. Given that, each user can implement his special process particularities. This special kinematics module of Simufact.forming provides an economic analysis of open die forging processes without user interaction and empowers the user to study a large number of variations to increase the process stability and to determine the main influencing factors on the final product properties.

The presented possibilities are transferable to ring rolling processes, in which starting from the ingot the upsetting, pre-forging and piercing are modelled prior the simulation of the ring rolling process in which the pre-form is rolled to its final shape. During these steps the microstructure is modelled. For the ring rolling process also all relevant process

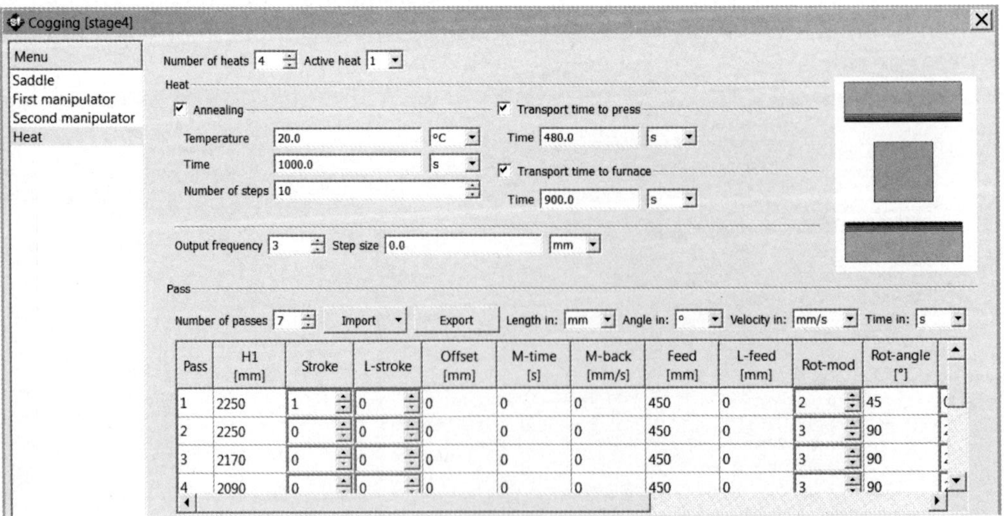

Fig. 8: Cogging kinematic for automatic control of the entire process in Simufact forming.

control mechanisms are implemented in the ring rolling module of Simufact.forming (Fig. 9). Based on the development of e.g. the ring diameter and force among others, the control parameters are automatically computed in the simulation program and used for the control of the rolls. The precise representation of the kinematics combined with the use of Hexahedral-elements with an elasto-plastic material law guarantee for a precise computation of the ring deformation and rotation and subsequently a precise computation of the residual stresses and the microstructure. This is particularly important for thin rings which can bulge and dent due to the

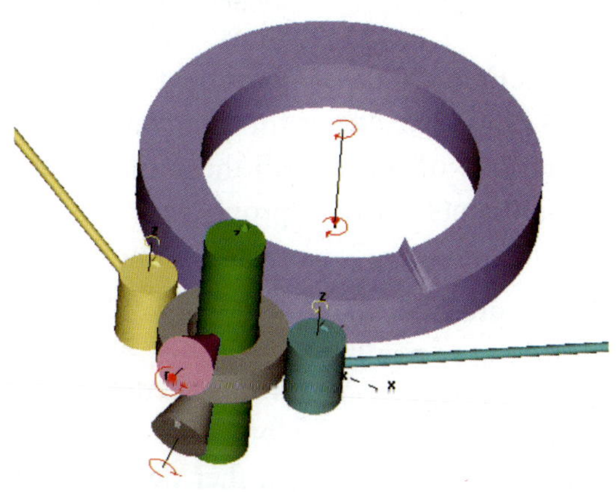

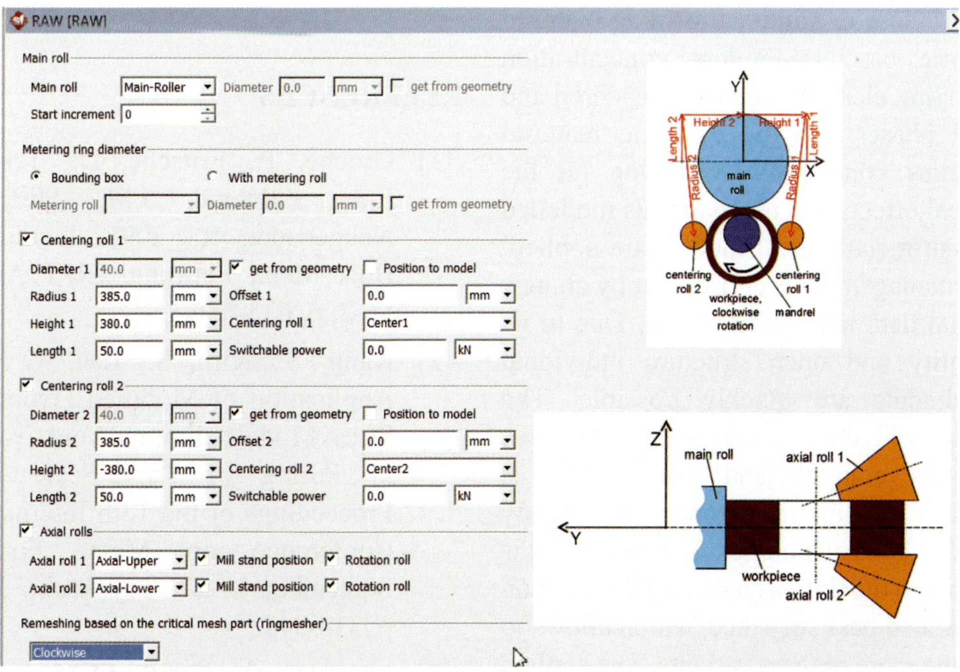

Fig. 9: Ring rolling module and automatic control in Simufact.forming.

remaining residual stresses as soon as the ring is released. After the cooling process the ring can be cut in Simufact.forming to its final geometry and then be subjet a heat treatment simulation. The so gained knowledge of the entire process chain can be used for further optimisations of the ring manufacturing.

CONCLUSIONS

In this paper the current developments of the simufact engineering GmbH in the field of process chain simulations of large incrementally formed components have been presented.

Newly developed interfaced provide the possibility to import simulation results from casting simulations to be used as initial conditions for open die forging simulations. With the new material data infrastructure in Simufact.forming material properties based on the local concentration of alloying elements can be considered and mixed phases and the resulting material properties computed. Depending on the physical effects and the materials modelled different metallurgical models are applied. Their management is carried out by an new material data management tool. Due to its flexibility and open structure individual amendments are quickly possible. The control of the cinematically complex processes cogging and ring rolling is provided by Simufact.forming by already available control modules. Through this it is possible to quickly set up a model based on the planned pass sequence, which allows to examine more process variants. The applied hexahedral element technology guarantees for the required highest result quality.

A target is to lay out manufacturing processes with the highest possible resource effectively and high process stability and optimised final product properties. This requires the understanding of the interactions of the individual manufacturing steps which allows to adjust product properties at an early manufacturing stage and to develop countermeasures for potential defects arising from previous steps. Thereby the milling and heat treatment can also be considered. Simufact. forming provides a common platform for all these manufacturing steps. The product properties are predicted by applicable material models, which are consequently further developed and extended. Due to its open and flexible structure and data base infrastructure Simufact.forming is prepared for future requirements and is continuously further developed according to the state of the science.

REFERENCES

[1] Groche, P., Fritsche, D., Tekkaya, E.A., Allwood, J.M., Hirt, G., Neugebauer, R., 2007. Incremental Bulk Metal Forming. CIRP Annals, 56-2, p. 635-656.

[2] Kang, S., Kim, S., Lee, Y., 2010. Application of Modified Hydrostatic Stress Model to Internal Void Crushing in Foring of Large-Scale Ingot. Proceedings of the 13th International Conference on Metal Forming, September 19-22, Toyohashi, Japan, p. 314-317.

[3] Mendoza, V., 2010. FEM Analysis of Defects and Microstructure Evolution during Hot Working of

Specialty Alloys. Proceedings of the 13th International Conference on Metal Forming, September 19-22, Toyohashi, Japan, p. 106-109.

[4] Schafstall, H., Barth, C., 2010. An innovate Approach to Automated Simulation of Full 3D Ring Rolling Processes and Other Incremental Forming Processes. Steel research international, vol. 81, no. 9, p. 202-205.

[5] Terhaar, J., Poppenhäger, J., Bokelmann, D., Schafstall, H., Kelkar, K., 2010. Considering the solidification structure of var ingots in the numerical simulation of the cogging process. 7th int. Symposium on Superalloy 718 and derivates, Oct. 10-13, 2010, Pittsburg, US.

[6] Utsunomiya, H., Yukimoto, T., Sakai, T., Suzuki, S., Nakajima, H., 2010. Pore Closure in Multi-Pass Cold Rolling of Lotus-Type Porous Copper. Steel research international, vol. 81, no. 9, p. 158-161.

[7] Xiu-quan, C., Yu-lin, H., Qin-xiang, X., Bei-gu, K., Wen-xuan, C., 2010. Closing of Central Cavity in Shaft Heavy Forging. Steel research international, vol. 81, no. 9, p. 330-333.

[8] Niyama, Uchida, Morikawa, Saito; A method of shrinkage prediction and its application to steel casting practice; AFS International Cast Metals Journal; 1982. Schafstall, H., Mc Bain, G.,Barth, C., Terhaar, J., Jaromileck, J., Sauermann, R.,; From cast ingots to the heat treated part - automated simulation of the process chain., Proceedings of the ICRF 1st International Conference on Ingot Casting, Rolling and Forging, 3.-6.06.2012, Aachen, Germany.

Thermo-Mechanical Axial-Torsion Testing to Assess Workability – Modeling Using Finite Element Method and Experimental Validation

Utpal Borah, Diptimayee Samantaray, Shaju K. Albert, A.K. Bhaduri and T. Jayakumar
Metallurgy & Materials Group, Indira Gandhi Center for Atomic Research, Kalpakkam-603102
Email: borah@igcar.gov.in

ABSTRACT

Owing to the inevitable multiaxial state of deformation kinematics and stresses in bulk metal forming processes, evaluation of workability calls for materials testing using combined axial and torsion tests. A high temperature combined axial and torsion testing system (ATTS) was developed in IGCAR for assessing workability of various materials under multi-axial state of stress and deformation conditions up to 1300 °C. In this paper, results from finite element modeling of axial torsion tests and seamless tube extrusion are presented. Combined torsion-compression tests would represent the state of stress and extent of deformation in tube extrusion more closely than axial torsion with tension or pure torsion tests.

KEYWORDS: *Axial-Torsion Testing, Finite Element Method, Tube Extrusion, Workability, Stress Triaxiality, Torsional Ductility.*

INTRODUCTION

In bulk metal forming processes, the billet experiences multi-axial state of stress and strain and workability of materials needs to be evaluated under multiaxial state of stress. Workability is generally defined as the ability to form a material within the load capacity of the available tooling and equipment without the introduction of failure or development of undesirable microstructure [1]. Evolution of microstructure during forming would depend on the kinematics and stress state in the material apart from temperature. Optimization of complex forming processes to produce defect-free parts with desirable microstructure calls for control of deformation parameters like strain rate, strain path, stress and temperature. Workability of the material is sometimes exceeded in the forming processes leading to undesirable flaws in the formed part and in some instances to fracture. Fracture controlled failure in relatively uniform deformation or shear localization controlled failure in localized failure plastic deformation may lead to such failures [2]. Hence, understanding of the behavior of materials under conditions similar to actual metal forming processes is essential for proper evaluation of workability.

Thermo-mechanical axial torsion testing imparts multi-axial state of stress and strain in the specimen up to a substantially large strain level. Numerical modeling for evaluation of the kinematics, stress state and temperature profile and history would assist in proper design of multi-axial deformation processing experiments to assess workability of materials. This work aims at evaluating the kinematics and stress state of viscoplastic axial-torsion deformation in both compression and tension modes usingFinite Element Method (FEM).

The next section discusses about thermo-mechanical axial torsion testing. Section 3 discusses about FEM modeling of seamless tube extrusion and axial torsion tests. Results are discussed in Section 4 followed by conclusions in Section 5.

THERMO-MECHANICAL AXIAL TORSION TESTING

A high temperature combined axial and torsion testing system (ATTS) was developed in Metal Forming & Tribology Programme (MFTP) in Indira Gandhi Center for Atomic Research (IGCAR) for assessing workability of various materials under multi-axial state of stress and deformation conditions up to 1300 °C. The machine has torque capacity of 1000 N-m up to maximum rotational speed of 850 RPM and axial load capacity of ±100 kN up to maximum linear speed of 1.5 m/s. The digitally controlled ATTS can carry out controlled thermo-mechanical tests under concurrent torsion and axial (tension/compression), sequential torsion and axial, pure torsion, compression and tension in twist, torque, axial load and stroke control. The ATTS is fitted with a 6 kW induction heating system along with a controlled water jet based specimen quenching system.

The machine holds the specimen in a hydraulic grip which can exert different clamping loads up to 120 kN on the specimen shoulder by varying the hydraulic fluid pressure. The machine is of modular design with independent actuators for torsion and axial loading of the specimen. The axial load is imparted using a double acting servo hydraulic actuator through the lower grip. A hydraulic motor mounted on the movable top crosshead imparts torsional load on the specimen through servo hydraulic valves up to 850 revolutions. This is a unique feature of the machine which enables imparting torsional load up to failure of the specimen thereby leading to evaluation of torsional ductility of materials. The angle of rotation is measured by an angular encoder with resolution of 0.036 degree. Conventional LVDT is used for measurement of axial stroke and a multi-axial load cell is used for measurement of torque and axial load simultaneously. Axial torsion tests were carried out on 316L stainless steel using this machine for validation of the FEM model for axial torsion tests.

THERMO-MECHANICAL FEM MODELING

Seamless Tube Extrusion

An axisymmetric FEM model was developed using the general purpose FEA code ABAQUS v6.10 and executed for extrusion of an annular billet of 172 mm outside diameter (OD), 80 mm inner diameter (ID) and 600 mm length to a seamless tube of 114 mm OD with 20 mm wall thickness (WT). The OD and WT reduction ratio were 1.51 and 2.3 respectively. As both the billet and tube are of circular shape, an axisymmetric FEM model would suffice for this process. Bilinear reduced integration elements were used for the billet. The die and mandrel were modeled as rigid bodies. Extrusion was carried out at ram speed of 250 mm/s. In view of inevitable severe distortion of the elements due to large deformation in extrusion, Arbitrary LagrangianEulerian (ALE) adaptive meshing was used to preserve the desired quality of the FE mesh. The coefficient of friction between the billet and tool surfaces was taken as 0.1. The mandrel was held fixed coaxially inside the die and downward stroke applied at the top surface of the billet thereby pushing the billet through the die.

Axial-Torsion Testing

Various authors have proposed multiple plasticity models for torsional deformation [2-7]. However, literature on numerical modeling of thermo-mechanical axial torsion deformation in steels is scarce. Although the geometry of the axial torsion specimen is axisymmetric, the process is not amenable to conventional axisymmetric FEM modeling due to the torsional loading. The generalized axisymmetric element formulation in ABAQUS allows for an additional degree of freedom to apply torsional loading to the axisymmetric elements about the axis [10, 11]. These elements in fully integrated bilinear form, designated as CGAX4, were used in the FE models of the following tests:

(i) Pure torsion at 1rpm speed: gauge diameter = 15 mm and gauge length =78.2 mm, fillet radius = 25 mm, one revolution.

(ii) Pure torsion at 1rpm speed: gauge diameter = 10 mm and gauge length =10 mm, fillet radius = 5 mm, one revolution.

(iii) Torsion at 1rpm with compression at 0.083 mm/s: gauge diameter = 10 mm and gauge length =10 mm, fillet radius = 5 mm, one revolution, 5 mm stroke.

(iv) Torsion at 1rpm with tension at 0.083 mm/s: gauge diameter = 10 mm and gauge length =10 mm, fillet radius = 5 mm, one revolution, 5 mm stroke.

Elasto-plastic FEM modeling was carried out for room temperature torsion testing of 316L Stainless Steel conforming to test case (i). The model was validated with experimental torsion test data generated by the ATTS. Subsequently, elasto-viscoplastic FEM modeling of axial-torsion tests were carried out using this validated FEM model for tests cases (ii) to (iv). Strain rate dependent material properties of 0.25C steel in austenitic state at 1100 deg C as given in Ref. [12] were used as constitutive data for the FE models.

RESULTS AND DISCUSSIONS

The FE mesh and deformed configurations of seamless tube extrusion, pure torsion, torsion with compression and torsion with tension are shown in Figs. 1, 2, 3 and 4 respectively. The stresses and deformation aremaximum in the conical die region during tube extrusion due to the normal and tangential forces exerted during conversion of billet into tube.The stresses get released as the tube exits the die. In pure torsion, the stresses and deformation are maximum at the surface of the specimen. However, imposition of axial stroke in the form of

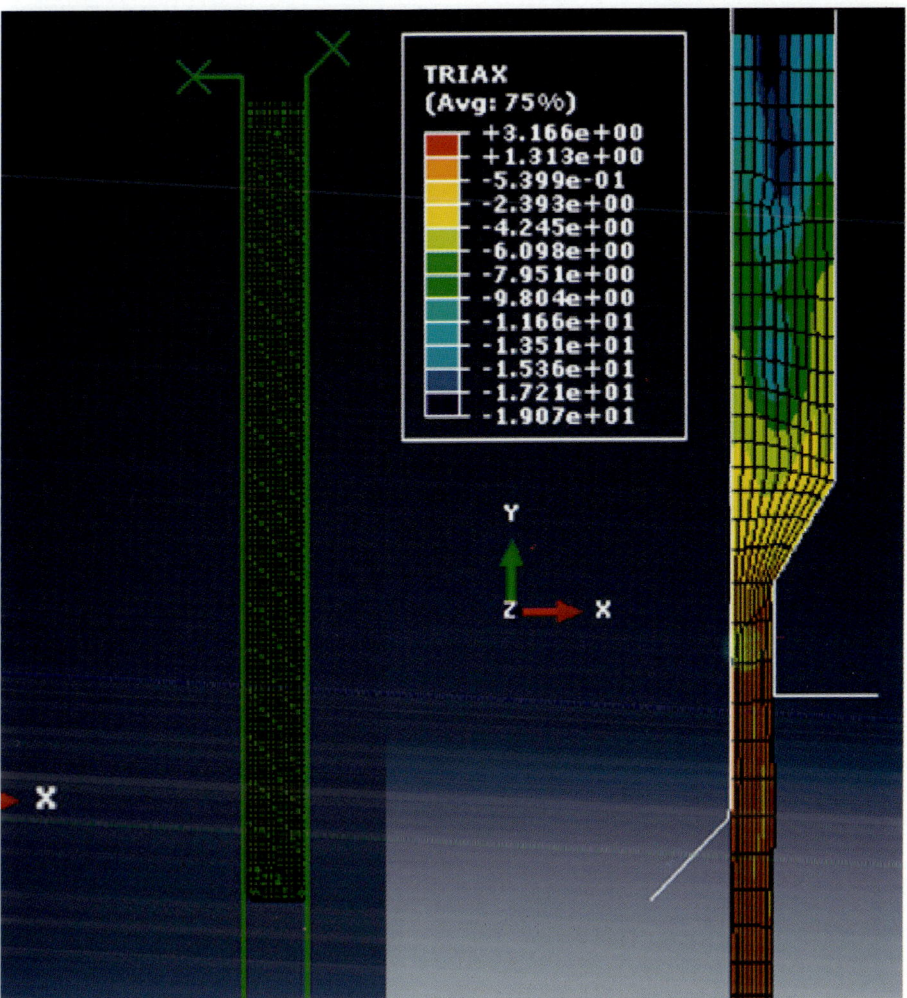

Fig. 1: Tube Extrusion – Stress Triaxiality distribution

compression or tension, substantially reduce the gradient of stresses and strains from centerline to the surface of the specimen.

Torque-Twist Curves in Axial Torsion

Comparison of torque-twist data from the FEM model for test case (i) with experimental data from room temperature torsion test of 316L stainless steel is shown in Fig. 5. The calculated torque-twist data matches very closely with the experimental data thereby validating the FEM model.

The torque-twist curves for the tests cases (ii) to (iv) generated by FEM are shown in Fig. 6. When simultaneous compression is imparted along with torsion, the torque increases as compared to that in pure torsion, indicating increase in material resistance. This is due to the triaxial compressive state of stress generated in the gauge section and

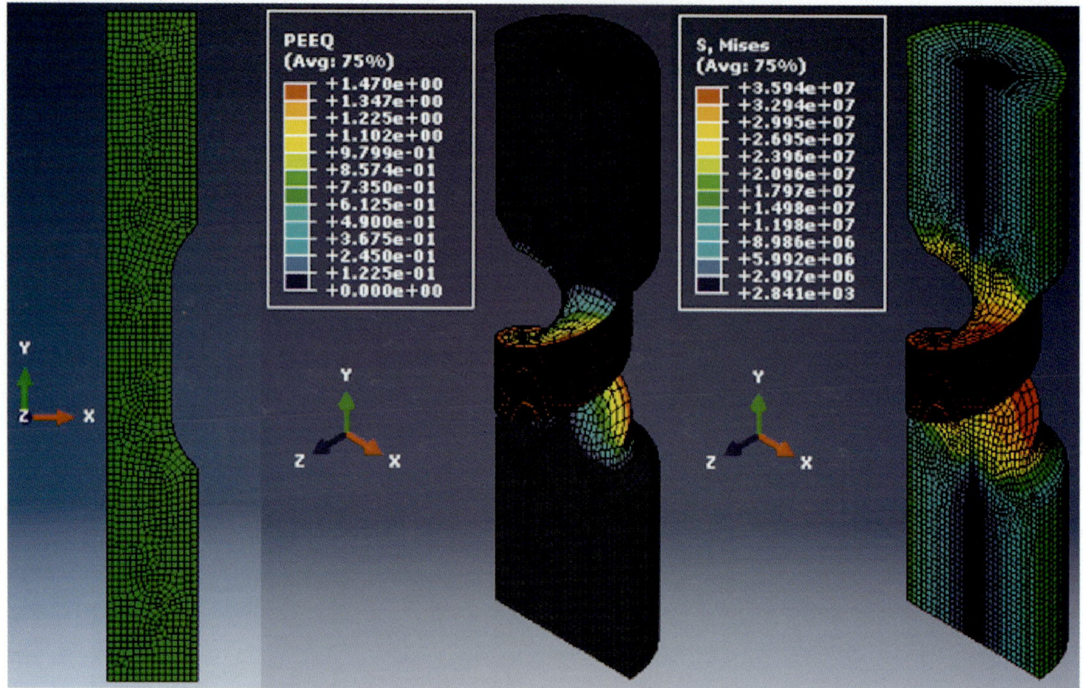

Fig. 2: Pure Torsion – FE Mesh, Equivalent Plastic Strain and von-Mises Stress distribution

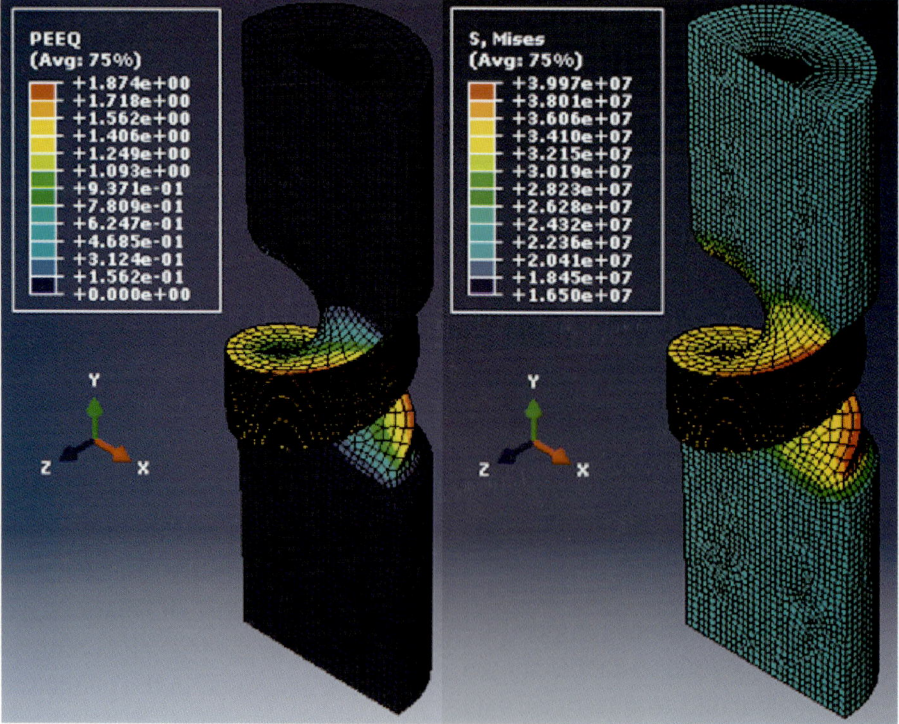

Fig. 3: Torsion with Compression – Equivalent Plastic Strain and von-Mises Stress distribution

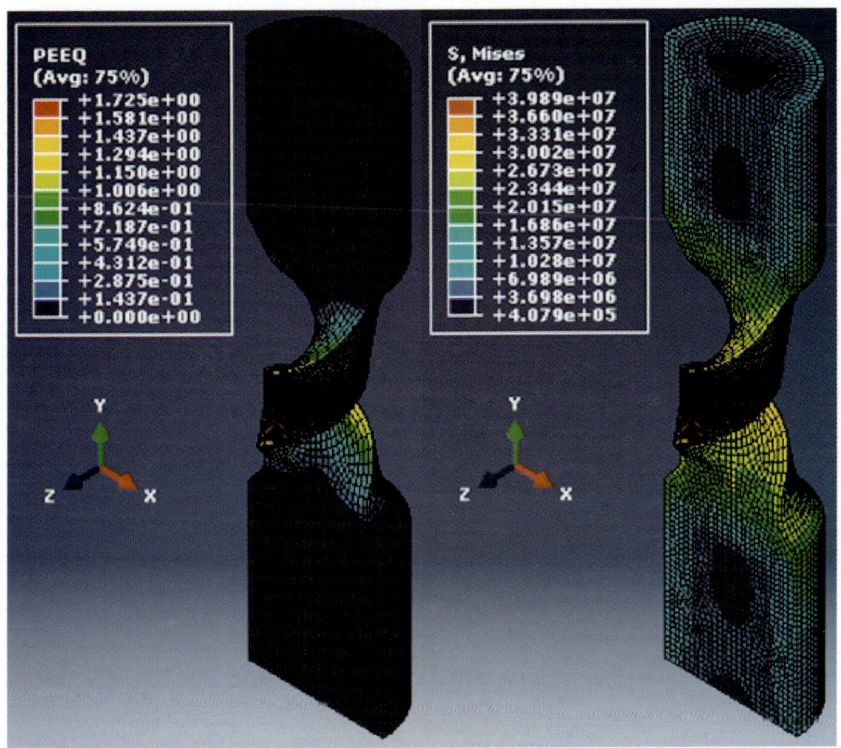

Fig. 4: Torsion with Tension – Equivalent Plastic Strain and von-Mises Stress distribution

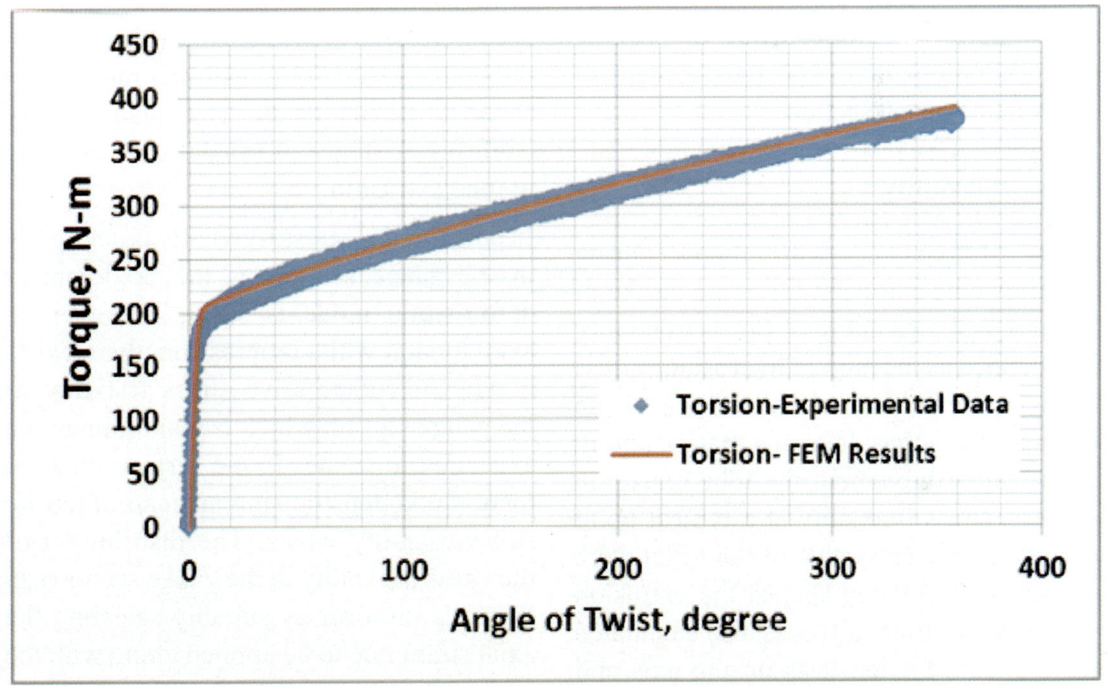

Fig. 5: Experimental Validation of Axial Torsion FEM Model

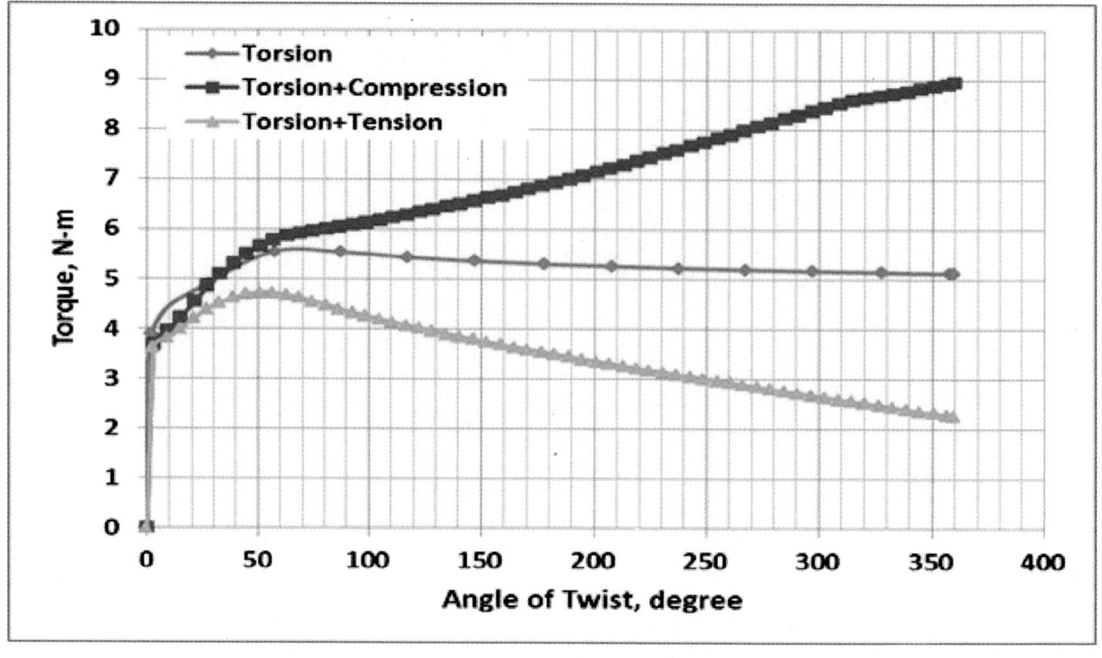

Fig. 6: Torque-Twist Curves in Axial-Torsion from FEM model

increase in diameter of the gauge section. However, the torque reduces substantially upon simultaneous application of tensile stroke along with torsion due to necking and generation of triaxial tensile state of stress in the gauge section.

Stress Triaxiality

Stress triaxiality distribution during seamless tube extrusion for a ram stroke of 500 mm is shown in Fig. 1. Stress triaxiality was defined as the ratio of hydrostatic stress to von-Mises effective stress. To evaluate the history of stress state of a material point in the billet throughout the tube extrusion process, stress triaxiality at a typical point near to the die entry side of the billet were extracted and plotted against the extrusion stroke. Stress triaxiality was also calculated for the axial torsion tests on the axis and surface along the centerline of the gauge section. The evolution of stress triaxiality in tube extrusion and axial torsion are shown in Fig. 7. In the extrusion, stress triaxiality is compressive in the die zone (identified by the dotted box) and transforms into tensile as the tube exits the die. This quantitatively confirms existence of triaxial state of stress in tube extrusion.

In pure torsion, stress triaxiality is negligible in the gauge section as the predominant deformation mode is shear. However, in axial torsion with compression, there exists a state of compressive stress triaxility in the gauge section which is maximum at the axis and propagates to the surface. In axial torsion with tension similar trend of tensile stress triaxility exists. The distribution of the stress triaxiality in the gauge section can be made uniform by suitably selecting the axial strain rate to be applied along with the torsion stroke.

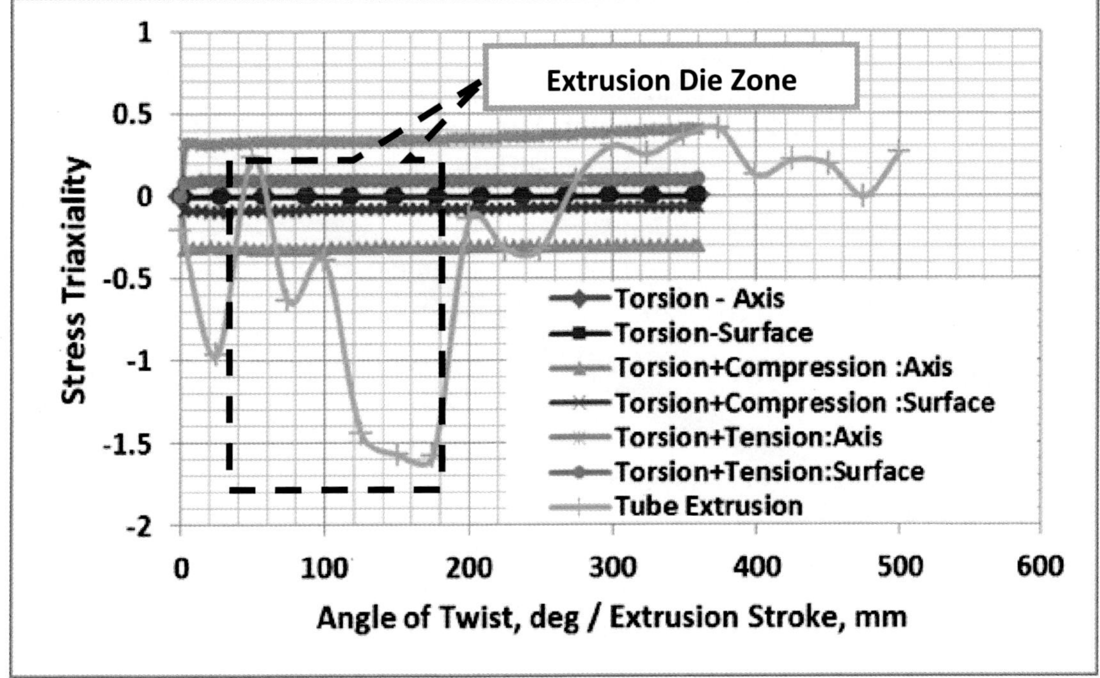

Fig. 7: Evolution of Stress Triaxiality in Seamless Tube Extrusion and Axial Torsion Testing

Thus, axial torsion tests would represent the state of stress and extent of deformation in tube extrusion more closely than pure torsion or axial tests.

Effective Strain Rate and Equivalent Plastic Strain in Axial Torsion

Evolution of effective strain rate and equivalent plastic strain in the gauge section during axial torsion tests are shown in Figs. 8 and 9 respectively. In pure torsion, there is nil equivalent plastic strain at the axis and it increases linearly with radius with the maximum occurring at the surface. Effective strain rate is also negligible at the axis and maximum at the surface. Thus a large gradient of kinematic parameters of deformation exists along with that of the stresses. This would lead to inhomogeneous distribution of various dynamic microstructural parameters.

In axial torsion with compression or tension strokes, the gradient of the kinematic and kinetic deformation parameters reduces drastically from the axis to the surface. This gradient in the gauge section can be eliminated by suitably selecting the axial strain rate to be applied along with the torsion stroke.

As the amount of twist increases, the equivalent plastic strain increases in axial torsion as compared to that in pure torsion. This phenomenon is more pronounced in axial torsion with tension and leads to reduction in failure shear strain. In room temperature axial torsion tests on stainless steel, the torsion stroke at failure was found to be maximum with compression

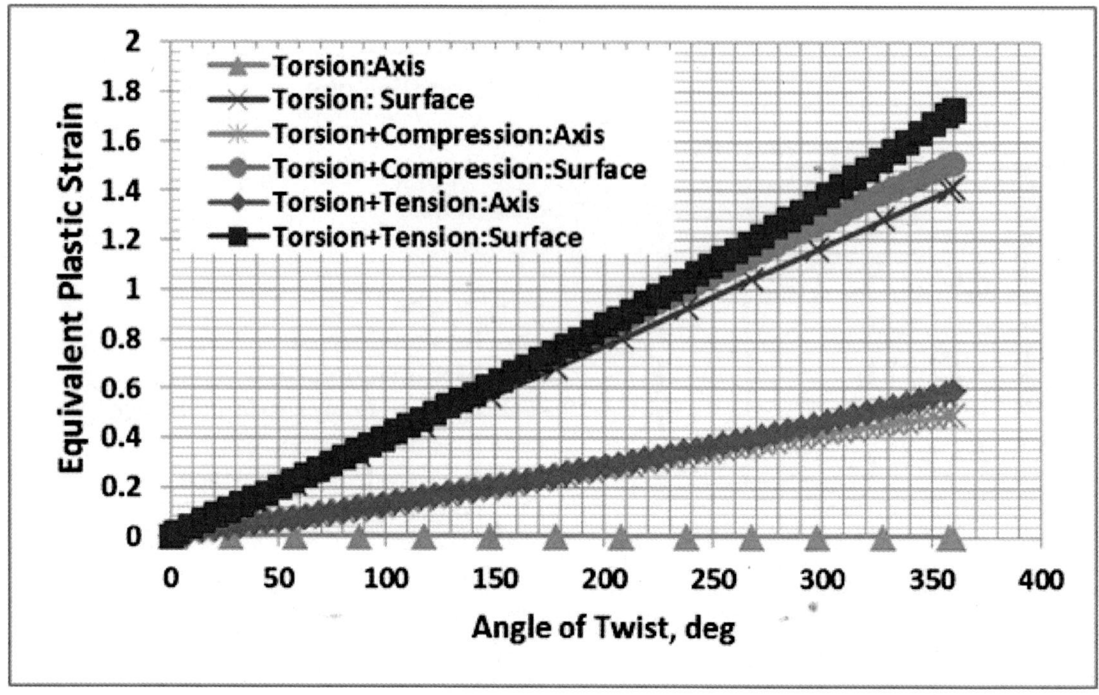

Fig. 8: Evolution of Equivalent Plastic Strain in Axial Torsion Testing

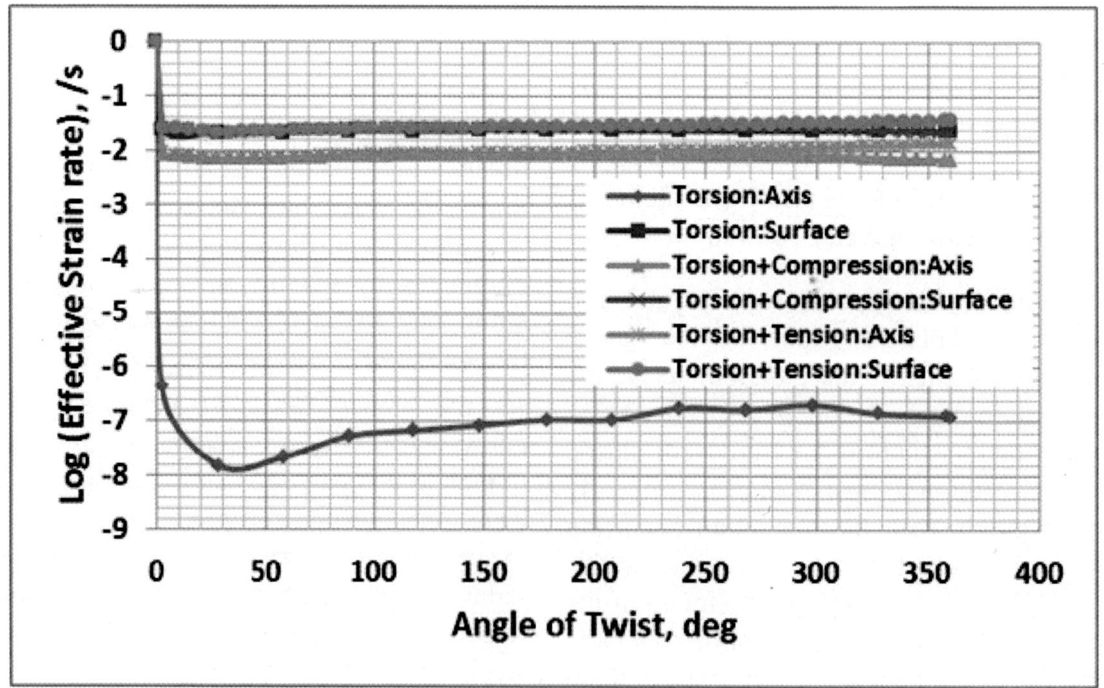

Fig. 9: Evolution of Equivalent Strain Rate in Axial Torsion Testing

followed by pure torsion and tension. The tensile stress triaxiality in the gauge section promotes necking thereby substantially preponing the failure event as compared to axial torsion with compression. Hence, axial torsion tests with compression would be more appropriate for evaluation of torsional ductility of materials.

CONCLUSIONS

Detailed FEM analysis of seamless tube extrusion and axial torsion tests were carried out in this work. The FEM model for axial torsion tests compared very closely with experimental results. In the die zone of tube extrusion compressive stress triaxiality exists which gets converted into tensile as the tube exits the die. Axial torsion tests with simultaneous compression would represent the state of stress and extent of deformation in tube extrusion more closely than axial torsion with tension or pure torsion tests.

REFERENCES

[1] G. E. Dieter, H. A. Kuhn, S. L. Semiatin (eds.), Handbook of Workability and Process Design, ASM International, 2003.

[2] E. Rauch, G. R. Canova, J. J. Jonas and S. L. Semiatin, Acta metal., Vol. 33, No. 3, pp 465-478, 1985.

[3] F. P. E. Dunne and I. Katramados, Int. J. of Mech. Sc., Vol. 40, No. 9, pp 901-912, 1998.

[4] Robert M. McMeeking, Int. J. Solids Structures, Vol. 18, No. 3, pp199-204, 1982.

[5] E. Lach& K. Pohlandt, J. os Mech. Work.Tech., 9, pp67-80, 1984.

[6] D. R. Barraclough, H. J. Whittaker, K. D. Nair & C. M. Sellers, J. of Test. &Eval., Vol 1, No. 3, pp 220-226, 1973.

[7] S. Khoddam& P. D. Hodgson, Mat & Des., 31, pp 2011-2017, 2010.

[8] P. Church & J. Manmohan, J. Phys. IV France, 7, pp C3-747-752, 1997.

[9] M. Zhou & M. P. Clode, Fin. Elem. In Anal.&Des., 31, pp 1-14, 1998.

[10] H. S. Kim, Mat. Sc. &Engg., A299, pp 305-308, 2001.(CGAX)

[11] D S Simulia Inc., Abaqus 6.10 Documentation, 2010.

[12] C. Rossard, Metaux-Corros. Ind., Vol. 35, pp 102-115, 140-153, 190-205. 1960.

Prediction of Roll Force during Hot Strip Rolling of CRNO Grades of Steel Using Mathematical-ANN Hybrid Model

S. Rath, P.P. Sengupta, V. Kumar, U. Bhaskar and R. Avtar
R&D Centre for Iron and Steel, Steel Authority of India Limited, Ranchi-834002, India
Email: srath@sail-rdcis.com

ABSTRACT

Prediction of roll force during hot rolling of CRNO grade of steel is highly challenging task to model developers compared that of other grades of steel. The major reason of this difficulty is the proper estimation of flow stress property of the material which shows some unexpected behavior during hot rolling. For most other grades of steel, flow stress generally increases with decrease of temperature for at a particular strain and strain rate. During experimentation for determination of flow stress of CRNO grade of steel in Gleeble-3500, a dynamic thermo mechanical simulator, it has been found that flow stress decreases with decreases in temperature for some temperature ranges at particular strain and strain rate. In literature, this reverse phenomenon has been attributed to the change of phase of the steel grade. This paper discusses the methodology of development, tunning and validation of a Mathematical-Artificial Neural Network (ANN) based hybrid model to predict roll force during hot strip rolling. Basic mathematical models have been developed using HSMM software to predict roll force. These models have been integrated with an ANN scheme and the hybrid model has been developed. The hybrid model has been trained with measured roll force data obtained from Hot Strip Mill of Rourkela Steel Plant through a PLC-based data acquisition system. The trained model has been validated with further data collected from the mill. It has been found that the hybrid model predicts roll force accurately. The paper also discusses the future use of the roll force model for optimizing reduction schedule to decrease the final output gauge of CRNO grade of steel coils from the mill.

KEYWORDS: *CRNO, ANN, Roll Force, Hot Strip Mill*

INTRODUCTION

Deformation characteristic of Cold Rolled Non-Oriented (CRNO) grade of steel during hot strip rolling is significantly different from other steel grades. During hot strip rolling, one can observed that if the temperature of rolled material increases then the roll force decreases and vice-versa. However, in case of CRNO grade of steel, the reverse phenomena occur at some stands. It is very difficult to design a proper drafting and speed schedule for CRNO grade due to these phenomena. Roll force plays a very important role for designing draft and speed schedule as allowable roll force is a mill structural limitation, torque and motor power are also related to roll force. This paper describes the development methodology of a hybrid model which predicts roll force accurately for CRNO grade of steel. A well-known computer simulation software HSMM has been used for simulating the hot strip rolling process of Rourkela Steel Plant which predicts roll force by mathematical model. The predicted roll force by HSMM has been passed to ANN scheme and the hybrid model has been formulated. It has been found that the hybrid model predicts roll force with high accuracy.

LITERATURE REVIEW

The objective of this literature review is three folds. Flow stress of steel grade is one of the most important material property requirements for prediction of roll force using mathematical model. So, the first objective is to find the flow stress value for some ranges to avoid repetition of experimentation in Gleeble. The second objective is to find some methodology that deal with complex combinations of chemical composition in the mathematical model. The third objective is to find out the methodology of hybrid modeling by other model developers of the world.

Serajzadeh and Taheri [1] investigated the effect of carbon and silicon on the flow stress of silicon steel. The paper gives the flows tress values of silicon steel at a low strain rate range ranging from 0.01 to 1.5 sec^{-1}. The flow stress range is 50-120 MPa at temperature 1000 ^{0}C. Sengupta et al. [2] performed experiments in Gleeble-3500 to determine the flow stress of CRNO steel at strain rate of 10 and 100 sec^{-1} at temperature range 850-1050 °C. Flow stress range is between 90-175 MPa in case of strain rate at 10 sec^{-1} and 200-275 MPa in the range in case of strain rate of 100 sec^{-1}. Calvillo et al. [3] gives flow stress values of low Carbon (0.003%) and high Silicon (2%) CRNO steel at temperature range of 800-1100 °C and strain rate at 0.01-2 sec^{-1}. The flow stress values changes from 20-85 MPa in this ranges. Zhou and Liu [4] also determined flow stress of silicon steel at strain rate 0.1, 1.0 and 8 sec^{-1} at temperature range 950-1200 °C. The flows tress values changes from 30-180 MPa in this range. It has been found from the above literature review that large amount of flow stress data are available in literature, which are very much helpful for model development. It has been also found that strain rate plays a very important role in flow stress. So, speed schedule has a significant contribution on roll force during hot strip rolling. Chemical composition of steel grade has also significant contribution on flow stress of CRNO steel.

Fischer and Schneider [5] described in detail the influence of deformation process on the improvement of non-oriented electrical steel. Hunady et al. [6] described a general equation for combining the effect of various chemical compositions on Silicon equivalent as:

$$Si_{eq} = [\%Si] + 2[\%Al] - 0.5[\%Mn] + 2.92[\%P] \quad (1)$$

Tang and Zhou [7] have also described a Artificial Neural Network (ANN) model for flow stress determination. They have used back-propagation algorithm in ANN to co-relate flow stress with strain, strain rate and temperature. Considering the advantages of the above works, ANN model has been formulated in the present work to model flow stress using ANN with back-propagation algorithm where flow stress has been related with strain, strain rate, temperature and Si_{eq} defined in Equation 1.

The third part of the literature review is to know the ANN based models developed for prediction of roll force during hot strip rolling. Son et al. [8] developed an on-line learning neural network for both long-term learning and short-term learning in order to improve the prediction of rolling force in hot-rolling mill. This paper shows that the predicted rolling force agrees with very close to the practical rolling force and the thickness error of the strip is considerably reduced. Lee and Lee [9] developed a corrective neural-network is trained to predict the rolling force for the first coil and the conventional learning output is used for the remaining coils of the lot. By doing so, the thickness error at the head-end part of the strip is considerably reduced. Moussaoui et al [10] developed a roll force based on ANN where flow stress is also considered as one of the inputs to the network. The model has been found to be very accurate. These works shows that ANN model is highly accurate for prediction of roll force during hot strip rolling. But, using the flow stress data makes the model more accurate.

METHODOLOGY OF MODELING AND SIMULATION

Development of Flow Stress Model

Flow stress is the most important material property which depends upon strain, strain rate, temperature and chemical composition. In the present study, the chemical composition has been grouped as Si_{eq} as described in Equation 1. Besides, the flow stress values obtained from literature, experiments have been conducted in Gleeble-3500 to determine flow stress values of CRNO steel. An ANN flow stress model using feed forward network and back-propagation algorithm has been developed. Input nodes of the neural network are strain, strain rate, temperature and Si_{eq}. Output node of the network is flow stress. The network has been trained using back-propagation algorithm. Neural Network Toolbox of MATLAB software has been used for training of the network.

Configuration of HSMM Software for HSM, RSP

Hot Strip Mill Model (HSMM) has been used for simulation of hot strip mill process. Department of Energy (DoE), USA and American Iron & Steel Institute (AISI) had jointly sponsored a project for development of an off-line model to achieve energy efficient optimization of flat rolling process. University of British Columbia (UBC) was the principal investigator in this project. Integ Process Group, USA was the software developer and commercial partner of the project. A software named Hot Strip Mill Model (HSMM) has been commercially developed from this project. SAIL-RDCIS has procured this software in 2006-07 from Integ Process Group under The entire mill line has been simulated with this software from furnace to coiler with customization of each stage.

The Hot Strip Mill (HSM), Rourkela Steel Plant (RSP) has been configured in this software. Two reheating furnaces, 2 nos reversing roughing stands, 1 no non-reversing stand, coilbox, 6 nos finishing stands, 4-bank laminar cooling section at runout tables and coilers were configured with required configuration data like distance between stations, roll diameter ranges, mill motor capacities, water flow rates at cooling sections etc.

Recording of rolling data for Model Calibration

An on-line data recording system has been installed in the Hot Strip Mill, Rourkela Steel Plant to record data of chemical composition, slab size, coil sizes and mill parameters like speed, temperature, roll gap and roll force.

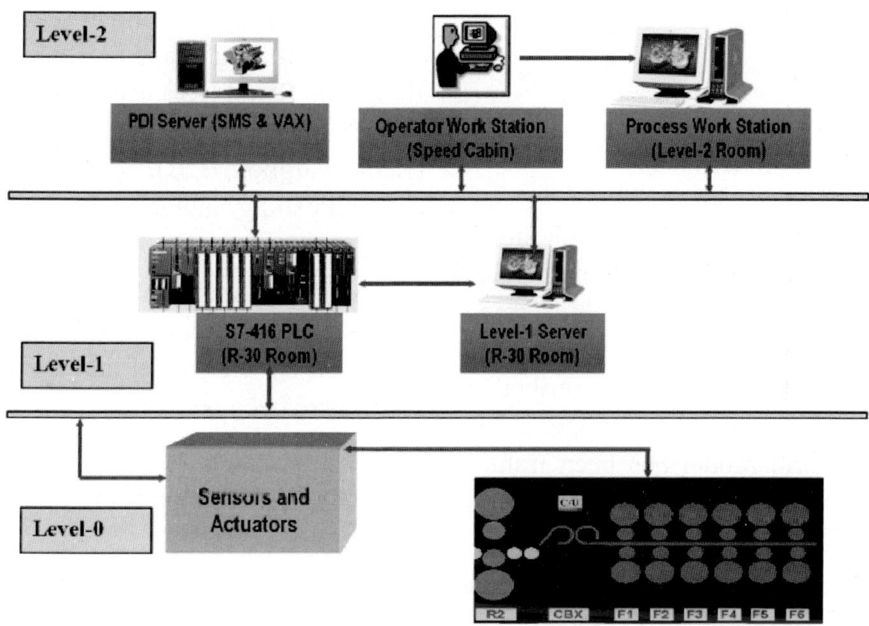

Fig. 1: Data recording system at HSM, RSP

The data recording system is shown in Figure 1. Primary data of each coil like chemical composition, slab and coil sizes and furnace tracking information is passed to data recording system (Process Work Station in Figure 1) using File Transfer Protocol (FTP). Data from mill PLC system is transferred to the data recording system using an OPC based software program developed for the purpose.

Simulation of CRNO rolling in HSMM

Both the primary data input obtained from the mill and recorded mill parameters have been given in the HSMM software for simulation of rolling of CRNO grades of steel. The model was calibrated with measured roll force data. The mill does not have load cells in roughing stands. So, model has been calibrated with roll force data obtained from finishing stands only.

Development of Hybrid Roll Force Model

A hybrid model scheme has been developed for prediction of roll force. The scheme is shown in Figure 2. This figure shows that roll force predicted by HSMM has been used as one of the input parameter to the hybrid model.

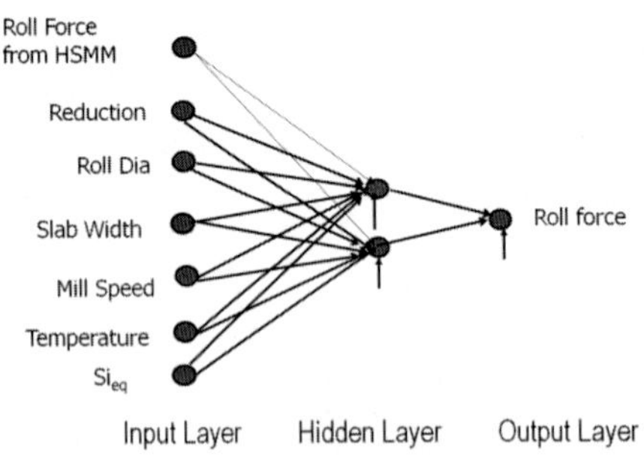

Fig. 2: Conceptual diagram of hybrid roll force model

The input nodes of hybrid roll force model are roll force predicted by mathematical model (in HSMM software), stand reduction, roll diameter, strip width, mill speed, temperature and Si_{eq} as described in Equation 1. The output of the model is roll force. The hybrid model has been trained using measured roll force.

RESULTS AND DISCUSSION

A screenshot of HSMM simulation is shown in Figure 3. The bottom portion of this figures shows that the representative diagram of each station at HSM, RSP which includes Reheating Furnaces (2 nos), Vertical RollingSstand (V0), Reversing Roughing Stand (R0), Reversing Roughing Stand (R1), Non-reversing Roughing Stand (R2), Coilbox (CB), Flying Crop Shear (FS), Finishing Stands (F1 to F6), Laminar Cooling Ssection (Bank1 to Bank-4).

Figure-3 also shows a typical result chart of HSMM simulation where the roll force predicted by the model and measured roll force have been compared.

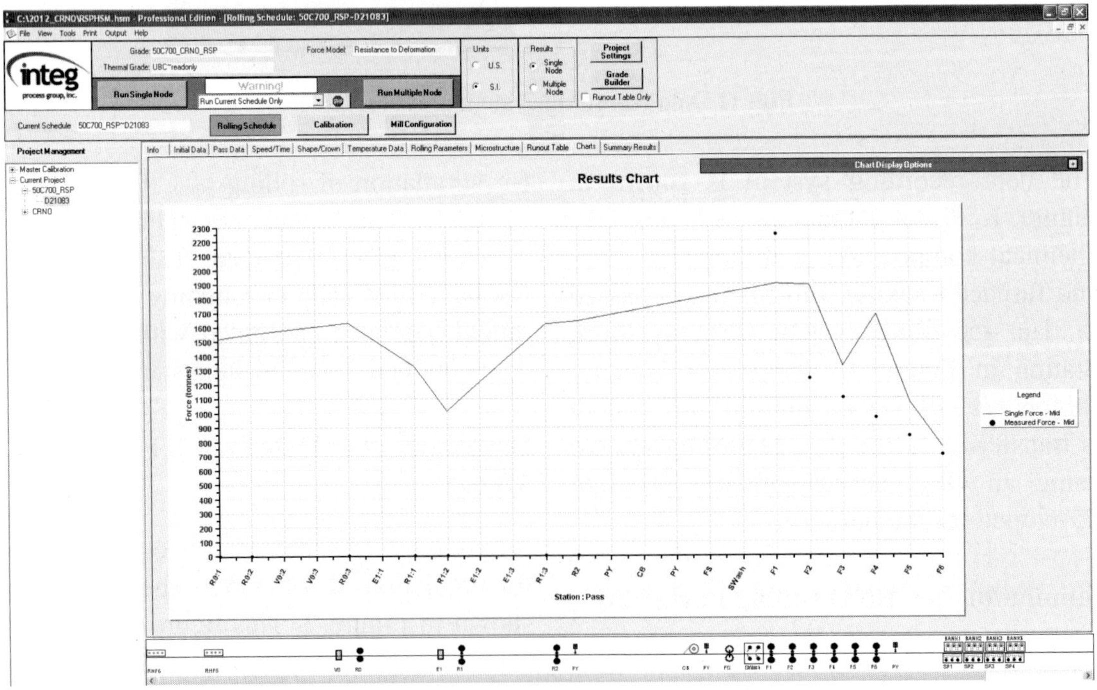

Fig. 3: Screenshot of HSMM with un-calibrated model

Roll forces predicted by HSMM at different stations of the mill is shown in the line plot where as the measured roll forces have been shown in dots.

There is a large mismatch between the predicted and measured roll force as shown in figure. Predicted roll force is much less than the measured roll force in stand F1 whereas the predicted roll force is much higher than the measured roll force in stands F2-F6. This is due to the traditional flow stress model inbuilt in the software.

The software has been calibrated for CRNO grade of steel using the hybrid flow stress model. Then the model was allowed to run again with new flow stress model. Result of calibrated model is shown in Figure 4.

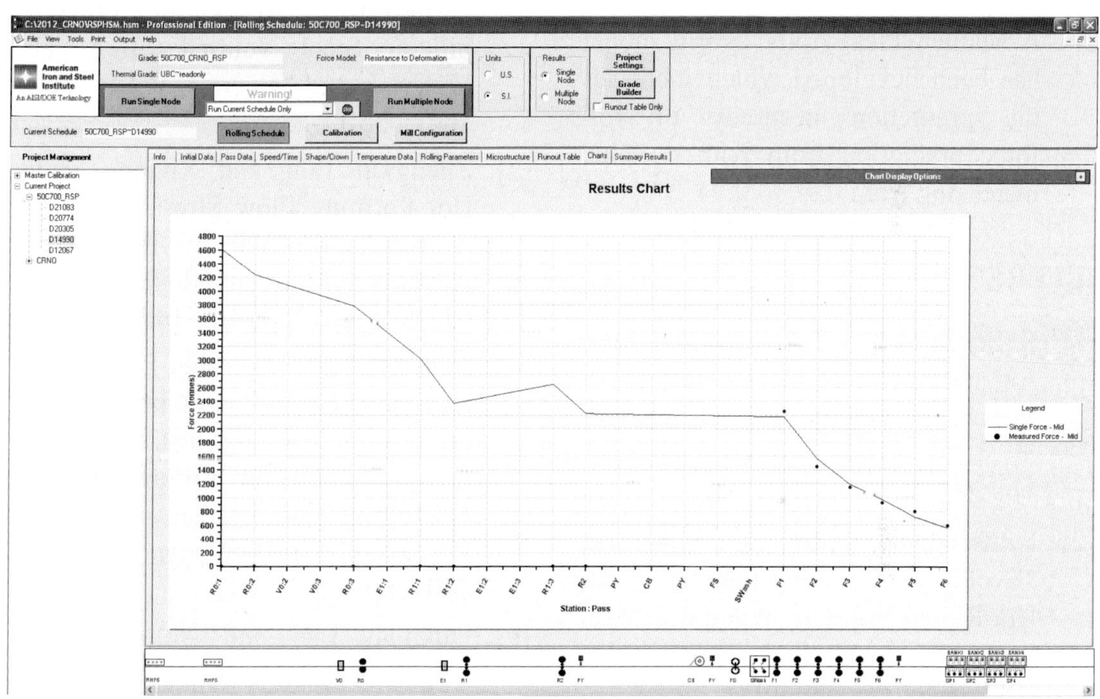

Fig. 4: Screenshot of HSMM with calibrated model

Figure-4 shows that there is a very good match between the predicted and measured roll force. Similar results have been obtained for more than 2000 coil data for which the model was simulated.

All the predicted roll force data have been used in the hybrid model described in Figure-2 and it has been found that the hybrid model has a better accuracy with r^2-value of 0.94 compared to 0.91 in HSMM model.

CONCLUSIONS

Following conclusions can be drawn from the literature review and above discussions:

- HSMM has been found to be an useful tool for simulation of hot rolling process

- Flow stress data of CRNO steel obtained from literature is useful for roll force prediction
- A factor Si_{eq} has been used to take care of complex combinations of chamical composition in the roll force model.
- Roll force prediction accuracy by HSMM has been improved after proper calibration using flow stress model.
- A Mathematical-ANN hybrid model has been developed which increases the prediction accuracy of roll force prediction from with r^2 value increasing from 0.91 to 0.94.

REFERENCES

[1] S. Serajzadeh and A. Karimi Taheri, An investigation on the effect of carbon and silicon on flow behavior of steel, Materials and Design, Vol. 23, 2002, pp. 271-276

[2] P.P. Sengupta, S. Rath and V. Kumar, "Evolution of Rolling Parameters for Hot Rolled Non-Orientated Electrical Steel", Proc., AISTech, Indianapolis, USA, 2007, pp. 489-493.

[3] Pablo R. Calvillo, Rafael Colás and Yvan Houbaert, Hot Deformation Analysis of a 2 wt.-% Si Electrical Steel by means of Torsion Tests, Advanced Materials Research, Vols. 15-17, 2007, pp 708-713

[4] Xudong Zhou and Xiangru Liu, Hot Forming Behavior and Flow Stress Model of Steel 50A1300, Advanced Materials Research Vols. 418-420, 2012, pp 1468-1471

[5] O. Fischer and J. Schneider, Influence of deformation process on the improvement of non-oriented electrical steel, Journal of Magnetism and Magnetic Materials 254–255, 2003, pp. 302–306

[6] J. Hunady, M. Cernik, E. J. Hilinski, M. Predmersky and A. Magurova, Influence of Chemistry and Hot Rolling Conditions on High Permeability Non-Grain Oriented Silicon Steel, Journal of Metals, Materials and Minerals, Vol.15 No.2, 2005, pp.17-23

[7] Zhengxin Tang and Xudong Zhou, Hot Forming Flow Stress Prediction of Steel 50A1300 by BPA ANN, Advanced Materials Research Vols. 532-533, 2012, pp 105-109

[8] Joon-Sik Son, Duk-Man Lee, Ill-Soo Kim and Seung-Gap Choi, A study on on-line learning neural network for prediction for rolling force in hot-rolling mill, Journal of Materials Processing Technology, 164–165, 2005, pp. 1612–1617

[9] Dukman Lee and Yongsug Lee, Application of neural-network for improving accuracy of roll-force model in hot-rolling mill, Control Engineering Practice, Vol. 10, 2002, pp. 473–478

[10] Abdelkrim Moussaoui, Yacine Selaimia and Hadj Ahmed Abbassi, Hybrid Hot Strip Rolling Force Prediction using a Bayesian Trained Artificial Neural Network and Analytical Models, American Journal of Applied Sciences 3 (6), 2006, pp. 1885-1889

Modeling the Effect of Ladle Bottom Design on Flow Characteristics and Refractory Wear in Gas Stirred Ladles

Rajeev Kumar Singh and Y.K. Khanna
R&D Centre for Iron and Steel, Steel Authority of India Limited, Ranchi, India-834002
Email: singhrajeev@sail-rdcis.com

ABSTRACT

With the ever increasing demand for better quality and lower cost of steel, secondary steelmaking has emerged as a critical step in the whole steelmaking process. It is being regularly used to achieve correct chemical composition and pouring temperature of the steel in ladle at the desired degree of cleanliness. In order to achieve the desired targets, stirring of the melt in the secondary steelmaking is practiced. Gas stirring has emerged as the most common means of stirring and is performed by passing inert gas through a number of porous plugs placed at the bottom of the ladle. The number of porous plugs, their position and the design of the ladle has a profound influence on its hydrodynamics and hence chemical efficiencies.

Traditionally, ladles in use have a flat bottom. During draining of ladle, a small amount of steel is left back in the ladle in order to avoid slag entrainment. In order to improve the metal yield during ladle drainage, newer practices involve using ladles having inclined or funnel shaped bottom. Use of inclined or funnel shaped bottom in the ladle is expected to influence the associated hydrodynamics, affecting stirring efficiency and hence metallurgical performance of the gas stirred ladles considerably.

The current study focuses on the fluid flow simulation of gas stirred ladles which have been fitted with different types of bottom geometries and compares them with the traditionally used flat bottom. Mathematical modeling has been carried out using the Discrete Phase Model available in CFD software FLUENT®. Investigation has been carried to study fluid flow, and hydrodynamic wear for different ladle bottom geometries at various flow rates.

Analysis of the results reveals that, ladle bottom design plays a significant role and influence hydrodynamic performance of the ladle. Inclined shaped bottom, while having a slightly lower mixing efficiency, compare well with the flat bottom ladle. This indicates that if there is a need to use a non-flat bottom ladle, inclined shaped bottom can be a good alternative albeit at a slightly lower efficiency.

KEYWORDS: *Ladle Design, Refractory Wear, Computational Fluid Dynamics, FLUENT®*

INTRODUCTION

For the steelmaking industry to remain competitive in the market, efforts must be made to reduce as well as modify harmful nonmetallic inclusions, minimize harmful impurities and achieve the optimum casting temperature, content of alloying elements, and homogeneity of the liquid steel.

Secondary steelmaking is a critical step in the steel production process between the primary processes and casting. Some elements are added and some have to be removed during secondary steelmaking in order to fine-tune the composition of the steel to meet specifications and customer's requirements. Temperature, internal quality and the inclusion content of the steel also have to be carefully controlled during secondary steelmaking operations.

Once steel of desired chemistry and temperature has been attained in the secondary steelmaking vessel, clean steel along with its slag cover (whose thickness varies from 0.05 to 0.20m) is taken for teeming into ingot mold or tundish. As the teeming process progresses, the interface that separates steel from the slag eventually deflect towards the drainage nozzle and adopts the form of a funnel. Funnel formation over the nozzle, leads to entrainment of a significant proportion of slag from ladle to tundish or ingot mold. Drainage is stopped when first traces of slag (onset of funnel formation) are detected in the nozzle. The height of the liquid left unteemed in the ladle has been found to be approximately equal to the diameter of the nozzle. Also, the inclination of the ladle bottom has no significant effect on the critical height of the funnel formation. As a consequence, the amount of steel left in the ladle can be strongly reduced by inclining the ladle floor[1]. Thus, though steel processing ladles are traditionally designed to have a flat bottom, in recent years[2], there has been growing interests to embody funnel shaped ladle bottom geometry, particularly in high capacity vessels. The primary reason for such a 'shift' is associated with operating benefits of reduced ladle slag carryover, minimal residual metal losses etc. Product quality and yield have both improved in plant by adapting funnel shaped ladle bottom geometry.

Any such design change, however, has the potential to influence the associated hydrodynamics and thereby influence the rates of various heat and mass transfer controlled processes (viz. melting, dissolution, mixing etc.) carried out in gas stirred ladles. Of specific concern is the influence of the ladle's bottom design on bulk mixing times as these are typically derived from the mixing characteristics of the slowly moving region, located towards the base of the ladle. Therefore, given the dependence of flow phenomena on vessel design, mixing times and correlations normally valid for flat bottom are likely to be ineffective, if extrapolated to different ladle bottom designs. In a recent article Mazumdar et. al.[3] have reported results from the physical modeling of mixing times in an aqueous model of gas stirred ladles with different types of bottom design. The present study aims to extend this work further through mathematical investigation and assess the impact of different types of bottom geometry on the fluid flow, mixing time and wear in a gas stirred ladle under a variety of operating conditions relevant to ladle metallurgy operations.

MATHEMATICAL MODEL

Mathematical simulation of fluid flow and mixing time has been performed for a laboratory scale model of a gas stirred ladle. The model configuration is same as the one used by Mazumdar *et. al.*[3]. This is to compare prediction with experimental mixing time in gas stirred ladles having different types of bottom designs (*viz.*, flat, curved and inclined). The mathematical analysis is carried out to investigate fluid flow, mixing time and wear profile as a function of different gas flow rates and plug configuration. Physical dimensions of the model and the material properties of the medium used for simulation are given in Table 1.

Table 1: Dimensions and operating parameters of the model ladle used in simulation

Diameter at the bottom and top	0.600m and 0.665m respectively
Height of model	0.700m
Bath depth	0.500m
Number of plugs	2 located at different locations, see Fig. 1
Gas flow rate	1.67, 2.0, 2.33, 2.50 and 3.33 $\times 10^{-3}$ m^3/s per plug
Water density	1000 kg $/m^3$
Water viscosity	0.001 kg/m.s
Air	1.17 kg/m^3

are located at 0.64R (where R is the radius at the bottom of the vessel). The included angle between the two plugs is 90^0. In Configuration-B, both the plugs are located at 0.5R, diagonally opposite from each other. Fig. 2 shows the sketch of the three different ladle designs along with the relevant dimensions which has been used in the present investigation.

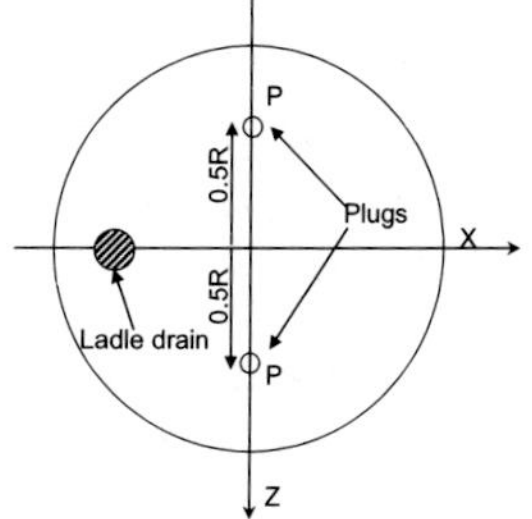

(a) Configuration-A: Two plugs located at 0.64R with an included angle of 900

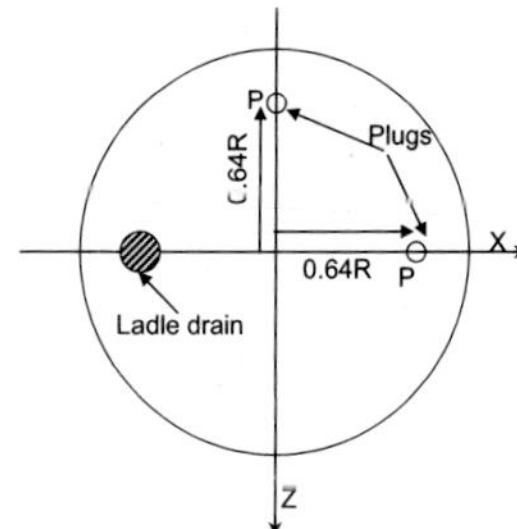

(b) Configuration-B: Two plugs located at 0.5R, diagonally opposite to each other

Fig. 1: Two different configurations used in the present investigation

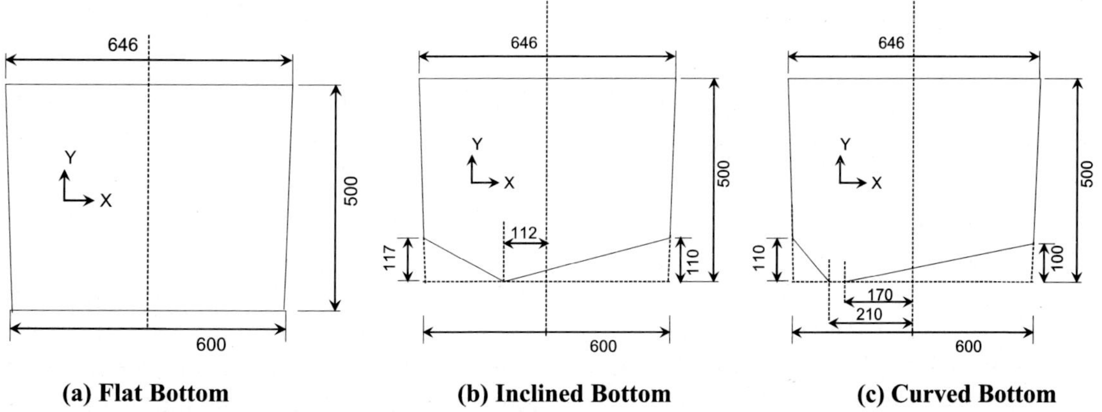

(a) Flat Bottom **(b) Inclined Bottom** **(c) Curved Bottom**

Fig. 2: A schematic of a ladle fitted with different types of bottom (All dimensions in mm)

In the present study, an unsteady, three-dimensional turbulent flow calculation scheme is first configured and the liquid or the continuous phase continuity and momentum conservation equations solved in conjunction with different bubble trajectory equations (*e.g.* force balance expressions along three coordinate directions) in a mutually coupled fashion using the Discrete Phase Model available in ANSYS FLUENT®.

The following assumptions has been applied in the discrete phase model, *viz.*,

- The flow system is essentially isothermal, turbulent, three dimensional and unsteady
- The presence of an upper buoyant phase is ignored and the bulk liquid-air interface is assumed to be flat and mobile. Only gas phase is allowed to escape from the top surface.
- Discrete mono size bubbles are assumed to form at the plug tip. Thus the size of the bubbles forming at the orifice is assumed to be known from available correlation reported in literature[4], e.g.,

$$d_b = 0.35\left(\frac{Q^2}{g}\right)^{0.2} \tag{1}$$

  Where d_b is bubble diameter, Q is the gas flow rate and g is acceleration due to gravity.

- The bubbles size is considered to remain constant during its rise through the liquid.
- Bubble-bubble interactions are ignored. Standard drag laws for spherical geometry are applied. The values of drag coefficient as given by Morsi and Alexander[4] is used in present study.
- Turbulence generated within the plume by the bubbles as well as the lateral lift forces are ignored.
- It is assumed that the bubbles have no direct impact on the generation or dissipation of turbulence in the continuous phase

At the onset of gas stirring, the rising gas imparts momentum to the surrounding stagnant liquid, thus setting up a motion in it. With time, as more and more gas bubbles rise through the fluid, the liquid velocity increases till a steady state is reached. The steady state is attained when the energy of the rising gas is balanced by the dissipation of the energy due to bubble slippage, liquid recirculation, wall friction and wave formation on the free surface. In order to capture the initial unsteadiness of the process, an unsteady state model is solved.

For the simulation of fluid flow in the system, the following boundary conditions are used.

- At the side wall and bottom wall, normal velocity components vanish and no-slip boundary conditions are employed.
- On the free surface of the liquid, tangential stress is zero and normal stress equals applied pressure, and no fluxes cross the surface. This is expressed as:

Both the flow and the turbulence equations are discretized using the Second Order Upwind scheme and default under relaxation parameters are used. All the initial values of flow variables and turbulence parameters set to zero and an appropriate time step is chosen for solution.

For the flat and inclined bottom ladle, the bottom of the ladle is paved with quadrilateral mesh whereas for the curved bottom, triangle shaped mesh is paved in the bottom. Then the domain is meshed with a cell size of 0.02m to create hexahedral or tri-wedge shaped volume elements. Care has been taken to keep the skewness level of the mesh below 0.8 to get a good mesh quality. This is, in particular, required for the curved geometry where due to curvature and the presence of drain opening a finer surface mesh for the bottom was required. Once the mesh is created, different boundary and continuum types are applied on the geometry and the mesh is exported for further processing. Fig. 3 shows the type of grid used for meshing on different types of ladle geometry.

The discrete phase model in conjunction with the flow and turbulence model is solved in an unsteady state to get a steady state velocity field. The volume average velocity was computationally monitored to know the instant when the flow had reached practically a steady state.

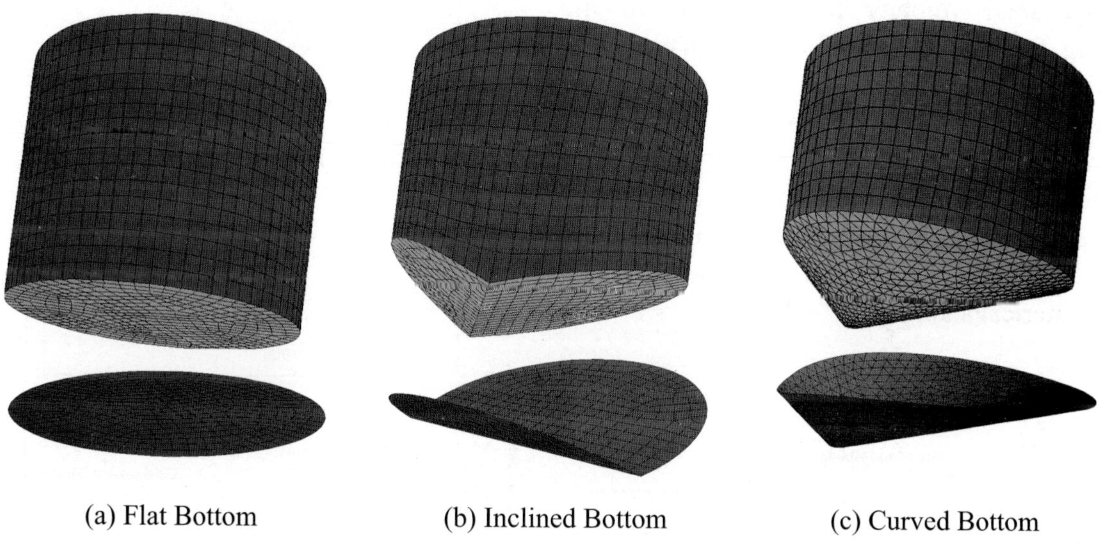

(a) Flat Bottom (b) Inclined Bottom (c) Curved Bottom

Fig. 3: Different types of mesh used for ladle of different types

RESULTS

Fig. 4 shows the axial velocity profile at the mid-height of the ladle for different grid sizes employed. From the figure it is clear that the varying grid size have practically no bearing on the axial velocity in the bulk fluid region. However, in the two phase plume region, the axial velocity is sensitive to grid size. Finer grids lead to higher velocity in the plume region. When the grid size has been decreased from 20mm to 10mm, there is no significant change in the axial velocity profile, indicating that 20mm grid is good enough to make the solution grid independent.

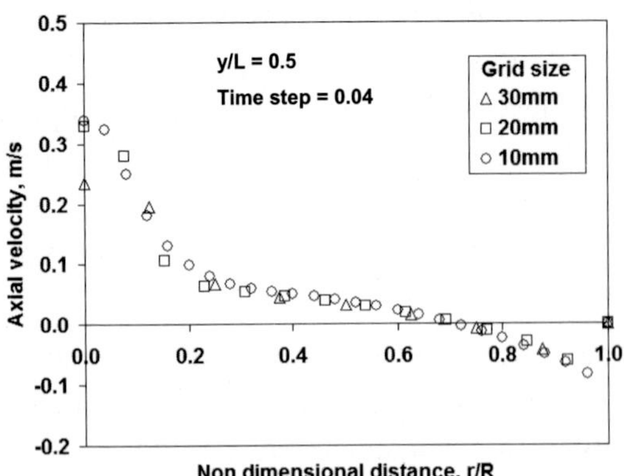

Fig. 4: Variation in axial velocity profile for different grid size

The effect of variation of time step on the axial velocity profile is shown in Fig. 5. It indicates that for the given grid size distribution, keeping all other parameters constant, the time step has practically no significant bearing on the fluid flow. However, smaller time steps lead to better convergence of numerical solution and hence are advantageous from the point of numerical stability.

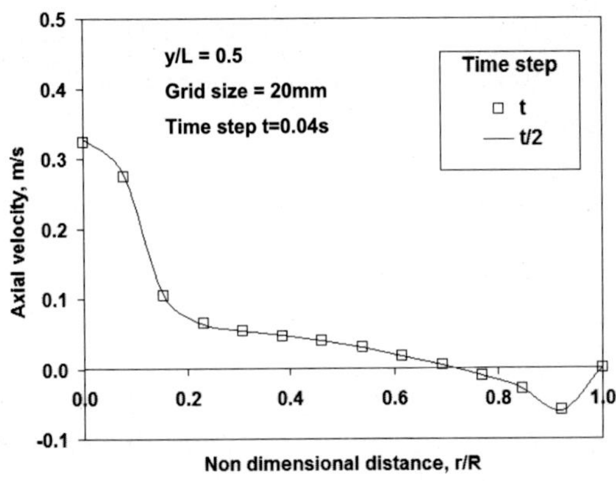

Fig. 5: Variation in axial velocity profile for different time steps

Fig. 6 and Fig. 7 show the effect of varying bubble diameter and the drag law as available in FLUENT®[5] on the axial velocity profile respectively. Varying bubble diameter, drag coefficient do not affect drastically the predicted axial liquid velocity component. However, the predictions within the gas plume region are very sensitive to the average bubble diameter and the drag coefficients used.

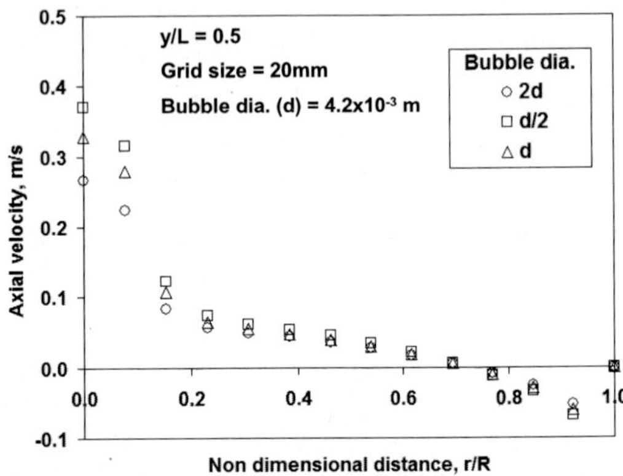

Fig. 6: Variation in axial velocity profile for different bubble diameters

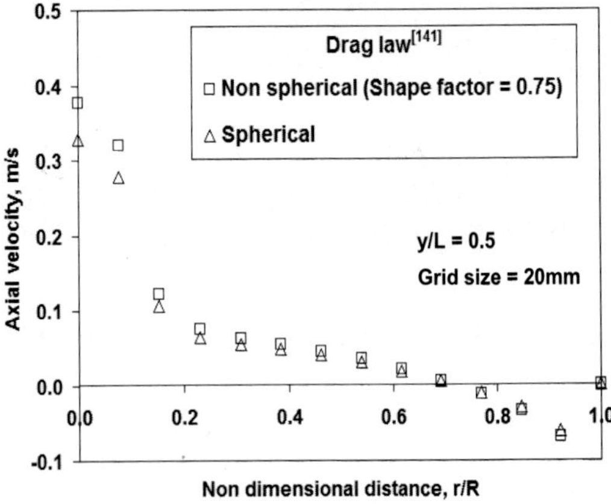

Fig. 7: Variation in axial velocity profile for different drag coefficient

Fig. 4 to Fig. 7 suggests that process and numerical parameters are of importance within the plume region only. However, in ladle metallurgy operations, the two phase plume region occupies only 3-4% of the domain space. This indicates that the variation in solution parameters such as grid distribution, time step, and bubble diameter as well as drag coefficient will have very little or no impact on the fluid flow in the bulk of liquid. In the present investigation, as we are primarily concerned with the bulk fluid flow, such variation can be neglected.

Before a detailed investigation of fluid flow profile in the ladle is carried out, the model employed is validated with data available from literature. For the purpose of validation, a two dimensional axi-symmetric discrete phase simulation for the ladle investigated by Sheng and Irons[6] has been performed.

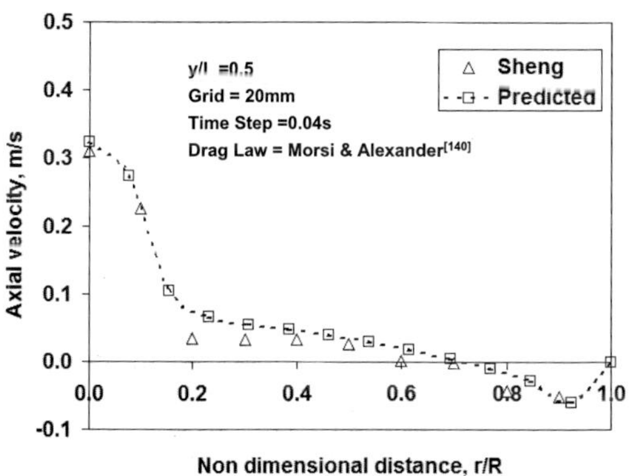

Fig. 8: Comparison of predicted and experimental axial velocity profile at mid height of the ladle[6]

Fig. 8 shows the comparison between the predicted and experimental axial velocity profile at the mid height of the ladle. It can be seen from the results that the discrete phase model employed in the current investigation predicts the fluid flow with reasonable accuracy. This suggests that the discrete phase model can be used for simulation of gas stirring of ladles with reasonable degree of success.

Effect of Bottom Design on Fluid Flow

Fig. 9(a), Fig. 10(a) and Fig. 11(a) show the stream traces in the planes bisecting the porous plugs for flat, inclined and curved bottom respectively. Fig. 9(a) clearly indicates the formation of two recirculating zone, the zone on the left, away from the plugs, being larger than the one on the right, which is nearer to the plugs. The two recirculating zones meet near the bottom where it forms a small recirculating zone. Flow profile for inclined bottom is shown in Fig. 10(a). The profile is similar to that of the flat bottom ladle, except for the fact that the recirculating zone near the bottom is shifted higher due to the inclination of the bottom. Fig. 11(a) shows the flow profile in the curved bottom ladle. Here the flow profile is completely different from that of the flat bottom or the inclined bottom. The flow forms a large recirculation loop in the left portion of the ladle. However, bulk of the recirculation is concentrated in the upper region of the ladle. The circulating flow behind the porous plug also does not reach up to the bottom, forming a recirculating zone near the surface. In such flow very slow mixing is likely to occur near the bottom, while the upper regions achieve faster mixing.

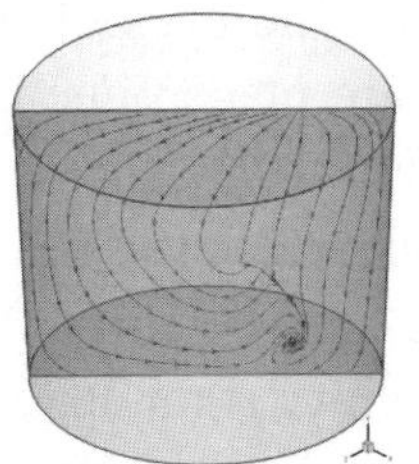

(a) Configuration–A

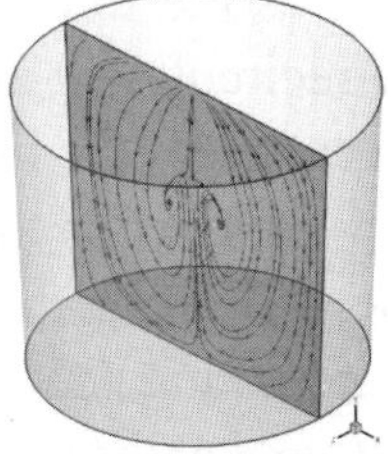

(b) Configuration–B

Fig. 9: Stream trace in the plane bisecting the porous plugs for flat bottom ladle.

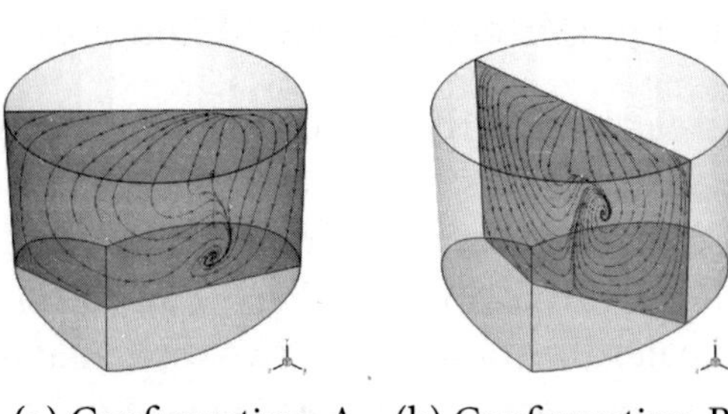

(a) Configuration–A (b) Configuration-B

Fig. 10: Stream trace in the plane bisecting the porous plugs for inclined bottom ladle.

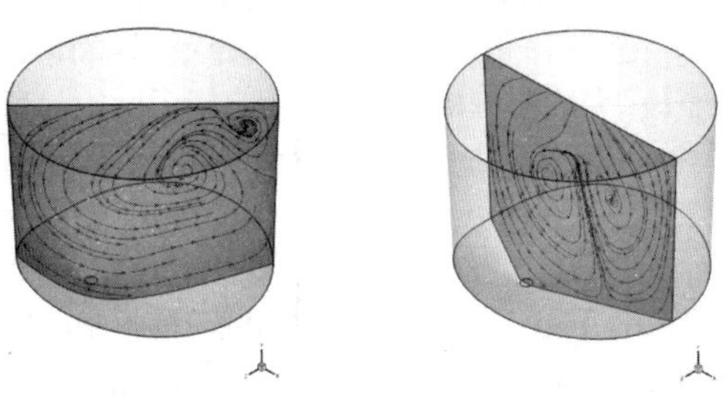

(a) Configuration–A (b) Configuration-B

Fig. 11: Stream trace in the plane bisecting the porous plugs for curved bottom ladle.

Fig. 9(b), Fig. 10(b) and Fig. 11(b) show the stream traces in the plane between the two diagonally located porous plugs. Fig. 9(b) shows the stream trace profile for a flat bottom ladle. Here, the fluid is divided almost equally into two recirculatory zones, which move to the bottom and then meet in the centre of the ladle from where it rises towards the free surface. The eyes of both the recirculating zones are placed just near the centre of the ladle. Fig. 10(b) shows the flow profile for the inclined bottom ladle. The fluid near the free surface forms two recirculating zones and move right up to the bottom finally rising up near the centre of the ladle. However, due to the inclination of the bottom, the eye of the recirculating zone is skewed slightly to one side where the inclination of the slope is lesser than the other side. Fig. 11(b) shows the flow profile for the same configuration in the curved bottom ladle. Here, it can be seen that fluid from the free surface bifurcates and moves towards the bottom where the two zones meet at a point away from the centre. Then, they move up towards the free surface, forming two recirculating zones, the eyes of which are placed near the centre of the ladle.

These flow profiles indicate that the bottom design of the ladle has profound influence on the overall fluid flow inside the ladle. The influence of the bottom design is severe in case of Configuration-A where the plugs are located in the adjacent quadrant.

Effect of Gas Flow Rate on Wall Shear Stress

Wall shear stress is caused due to change in the velocity gradient near the walls and is to some extent responsible for the hydrodynamic wear of the ladle. As the fluid flow profile in the ladle is affected by different variables such as gas flow rate, porous plug configuration and ladle bottom designs, similar behaviour can be expected with wall shear stress also. The effect of different variables on the hydrodynamic wear in the ladle has been analysed and presented in the subsequent sections.

Fig. 12 and Fig. 13 show the area weighted average and maximum shear stress on the side walls and the bottom of the ladle respectively. From these figures it is evident that with the increase in gas flow rate, the average as well as the maximum shear stress on the side and bottom walls of the ladle increases. This implies that the hydrodynamic wear of the ladle will increase with increasing gas flow rate. The figures also indicate that the shear stress for the flat and inclined bottom is very similar though for curved bottom the shear stress values are significantly different.

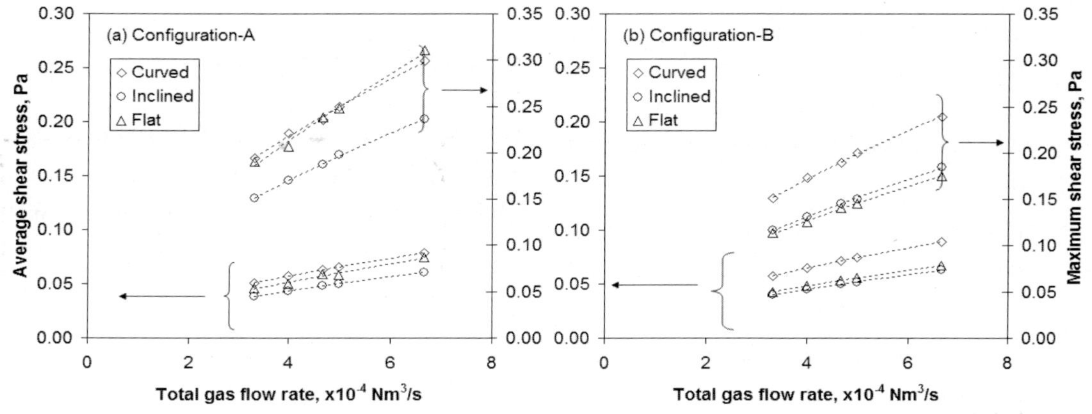

Fig. 12: Average and maximum shear stress on the side wall of the ladle for different configuration

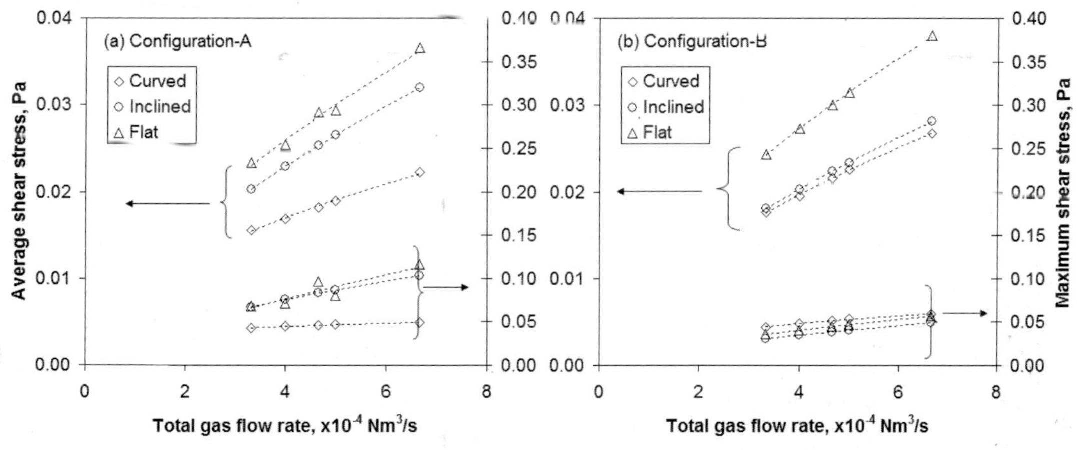

Fig. 13: Average and maximum shear stress on the bottom of the ladle for different configuration

CONCLUSION

Mathematical modeling of gas stirring of steel ladles has been found to be a useful tool to optimize the performance of the secondary steelmaking units. In the present work Discrete Phase Model (DPM) as available in the CFD software FLUENT® has been adopted to calculate the fluid flow in gas stirred ladles.

Different types of ladle bottoms have been investigated for their performance with regard to fluid flow, shear stress and mixing behavior.

- Discrete Phase Model as available in FLUENT® can be suitably modified to simulate gas injection operation in ladles.
- With increase in gas flow rate, the recirculation velocity inside the ladle as well as wall shear stress increases whereas the mixing time of the vessels decreases.
- Different porous plug configuration has profound influence on the fluid flow profile in the vessel.
- Different types of ladle bottom have significant impact on the fluid flow inside the ladles. Ladles having curved shaped bottom have very weak flow near the bottom of the vessel causing large dead zone regions.
- Shear stress is more distributed in case of Configuration-B (plugs located in diagonal quadrant) but it is not so in case of Configuration-A (plugs located in adjacent quadrant) where there is much higher localized wear.
- In the curved bottom ladle, the bottom shear stress is lowest further confirming that flow is weakest near the bottom in these ladles.

From the fluid flow profile, shear stress distribution pattern analysis it can be finally concluded that, *if ever there is a need for different kind of bottom design to improve the metal yield, then an inclined bottom ladle with porous plugs located diagonally across can be used.* These types of ladle bottom have very similar flow profile, and mixing time in comparison to the traditional flat bottom ladle, but with a significant reduction in the left over metal in the ladle.

ACKNOWLEDGEMENT

The authors are thankful to the management of RDCIS, SAIL for their permission and encouragement. The principal author is also very thankful to Prof. Dipak Mazumdar of Indian Institute of Technology, Kanpur, India under whose guidance is study was carried out.

REFERENCES

[1] G.M. Mazzaferro, M. Piva, S.P. Ferro, P. Bissio, M. Iglesias, A. Calvo and M.B. Goldschmit, Ironmaking & Steelmaking, Vol. 31, 2004, pp. 503-508.

[2] R. Jouhan and L. Heaslip, European Patent No. 1474257, 2003.

[3] N. Mazumdar, A. Mahadevan, M. Madan and D. Mazumdar, ISIJ Int., Vol. 45, 2005, pp. 1940-1942.

[4] S.T. Johansen, F. Boysan, Metall. Trans. B., Vol. 19B, 1988, pp. 755-764.

[5] Fluent 6.1 User's Guide, Fluent Inc., Lebanon(NH), USA, Feb 2003, Ch. 21, pp. 5-6.

[6] Y.Y. Sheng and G.A. Irons, Metall. Trans. B, Vol. 24B, 1993, pp. 695-705.

Parametric Optimization of Closed Die Forging Process Using FEM and Taguchi Method

Md. Israr Equbal, Rajkumar Ohdar* and P. Talukdar
Department of Forge Technology
National Institute of Foundry and Forge Technology
Ranchi-834003, India
Email: israr_equbal@yahoo.co.in, rkohdar@yahoo.com*, pinaki.t@rediffmail.com

ABSTRACT

In any forging process the performance of the die and product quality is heavily dependent on various parameters which can be broadly divided into two groups that are design parameter and process parameter. Design parameters such as flash thickness, flash width, corner radius, fillet radius, draft etc. represents the geometrical aspect of the die and are important for die modeling stage. Whereas process parameters like billet temperature, die temperature, friction coefficient, lubricant, die velocity etc. are related to the forging process condition. In this study finite element based DEFORM in conjunction with Taguchi method is used to determine the optimum values of design and process parameter for the closed die forging of Automotive part. Analysis of variance (ANOVA) is employed to determine significant parameters. The results shows that at optimum factor level setting for minimizing the forging load complete die filling is achieved with less defects. This is useful for improving the die life and product quality.

KEYWORDS: *Finite Element Analysis, Taguchi Method, Simulation, Optimization, Anova*

INTRODUCTION

Because of their advantages, such as high productivity, small finishing allowances and the very good service properties of the finished products, forging processes belong to the most commonly used product manufacturing technologies [1]. The degree of the structural reliability and high strength to weight ratio achieved in the forging process, make the forgings the best choice for high performance, high strength, high reliability and long term usage applications, where tension, stress, load, and human safety are critical considerations. Forgings are widely used in the automotive industry, defense industry, marine industry and aerospace industry, agricultural machinery, off-highway and railroad equipment, valves, fittings, petrochemical applications, industrial hardware and hand tools. Parts such as connecting rods, piston rods, crankshaft in a car, landing gear supports, wheels, propeller blades in an aircraft etc. are produced by forging process [2]. Forging process design in industries often depends on experience and the trial-and-error method, which results in a great waste of time and cost. Furthermore, the design quality of the forging dies and process conditions clearly affects the production efficiency directly. The main goals of optimal process design are to reduce forging load, minimize stress, control the material microstructure, and obtain a sound product with no defects. Therefore, it is very important to increase the die life and improve the product quality by means of optimizing die geometry design, proper selection of the die material, and optimum control of the process parameters.

FACTORS AFFECTING CLOSED DIE FORGING

In closed-die forging, two or more dies are moved to each other to form a metal billet that has a relatively simple geometry to obtain a complex shape. In a common multistage forging process, the part is first forged in a set of busting dies then moved to one or more sets of blocking dies and finally forged in finisher dies. Finisher dies are used to enhance geometrical detail without significant material flow. To improve the efficiency of forging processes, it is necessary to have a complete process understanding and model. To this end, a great deal of research has been performed in order to quantify the effect of various finisher die parameters to forging process. For example Santos et al. [3] optimize the die design by taking into consideration shape and size of the initial billet and predict the forces needed as well as the defect occurrences. Lapovok [4] study the extension of die life of hot forging by minimizing the damage accumulation and development of optimal preform design. Ab-Kadir A. R. et al. [5] studied the effect of corner radius and friction parameters on the optimization of the cold forging die design by conducting finite element (FE) simulation. Satish et al. [6] uses FEM-based computer simulation to optimize the design parameters and input billet cross-section for front axle beam. Gangopadhyay et al. [7] uses Three-dimensional finite element analysis DEFORM 3D software on multistage hot forming of railway wheels involving the processes of upsetting, forging, and punching of wheels. Srivastava et al. [8] studied the effect of billet temperature, die velocity and coefficient of friction on forging load and strain rate. Saniee and Hosseini [9] studied the effects of flash allowance and bar size on the die filling and required load. These works reveal that Die design, material behavior during deformation, as well as friction, and material flow characteristics in a die cavity, are important in design consideration. Also important is the proper selection of die materials, temperature, speed, lubricant and equipment. There are many parameters that can influence the performance of the die and hence product quality is heavily dependent on various parameters which can be broadly divided into two groups that are design parameter and process parameter. Design parameters such as flash thickness, flash width, corner radius, fillet radius, draft etc. represents the geometrical aspect of the die and are important for die modeling stage. Whereas process parameters like billet temperature, die temperature, friction coefficient, die velocity etc. are related to the forging process condition. The present work concentrates on the study of effect of design parameters viz. flash and process parameters viz. billet temperature, die temperature and friction coefficient on forging load and die filling. During the die design and process planning stage it is necessary to estimate forging load to avoid unexpected die failure and inadequate die filling. Overloading can cause premature die failure because it crosses the strength of die material. Spring saddle of an automobile is taken as an example product and Taguchi experimental method is utilized to design the parameter combinations so as to identify the relative influence of each parameter on the forging load and die filling. As the real-time forging is time consuming and costly therefore FEM based DEFORM™ 3D software is used for simulating forging process. For simulation purpose 3D (three-dimensional) solid model of spring saddle and its top and bottom die are modelled in CATIA V5 software as shown in Fig. 1 and Fig. 2 (a) and 2 (b). The material of Spring Saddle is AISI 1035 alloy steel, whereas Tool Steel DIN 1.2714 is used as die material and their properties is given in Table 1.

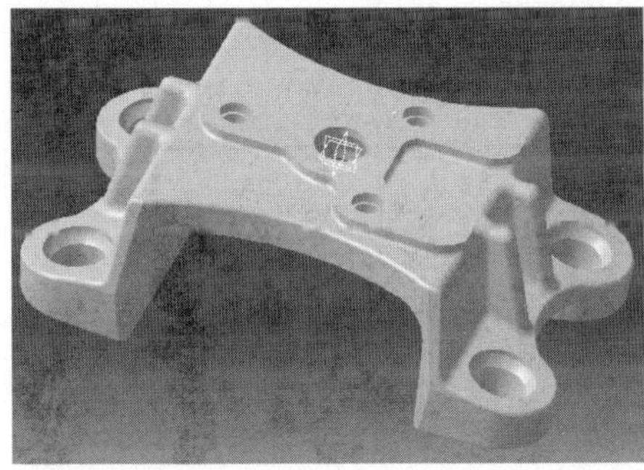

Fig. 1: 3D CAD model of spring saddle

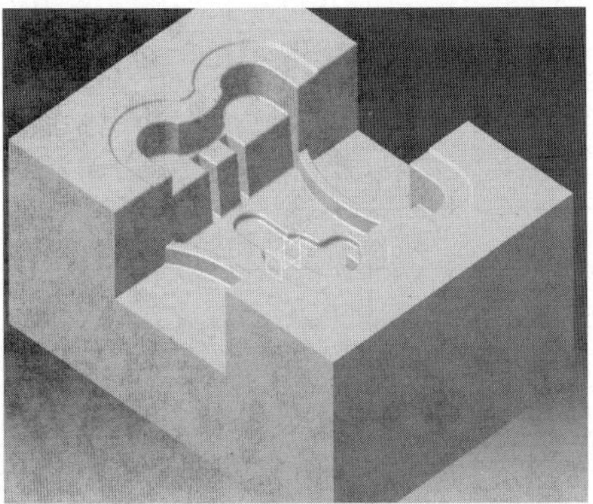

Fig. 2(a): 3D CAD model of Bottom Die

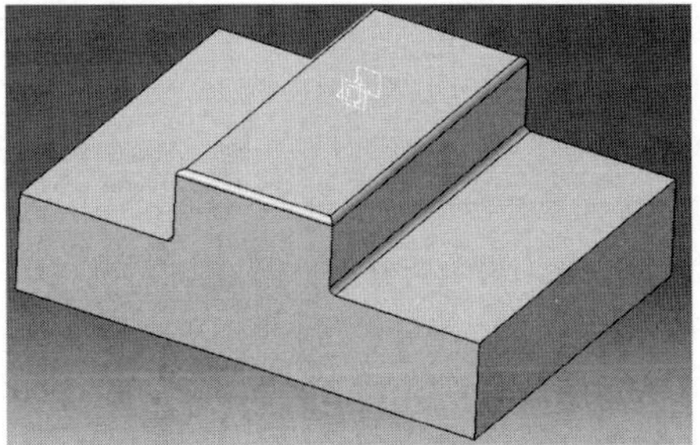

Fig. 2(b): 3D CAD model of Top Die

Table 1: Material Properties

Property	Component	Die	Units
Density	7.7-8.03	7.86	(x1000 kg/m^3)
Poisson's Ratio	0.27-0.3	0.27-0.3	
Elastic modulus	190-210	190-210	GPa
Tensile strength	485	655	MPa
Yield strength	415	380	MPa
Hardness	143	93	HB

TAGUCHI METHOD

Although the simulations based on the FEM method are capable of predicting many variables such as forging load, strain, strain rate and stress distributions more accurately, but those are time consuming and computationally expensive for a completely optimization procedure. Therefore in this work, statistical design of experiments is used to get information such as main effects of design parameters from the minimum number of experiments or, in our case, the minimum number of simulations. The Taguchi method, which is one of the fractional factorial designs, utilizes orthogonal arrays (OAs) to study a large number of variables with a small number of experiments or, in this study, the minimum number of simulations. Furthermore, the Taguchi method can be used to determine the values of the processing parameters optimizing the quality characteristics. Also analysis of variance (ANOVA) can be applied to investigate which parameters influence the closed die forging process most significantly. Taguchi method is employed using the following procedure:

(i) Select the factors affecting on the forging load and die filling.

(ii) Select the appropriate levels that each factor may hold within the simulation.

(iii) Select the orthogonal array to determine the number of simulation to run.

(iv) Conduct the simulations.

(v) Analyze the results; determine the optimum factor level combination.

(vi) Predict the optimum performance and influence of factors.

(vii) Confirm the experimental design.

The tool used in the Taguchi method is the orthogonal array (OA). OA is the matrix of numbers arranged in columns and rows [10]. The Taguchi method employs a generic signal-to-noise (S/N) ratio to quantify the present variation. These S/N ratios are meant to be used as measures of the effect of noise factors on performance characteristics. S/N ratios take into account both amount of variability in the response data and closeness of the average response to target. There are several S/N ratios available depending on type of characteristics: smaller is better, nominal is best (NB) and larger is better. Table 2 shows four factors and three levels used in the experiment. If three levels were assigned to each of these factors and a factorial experimental design was employed using each of these values, number of permutations would be 3^4. The fractional factorial design reduced the number of experiments to 9. The orthogonal array of L9 type was used and is represented in Table 3. This design requires sixteen experiments with five parameters at two levels of each. The interactions were neglected.

Table 2: Factors and Factor levels

Factors	Factor Levels			Unit
	1	2	3	
Flash thickness (A)	2	3	4	mm
Billet temperature (B)	1000	1100	1200	^{0}C
Die temperature (C)	200	250	300	^{0}C
Friction coefficient (D)	0.25	0.3	0.7	-

Table 3: L9 Orthogonal array

Trial	A	B	C	D
1	2	1000	200	0.25
2	2	1100	250	0.30
3	2	1200	300	0.70
4	3	1000	250	0.70
5	3	1100	300	0.25
6	3	1200	200	0.30
7	4	1000	300	0.30
8	4	1100	200	0.70
9	4	1200	250	0.25

FINITE ELEMENT METHOD

Forging die design and process design traditionally has relied mostly on empirical knowledge and trial-and error experience of the designer, which requires much time and money. However, the adopted simulation method by FEM used here can effectively evaluate the influence and the significance of process parameters, and is helpful for different aspects, including shortening research and development and production lead time, improving the product quality, and reducing production cost. The forging process was modelled using the FEM software package DEFORM-3D, which was designed for metal forming calculations.

PARAMETRIC DIE DESIGN

For investigation of this research an automobile part spring saddle has been considered. The dies were drawn using CATIA Software prior to actual simulation in Deform- 3D. After construction of the dies, the Catia files were saved as Stereo Lithography (.stl format) files to be imported into Deform-3D models which insure accurate integration. For Taguchi method simulation it is necessary to consider several geometric parameters in finisher die that affect the output factors so parametric 3D modeling has been carried out with this software. See Fig. 3.

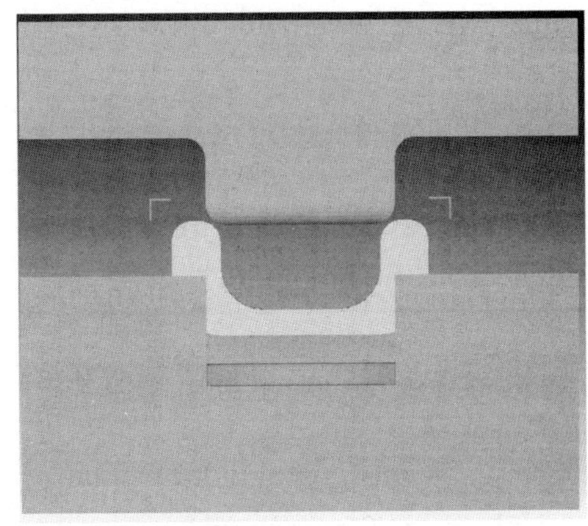

Fig. 3: Parametric 3D Modelling of spring saddle Finisher Die

SIMULATION CONDITIONS

The numerical analysis was carried out using Deform-3D. This analysis is performed to ensure die cavity filling and suitable forging load. There are some assumed models in this paper, including that the plastic material model and the rigid die model are assumed. The velocity of the moving ram is kept constant. The strain rate at the first of deformation is equal to $2S^{-1}$. All the simulations started with 45000 elements. Table 4 shows operations parameters assigned to complete the simulation.

Table 4: Operation parameters assigned to complete the simulation.

Forging Equipment	Mechanical press
Press capacity	2500Ton
No. of elements	45000
Mesh type	Tetrahedral
Die displacement	0.5 mm
Simulation mode	Isothermal
Simulation step	100
Step increment	2
Primary die	Top die
Environment Temperature	20^0C

Parameters for fem simulation are constructed in Table 5.

Table 5: Simulation Parameters and results

Trial	Factors				FEM Observations		S/N ratio
	A	B	C	D	Forging Load (Tons)	Die filling	
1	2	1000	200	0.25	1870	yes	-5.24902
2	2	1100	250	0.30	1260	yes	-1.79810
3	2	1200	300	0.70	757	No	2.61537
4	3	1000	250	0.70	2420	yes	-7.49497
5	3	1100	300	0.25	1170	yes	-1.21396
6	3	1200	200	0.30	647	No	4.01319
7	4	1000	300	0.30	926	No	0.81917
8	4	1100	200	0.70	935	No	0.08730
9	4	1200	250	0.25	355	No	9.11864

RESULTS AND DISCUSSION

The simulation results of Taguchi's L9 orthogonal arrays shows in Table 5. In the third, sixth, seventh, eighth and ninth simulation under-filling in the dies has been observed, which is shown in fig. 4. Out of these simulations third and sixth have given under-filling but amount of under-filling is less compared to others. In the simulation seventh, eighth and ninth, it has been observed that material is going out in the form of flash without filling die cavity completely. Simulation no. one, two, fourth and fifth gives complete die filling. Out of which fifth simulation has given complete die filling (Fig. 5) with minimum forging load of 1170 Ton as given by load stroke curve in Fig. 6. At this load not only the complete die filling is achieved but also uniform effective strain rate is developed within the part (Fig. 7). This increasing the

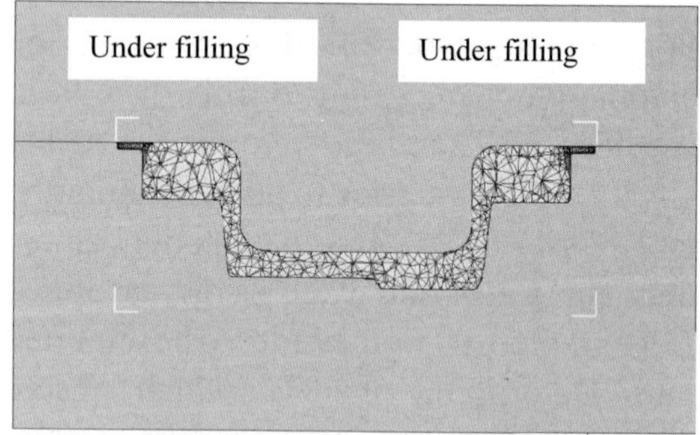

Fig. 4: Uunder filling in die

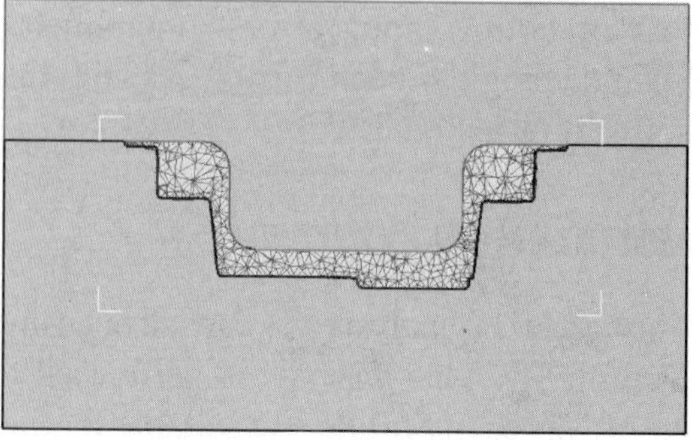

Fig. 5: Complete die filling

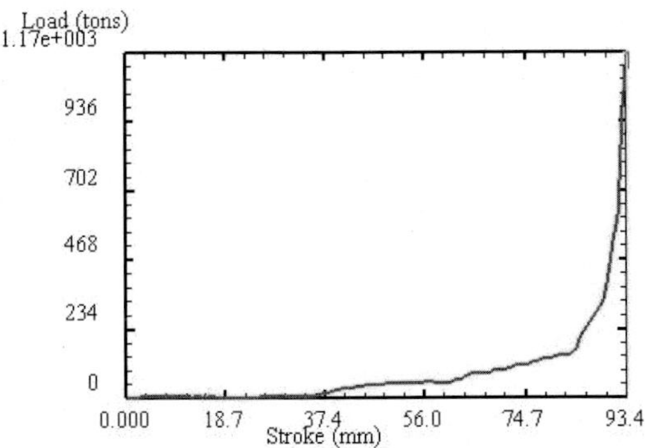

Fig. 6: Load- Stroke curve

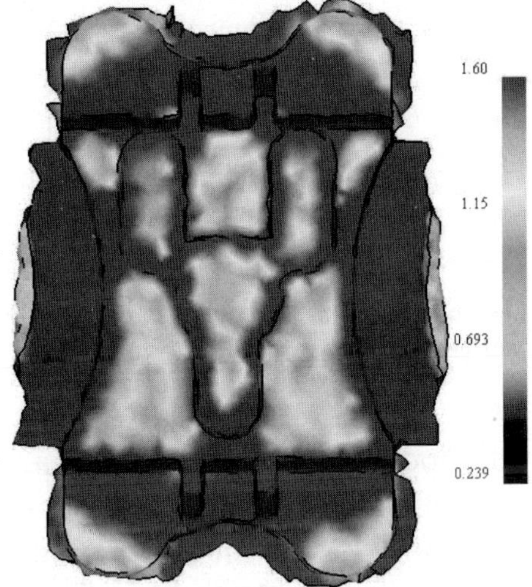

Fig. 7: Material Flow and Effective strain rate distribution in forging of Spring Saddle

die life and reduces the within part property variation. From the above discussion it is clear that die filling has been achieved in first, second, fourth and fifth simulation. Out of these the load requirement is least for fifth simulation. In first and fourth simulation the load is very high whereas for second simulation the load is slightly higher than the fifth simulation.

ANALYSIS OF RESULTS

Data analysis is made using MINITAB R14 software at 95% of confidence. Relative influence of each factor is determined by analysis of variance method (ANOVA) presented in Table 5. ANOVA depicts that the process parameter namely Flash thickness and Billet Temperature are the most significant parameters affecting the Forging load. Adequacy of analysis is carried out using Anderson–Darling (AD) test and results are illustrated in Fig. 8, which shows that data follows normal distribution and develop procedure is suitable enough to explore the design space.

Table 5: ANOVA Table for forging load

Factor	Degree of freedom	Sum of |square	Variance	% contribution	F-value
Flash Thick.	2	45.73	22.865	24.34	3.44
Billet Temp.	2	128.88	64.44	68.59	9.7
Die Temp.	–	1.903*	–	–	–
Friction Coeff.	–	11.376*	–	–	–
Error		13.279	6.639		
Total		187.889			

*pooled

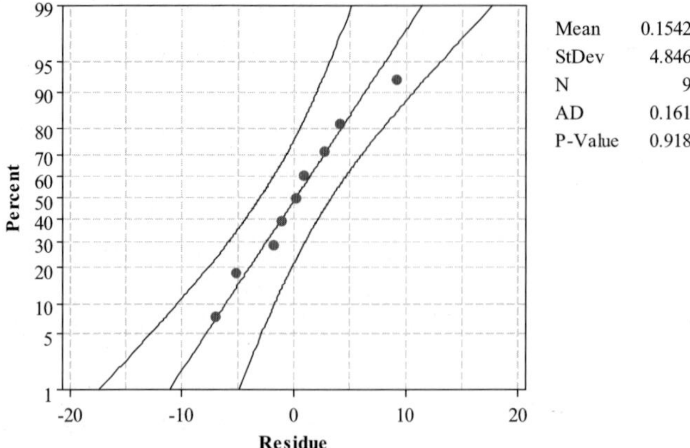

Fig. 8: Normal probability plot of residue at 95% of confidence

CONCLUSION

In this study finite element based DEFORM in conjunction with Taguchi method is used to determine the optimum values of design and process parameter for the closed die forging of spring saddle. Analysis of variance (ANOVA) is employed to determine significant parameters. The most influence factors over forging load are flash thickness and billet temperature.

ACKNOWLEDGEMENT

We are thankful to the R.S. Behera group of industry, Bhubaneswar, Orissa for his valuable inputs and supports. We are also thankful to Prof. S.N.Sinha, Director, NIFFT for supporting us in preparation of this manuscript.

REFRENCES

[1] V. Vazquez and T. Altan, Die design for flashless forging of complex parts, Journal of Materials Processing Technology, Vol. 98, 2000, P. 81-89.

[2] Tunc, Mehmet, Computerized Cost Estimation for Forging Industry, Master's Thesis, The Graduate School of Natural and Applied Sciences of The Middle East Technical University, 2003.

[3] A.D. Santos, J.F. Duarte, A. Reis, B. da Rocha, R. Neto and R. Paiva, The use of finite element simulation for optimization of metal forming and tool design, J. Mater. Proc. Technol., Vol. 119, 2001, P. 152-157.

[4] R. Lapovok, Improvement of die life by minimization of damage accumulation and optimization of preform design, J. Mater. Proc. Technol., Vol. 80-81, 1998, P. 608-612.

[5] A.R. AbKadir, A.R. Othman, Z. Samad, K. Hussain, M.T. Abdullah, and A.B. Abdullah, Effect of corner radius and friction parameters on the optimization of the cold forging die design, Journal of Applied Sciences, Vol. 7, 2007, P. 777-784.

[6] G.D. Satish, N. K. Singh and R.K. Ohdar, Preform optimization of pad section of front axle beam using DEFORM. Journal of materials processing technology, Vol. 203, 2007, P. 102-106.

[7] T. Gangopadhyay, R.K. Ohdar, D. K. Pratihar and I. Basak, Three-dimensional finite element analysis of multi-stage hot forming of railway wheels. International Journal of Advanced Manufacturing Technology, Vol. 53, 2011, P. 301–312.

[8] F. F. Saniee, and A.H. Hosseini, The effects of flash allowance and bar size on forming load and metal flow in closed die forging, Journal of Materials Processing Technology, Vol. 177, 2006, P. 261–265.

[9] M. Sanjari, A. Karimi Taheri and M. R. Movahedi, An optimization method for radial forging process using ANN and Taguchi method, International Journal of Advanced Manufacturing Technology 2008.

SECTION VIII

Advances in Materials Characterization

Corrosion Characterization of Advanced Steels for Use in the Oil & Gas Industry

Brajendra Mishra
Professor & Director, NSF Center for Resource Recovery & Recycling
Metallurgical & Materials Engineering
Colorado School of Mines, Golden, Colorado, USA
Email: bmishra@mines.edu

ABSTRACT

Drill pipe steels are in contact with CO_2 environments depending on the oil and gas field. These steels have to be resistant to various in-service conditions including aggressive environments containing CO_2, H_2S, O_2, and chlorides, in addition to static and dynamic mechanical stresses. In this respect stress corrosion cracking susceptibility of two grades of drill pipe steel in CO_2 environment have been studied simulating the bottom hole oil and gas well conditions. SSRT results show that SCC susceptibility or loss of ductility changes with temperature and increasing temperature increases the loss of ductility. Optical and electron microscopy shows presence of cracks which are the cause of ductility loss. No $FeCO_3$ is observed below 100ºC, and density of $FeCO_3$ is higher in grip section than gauge length and this is due to strain disturbance of growth of iron carbonate crystals. Material selection for down hole in CO_2 containing environments needs has been reviewed and probability of SCC occurrence in higher temperatures has been considered.

In another critical application, during oil and gas operations, steel pipeline networks are subjected to different corrosion deterioration mechanisms, one of which is microbiologically influenced corrosion (MIC) that results from accelerated deterioration caused by different microbial activities present in hydrocarbon systems. Bacterial adhesion is a detrimental step in the MIC process. The MIC process starts with the attachment of planktonic microorganisms to metal surface that lead to the formation of the biofilm and subsequently results in metal deterioration. The tendency of a bacterium to adhere to the metal surface can be evaluated using thermodynamics approaches via interaction energies. Thermodynamic and surface energy approaches of bacterial adhesion will be reviewed. Also, the factors affecting bacterial adhesion to the metal surface will be presented. In addition, the subsequent physical-chemical interaction between the biofilm and substratum and its implication for MIC in pipeline systems will be discussed.

KEYWORDS: *Steel, Corrosion, Microbiological, Transport, Sweet Corrosion, Drill Pipe*

(A) SWEET CORROSION OF DRILL PIPE STEELS

Corrosion is one of the most important continuous problems of oil and gas production. Several forms of corrosion occur in oilfield and CO2 corrosion or — sweet corrosion is one of the most common forms of attack which hydrocarbon production industry is faced with[1]. Drill pipes may be in contact with wet CO2, H2S, O2 and chlorides depending on the oil and gas field. In addition, through drilling action these steels bear static and dynamic mechanical stresses. These steels have to be resistant to these aggressive environments if they are to be viable[2].

Uniform corrosion of steels in aqueous environments containing CO2 has been extensively studied in the literature, and corrosion mechanisms are now very well defined and are already incorporated in corrosion prediction models[3-5]. Also extensive studies have been carried out on the sulfide stress corrosion cracking (SSC) behavior of high strength low alloy (HSLA) carbon steels [6-11]. As the depletion of shallow oil and gas wells forced the industry in moving toward more aggressive and deeper and sourer environments particularly for natural gas extraction, in these environments the probability of SSC occurrence is increased which restricts the utilization of service strength of steels used for casing, tubing, and tool joints. Thus, the need for new grades of HSLA steels having higher mechanical properties with good SSC resistance has grown[6]. It has been mentioned that high tensile strength, high H2S concentrations, low pH, high total pressure, high chloride concentration, lower temperature, high hardness of steel promote SSC and this phenomena is closely related to hydrogen embrittlement as a result of hydrogen ingress into the stresses steel[8,9]. However, there is less information about the stress corrosion cracking (SCC) susceptibility of these steels in environments containing CO2 as they are subjected to static and dynamic mechanical stresses along with wet CO2 gas; Perhaps, this phenomena has somewhat been neglected in favor of H2S corrosion and sulfide stress cracking. Additional research is required to fully define the environmental-material relationship of high strength low alloy drill pipe steels in sweet high temperature, high pressure wells[10].

Intergranular and transgranular stress corrosion cracking of pipeline steels from the soil side have been studied widely[12-16]. Parkins and his research collaborators showed[12] the occurrence of intergranular stress corrosion cracking (IGSCC) due to the presence of relatively concentrated solution of carbonate-bicarbonate. Also their works[13,14] demonstrated happening of transgranular stress corrosion cracking (TGSCC) that produced by a dilute, lower pH solution of carbonate-bicarbonate. Mechanism for IGSCC was described as preferential dissolution at grain boundaries and repeated rupture of the film formed at the crack tip in which crack sides were not attacked. In contrast, TGSCC is associated with dissolution of crack tip and crack sides which accompanied with hydrogen ingress into the steel. Moreover, Colwell et al.[15] have proposed a mechanism for TGSCC of pipe line steels. According to their studies, iron carbonate found in the vicinity of cracks under disbonded coating in the field is formed from bicarbonate which releases its hydrogen. Hydrogen that diffuse into the steel and lead to embrittlement results in TGSCC.

To these ends, stress corrosion cracking susceptibility of API S-135 grade drill pipe steel in CO2 containing environment was studied, simulating the bottomhole oil and gas well conditions to advance the present understanding of the prediction of materials performance.

EXPERIMENTAL PROCEDURE

Slow strain rate test (SSRT) on API S-135 grade drill pipe steel were carried out using a constant extension rate test (CERT) machine that was activated by a ¼ HP 1725 RPM 0-90 VDC permanent magnet motor. The geometry of the SSRT specimens is shown in Figure 1. The composition and mechanical properties of the API S-135 steel are listed in Table I.

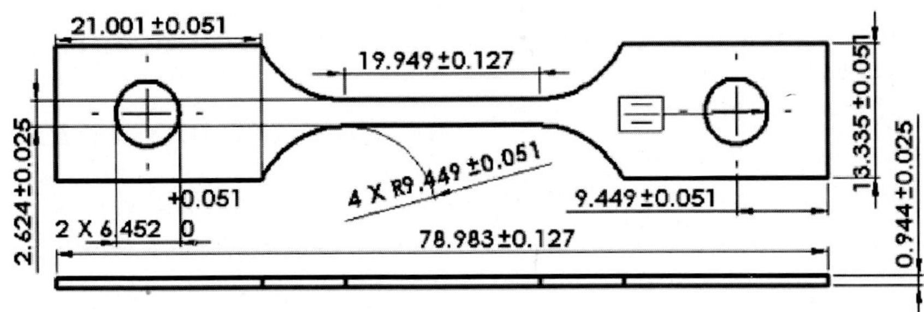

Table I: Mechanical properties & chemical composition of drill pipe steels[2].

	Yield Strength					Tensile Strength				
Grade	Min			Max				Min		
	psi		MPa	psi		MPa	psi		MPa	
G-105	105,000		724	135,000		931	115,000		793	
S-135	135,000		931	165,000		1138	145,000		1000	
Element	**C**	**Mn**	**Si**	**Ni**	**Cr**	**Mo**	**Al**	**P**	**S**	**Fe**
G-105	0.27	1.33	0.31	0.03	0.92	0.28	0.11	0.013	0.006	Bal.
S-135	0.25	0.8	0.27	0.02	1.26	0.69	0.03	0.008	0.002	Bal.

SSRTs were performed to understand the effect of temperature on the SCC susceptibility of API S-135 drill pipe steel. Experiments were performed in a one liter Inconel 600 autoclave at different temperatures in CO2 saturated deionized and ultra filtered (DI.U.F) water. The SSRTs were carried out in air and in the corrosive solution at strain rate of 1x 10^{-6} s^{-1} at 25, 50, 75, 100, 125, 150, and 175°C. Every experiment was repeated twice and three times in some cases. To perform the experiments in corrosive environment, 850 ml DI.U.F water was poured inside the autoclave and after tightening the autoclave; CO2 was purged one hour to deaerate and saturate the solution and finally heated to the desired test temperature using a proportional integral derivative (PID) temperature controller. As soon as temperature reached the desired temperature SSRT was started. To calculate the SCC susceptibility of S-135 drill pipe steel at different temperatures the plastic strain to failure (EP) and reduction in area (RA) were determined from the stress-strain curve and broken specimens respectively. Plastic strain to failure was obtained by subtracting the elastic strain at failure from the total strain at failure. As aforementioned CERT apparatus measures the crosshead displacement instead of the direct gauge section measurement, to avoid displacement contribution from shoulders of test specimens and from the load train this parameter is used. Reduction in area was measured from the difference between the initial area of specimens at the middle of gauge length and reduced area after fracture which was measured using optical microscopy. The ratio of Plastic strain to failure and reduction in area at corrosive environment (EPE and RAE) to inert (air) environment (EPA and RAA) determined as ductility ratio (EPR% and RAR%) which is a measure of SCC susceptibility. The closer the ductility ratio is to 100, the higher is the resistance to environmental cracking[16].

Phase analysis of corrosion products was obtained using X-ray diffraction (XRD) method with a Philips PW 3040/60 spectrometer using Cu Kα radiation. FEI Quanta 600i scanning electron microscopy (SEM) was used to study the morphology of corrosion product layer.

RESULTS AND DISCUSSION

(i) *SCC Susceptibility of S-135 Grade Drill Pipe Steel at Different Temperatures*

Ductility ratios (EPR% and RAR%) curves for API S-135 grade drill pipe steel measured in SSRT method in DI.U.F. water saturated with CO2 are shown in Figures 2 and 3. The test results are summarized in Table II.

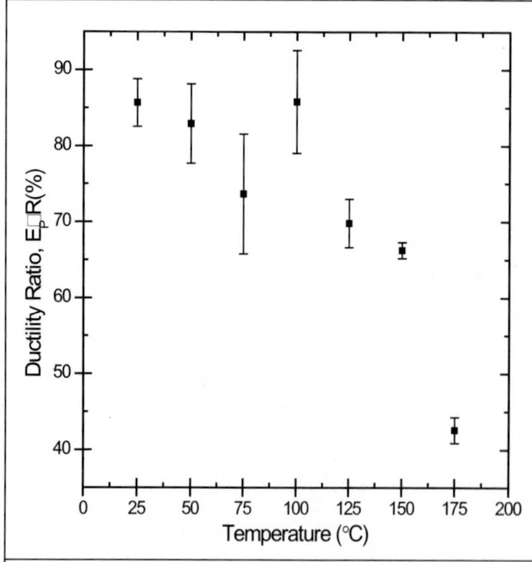

Fig. 2: Ductility ratio, E_pR, in the CO_2 saturated solution and air for the S-135 grade drill pipe steel at different temperatures & 10^{-6} s^{-1} strain rate.

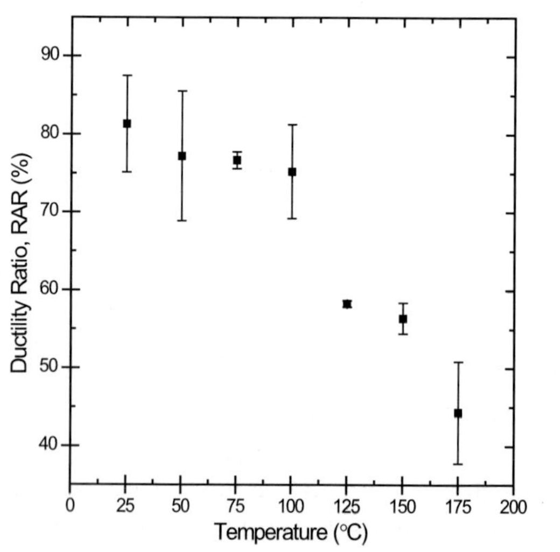

Fig 3: Ductility ratio, RAR, in the CO_2 saturated solution and for the S-135 grade drill pipe steel at different temperatures and 10^{-6} s^{-1} strain rate.

As Figures 2 and 3 show, the ductility of the specimens decreased in solution due to corrosivity at 25, 50, and 75 ^{0}C. An increase from 75 to 100 ^{0}C for EPR was observed but again increasing temperature from 125 to 175^{0}C caused reduction in ductility ratios. The first decrease is attributed to the increase in corrosion rate and reduction in the cross section of tensile specimen. Since corrosivity and temperature both playing competing role on ductility ratio, at 100^{0}C the influence of temperature becomes predominant and this governance cause an increase in ductility ratio. Ductility ratios in Figures 2 and 3 for 100^{0}C show that longitudinal strain is more significant than transverse strain. The second decrease from 125 to 175^{0}C is due to the formation of protective corrosion products which passivate the steel from further uniform corrosion and leads to localized corrosion, initiation of cracks and finally stress corrosion cracking. SEM studies of fracture morphology of the specimens were carried out to investigate this behavior. Figure 4 shows cracking and fracture morphology (side view) of S-135 grade drill pipe steel tested in CO2 saturated DI water at different temperatures with a strain rate of 10^{-6} s^{-1}. Results indicated that up to 75^{0}C no crack was observed on the specimen, but at 100 ^{0}C cracks are developed only in necking area and extended to the gauge length at 125, 150, and 175^{0}C. In addition, increasing temperature increased the depth of the cracks as well as cracks size. In the next phase of this study the crack formation mechanism and type of stress corrosion cracking (IGSCC or TGSCC) will be investigated and published in a little while.

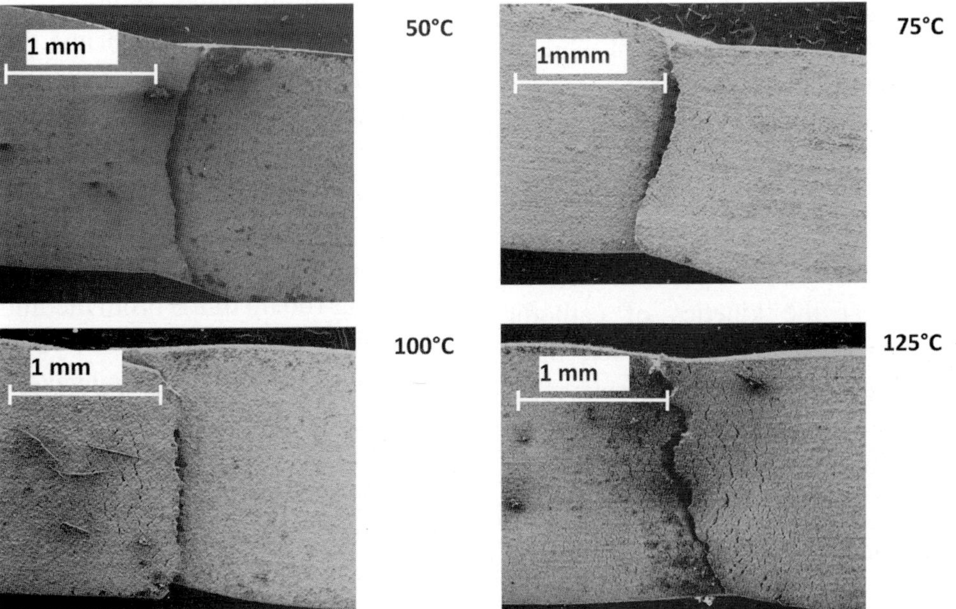

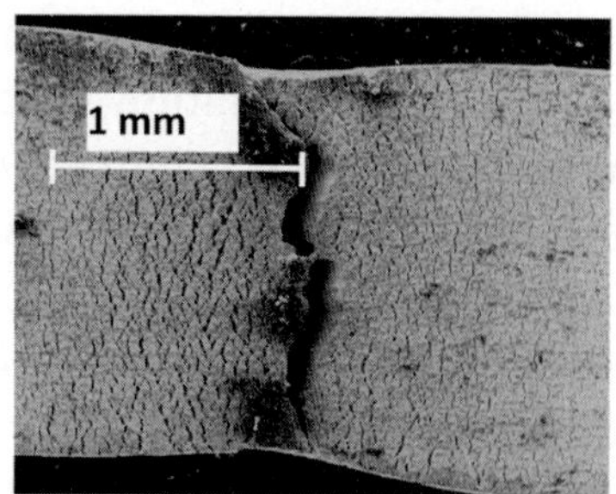

Fig. 4: *F*racture morphologies of S -135 grade drill pipe steel tested in CO_2 saturated DI water at different temperatures with a strain rate of 10^{-6} s^{-1}.

B. MICROBIOLOGICALLY INDUCED CORROSION OF PIPELINE STEELS

Microorganisms that are present in oil reservoirs are able to induce localized changes in the aqueous environment (e.g., alter the concentration of the electrolyte components pH and oxygen concentration) leading to localized corrosion known as microbiologically influenced corrosion (MIC). Microbial activities are responsible for approximately 20% of the total corrosion cost in the oil and gas industry, of which a significant part is due to anaerobic corrosion influenced by sulfate reducing bacteria (SRB) and aerobic corrosion influenced by iron-reducing and oxidizing bacteria (IRB/IOB)[17]. The metabolic by-products of these microorganisms found in biofilms on steel surfaces affect the kinetics of cathodic and/or anodic reactions. Moreover, these metabolic activities can considerably modify the chemistry of any protective layers, leading to either acceleration or inhibition of localized corrosion[17].

The objective of this study is to investigate the impact of environmental aerobic bacteria (cultivated from oil field samples rather than obtained from a culture collection) on the corrosion behavior of low alloy high strength (API 5L X80) linepipe steel. The IRB consortium used in this study was cultivated from a sour oil well in Louisiana, USA.

EXPERIMENTAL METHODS

Organisms and Testing Medium

The IRB consortium used in this study was cultivated from water samples obtained from a sour oil well located in Louisiana, USA. The water samples were collected and bottled at the wellhead from an approximate depth of 2200 ft. as described by NACE Standard TM0194[18]. For IRB isolation, one milliliter of the water sample was transferred to a 250 ml of Erlenmeyer flask containing 100 ml of LB broth medium. The LB medium was composed of tryptone (10 g), sodium chloride (10 g), and yeast extract (5.0 g) added to one liter of distilled water. The pH of the medium was adjusted to 7.2 using 5M sodium hydroxide and autoclaved for 20 min. The bacteria were incubated at 30 °C in a rotary shaker at 150 rpm until turbid growth was observed.

Identification of the iron-reducing consortium

DNA was extracted from cultivars using the MoBio Powersoil DNA extraction kit. 16S rRNA gene amplification was carried out using the 'universal' polymerase chain reaction (PCR) primers 515F and 1391R. PCR, cloning and transformation were then carried out as described by Sahl et al.[19]. Unique restriction fragment length polymorphisms (RFLP) were sequenced on an ABI 3730 DNA sequencer at Davis Sequencing, Inc. (Davis, CA). Sanger reads were called with PHRED via Xplorseq[20]. Sequences were compared to the GenBank database via BLAST (http://blast.ncbi.nlm.nih.gov/Blast.cgi).

Specimen preparation

Steel coupons (10 mm x 10 mm x 5 mm) were cut from a 6-inch (12.5 mm) section of API-5L X52 carbon steel pipe with chemical composition (wt.%): 0.07C, 1.05Mn, 0.008S, 0.03Cr, 0.008P and 0.195Si. The coupons were embedded in a mold of non-conducting epoxy resin, leaving an exposed area of 100 mm^2. For electrical connection, a copper wire was soldered at the rear of the coupons. The coupons were polished with progressively finer sand paper to a final grit size of 600 microns. After polishing, the coupons were rinsed with distilled water, ultrasonically degreased in 100% acetone followed by 100% ethanol and sterilized by exposure to 100% ethanol for 24 hours.

Electrochemical Tests and Surface analysis of the coupons exposed to IRB

The electrochemical measurements and subsequent surface analysis were performed per the procedure described by Al-Abbas et al.[21] elsewhere.

RESULTS AND DISCUSSION

Identification of the iron-reducing consortium

16S rRNA gene analysis indicated that the mixed bacterial culture consortium contained two phylotypes: members of the *Proteobacteria* (*Shewanella oneidensis sp.*) and *Firmicutes* (*Brevibacillus sp.*). *Shewanella oneidensis* are facultative anaerobic iron-reducing bacteria that are capable of diverse metabolisms. They are able to reduce ferric iron and sulfite, oxidize hydrogen gas, and produce sulfide. It has been reported that these bacteria might be involved in biocorrosion whereas some studies suggest that *Shewanella oneidensis* may have an inhibitory effect towards corrosion[22]. *Brevibacillus* are aerobic spore-forming microbes. They have the ability to degrade hydrocarbons and plastics. They were used to treat part of the Daqing oil field, in which the oil production increased by 165% for about 200 days[23].

Morphology and composition of interfacial surfaces

The morphology observations of corrosion products of API X52 carbon steel exposed to sterilized control medium (abiotic) and inoculated medium (biotic) are shown in Figure 5A and 5B, respectively. For the abiotic system, there is one homogenous layer of corrosion product (Figure 5A) formed on the surface with some deposited salt crystals. Quantitative EDS analysis, spectra not shown here, revealed peaks for oxygen, iron, chloride sodium and carbon that accumulated from the growth medium. The EDS analysis suggests that the corrosion layer is composed of iron oxides mixed with sodium chlorides and organic compounds. There is significant difference in the characteristics of the layer that developed in the presence of the IRB microbial consortium (biotic) as shown in Figure 5B. The significant amount of products observed in the biotic system is most likely due to the production of a biofilm matrix. Figure 5C displays higher magnifications of the biofilm structure that confirms bacterial attachment to the metal surface. In the case of the biotic condition, the EDS analysis revealed peaks for nitrogen and phosphorus in addition to oxygen, iron, chloride, sodium and carbon. The presence of the nitrogen and phosphorus may be due to the microbial production of extracellular polymer substance (EPS)[24].

IRB cells attached to the substrate produce EPS, which results in biofilm formation. The biofilm has a heterogeneous morphology and thickness as shown in Figure 5B. The rod-shaped bacteria occupied a small volume fraction as compared to the precipitated corrosion products and EPS. The EPS and corrosion products usually occupy 75-95 % of biofilm volume, while 5-25% is occupied by the cells[25].

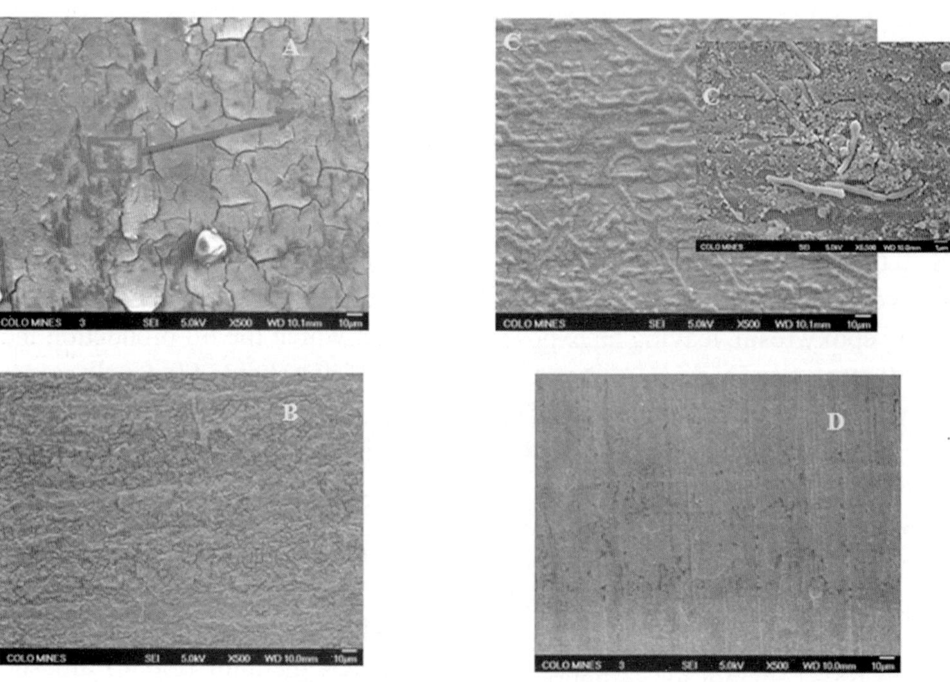

Fig. 5: FESEM for the API 5L X52 surface: (A) FESEM image of carbon steel exposed to sterile LB medium nutrients, at 500X. (B). FESEM Image of carbon steel exposed to LB medium inoculated with IRB at 500X. (C) Higher magnification (X6500.) of image (A) shows the bacteria cells. (D) Clean metallic surface removed from biotic system at 500X. (F) Clean metallic surface removed from biotic system.

The coupons immersed in the abiotic medium exhibit aggressive and elongated pitting (Figure 5C) in comparison with the biotic coupons that still show the polishing marks along with shallow pitting (Figure 5D). The protective effects of the biofilm in the case of biotic condition are attributed to the reduction of ferric ions to ferrous ions and increased consumption of oxygen by *Shewanella oneidensis sp* respiration. Dubiel et al.[26] reported that *Shewanella oneidensis sp.* inhibit corrosion on stainless steel surfaces in a medium containing yeast extract and peptone. Interestingly, it has been proposed that *Shewanella oneidensis sp.* be used to control corrosion in pipeline systems. *Shewanella oneidensis sp.* may colonize the metal surface and consume oxygen molecules adjacent to the metal surface by aerobic respiration. As oxygen is depleted, the bacteria turn to Fe^{3+} anaerobic respiration and produced Fe^{2+} ions diffuse into the bulk fluid. In static environments, this process will create a chemical shield that reduces oxygen diffusion, which in turn, inhibits the cathodic reaction due to lower oxygen availability. By electrochemical reaction Fe^{2+} is oxidized to Fe^{3+} and again reduced by bacterial respiration. Consequently, corrosion will be inhibited[27].

Polarization resistance/ Corrosion Rate

The polarization resistance (Rp) variations for the biotic and abiotic systems are shown in Figure 6A. The LPR as a function of time data revealed that in the biotic medium a substantial increase of polarization resistance (Rp) to 7000 $\Omega.cm^2$ was followed by a decrease to 5000 $\Omega.cm^2$ at 100 hours, which then remained stable at approximately 6500 $\Omega.cm^2$ throughout the period of exposure. The polarization resistance is inversely proportional to the corrosion current, which means lower corrosion rate at higher Rp resistance. This trend confirms the corrosion inhibition effects induced by the activity of the bacterial consortium. On the other hand, in the abiotic medium, the Rp trend remained more or less steady at approximately 1500 $\Omega.cm^2$ over the entire period. When comparing the two Rp trends for the biotic and abiotic medium, the Rp values for the abiotic medium are significantly lower than those for the biotic medium. This observation suggests that the corrosion rate is substantially higher in the abiotic medium. The corrosion rate plots over time for the biotic and abiotic systems are shown in Figure 6B. The corrosion rate for the abiotic medium reached a maximum value of 20 mpy after 200 hours whereas the corrosion rate for the biotic system was around 2 mpy over the experimental period.

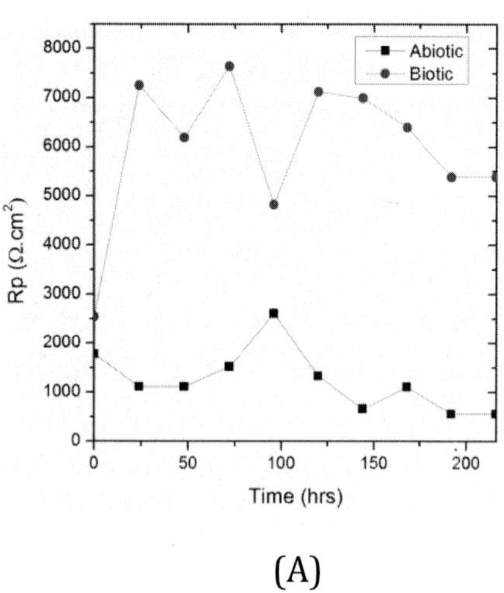

(A)

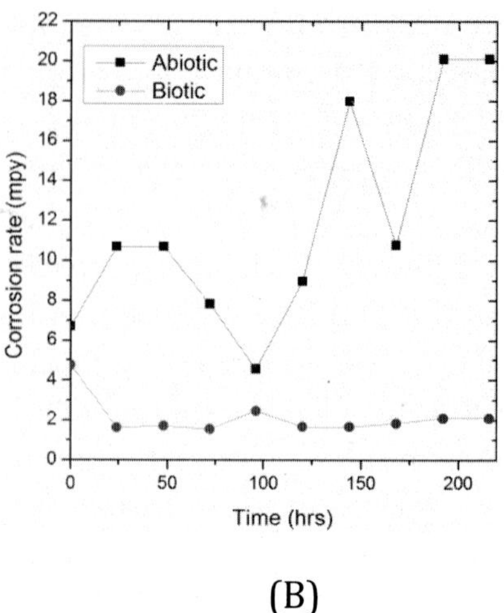

(B)

Fig. 6 : (A) Polarization resistance, (B) Corrosion rates variations for both biotic and abiotic media.

Both Rp and corrosion rate data confirm the corrosion inhibition properties of the bacterial consortium. This inhibition cannot be explained by a single mechanism[27]. It might be attributed to the oxygen depletion induced by bacterial respiration[27]. The change in oxygen concentration caused by microbial respiration can decrease the cathodic reaction rate by reducing the amount of reactants available for the cathodic reaction. A similar inhibitory mechanism for steel in the presence of two marine isolates in saline media was reported by Pedersen and Hermansson[28]. It is also possible that the biofilm developed by the bacteria contributes to the low corrosion in the biotic medium. There have been a few reports of corrosion inhibition in the presence of biofilms[17]. Biofilms and attached cells may form a diffusion barrier hindering the reactants diffusion to and from the metal surface[29]. Another mechanism that might be responsible for the corrosion inhibition is related to the generation of a cathodic protection current by the bacterial consortium. It was reported that *Shewanella oneidensis sp.* produced electrically conductive pilus-like appendages called bacterial nanowires that can transfer electrons to the electrode. The capability of *Shewanella oneidensis sp.* to transfer electrons from organic sources to electrodes without intervening catalysts serves as the basis for electricity production in microbial fuel cells[30]. Moreover, *Brevibacillus sp.* has been investigated in microbial fuel cells for their capacity to generate electrons[31]. Interestingly, previous studies show the capacity of *Brevibacillus sp.* to produce electricity increases in the presence of *Pseudomonas sp.* due to the metabolites produced by *Pseudomonas sp.*, which enable *Brevibacillus sp.* to achieve extracellular electron transfer[31.] Based on these facts, it is possible, that *Brevibacillus sp.* and *Shewanella oneidensis sp.*, through their complimentary metabolisms, produce cathodic currents that result in corrosion inhibition. The increase in corrosion rate noticed in the abiotic medium is possibly due to the effect of a formation of a mixed non-protective layer of sodium chloride, iron oxides and carbon-based compounds on the electrode surface[25].

Electrical Impedance Spectroscopy (EIS) results

Figure 7A displays the Nyquist plots for the carbon steel coupon exposed to the abiotic medium over time. The steady state was reached at 120 hours. At low frequencies (LF), shown in Figure 7A, the magnitude of the capacitive loop represented by the semicircle diameter decreased with time. These LF magnitudes represent the change in charge transfer resistance (Rct) that describes the evolution of the anodic reaction that is controlled by charge transfer processes[25]. The decrease of Rct with time indicates an increase in corrosion rate, possibly due to the effect of a formation of a mixed layer of sodium chloride, sulfide, potassium and carbon-based compounds on the electrode surface. Moreover the presence of oxygen in the system enhances the cathodic reactions that drive the anodic dissolution of the metal[25]. The formation of a corrosion product layer was confirmed by the phase angle spectra (Figure 7B) that shows two time constants at 10 Hz and 0.1 Hz medium frequencies, as indicated by the arrows in Figure 3B.

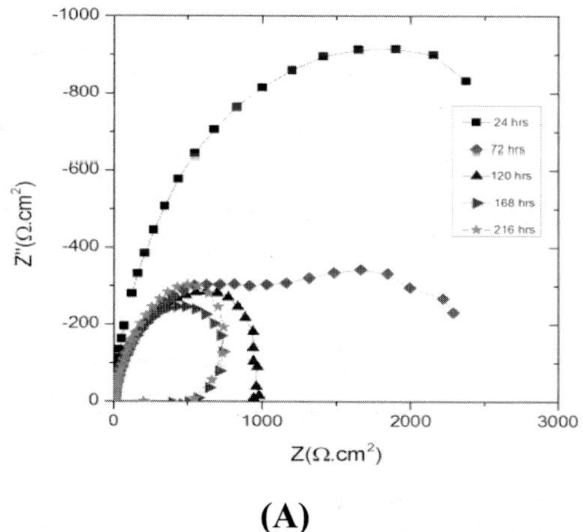

(A)

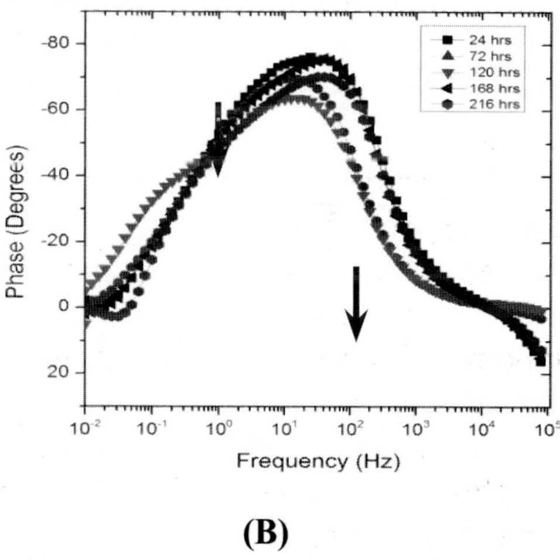

(B)

Fig. 7: EIS date for abiotic culture medium; (A) Nyquist Plots (B) Phase angle plots.

When the carbon steel was exposed to the inoculated culture medium (biotic), the EIS spectra varied significantly with exposure time as shown in Figure 8A. The low frequency (LF) magnitude, represented by the semicircle diameter, significantly increased with time indicating a decrease in corrosion rate and increase in charge transfer resistance (Rct) as supported by Figure 8A.

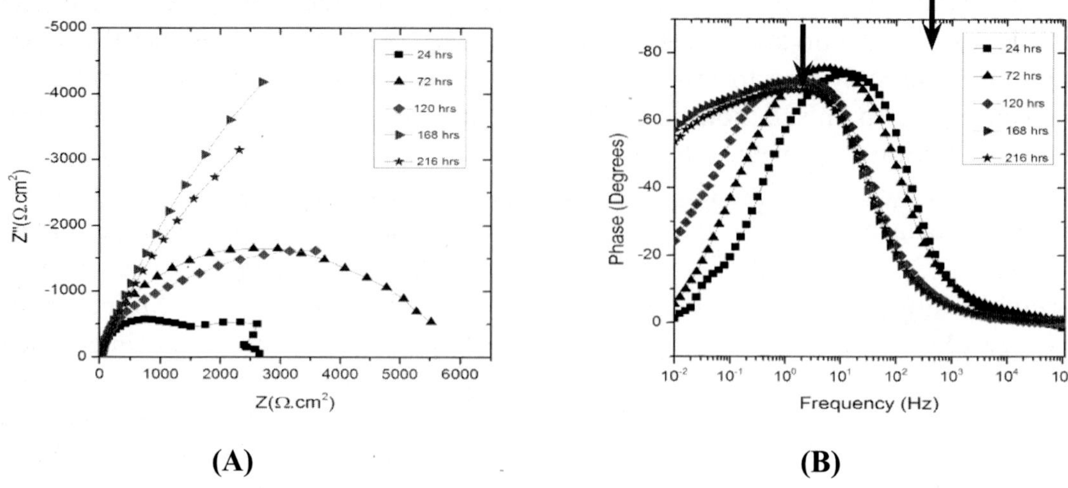

Fig. 8: EIS date for inoculated biotic medium; (A) Nyquist Plots (B) Phase angle plots.

For the first 72 hours, the medium frequency (10 Hz) response presented in the phase diagram in Figure 8A shows one time constant that indicates an activation control process. This behavior is attributed to the formation of an unstable conditioning layer based on a mixture of inorganic/ organic compounds[21]. However, when stable biofilm formation was achieved at 120 hours and when steady state is reached, mass transfer limitations overcome the interfacial activation, which is reflected in a change from a semicircle behavior to a straight line shown in Figure 8A. It is speculated that the formation of an adherent biofilm influenced the mass transfer processes in the electrochemical cell[25]. The formation of biofilm is confirmed by the two time constants indicated by arrows in Figure 8B. The EIS spectra affirm the inhibitory effect of the bacterial activities and subsequent biofilm formation described in the previous sections of this paper. Potentially the IRB consortium could be use as a strategy to combat corrosion especially in aerobic environment.

CONCLUSIONS

- Fracture surface studies revealed the occurrence of stress corrosion cracks when conducting SSRTs on API S-135 grade drill pipe steel in CO2 saturated environment at temperatures above 100^0 C.
- SEM and XRD studies showed the formation of protective rhombohedral FeCO3 at 100^0C and above temperatures on the surface passivating the steel which later leads to localized corrosion and initiation of cracks. Some of the cracks formed on the surface may become inactive, but some may be reactivated by coalescence with later nucleated cracks and perhaps by stress condition variation.

- At lower temperatures cracks did not develop in gauge length meaning that protective corrosion products were not formed on the surface and uniform corrosion of steel occurred.
- SCC occurs in the absence of uniform corrosion. Creation of favorite condition for protective scale formation at 100^{0}C and above prevents uniform corrosion which leads to initiation of cracks from susceptible areas. Increasing temperature increases the kinetics of reactions and faster initiation and propagation and higher loss of ductility.
- In the microbiologically influenced corrosion (MIC) study of API 5L grade X52 carbon steel coupons by an iron-reducing consortium cultivated from a sour oil well produced water was investigated. Interestingly, IRB metabolic reactions and biofilm formation inhibit the corrosion process. The maximum corrosion rate in the biotic system was 2 mpy while it was 20 mpy in the abiotic solution. The corrosion inhibition was attributed to oxygen depletion induced by IRB bacteria respiration and the protective affect of the biofilm that formed on the metal surface.

REFERENCES

[1] M. B. Kermani, and A. Morshed, – Carbon dioxide corrosion in oil and gas production-A Compendium, Corrosion **59**, 659 (2003)

[2] J. Mougin, M.S. Cayard, R.D. Kane, B. Ghys, and C.Pichard, –Sulfide stress cracking and corrosion fatigue of steels dedicated to bottom hole assembly components, NACE, paper 05085, 2005.

[3] B. Mishra, S. Al-Hassan, D.L. Olson, and M.M. Salama, –Development of a predictive model for activation-controlled corrosion of steel in solutions containing carbon dioxide□, Corrosion, Corrosion **53**, 852 (1997)

[4] B. Mishra, D.L. Olson, S. Al-Hassan, and M.M. Salama, –Effect of Microstructure on Corrosion of Steels in Aqueous Solutions Containing CO2□ Corrosion, **54**, 480 (1998).

[5] C. De Dewaard and D. E. Milliams, – Carbonic Acid Corrosion of Steel□, Corrosion, **31**, 177 (1975).

[6] J. C. Charbonnier, H. Margot-Marette, A. M. Brass, and M. Aucouturier, –Sulfide Stress Cracking of High Strength Modified Cr-Mo Steels, Metallurgical Transaction A, Vol. 16A, May 1985, p 935.

[7] R. Brett Chandler, Michael J. Jellison, James W. Skogsberg, Tom Moore, –Advanced drill string metallurgy provides enabling technology for critical sour drilling, Paper No. 02056, NACE 2002.

[8] R. D. Kane and M. S. Cayard, – Roles of H2S in the behavior of engineering alloys: A review of literature and experience, Paper No. 274 NACE 1998.

[9] M. Elboujdaini and R. V. Revie, —Metallurgical factors in stress

corrosion cracking (SCC) and hydrogen induced cracking (HIC), J Solid State Electrochem (2009) 13: 1091-1099.

[10] A Robert Mack, —Stress corrosion cracking of high strength steels in aqueous solutions containing CO2-effects of yield strength, dissolved oxygen, and temperature, NACE, paper 01076, 2001.

[11] Byoung-Ho Choi and Alexander Chudnovsky, —Observation and modeling of stress corrosion cracking in high pressure gas pipe steel, Metallurgical and Materials Transaction A, 13 August 2010.

[12] R. N. Parkins, —Mechanistic Aspect of intergranular stress corrosion cracking of ferritic steels, Corrosion, **52**, 363 (1996).

[13] R. N. Parkins, W. K. Blanchard Jr., and B.S. Delanty, —Transgranular stress corrosion cracking of high-pressure pipelines in contact with solutions of near neutral pH, Corrosion, Vol. **50**, 395, (1994).

[14] J. A. Colwell, B. N. Leis, P. M. Singh, —Recent developments in characterizing the mechanism of near neutral pH SCC, NACE, paper 05161, 2005.

[15] B. Gu, W. Z. Yu, J. L. Luo, X. Mao, —Transgranular stress corrosion cracking of X-80 and X-52 pipeline steels in dilute aqueous solution with near-neutral pH, Corrosion, Vol. **55**, No. 3, p 312.

[16] National Association of Corrosion Engineers standard —Standard Test Method Slow Strain Rate Test Method for Screening Corrosion-Resistant Alloys (CRAs) for Stress Corrosion Cracking in Sour Oilfield Service, TM0198-2004.

[17] B.J. Little and J.S. Lee. 2007. *Microbiologically Influenced Corrosion*. John Wiley & Sons Inc., Hoboken, NJ, USA; pp: 1-50.

[18] NACE Standard TM0194. 2004. "Field Monitoring of Bacterial Growth in Oil and Gas Systems". Published by NACE, Houston, Texas.

[19] W. Sahl, F. Nathaniel, H.J. Kirk, W. David, W.C. Stone and J. R. Spear. 2010. "Novel Microbial Diversity Retrieved by Autonomous Robotic Exploration of the World's Deepest Vertical Phreatic Sinkhole" Astrobiology, 10(2):201–13.

[20] D. Frank, 2008. XplorSeq: a software environment for integrated management and phylogenetic analysis of metagenomic sequence data. *BMC Bioinformatics* 9: 9–420.

[21] F.M. Alabbas, A.B. Gavanluei and J.R. Spear et al., 2011. "Effects of Sulfate Reducing Bacteria on the Corrosion of X-65 Pipeline Carbon Steel" Paper # C2012–0001140, Proceeding NACE 2012 Conference.

[22] M.T. Lutterbach, L. Contador, A. Oliveira and M. Galvão . 2009. "Iron Sulfide Production by *Shewanella* Strain Isolated From Black Powder" Paper # 09391, Proceeding NACE 2009 Conference.

[23] N. Youssef, M.S. Elshahed and M.J. McInerney.2009. "Microbial

Processes in Oil Fields: Culprits, Problems, and Opportunities", *Adv. Appl. Microbiol.* 66:141–251.

[24] C. Shobhana, G. Gunasekaran and Pradeep Kumar. 2005. "Corrosion Inhibition of Mild Steel by Aerobic Biofilm" *Electrochimica Acta*, 50 (24):4655–4665.

[25] H. Castaneda and X.D. Benetton. 2008. "SRB-Biofilm Influenced in Active Corrosion Sites Formed at the Steel-Electrolyte Interface When Exposed to Artificial Seawater Conditions" *Corrosion Science* 50:1169–1183.

[26] M. Dubiel, C.H. Hsu, C.C. Chien, F. Mansfield and D.K. Newman. 2002. "Microbial Iron Respiration Can Protect Steel From Corrosion" *Applied and Environmental Microbiology* 68, 1440–1445.

[27] L.K.Herrera and H.A. Videla. 2009. "Role of Iron-Reducing Bacteria in Corrosion and Protection of Carbon Steel" *International Biodeterioration & Biodegradation* 63: 891–895.

[28] A. Pedersen, M. Hermansson.1991. "Inhibition of Metal Corrosion by Bacteria" *Biofouling* 3:1–11.

[29] H.A. Videla and L. K. Herrera. 2009." Understanding Microbial Inhibition of Corrosion. A Comprehensive Overview" *International Biodeterioration & Biodegradation* 63:896–900.

[30] Y.A. Gorby, Y. Svetlana and McLean J. S. et al. 2006. "Electrically Conductive Bacterial Nanowires Produced by *Shewanella oneidensis* Strain MR-1and Other Microorganisms" *PNAS* 103: 11358-11363.

[31] T.H. Pham, Boon N. and Aelterman P. et al. 2008. "Metabolites Produced by *Pseudomonas sp.* Enable a Gram Positive Bacterium to Achieve Extracellular Electron Transfer" *Appl Microbiol Biotechnol* 77:1119–1129.

Miniature Specimen Technique for Assessing the Microstructural Degradation in 2.25Cr-1Mo Steel

B.B. Jha, T.K. Sahoo, D. Tripathy and B.K. Mishra
Surface Engineering Department
CSIR-Institute of Minerals and Materials Technology
Bhubaneswar, Odisha, India
Email: bbjha59@yahoo.com

ABSTRACT

Miniature specimen technique viz. small punch tests (SPT) have been carried out at room temperature to correlate the microstructural degradation of 2.25Cr-1Mo steel with that of SPT parameters. Microstructural degradation of this steel has been introduced as a result of thermal ageing corresponding to Larson-Miller parameters (LMP) values of 33,012, 35,402, 37,846 and 38,374. SPT parameters viz. total area and area under the region of plastic instability of load-displacement curve have been found to decrease with increase in LMP values. Strength parameter viz. UTS obtained using uniaxial tensile test has also been found to decrease with increase in LMP values. Results indicated that miniature specimen technique viz. small punch test could be successfully used to assess the degradation of microstructure in 2.25Cr-1Mo steel generated due to their exposure at high temperatures.

KEYWORDS: *Small Punch Test, 2.25Cr-1Mo Steel, Microstructural Degradation, Larson-Miller Parameters, Plastic Instability*

INTRODUCTION

Ferritic 2.25Cr-1Mo steel is a popular engineering alloy used for high temperature application such as heat exchanger, steam receivers, high temperature headers and pipings in various power generating units. During service, components made up of this steel suffer material degradation in terms of decrease in strength as a result of change in microstructure due to long term exposure at high temperatures. Therefore, in order to estimate the exhausted life of components, the determination of their material properties in their current state of damage is highly essential. This needs removal of metal samples from the service exposed components without jeopardising their service suitability. Miniature specimen technique has the potential to evaluate this degradation trend without affecting their service-worthiness and hence are gaining importance in recent days [1-3]. This technique is also able to characterise the mechanical behaviour of materials using a greatly reduced volume of material. As a result, it has greater significance in evaluating the degradation of costly and critical components like turbine rotor, disk, shaft, casing etc.

The small punch (SP) testing method is a miniaturised testing technique based on the determination of the curve of load vs displacement for a small disc shaped specimen of size about 8 to 10mm dia and the thickness of the disc could vary from 0.25mm to .05mm [4]. A typical SP load-displacement curve shows three regions viz. elastic bending, plastic bending and plastic instability as shown in Fig. 1.

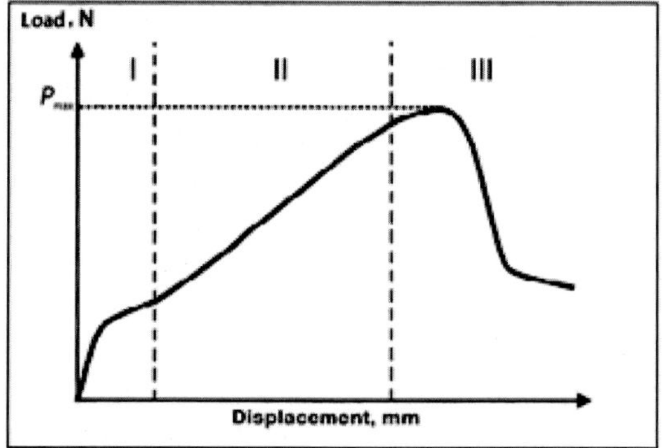

Fig. 1: A schematic of small punch load-displacement curve showing different regions viz. (I) Elastic bending, (II) Plastic bending and (III) Plastic instability.

The deformation mechanisms with each regions have studied in detail [1-3,5] and correlated with different parameters of damage [5]. Though large number of attempts have been made to correlate the radiation damage, ductile - brittle transition temperature and toughness properties of materials with that of different parameters of SP test [1-5]; this method has not found its full application in assessing the microstructural degradation of materials taking place at high temperatures. Since the microstructural deterioration in high temperature metals and alloys acts as a precursor for major damage mechanisms, it is very appropriate to correlate these deteriorations with SP parameters. This correlation would be very useful not only to predict the exhausted life of components but also help the utility to plan their run / repair / replacement schedule. In the present paper, therefore, an attempt has been made to correlate the microstructure degradation of 2.25Cr-Mo steel with that of the parameters of small punch testing. Small punch test for various degraded samples of 2.25Cr-1Mo steel has been carried out and the results are correlated with that of microstructural degradation. Results indicate that small punch testing could be successfully used to assess the degradation of microstructure in this steel resulted due to their exposure at high temperature service environment.

EXPERIMENTAL

Materials

2.25Cr–1Mo ferritic steel having nominal chemical composition (wt.%) as C-0.07, Mn-0.42, Si-0.019, S-0.025, Cr-2.28, P-0.02, Mo-0.98, Ni-0.09, Fe-balance, has been used in the present investigation.

Thermal Ageing

The starting material used in this study was in the form of 12 mm thick plate. From this plate strips of 12 mm x 12 mm x 250 mm were cut using a power hacksaw and then machined to a size of 10 x 10 x 80 mm dimension with rolling directions parallel to the length of the strips. These strips were then sealed in evacuated quartz tubes and given annealing treatment at 1223 K for two hours. The temperature of the furnace was maintained, within an accuracy of $\pm$ 1.0 K.

In order to study the microstructural degradation, these annealed strips were subjected to ageing treatment at 873 K/10 hrs, 923K/20 hrs, 973K/40 hrs and 973K for 80 hrs using a tubular furnace. These treatments correspond to Larson-Miller Parameter (LMP) values of 33,012, 35,402, 37,846 and 38,374 respectively. Scales of the heat treated samples were then removed by coarse polishing on silicon carbide papers.

Optical Metallography and SEM

Cubes of size 10 x 10 x 50 mm were cut from the heat treated samples, annealed and thermally aged to various degrees. These specimens were suitably mounted in a cold setting resin and were ground using successive grades of silicon carbide papers and then polished mechanically using 0.25 micron diamond paste for final scratch free polishing. Chemical etching was then done using a solution of 2 ml HNO_3 in 100 ml methanol (i.e., 2% Nital) maximum for 10 seconds.

Chemically etched and dried specimens were carried out using Nikon make metallurgical microscope (Model LV-150). Complete scanning was done at the lower magnification to know the representative microstructural features. These representative microstructural features were then captured at higher magnification and presented as observations. All the fractured samples after SP testing were examined using SEM (Hitachi Model S-3400N) for studying the morphology of fracture surfaces.

Transmission Electron Microscopy (TEM)

For transmission electron microscopy (TEM) a very thin specimens of thickness about 0.5 mm were cut from various heat treated samples using electric discharge machining (EDM). These specimens were further thinned using fine grade silicon carbide papers up to a thickness of about 0.2 mm and then a disc of size 3 mm diameter was cut from these thinned specimens. These 3 mm diameter discs were further thinned by polishing both sides using silicon carbide papers and emery papers up to a thickness of about 80 microns. The thickness of these discs was further reduced at the central portion using dimpler (M/s Gattan make) to a thickness of about 30 microns. Dimpled discs were then subjected to ion milling till a fine perforated hole is made at the centre of the disc enabling the adjacent area of the hole to be electron transparent.

Small Punch Test Procedure

The sample of size 8mm diameter and with 0.5mm thickness were cut from various heat treated /thermally aged material samples using electric discharge machining (EDM) and polished with1000 & 1200 grit paper to obtain fine surface finish. The experimental setup consists of miniature specimen, specimen holder, plunger (shown in Fig. 2) were clamped in screw and the punch with hemispherical headed tip of diameter of 2.5 mm pushed against the specimen under a constant punch displacement rate of 0.05 mm/min upto a 40% drop of peak load.

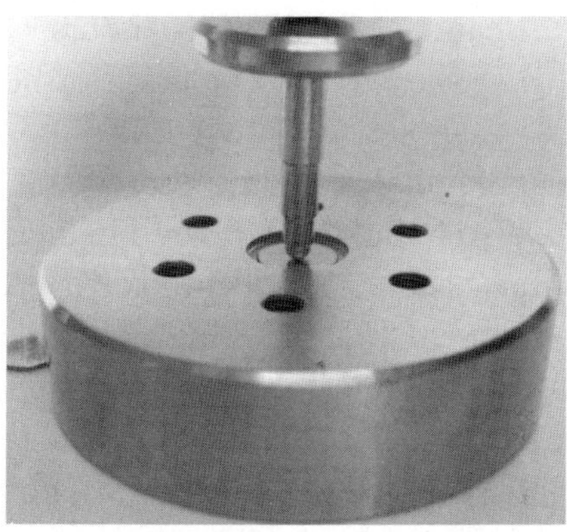

Fig. 2: Small punch test assembly showing plunger, specimen holder and disc specimen.

The test was done with INSTRON 8801 universal testing machine. After each test, the specimens were unloaded and the areas under the load- displacement curve for all samples (up to fracture load) were obtained. Numerical data for load-displacement curve has also been used to regenerate the load- displacement curve using MATLAB software. This helped in calculating the area under different regions of the curve viz. elastic bending, plastic bending and plastic instability.

RESULTS AND DISCUSSIONS

Microstructural Features

Starting microstructure for this material in annealed condition has a ferrite-bainite mixture as shown in Fig. 3(a). The average initial grain size was measure and found to be 35 μm. Thermal ageing at 873K/10 hours corresponding to Larson Miller Parameter (LMP) values of 33,012 shows first sign of carbide spheroidisation as indicated in Fig. 3(b). Figure 3(c) shows the optical micrograph tempered at 923K/20 hrs corresponding to LMP value of 35,402.

Appreciable spheroidisation of bainite phase could be noted in this micrograph. Ferrite grain boundaries have also become comparatively more defined as compared to Fig. 3(b) indicating that some precipitation at grain boundary has also taken place. An increase in ageing temperature resulted in a progressive increase in the degree of spheroidisation as well as in the density of precipitation. Figure 3(d) shows the optical micrograph of aged specimen corresponding to LMP value of 37,846, which shows that spheroidisation is almost complete but the carbides are still grouped in banitic phase. When ageing was done at 973K/80 hours corresponding to LMP values of 38,374, it

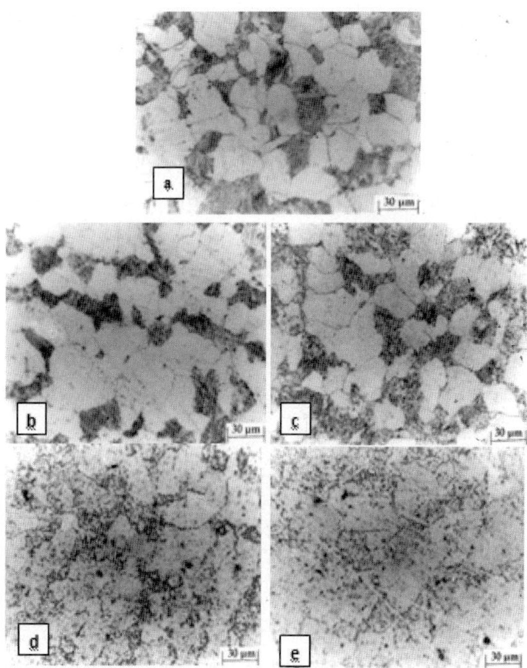

Fig. 3: Optical micrographs of 2.25Cr-1Mo steel a) Annealed and thermally aged having LMP values (b) 33,012 (c) 35,402 (d) 37,846 and (e) 38,374.

is found that spheroidisation of Fe-carbides is complete, carbides are dispersed leaving little trace of original bainitic area as shown in Fig. 3(e).

Spheroidisation of the carbides in this steel occurs over time when heated at various temperatures because spheroidised microstructure is the most stable microstructure. A platelet or blade of Fe-carbides has a large surface to volume ratio and thus, a higher surface energy. This high surface energy can be reduced by changing its shape from a blade or plate to a sphere. Further, reduction in surface energy can be achieved by decreasing the number of iron carbide particles from a large number of small particles to small number of large particles. Therefore, the first stage is the breakdown of platelets or blades of iron carbides. This is followed by appreciable breakdown of iron carbide platelets and also precipitation at grain boundaries. Further changes in microstructure shows a complete breakdown of platelets but they are still grouped in bainitic phase. Lastly, iron carbide particles coalesce to form larger particles of iron carbides. Figure 3 shows this progressive change in microstructure as a result of various degrees of ageing in this 2.25Cr-1Mo steel.

Transmission electron micrograph (TEM) of annealed sample (Fig. 4(a)) showed fine precipitation of lenticular shaped carbides uniformly distributed within the grains of the bainitic matrix. The shape of carbides is mostly lenticular aligned parallel to each other along their axes as shown in Fig. 4(a). These carbides are identified as M_3C type[6]. Transmission electron micrograph of the samples corresponding to the treatment of 873K/10 hours (LMP = 33,012) showed the evolution of needle type of precipitates as shown in Fig. 4(b). These precipitates having size about 300 nm are also found to be ordered in the same direction. These are identified as M_7C_3 type carbides [6]. Transmission electron micrograph of the samples corresponding to the treatment of 923K/20 hours (LMP = 35,402) showed precipitation of globular carbides (size about 500 nm) within the bainitic matrix (Fig. 4(c)) and identified as $M_{23}C_6$ type carbides. These carbides at this stage rich in iron with considerable chromium content [6]. Transmission electron micrograph of specimen corresponding to the treatment of 973K/40

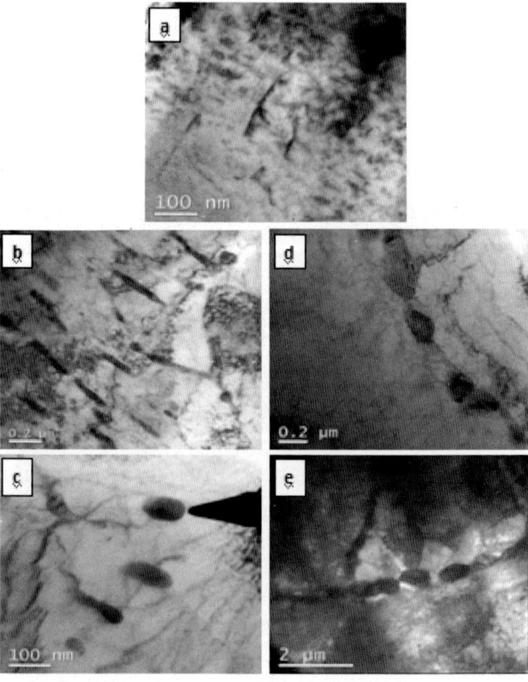

Fig. 4: Transmission Electron Micrographs of 2.25Cr-1Mo steel a) Annealed and thermally aged having LMP values (b) 33,012 (c) 35,402 (d) 37,846 and (e) 38,374.

hours (LMP = 37,846) and 973K/80 hours (LMP = 38,374) are shown in Fig. 4(d) and 4(e) respectively. It is observed that by this treatment the precipitates have coarsen and they are aligned at grain boundaries. Their size has also increased (about one micron). Some plate shaped precipitates have also been observed (Fig. 4(d)) as a result of this treatment. These carbides are identified as $M_{23}C_6$ rich in chromium and plate shaped carbides are M_6C type rich in Mo content [6].

Apart from changes in microstructure which take place due to spheroidisation, thermal exposure of this steel also contributes to complex ageing phenomena which are governed by complicated precipitation process that occur in this variety of steel. It has been reported that ageing at lower temperature and lesser time produces a microstructure consists of Fe_3C, epsilon carbides and fine Mo_2C carbides [6]. The Mo_2C forms a high density of fine needles or platelets and thus, contributes to strengthening by dispersion hardening [7]. As the degree of ageing is increased, the unstable Mo_2C carbides eventually transforms to larger globular particles of various carbides [7,8]. Not only that the amount of precipitates increased as a result of increase in the degree of ageing but the size and shape of carbides also changed with the severity of ageing. These observations clearly suggest that coarsening of precipitates take place as a result of increase in the severity of ageing treatment. Present results on sequence of precipitation and coarsening of precipitates are found to be consistent with the reported observation in the literature [7,8].

Small Punch Test

In order to examine the influence of ageing treatment on the behaviour of materials under small punch test, these aged samples were tested using the fixture assembly (Fig. 2) under the punch displacement rate of 0.05 mm/min. Figure 5 shows the load vs deflection curve of all the samples i.e. annealed and thermally aged at different values of LMP. These plots are generated using MATLAB software from the data obtained from SPT test. Typical three regions viz elastic bending, plastic bending and plastic instability is clearly depicted in all these plots. Table 1 indicates the variation of total area and the area under the region of plastic instability of load displacement curve with that of L-M parameters. These

Table 1: Area under the load-displacement curve and ultimate tensile strength (UTS) values of annealed and thermally aged 2.25Cr-1Mo steel.

Sample heat treatment	Area under the curve (N.mm)			
	LMP Values	Area under region III	Total area	UTS (MPa)
1223 K/2hrs/FC	Virgin	411.07	3405.9	601
873 K/10hrs/FC	33,012	385.24	2378.4	555
923 K/20hrs/FC	35,402	354.28	2301.11	548
973 K/40hrs/FC	37,846	299.91	2035.7	468
973 K/80hrs/FC	38,374	303.28	1698	452

variations are depicted clearly in Fig. 6 for the total area and in Fig. 7 for the area under the region of plastic instability. It is clearly indicated in both the figure that the total area and area under the region of plastic instability decreases with increase in L-M parameters. Also for annealed sample, the total area and area under plastic instability region has been found to be higher in comparison to thermally aged samples of various degrees.

Ferritic 2.25Cr-1Mo steel derives its strength mainly from two sources, (i) solid solution strengthening of matrix ferrite by carbon, molybdenum and chromium and (ii) precipitation hardening by carbides. Although number of different types of carbides may be present (viz., Fe_3C, epsilon carbides, etc.) in annealed material, but the principal carbide phase responsible for its strength is fine dispersion of Mo_2C carbides [7,9]. The initial microstructure of this steel in annealed condition consists of bainite and ferrite (Fig. 3(a) containing Fe_3C (Fig. 4(a)). After thermal ageing fine precipitation of other carbide phases like M_7C_3, $M_{23}C_6$ and M_6C where *M* is mostly molybdenum and chromium takes place. These precipitates are shown in transmission electron micrographs of aged specimens (Fig. 4(b-e)). This precipitation sequence is in line with the observations of other workers [7,9]. Carbides like M_3C which are found in annealed material are of lenticular shape (Fig. 4(a)) and are coherent with the matrix. As a result, their presence in the microstructure contributes to the strength significantly. After thermal ageing unstable carbides transforms to larger globular carbides of M_7C_3 , $M_{23}C_6$ and M_6C (M is mostly Cr & Mo) which are not coherent and are also of large in shape and size. For these reasons, such carbides are reported to have little strengthening effects [8]. Table 1 shows the variation of total area and area under the region of plastic instability including ultimate tensile strength as a function of L-M parameters. All these parameters show a decreasing trend with increase in the value of LMP. This decrease could be attributable to the increased tendency towards carbides precipitation and precipitate coarsening. Also the precipitation of carbide phase (Fig. 4(b-e)) results in depletion of Cr, Mo and C from the matrix. From this investigation it appears that decrease in solid solution strengthening dominates over increase in strength due to precipitation although generally the precipitation of carbide increases strength if precipitates are coherent. Coarsening of carbide particles (Fig. 4(b-e)) along with change of shape of iron carbides from platelet to spheres (Fig. 3(b-e)) contribute to decrease in strength parameters of this steel including decrease in total area and area under the region of plastic instability (Fig. 6 & 7) as a result of increase in LMP values. Apart from nullifying the effect of coherency hardening and order hardening, coarsening of precipitates (Fig. 4(b-e)) also facilitates dislocations to bypass them either by cross-slip or Orowan looping [10]. This could be the possible reason of decrease in total area and area under the region of plastic instability with an increase in LMP values.

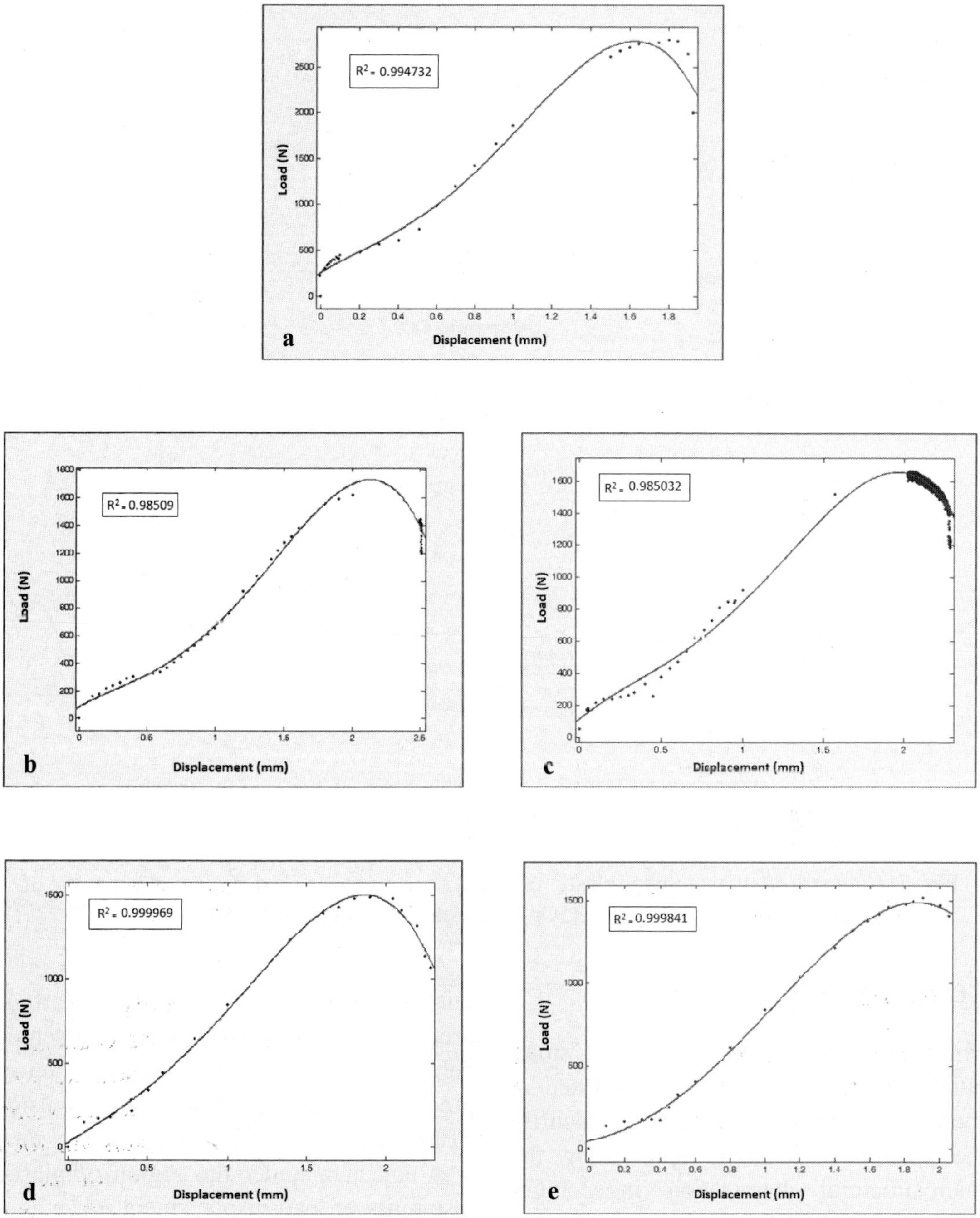

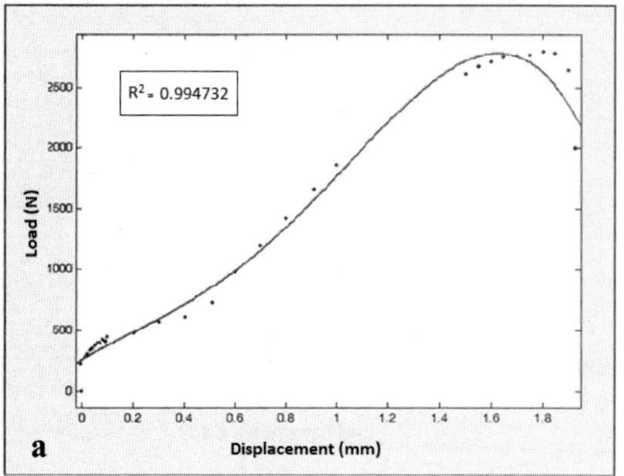

Fig. 5: Compressive load vs central deflection curve of 2.25Cr-1Mo steel a) Annealed and thermally aged having LMP values (b) 33,012 (c) 35,402 (d) 37,846 and (e) 38,374 (through MATLAB Software).

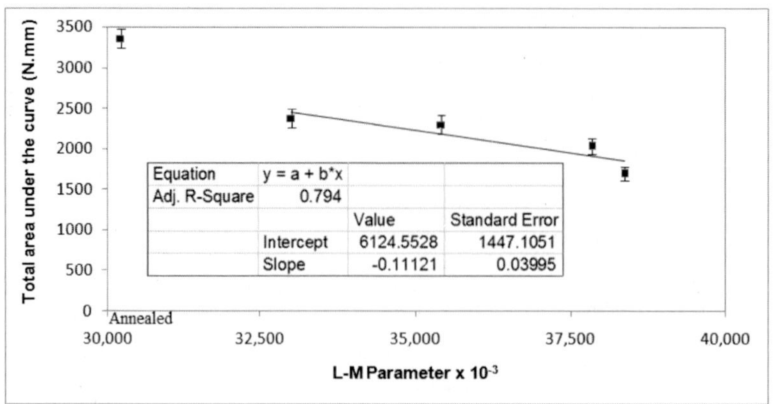

Fig. 6: Comparison of total area under the load-displacement curve of annealed and thermally aged 2.25Cr-1Mo steel as a function of L-M parameters.

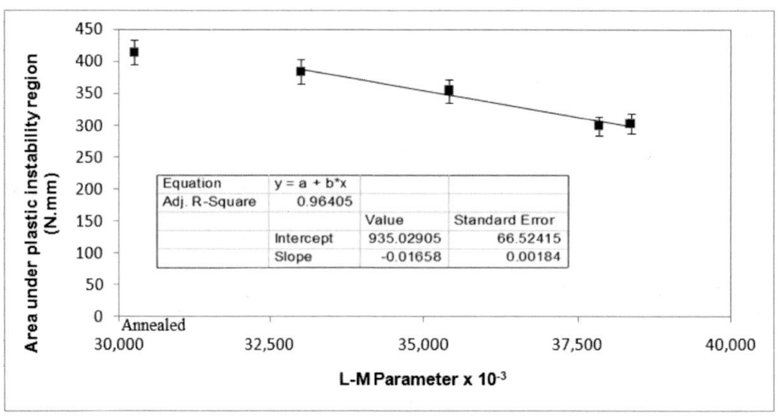

Fig. 7: Comparison of area under plastic instability region of the load-displacement curve of annealed and thermally aged 2.25Cr-1Mo steel as a function of L-M parameters.

CONCLUSIONS

Miniature specimen technique viz. small punch test (SPT) has been performed at room temperature in order to identify the parameters of SPT representing the microstructural degradation in 2.25Cr-1Mo steel. Microstructural degradation in this steel has been introduced as a result of thermal ageing corresponding to LMP values of 33,012, 35,402, 37,846 and 38,374 . Optical and transmission electron microscopy were used to identify the carbide precipitates and examine their coarsening behaviour. Shape, size and distribution of precipitates have also been studied using TEM. Two of the SPT parameters viz. total area and area under the region of plastic instability of load-displacement curve have been found to decrease with increase in L-M parameters. Coarsening of carbide particles along with change of shape of iron carbides from platelets to spheres could be the possible reason for decrease in these SPT parameters with increase LMP values. Our result indicated that small punch test could be successfully used to assess the degradation of microstructure in 2.25Cr-1Mo steel generated due to their exposure at high temperature service environment. This is a step forward towards the application of this technique for failure analysis and life extension studies.

REFERENCES

[1] Kundan Kumar, K. Madhusoodanan and B.B. Rupani, Conference on pressure vessel and piping, (OPE-2006) Chennai, 2006, p. 92.

[2] I. Klevtsov, A. Dedov, Molodtsov, 6th International DAAAM Baltic Conference Industrial engineering, Tallinn, Estonia, 2008.

[3] T. H. Hyde and W. Sun and J.A. Williams, International Materials Reviews, Vol. 52 2007, p.213.

[4] K.V. Kasiviswanathan, S.K. Hotta, C.K. Mukhopadhyay and Baldev Raj, American Society for Testing and Materials, ASTM STP - 1329, Philadelphia, 1998, p. 523.

[5] J. Foulds and R. Viswanathan, Journal of Engineering Materials and Technology, Vol. 116, 1994, p.457.

[6] B.B.Jha, B.K.Mishra, B. Satpati and S.N.Ojha, Material Science – Poland, Vol. 28, 2010, p. 335.

[7] Vaidehi Ganesan and P. Rodriguez,

[8] Proc. of Int. Symp. on Materials aging and Life Management, Kalpakkam, India, 2000, p.293.

[9] V. Jayan, M.Y. Khan and M. Husain, Materials Letters, Vol. 58, 2004, p.2569.

[10] S. Saroja, P. Parameswaran, M. Vijayalaxmi and V.S.Raghunathan, Acta. Metall. Mater., Vol. 43, 1995, p.2985.

[11] E. Orowan, Institution of Metals, London, 1948, p.451.

Fatigue Resistance of Structural Alloys

Vakil Singh
Center of Advanced Study, Department of Metallurgical Engineering, Institute of Technology, Banaras Hindu University, Varanasi-221005.

ABSTRACT

Fatigue resistance of structural alloys is characterized in terms of appropriate parameters, depending upon the nature of cyclic loading, high cycle and low cycle fatigue. Resistance of a structural component under cyclic loading comprises of two components, namely, the resistance against crack initiation and resistance against crack propagation. While there is dominant role of the process of crack initiation in high cycle fatigue, there is major role of crack propagation in low cycle fatigue. Invariably fatigue cracks initiate at surface, hence there is growing concern for surface modification of structural components to increase the resistance of material against crack initiation. It largely depends on the nature of plastic deformation, occurring by planar slip or wavy slip, stability of precipitates in age hardening alloys and the environment. Several important structural alloys exhibit dual slope in Coffin-Manson relationship between plastic strain amplitude and fatigue life. It is a very important aspect of low cycle fatigue resistance, both from the design as well as from application point of view of the component. The role of surface modification, through a novel technique of surface nanocrystallization in improving fatigue life, is being widely acknowledged. Some important structural alloys like Zircaloy-2 is observed to exhibit pronounced secondary hardening, in later stage of strain controlled cycling, even at ambient temperature. The magnitude of secondary hardening is observed to be dependent both on strain amplitude as well as the strain rate. Dynamic strain aging is also found to play important role on LCF resistance of several structural alloys.

KEYWORDS: *Inconel 718, Nimonic PE16, Timetal 834, 9Cr-1Mo Steel, Zirconium Alloys, Low Cycle Fatigue.*

INTRODUCTION

Fatigue is an important material property for the integrity, efficient and safe operation of structural components, subjected to repeated loads, at room temperature as well as elevated temperatures. The important structural components where fatigue play vital role are turbine discs of high temperature gas turbine engine, air craft blades, ship propellers, and heat transfer piping system, and fuel tube claddings of nuclear reactors.

The present paper describes different aspects of low cycle fatigue behavior of important structural materials; like precipitation hardened structural alloys superalloys, aluminum alloys, titanium alloys, steels and zirconium alloys.

Superalloys are one of the oldest and widely used high temperature structural alloys as turbine blades, compression discs and shafts for high performance gas turbine engines, where low cycle fatigue plays important role, because of their superior high temperature tensile, fatigue, creep strength, and oxidation resistance in comparison with the other structural alloys. Alloy IN718 is a precipitation-hardenable nickel-iron base superalloy containing significant amounts of chromium, niobium and molybdenum, and lesser amounts of aluminum and titanium. The main strengthening phase γ'' is a metastable phase based on the Ni3Nb stoichiometry [1]. Alloy IN718 offers exceptionally high strength potential for applications from cryogenic temperatures to 650°C and excellent weldability due to slow precipitation of the γ'' phase in the γ matrix. Alloy IN 718 is one of those alloys that exhibit planar slip behavior [2, 3]. However, one of the main draw back of superalloys is their low specific strength which limits the use of these alloys in new generation high performance gas turbine engines. In this respect titanium alloys have attracted the designers because of their high specific strength compared with the widely used superalloys and good high temperature tensile, fatigue strength and superior oxidation resistance.

Alloy Timetal 834 is a near α-titanium alloy designed for high temperature applications as compressor disc and blade of advanced gas turbines of jet engines, as substitute of heavy nickel base superalloys. It exhibits a wide variety of microstructures depending on the solution treatment temperature and the subsequent rate of cooling [4, 5]. A good combination of high tensile strength, creep strength, low cycle fatigue (LCF) life, and resistance to fatigue crack propagation is observed in a dual phase microstructure, consisting of 15 vol.% fine-grained primary a (Pα) in a matrix of relatively fine-grained transformed β with a lamellar structure [6].

Stainless steels are another class of ferrous structural alloys; widely used in nuclear, aerospace, and marine industry because of their corrosion resistant against a wide variety of environments, god mechanical strength and low cost, compared with other structural materials mentioned above. Austenitic stainless steels (types 304 and 316) are currently being used in liquid metal cooled fast breeder reactors (LMFBRs) as cladding and structural materials. Recently, ferritic martensitic steel (modified 9Cr–1Mo) usually designated as grade 91 steel, is extensively used as a structural material at elevated temperatures in fossil-fired power plants. This is also chosen material for the tubing in the reheater and superheater portions and as a thick-section tube plate material in the steam generators of liquid metal cooled fast breeder reactors (LMFBRs) due to its higher thermal conductivity and lower thermal expansion coefficient, coupled with better resistance to stress corrosion cracking, and microstructural stability over very long periods of exposure at high temperature water–steam systems, compared with austenitic stainless steels [7,8]. The heat exchanger components of LMFBRs undergo thermally induced strain cycling during start-up and shut-down or during variation in the operating conditions. Thermal fluctuations in the liquid coolant results in development of thermal gradient from the surface to core of the component across the wall thickness. Such thermal gradients occurring cyclically during operation result in thermomechanical fatigue (TMF). Thermomechanical fatigue (TMF) is a widespread cause of failure in components of industrial machinery and structures exposed to thermal and mechanical loads that change with time.

Zirconium alloys are highly important material in nuclear energy engineering, used as structural material for core components of thermal neutron reactors like fuel claddings, pressure tube. Zirconium alloys are known for their inherent property of low neutron absorption cross section along with good resistance to corrosion against high temperature water-steam environment, good thermal conductivity and acceptable mechanical strength and ductility, under nuclear reactor operating temperatures conditions. Zirconium alloy core components experience plastic strain cycles due to power oscillations in the reactors and it leads to their eventual failure resulting from low-cycle fatigue. The cladding also undergoes cyclic deformation due to various kinds of external factors, such as coolant temperature, pressure and coolant flow rate. The main zirconium alloys used are Zircaloy-2, Zircaloy-4 and Zr-2.5%Nb alloy. The main alloying elements are Sn, Fe, Cr, Ni, Nb, N, and there are trace elements like O, Hf, S, P, C, etc. in the ppm level. Zircaloy-2 is an alpha base alloy containing Sn, Fe, Cr, Ni as major alloying elements. Zircaloy-4 is also an alpha base alloy containing the same alloying elements as Zircloy-2 except Ni which is found to cause hydrogen embrittlement forming nickel hydrides.

LOW CYCLE FATIGUE OF STRUCTURAL MATERIALS

Superalloys

Low cycle fatigue resistance of a material is displayed by the Coffin Manson (C-M) equation which is a relation between the plastic strain amplitude and number of reversals to failure. In general this gives a linear single slope on log-log scale for most of the materials. However, Superalloys Inconel 718, Nimonic PE16 alloys in the peak aged condition are found to exhibit dual slope in the C-M plot (Fig.1) [9-11]. The dominant mode of deformation in alloy 718 has been identified as planar slip, as shown in Fig.2. It may be seen that the at low strain amplitudes, the degree of strain homogeneity increases with increasing γ' particle size, as seen by a progressive increase in the density of deformation bands and also shown by the increase in the number of grains containing deformation bands. It has been considered as the cause of dual slope in the Coffin– Manson (C-M) relationship of this alloy [9-11]. Various other mechanisms have also been proposed to explain the bilinear C-M relationship. Lerch and Gerold [12] found the dual slope C-M behavior in Nimonic 80A at room temperature. They observed that slip bands were formed during cycling and their number increased with increasing strain amplitude till the grains were filled with slip bands. Thereafter, further strain could only be accommodated through other mechanisms like micro-twinning and grain rotation. This change in deformation mode with increase in strain amplitude was considered to lead to a change in slope in the C-M plot. According to Singh et al [9] the dual slope in the C-M plot obtained in the alloy Nimonic PE16 at room temperature is associated with the change in deformation from single slip to multiple slip. Mediratta et al [13] investigated the LCF behavior of a series of dual slope steels. They noticed that the cyclic strain-hardening exponent (n') was substantially higher in the high strain region than in the low strain region.

Once a set of slip bands form in the favorably oriented regions, deformation resulting from the subsequent cycles continues to be accommodated in the same bands, particularly in the age-hardened condition due to progressive shearing of precipitates.

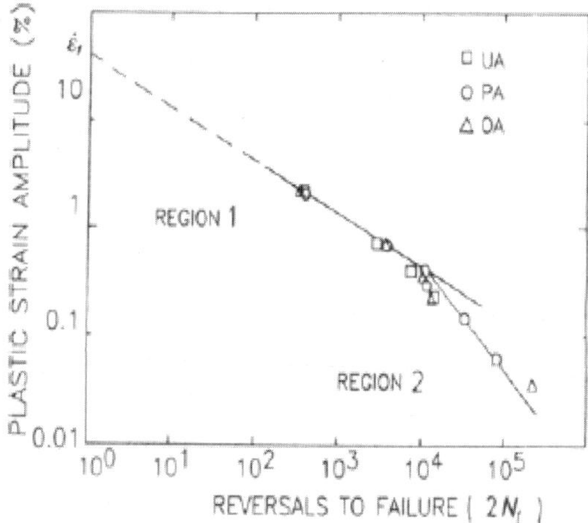

Fig. 1: Coffin-Manson plots for NIMONIC PE16 of the UA, PA, and OA materials.

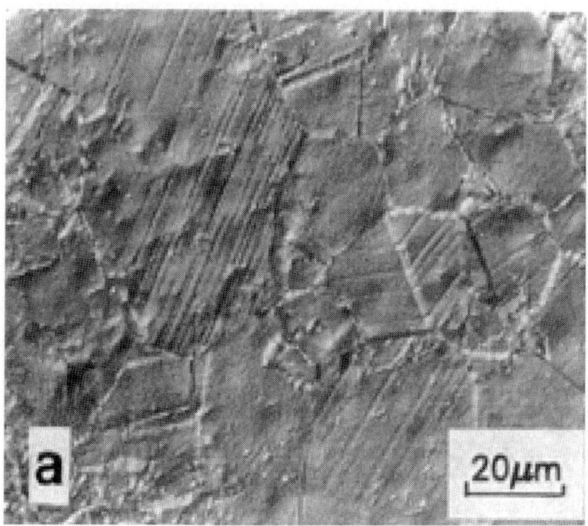

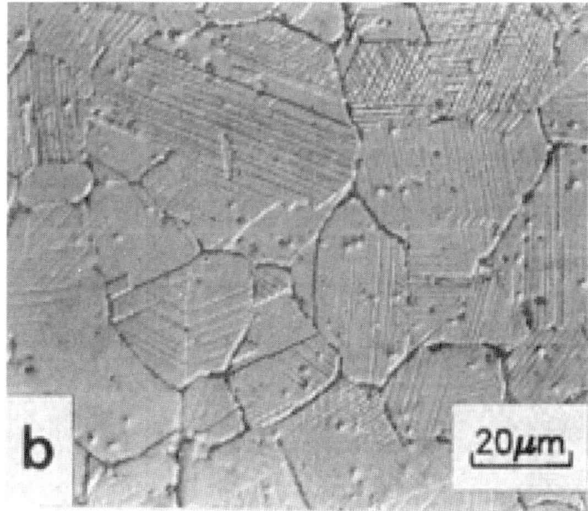

Fig. 2: Optical micrographs showing slip bands in the differently heat-treated specimens of NIMONIC PE 16 tested in LCF at Ae,/2 = 0.60 pet: (a) UA, (b) PA

Therefore, it was proposed to harden the regions of soft orientations of the peak-aged (PA) blanks by cold rolling to cause activation and intersection of different slip systems. Thus, the initial regions of soft orientations are hardened and the process of crack initiation during cyclic loading is delayed. Fig. 3(a) show Coffin-Manson plot for super alloy Inconel 718 in peak aged condition and peak age with 5% cold work, and Fig. 3b show cold work with stress relieving treatment. The stress relieving treatment was carried out at 550°C for 1.5hrs followed by cooling in air. It may be seen that the bilinear in the Coffin Manson plot is eliminated after cold working and the linearity of slope is retained even after stress relieving treatment.

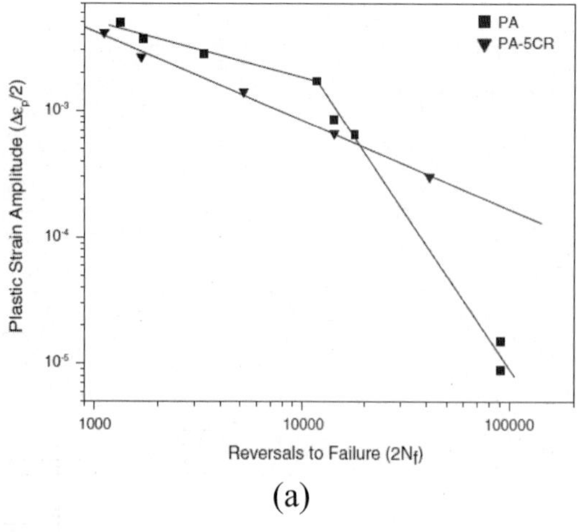

(a)

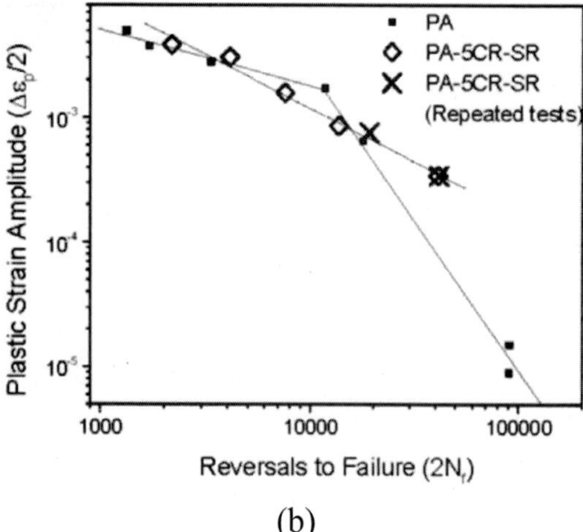

(b)

Fig. 3: C-M plot for IN718 (a) peak aged and peak aged with 5% cold work (b) peak age and peak age+5CR+stress relieving.

TITANIUM ALLOYS

Low cycle fatigue behavior of the Ti alloy Timetal 834 alloy, solution annealed in the α+β phase field at 1020°Cfor 2 h, cooled in air (ST-AC-A), and subsequently aged at 700°C for 2h had been investigated at room temperature [14] and 600°C[15]. It was observed that at room temperature there was continuous cyclic softening, starting from the first cycle to the last cycle of failure with increased degree of softening, at greater than 1% strain amplitude [Fig. 4]. On the other hand alloy tested at 600°C exhibited rapid cyclic softening in the initial stage (<10% of fatigue) and saturation of stress for the major fraction of fatigue life. However, the rate as well as degree of softening was found to decrease with increase in strain amplitude at elevated temperature 600°C [15]. The softening at room temperature, tested at higher strains was attributed to complete shearing of the ordered Ti3Al precipitates, with increase in the number of slip bands and glide dislocations. Whereas at high temperature, the initial softening occurs due to planarity of slip, shearing of the ordered Ti3Al precipitates and coarsening of silicides. The saturation in cyclic stress results from balance between the softening and hardening effects. Titanium alloys are also found to exhibit dual slope in the C-M plot in the (α + β) ST-AC-A condition (Fig. 5) during LCF tested at room temperature [14]. The fatigue life, in the region of low strain amplitude, is much lower than that expected from extrapolation of the data of high strain amplitude. However, only single slope is exhibited at elevated temperature of 600°C. Also fatigue life at 600°C was considerably higher than that at room temperature, due to increased homogeneity of deformation and higher stability of Ti3Al precipitates, at the elevated temperature of 600°C. Fatigue life is markedly enhanced at low strain amplitude due to cold rolling, and consequently, the dual slope is eliminated from the C-M plot as shown in the Fig.5. Fracture behavior of the specimens tested in LCF at room temperature and 600°C is shown in Fig. 6.

While the facture surface of the specimen tested at room temperature is quite rough (Fig. 6a), distinct fatigue striations may be seen in case of the specimen tested at 600°C (Fig. 6b). The striations in Fig. b are quite close to the crack initiation site.

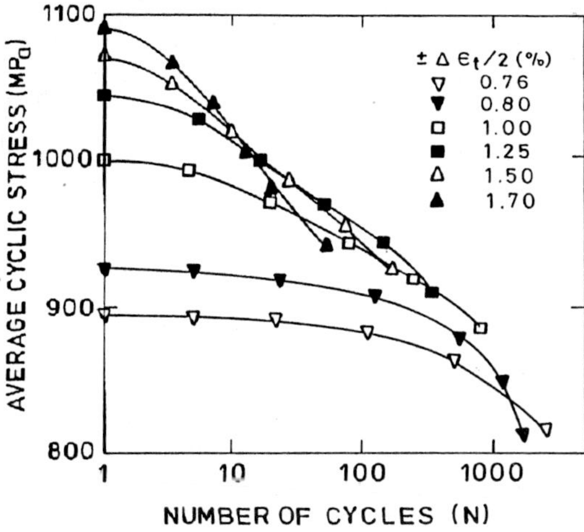

Fig. 4: Cyclic stress response curves for the Ti alloy Timetal 834 in the ST-A condition at room temperature.

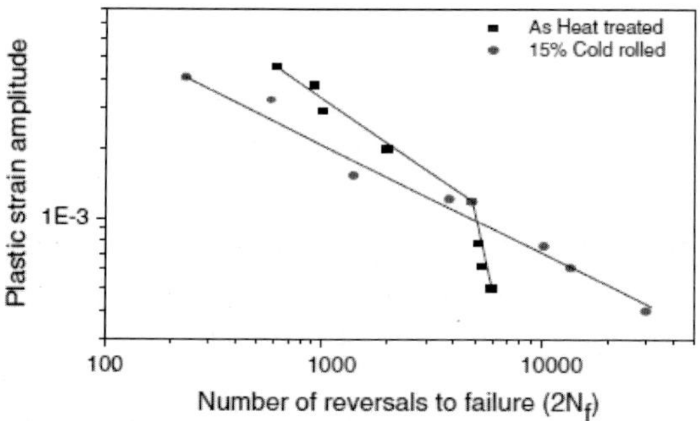

Fig. 5: Coffin-Manson plot for Timetal 834 LCF tested in ST-AC-A heat treatment condition and 15% cold work condition.

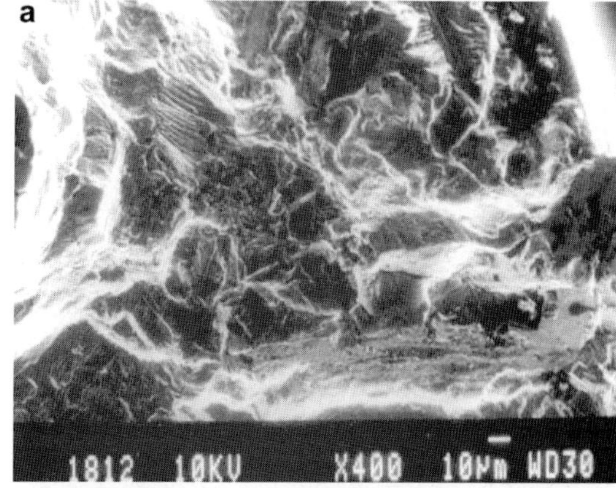

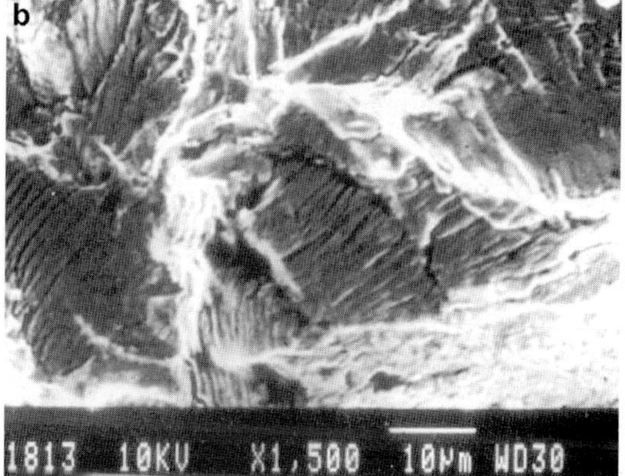

Fig. 6: SEM fractographs showing fracture behavior in LCF at: (a) room temperature and (b) 600°C.

In view of the service conditions, structural components subjected to cyclic loading at elevated temperature should also consider the creep-fatigue interaction behavior of the components. It has been established that tensile holding in LCF testing at elevated temperature was more detrimental than tensile-compression and only compression holding. Figure 7 shows the typical wave forms used in LCF testing to know the creep-fatigue interaction.

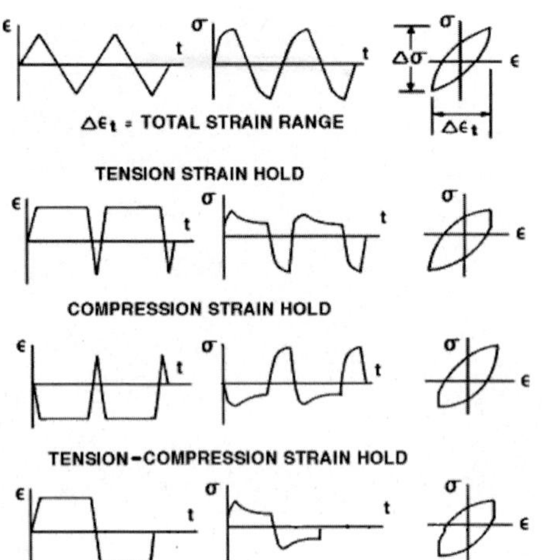

Fig. 7: Typical wave forms employed LCF testing in studying creep-fatigue interaction.

Effect of hold period on low cycle fatigue behavior of the Ti alloy Timetal 834 has been investigated at room temperature [16] at two different strain amplitude of ±1% and ±1.25% and hold periods of 60 sec, 180 sec and 300sec in tension and compression strain cycles. Fatigue life was reduced with increasing hold period. The deleterious effect of holding on fatigue life was relatively more for compressive than the tensile at room temperature.

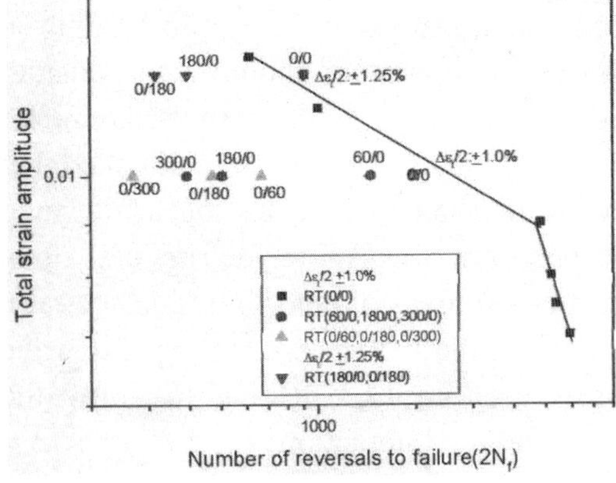

Fig. 8: Effect of hold period on fatigue life at total strain amplitudes ±1% and ±1.25% tested at room temperature.

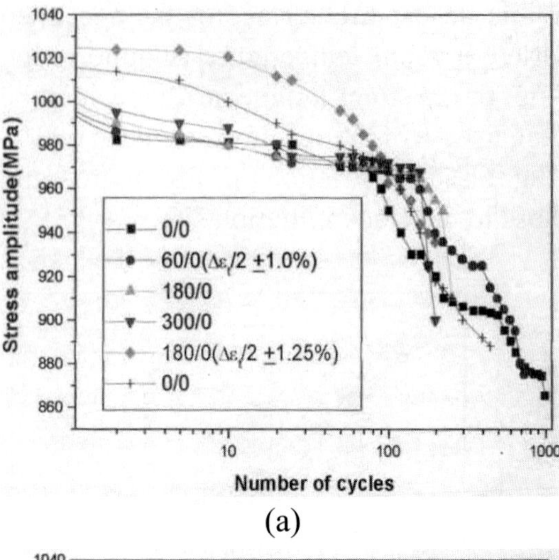

(a)

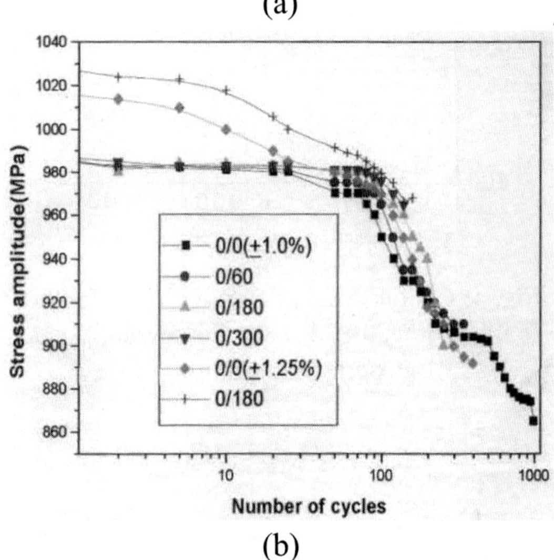

(b)

Fig. 9: Cyclic stress response at different hold periods tested at RT (a) Tensile hold (b) compression hold

The decrease in fatigue lives at room temperature with introduction of hold periods is attributed to stress induced diffusion of hydrogen at the crack tip and the embrittling from of titanium hydride precipitating. The cyclic stress response showed softening in both tensile and compression holds at high strain amplitude of 1.25%. However, at strain amplitude of 1%, in the tensile hold, there is initial softening for 20 then near stabilization followed by softening till failure. In the compression hold cycle the alloy showed stabilized stress response up to 100 cycles followed by softening till failure.

STAINLESS STEELS

Comparative study of isothermal and thermo-mechanical fatigue has been carried out on 316L N stainless steel under IP and OP loading conditions. Temperature intervals (ΔT) of 200–300°C were employed and tests were carried out in the range, 300–650°C [17]. It was observed that OP TMF yielded lower lives compared to IP TMF in the sub-creep regimes, whereas the reverse was found true in the creep temperature domain. Isothermal cycling at the peak temperature of TMF yielded lower lives compared to both IP and OP cycling. Depending on the testing conditions, nitrogen-alloyed 316L stainless steel shows DSA within a wide temperature range of 250–600°C, during low cycle fatigue, it was marked by pronounced cyclic hardening and serrated flow [18–21]. The difference in fatigue lives has been attributed to DSA, carbide precipitation, coupled with occasional grain boundary serrations.

Thermomechanical fatigue (TMF) and isothermal fatigue (IF) behavior of modified 9Cr–1Mo ferritic martensitic steel had been studied at different temperatures between 300–650°C at constant strain amplitude of 0.4% at strain rate 1.2×10^{-4} s^{-1}[22]. Fatigue life sequence was found to be IF < OP (TMF) < IP (TMF) (Fig.10). However, the difference in lives was reduced with increasing the Tmax of TMF test temperature in IF cycling. The net tensile hysteresis energy is a well-known parameter, used to characterize and quantify the extent of fatigue damage, and is usually represented by the area of the tensile portion of the stress–strain hysteresis loops. The variation of the hysteresis energy with the strain–temperature phasing at different temperature ranges, is presented in Fig. 11. A consistently greater hysteresis energy for the case of OP TMF, in comparison with that for the IP TMF can be noted, which could also be construed as an additional contributing factor for the lower lives obtained under OP TMF cycling conditions.

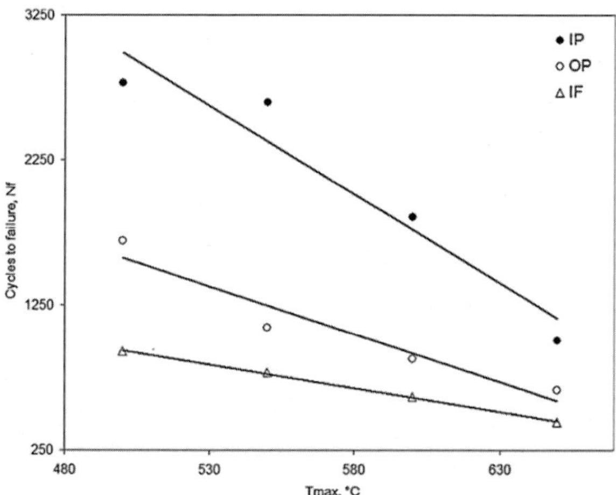

Fig. 10: A comparative plot of the cyclic lives obtained under TMF (IP, OP) and IF cycling at T_{max}.

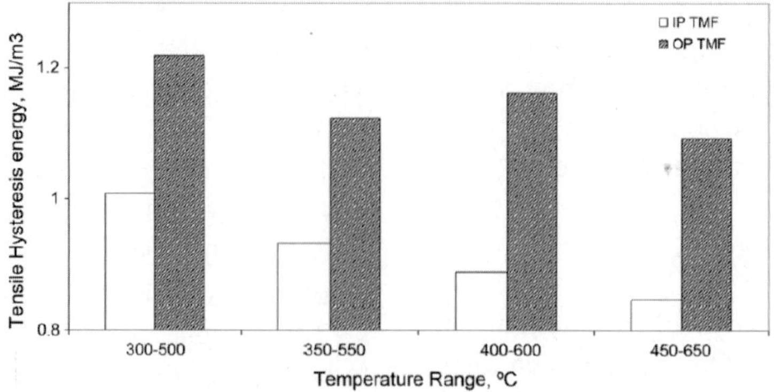

Fig. 11: Half-life hysteresis energy as a function of strain–temperature phasing at different temperature ranges.

Dynamic strain aging has been observed in the temperature range between 300-400°C in the IP and OP TMF loading. The behavior can be seen by serrations in the hysteresis loops shown in the Fig. 12. Investigation by Choudhary et al. [23] on the serrated yielding in a plain 9Cr-1Mo steel showed that dynamic strain ageing (DSA) occurs in the temperature range, 250-400◦C during tensile deformation. Interaction of dislocations with carbon atoms in the above temperature range, as reflected by the observed activation energy of 83 kJ/mol, was found to be responsible for DSA.

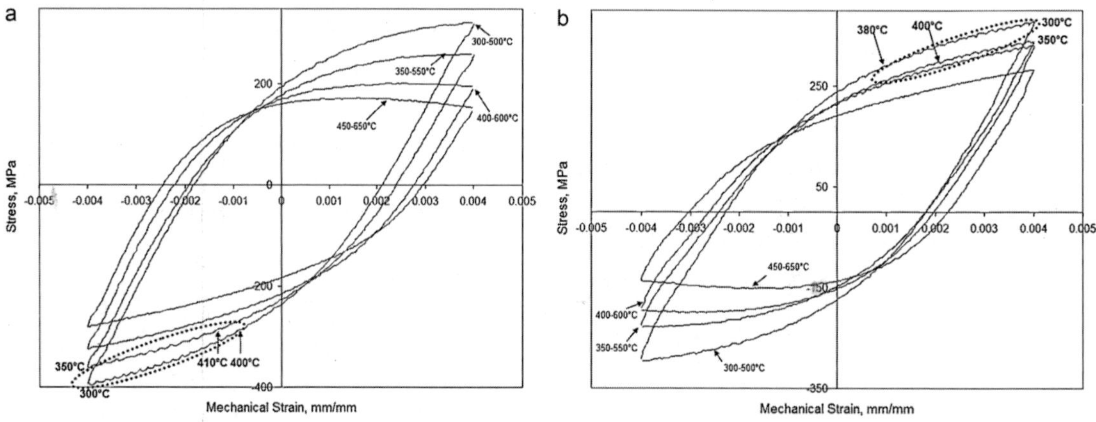

Fig. 12: Half-life stress–strain hysteresis loops under (a) IP and (b) OP TMF tests at different temperature ranges.

ZIRCONIUM ALLOYS

Low cycle fatigue behavior of zirconium and its alloys had been reported at different strain rates, heat treated conditions, from RT to 600°C [24-27]. Total strain amplitude controlled tests carried out in air and vacuum for the Zr and Zircaloy 2 showed increase in fatigue life up to 300°C and decrease in life above this temperatures [25]. The higher life was attributed to fatigue strengthening of the Zr and Zircaloy-2 at elevated temperature up to 300°C due to DSA. Prior cold working (~ 15%) was found to be detrimental to fatigue life of both Zr and Zircaloy2. Microstructural investigation of near fracture surfaces, the internal sections, revealed twinned regions in both Zr and Zircaloy-2, up to 500°C. From the shape of the twins it was deduced that the twinning was mainly of {1121} type (parallel sided) and {1012} type (lenticular shaped). At higher temperatures the slip traces were localized to form slip bands at ~ 400°C and even at higher temperatures finely dispersed slip traces were seen.

The cyclic stress response of annealed zircaloy-4 at room temperature showed unusual softening at low strain ranges and hardening at high strain ranges (Fig. 13). Usually in the annealed condition most of the materials exhibit cyclic hardening due to creation of new dislocations and softening in the hardened condition due to annihilation of dislocations and formation of more stable structures of dislocation cells. Softening at low strain amplitudes has been attributed to dislocation rearrangement [28], texture rotation [29, 30] and decrease in friction stress resulting from unlocking of dislocations from oxygen interstitial solute atoms [31].

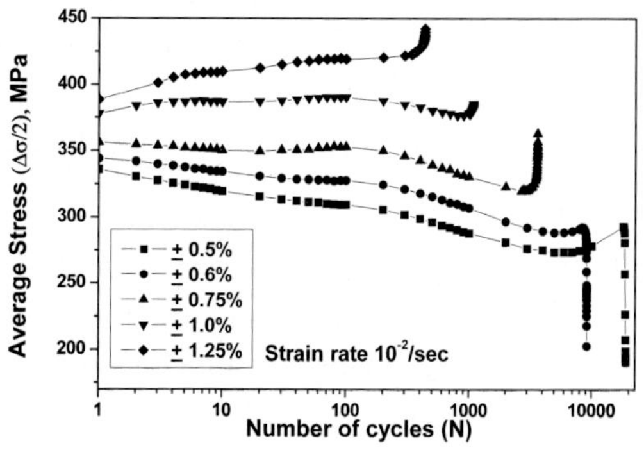

Fig. 13: Stress response of Zircaloy-2 during LCF at room temperature, at strain rate 10^{-2} s^{-1}

It may also be seen from the Fig. 13 that there is secondary cyclic hardening exhibited at all the strain rates investigated. However, thorough furhter investigation is required to explain the possible mechanism responsible for this.

ENHANCEMENT OF FATIGUE RESISTANCE BY SURFACE NANOCRYSTALLIZATION

As discussed above the majority of fatigue life cycles under cyclic stress control loading is spent in the process of initiation of fatigue crack. The conventional technique of enhancing high cycle fatigue of a structural component is by shot peening in which surface compressive stress are induced to enhance resistance against crack initiation. Surface nanocrystalization is an advanced technique of surface modification and it has attracted many researchers because of combined effect of surface compressive residual stress and surface nanostructure surface.

SUMMERY

- Nickel base Superalloys PE16 and IN718, Titanium alloy Timetal 834 have been found to exhibit bilinear slope in Coffin-Manson plot in peak aging and (α + β) ST-AC-A heat treated condition respectively.
- Cold working after peak aging eliminates bilinear behavior in C-M plot.
- Cyclic hardening is exhibited in annealed Zircaloy-2 from beginning to failure, at higher strain amplitudes

whereas cyclic softening is exhibited at lower strain amplitudes.

- Dynamic strain aging has been observed in the modified 9Cr-1Mo steel in the temperature range 300-400°C and it affects its LCF behavior.

REFERENCES

[1] E.E. Brown, D.R. Muzyka, In Superalloys II, John Wiley, New York, NY, (1987) 165.

[2] D.F. Paulonis, J.M. Oblak, D.S. Duvall, Trans. ASM, 62 (1969) 611.

[3] L Xiao, D.L. Chen, M.C. Chaturvedi, Metall. Mater.Trans. A, 35A (2004) 3477.

[4] B. Borchert and M.A. Daeubler: 6th World Conf. on Titanium, P. Lacombe, R. Tricot, and G. Beranger, eds. Les Editions de Physione, Les Ulis, France, 1988, pp. 467-72.

[5] C. Andres, A. Gysler, G. Lutjering: Z. Metallkd., 1997, vol. 88, pp. 197-202

[6] D.F. Neal, in: G. Lutjering, U. Zwicker, W. Bunk (Eds.), Titanium Science and Technology: Proc. Fifth Int. Conf. on Titanium Congress-Centre, Munich, FRG, September 10–14, 1984, DGM-Oberursel (1985) p. 2419.

[7] H. Haneda, et al., Proc. Int. Conf. on Advances in Materials Technology for Fossil Power Plants, ASM International, Chicago, 1987, pp. 231–241.

[8] P.J. Ennis, A. Czyrska-Filemonowicz, Sadhana 28 (2003) 709–730.

[9] V. Singh, M. Sundararaman, W. Chen, R.P. Wahl, Metallurgical Transactions A Volume 22a, February 1991—499

[10] K.V.U. Praveen, Vakil Singh, Materials Science and Engineering A 485 (2008) 352–358

[11] K.V.U. Praveen, G.V.S. Sastry, V. Singh: Trans. IIM, 2004, vol. 57 (6), pp. 623–30.

[12] B.A. Lerch and V. Gerlod, Acta Metall., 1985, 33, 1709

[13] S.R. Mediratta, V. Ramaswamy, P. Rama Rao, Scripta Metall., 20(1986)555

[14] Singh Nidhi, Gauthama, Singh V., Mater Sci Eng A 325(2002)324.

[15] Nidhi Singh, Gouthama 1, Vakil Singh, International Journal of Fatigue 29 (2007) 843–851

[16] Sai Srinadh, Phd thesis, Microstructure: tensile, low cycle fatigue, and oxidation behaviour of the near α-Ti alloy IMI834, Banaras Hindu University, 2005.

[17] A. Nageshaa, R. Kannan, P. Parameswaran, R. Sandhya, K. Bhanu Sankara Rao, Vakil Singh, Materials Science and Engineering A 527 (2010) 5969–597

[18] V.S. Srinivasan, R. Sandhya, K. Bhanu Sankara Rao, S.L. Mannan, K.S. Raghavan, Int. J. Fatigue 13 (6) (1991) 471.

[19] S.G. Hong, S.B. Lee, J. Nucl. Mater. 328 (2004) 232–242.

[20] V.S. Srinivasan, R. Sandhya, M. Valsan, K. Bhanu Sankara Rao, S.L. Mannan, Scripta Mater. 37 (1997) 1593–1598.

[21] S.G. Hong, K.O. Lee, S.B. Lee, Int. J. Fatigue 27 (2005) 1420–1424.

[22] S.G. Hong, S.B. Lee, J. Nucl. Mater. 340 (2005) 307–314.

[23] A. Nagesha, R. Kannan, G.V.S. Sastry, R. Sandhya, Vakil Singh, K. Bhanu Sankara Rao, M.D. Mathew, Materials Science and Engineering A 554 (2012) 95– 104

[24] A.F. Armas, I. Armas-Alvarez, ASTM STP 939, (1987) 617.

[25] K.U. Snowden, P.A. Stathers, J. Nuc. Mat. 67 (1977) 215.

[26] A.F. Armas., I. Alvarez-Armas, G. Mansilla, J. Mat. Sci., 27 (1992) 1307.

[27] A.F. Armas, I. Alvarez-Armas, G. Moscato, Scripta. Met., 34 (1996) 281.

[28] J.I. Dickson, L. Hanfield,G. L'Esperance, Mater. Sci.Eng, 60(1983) L3.

[29] H.I. Choi, D.H. Lee, S.I. Kwun, Scripta Mater, 38(1998)565.

[30] D.H. Lee, S.I. Kwun, Scripta Metall Mater, 31(1994)1475

[31] A.F. Armas, S. Herenu, R. Bolmaro, I. Alvarez-Armas, Journal of Nuclear Materials 326 (2004) 195–200.

Effect of Plasma Nitriding on Fatigue Characteristics of AISI 4330 Low Alloy Steel

Rajkumar Singh[1], Jagdish Sherkar[1], Ram Chandra Prasad[2], Ganesh Jawale[1] and Raju Kadam[1]
[1]Bharat Forge Limited, Mundhwa, Pune-411 036, India
[2]Department of Metallurgical Engineering and Materials Science, Indian Institute of Technology Bombay, Mumbai-400 076, India
Email: RajkumarSingh@bharatforge.com, jagdish.sherkar@gmail.com, rcp@iitb.ac.in, GaneshJawale@bharatforge.com, raju.kadam@bharatforge.com

ABSTRACT

The plasma nitriding technique has been used to improve surface hardness, fatigue strength, corrosion resistance and wear resistance of materials, especially iron-based alloys. In this study the fatigue properties of plasma nitrided quenched and tempered AISI 4330 low alloy steel were investigated by varying different process parameters, like, time and temperature. Steel components made of AISI-4330 low alloy steel are used in oil field industries such as crank shafts and fluid end parts. Rotating and bending fatigue tests has been carried out ,it has been found that the plasma nitriding surface treatment improves the fatigue strength of AISI 4330 low alloy steel by 90% depending on nitriding time and temperature. The fractured surfaces of nitrided and without nitrided samples were characterized by scanning electron microscopy (SEM). It was determined that, the subsurface 'fish eye' type fatigue crack initiation mechanism in ion nitrided AISI 4330 steel.

KEYWORDS: *DC Pulse, Plasma Nitriding, Fatigue Strength, Surface Hardness, AISI 4330 Low Alloy Steel.*

INTRODUCTION

Plasma or ion nitriding is a surface treatment process widely used to improve surface hardness, fatigue strength, corrosion resistance and wear resistance. The technology is based on glow discharge technology. In this process nascent nitrogen is introduced to the surface of a metal part for subsequent diffusion into the material. In a vacuum environment, high-voltage electrical energy is used to form plasma, through which nitrogen ions are accelerated to impinge on the work piece. During plasma nitriding, hydrogen and nitrogen ions sputtering action on surface takes place. Because of this, high compressive residual stresses are introduced in the surface region, which increase the surface hardness. In the process precipitate of nitrides form on the surface which causes remarkable improvement in fatigue properties [1]. Diffused nitrogen into a steel surface is combined with alloying elements to form a fine dispersion of alloy nitrides. As a result, a thin iron nitride layer, which consists of γ'-Fe4N and ε-Fe2-3N intermetallic phases, referred to as the 'white layer' is produced on the surface together with a relatively thick and strong diffusion zone in the subsurface of a steel component, which gradually reduces the hardness and nitrogen concentration towards the core, resulting in a diffuse case–core interface [1].A significant advantage of plasma nitriding is that it causes less distortion than hardening and carburising, because of the relatively low processing temperature, usually below 570°C. The minor probability of surface distortion and dimensional variations makes it possible to reduce or even eliminate the need for expensive finishing operation before component is put in to the service [9]. A diffused layer, which is formed during ion nitriding process, causes compressive residual stresses on or near the surface of the component, as well as low tensile residual stress in the core. Both increased residual stress and surface hardness result in improvements in the fatigue life because the thin, hard layer prevents plastic flow. This means that plastic deformation, which is the source of crack initiation, probably appears in the subsurface. The suppression of slip band formation at the surface delays crack initiation and propagation, thus being responsible for the improvement of fatigue life [8]. The effect of increasing the case depth on the fatigue limit can be viewed as effectively moving the fatigue crack initiation site further into the core. There have been studies of the effects of plasma nitriding on the rotating bending fatigue strength of AISI 1045, AISI 4140, AISI 5140, EN40B, 34CrNiMo6 and AISI 4340. However, there has not been any information about fatigue behaviour of plasma nitrided AISI 4330.In the present research AISI 4330 low-alloy steel was plasma nitrided under different conditions including time (6, 9,12 and 16 h) and temperature (500°C and 520°C and their effect on the rotating bending fatigue strength was investigated.

EXPERIMENTAL WORK

A standard medium carbon, CrMoNi low alloy AISI 4330 steel, used in the production of crankshafts and fluid end jobs of pump, has been used in this study.

Table 1: Chemical composition of AISI 4330 low alloy steel

ELEMENTS	WEIGHT %
Carbon (C)	0.30
Manganese (Mn)	0.45
Silicon (Si)	0.15
Sulphur (S)	0.010 max
Phosphorus (P)	0.015 max
Chromium (Cr)	1.20
Nickel (Ni)	2.60
Molybdenum (Mo)	0.50
Vanadium (V)	0.12
Copper (Cu)	0.25 max
Aluminium (Al)	0.035 max
Tin (Sn)	0.015 max
Antimony (Sb)	0.010 max
Arsenic (As)	0.015 max
Cobalt (Co)	0.025 max
Lead (Pb)	0.010 max
Hydrogen (H)	2 ppm max
Iron (Fe)	Balance

The chemical composition of the steel is given in Table 1. The received material undergoes heat treatment process is shown in Table 2.The hourglass shaped rotating bending fatigue test specimens were machined according to DIN 50113 Fig. 1 at CNC turning lathe. The surface of the gage length section was ground through 200,

400, 600 and 1200 grades emery papers .The surface roughness within the gauge length of the samples was maintained below approximately 0.3 μm.

Table 2: Heat Treatment process and parameters

Parameters	Normalizing	Hardening	Tempering	Double tempering
Temperature	900°C	850°C	630°C	580°C
Socking time	4hrs.	5hrs	10 hrs	8hrs
Cooling	Air cooling	Oil quench	Furnace cool to 425°C then Air cool	Air cool

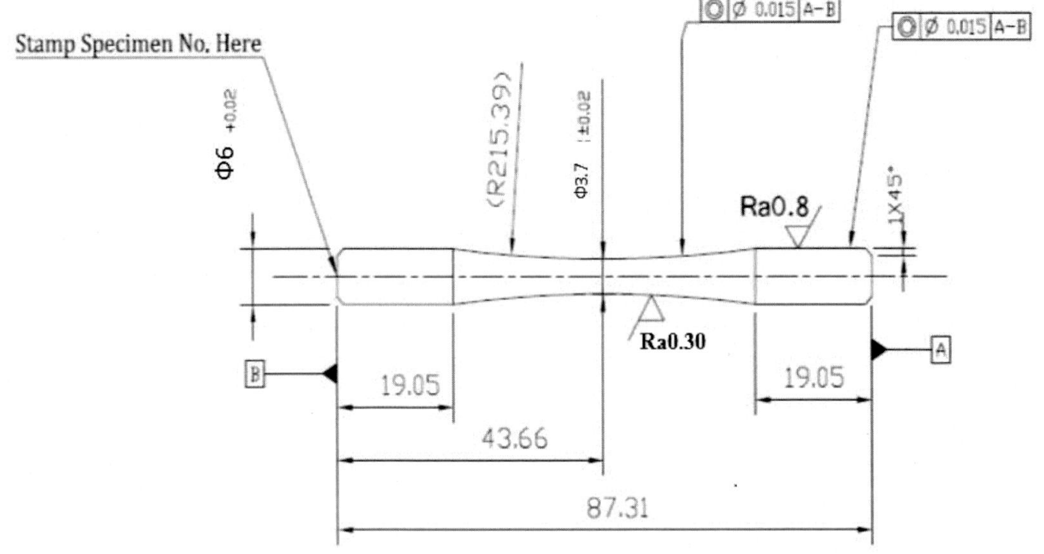

Fig. 1: Fatigue test specimen

Plasma nitriding was done at industrial ion nitriding DC pulse control furnace. Specimens were connected to cathode. Specimens were plasma nitride for 6,9,12 and 16 hours at 500°C and 520°C temperatures by using separate hydrogen and nitrogen gas. The process time was considered after treatment temperature was reached. After ion nitriding, the specimens were cooled to room temperature in the vacuum chamber.

Fatigue strength was determined by using R.R. Moore rotating bending fatigue test machine. Rotating bending fatigue test were conducted at KCTI, R & D, Centre Pune, India by using a sinusoidal load of 4500 rpm in laboratory air atmosphere and load ratio is R= −1. Specimens ranging from 16 to 18 numbers were used to obtain the fatigue strength for each nitriding parameter. The specimens were tested for different stress levels. The test started at stresses half of the ultimate tensile strength. The tensile test for specimen without nitriding AISI 4330 material was also conducted. From tensile test the ultimate tensile strength of AISI 4330 low alloy steel is 864MPa. In each fatigue test number of cycle to fatigue failure was noted. The test was stopped if no failure occurred after cycles and two specimens were tested for each stress levels. After all fatigue tests statistical analysis was done by using eight stress levels to find out fatigue strength. To determine diffused layer case depth for each nitriding parameter, microhardness measurement was taken by using 200gm load. By using optical microscope compound layer thickness was determined for each nitriding parameter. The fracture surfaces of specimens were examined under optical microscopy and scanning electron microscopy (SEM) in order to evaluate the crack initiation and growth characteristics of the materials in the core and nitrided case regions of specimens.

RESULTS AND DISCUSSION

Fig. 2 shows the microstructure of AISI 4330 low alloy steel surface plasma nitrided for 6 and 9 hours and at 520°C temperature and at 500X magnification. It is observed that the microstructure consists of an internal core of tempered martensite and nitrided layer in the external surface. The nitrided layer consists of a nitrogen diffusion zone with fine nitride precipitates and a compound (white) layer in the outer part. As plasma nitride time increases the diffuse nitride layer and compound layer thickness increase. The compound layer thickness, surface hardness, fatigue strength of plasma nitrided specimen are given in Table 3. The fatigue strength improvement is because of plasma nitriding treatment. As plasma nitrding time and temperature increases the case depth and hardness increases which caused higher fatigue resistance. An improvement in fatigue strength of 90% was noticed for sample nitride for 16 hours-at 520°Ctemperature.

White layer

(a)

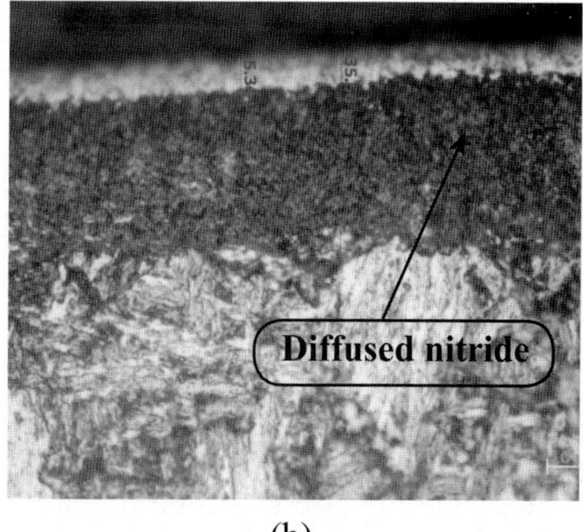

(b)

Fig. 2: The microstructure of the plasma nitride samples at (a) 6hrs and 520°C (500X) (b) 9hrs and 520°C (500X).

Table 3: Structural and fatigue properties of plasma nitrided AISI 4330 low alloy steels

Nitriding Parameters	Surface Hardness Hrc@500gm	Compound Layer, (μm)	Fatigue Strength, (σ MPa)	% Improvement
without plasma Nitriding	32		**510**	
6h,500°C ion Nitriding	57	2.5	**830**	**64**
9h,500°C ion Nitriding	61	4.2	**905**	**79**
12h,500°C ion Nitriding	62	5.1	**912**	**81**
16h,500°C ion Nitriding	62.8	6.2	**945**	**85**
6h,520°C ion Nitriding	57.8	3.04	**840**	**66**
9h,520°C ion Nitriding	62	6.2	**920**	**82**
12h,520°C ion Nitriding	64	7.6	**935**	**86**
16h,520°C ion Nitriding	61	8.4	**955**	**90**

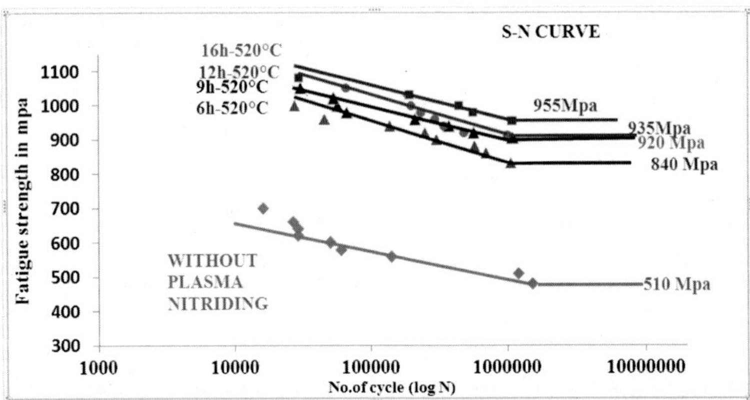

Fig. 3: S-N curve of plasma nitrided AISI4330 low alloy steel for 520°C temp.

The S-N curves obtained from rotating and bending fatigue tests are given in Fig.3. Fatigue properties obtained are different for different nitriding parameters. The crack initiation site moves further into core or below case-core interface zone, this is due to increase in case depth. This means that a greater bending stress will be required at the surface to create a sufficient high level of stress at the case-core interface to initiate failure [6]. Scanning electron microscope (SEM) examinations of the fracture surface of fatigue plasma nitrided samples were done. It shows that the fracture starts below the case-core interface zone. All the fractured specimens were observed to have failed by 'fish eye' phenomenon with fatigue crack originating from below case-core interface region i.e. high-cycle/low stress region. The fish eye cracks are circular in shape which is direct consequences of pure rotating bending type of cyclic stressing [7].

The images of SEM fractograph of typical fish eye crack initiation are given in Fig.4.

Fig.5 shows High resolution Transmission Electron Microscopic (TEM) image of plasma nitrided (16 hours and 520°C) surface which consists of γ′-Fe4N particles. Plasma nitrided cross section had diameter of 3mm and was polishing carried out up to 80μ followed by jet polishing up to 1 nm to ravel nitride layer in the specimen. The sample preparation is the most critical part in TEM. The TEM analysis focused on examination of particle size of γ′-Fe4N and morphology of nitride layer.

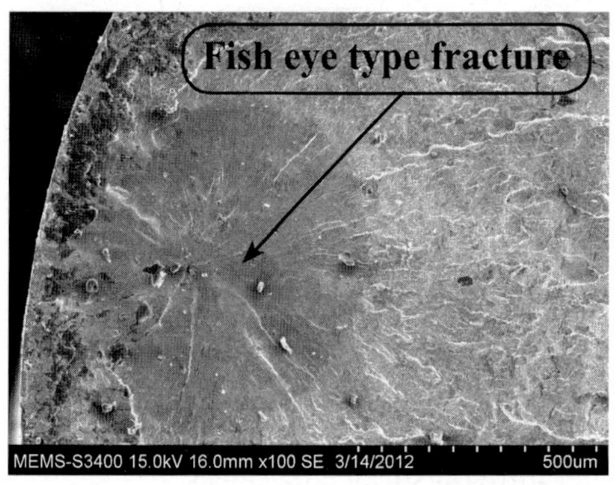

Fig. 4: SEM fractograph of nitrided sample
(Fish eye type crack initiation below case-core interface)

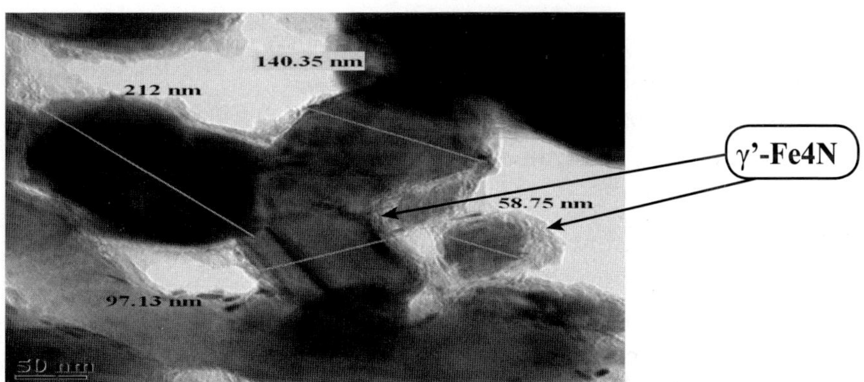

Fig. 5: TEM image of nitride layer

The surface hardness is a strong function of a temperature [5]. The hardness decreases can be explained in terms of precipitate size and density. Although the surface hardness of nitrided specimen increases with increase time and temperature, the reduced surface hardness was observed at longer nitriding time because of the formation of larger nitride precipitates which are less effective in increasing hardness than smaller ones [4]. Fig.6 shows the relation between hardness and nitriding time. It shows at lower nitriding time 6h and temperature 500°C, precipitate density are low, therefore the hardness is also low (57.8 HRC). At higher treatment time of 16h and temperature 520°C, precipitate size and density are more, therefore hardness start decreasing. The maximum hardness is obtained at temperature of 520°C and a time of 12 hours because of optimum precipitates size and density.

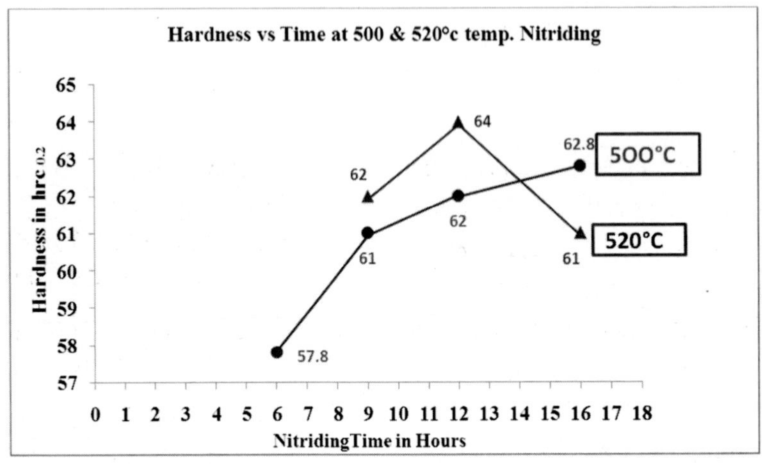

Fig. 6: Effect of nitriding time and temperature on surface hardness

CONCLUSIONS

I. The case depth of diffused nitrides and thickness of compound layer increased with increasing in plasma nitriding time and temperature (500°C to 520°C).

II. The surface hardness of AISI 4330 low alloy steel increased with increase in plasma nitriding time and temperature. But if nitriding time and temperature increases further, significant loss of surface hardness was observed as a result of the formation of large nitride precipitates.

III. By Plasma nitriding process we can increase the fatigue strength of AISI 4330 material by 90%. The improvement of 90% in fatigue strength was achieved for 16 hrs-520°C plasma nitriding specimens.

IV. The main characteristics of fractured plasma nitrided specimens are fish eye crack initiation below case-core interface zone.

REFERENCES

[1] Sule Yildiz Sirin, Kahraman Sirin, Erdinc Kaluc,"Effect of the ion nitriding surface hardening process on fatigue behaviour of AISI 4340 steel", Materials Characterization 59 (2008) P. 351-358.

[2] James M. O'Brien, O'Brien & Associates; Dan Goodman, "ASM Handbook, Heat Treating, vol.4", American Society for Metals,

Materials Park, Ohio, 1991. P. 944-953.

[3] I.M. Hutching, "Tribology: Friction and wear of engineering materials", chapter 5, pp. 77-78, 1992.

[4] G. Castro, A. Fernandez-Vicente, J. Cid, "Influence of the nitriding time in the wear behavior of an AISI H13 steel during a crankshaft forging process", Wear 263 (2007) P.1375-1385.

[5] Karamış MB, Staines AM, "An evaluation of the response of 722M24 steel to high-temperature plasma nitriding treatments", Heat Treatment Met 1989; 3:79–82.

[6] Alsaran A, Çelik A, "Structural characterization of ion-nitrided AISI 5140 low-alloy steel", Materials Characteristics 2001; 47:207–13.

[7] Bell T, Loh NL, "The fatigue characteristics of plasma nitrided three-pct. Cr–Mo steel", J Heat Treatment 1982; 2/3:232–7.

[8] Genel K, DemirkolM, Çapa M, "Effect of ion nitriding on fatigue behaviour of AISI 4140 steel", Material Science Eng A 2000; 279:207–16.

[9] Çelik A, Karadeniz S, "Investigation of compound layer formed during ion nitriding of AISI 4140 steel", Surface Coat Technology 1996; 80:283–6.

[10] J.L. Arques, J.M. Prado, "Wear 100", (1985) 325-399.

[11] Mittemeijer EJ, "Fatigue of case-hardened steels; role of residual macro and microstresses", Journal of Heat Trcat 1983; 3/2:114–9.

Characterization of Modified Steel Surfaces with Scanning Auger and Scanning Microprobe XPS

J.S. Hammond, D.F. Paul, S. Alnabulsi, K.W. Bomben and W. Betz
Physical Electronics USA, 18725 Lake Drive East, Chanhassen, MN, 55317, USA

ABSTRACT

The imaging surface analysis techniques of scanning Auger and scanning X-ray photoelectron spectroscopy (XPS) have been used to characterize modified steel surfaces. The interior surfaces of ductile iron as prepared by *in situ* fracture analysis have been characterized by scanning Auger to reveal the interfacial segregation of minor constituents to the boundaries of graphite nodules with nanoscale resolution. Field emission scanning Auger has also been used to reveal the surface modification of steel surfaces prepared under different annealing atmospheres. Monochromated scanning XPS has been used to elucidate the chemical modifications within tribological wear tracks of steel surfaces. These three examples validate the capabilities of scanning Auger and scanning X-ray photoelectron spectroscopy (XPS) to provide nanoscale surface sensitivity with nanometer and micron scale spatial resolution imaging capabilities for steel samples.

KEYWORDS: *Auger, XPS, Facture Analysis, Annealed Steel*

INTRODUCTION

The modification of steel surfaces has important benefits for commercial products. The surface analysis techniques of scanning Auger and X-ray photoelectron spectroscopy (XPS) have become powerful tools for the characterization of modified steel surfaces. Recently, the improvements in field emission electron based scanning Auger instrumentation allow the elemental and chemical analysis of features with spatial resolution down to 10 nm. In addition, the unique scanning x-ray microprobe XPS can provide even more complex chemical surface characterization with a 10 micron spatial resolution.

Three surface analysis examples will be briefly discussed. The first will be the interior surfaces of ductile iron as prepared by *in situ* fracture analysis with scanning Auger. The second will be the scanning Auger characterization of steel surfaces prepared under different annealing atmospheres. The third will be the scanning XPS microprobe characterization of tribology wear tracks on steel. These examples will illustrate how the dramatic 10 fold improvements in spatial resolution and the utilization of scanning microprobe analytical protocols offer powerful tools for developing new steel products.

SEGREGATION OF IMPURITIES TO INTERPHASE BOUNDARIES IN DUCTILE IRON

Grain boundary embrittlement and high temperature grain boundary instability are important factors in the quest for higher performance metals for many industrial applications. Of particular importance is the composition of interphase boundaries (the interface between separate phases) in these metals. Scanning Auger spectroscopy combined with *in situ* facture analysis provides a unique technique to expose the surfaces of the interphase boundaries and provide compositional analysis and elemental mapping of these surfaces. An earlier study provided a wealth of information about the effects of different additives on the eutectic transformation of ductile iron [1]. This earlier study utilized *in situ* fracture analysis with a PHI 590 Scanning Auger with a spatial resolution of 200 nm. The dramatic improvement in spatial resolution afforded with the Field Emission electron gun column of the PHI *700Xi* Scanning Auger Nanoprobe can now provide a spatial resolution down to 8 nm. The PHI *700Xi* utilizes a unique CMA for detection of Auger electrons excited by the field emission electron source which can achieve a beam size as small as 0.006 μm. The field emission electron source is coaxial with the CMA. The CMA provides a 360° azimuthal collection angle around the field emission source. The CMA therefore collects Auger electrons from any sample surface that is line of sight to the incident electron source. The result of this geometry is that the Auger sensitivity remains constant over a very wide range on emission angles. This paper illustrates the improved imaging for the analysis of the interphase boundaries and nanostructures from in situ fractured metallurgical samples.

A ductile iron sample with 0.11% Sn was fractured *in situ* and analyzed in a PHI *700Xi* Scanning Auger Nanoprobe. For higher sensitivity, all analyses were performed with a 20 kV beam at 10 nA with a resulting beam size of 12 nm. Auger spectra and maps were acquired with a Cylindrical Mirror Analyzer (CMA) with an eight channel detector operating at 0.5% energy resolution. Ultrahigh vacuum conditions (less than 7 x 10^{-8} Pa or 5 x 10^{-10} Torr) were maintained during the fracture and analysis to prevent contamination of the freshly exposed surfaces. A secondary electron image (SEI) shows the fracture occurs along the graphite/metal interface with extremely high topography. Overlay Auger maps, acquired without any observed topographical shadowing, clearly show that a crater from which a ~ 40 μm graphite nodule was removed has a surface composition of Sn that has segregated to the boundary of the graphite/iron nodule that was removed from the crater by the fracture process. A carbon Auger image confirms the presence of graphite/iron nodules. The Auger spectrum acquired in the area of the iron crater confirms that the segregation of Sn as well as Si, Mg and O occurs at the interphase boundary with the graphite nodules.

The Auger color overlay maps also show ~ 500 nm regions of high intensity Ca. Auger maps acquired with a 20 kV, 1 nA electron beam, to achieve the highest spatial resolution of 8 nm, produced Auger overlay maps of Ca, Mg and Ti in addition to the Fe matrix.

Six areas of higher Mg intensity with a feature size smaller than 50 nm are observed in both the Mg and Fe maps. The S map shows S to be coincident with Ca. To confirm the elemental compositions of the three phases observed in the Auger maps, multiple Auger point survey spectra were acquired from each of the three phases. The phase associated with high Ca also contained higher amounts of S, Mg, and O. The phase with highest Mg contained very high O, and low amounts of Ca and Ti. The phase with the highest Ti had very high intensity C as well as lower intensity O, Ca, S and Mg. Fe peaks, potentially from the matrix, were also observed in the spectra from the three phases.

The PHI *700Xi* Scanning Auger NanoProbe combined with an *in situ* fracture stage is a unique instrument for the analysis of interface boundaries in iron. The 8 nm spatial resolution of this state-of-the art scanning Auger facilitates comprehensive analysis of the smallest features. This spatial resolution, when

combined with software controlled image registration, allows high sensitivity Auger mapping of multiple phases observed within 500 nm craters. The CMA allows full Auger mapping of all nodules and craters produced during the fracture analysis without any observed topographical shadowing. The Auger analysis confirms that the graphite nodules separate at the interface with the interphase boundaries that form as a result of segregation of minor constituents in the iron such as Sn, Ca, S, Mg, O and Ti.

SURFACE COMPOSITION OF ANNEALED AND HOT DIP GALVANIZED STEEL

To reduce vehicle weight, the automobile industry is interested in high strength steel alloys that could provide lighter body components. These complex steels typically contain alloying agents including Mn, B, and Si that may segregate to the surface during annealing of the steel. The composition of the atmosphere used during the annealing process impacts surface oxidation and segregation processes. Since the steel will be galvanized to provide corrosion protection, a surface composition must be provided that promotes good wettability for the hot dip galvanizing material. The PHI *700Xi* scanning Auger can be used to study segregation and phase separation before and after the galvanizing process.

Two annealed samples were supplied by the Max Planck Institute in Germany. The first sample had a bulk composition of Fe – 1 wt % Mn – 0.01 wt % B. This sample was annealed at 800°C in 5% H_2N_2 atmosphere with -40°C dew point. The second sample had a bulk composition of Fe – 5 wt % Mn – 3 wt % Si. This sample was annealed at 800°C in 5% H_2N_2 atmosphere with -80°C dew point. The third sample was a hot dipped galvanized steel sample with non-uniform Zn and Si.

The SEI images of the first sample revealed a surface with highly topographic features with irregular shapes varying in dimensions from 50 nm to 1 micron. Auger spectra identified the presence of C, O, Si Fe, Mn, B and N on the surface. Auger maps of these elements identified at least 4 different phases on the surface. In addition to the Fe matrix, surface regions with high concentrations of C and O with low amounts of Mn, N and Fe were observed. In addition a phase rich in Fe, Mn, B, Si and O was observed as well as a phase with just Fe, Si and O. The Auger mapping was obtained without any observed topographical shadowing.

The second sample also showed a SEI with highly topographic features with irregular shapes varying in dimensions from 100 nm to 2 microns. Auger spectra identified the presence of C, O, Si Fe and Mn on the surface. The Auger maps showed the presence of three phases: one phase is rich in Si oxide which has sphere like surface structures; one phase has structures rich in Mn oxide, with the remaining areas of the surface with the expected Fe matrix phase.

These results demonstrate how the annealing process can dramatically alter the morphology and surface composition of the iron samples. The use of a field emission electron gun coaxial to the Cylindrical Mirror Analyzer facilitates the analysis of features as small as 10 nm without any observed topographical shadowing in the Auger mapping. It is expected that both the surface composition as well as the surface morphology can influence the results of a hot dipped galvanizing process.

The surface chemistry and surface morphology of third sample was examined following a hot dipped galvanizing process. The SEI showed a surface with striations whose width appeared to be less than 100 nm. After the sputter ion removal of a very thin surface contamination, the Auger spectra showed that the surface included O, Zn, Al and Si. Multipoint quantitative elemental analysis indicated that the striations were enriched in Zn and O. Auger mapping and Auger line scans indicated that the striations have an enrichment of Zn and O and that the striations have a width of ~ 50 to 80 nm.

These results indicate that the hot dipped galvanizing process produces and non-uniform coating with phases that have dimensions of 10's of nm's. This non-uniform surface chemistry and surface morphology would be expected to influence the adhesion and corrosion protection characteristics of this galvanizing coating.

TRIBOLOGY WEAR TRACKS ON STEEL

Tribology and corrosion studies of steel surfaces are highly relevant to many commercial products. In both of these areas of research, the chemical surface characterization of micro-areas of steel surfaces is highly desirable. In this study, a wear track of MoS_2 on a steel surface was characterized by a scanning x-ray monochromated XPS. The x-ray source utilizes a focused electron beam that can be electronically raster scanned upon an Al anode for x-ray generation and an ellipsoidally shaped quartz crystal monochromator that refocuses the x-ray beam onto the sample surface. Therefore, when the electron beam is scanned on the anode surface the refocused x-ray beam is scanned on the sample surface. This scanning x-ray source provides high performance large-area spectroscopy, x-ray beam induced secondary electron imaging for rapid and confident location of small sample features, unique multi-point micro-area spectroscopy performance, chemical state XPS imaging, and multi-point high performance sputter depth profiling with x-ray beams from less than 10 μm to 300 μm in diameter.

The scanning x-ray induced secondary electron image (SXI) identified parallel 15 micron and 25 micron tracks separated by 10 microns. XPS survey spectra, XPS chemical state imaging and XPS chemical state line scans confirmed the presence of MoS_2 tracks on the surface of the steel.

Multi-point sputter depth profiles on and off the track confirm the presence of the chemistry of MoS_2 mixed with iron oxide and iron hydroxide in the wear track. Underneath the wear track, the depth profile confirmed the presence of elemental iron. The depth profile at the point adjacent to the wear track confirmed the presence of an iron oxide and iron hydroxide layer covering the elemental iron.

SUMMARY

The electron spectroscopy techniques of field emission scanning Auger and monochromated scanning XPS have been shown to be unique and extremely valuable tools for the characterization of the surface chemistry of modified steel surfaces. The surfaces can be exposed by in situ fracture analysis in a scanning Auger or by direct insertion of samples that have been surface modified external to the instruments. Both of these techniques have a surface sensitivity of 1 to 5 nm. Scanning Auger has an ultimate spatial resolution less than 10 nm while scanning XPS has an ultimate spatial resolution less than 10 microns. Scanning Auger with a field emission electron gun with a coaxial Cylindrical Mirror Analyzer can provide compositional mapping of *in situ* fracture analysis samples to elucidate the fracture mechanisms of iron and other high strength metals as a function of trace elements in the metallic matrix and the heat treatment of the metals before fracture analysis. Scanning Auger can also image the phases that are formed as a function of annealing conditions and bulk metallic composition.

Monochromated scanning XPS can provide chemical state analysis, chemical state imaging and chemical state depth profiling on tribology samples as well as corrosion and other surface modified samples with a spatial resolution of 10 microns. Based on the differences in ultimate spatial resolution as well as the quantitative chemical state analysis of XPS, the scanning Auger and monochromated scanning XPS techniques provide powerful and synergistic tools to characterize many types of steel research problems.

REFERENCE

[1] W. C Johnson and B. V. Kovacs, *Metallurgical Transactions A,* Vol. 9A, 1978, p.219-229.

Identifying Boron Distribution in a Nb-B Containing Structural Steel

Balbir Singh and Gadadhar Sahoo
R&D Centre for Iron and Steel, Steel Authority of India Limited
Ranchi-834002
Email: bsingh@sail-rdcis.com

ABSTRACT

The C-Mn-Nb-B steel, which was made in LD converter and subsequently rolled to 8 mm diameter wire rods, was used as experimental material to identify boron distribution in as-rolled state. When examined under optical microscope, the steel was found to have polygonal ferrite with 10-15% degenerated pearlite in the microstructure. Being boron as a lower atomic number element, its presence at different locations was identified by employing Auger Electron Spectroscopy. During the study, elemental distribution spectra were obtained at the grain boundary, within ferrite grain and the pearlitic regions. On superimposing these spectra at the optical image, it was revealed that boron rich compounds were present in the pearlitic regions as well as at grain boundaries. In order to understand the kinetics of boron segregation and the type of precipitates at different locations, differential thermal analysis (DTA) was carried out on powder samples of the steel for which transformation temperatures were calculated through empirical relationships. During DTA study, enthalpy and latent heat were calculated, which helped in inferring that $M_{23}(CB)_6$ precipitate is the most probable boron compound present at the grain boundaries.

KEYWORDS: *Boron Distribution, Hardening, Precipitation, Structural Steel.*

INTRODUCTION

Boron is added to structural steels to improve their hardenability and is effective when added in very small amounts; the optimum being in the range of 5 to 40 ppm. Since it enhances the hardenability effect of other alloying elements, it is called alloy intensifier. It has been found to be very effective when added to low carbon alloy steels, although beneficial effect is reduced at higher carbon contents. The largest effect of boron on hardenability occurs at carbon content of about 0.10 wt.%. In addition to its beneficial effects boron appears to be detrimental to low temperature toughness properties of steels.

The mechanism through which B additions enhance hardenability needs to be understood as its enunciation may greatly help in understanding the different effects, its additions produce. Despite a relatively larger atomic size B dissolves interstitially in steels. Therefore, B atoms have an automatic tendency to migrate to open spaces such as prior austenite grain boundaries to reduce strain. Due to this they temporarily block the nucleation of equilibrium transformation products - a pre-requisite so important to enhancing hardenability in general (both the bainitic and martensitic hardenability). As the steel is being continuously cooled the final microstructure to form would thus either be bainite or martensite depending upon the alloy design. Boron is detrimental to ambient and low temperature toughness because of the embrittling effect of B atoms present along the prior austenite grain boundaries. Another feature arising out of the presence of boron in steels is that banded structure is produced. This also has an adverse effect on toughness. It is for this reason that it is often suggested that B addition would prove most efficient when added to a low carbon iron base as the embrittling effect of higher carbon content is at a minimum.

Some studies [1] have demonstrated that in Nb containing B treated steels, Nb also segregates temporarily at prior austenite grain boundaries. Thus, both the atoms when present at austenite grain boundaries retard the nucleation and development of new grains of equilibrium transformation product takes place [2]. Depending upon the effects of these two atoms on grain boundary energy and on the velocity of moving boundaries it was observed experimentally that their retarding effect on recrystallisation is more when added jointly compared to their separate effects [3]. When added in combination, Nb and B have a synergistic effect on the recrystallisation kinetics of austenite which has been attributed to the formation of (Nb,B) complexes at the newly formed grain boundaries. This increases the solute drag force of Nb atoms and in turn, decreases the boundary velocity and slows down the recrystallisation kinetics of B modified Nb containing steels. In an alloy similar to the one under study, Tomehiro et al. [4], however, observed the recrystallisation temperature of about 980°C. The studies have also been confirmed that presence of solutes like Nb and B in the austenite retards austenite grain growth even after recrystallisation. This directly influences the kinetics of austenite to ferrite transformation when steels are cooled continuously below Ar3 temperature.

In view of the above mentioned segregation patterns of Nb and B atoms, present study has been carried out to identify experimentally the distrution of their compounds at various locations in a structural steel. For this purpose the microstructure of the as-hot-rolled Nb-B containing steel was studied under optical microscope and Auger Electron Spectroscope. The study was supported by differential thermal analysis and empirical calculation of critical/ transformation temperatures.

EXPERIMENTAL MATERIAL

The present study was conducted on specimens drawn from 8 mm diameter wire rods of C-Mn-Nb-B steel. Prior to hot rolling, the steel was made in LD converter and cast into ingots for further rolling to 85 mm square billets in continuous billet mill. After rolling the billets were cut to 12 meter standard length and allowed to cool in cooling bed. The billets prior to being further rolled in wire rod mill were charged into the reheating furnace wherein a temperature of 1170°C was maintained at the front end and 1220°C at the back end. After soaking for 1-1½ hour, the billets were rolled to 8 mm diameter wire rods with a cumulative reduction of about 99 %. Rolling was initiated in the range of 1000-1150°C and completed in 8 passes to attain a finishing temperature of 950-1000°C. Just after finish rolling the wire rods were coiled and allowed to cool on hook conveyors. Samples for various tests were taken after the coils had cooled to room temperature. The chemical composition of the rolled product was analysed as given in Table 1.

Table 1: Chemical composition of microalloyed steel wire rod

C	0.13	S	0.025
Mn	1.05	P	0.036
Si	0.05	B	0.004
Nb	0.023	N	0.0070
Al	0.04	-	-

The mechanical properties (hardness and tensile strength) of the product conformed to ISO:898-I specification, which are presented in Table 2.

Table 2: Mechanical properties of as-hot-rolled microalloyed steel wire rod

Yield Strength	360 MPa
Ultimate Tensile Strength	550 MPa
Uniform Elongation	17.4 %
Total Elongation	35 %
Reduction in Area	70 %
Hardnes (Hv20)	189

EXPERIMENTAL PROCEDURE

Optical Metallography

Optical metallography was carried out on 8 mm diameter as-rolled specimens. About 10 mm long pecimens were cut in transverse direction, ground and polished in usual manner on amery paper of grits up to 600. Final polishing was done on a wheel cloth with 0.1μ alumina solution as abrasive. After polishing, the specimen were rinsed in soap solution thoroughly and dried before further etching with freshly prepared 2% nital solution. The specimens thus prepared, were observed under Neophot-30 optical microscope and microstructures photographed.

Auger Electron Spectroscopy

The experimental steel was also investigated under Auger Electron Spectroscope (AES) to identify locations of B segregation. For this study about 2mm thick disc specimens were cut from 8mm diameter wire rod and polished in the usual manner. The specimens were carefully etched with freshly prepared 2% nital solution and subsequently cleaned in acetone by an ultrasonic cleaner. After drying, the freshly prepared specimen was loaded into the vacuum chamber of AES maintained at a pressure of 1.8×10^{-6} Pa. The specimen was exposed to primary electron beam at an accelerating voltage of 10 kV and primary electron beam current of around 10^{-7} Amps.

Segregation behaviour of various elements present in the alloy was studied by selecting different locations, such as pearlite-ferrite grain boundary, ferrite-ferrite grain boundary, within pearlitic regions and ferrite grains. During the experiment 10 sweeps were taken at each location and the average concentration calculated through dedicated computer software. Kinetic energy versus Intensity plots were obtained on X-Y recorder through the same software. For each location relative atomic concentrations of all detectable elements were determined based on the peak height after correcting for sensitivity (5).

Differential Thermal Analysis

The Differential Thermal Analysis (DTA) was conducted on the microalloyed steel wire rods on NETZSCH Simultaneous Thermal Analyser STA-409 system. About 45 mg steel sample was used in powder form which was heated in air at a controlled rate in dry alumina crucible. For this study Kaolin was used as a reference.

A heating rate of 10°C/minute was selected during the test. The start temperature for the alloy was 21°C while end temperature was set to 1200°C. The sampling time, the acquisition rate and the total time taken for the test were 3 seconds, 2 points/°K and 2 hours 6 minutes, respectively. Range for both the DTA and the DDTA were fixed at 200 °V. During an experiment, data were recorded continuously and analysed by NETZSCH DATA ACQUISITION SYSTEM. The DTA and DDTA values were plotted against temperature or time through the acquisition system itself.

Calculated Transformation Temperatures

Using the empirical relationships proposed by Andrews [6], valid for low carbon and medium carbon structural steels, the critical / transformation temperatures were worked out as follows:

Ac_1 (°C) = 723 -10.7Mn -16.9Ni +29.7Si +16.9Cr +290As +6.38 W

Ac_3 (°C) = 910 -203C1/2 -15.2Ni +44.7Si +104V +31.5Mo +13.1W

M_S (°C) = 512-453C+217(C)2-71.5(C) (Mn) -67.6(C)(Cr)-16.9Ni+15Cr-9.5Mo

M_{10}(°C) = Ms - 10 [error ± 3]

M_{50}(°C) = Ms - 47 [error ± 6]

M_{90}(°C) = Ms - 103 [error ± 12]

M_f (°C) = Ms - 215 [error ± 15]

Substituting for the elements present in the experimental steel, critical/ transformation temperatures were estimated to be:

Ac_1 = 713°C, Ac_3 = 839°C

M_s = 447°C, M_{10} = 437°C

M_{50} = 400°C, M_{90} = 344°C

M_f = 232°C

RESULTS

Optical Microscopy

Representative optical photomicrograph of as-rolled wire rod is depicted in Fig.1. The figure revealed that microstructure of experimental steel comprised polygonal ferrite and pearlite. Volume fraction of pearlite, as measured at 10 different locations on the specimen, varied between 12 to 15 percent. The average ferrite grain size measured using the line intersection method was found to be ~12 μm. It is also seen from the figure that the ferrite had a

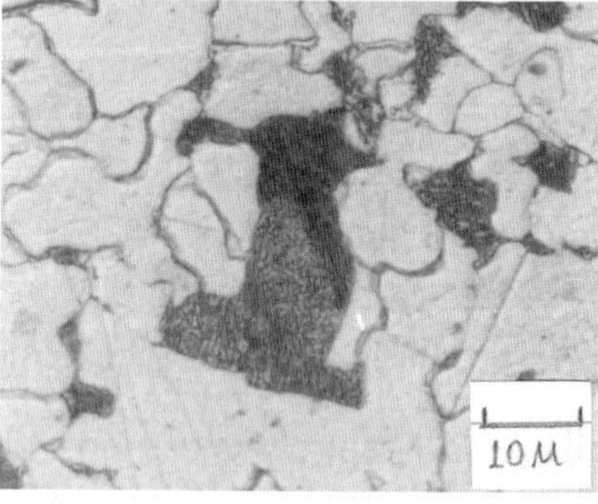

Fig. 1: Microstructure of Nb-B steel

mixed grain structure with about 25% of the grains being finer and in the size range of 8-10μm. Pearlite in most locations was non-lamellar and comprised degenerated morphology.

Auger Electron Spectroscopy

Auger electron spectroscopic maps presented in Figs. 2 (a & b) showed that boron rich regions existed in the steel microstructure. On comparing this map with the optical microphotograph it was revealed that the boron rich pockets existed within the pearlitic colonies. Besides this, boron was also found segregated along the grain boundaries. Line scan of boron duly supports these observations (Fig.3). Auger spectra taken at ferrite-pearlite interface (Fig.4) showed a feeble B-intensity peak at 179 ev kinetic energy. Before taking the spectra, the specimen was argon etched for three minutes to remove any adsorbed oxygen within the surface. Ion current was maintained as 20 mA at an accelerating voltage of 3 kV. Sample tilt was kept ~40° and aperture size was medium. Elemental distribution spectra obtained inside the ferrite grains (Fig.5) did not show even a

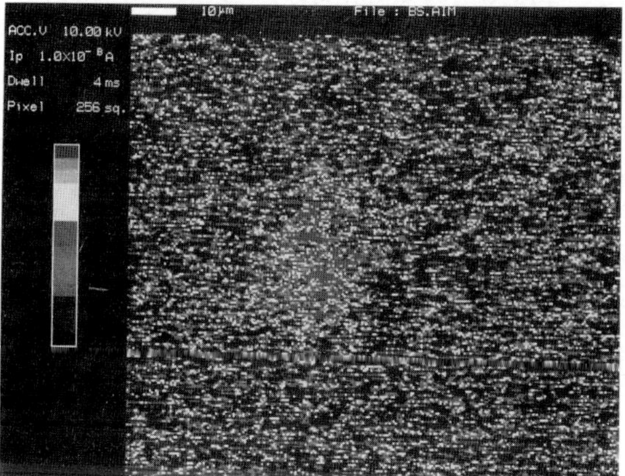

Fig. 2(a): AES map showing B distribution

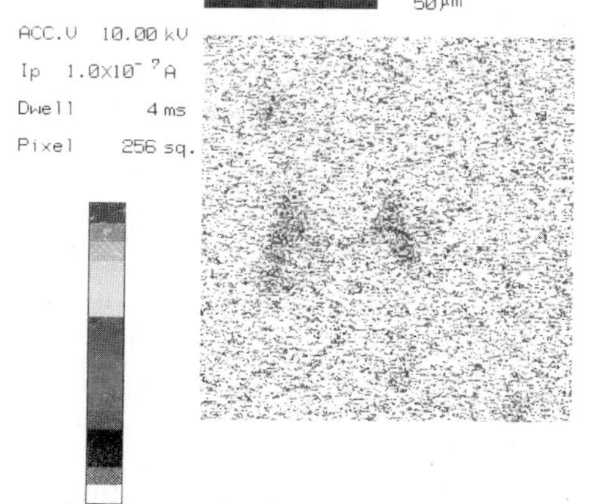

Fig. 2(b): AES map showing B distribution

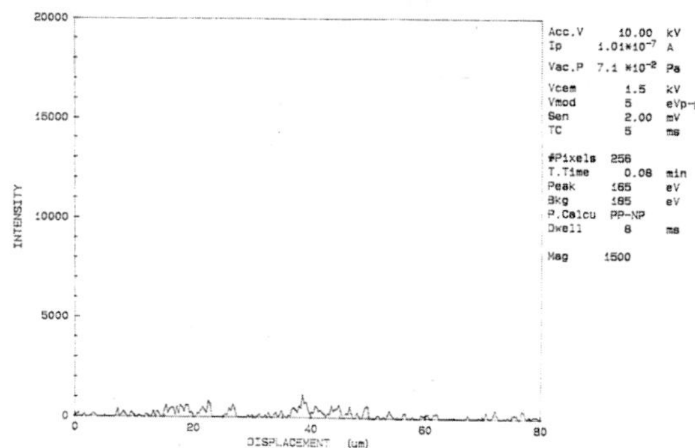

Fig. 3: AES Line scan showing B peaks at different locations

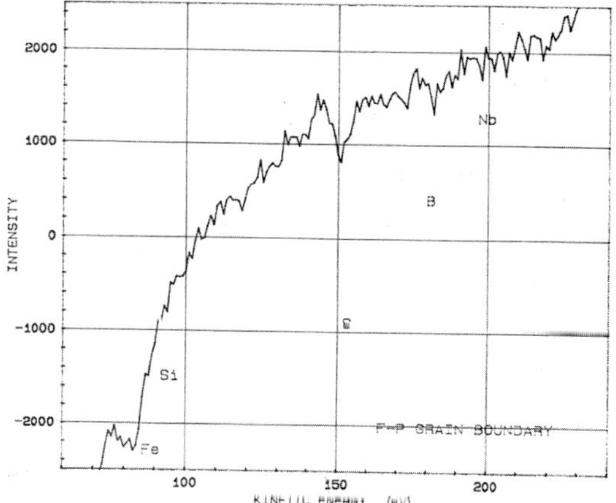

Fig. 4: AES point analysis showing B peaks at ferrite-pearlite grain boundary

feeble peak of B at 179ev kinetic energy, thereby confirming its absence at this location.

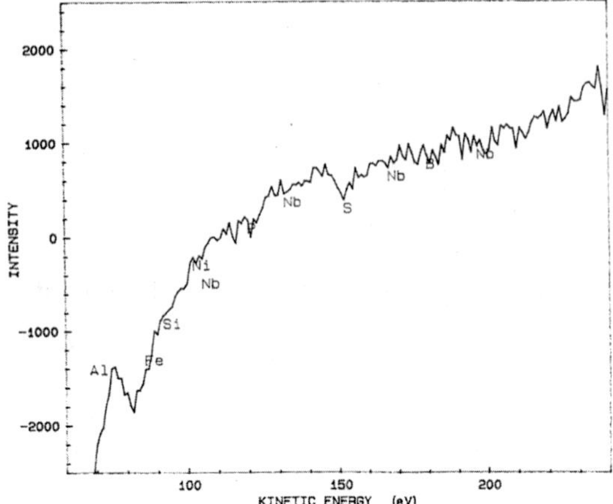

Fig. 5: AES point analysis showing no B peak within ferrite grain

Thermal Analyses

Studies on thermal analyses of microalloyed steel comprised an assessment of critical/ transformation temperatures through differential thermal analysis, and calculation of enthalpy of transformations. DTA versus temperature plots along with DDTA are depicted in Fig. 6. The results obtained on analysing the data are presented below:

Critical/Transformation Temperatures

i) First deviation in the DTA curve, was observed at 430°C. This, most probably, corresponds to relaxation of stresses introduced during machining of the specimen to powder form. The said process apparently got completed by ~610°C.

ii) First phase transformation took place at 715°C which is close to the ferrite→austenite transformation start temperature (Ac_1:713°C) as calculated empirically. The transformation reached to its peak at 734°C followed by pearlite→austenite transformation at 745*C. Subsequently, this transformation got completed at ~768°C. Between 715-768°C, with a peak at 734°C, the enthalpy (ΔH) of the transformation and latent heat (ΔC_p) were estimated as -9.4 J/g and -0.232 J/g°K, respectively. The negative values of ΔH and ΔC_p indicated that ferrite to austenite and cementite to austenite transformations are exothermic in nature.

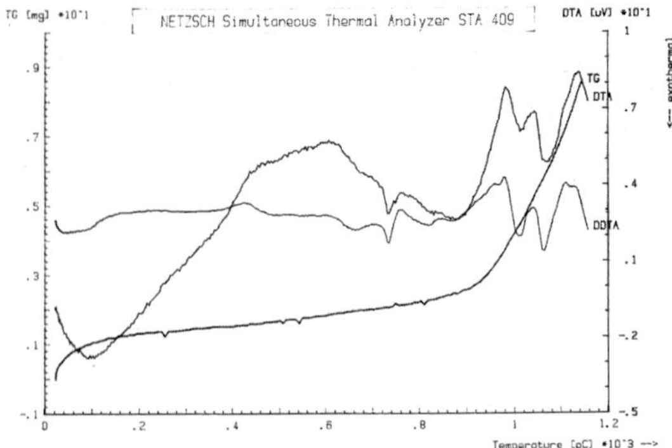

Fig. 6: DTA and DDTA plots showing transformation temperatures

iii) Second transformation was observed between 980-1044°C with a peak at 1014°C. Enthalpy and latent heat calculated for this peak were -48.6 J/g and -0.583 J/g°K, respectively. This exothermic energy was liberated, in all probability, due to the dissolution of $M_{23}(C,B)_6$ precipitate in the aforementioned temperature range [7,8].

iv) The above transformation was followed by another transformation between 1044°-1140°C. Enthalpy and latent heat, calculated for a peak at 1074°C, were found to be -58.4 J/g and -0.692 J/g°K, respectively. The energy values, thus attained in all probabilities represent the dissolution of NbC precipitate into austenite [7,8].

DISCUSSION

The microstructure of the material in the hot rolled condition followed by air cooling comprised ferrite+degenerated pearlite with an average ferrite grain size of ~12 μm. The important feature is definitely the presence of degenerated pearlite. It is evident that its formation has been largely brought about by the presence of boron. This observation is based on the premise that in an equivalent B-free steel, no degenerated pearlite forms. It would therefore be necessary to arrive at an indepth understanding of the effect of Nb & B both individually and when acting in tandem on the overall transformation behaviour of the steel under investigation. Without going into details, it would not be out of place to suggest that the formation of degenerated pearlite is perhaps an out come of; i) the likelihood of forming bainite type structure at relatively faster cooling rates due to the presence of B, and ii) the likely effect of Nb & B in influencing the carbide formation step constituting pearlite.

During reheating of the billets at 1200°C, for 1-1½ hours, carbides and other similar compounds mostly dissolve, thereby ensuring that alloying elements such as Nb, B, Al, C go into solution. In so far as the solubility of these elements at high temperatures is concerned, it mainly depends upon the atomic size and the affinity amongst interacting species. Empirical relationships exist for computing solubility at a given heat treating temperature. For the experimental steel, the amount of Nb dissolved in equilibrium with C, could be calculated by using the relationship [9] :

$$\log [\,Nb\%\,][\,C\%\,] = -7900/T + 3.42.$$

On substituting for C (as 0.13%), ~0.088% Nb could be dissolved at 1473°K. Irvine et al. indicated that N also contributes to carbon equivalence. For such steels (C:N>10) they proposed the relationship:

$$\log [Nb]\,[C + 12\,N/14] = -6770/T + 2.26$$

Since the experimental steel contains C and N in the ratio of ~18.57, the amount of soluble Nb that can be accommodated in equilibrium with 0.13%C and 70 ppm N at 1473°K is ~0.034%. The amounts of Nb calculated through the aforementioned relations, namely 0.088 & 0.034%, are higher than the one present in the experimental steel (0.023%). It would not be incorrect to infer, therefore, that most of the Nb present in the experimental steel is in the dissolved state on reheating the billets at 1200°C.

It would now be pertinent to examine the effect of Nb and other alloying elements in influencing the processing schedule. While little effect has been observed on reheating at 1200°C, ease of deformation is influenced at lower temperatures [10]. During hot deformation, excess dislocations being generated anneal out, thereby leading to the attainment of a sub-grain structure. At rolling temperatures of 1050-1150°C, in combination with larger strains generated due to higher speed of rolling, critical strain is easily reached and this causes austenite to recrystallise. In roughing stands of wire rod mill rolling speed is comparatively low but temperature of about 1150°C is good enough for inducing recrystallisation. Similarly, at the finishing stage, despite the temperature being lower (1075°C), higher strains generated due to higher rolling speeds induce recrystallisation to occur.

The occurrence of recrystallisation during hot working is thus a fundamental happening. Presence of alloying elements would, however, influence the recrystallisation kinetics of austenite and it is quite likely that under certain conditions it may not go to completion. The presence of alloying elements, in general, raises the recrystallisation temperature. Nb and B are known to influence it markedly through their precipitate forming tendency and the resultant well known pinning and solute drag effects. In order that maximum benefits are derived from these effects, it would be most useful to have Nb and B in solution before starting rolling. This is important because the basic condition for obtaining grain refinement during processing is that reprecipitation of the grain refining precipitates (in the present instance Nb bearing compounds) and recovery/ recrystallisation should occur simultaneously.

The mechanism suggested envisages that as rolling is initiated, the fibered structure formed due to deformation recovers/ recrystallises. At the same time reprecipitation of the Nb bearing compounds begins, perhaps on the sub-grain boundaries formed due to recovery. This is useful in pinning sub-boundaries. As the number of passes increase the aforementioned steps are repeated. Despite a likely fall in temperature as the rolling proceeds, the Nb bearing precipitates coarsen since the absolute value of temperature is high. The precipitate in its coarsened form is useful in restricting grain boundary migration. By the time the temperature reaches 950-1000°C most of the grain boundaries/ sub-boundaries are expected to be pinned and as such it is most useful to terminate rolling at this stage. Consequently it would be possible to retain the microstructure attained at the finishing stage down to room temperature without the necessity of having to alter the cooling rate. Thus, it is clear that the fraction of Nb that has separated as a precipitate in austenite is contributing to grain refining during controlled processing.

Regarding the nature of the precipitating Nb compounds, Nordberg and Aaronson [11] have demonstrated through thermodynamical calculations that it forms NbC rather than Nb(C,N) in the alloys containing C:N≥5. In the material under study the ratio happens to be ~19. Therefore, presence of Nb in the form of NbC was most probable. Moreover, since C in solution was surplus, the amount left over after combining with Nb, could be effectively fixed with boron. Studies [8,12] have revealed that if C and B are present in solution near Ac_3 temperature they combine and form $M_{23}(C,B)_6$ type of precipitates at austenite grain boundaries and grain interiors. Since Nb has already been fixed as NbC, iron atoms are thermodynamically the next option to adopt the atomic configuration of $M_{23}(C,B)_6$. Therefore, $M_{23}(C,B)_6$ is the favoured precipitate to consume the available boron. Since Nb is effectively consumed in forming NbC, it is no longer effective in suppressing the $\gamma \rightarrow \alpha$ transformation. Therefore, on cooling at slower rate of about 1°C/sec, as used in industrial wire rod mills, lower transformation products such as bainite and martensite did not form since neither Nb nor B is able to contribute to such an happening. At such a low cooling rate ferrite did not experience much resistance to nucleate at austenite grain boundaries and other imperfections. Due to its low solubility in ferrite, carbon was rejected during ferrite grain growth and this finally resulted into the formation of pearlitic microstructure. It is interesting to note that inspite of low cooling rate in the wire rod mill, formation of lamellar morphology of carbides (cementite) could be avoided. This was made possible due to the synergistic effect of Nb and B, exerted on the kinetics of $\gamma \rightarrow \alpha$ phase transformation. The non-lamellar morphology of degenerated pearlite is some what similar to that of acicular ferrite having broken chains of cementite inside ferrite matrix. As suggested in the literature acicular ferrite has also been described as a form of upper bainite. It may, therefore, be appropriate to mention that degenerated pearlite is a microstructure which has its characteristics between lamellar pearlite and upper bainite and hence, in terms of properties, is expected to perform better than lamellar carbides. In the light of the above mentioned analysis the properties of the as rolled material appear to be reasonable with total elongation and reduction in area values better than what can be normally expected.

It is known that on cooling the recrystallised austenite from finishing temperatures of about 950-1000°C to room temperature compounds of Al, N, Nb, B, C etc. start precipitating out after the $\gamma \rightarrow \alpha$ transformation. The extent of precipitation depends upon their volume fraction, affinity towards each other and solubility in α (which is negligible). As far as N is concerned, it forms AlN rather than BN if Al:N exceeds the stoichiometric ratio of 1.93 [8,13,14]. In the experimental alloy, therefore, precipitation of AlN was expected since Al:N ratio was about 5.7. Similarly, Nb will precipitate as NbC and along with AlN will precipitation harden ferrite. Coming to the role of boron, Cameron and Morral [15] have reported that the solubility of B in austenite and ferrite phases is extremely small as based on the Fe-C-B system. On the other hand, cementite can take up a large amount of B as a substitute for carbon forming $Fe_3(C,B)$ precipitates consisting of Fe_3C and Fe_3B components. Perez et al. [14,16] detected three types of boron precipitates through high resolution electron microscopy. These precipitates are FeB, $M_{23}(C,B)_6$ and B_4C type in boron microalloyed steel. The aforementioned information is likely to prove useful in deciding how B would be distributed in the experimental material.

A study of the relevent portion of the Fe-C-B diagram revealed that for a composition containing 0.13 wt.% carbon and 10 ppm B precipitation of $M_{23}(C,B)_6$ occurred in the temperature range ~740-770°C [16]. It was further proposed by Watanabe et al. [8,16] that precipitation of $M_{23}(C,B)_6$ mostly takes place between Ac_3 and Ac_1 temperatures. For the experimental steel they are 839°C and 713°C, respectively and within this temperature range only about 10 ppm boron would participate in the formation of $M_{23}(C,B)_6$. Evidently, most of it (~30 ppm) is likely to precipitate with cementite in the form of boron containing compounds such as FeB/Fe_2B/B_4C or a combination of these during $\gamma \rightarrow \alpha$+pearlite transformation. This perhaps accounts for the formation of boron rich regions inside pearlitic colonies and comparatively, lower density of $M_{23}(C,B)_6$ precipitates at ferrite grain boundaries at slower cooling rates (Figs. 2a & b). As the cooling rate is raised a part of this 30 ppm boron will effectively contribute towards the formation of intermediate or lower temperature transformation products.

Having thus explained B segregation at the boundaries and reasons for its presence in larger amounts within the pearlitic regions, it is now necessary to comment upon its presence/ segregation at the ferrite-pearlite interface and its absence inside ferrite grains. The former is consistent with the general understanding that boron would be found in close proximity of the carbide phase. The latter observation agrees well with the known tendency of B to segregate to prior boundaries.

Despite a relatively lower Nb content, the ferrite grain size obtained in the as hot rolled condition (~ASTM;9) is comparable to that obtained in microalloyed HSLA steel. This is perhaps on account of the presence of Al (0.04%) which may be contributing to grain refining/ precipitation hardening due to the formation of AlN. A yet another aspect which could have contributed indirectly is the Nb-B synergism, thereby facilitating the fixing up of extra carbon (beyond that which combines with 0.023% Nb).

It is well understood that for attaining desired properties in steels transformation temperatures and cooling rate play an important role since they ultimately control the end microstructure in materials. Accurate measurement of these parameters is, therefore, necessary. For this reason, transformation temperatures were calculated for the chemistry using Andrews [6] empirical relations and confirmed through differential thermal analysis. This validates the relevance of the empirical formulations and justifies calculation of temperatures such as M_s, M_f etc. using them.

The aforementioned critical/ transformation temperatures alone did not fully reveal the transformations occuring in the experimental material. DTA, besides helping in ascertaining the additional information, was also useful in confirming either fully or partly the information thus far obtained. It was revealed that the energy peaks at ~1014°C and 1074°C (latent heats of about -0.583 j/g°K and about -0.692 j/g °K, respectively) denoted dissolution/ precipitation of $Fe_{23}(C,B)_6$ and NbC precipitates [7,8], respectively. Thus DTA proved helpful in arriving at a more detailed understanding of the transformation behaviour as it was able to detect transformations not discernible with the help of dilatometry.

On analysing tensile results it was revealed that above referred microstructures helped in attaining a total elongation of 35%, uniform elongation of 17% and reduction in area of 70% in the experimental steel. The high value of elongation along with large reduction in area values indicated that the steel is adequately formable for sustaining large deformations imparted during cold heading operations carried out while forming the components.

CONCLUSIONS

Based upon the results obtained through optical microscopy, Auger electron spectroscopy and differential thermal analysis and subsequent discussions thereon, following inferences may be arrived at:

1. In as rolled condition the wire rods contained about 12 to 15 % degenerate pearlite the balance being polygonal ferrite. The average ferrite grain size was found to be ~12 μm.
2. Auger electron spectroscopy revealed that B was segregated within pearlitic colonies as well as at the grain boundaries. Through DTA analysis and enthalpy calculations it emerged that B is likely to be present as $M_{23}(C,B)_6$.
3. Nb was found to be distributed throughout the matrix, most likely as NbC.

ACKNOWLEDGEMENTS

Authors are thankful to the management of RDCIS for permitting presentation of this paper in the international conference SIMPRO'2012.

REFERENCES

[1] R. Herschitz and D.N. Seidman, Acta Met. 33, 1547 (1985).

[2] M.J. Luton, R. Dorvel and R.A. Petkovic, Met. Trans., 11A, 411 (1980).

[3] X.L. He, M. Djahazi, J.J. Jonas and Jackman, Acta Met., Vol. 39, No. 10 (1991), P. 2295.

[4] H. Tomehiro, M. Murata, R. Habu and M. Naguma, Trans. ISIJ 27, (1987), P. 120.

[5] Auger Electron Specroscope Manual.

[6] KW Andrews, JISI, July 1965, P.721.

[7] S. Watanabe et al., Trans. ISIJ,Vol.23, (1983), P.31.

[8] S. Watanabe and H. Ohtani, Trans. ISIJ, Vol.23, (1983), P.38.

[9] K.J. Irvine, F.B. Pickering and T. Gladman, JISI, 205, (1967), P. 161.

[10] D.Zhang, A.J. Baker and P.R. beeley, Material Science and Technology, Vol.10, Oct (1994), P.903.

[11] H. Nordberg and B. Aronsson, JISI, 206, (1968), P. 1263.

[12] Y. Omori, Trans. ISIJ, No.11, P.339.

[13] S. Watanabe, H. Ohtani and T. Kunitake, Trans. ISIJ, Vol.23, (1983), P.121.

[14] K. Sugimoto et al., Trans. ISIJ, Vol.26, (1986), P.329.

[15] T.B. Cameron and J.E. Morall, Met. Trans., Vol.17A, Aug. (1986), P.1481.

[16] R. Perez, A. Garcia and J.A. Jaurez, Scripta Met., Vol.28, (1993), P931.

Ultrafast, Complete Inclusion Characterization for Steel Mills: Advances in Turnkey, Automated Scanning Electron Microscope

Vishal L. Patel, Timothy J. Drake and Susan Benes
ASPEX Corporation an FEI Company, Delmont, PA USA
Email: Vishal.Patel@fei.com

ABSTRACT

For many years, when microscopic analysis is mentioned in steel mills, a natural reaction is to assume that the familiar light-optical microscope is the obvious candidate. Optical microscopy has been widely used in the steel industry to monitor the cleanliness of the steel via procedures defined by the ASTM, JIS, DIN and ISO organizations. Even the superior imaging microscopes such as Scanning Electron Microscopes (SEMs) are often impractical to do routine inspection work. In this day and age, more and more steel makers are focusing their attention on cost savings and improving efficiency. One realization that has occurred is that, in order to improve the process and reduce waste, a better understanding and control of inclusions is necessary. Automated SEMs equipped with inclusion characterization software that can provide the size, shape, elemental composition, and image, have been widely used in recent years at steel mills for routine investigations of non-metallic inclusions and their effect on the production of steel. Data presented from a Metals Quality Analyzer will demonstrate the distinct differences between the samples collected at a various stages of steel making. Savings up to $1,000,000 in lost revenue due to the presence of inclusions, clearly establishes the automated SEM as a complete inclusion characterization solution for the steel mills over optical inclusion identification or any other methods.

KEYWORDS: *Automated SEM/EDX, Metals Quality Analyzer (MQA™), Inclusion Analysis, Cost savings, OmegaMax*

INTRODUCTION

As the global demand and production for steel increases every year, the focus on cost savings and improving the efficiency for the steel manufacturer is also on the rise. In particular, to improve the process and reduce waste, a better understanding and control of non-metallic inclusions is necessary. In order to achieve these goals, microscopic analysis is one of the methods that have been widely used by the industry to monitor the non-metallic inclusions.

When microscopic analysis is mentioned, particularly in the context of a routine industrial-type process, a natural reaction is to assume that the familiar light-optical microscope is the obvious candidate. Even those aware of the superior imaging capabilities and functionalities of the electron beam instruments may assume that such instruments are impractical and too 'high-end' to do routine inspection work. This perception of electron microscopes is warranted with most SEMs on the market today; however, SEM/EDX products with automated inclusion analysis software are at the fore-front to understand the process control and the cost savings.

SEM/EDX equipped with the Metals Quality Analyzer (MQA™) software isin steel mills where the operators of these analyzers are not SEM experts, but rather highly trained metallurgists troubleshooting the day-to-day concerns of a steel making facility. In addition, the SEM with MQA™ is designed to minimize downtime, function in a wide range of ideal operating conditions, and is a fully automated and integrated software package. All of these features combined are at the core of the automated SEM/EDX technology and will provide a vast amount of knowledge in comparison to the typical light microscopy analysis.

It is the intent of this discussion to outline the major differences and advantagesthat SEM equipped with the Metals Quality Analyzer™ (MQA)can offer steel producers over an optical microscope. In addition, there have been recent improvements in spectroscopy that have dramatically reduced analysis speeds for a typical analysis.

THE FUNDAMENTALS

The full appreciation of the difference between optically analyzing inclusions and using the Automated SEM/EDX starts with the fundamental principles of the operations of the system.

Optical Microscopy

Figure 1 schematically illustrates the general principle of an optical microscope configured for inclusion analysis. The inclusion sample is mounted on a motorized stage and illuminated by a diffuse source of light. The image of a tiny portion (field) of the specimen is projected via the lens system onto the sensor array such that each 'pixel' (picture element) is recorded simultaneously as a 'color' value (RGB intensity value). The spatial area of the sample viewed in each field is dictated by the strength of the lens system (~100X) and the size of the sensor array.

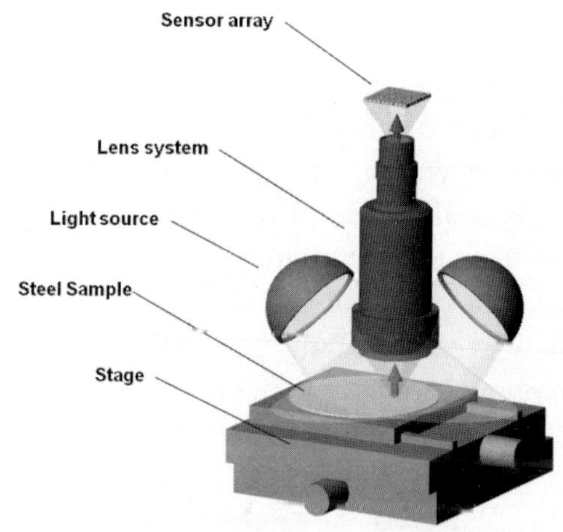

Fig. 1: Schematic illustration of an optical microscope.

The sensor array is read into a computer memory (not shown) where the image is processed by software to recognize inclusion boundaries and extract size and shape parameters. When analysis of one field of inclusions is completed, the motorized stage shifts another field into view. Stringers are identified according to their 'color' value. For example, Type A inclusions appear brighter than Type B inclusions, etc. For optical analysis, no compositional information is directly obtained from the sample; it is an indirect,inferredidentity by brightness.

The SEM

The principle of the SEM is schematically illustrated in Figure 2. The electron optics 'column' produces a highly focused beam of electrons that impinges on a point of the steel specimen. The interaction of the beam with the specimen induces a signal that is sensed by the detector. In the case of inclusion analysis, we utilize atomic number as the contrast mechanism. The sensed intensity of the response varies as the beam is sequentially scanned over consecutive points of the specimen and this variation of intensity is recorded in the computer memory as a gray-scaleimage. The gray-scale value of heavy elements is higher than that of lighter elements. The spatial area (mm^2) of the field collected is dictated by the spacing between adjacent beam points on the specimen.

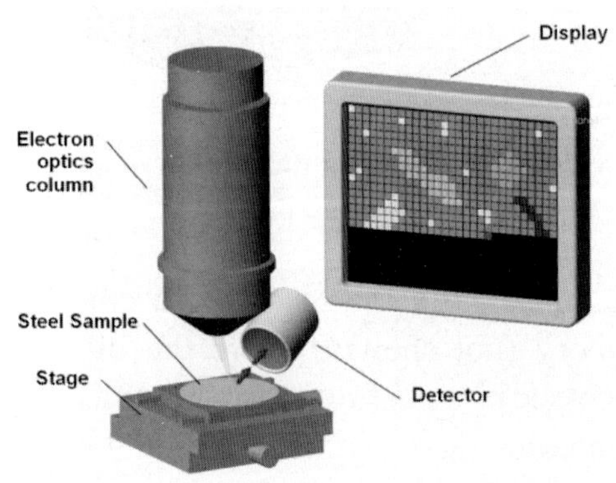

Fig. 2: Schematic illustration of an electron beam analyzer.

Electronic displacement of the beam also allows multiple particle fields to be analyzed without moving the stage; stage motion is reserved for large field displacements. Since a focused electron beam can be produced only within a relatively high vacuum, the beam, specimen, and detector are housed in a vacuum enclosure (not shown).

Automated Analysis with a SEM/EDX

In comparison to an optical microscope, the automated inclusion SEM dynamically scans the steel sample. Rather than capturing a high-resolution image of the field, the system movesthe beam across the field of view in an array of fairly coarse steps (~1-2 microns in size – See Figure 3-1.). At each point, the brightness of the signal is recorded. If the signal is dark enough to indicate that an inclusion is present at that position, then the software initiates aninclusion sizing sequence (See Figure 3-2.).

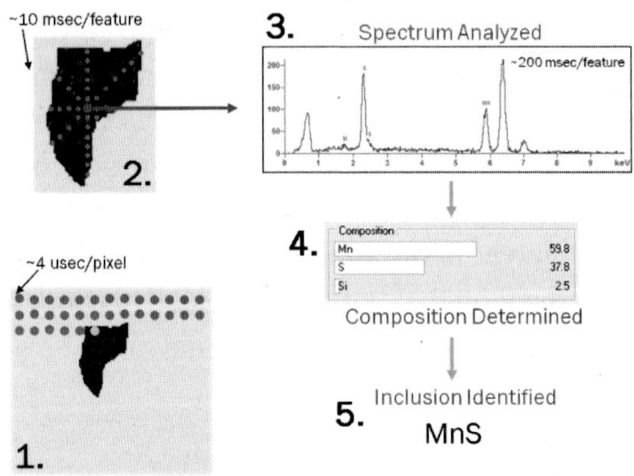

Fig. 3: Illustration of Automated Feature Analysis™ on an MQA.

Once the "coarse" scanning (indicated by the dots) identifies an inclusion (green dot), the center of inclusion is identified and 'chords' are drawn with the beam on the inclusion to define the particles size and shape (Figure 3-2). The major reason for an improvement in speed is because the instrument only spends time collecting detailed data where particles are known to be present, rather than spending time capturing and analyzing vast numbers of "empty" pixels. Since there is almost always much more empty space on the specimen then space occupied by inclusions, the result is a large speed advantage in searching, thus decreasing dramatically the overall analysis time.

THE DIFFERENCES

Contrast Mechanism

The information in an image comes from the various contrast mechanisms by which a subject feature interacts with the illumination source to produce variations in the reflected signal. The contrast mechanism in light optical microscope is drastically different from that of the automated SEM/EDX. For light optical microscopes, the interaction of light with inclusions is different than that of polished steel, so that under an optical microscope inclusions appear darker than the steel. Similarly, scratches on the surface are often detected as inclusions due to darker appearance. Algorithms have been created to alleviate these issues, but not all. Occurrences of falsely detected inclusions routinely occur.

By way of contrast, electron-beam interaction mechanisms tend to provide simpler and more useful means of inclusion discrimination. The signal that is most commonly used for producing SEM images (micrographs) is the secondary electron response, which is very sensitive to surface detail with its realistic shadowing a pleasing 'three dimensional' appearance is obtained. However, the less familiar phenomenon of backscattered electrons provides a contrast mechanism that is particularly useful in automated inclusion applications (Figure 4).

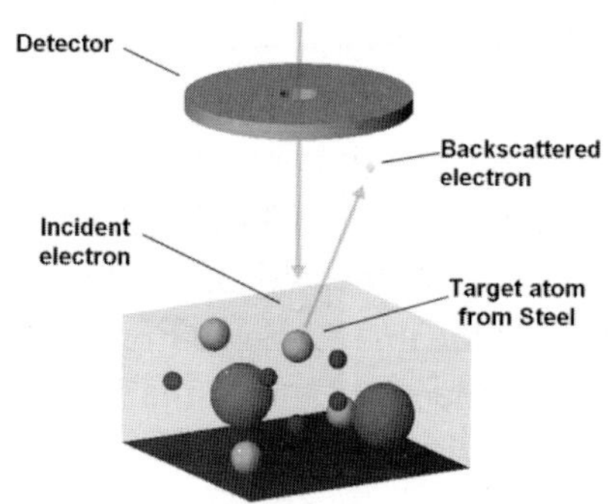

Fig. 4: Backscatter electron contrast mechanism.

The backscatter contrast mechanism involves the elastic scattering of incident electrons off of the atoms of the specimen. Since even the lightest atoms are orders of magnitude more massive than electrons, the effect is like a ping-pong ball bouncing off a bowling ball – the electron recoils with nearly all of its initial energy. An annular detector element arrayed symmetrically around the incident beam intercepts these backscattered electrons. Though the illustration in Figure 4 (not to Scale), it does correctly suggest that the probability of a backscattered event will be proportional to the average size of the atoms in the specimen material – hence atomic number contrasting.

During a standard inclusion analysis, the magnification is between 200X and 450X (Based on standard image display sizes for the SEM with MQA™). Figure 5 powerfully demonstrates the advantage of backscattered electron contrast mechanism for automatically characterizing inclusions in steel. Figure 5a represents the Backscatter

Electron Detector (BSED) image of the analyzer, while Figure 5b represents a false colorization of the BSED signal of Figure 5a based on the detection thresholds for inclusions. There are two important features to realize for both images: (1) the inclusions are the darker regions of 5a, and (2) there is a scratch (~100 microns wide) in the field of view located in approximately the center of the image. From Figure 5b, it is noticed that there are some smaller features that are imbedded in the scratches that are detected as features.This is *not* a concern for the inclusion analysis results because of the additional discrimination that an electron beam provides. Backscatter contrasting provides the SEM with a unique way to find and locate inclusions. In addition, it provides the analyzer with a chemically dependent mechanism of differentiating inclusion types. For example, sulfides have heavier average atomic numbers than oxides, so as a result the instrument can easily differentiate them.

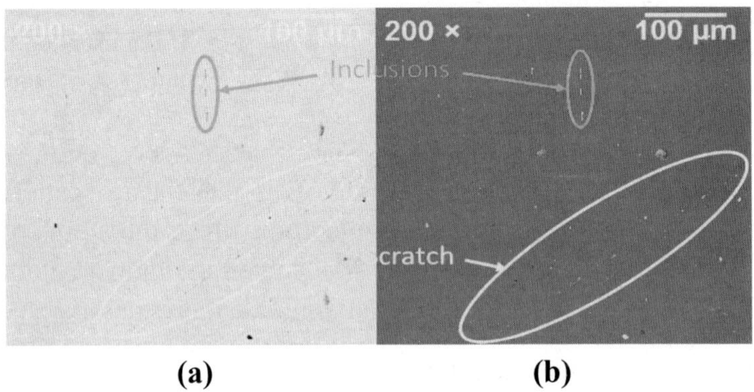

Fig. 5: Illustration of the contrast mechanism and the detection of inclusions.

Sample Preparation

Optical microscopes require 'pristine' polishing in order to the detection of scratches on the surface. Typically, <1 μm diamond finish is required. By comparison, the MQA™ will analyze steel samples that have been polished using a 3 μm diamond finish. This saves valuable time during the sample preparation step. In addition, it is impossible to analyze samples with numerous scratches such as shown in Figure 6 & 7. The sample was polished to a 1um finish and then purposely scratched to determine the ability of the MQA™and the optical microscope to analyze samples with a significant amount of scratches on the surface. The sample was impossible to analyze using an optical microscope due to the inability to determine the scratches from the inclusions. The same sample was analyzed and compared using the SEMto a sample from the same heat and almost identical location; the ASTM E2142 ratings were almost identical. The only difference reported was in the scratched (damaged) sample; one additional Type B stringer was found that had a severity rating of 1.0 whereas the other sample wasrated 0.5. This difference could be attributed to a different sampling location. In addition, if a comparison is done between an optical

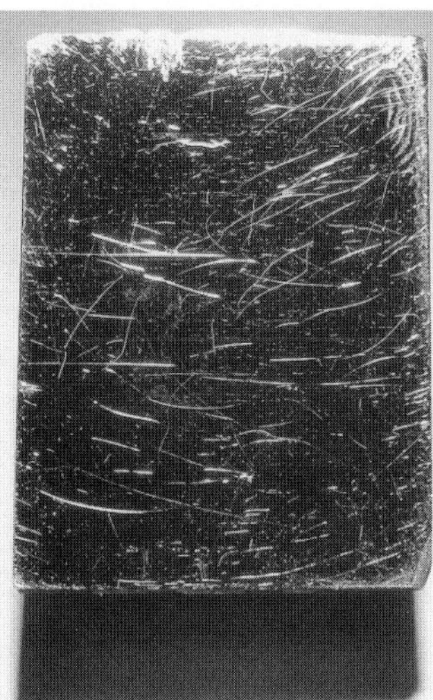

Fig. 6: Optical image of scratches on the surface of steel sample. This sample was impossible to analyze in an optical microscope, while the SEM with MQA had no difficulty in analyzing it and producing excellent results as compared to a non-damaged surface.

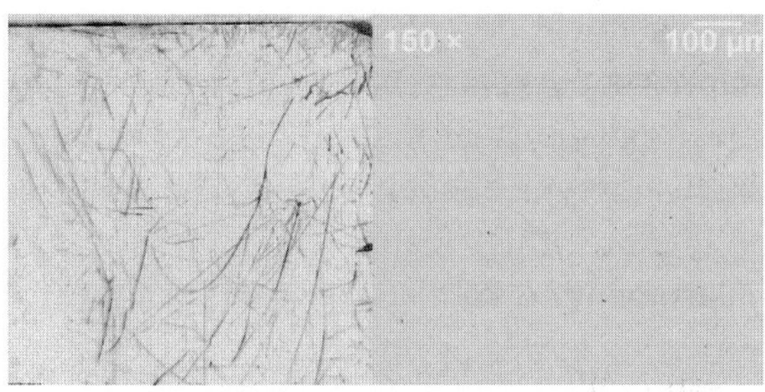

Fig. 7: Comparison of surface defects with an optical camera (a. ~10X) and with the MQA (b.). Using atomic number contrasting dramatically minimizes the effects of surface deflects.

micrograph and the SEM, it is quite apparent that surface quality is not as important for theSEM. The scratches are almost undetectablebecause although the surface has groves in it, the material (i.e. steel) is still the same atomic number regardless of the damage. As a result, the instrument does not detect the scratch as an inclusion (like an optical system would do) and more or less ignores it all together.

Chemistry Contrasting

Comparing the compositional analysis capabilities of the electron-beam-based inclusion analyzer with its light-optical counterpart is somewhat of an easy comparison. Modern-day light microscopy has no capacity to conduct compositional analysis. Light microscope systems can be programmed to make use of color cues, but this is both inexact and unreliable way to characterize composition. Therefore, light optical inclusion analysis relies only on the shape of the inclusions for classification, – not chemical information.

Electron beam inclusion analyzers have for the past fifty years routinely analyzed the x-ray emissions that are produced by the beam's interaction with a steel specimen to obtain accurate compositional information.

The material processes that produce visible light involve molecular interactions. By contrast, X-ray production derives from atomic effects. X-ray production, since it occurs deep in the atom where it is not affected by factors such as temperature, pressure, and chemical bonding, is characterized by simple and extremely consistent emission patterns (The energy of the X-ray corresponds directly to the elements of the inclusion).

Since every element above lithium produces characteristic X-rays when excited by energetic electrons, the detection of such emissions provides a simple and unambiguous way of identifying which elements are present within an inclusion and the relative abundance of each.

Consequently, during automated inclusion analysis with the SEM, the final step involved in classifying aninclusion is to collect its x-ray spectrum. The inclusion identity is then truly known and placed in an appropriate material category, such as a spinel or calcium aluminate, for example. Such analysis provides a definitive identification of the elemental composition of the inclusion material where light optical systems cannot.

The ability of the SEM to perform X-ray analysis on features that are detected also provides the instrument an added discrimination mechanism. The discrimination of inclusions and steel not only rely on the differences in their atomic numbers but also their chemical signatures. For example, if the beam sweeps across steel and the video level drops due to an interaction with a feature that is darker (or brighter) than the background steel matrix, the instrument believes that this is a feature of interest to the user and that it should perform an additional screening of it. The additional screening will determine if it is one of four items; surface defect (scratches), debris, porosity, or an inclusion. If the feature is either surface defects, debris, or porosity, the spectrum that is acquired by the SEM will obtain no chemical information relevant to what is found in inclusions – mainly, Al, Si, Mg, Mn, S, Ti, Zr, etc. Figure 8 illustrates this point very precisely. In the image, a scratch exists on the surface; and upon the edges of the surface defect, there is a residual contamination from the device used to make the defect (right image). If the SEM scans this surface as is indicated in Figure 5, this contamination would trigger the instrument to analyze it

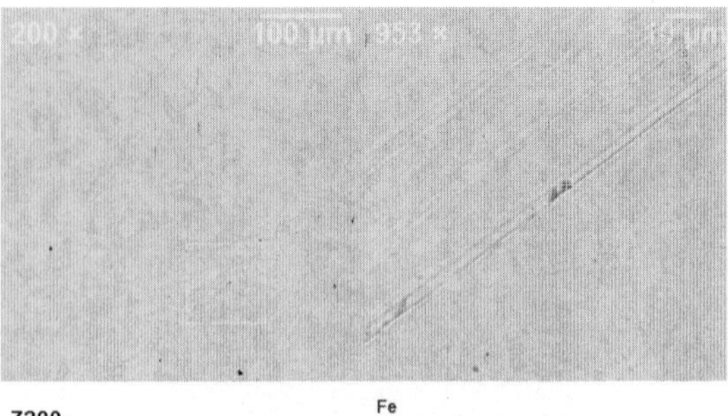

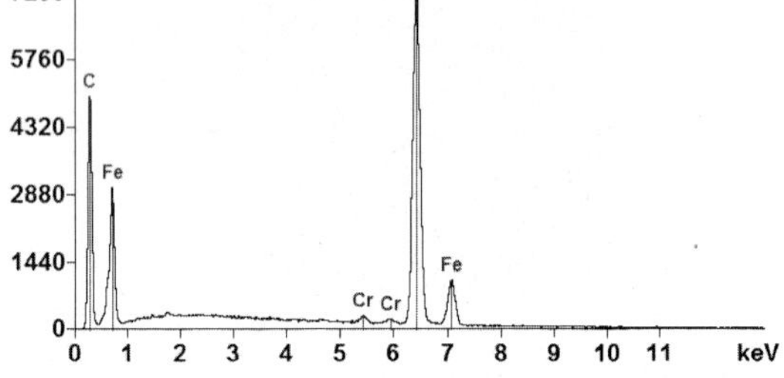

Fig. 8: Image and Spectrum from a contamination on the surface of a polished steel sample.

and determine if it is an inclusion. Based on the chemical composition, we see that the debris is carbon (C). The Iron (Fe) peaks are from the steel matrix). The absence of inclusion type elements allows the analyzer to ignore this feature. Conversely, the instrument could also track this feature for reporting at a later time.

For rapid and accurateidentification of nonmetallic inclusions, the capability to perform rapid X-ray analysis on inclusions with spatial dimensions down into the sub-micron range establishes electron-beam analysis as the premier technique for inclusion classification.

Inclusion Sizing and Detection

Measuring systems critical to process evaluations are typically scrutinized for theirinherent capabilities to accurately and precisely measure objects. In the realm of inclusion analysis, this is often overlooked. Engineers are used to running samples using optical microscopes and a critical assumption is made that they are accurately analyzing a sample. Over the last decade or so, scientists and engineers have become became more conscience to quality and have directed their focus on metrics to evaluate procedures that are routine. The main goal of these projects revolves around streamlining the process and reducing variations that occur.

A routine approach to evaluate any analytical tool is performing multiple analyses on several samples to determine an effective standard deviation of the measurement. This informs the user of how well theequipment is calibrated or the gage of the analysis. The gage is a term used to indicatethe ability of procedure to measure an object repeatedly and reproduce that result.For light optical microscopes, the maximum resolving power is typically only 1 micron. By definition, the resolution of a microscope is determined by the smallest measurement that can be acquired. In an optical microscope, this is the size of one pixel on the sensor array.At 100X, 1 micron represents the extreme capabilities for light optical instruments. This is mainly due to the diffraction limitations of light. For inclusion analysis, the limitation for most analyses is 2 micron; so at best there are only two pixels that represent one inclusion. The inclusions in this size range are usually the most important inclusions for obtaining information about the process. Knowing this the resolving power of light optical systems is limited in the most important region – less than 5 microns. Subsequently, studies have shown that the reproducibility of light optical instruments range significantly. Reasons for such a poor precision are: (1) the accuracy (i.e. the true measure of an object) is poor because it is operating at its limit of resolution – 1 pixel is 1 um, so there is only a 50% chance to accurately measure a feature 1 micron in size; (2) false negative results increase dramatically as you approach the 1 micron=1 pixel resolution limit; and finally (3) the depth of focus is poor, so the reliability of the measurement is also poor.

In the case of the MQA, the useful magnification that can be achieved by a SEM is defined by the size of the focused spot projected on the specimen, and this too is ultimately limited by the diffraction of the focused electron beam. The ultimate resolving power is 7 nm for imaging, and during automation it can be as lowas ~0.030 nm. This simple fact alone means that the reliability of measurements is better by several orders of magnitude. Simply stated, there are more pixels for the smallest of features and as a consequence, a SEM begins to 'run out of steam' for practical imaging of inclusions below about 200 nanometers, whereas an optical microscope has difficulty imaging inclusions below the 5 micron size range.

Because of the drastic difference in resolving mechanisms and capabilities, the SEM provides a vastly superior reproducibility and reliability in its measurements. Figure 9 plots the results from a study where the size of the features were analyzed repeatedly for 3 samples, ran 3 times by 3 different operators. The results illustrate that the %Relative Standard Deviation isonly ~4% of the measurement for inclusions smaller than 10 microns. This high level of precision in the inclusion analysis industry was almost unheard of until the automated inclusion analyzers became available.

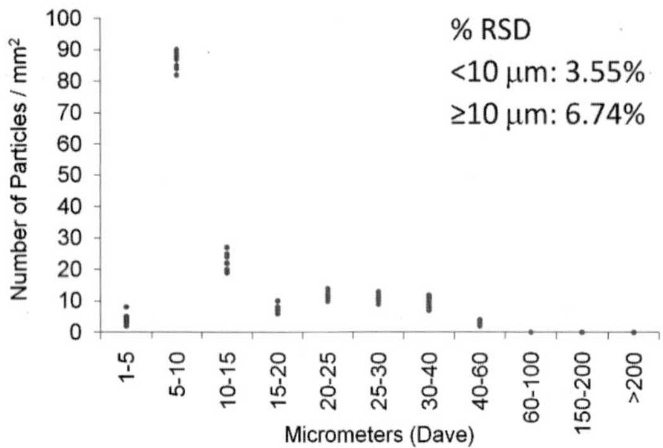

Fig. 9: Percent Relative Standard Deviation (%RSD) for measuring sizes of features.

International Inclusion Rating Standard

International Inclusion ratings are used the World over to standardize the classification of an inclusions. It becomes critical to understand the differences in how the ratings are determined via optical microscopy vs. SEM.

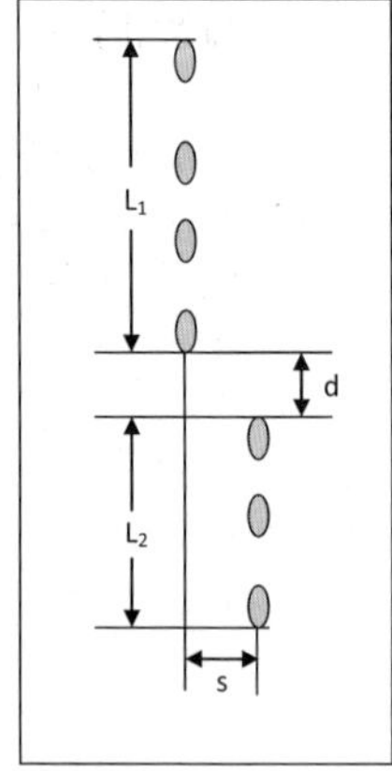

Fig. 10: Stringer length schematic.

$$Stringer\ Length\ (SL) = L_2 + L_1 + d$$
$$d \leq 40\ microns$$
$$s \leq 15\ microns$$

Although the structure and details of the algorithms are almost identical for both a light optical microscope and the MQA, there are some significant fundamental differences between the two types of instruments – specifically in the areas of effectively measuring stringers and very long inclusions.

An optical microscope, by design, requires the acquisition of a field of view image typically 0.5mm² in size. This 'static' image is then analyzed to determine the size of the features that are within that image. Figure 10 outlines one example of how stringer lengths are determined.

Based on these measurements, the requirement of a static image for an optical microscope inherently restricts its analysis to inclusions within one field. In any automated inclusion analysis, there are upwards of 60 fields and each field is rated based on its severity of cleanliness, or inclusion content. If for example (Figure 11), an inclusion (1) extends across two fields (F1 and F2); an optical instrument would use the two static images of the field as independent images. Consequently, the inclusion is counted twice, or not at all – once in field F1 and again in field F2. Each time it is counted, the inclusion will also be incorrectly sized as smaller than it actually is. In another example, if a stringer (2) is present and extends through multiple fields (F2 and F3), an optical microscope will split the stringer into two parts and count them separately – one 2-piece component in F2 and one inclusion in F3.

The biggest virtue of the electron-beam imaging system is the speed and dynamic range over which it can be adjusted. Even sophisticated optical 'zoom' lenses struggle

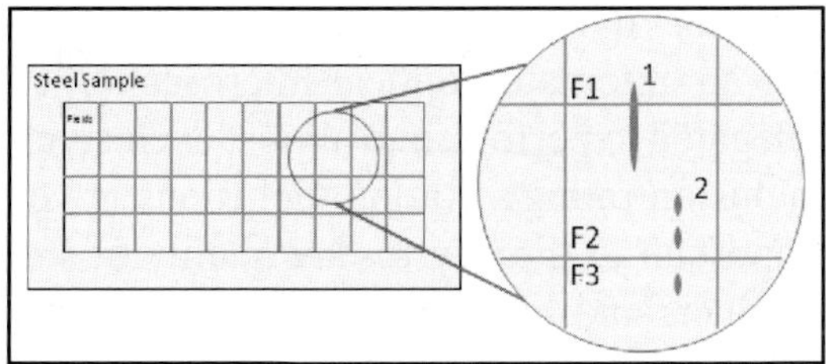

Fig. 11: Schematic of inclusion analysis fields.

to approach a factor of ten in magnification change, whereas any SEM can 'zoom' continuously through at least four orders of magnification change. This can be accomplished virtually easily, without any kind of mechanical adjustment.As a consequence, a truly optimized electron-beam inclusion imaging system, such as MQA, can dynamically adjust to imaging requirements in ways that no optical system can approach.

The MQA does not acquire a staticimage; it rather dynamically controls the beam to search for inclusions. The rating software used by SEM then overlays the fields to the entire x,y locations of the inclusions that were detected. By contrast, the MQA uses the entire array to determine the size of the inclusions. This eliminates the discrepancies seen via optical microscopy as shown in Figure 10. Consequently, inclusion 1 and 2 in Figure 11 can be analyzed correctly.

In Figure 11 - inclusion 1, the MQA uses the electron beam to extend past the border of its fields to analyze the full length of an inclusion regardless of its overlap with multiple fields. The MQA does so with the same precision, and the analyzer even knows which field the inclusion is contained in.

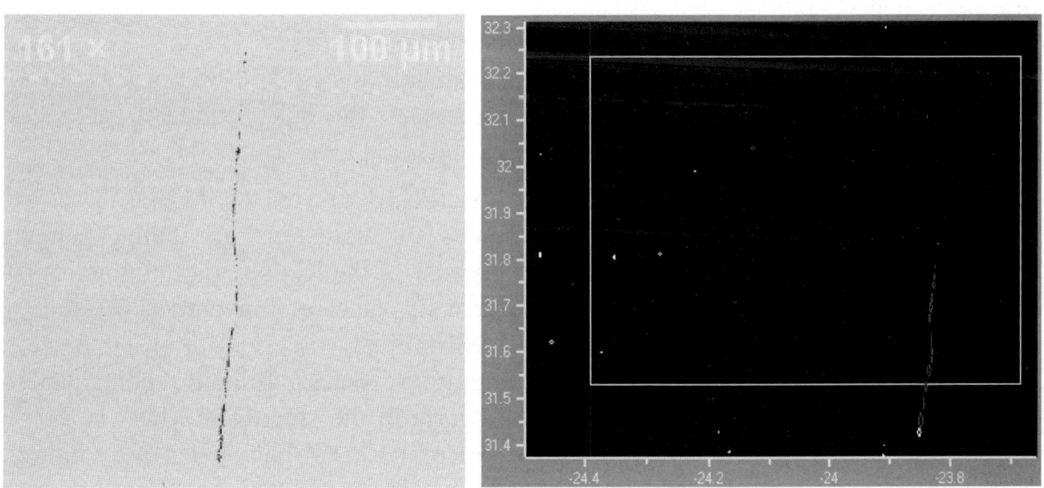

Fig. 12: Long stringer. Left – SEM BSED image, Right – MQA Rating software analysis of the stringer.

Figure 12 illustrates an actual analysis of a stringer that extends across two fields. The inclusions highlighted in RED indicate the total length of the inclusion (*SL*). It is important to note that the MQA also identified a unique chemical phase on the end of the inclusion (shown in WHITE). Rather than being part of the Al-Type inclusion (RED), this unique phase (WHITE dot on the end) was a Si-Type inclusion. In an optical instrument, this inclusion would identify this stringer most likely as a single, longer inclusion, which is inaccurate based on itschemical composition.

Process Related Analysis

The most powerful advantage of the SEM equipped with MQA relates to the overall superior dynamic range of applications. For an automated optical microscope, it is relegated to performing, almost exclusively, inclusion ratings. Rating inclusions is very important to the day-to-day quality of steel products and building customer assurance. What is often overlooked, and is gaining tremendous attention recently, is how quality can be directly impacted by the process further upstream. Figure 13 outlines a basic overview of the steel process. While automated optical microscopes almost exclusively analyze samples from the rolling/finishing side of the process, the MQA adds the flexibility of running process chemistry samples as well (Figure 14). The primary reason optical microscopes are not used for process analysis is because >90% of the inclusions are <5 microns in size. The information revealed by these types of 'chem- samples' (Figure 14) provides invaluable insight into the process of steel making at the ladle, tundish, furnace and caster.

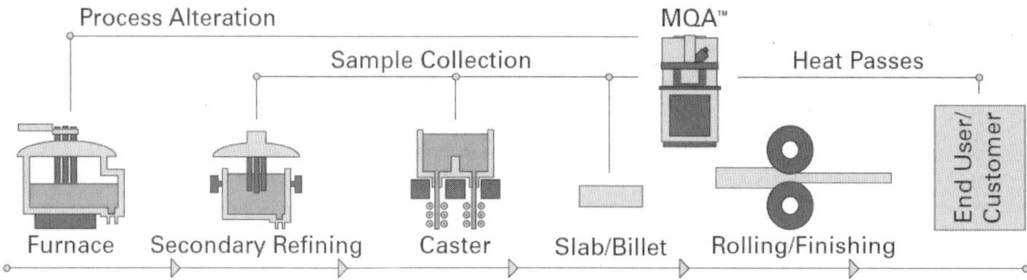

Fig. 13: Overview of Steel Making process.

Specifically, if you would like to see what types of inclusions are being generated because of the Al deoxidation practices you have in place, or how much calcium you need to treat your steel in order to prevent clogging issues? Imagine being able to fingerprint the steel. The MQA is the only tool of its kind that opens the window to these elusive questions. The automated reporting software can be customized to report almost any inclusion parameter. Figure 15 is an example report generated from the analysis of a chem-sample. When it comes to addressing the needs of the steel industry, the MQA was designed to meet those needs – and increase the number ofuseful applications of inclusion analysis.

Fig. 14: 'Chem-Samples' from the secondary refining ladle.

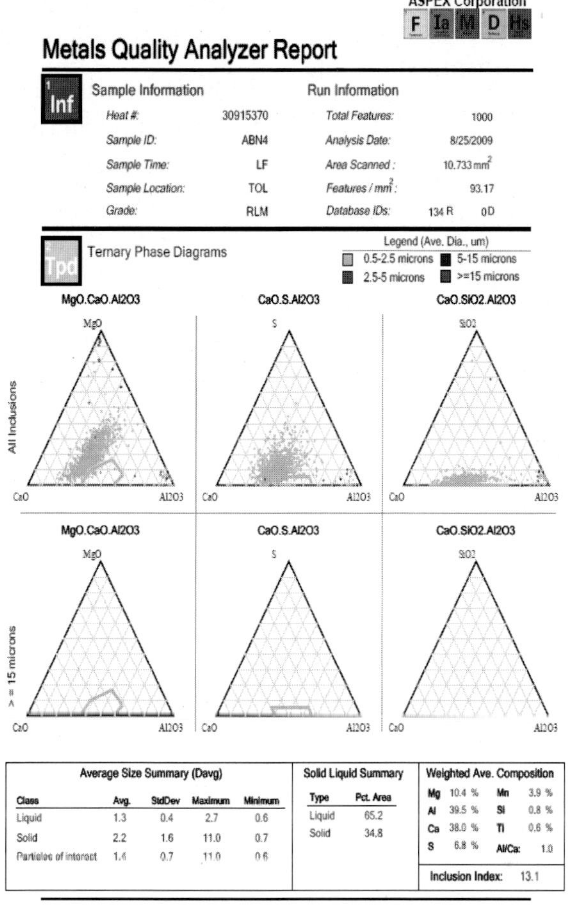

ASPEX Corporation

F Ia M D Hs

Metals Quality Analyzer Report

Inf

Sample Information		Run Information	
Heat #:	30915370	*Total Features:*	1000
Sample ID:	ABN4	*Analysis Date:*	8/25/2009
Sample Time:	LF	*Area Scanned :*	10.733 mm^2
Sample Location:	TOL	*Features / mm^2 :*	93.17
Grade:	RLM	*Database IDs:*	134 R 0 D

Tpd Ternary Phase Diagrams

Legend (Ave. Dia., um)
0.5-2.5 microns 5-15 microns
2.5-5 microns >=15 microns

Average Size Summary (Davg)

Class	Avg.	StdDev	Maximum	Minimum
Liquid	1.3	0.4	2.7	0.6
Solid	2.2	1.6	11.0	0.7
Particles of interest	1.4	0.7	11.0	0.6

Solid Liquid Summary

Type	Pct. Area
Liquid	65.2
Solid	34.8

Weighted Ave. Composition

Mg	10.4 %	Mn	3.9 %
Al	39.5 %	Si	0.8 %
Ca	38.0 %	Ti	0.6 %
S	6.8 %	Al/Ca:	1.0

Inclusion Index: 13.1

Fig. 15: Processed Chem-Sample analysis for inclusions >1 micron in size.

The SEM/EDX with the leatest Silicon Drift Detector (SDD) techology such as the OmegaMax^{TM} detector maximizes the effectiveness of EDX sensors in a SEM, resulting in dramatic gains in sesitivity and speed. The pahse ternary diagrams in Figure16 illustrate the tremendous inmprovement in x-ray data collection efficiency that can be realized with the OmegaMax.

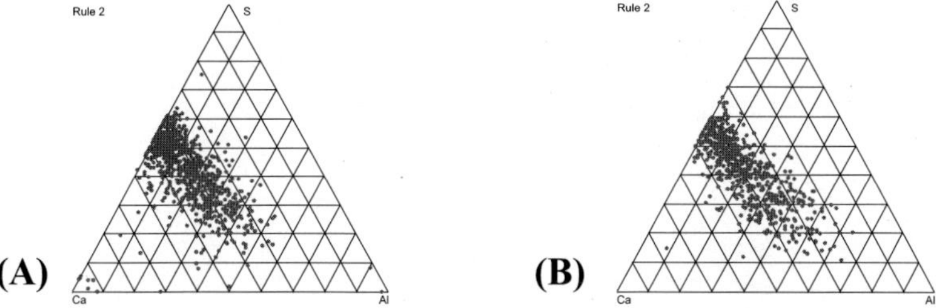

Fig. 16: (A) Data collected on standard 10mm^2 SDD at 1000ms acquisition. (B) Data collect on OmegaMax 5mm^2 at 200ms acquisition. Identical data collected 5 times faster.

CONCLUSION

This perception can be attributed to the fact that basic optical microscopes are fundamentally simple mechanical devices with centuries of evolution behind them, whereas the SEM is a relatively recent electronic development which has not, until recently, significantly evolved from its research-laboratory roots.

When comparing the fundamental operation of optical microscopes vs. automated SEM one clearly understands the unique advantages the SEM brings to automated inclusion analyses.

Fatigue Failure of an Expansion Bellow of a Blast Furnace due to Improper Welding: A Case Study

Pankhuri Sinha, Hrishikesh R. Jugade, Goutam Mukhopadhyay and Sandip Bhattacharyya
R&D and Scientific Services, Tata Steel Limited, Jamshedpur-831 001, India
Email: pankhuri.sinha@tatasteel.com

ABSTRACT

Failure analysis of an expansion bellow of a blast furnace has been presented. Expansion bellows are the wavy or convoluted tubes which are designed for high flexibility when subjected to loads, internal pressure or bending moments. The base metal of expansion bellows is specified considering the movement, pressure, temperature, service life and corrosion resistance required for the particular application.

One of the expansion bellows used in the cold blast transmission line of a blast furnace failed catastrophically within a year of its installation. Failure analysis of the failed expansion bellow includes visual observation, macro- and micro-structural analysis, fractography, chemical testing and hardness measurements. The location of primary failure of the component was identified as the weld joint of the bellow and hinge.

Visual observation of the fracture surface of component reveals beach marks indicating fatigue mode of crack propagation. Macro- and micro-structural examination across the location of failure reveals regions of under-fill indicating improper welding. These under filled regions have acted as stress concentrators resulting in easy crack generation sites. Once the crack nucleated at under filled region, it propagated while the normal functioning of the bellow, resulting in its failure within a year of operation.

This paper presents the detailed root cause failure analysis of the expansion bellow revealing weld defect due to improper welding practice as the cause of failure.

KEYWORDS: *Expansion Bellow, Fatigue, Improper Welding.*

INTRODUCTION

Blast furnace that is used for iron-making is basically a counter-current reactor where iron ore is processed at ambient temperature, in presence of coke and other carbonaceous material as fuel, to produce liquid iron. The raw materials, iron ore and coke, are charged at the top of the furnace while pre-heated air or blast enters through the lower part of the furnace. Atmospheric air which is sucked in at the blower house moves through the cold blast transmission line into the stove where it is pre-heated prior to being forced into the furnace at a high pressure. The product of blast furnace i.e. liquid iron is chiefly used as raw material for steelmaking.

The cold blast transmission line of a blast furnace is typically a large diameter pipe with a number of thermal expansion components. All materials get subjected to dimensional variations on thermal exposures. In case of piping systems these dimensional changes due to thermal variations can produce excessive stresses, either tensile or compressive, throughout the piping system and at certain fixed points such as the joints within the system. A practical and economic means of achieving the piping system flexibility in a compact design is through the application of expansion joints/ bellows. Expansion bellows are the wavy or convoluted tubes which are designed for high flexibility when subjected to loads, internal pressure or bending moments.

One of the expansion bellows used in the cold blast transmission line of a blast furnace and having a diameter of 1800 mm, failed catastrophically within a year of installation. It was a hinge type expansion bellow. It consisted of two hinges welded to the body of the bellow at diagonally opposite positions as illustrated in Fig. 1. These hinges were fixed to an adjacent support. Function of the hinges was to restrict the movement of the bellow to a particular plane, i.e. the axial plane. The failure resulted in complete removal of the hinge from the set-up resulting in catastrophic breakdown of the system.

The failure adversely affected the entire operation of the blast furnace leading to unplanned outage of the furnace along with damage to the associated equipments. Primary aim of this investigation is to find out the root cause of failure of the component and to suggest preventive measures so as to avert such failures in future.

Fig. 1: Overall view of the failed expansion bellow

INVESTIGATION

Visual Observation

General view of the failed expansion bellow is shown in Fig. 2a. Convolute of the bellow got flattened and damaged as a result of failure which can be seen in Fig. 2a. The location of primary failure i.e. crack initiation is identified to be the weld joint between the bellow and the hinge as depicted in Fig. 2b. Visual observation of the location of failure reveals cracks originating at the weld zone as shown in Fig. 2c. Fracture surface of the failed component is shown in Fig. 2d while its closer view is shown in Fig. 2e. The fracture surface reveals different color layers separated by well distinguished boundaries or beach marks. This signifies progressive crack propagation indicating fatigue failure of the component.

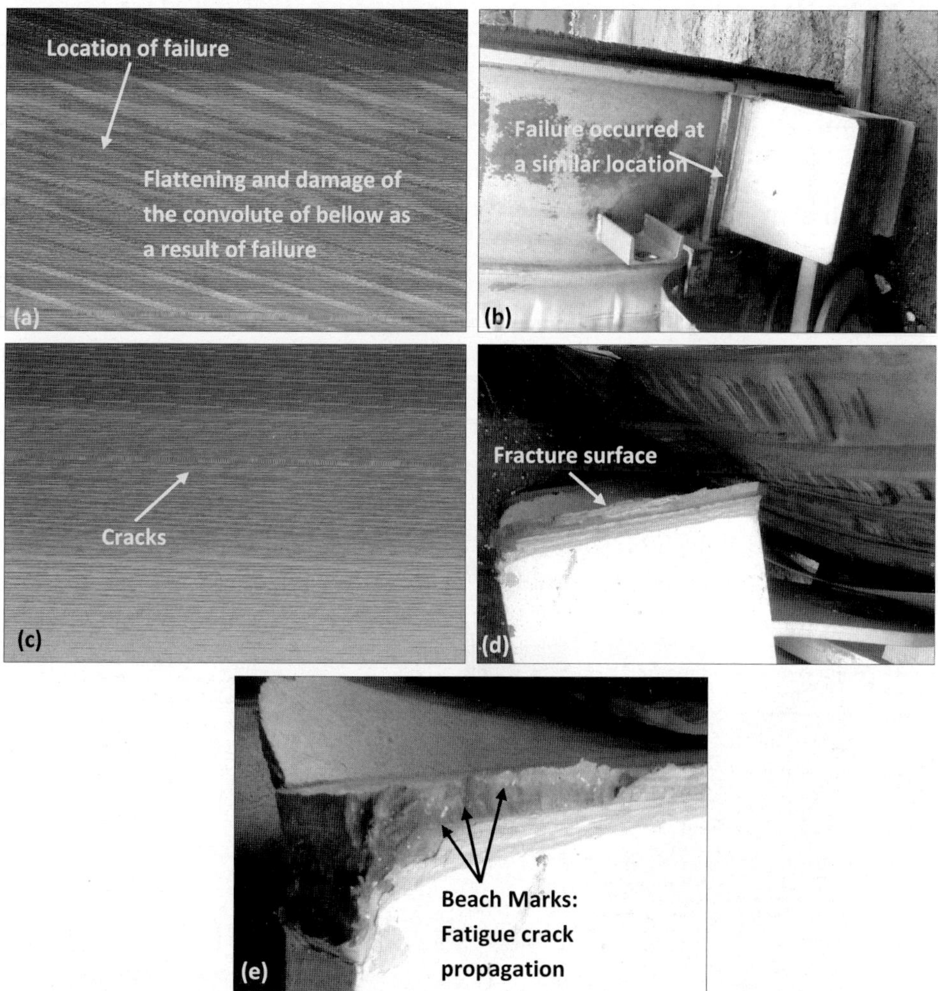

Fig. 2: (a) General view of the failed expansion bellow showing location of failure; (b) Weld joint between the bellow and hinge, failure/crack initiation occurred at a similar joint. (c) Cracks at the weld joint between the bellow and the hinge; (d) Fracture surface of the failed component; and (e) Closer view of the fracture surface showing signature of fatigue.

Chemical Analysis

Chemical analysis of the samples of bellow and hinge is carried out using wet analysis; carbon (C) and sulfur (S) contents are determined using combustion infrared technique. The obtained chemical analysis is compiled in Table 1. It is revealed that the bellow as well as the hinge is made up of same material which is Fe 410 grade, IS 2062, of hot rolled low, medium and high tensile structural steel [1].

Table 1: Chemical analysis (wt.%) of the bellow sample.

Sample	C	Mn	S	P	Si	V	Al	Ti	Cr	Mo	Cu	Ni	Nb
Bellow and Hinge	0.15	1.08	0.0139	0.028	0.215	0.0045	0.044	0.0034	0.051	0.0029	0.0145	0.017	0.0021
Fe 410 (IS 2062)	**0.23 max**	**1.50 max**	**0.045 max**	**0.045 max**	**0.40 max**								

Macro-structural examination

A number of samples are cut across the weld zone between the bellow and the hinge to examine the nature and soundness of welding. Digital photographs of the as received samples as well as the macro dual zoom images of the etched micro-samples (micro-samples preparation is described in the subsequent section) are shown in Fig. 3. Digital photographs of the samples reveal regions of under-fill at the root of the weld as shown in Fig. 3a and 3b indicating incomplete depth of penetration of the weld. The macro dual zoom images of the etched micro-sample as captured by an optical microscope (LEICA DMRX) are shown in Fig. 3c and 3d. These also expose the under filled regions at various locations of welding. A crack initiating from the under filled region and propagating through the HAZ is revealed as shown in Fig. 3d.

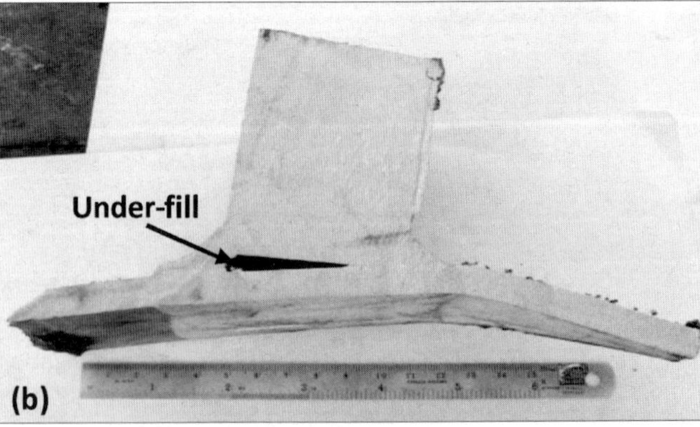

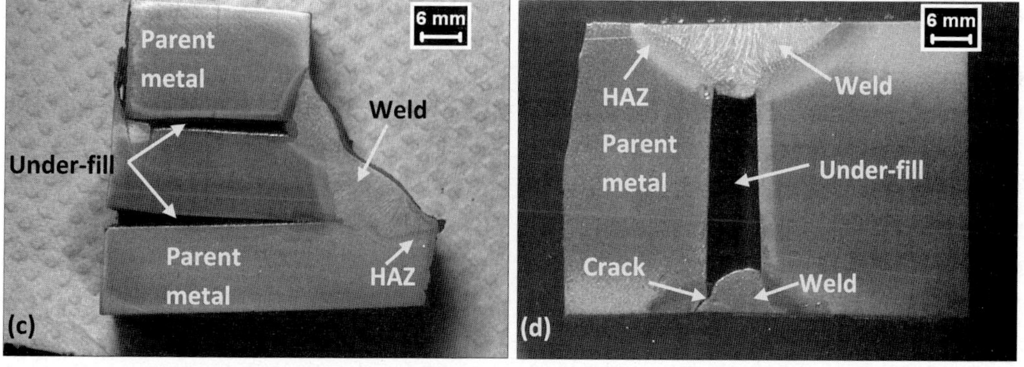

Fig. 3: (a and b) Digital photographs of the sample across the weld joints; (c and d) Macro-dual zoom images of the etched micro-samples.

Micro-structural examination

Micro-specimens are prepared from the samples collected across the weld zone between the bellow and the hinge for microstructural examination so as to explore the presence of abnormality, if any. The specimens are mounted with resin, ground and polished using standard metallographic techniques. The polished specimen is etched with 3% Nital solution to reveal the microstructures. The optical microstructure of the parent material, heat affected zone or HAZ and the weld bead as revealed in an optical microscope (LEICA DMRX) is shown in Fig.4. While the parent material exhibits a banded ferrite-pearlite matrix (Fig. 4a), HAZ consists of a predominantly ferrite matrix (Fig. 4b). The weld zone has elongated grains with widmanstatten ferrite at the grain boundaries (Fig. 4c).

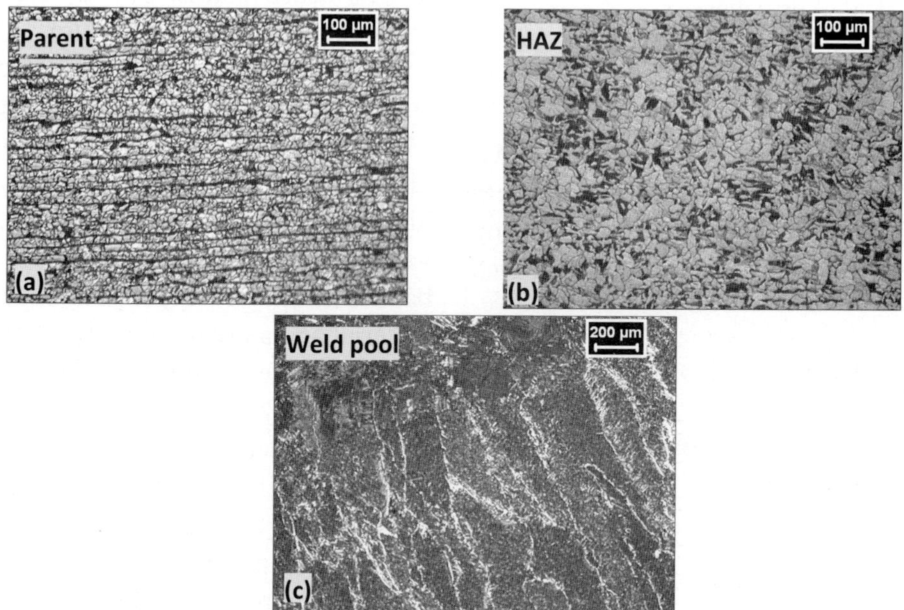

Fig. 4: Optical microstructures of (a) Parent material; (b) HAZ and (c) Weld bead.

Fractography

A specimen containing a portion of fracture surface is cut for fractographic examination. The specimen is firstly cleaned with tartaric acid followed by ultrasonic cleaning. It is thereafter observed in a scanning electron microscope (INCA 7426) operated at an accelerating voltage of 15.0 kV. Striations are clearly observed at a magnification of 4000X, as shown in Fig. 5, confirming fatigue.

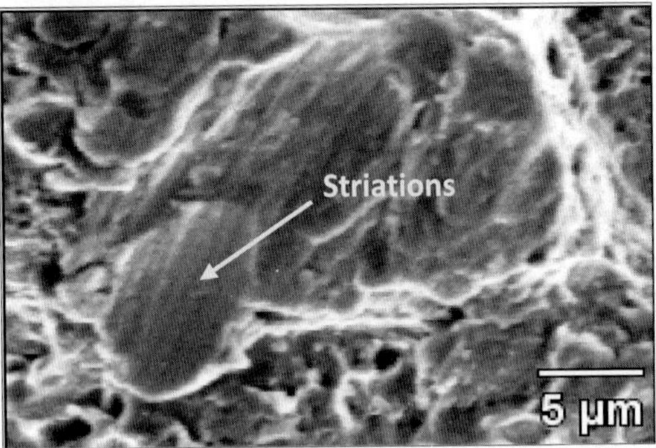

Fig. 5: Fractographic image showing striations.

Measurement of hardness

Hardness values of the parent material, HAZ and weld bead are measured with the help of a Vickers hardness testing machine using 30 kg weight. The results are summarized in Table 2.

Table 2: Hardness values of the samples.

Sample	Hardness (HV 30)
Parent metal	176, 180
HAZ	277, 283
Weld bead	248, 249

THE ANALYSIS

An expansion bellow used in the cold blast transmission line of a blast furnace failed disastrously in service within a year of installation. It was a hinged type bellow and had two hinges welded to it at diagonally opposite positions as shown in Fig. 1a. These hinges were used to fix the bellow to an adjacent support so as to restrict its operation in the axial plane which was the operational requirement. The component was designed to operate at temperatures within 300°C and pressures within 6 bar. At the time of failure, the operational parameters were reported to be well within the specified limit. The measured operational temperature and pressure were 240°C and 4.5 bar respectively.

Visual observation of the location of failure reveals that failure initiated at the weld joint between the bellow and the hinge as shown in Fig. 2b. Fracture surface of the component reveals beach marks as illustrated in Fig. 2d indicating fatigue failure. Chemical analysis of the samples reveals the bellow and hinge material to be Fe 410 grade of IS 2062 of hot rolled low, medium and high tensile structural steel as tabulated in Table. 1. Optical microstructures of the various zones as shown in Fig. 4 reveals a banded ferrite-pearlite matrix of the parent material and a predominantly ferrite matrix of HAZ. Weld zone exhibits elongated grains with widmanstatten ferrite which is typical of a weld pool. Measured hardness values which are tabulated in Table. 2 are found to be satisfactory for the obtained composition and microstructure as well as the application. It could be inferred that there were no material related abnormality/

inferiority in the component. Fractography reveals striations as shown in Fig. 5, confirming fatigue failure of the component.

Manual metal arc welding (MMAW) process had been employed to weld the hinges to the bellow. In a MMAW coalescence of metals is produced by heat which comes from an electric arc that is maintained between the tip of a covered electrode and the surface of the base metal in the joint being weld [2], which is the bellow in this case. To examine the nature and soundness of weld, a number of samples are cut across the weld region of the bellow and the hinge. Digital images of the as-received samples and macro-dual zoom images of the etched micro-samples (Fig. 4) reveal regions of under-fill at the root of the weld. This indicates incomplete penetration since the weld bead did not start at the root of the weld groove. Such incomplete penetration is detrimental as it forms channel and crevices in the root of the weld which acts as stress concentration site (notch effect). Crack can easily nucleate at these sites as in the present case which is highlighted in Fig. 3d. This type of defects occurs when the welding procedures are not strictly followed. Most probable causes of such defects include the current setting, arc length and electrode angle manipulation, etc. [3].

Above observations suggest that the fracture occurred by cyclic crack growth that initiated at the root of the weld, due to incomplete penetration of the weld bead, resulting in regions of under-fill. These under filled regions provided notch effect thereby acting as easy crack nucleation site. Once the crack initiated at these stress concentration sites, it propagated under normal operating condition resulting in failure of the component within a year of service.

CONCLUSION

From the investigation it is inferred that the component failed due to poor workmanship in welding.

RECOMMENDATIONS

To avoid such failures, proper and stringent welding practice must be ensured. The welded structure must be subjected to non destructive evaluation techniques of flaw detection like ultrasonic testing or radiography prior to installation in service.

REFERENCES

[1] Indian Standard for hot rolled low, medium and high tensile structural steel; IS 2062; 2006.

[2] The metallurgy of welding; D Sefferian; Champman and Hall; London; 1962.

[3] Pipe welding procedures; Rampaul; Industrial Press; 2003; pp. 3141-3148.

SECTION IX

Process Optimization

Improvement in Mechanical Property Distribution of High Strength Steel

Manashi Adhikary, Nikhil Ramakrishnan, Sundara Bharathy, S. Ray,
Avtar Singh, Vinay Mahashabde and T. Venugopalan
TATA Steel
Email: manashi.adhikary@tatasteel.com

ABSTRACT

With India emerging as a major hub for automotive companies in recent times, demand of locally sourced High Strength Steel meeting global standards is on the rise. With current product portfolios in India, most Indian Steel companies are manufacturing high strength steel at ~440 MPa tensile strength. However in India, currently these steels are produced through batch annealing route rather than conventional continuous annealing route. The challenge lies therefore to produce high strength steels with consistent mechanical properties. High strength steels, mostly are micro-alloyed, needs special process control especially through batch annealing route to achieve consistency in mechanical properties for intra and inter coils.

This paper illustrates the effects of thermo-mechanical and thermal parameters affecting the microstructure and therefore the end properties of high strength steel. The work done in one of the 440 MPa high strength steel produced through batch annealing route will also be explained. While the development of the grade was completed, the yields were lower to the tune of 60-65%. Even at these yields, there were associated bottlenecks in the process. Improving the consistency in mechanical properties and thereby improving the yields while debottlenecking the process constraints was the motivation of the work.

Microstructure characterization of hot rolled coil ends *vis-a-vis* mechanical properties, experiments with hot rolling cooling strategy, assessment of properties under different annealing cycle and position of coil in the BAF, studying the impact of hot strip mill process parameters, were carried out to arrive at the optimized process norms for the grade and thereby improve mechanical properties. With the current work, the authors could establish a yield to the tune of ~80-85%.

KEYWORDS: *440 MPa, High Strength, Batch Annealing, Microstructure*

INTRODUCTION

High strength micro-alloyed 440 MPa steel is one of the cold rolled product variant produced in TATA Steel. Achieving uniform strength through batch annealing route in this grade was a challenge. The authors looked for correlation with microstructures and end properties. It was observed that obtaining uniformity in microstructure in addition to rationalization of Annealing cycle and chemistry was the key success factor.

The work was initiated based on the observation of cold rolled coils having a gradual difference in mechanical properties at Head end (HE) and Tail end (TE) of the finished coils. Sometimes the Tail end of the coil could not reach the minimum of the specification of 440 Mpa tensile. This led to subsequent salvaging of the coils and hence internal rejections. The authors looked for the reason behind such difference in properties and the details of the work is explained further below.

EXPERIMENTAL WORK

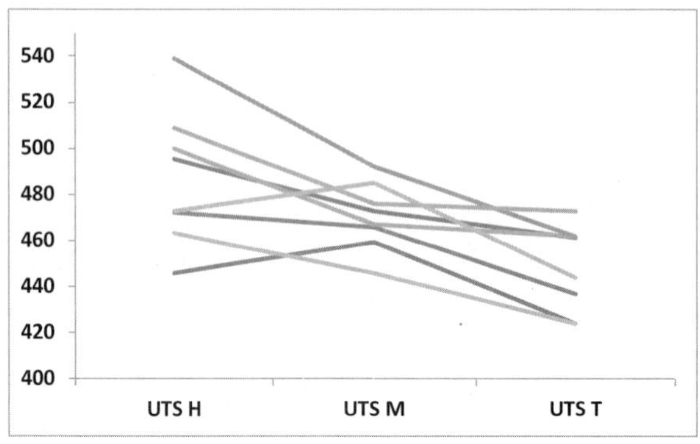

Fig. 1: Variation in tensile strength at Head (H, Middle (M) and Tail (T) end of the coils

The observation of test results of the cold rolled coils showed difference between Head and Tail end of the coils. The above graph indicates the tensile property at the three locations along the length of a coil (eight numbers of coil shown as illustration), against the target of minimum 440 MPa tensile strength (Fig. 1).

Microstructure study of few coils did not reveal any perceptible difference except in grain size difference to the tune of few 10-20 microns. See Fig 2

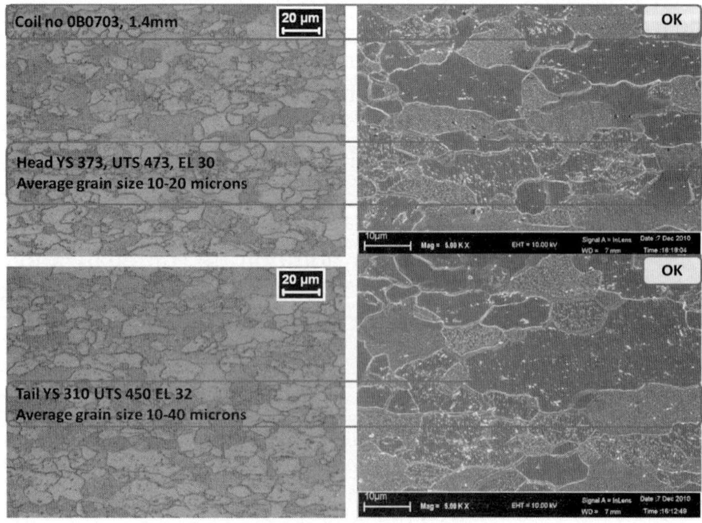

Fig. 2: Microstructure at HE and TE of a coil showing minor grain size variation of 10-20 microns

Effect of coil thickness

The rejection on account of property issues were mostly found in thinner coils (Table 1)

Table 1: % Failed coils thickness wise due to property variation

Thickness	% Failure
0.8	67
0.9	0
1	54
1.2	28
1.4	15
1.6	0
1.8	0
2	0

This was correlated with the difference in hot and cold spot temperatures (Delta -T) that these coils were subjected to.

It was found that

Delta - T >100 deg – Higher % failures

Delta - T <100 deg – Lower % failures

This Delta - T has a direct correlation with uniformity of properties within a coil. Lower the Delta - T during batch annealing lower is the difference in properties within a coil.

Effect of coil position in a stack

Table 2: Thickness wise mapping of failure at TE of the coil due to property lower than 440 MPa vs Coil position in Batch annealing furnace

Failure/ TE	0.784		1	1.176		1.372	1.4		1.6			2	Total
Position 1			2	1					1			1	5
Position 2			1				1						2
Position 3			1			1							2
Position 4	3		1										4
Position 5				1									1
Total	3		5	2		1	1		1			1	14

Processed	0.784	0.98	1	1.176	1.2	1.372	1.4	1.568	1.6	1.764	1.96	2	Total
Position 1			2	4	2	1	3	7	1	1	1	2	24
Position 2			3	4	4		5	4		1			21
Position 3	2	4	8	2	6	3	2	4	2				33
Position 4	4	7	8		5	3	3	2	2				34
Position 5			6	1	2		2				1		12
Total	6	11	27	11	19	7	15	17	5	2	2	2	124

% Failure	0.784	0.98	1	1.176	1.2	1.372	1.4	1.568	1.6	1.764	1.96	2	
Position 1			100	25	0	0	0	0	100	0	0	50	21
Position 2			33	0	0		20	0		0			10
Position 3	0	0	13	0	0	33	0	0	0				6
Position 4	75	0	13		0	0	0	0	0				12
Position 5			0	100	0		0				0		8

Analysis of failure of the coils due to lower properties at TE was done w.r.t coil position within a stack (Table 2). Although there was no strong correlation between thickness and position, Position- 1 seemed to be critical for 1mm. Position- 1 is the bottom- most coil in an annealing stack.

Subsequent trial in thinner section (<=1mm) with all stack position and reduced Delta- T (<100 deg c)

A trial was initiated with thinner section (<=1mm) to understand the effect of reduced Delta- T and stack positions. Following were the results obtained (Table 3)

Table. 3: Trial in 1mm with 90 deg Delta - T

Coil	Sec	Width	Position	YS H	YS M	YS T	UTS H	UTS M	UTS T	EL H	EL M	EL T
324040	1	1200	3	329	340	340	458	460	449	34	32	33
324030	1	1200	4	360	319	338	469	447	448	30	34	35
351150	1	1200	5	334	344	339	458	459	454	31	31	30
366240	1	1200	3	335	345	342	455	463	457	30	32	32
351160	1	1200	2*	325	324	330	444	440	444	35	34	34

Except one coil at the boundary of tensile, the results were favorable. The results indicate the H,M, T end values w.r.t YS, UTS and El

Microstructure of the above study in comparison to normal annealing cycle is shown in Fig. 3-6 for two cold rolled coils

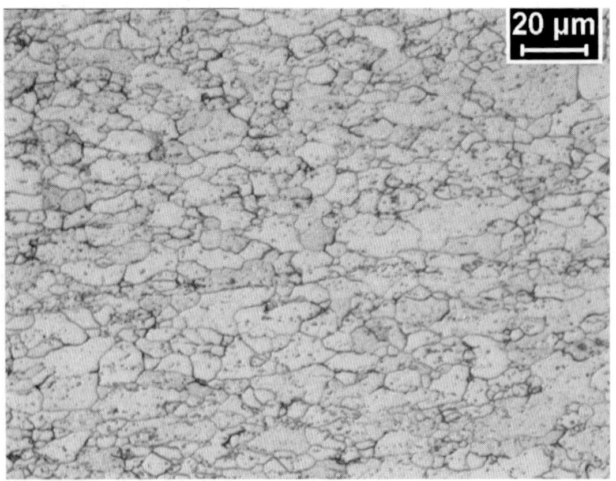

Fig. 3: HE with 110 deg Delta - T (coil no 032399)

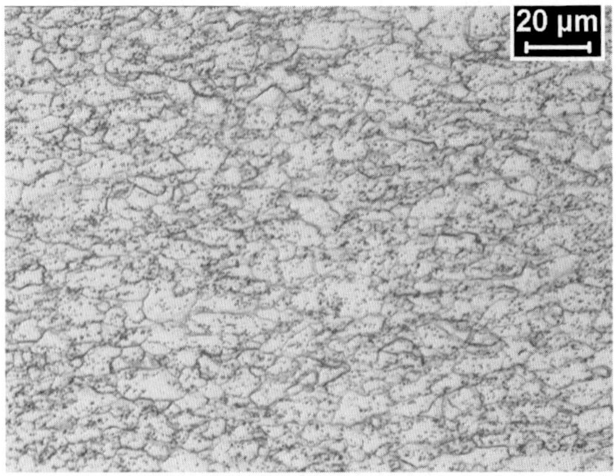

Fig 4: HE with 90 deg Delta - T (coil no 032403)

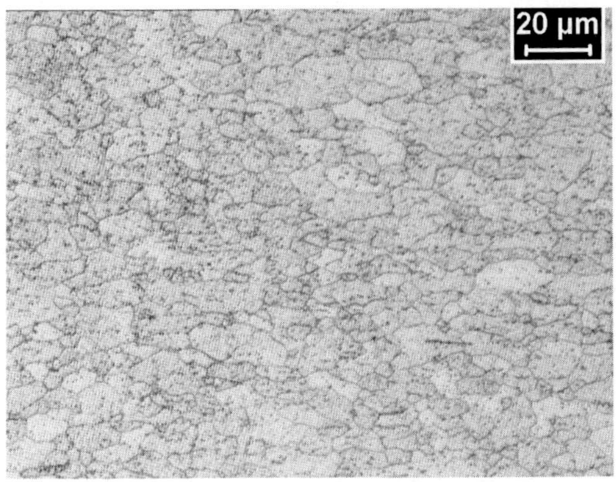

Fig 5: TE with 110 deg Delta- T (coil no 032399)

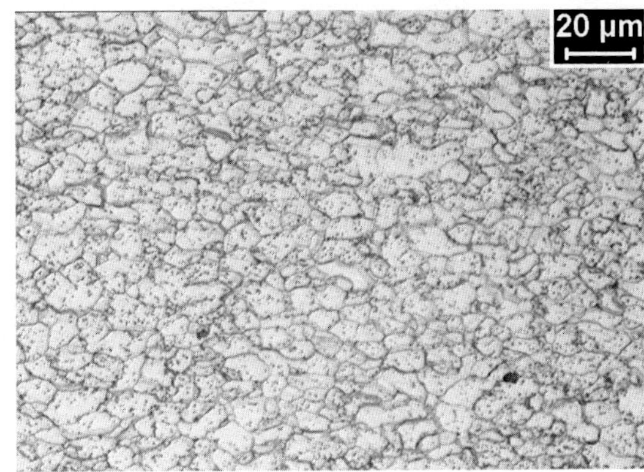

Fig 6: TE with 90 deg Delta - T (coil no 032403)

Grain size in 110 deg Delta - T was found to vary marginally more from head to tail end w.r.t to 90 deg Delta - T

For other thicknesses, the difference in body and core temperatures were analyzed through a 'HEAT MOD' simulator in BAF and no perceptible difference were obtained w.r.t to coil position; however there were few favorable stack positions w.r.t thickness (Table. 3). Trials were carried out in these favorable stack positions. Later the same was extended to a full stack of the grade through extensive trials and property analysis.

Table. 3: Favorable stack positions w.r.t thickness of coil

Thickness (mm)	Favorable Stack positions	Coils investigated
0.9	4	2
1.2	1,2,3	11
1.4	1,4	8
1.6	4	2
1.8	1	1

(Criteria for OK coils: UTS >=440 MPa)

Effect of input properties from Hot rolled coil

The phenomenon of drop in tensile was further correlated with input hot rolled coil properties. It was observed that the tail end of HR coil which was having highest tensile strength was prone to the maximum drop in tensile post cold rolling and annealing (Fig. 7). This was mapped w.r.t to the process of unwinding and rewinding of the coils beginning hot rolling (Table 4)

Table. 4: Mapping of coil ends of CR with upstream

Coil	HR Coil	PLTCM Exit	ECL Exit	BAF Exit	SPM Exit	RCL exit CR sample
Outer Dia	Tail/OD	Head/ID	Tail/OD	Tail/OD	Head/ID	Tail
Inner Dia	Head/ID	Tail/OD	Head/ID	Head/ID	Tail/OD	Head

An intermediate recoiling process was incorporated on a trial basis to understand the effect in the tensile variation from head to tail end of the coils (Table. 5).

Table. 5: Introduction of Intermediate coiling step

Coil	HR Coil	CDL exit	PLTCM Exit	ECL Exit	BAF Exit	SPM Exit	RCL exit CR sample
Outer Dia	Tail/ OD	Head/ ID	Tail/OD	Head/ ID	Head /ID	Tail/ OD	Head
Inner Dia	Head /ID	Tail/OD	Head/ ID	Tail/OD	Tail/ OD	Head /ID	Tail

All coils in-spite of intermediate recoiling showed the same orientation of slope, however the steepness decreased and except one coil (TE) (30 coils were tried), all coils were well above 440 MPa tensile strength (see two such graphs below, Fig. 8 and Fig. 9). This indicated that both Hot rolled as well as annealing was the reason of the variation of the tensile strength within a coil.

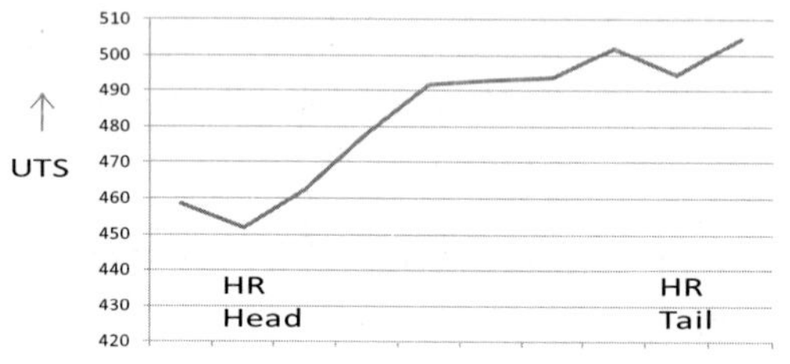

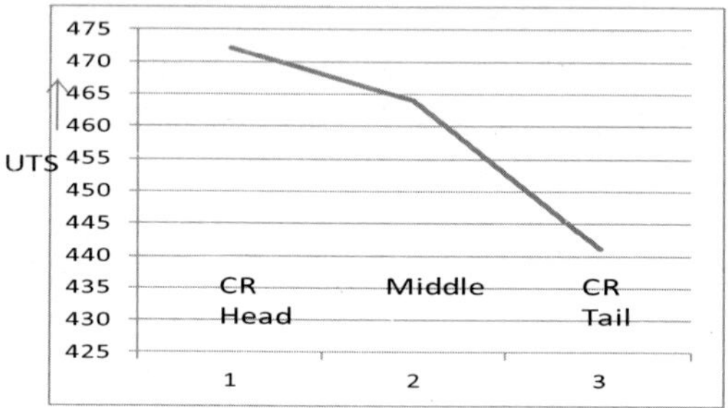

Fig. 7: Mapping of cold rolled coil tensile strength variation w.r.t hot rolled tensile strength variation

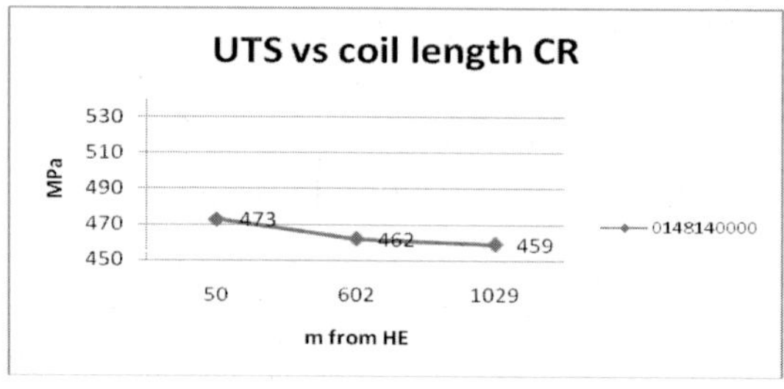

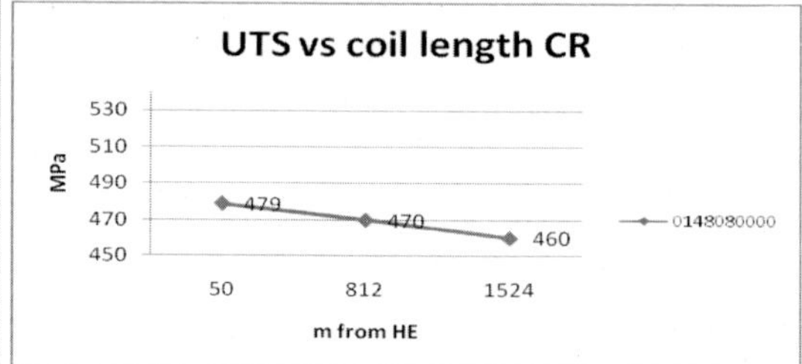

Fig. 8 and Fig. 9: Result of introduction of intermediate coiling step on tensile strength variation

Microstructure study of hot rolled coils from coil ends revealed the following

Microstructure vs HR properties of Hot rolled head end (Fig. 10)

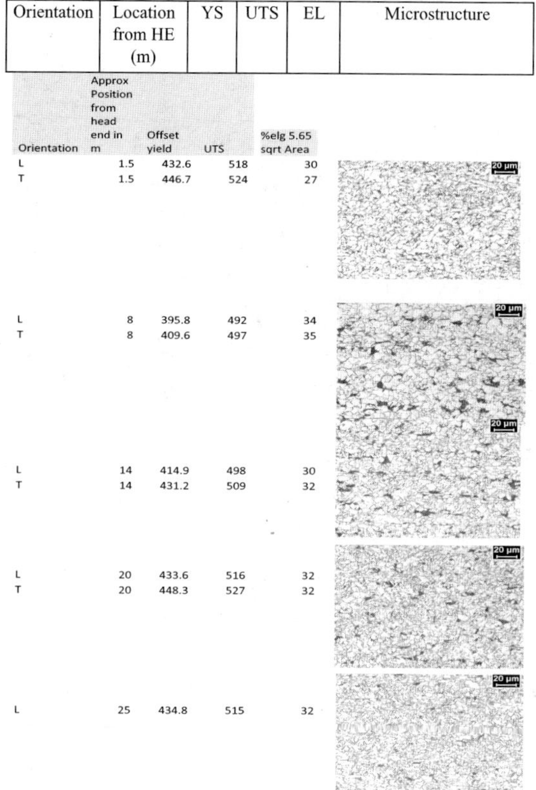

Orientation	Location from HE (m)	YS	UTS	EL	Microstructure
Orientation	Approx Position from head end in m	Offset yield	UTS	%elg 5.65 sqrt Area	
L	1.5	432.6	518	30	
T	1.5	446.7	524	27	
L	8	395.8	492	34	
T	8	409.6	497	35	
L	14	414.9	498	30	
T	14	431.2	509	32	
L	20	433.6	516	32	
T	20	448.3	527	32	
L	25	434.8	515	32	

Fig. 10: Microstructure of hot rolled coil ends at different lengths from HE of the coil showing mechanical properties and microstructure differences

Perceptible difference in microstructure was observed lengthwise in hot rolled coils. Illustrative micrographs shown for HE and TE of Hot rolled coil at every ~5m of the coil ends.

The coiling temperature at either ends of hot rolled coil was as below (Fig. 11)

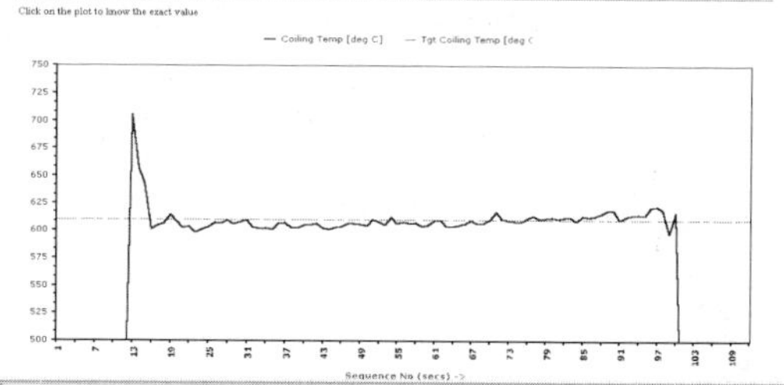

Fig. 11: Coiling Temperature pattern from HE to TE of hot rolled coil

Microstructure vs HR properties of hot rolled tail end (Fig. 12)

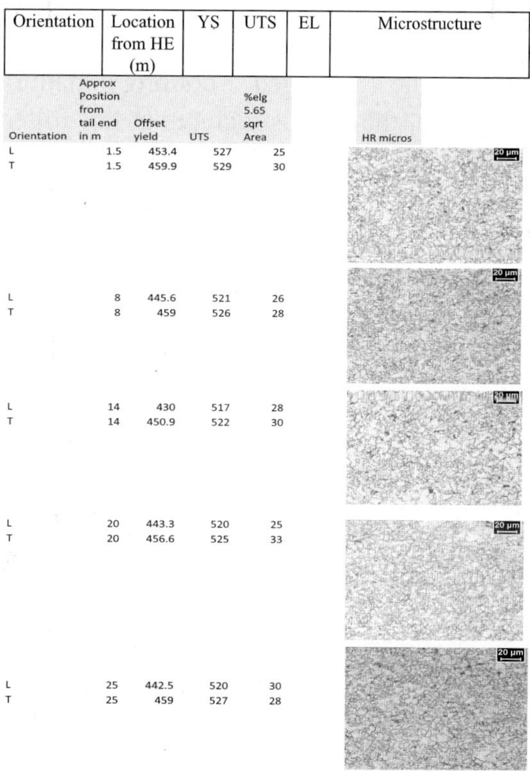

Orientation	Location from HE (m)	YS	UTS	EL	Microstructure
Orientation	Approx Position from tail end in m	Offset yield	UTS	%elg 5.65 sqrt Area	HR micros
L	1.5	453.4	527	25	
T	1.5	459.9	529	30	
L	8	445.6	521	26	
T	8	459	526	28	
L	14	430	517	28	
T	14	450.9	522	30	
L	20	443.3	520	25	
T	20	456.6	525	33	
L	25	442.5	520	30	
T	25	459	527	28	

Fig 12: Microstructure of hot rolled coil ends at different lengths from TE of the coil showing mechanical properties and microstructure differences

Trial with changed pattern (U type) of coiling temperature

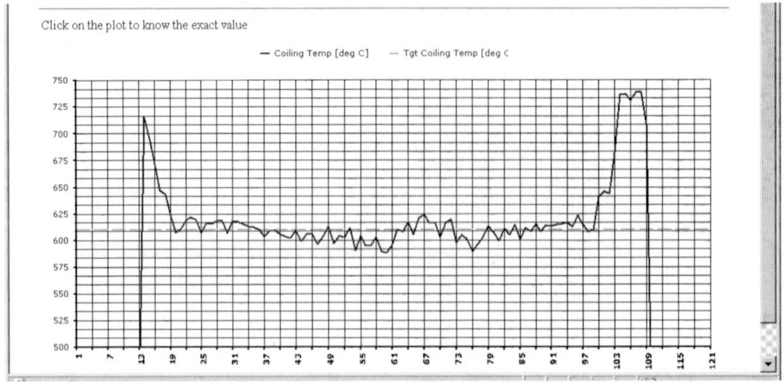

Fig 13: Trial with changed coiling temperature pattern from HE to TE of hot rolled coil

With the changed cooling pattern (Fig. 13), the microstructures at hot rolled stage at either ends was found to be similar.

With both the changes of reduction of Delta - T and change in coiling temperature pattern, the % of coils with >20 MPa within coil variation reduced from 20% to 11% (Fig. 14)

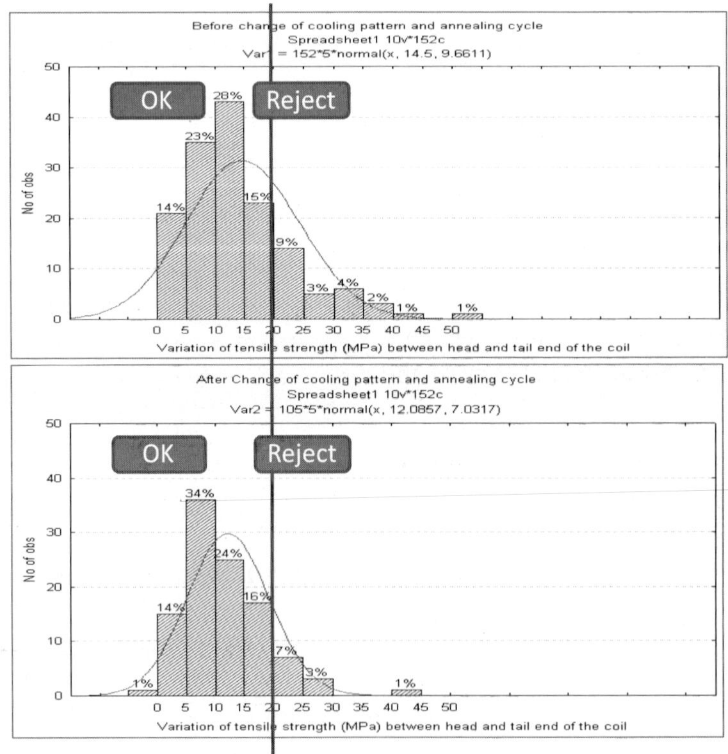

Fig. 14: Effect of change in coiling pattern and Delta –T on entire range of coil

Effect of other hot rolled parameters on cold rolled tensile properties

It was also observed during cold rolled coil property analysis that lower slab retention time and higher finish rolling temperatures also led to higher variation of tensile in cold rolled coils in-spite of 'No or Less' correlation among hot rolled coil properties.

CONCLUSION

The above study indicates that consistency in mechanical properties in high strength steels (440 MPa) is dependent majorly on the following factors:

1. Higher difference in hot spot and cold spot temperature during annealing impacts the microstructure and leads to higher within coil variation.
2. Hot rolled coil microstructure has an impact on the cold rolled coil microstructure.
3. Coiling temperature pattern post hot rolling impacts the final cold rolled and annealed coil properties.
4. Hot rolling parameters like furnace retention time and temperature and finish rolling temperature may have an impact on the end properties in cold rolled coil.

ACKNOWLEDGEMENTS

The authors would like to specially thank LD2#SC, HSM, CRM including BAF operations for their agreement with the process and product design. Special acknowledgement is rendered to Mr Kishore K Ojha of BAF, Cold Rolling Mill and Ms Arthita Dey of Metallography Laboratory R&D Scientific Services for their continued support on the annealing trials and microstructure front.

REFERENCES

[1] *Cold rolled HSLA sheet and strip products*, Bernhard Engl, ThyssenKrupp Stahl AG, Dortmund, Germany

[2] *The challenge to produce consistent mechanical properties in Nb- HSLA strip steels*, J K Patel, B Wilshire

Determination of Hot Rolling Parameters for CRNO Steel at Rourkela Steel Plant

P.P. Sengupta, S. Rath and Vinod Kumar
Research & Development Centre for Iron & Steel, Steel Authority of India Limited, Ranchi
Email: ppsengupta@sail-rdcis.com

ABSTRACT

Investigations on the process parameters of M47 grade rolling at Hot Strip Mill (HSM) backed up by theoretical considerations and laboratory analysis have helped to identify the process parameters responsible for generation of high gauge and rundown. Thermo mechanical study revealed that output gauge of M47 grade Hot Rolled Non-oriented (HRNO) steel is dependent on temperature of hot working and requires a temperature range of 870 – 900C for its deformation in the Finishing Stands. Modification and stabilization of process parameters led to reduction of gauge of M 47 grade coils by 57% (output gauge for M47 grade coils ≤2.5mm increased from 25% to 82%). This in turn led to decrease in no. of passes in 4-Hi Reversing Cold Mill thus increasing Cold Rolled Non-oriented (CRNO) productivity by 8% (from 14.8 to 16.0 t/hr)

KEYWORDS: *CRNO Steel, Hot Rolling, Flow Stress, Deformation Characteristics.*

INTRODUCTION

CRNO steel finds wide applications in electrical machineries like motors, generators, refrigerator compressors and small transformers. Rourkela Steel Plant (RSP) is the prime producer of different grades of CRNO steel in India with Si ranging from 1.0 to 2.6%. Selection of composition & hot rolling parameters are of vital importance in production of these steels at HSM. The steel is produced through concast route. Hot rolling of the concast slabs is done in a 1700 mm HSM which is equipped with 3 numbers of roughing stands (two of reversing type and one of non-reversing type) and 6 numbers of finishing stands. A typical problem associated with hot rolling of M 47 grade CRNO steel having Si in the range of 1.2 to 1.4% was generation of higher mill load and gauge. The higher gauge (usually of ≥2.5 mm) resulted in more number of passes in the subsequent cold rolling stage leading to loss in productivity. The desired HR thickness for this grade is 2.5mm, which is to be reduced to 0.5mm during cold rolling.

It was observed that rolling of M47 grade HRNO steel in the finishing stands of HSM was characterised by generation of high mill loads, which prevents the operator for aiming at higher reduction. This restriction in degree of reduction particularly in the first three finishing stand results in higher gauge of the HR coils. It was essential therefore to examine the flow stress characteristic of the material under actual rolling conditions.

Researchers have systematically investigated the behaviour of Si steel under conditions of hot rolling. Rise of cumulative work hardening during hot rolling can be attributed to increase in its strain rate [1]. Also % C in steel and temperature of deformation sensitively influence the resistance to deformation of this grade of steel [2]. An empirical relationship was developed between chemical composition and hot rolling parameters whereby ductility sufficient for processing this steel in the cold rolling mill can be predicted [3].

Examination of Fe-Si Phase diagram for 0.01 to 0.02 weight percent Carbon as shown in Fig.1 revealed that the HRNO grade steel investigated presently becomes completely austenitic between the temperature of 950 and 1325C and also a narrow band of $\alpha+\gamma$ extends up to 25C beyond. Hot working in this range of temperature would therefore be difficult and would require extra precaution.

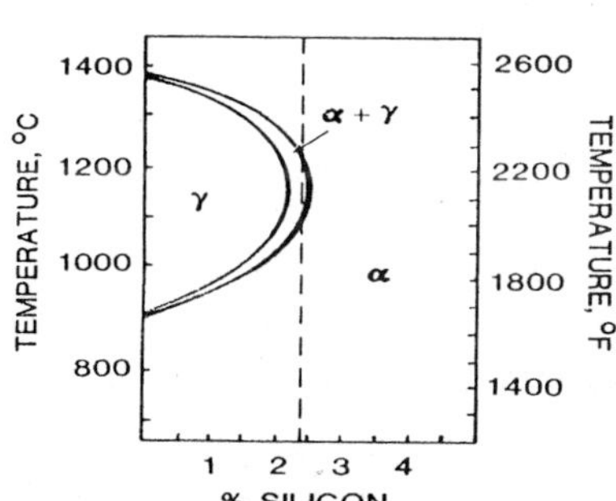

Fig. 1: Fe-Si Phase Diagram

With the information available it was felt necessary to carry out a thermo-mechanical simulation study for M47 grade HRNO steel and formulate the thermal and reduction regimes necessary during its actual hot rolling in the mill.

EXPERIMENTAL WORK

Steel Chemistry

The chemical composition of the steel used for this study is given in Table 1.

Table 1: Chemistry of M47 grade steel

%C	%Mn	%P	%S	%Si	%Al
0.026	0.210	0.018	0.008	1.28	0.06

To understand the behaviour of this steel during its actual hot working conditions at finishing stands of Hot strip Mill, RSP, the flow characteristics of the steel was to be determined. The objective of the study was to formulate thermal and reduction regimes for this steel so as to achieve reduced gauge of output coils.

Samples were prepared from M47 grade steel on the event of its removal from the roughing stand R2. The flow stress of the material was measured at different temperatures. Strain rates of 10/sec and 100/s were used, the former for rolling at F1 stand and the latter for F2 - F6 stands based on theoretical calculations.

Experimental Techniques

Mean flow stress of the steel was determined by flow stress compression testing with the help of a Thermo-mechanical Simulator (Gleeble-3500C). Cylindrical specimens of 10 mm diameter and 12 mm height were used for this purpose. The compression testing was done in the temperature range of 850 C to 1050 C at an interval of 50 C. Additional test was carried out at 925 C to find out the minima in the flow stress vs.

temperature plot. All the specimens were heated to the test temperature at a heating rate of 5C/sec and were soaked for 2 minutes before deforming them at a strain rate of 10/s and 100/s. Mean flow stress was determined by integration method from the stress-strain plot obtained from flow stress compression testing.

The mean flow stress values obtained from the stress-strain plot obtained by flow stress compression testing were plotted against the test temperature which revealed the following:

(a) Higher mean flow stress at higher strain rate

(b) Minima at around 950C.

Higher mean flow stress at higher strain rate is obvious. However, minima at around 950C can be explained in the light of the phase transformation taking place at this temperature.

The flow stress curves for HRNO M47 steel at strain rates of 10/sec and 100/sec were drawn separately and given in Fig. 2 & 3.

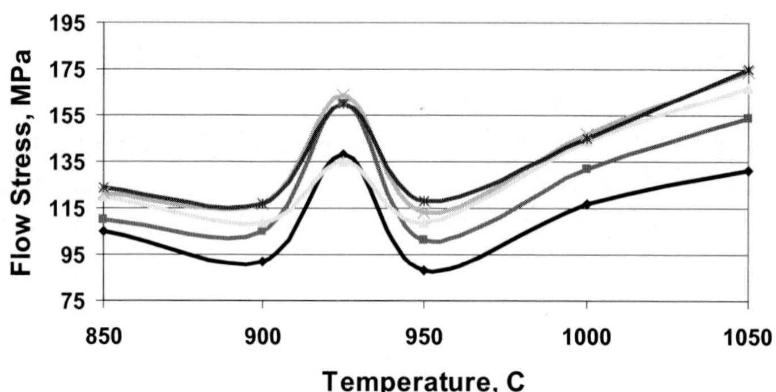

Fig. 2: Flow Stress Curve at 10/sec

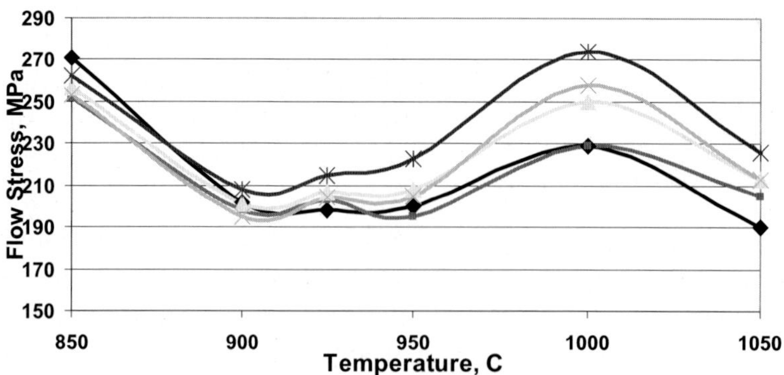

Fig. 3: Flow Stress Curve at 100/sec

Since flow stress is a function of strain, strain rate and deformation temperature, it was evident that at lower strain and strain rate as encountered in Stand F1, the deformation temperature should not be allowed to exceed 900C. For strain rates encountered at Stands F2 to F6, the deformation temperature should be 870 – 900C such that lower flow stress values leading to lower work hardening of the material would be achieved. It was also evident that higher deformation should be aimed at the 1st three stands, F1, F2 and F3 where stock temperature is in the vicinity of 900C and lower flow stress is encountered.

Based upon the above findings the process parameters for M47 rolling at Hot Strip Mill was fixed. Also reasons for higher output gauge of M47 was evident from this study. It was concluded that the higher output gauge of M47 grade coils resulted from increased flow stress on the event of working at a higher temperature range.

Also the micro structure of deformed material for M47 grade at strain of 0.5, strain rate 100/sec and temperature of 950 and 900 C as shown in Fig 4 revealed recrystallised fine grains at 900C

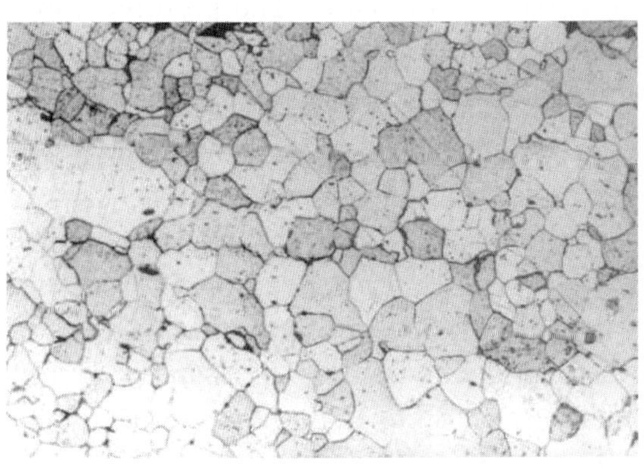

a) Temp. of deformation: 950C

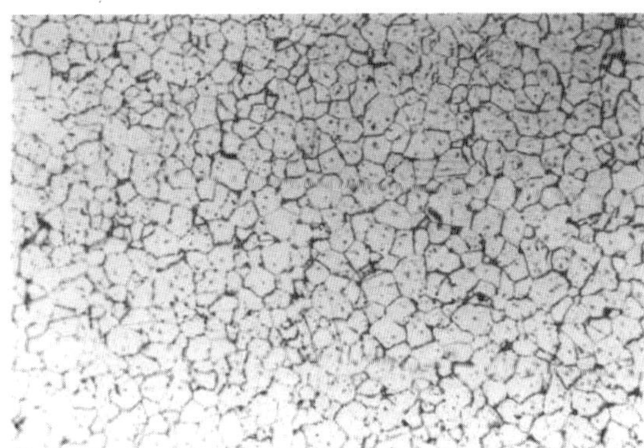

b) Temp. of deformation: 900C

Fig. 4: Micro structure of deformed M47 material

This increase in flow stress at higher temperature is due to hot working either in the α-phase or in the two-phase $\alpha+\gamma$ region. It is apparent that some sort of strengthening process sets up at higher range of temperature and probably onset of strain induced phase transformation of ferrite to hard austenite along ferrite- ferrite grain boundaries is responsible for such hardening at $\geq$ 950C.

Extensive trials were taken up with a view to formulate the thermal and reduction regimes during M47 grade processing at HSM. It was observed that working at lower range of temperature was instrumental in achieving the desired gauge and rundown.

Based upon these observations and substantiating the experience of thermo mechanical study conducted, the following range of temperature were recommended for M47 rolling which resulted in lower gauge and rundown.

Furnace drop-out:	1180 – 1200C
Ro entry:	1120 – 1140C
R1 entry:	1060 – 1080C
R1 exit:	1020 – 1040C
R2 exit:	990 – 1010C
F1 entry:	900 – 920C
Finishing:	860 – 880C
Coiling:	680 – 700C

Also based on follow up of different campaigns of M47 grade rolling in HSM, the following reduction regimes were recommended to achieve the targeted gauge of output coils. The transfer thickness from R2 was reduced from 23mm to 21mm thus creating a possibility of further reduction of gauge in finishing stands. The higher reductions employed in the 1st three finishing stands were also instrumental in achieving the targeted gauge.

Ro reduction: 3 pass
210 – 165 – 130 –100 mm

R1 reduction: 5 pass
100 – 86 – 70 – 56 – 42 – 30 mm

R2 reduction:
30 – 21.5 / 21.0 mm

Finishing:
21.0 – 12.5 – 7.6 – 4.6 – 3.4 – 2.6 – 2.2 mm

% Reduction
40.5 39.2 39.5 26.1 23.5 15.4

During processing of the coils, the finishing mill descaling sprays were not operated, as it would have unnecessarily brought down the temperature of the incoming bar to the finishing stands. The coil box, which is an important tool for homogenization of bar temperature, was not used as working in the finishing stands has been aimed at lower range of temperature. At this the range of temperature zoom rolling was practiced, as mill acceleration promoted uniform temperature and consequently uniform gauge along the length of the coil.

RESULTS AND DISCUSSIONS

Investigations on the process parameters of M47 grade rolling at HSM backed up by theoretical considerations and laboratory analysis have helped to identify the process parameters which are responsible for generation of high gauge and rundown. Modification and stabilization of these parameters have helped to reduce gauge and rundown in M47 coils. The modifications/ actions carried out to achieve lower gauge and rundown are discussed below.

First and foremost, the chemistry of M47 grade steel particularly with respect to %C was monitored for every heat as higher amount of C lead to generation of excessive mill loads during processing of this steel. C was restricted within ≤ 0.03% and Si within the range of 1.2 to 1.5% and other elements within the stipulated norms.

The soaking of slabs plays a vital role in achieving the targeted gauge and rundown. Base on the trials conducted at HSM a retention time of 4 hours minimum was kept which ensured uniform soaking of slabs. The drawal rate from the furnace was adjusted such that this retention time was maintained. With attainment of uniform soaking, the mill loads were kept within the stipulated norms and temperature rundown of slabs was also minimized.

Another important aspect of hot rolling of M47 grade steel was working at lower temperature range, which was verified by actual measurements of temperature at different stages of processing at HSM. Thermo-mechanical simulation study conducted had given an insight on the temperature of deformation necessary to achieve lower flow stress i.e., lower roll force in the mill. Working at lower temperature was beneficial to achieve lower output gauge. The temperature was kept at 870 – 900C such that lower flow stress values leading to lower work hardening of the material could be achieved. By practice of zoom rolling, mill acceleration could be increased which led to constant temperature and uniform gauge throughout the length of the coil. The reduction regimes were studied and its optimization was done. The output gauge from R2 was reduced from 23mm to 21mm which helped to achieve lower thickness of output coils. From the thermo-mechanical study it was also evident that higher deformation ought to be given at temperature in the vicinity of 900C. Therefore maximum drafting was aimed at the 1st three stands, F1, F2 and F3 to achieve the desired gauge.

As a result of modification and stabilization of the above strategies, the gauge and rundown in HRNO M47 grade coils could be reduced. by 57% (output gauge for M47 grade coils ≤2.5mm increased from 25% to 82%). Also rundown ≥ 0.15mm in M47 grade coils reduced from 30% to 18%. These in turn led to decrease in no. of passes in 4-Hi Reversing Cold Mill thus increasing CRNO productivity by 8% (from 14.8 to 16.0 t/hr)

CONCLUSIONS

Investigations and analysis of parameters affecting gauge and rundown of HRNO M47 grade coils at HSM have revealed the following:

- Chemistry of slab particularly with respect to %C and %Si influences the rolling load in finishing stands. Lower C (≤0.025%) and higher Si (≥1.2%) are pre-requisites to achieve lower output gauge.
- Soaking of slabs has an important bearing on the final gauge and rundown. A properly soaked slab of 4 hrs min. residence time and with minimum thermal gradient from its head to tail end leads to lower mill load and less rundown in the output coils.
- Gauge of M47 grade coils is dependent on the temperature of hot working and should be performed at temperature region of 860 to 900C in the finishing stands.

ACKNOWLEDGEMENT

The authors are thankful to the management of SAIL, RDCIS for giving permission to present this paper

REFERENCE

[1] G.P.Verma, Transactions of IIM, Vol.34, No.4. 1981, p 11

[2] G.K.Schoder, Handbook of Electrical Resistivity of Binary Metallic alloys, CRC Press, 1983, p.205

[3] T.W.Smythe, BHP Technical Report, Vol.25, No.2, 1982, p125

Optimization of Process Parameters in Bar and Rod Rolling Using Gleeble Simulation

Pranav Kumar Tripathi, D. Satish Kumar, H.A. Lanjewar, S. Manjini and Kamleshwar Prasad*
JSW Steel Ltd, Vijayanagar Works, Toranagallu, KA, India
*JSW Steel Ltd., Salem Works, Salem, TN, India
Email: pranavkumar.tripathi@jsw.in

ABSTRACT

In modern hot rolling of steel, the rolling speeds have been continuously increased in addition to increased alloy additions. To achieve the desired properties, high speed rolling processes must be simulated to study the behavior change of various alloy steels under high strain rate hot deformation conditions. Thermo-mechanical simulations using Gleeble - 3800 was used to simulate bar and rod rolling for optimizing rolling parameters in specific Fe-500 and high alloy steels respectively. Property variation with varying carbon in Fe-500 grade steel was investigated through Hot Axisymmetric Compression tests to measure the Flow stress of the bar steel samples at varying temperatures and strain rates. Thermo-mechanical simulation studies helped in optimizing the reheating furnace temperatures in bar mill leading to significant energy savings. Variation in flow behavior of billets of high Cr-Mo steel between surface and centre due to enhanced piping and segregation was leading to centre cracks. Change in high temperature plastic deformation behavior has been investigated using hot deformation studies simulating large diameter bar rolling in horizontal vertical continuous rolling mill. Centre and surface samples were deformed at different temperatures and strain rates to deduce optimum rolling schedules. The optimum combination of bloom heating temperature profile, strain, strain rates and temperature in roughing and finishing strands lead to defect free rolling.

KEYWORDS: *Thermo-mechanical Simulation, Gleeble, Flow Stress, Rod Mill, Segregation, Centre Cracks.*

INTRODUCTION

JSW Steel, Vijayanagar Works is presently a 10Mtpa integrated steel plant having two modules of COREX C-2000 to produce 1.6 Mtpa and four Blast furnaces to produce 8.4 Mtpa of hot metal. Two steel melt shops comprise seven Basic oxygen furnaces, six Ladle heating furnaces, two RH degasser and six single strand slab casters and an eight strand billet caster. JSW Vijayanagar works also comprises of two Hot strip mills to produce 3.2 Mtpa of HR Coils and 1.1 Mtpa of CR coils. Recently, bar rod mill and wire rod mills have been commissioned having 1 Mtpa and 0.9 Mtpa capacities respectively. The product range comprises of low carbon aluminum killed (LCAK), tube, structural, line pipe, micro-alloyed, and a range of medium carbon steels. With the introduction of hot strip mill - II JSW steel has started production of high strength line pipe steels (X80 and above) and dual phase steels. Within a very short span of its commissioning WRM has produced earthquake and corrosion resistance steels with innovative heat treatment techniques.

With increasing stringent requirements, extensive work has been carried out at JSW Steel, Vijayanagar works to develop higher grade steels, to meet the growing needs of customers in their endeavor to reduce costs and make high end products. The successful development and production of these strength steels requires a close control of process parameters and alloy designing. To meet these demands and accelerate the process of new product development and customization, physical and mathematical modeling approach is regularly applied.

GLEEBLE SIMULATION

Physical and Mathematical modeling has become an inevitable tool in the present day research for design development and control of steel processing. Owning to low cost and inherent limitation of experimentation in both Physical simulation and mathematical simulation, both methodologies have become interdependent. Relevant data for mathematical simulation in now economically obtained from physical simulation tests. Numerous works [1-12] on physical simulation have been reported for various processing operations using GLEEBLE which has been extensively used for simulating continuous casting, rolling, forging, heat treatment, welding, annealing etc. GLEEBLE provides a platform for testing samples and determine its limits of mechanical properties under various conditions. This has also been widely used for validating mathematical models. Recently installed GLEEBLE-3800 at R&D JSW Vijayanagar works has two mobile units, the pocket jaw and the hydrawedge. The machine is playing a significant role in product development and process improvement. With its continuous usage, it is JSW is working towards shorten the time of new product development along with solving many perennial issues in some of the critical grades. In the present paper two such cases (a) Bar rolling and (b) Rod Rolling process have been simulated using the Gleeble 3800 to develop or optimize the process parameters.

(A) Optimization of Process Parameters in Bar Rolling Using Gleeble Simulation

The bar mill at JSW is designed for full product range of Thermo-Mechanically Treated Re-Bars achieved with HYQST (High Yield strength Quenched and Self Tempered) Technology for sizes ranging from 8mm to 40mm diameter. It has a rated capacity of 1 Mtpa and mill speed upto 38 m/sec. It employs the latest proven rolling mill technology including automatic control of the rolling process to provide the maximum processing flexibility and operating efficiency to meet present and future market demands. Automatic cascade, cobble detection, rolling mill monitoring and billet tracking ensures a high degree of process efficiency. It is equipped with a walking beam type reheating furnace with computer controlled combustion system capable of 160 t/h cold charge and 190 t/h of hot charged billets of 165*165*12000 mm size. The product mix consists of TMT rebars of Fe 415, Fe500/550 grades as per IS1786 (2008) specifications with cross-section from 8 mm to 32.

At JSW Steel, processing of reinforcing steel, involves hot rolling followed by controlled cooling (TMT/HYQST process).

The practice is finishing online heat treatment while rolling, known as thermo-mechanically treated (TMT/HYQST) process. After leaving the last stand of the rolling mill, the bars are quenched (rapid cooling) in water from final rolling temperature of about 1050° C. The quenching is partial, only until a surface layer has been transferred from austenite (a steel phase stable only at very high temperature) to martensite (stable at temperatures below 350° C). This controlled quenching is achieved in one or more online water cooling devices through the steel passes at a very high speed before reaching the cooling bed. As the quenching is only partial, a part of the original heat remains in the core of the steel and on the cooling bed, this migrates towards the surface, this results in an automatic self tempering process where the surface layer of martensite is tempered, this "tempering temperature" (or equalization temperature) refers to the maximum temperature attained by the bar surface of the quenching. Tempering enables a partial diffusion of carbon out of the extremely brittle but strong martensite, thus relieving the inherent stress locked in during the sudden quenching of the red hot steel in cold water. The resulting tempered martensite shows improved deformability compared to the as-quenched martensite. The core of the heat treatment bars consists of ferrite and pearlite - more ductile but less strong than martensite. Computerized process control is used to dynamically adjust the parameters depending on the chemical composition of the steel, the desired grade and size of the reinforcing bar. After the commissioning of the bar mill, over a period of time with increased production, mechanical properties started deteriorating in the Fe-500 grade. This problem was attributed to the insufficient quenching leading to much smaller martensite layer. As a preliminary solution, the carbon percentages were increased from 0.15-0.16% to 0.22%. It was conceived that increased carbon content will result in higher martensitic formation at the periphery and improve the mechanical properties. However, it was assumed that increased carbon content will increase the flow stress which would result in higher mill loads. To counter mill load issue, the reheating furnace exit temperature was increased. Due to higher RF exit temperature, the final temperature obtained after quenching were still higher which hindered proper microstructure development. Moreover, due to increased RF exit temperature, fuel consumption also increased which lead to overall higher operational costs.

EXPERIMENTS

Fe-500 is one of the most common billet grades at BRM. However, there has been a significant variation in the operating parameters (re-heating temperatures) for different carbon ranges in this grade which results in variation in properties. Specifications of Fe-500 grade, are as shown in Table 1.

Table 1: Fe-500 (IS: 1786) chemical composition

Chemical Composition (%)	Grade Fe – 500, (IS: 1786)
Proof Stress (Min.)	500 MPa
Tensile Strength (Min.)	545 MPa
Elongation (Min.)	12%
Bend Test (Min.)	Upto 22 mm-4D

To understand the effect of temperature and chemistry on the mechanical properties in the above grade, gleeble experiments were conducted. To study the effect of carbon content on flow stress, three set of samples with different carbon concentrations (Table 2) were tested at different temperatures and strain rates.

Table 2: Samples carbon range

Composition	Samples		
	A	B	C
Carbon	0.244	0.164	0.155

Single hit deformation tests for each set of sample were conducted at lower (10/s) and higher (50/s) strain rates using Hydrawedge module. Each set of samples was deformed at 800, 850, 900, 950 and 1000ºC and flow stress values were obtained. The program cycle is shown in the fig. 1. The deformed samples were also analyzed for microstructure and hardness values.

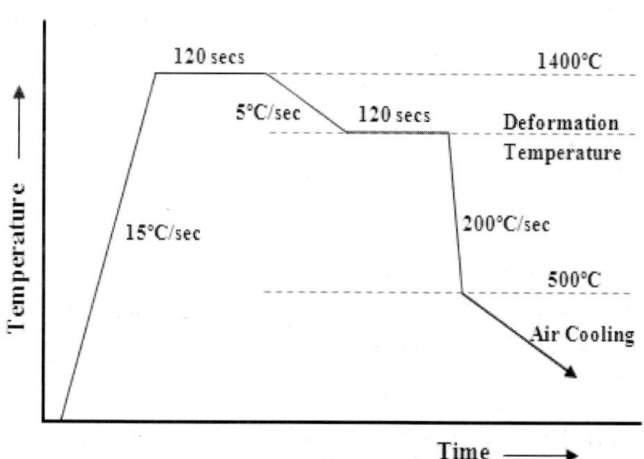

Fig. 1: Single Hit Program Cycle

RESULTS AND DISCUSSION

The samples were deformed at different temperatures and strain rates followed by faster cooling. Fig. 2 shows the typical cooling curve obtained in one of the cases. High cooling rates of (200º C) were obtained through mist quenching using the quench assembly.

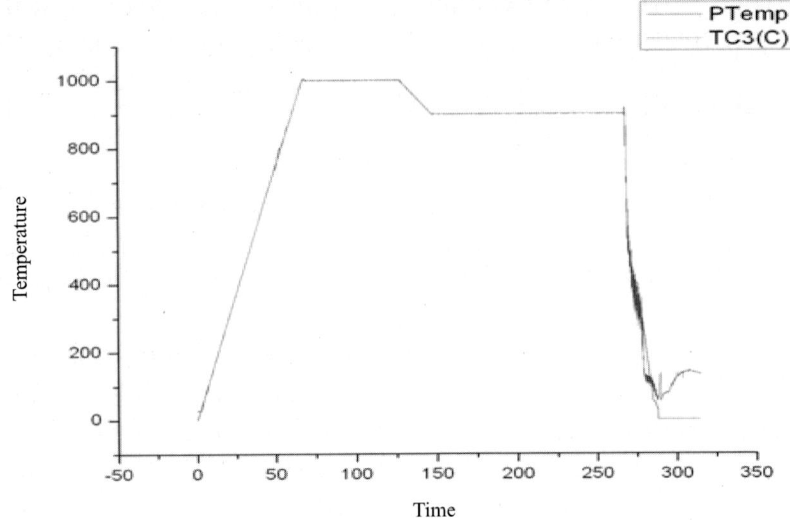

Fig. 2: Typical Cooling Curve (900°C, Strain Rate – 10, 0.244% C).

Fig. 3 shows the typical Stress stain curve obtained for single hit deformation cycle in one of the cases during the experiments.

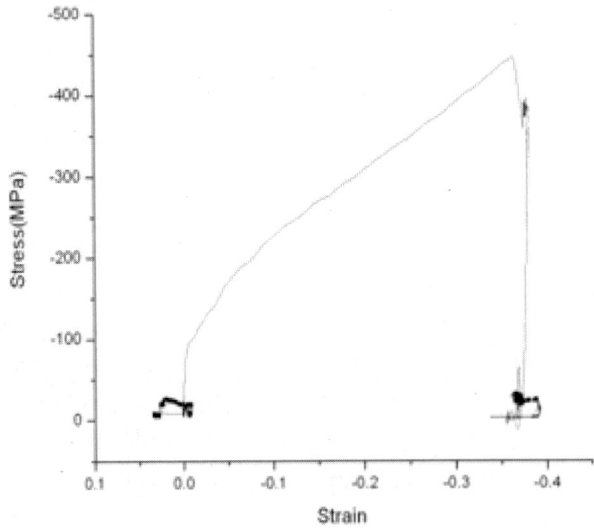

Fig. 3: Typical Stress Strain Curve (800°C, Strain Rate – 10, 0.244% C).

Fig. 4 (a) and (b) compare the flow stress values at different temperatures for selected carbon range samples at lower (10 s^{-1}) and higher (50 s^{-1}) strain rates respectively. The flow stress values obtained from single hit deformation tests revealed that the flow stress values decrease with increasing deformation temperatures. However, there is no appreciable change in the flow stress with carbon levels at all temperatures at particular strain rates. This concludes that increasing the carbon levels will not increase the mill load. Hence increasing the carbon concentration should not be followed or manifested by increasing reheating furnace exit temperature to reduce flow stress and therefore, mill load.

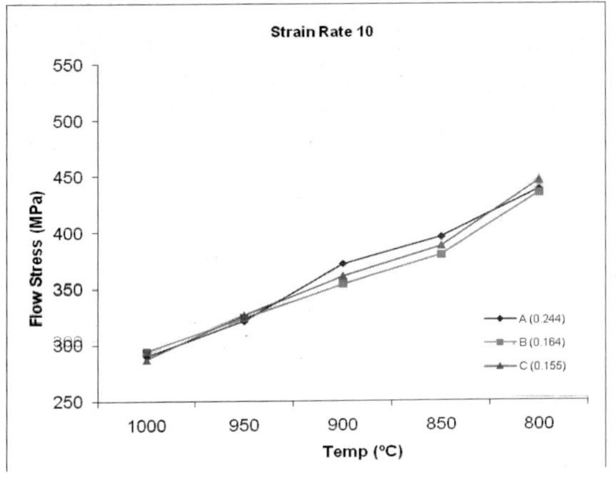

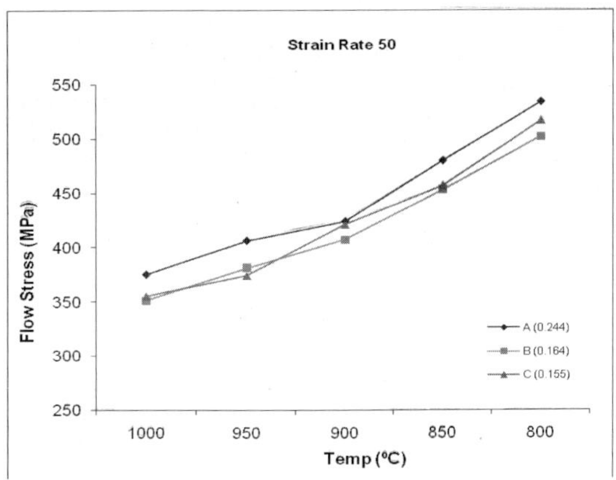

Fig. 4: Comparison of Flow stress values at different temperatures and strain rates

The tested samples were analyzed for microstructure and hardness (centre and periphery). Fig. 5 shows the microstructure at the centre (a) and periphery (b) of one of the samples. Microstructure clearly reveals the ferrite-pearlite structure at the centre and martensitic structure at the periphery as is required in a standard TMT bar sample. Fig. 6 shows the comparison of hardness values at centre and periphery at different strain rates for all the 3 carbon range samples tested at the plant operating temperatures. Higher hardness values are obtained at the periphery due to higher quenching at the surface which matches well with the microstructure. It is therefore concluded, a lower temperature of rolled bars during quenching stages will help in developing optimized martensitic structure development. It will also help in energy and cost savings. As per the recommendations on the basis of trials conducted on Gleeble, the reheating furnace exit temperatures for Fe-500 grades have been lowered from 1250°C to 1200°C.

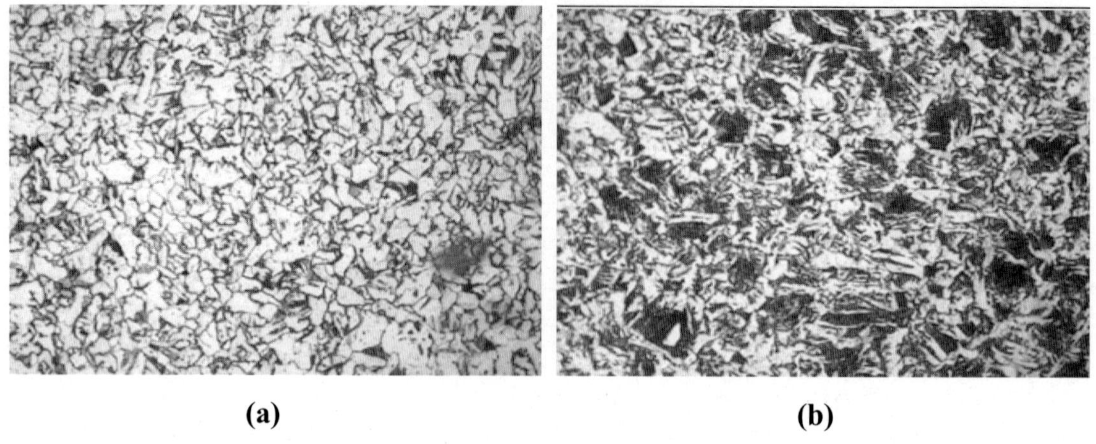

(a) (b)

Fig. 5: Microstructure of the Gleeble tested samples at (a) centre (b) periphery

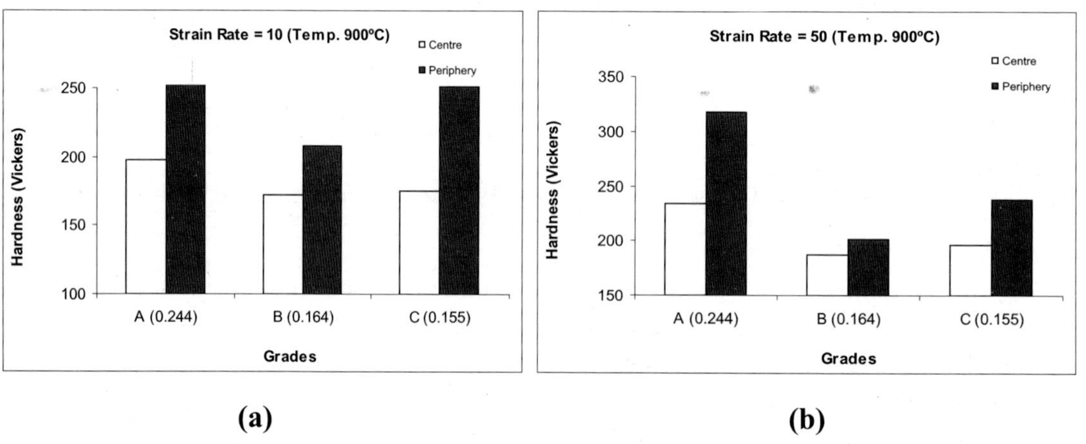

(a) (b)

Fig. 6: Comparison of hardness values at centre and periphery for (a) low and (b) high strain rates

(B) Optimization if Process Parameters in Rod Rolling Using Gleeble Simulation

JSW Salem Works is the largest integrated special steel complex in India with 1 MTPA capacity. Wire rods, special steel, RCS, rounds and spring steel flats are manufactured at JSW Salem works. 42CrMo4 and 34CrMo4 are two important grades for automotive applications being rolled at JSW Salem and reported for piping defects. On microscopic examinations of billets, both 42CrMo4 and 34CrMo4 grades were found to be undergoing piping and segregation near the center as a result of solidification during continuous casting. 42CrMo4 grade was developing cracks near the centre of the bar after rolling operation, whereas 34CrMo4 grade was showing normal behavior after finish rolling. It is worth noting that a 4 pass reduction profile was being followed in case of 42CrMo4 grade as compared to 8 pass reduction in 34CrMo4 grade in the finishing mill.

The diameter of the deformed blooms after roughing mill for both 42CrMo4 and 34CrMo4 grades was approx 190 mm. This diameter was reduced to 130 mm for 42CrMo4 in 4 passes, and to 90 mm for 34CrMo4 in 8 passes. It was understood that to avoid crack generation after rolling, the piping should get welded and also the flow stress applied should be enough to roll the segregations at the centre, evenly. Such a scenario will help in bringing down the instances of internal cracks. Hence, with an objective of comparing the flow stress of both the grades at different strain rates and temperatures, single hit deformation studies were performed on Gleeble. Chemical composition of both the grades is shown in Table 3.

Table 3: Chemical composition of 42CrMo4 and 34CrMo4

	C	Mn	Si	S	P	Al	Cr	Mo	Ni	N
42CrMo4	0.42	0.75	0.26	0.022	0.016	0.024	1.17	0.26	0.016	0.0075
34CrMo4	0.34	0.76	0.22	0.012	0.016	0.020	1.10	0.21	0.018	0.0075

Samples were taken from these rods from centre and periphery and analyzed for hardness values. It was found that the hardness at the centre was much higher than that at the surface for 42CrMo4 grade (Table 4). It is normally observed that the centre of a rod generally has lower hardness than the surface due to higher quenching rates obtained at the outer surface and this was quite contrary to the findings in present case. The hardness was also measured at the centre of 34CrMo4 grade rods and was found to be lesser than 42CrMo4 grade rods.

Table 4: Comparison of average microhardness values

	34CrMo4 Centre	**42CrMo4 Surface**	**42CrMo4 Centre**
Microhard-ness (HV) 300 gm	188	196	226
HRB	90	92	97

EXPERIMENTS

To understand the effect of temperature and strain rates on the flow stress in the above grades, single hit axisymmetric deformations were performed using Gleeble at different finishing temperatures and strain rates for a fixed amount of strain followed with air cooling after the deformation. Single hit deformation tests for each set of sample were conducted at strain rates of 0.1, 0.5, 1.5 and 5 for a fixed amount of strain (0.26) using hydrawedge module. Each set of samples was deformed at 950, 1000 and 1050°C and flow stress values were obtained. The typical program cycle is shown in the fig. 7. The deformed samples were also analyzed for hardness values.

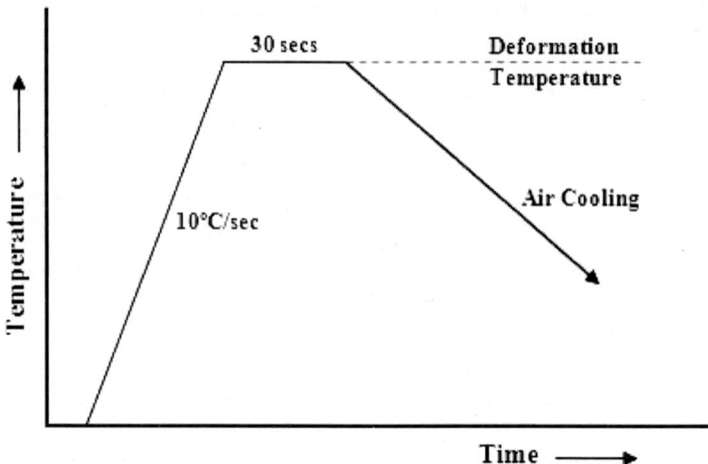

Fig. 7: Single Hit Program Cycle

RESULTS AND DISCUSSION

Fig. 8 shows the typical cooling curve obtained in one of the cases. Fig. 9 shows the typical flowstress curve obtained in one of the cases during the experiments.

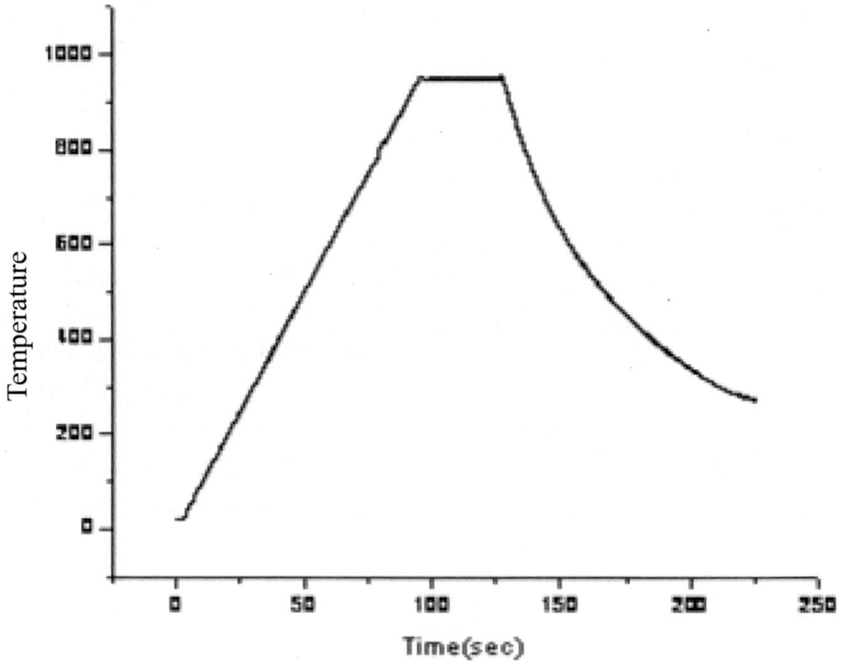

Fig. 8: Typical Cooling curve

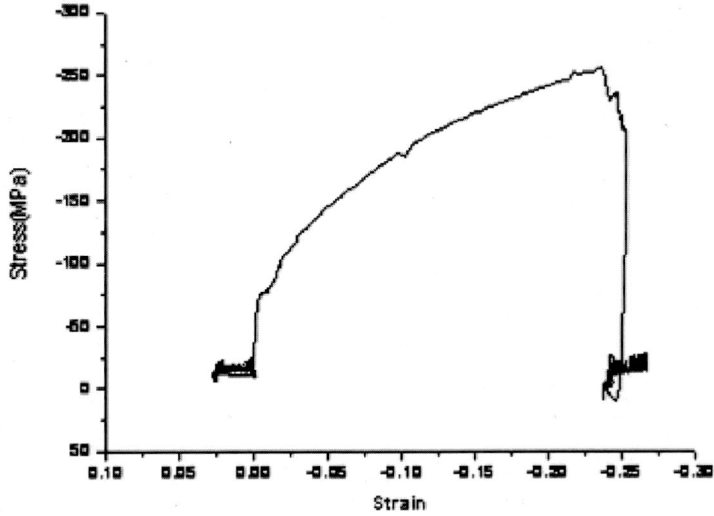

Fig. 9: Typical Stress strain curve

Fig. 10 (a, b) compare the flow stress values for 42CrMo4 grade at surface and centre locations respectively at different temperatures with varying strain rates. Fig. 10 (c) shows the corresponding flow stress values at centre location for 34CrMo4 grade.

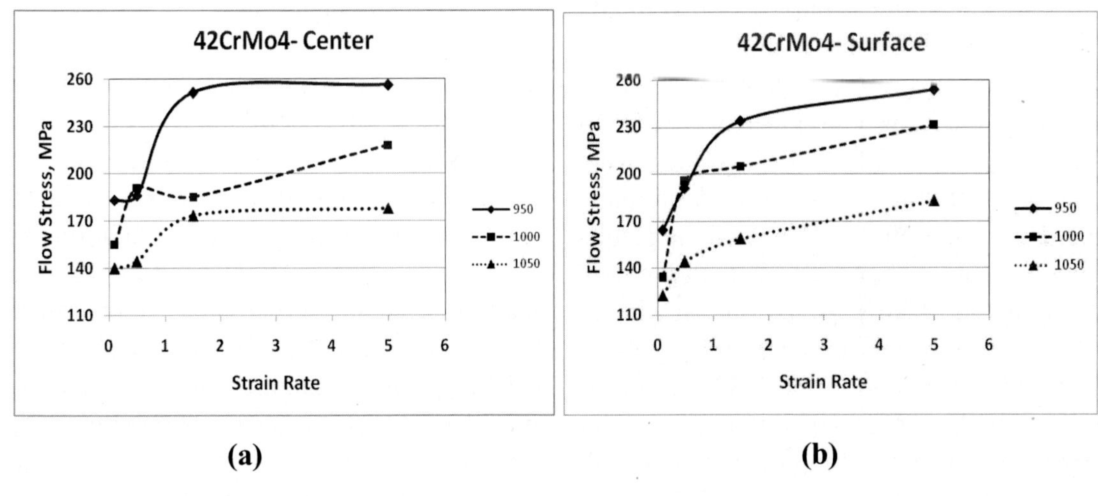

(a) **(b)**

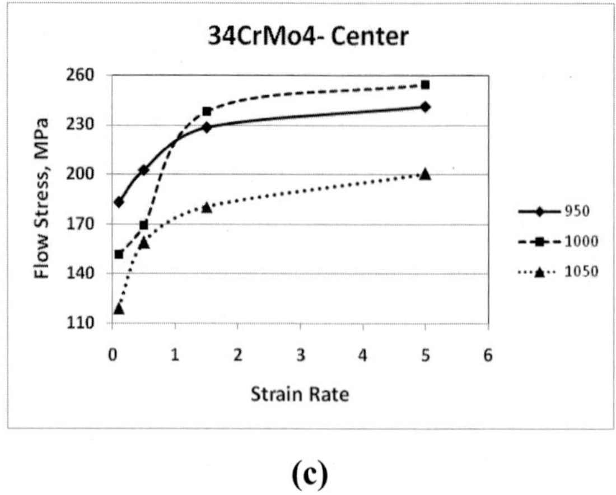

(c)

Fig. 10: Comparison of Flow stress values at different temperatures and strain rate

From the above figures, with few deviations it can be clearly deduced that the flow stress increases with increasing strain rate and decreases with increasing deformation temperature which matches well with the established literature. Though the flow stress of the centre samples is marginally higher than the surface samples, there is no significant difference between the flow stress of the centre samples of 42CrMo4 and 34CrMo4. The tested samples were analyzed for microstructure and hardness (centre and periphery). Based on the results obtained, it is believed that the centre cracking in 42CrMo4 is due to lower reduction ratio. It was therefore suggested that reduction ratio of 42CrMo4 can be increased by increasing the strain per strand, either in roughing or finishing mill. Increasing the strain will definitely increase the strain rate, which however is not expected to increase the present flow stress of the material significantly as shown from the Gleeble tests. Thus, centre cracking in 42CrMo4 grade can be eliminated by increasing the strain while keeping other rolling parameters unchanged without any appreciable increment in the mill load. The same has been practiced at JSW Salem and has shown good results.

CONCLUSIONS

GLEEBLE 3800 system at JSW Vijayanagar works has been a useful tool for simulating various operations. Issues in bar and rod rolling has been successfully addressed through thermo-mechanical simulation. In BRM simulation studies, relation between carbon content, processing temperature and strain rates has been established for effective determination of the thermo-mechanical processing parameters to achieve the desired microstructure. The reheating furnace exit temperature at BRM has been reduced by almost 50ºC leading to huge energy savings and better process economics. The rod mill simulation studies for JSW ISPAT addressed the issue of centre cracking in 42CrMo4 through flow stress measurement studies. Change in reduction ratio by increasing the strain per strand, resolved the problem significantly.

REFERENCES

[1] Hirowo G. Suzuki, National Research Institute for Metals, 1-2-1 Sengen, Tsukuba 305, Japan, pp. 27-39.

[2] K. Airaksinen, L.P. Karjalainen and D. Porter, 2nd ESAFORM Conference on Material Forming, Guimaraes, Portugal (1999), pp. 555-558.

[3] Yoji Hirana and Takeyoshi Doko, Furukawa Review, No. 13, 1994, pp. 163-168.

[4] J. L. Lee, Y. T. Pan and Y. S. Hwang, 37th MWSP Conf. Proc. ISS, Vol. XXXIII. 1996, pp. 11-21.

[5] A. Kumar, C. McCulloch, E. B. Hawbolt, and I. V. Samarasekera, Materials Science and Tech., April 1991 Vol. 7, pp. 360-368.

[6] H. S. Ferguson, International Symposium on Physical Simulation (7th ISPS) National Research Institute for Metals, Tsukuba, Japan, Jan'1997, pp. 1-26.

[7] Hirowo G. Suzuki, International Symposium on Physical Simulation (7th ISPS) National Research Institute for Metals, Tsukuba, Japan, Jan'1997, pp. 27-40.

[8] Y. Watanabe and A. Yoshie, International Symposium on Physical Simulation (7th ISPS) National Research Institute for Metals, Tsukuba, Japan, Jan'1997, pp. 193-200.

[9] J. J. Jonas, International Symposium on Physical Simulation (7th ISPS) National Research Institute for Metals, Tsukuba, Japan, Jan'1997, pp. 127-140.

[10] S. Kleber, K. Kerschbaumer and B. Buchmayr, International Symposium on Physical Simulation (7th ISPS) National Research Institute for Metals, Tsukuba, Japan, Jan'1997, pp. 259-264.

[11] Bainian Qian, Jingli Li and Dehua Yu, International Symposium on Physical Simulation (7th ISPS) National Research Institute for Metals, Tsukuba, Japan, Jan'1997, pp. 443-448.

[12] Shujun Jiao, Ru Zheng, Xianjin Wang, Yan Zhang, Beihua Feng, Shiyan Wang, Xuegang Zhu, International Symposium on Physical Simulation (7th ISPS) National Research Institute for Metals, Tsukuba, Japan, Jan'1997, pp. 523-528.

Minimization of Cracking of Wheel Rims Manufacturing from Fe 410 Grade Structural Steel

Arthita Dey, Tanmay Bhattacharyya, G. Senthil Kumar and Sandip Bhattacharyya
R&D and Scientific Services, Tata Steel Limited, Jamshedpur-831 001, India
Email: arthita@tatasteel.com

ABSTRACT

Automotive industries demands wheels without any defect. However, process as well as material abnormalities affects the quality of wheels due to different reasons. Wheels for heavy and commercial vehicles are manufactured by cold forming of slit hot rolled low carbon structural steel with or without micro alloying. Cracking of wheels during manufacturing is a major problem at the production line. Hot rolled structural steel grade, Fe 410, is a common material for wheel rim manufacturing for its good formability. This hot rolled grade with thickness ranging from 5 to 6 mm is used to manufacture wheel rim for automobiles. These materials pass through flush butt welding at the customer's end during manufacturing the final product.

For past 7-8 years customers were facing problems of cracking during wheel rim processing. There was about 40-50% rejection in this particular grade due to cracking. Few trials were conducted in the past involving changing the operating parameters during hot rolling, casting temperature etc. However, these modifications in the past did not have much effect on the reduction of customer complaints. Multiple numbers of samples with different failure modes were collected from customer's end for analyzing the root cause.

The investigation of the failed wheel rims involved visual observation, mechanical and chemical testing with microstructural characterization. It was observed that use of peritectic grade steels, centerline segregation and improper welding are the prime causes of the cracking of this grade. To mitigate the problem the steel chemistry and the welding process parameters were modified. Finally the rejection has been reduced to 2.5%.

The paper describes the causes of the rim failure manufactured from Fe 410 grade along with the probable solution to eliminate the problems.

KEYWORDS: *Wheel Rim, Fe 410, Centreline Segregation, Flush Butt Welding*

INTRODUCTION

Low Carbon and micro alloyed hot rolled strips with different thickness are used to manufacture wheel rims for light and medium commercial vehicles. Wheel rims are applied either for tube wheels or for tubeless wheels. The cross sectional view of the assembly of a wheel rim and a wheel disc has been illustrated in Fig. 1. Wheel Disc and Rim are produced in separate production lines and subsequently welded together in an assembly line. The wheel disc is produced from coil stock in a deep-

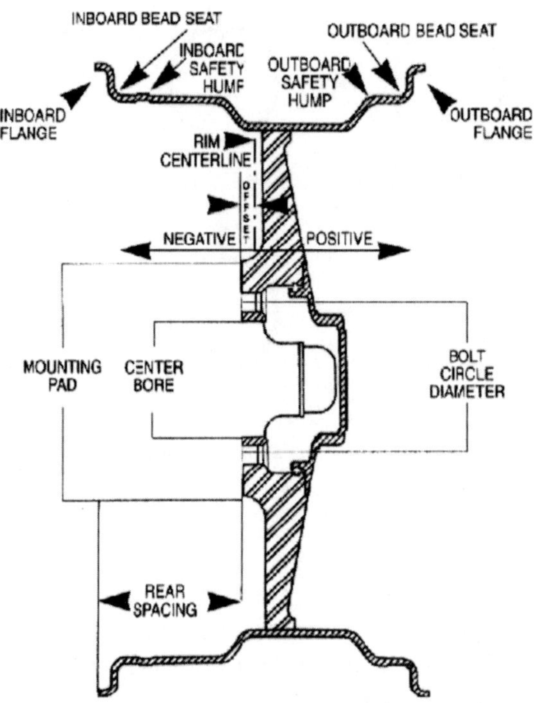

Fig. 1: Illustrative design showing cross sectional view of wheel rim and wheel disc [2].

drawing process using a progressive press (shaping and punching). The associated wheel rim is also produced from a sheet metal coil. The coil is unwound and cut to length, and subsequently pre-bent (round bending) and welded into a circular blank with the help of flush butt welding (Fig. 2).

In the process upset force is applied at some point near the end of the heating cycle to bring the parts into intimate contact. The flashing action caused by the high current densities at small contact points between the workpieces forcibly expels material from the joint as the workpieces are slowly moved together. The weld is completed as a forge welded joint by rapidly upsetting the workpieces. Then the circular blank formed by the process is profiled using roll forming stands and is shaped into the typical wheel rim profile. In the final assembly line the disc is pressed into the wheel rim, welded, checked and subsequently dip-primed. The process is shown schematically in Fig. 3[1].

Cracking of wheel rim is a major cause of concern in manufacturing and forming of the strips. Generally cracking and sometimes thinning occurs along the weld during flaring operation. These cracks originate from inherent material properties of the parent strip arising from inappropriate chemistry, the improper processing during welding and inferior mechanical properties from intermediate processing operations [3]. Tata Steel Limited (TSL) supplies 6.20mm thick steel plate (Grade: Fe-410) to a leading wheel producer for rim

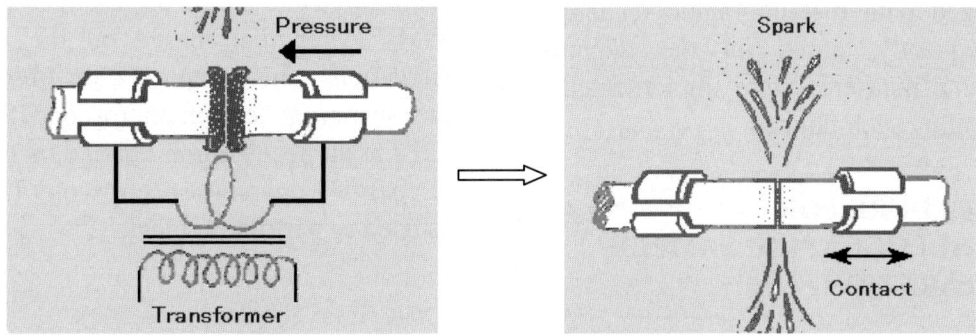

Fig. 2: Schematic of Flush butt Welding Process

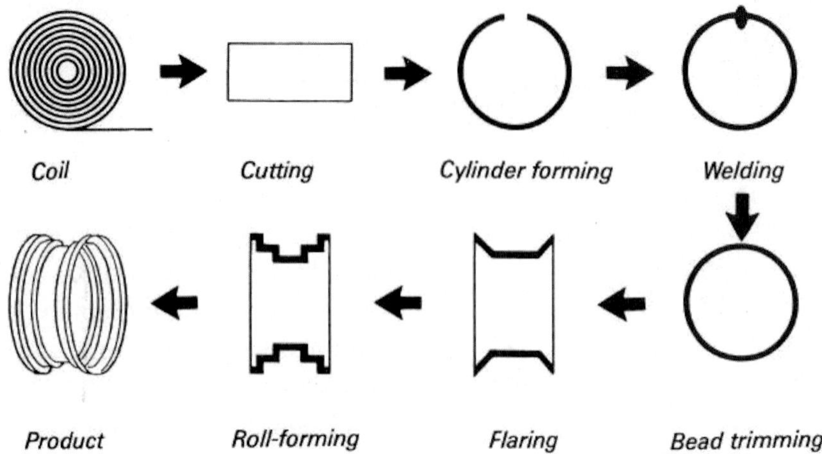

Fig. 3: Wheel manufacturing process [1]

manufacturing of wheel which was used for passenger movement in a high speed vehicle. There was 40-50% rejection of TSL supplied material in the year of 2007 and 2008. Over the periods different measures have been taken which has reduced the rejection to around 2.5% at customer end.

The present study deals with the failures of wheel rims which caused huge rejection of TSL's material in the 2007-2008 and also the experiences of the journey made to reduce the rejection from about 50% to 2.5%. Seven wheel rims of same grade, diameter and thickness were selected for the study. The root cause of the failures was analyzed with metallographic techniques. Subsequently, based on the diagnosis, remedial measures were implemented for preventing such failures.

PROCEDURE FOR METALLOGRAPHIC SAMPLE PREPARATION

The current study was carried out with seven different wheel rims, five were cracked from the weld and rests two were of good quality. All the rims were manufactured from 6.2 mm thick Hot Rolled (HR) plate with Fe 410 grade structural steel. The metallographic preparation of the samples was carried out by sample sectioning, mounting, grinding and polishing in the usual sequence. Sample sectioning was carried out by use of an abrasive cutter. Then the samples were prepared by hot mounting method in a conducting resin and polished up to 1200 grit emery paper followed by polishing in diamond paste of up to 3 µm. After preparing the scratch free sample, they were thoroughly washed with alcohol and observed under light optical microscope (LOM). After that, based on the LOM observations, the samples were observed under scanning electron microscope (SEM-JEOL) at an acceleration voltage of 15 kV and energy dispersive spectroscopy (EDS) was used as per requirement.

OBSERVATIONS

Visual Observations

Visual observation of all the samples under study revealed cracking in the weld (Figs. 4a to 4e), the bluish region due to heat effect during welding, clearly confirms the location of the weld in the cut rim sample referred to the laboratory. Those of good samples do not contain any defect (Figs. 5a and 5b). Multiple microsamples were prepared across the crack and crack tip

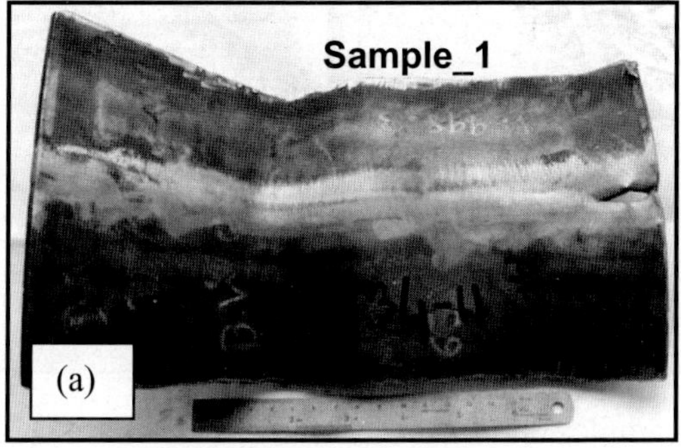

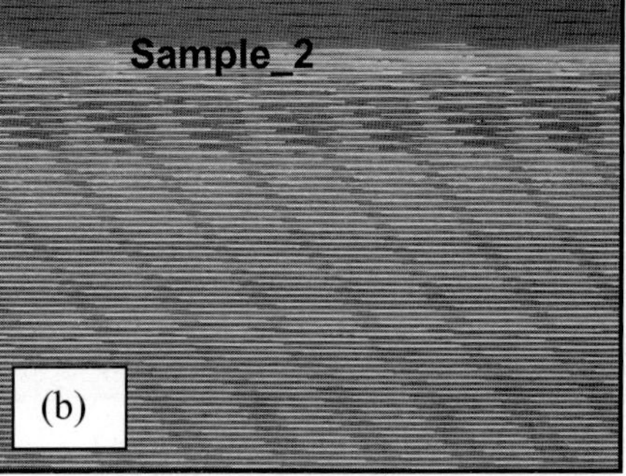

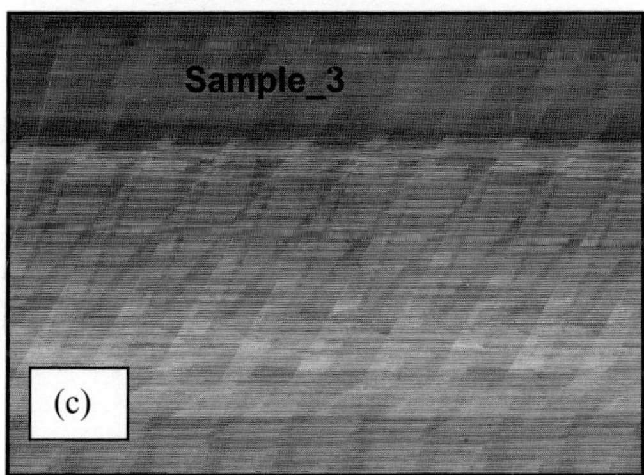

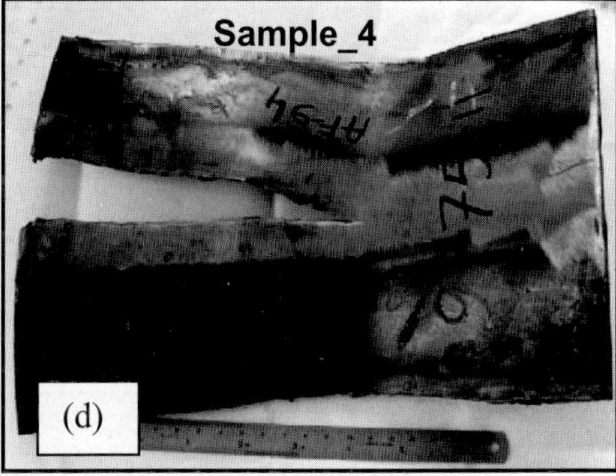

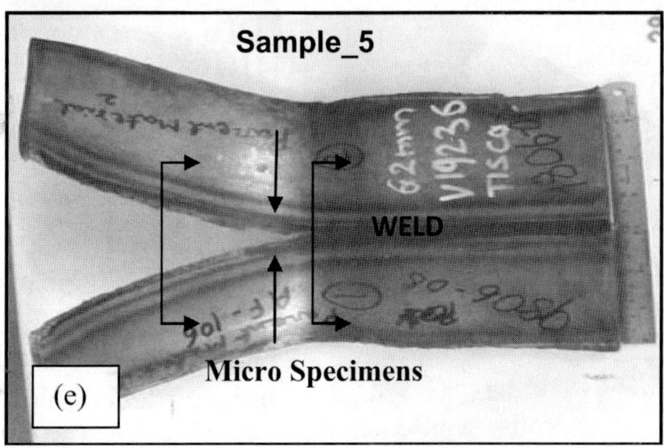

Figs. 4: (a) to (e) Cracked samples under current study

of the samples under study (Fig. 4e). The stereoscopic observation of one of such typical failure revealed necking at the crack tip followed by cracking. Both the necking and cracking were associated with the weld (Figs. 6a and 6b).

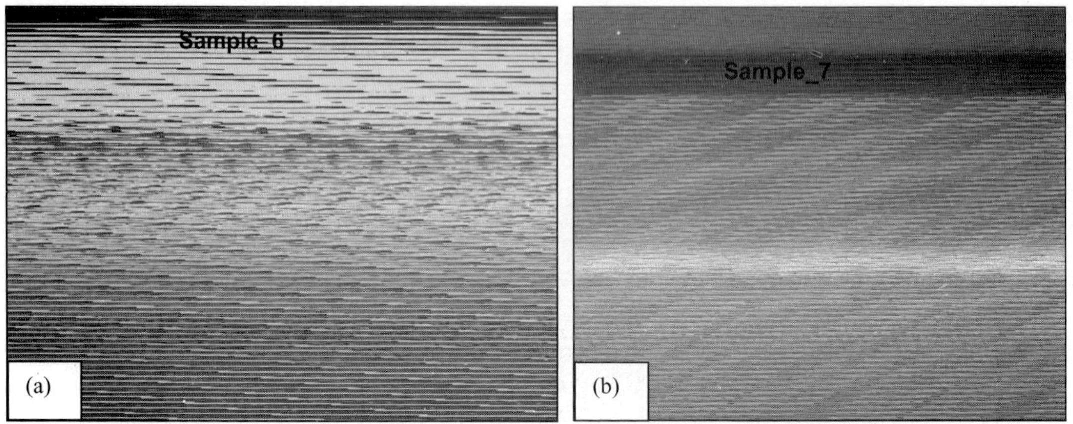

Figs. 5: (a) and (b) Cracked samples under current study

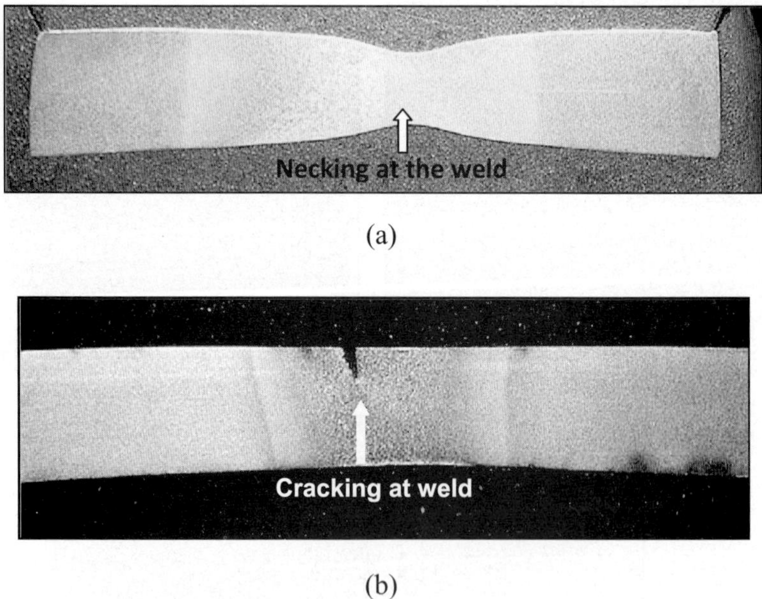

(a)

(b)

Figs. 6 (a) and (b): Stereoscopic observation of micro-sample

Chemical analysis

Chemical analysis of samples was carried out using X-ray fluorescence spectroscopy (XRF) except C and S. The carbon (C) and (S) content were analyzed by combustion infrared technique. The chemical analysis has been compiled in Table 1. All the materials conformed to Fe 410 grade steel.

Table 1: %Chemical composition of the samples under study

	C	Mn	S	P	Si	Al	Mn/Si	Ceq
Sample_1	0.082	0.88	0.003	0.015	0.065	0.025	13.53846	0.15
Sample_2	0.08	0.81	0.008	0.017	0.059	0.04	13.72881	0.15
Sample_3	0.08	0.8	0.005	0.01	0.064	0.035	12.5	0.15
Sample_4	0.079	0.8	0.006	0.017	0.065	0.042	12.30769	0.15
Sample_5	0.08	0.86	0.005	0.015	0.07	0.033	12.28	0.15
Sample_6	0.07	0.78	0.004	0.017	0.09	0.04	8.666667	0.13
Sample_7	0.06	0.77	0.005	0.0148	0.09	0.05	8.555556	0.12
Fe 410 Specification	0.18 max	1.25 max	.015 max	.020 max	0.10 max	.02 / .06		

Microstructural Examination

Unetched micro structure across the crack revealed entrapment of non metallic inclusions as shown in Figs. 7a to 7d. Etched microstructure showed mixture of altriomorphic ferrite, primary ferrite PF(G) and acicular ferrite AF around the inclusions

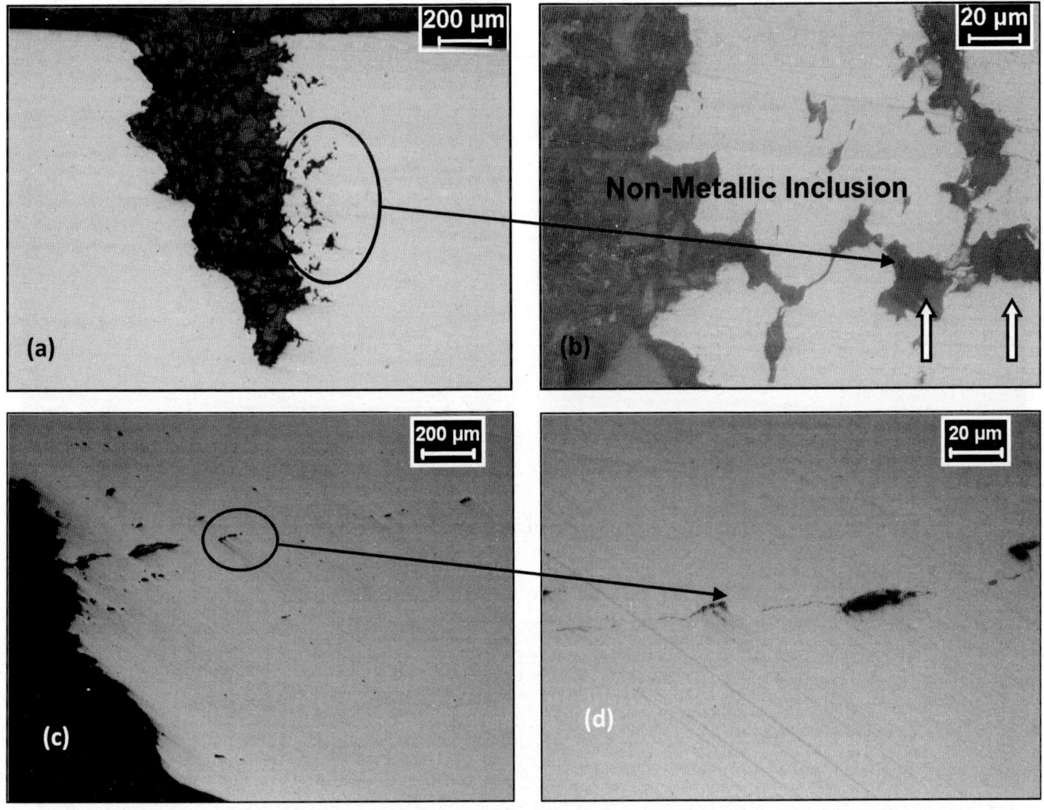

Figs. 7 (a) to (d): Unetched sample showing entrapment of non metallic inclusion at the crack location.

confirming their presence inside the weld (Figs. 8a to 8d) [4]. The microstructure observation across the weld of the cracked rim revealed mixture of primary ferrite PF(G), ferrite with aligned second phase FS(A) and acicular ferrite AF at the [5]. The HAZ revealed fine ferritic matrix of average grain size 8 ASTM with pearlite at the grain boundaries and parent revealed ferritic matrix with average grain size of 9 ASTM (Figs. 9 and 10). Microstructure of the good weld revealed mixture of primary ferrite PF(G), widemanstratten ferrite WF and acicular ferrite AF. That of the HAZ showed ferrite pearlite structure and matrix was predominantly ferritic (Figs. 11 and 12). The average ferrite grain size of HAZ as well as matrix was about 9 ASTM.

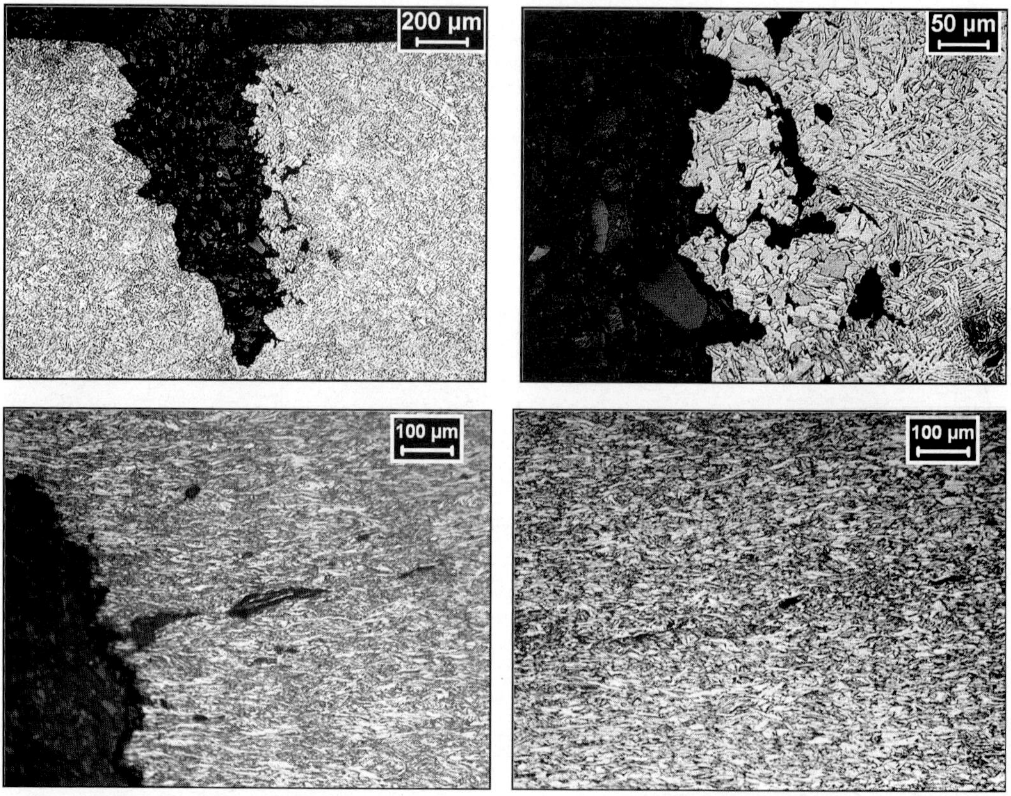

Figs. 8 (a) to (d): Etched sample showing cracking at the weld.

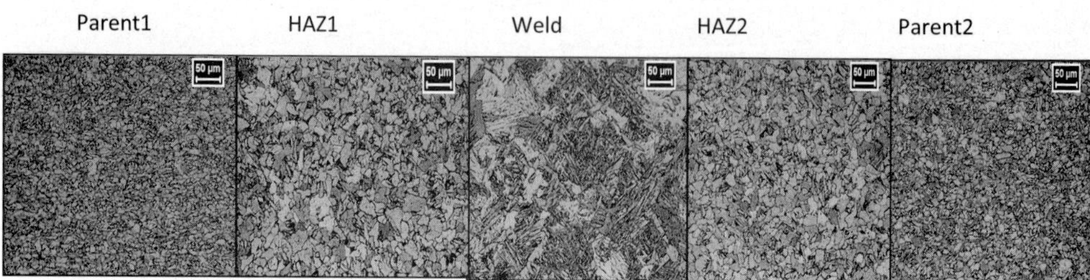

Fig. 9: Micro structure across the weld of a cracked sample

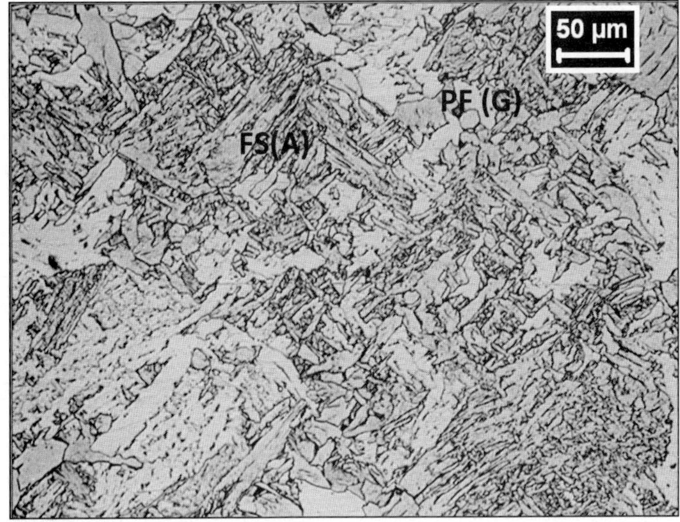

Fig. 10: Microstructure of weld of cracked rim

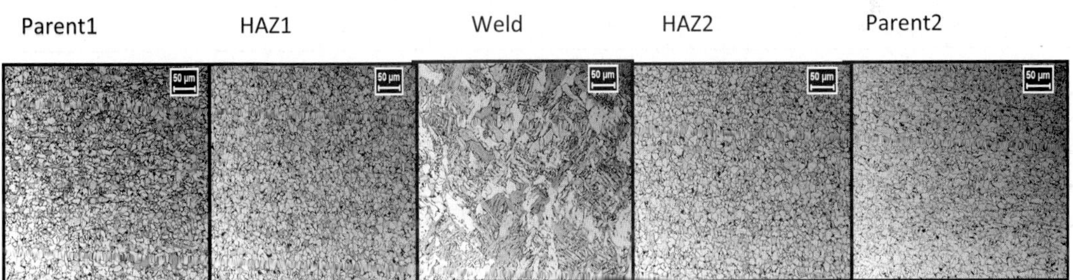

Fig. 11: Micro structure across the weld of a good sample

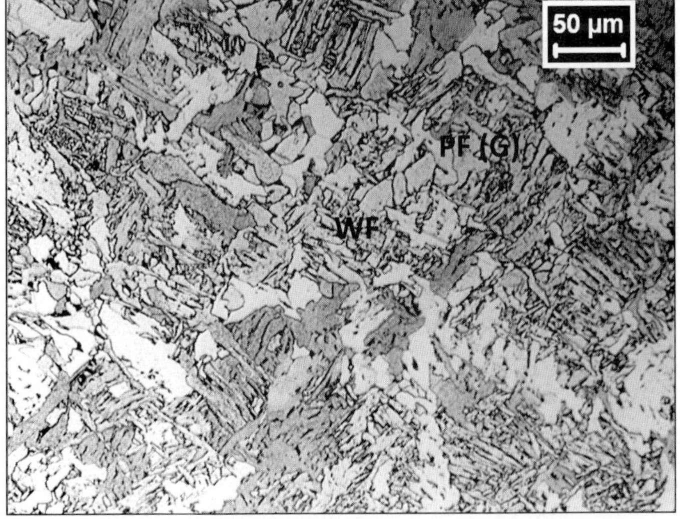

Fig. 12: Microstructure of weld of un-cracked rim

Physical Property Testing and Hardness Survey

Evaluation of physical properties such as, yield strength (YS), ultimate tenise strength (UTS) and % elongation were carried out on 100 ton UTM. Table 2 depicts that the properties analyzed for the input material were within specification. Specimen across the weld region was prepared for taking hardness profile across the weld. Hardness values were measured on HV scale using Vickers Hardness testing machine at a load of 30 kg as shown in Fig. 13. The hardness survey revealed uniform hardness all over the weld, HAZ and matrix.

Table 2: Mechanical properties of the sample under study

	Yield Strength (Mpa)	Ultimate Tensile Strength (Mpa)	%El
Sample_1	360	465	29
Sample_2	362	467	27
Sample_3	386	452	38
Sample_4	372	447	33
Sample_5	412	466	27
Sample_6	366	469	29
Sample_7	370	442	31
Fe 410 Specification	**250 min**	**410 min**	**23 min**

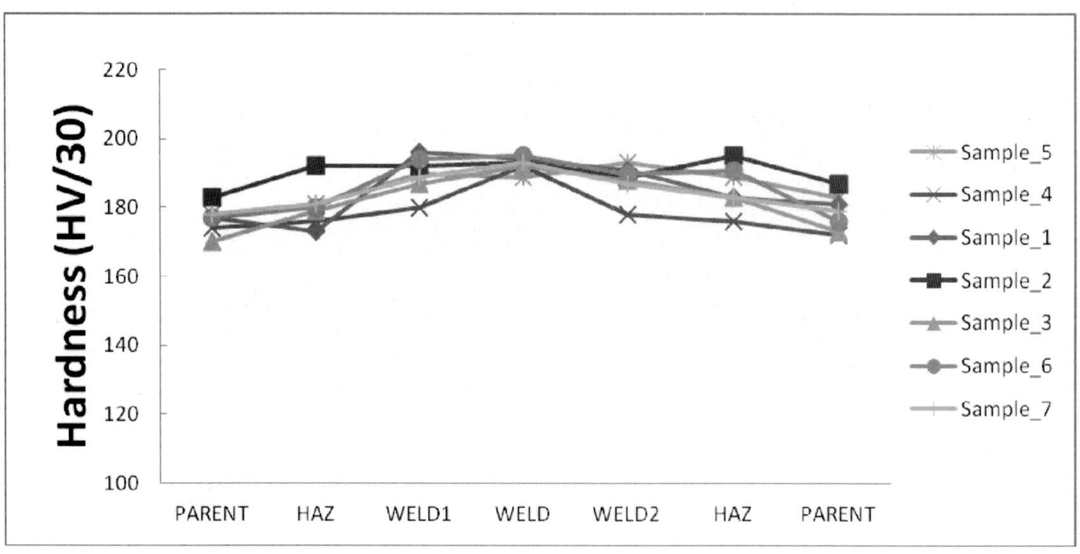

Fig. 13: Hardness Survey across the weld

SEM AND EDS ANALYSIS

The fracture face of the cracked wheel rims were ultrasonically cleaned and examined under Scanning Electron Microscope. Fractography study revealed ductile fracture with presence of inclusions (Fig. 14). The EDS analysis of the non metallic inclusion close to the weld revealed peaks of O, Al, Si Mn, S and Fe (Figs. 15 (a and b); Tables 3 and 4) indicating alumino-silicate type non metallic inclusion associated with manganese sulphide inclusion.

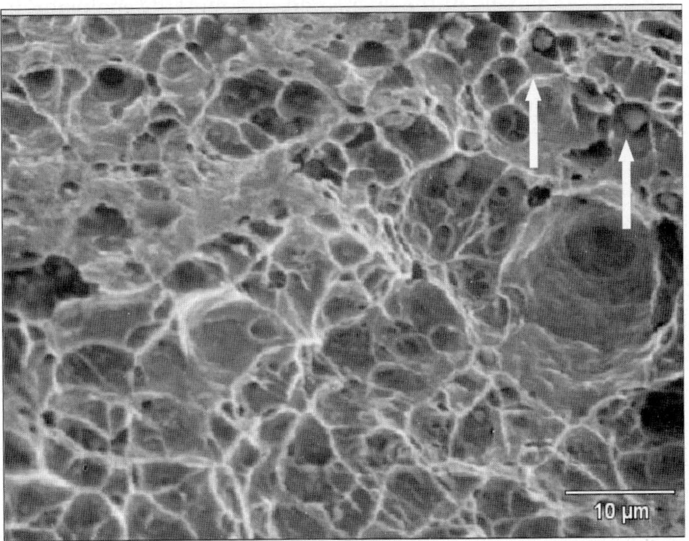

Fig. 14: Fractography of cracked region showing necking

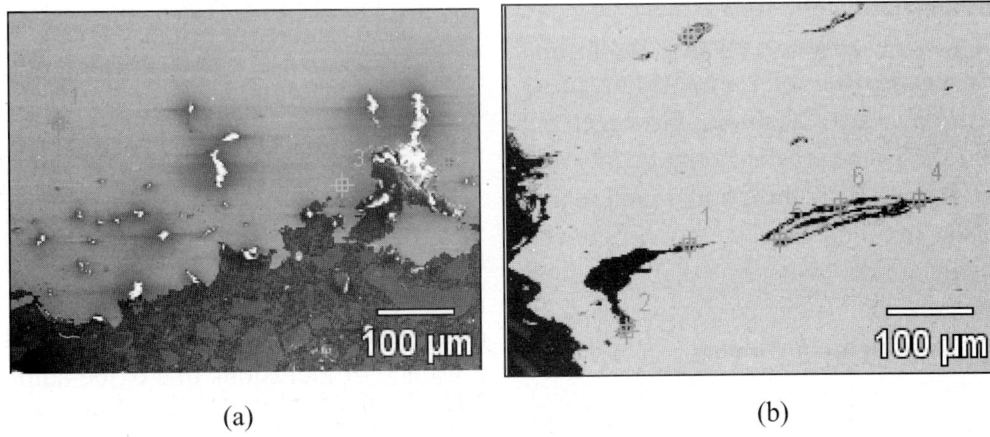

(a) (b)

Figs. 15 (a and b): Location of EDS analysis of the suspected non metallic inclusion

Table 3: Weight % analysis of the suspected non metallic inclusion in Fig. 15a

	O	*Al*	*Si*	*Mn*	*Fe*
Point_1	47.73	18.21	9.48	17.68	6.91
Point_2	49.74	10.43	13.49	24.22	2.12
Point_3	49.30	10.53	13.16	24.16	2.86

Table 4: Weight % analysis of the suspected non metallic inclusion in Fig. 15b

	O	*Si*	*S*	*Al*	*Ca*	*Mn*	*Fe*
Point_1	16.115	0.348	0.371	8.621		2.430	72.115
Point_2	11.683		0.586			0.777	86.954
Point_3						1.356	98.644
Point_4			8.796	9.004		14.353	67.846
Point_5	12.189	0.806	1.879	3.082	1.766	1.420	78.858
Point_6	15.664	2.135	13.234	5.106		16.998	46.863

DISCUSSION

The nature of failure of all the samples under study showed that the wheel rims failed from the weld. To diagnose the reasons for failure, chemical, physical and metallurgical properties of good samples were compared with failed material. The chemical composition of all the samples was within specification of Carbon-Manganese Fe 410 grade steel (Table1). However the Mn/Si ratio of the cracked components was higher than that of the good samples and chemistry of all the cracked wheels showed peritectic grade, whereas that of good sample revealed non peritectic. Usually steel with carbon equivalent 0.15 to 0.25% are known as peritectic grade chemistry [6]. The mechanical properties of all the samples conformed to the specification and the hardness survey along the weld found to be comparable in both the cracked sample as well as good sample. This confirms that in cracked sample the weld is satisfactory.

Metallurgical study revealed presence of non metallic inclusion at the defect area (Figs. 7a to 7d). On etching the same the inclusions were observed in the weld (Figs. 8a to 8d). The major micro-constituents of the steel weld were mixture of primary ferrite PF(G) and widemanstratten ferrite [5]. Interiors of the grains were transformed wholly to fine acicular ferrite AF. Incase of the cracked sample some lamellar ferrites aligned with second phases were observed (Figs. 10 and 12). The nature of the transformation product was influenced by the size and distribution of the nucleating particles. In presence of inclusions like oxide-sulphide-sillicate the widemanstratten ferrites take the form of a side plates, which are nucleated at the austenite grain boundaries and form lamellar ferrite with aligned second phase

FS(A)[5]. As a result lamellar ferrites were observed in the cracked sample, which were not found in the weld of good sample (Fig. 12). The marginal grain size difference between the HAZ of defective weld and the good weld was not of much importance as the hardness variation of all the welds was comparable. The EDS analysis of the non metallic inclusion close to the weld revealed alumino-silicate type non metallic inclusion associated with manganese sulphide inclusion (Figs. 15 (a and b); Tables 3 and 4). These are oxides which were formed in the sheet edge in the flashing process and remain in the weld surface without being ejected during upsetting. All the inclusions were lie beneath the fractured surface, so the inclusions were observed inside the dimples in the current study (Fig. 14). In principle the steel's chemical composition determines the presence of weld surface defects. In case of mild steel the rate of susceptibility of cracking increases with increasing Si, Mn and Al content in steel [7]. Basically Si, Mn and Al affect the formation of oxides in the steel, which originates cracking to the surrounding parts. The current study shows that high Mn/Si ratio around 12 to 13 caused more oxide formation at the weld, resulting in cracking during flaring. Lowering the same in the level of 8 has resulted in better result by reducing the rejection from 40 to 50% to 2.5%. Literature study reveals that in carbon manganese steel optimum impact toughness was achieved at Mn:Si ratio of 8 to 9 [9]. Mn:Si ratio affects mechanical properties through its influence on inclusion content[10]. Lowering the ratio by increasing silicon content results in a molten manganese silicate, this helps in welding [8]. The ratio outside this range forms solid oxide inclusions which got entrapped into the material resulting cracking. The cracking can further be decreased by use of the non peritectic chemistry as shown in (Table 1). In peritectic grade the ductility of the steel is lowest in the peritectic range due to the sudden contraction as a result of the peritectic reaction [6]. This leads to cracking during welding. The effect can be overcome by the use of non peritectic chemistry and susceptibility of cracking will be minimized.

CONCLUSION

Failure of the wheel rim occurred at the weld during flaring because of presence of non metallic inclusions. The higher Mn and Si ratio in the steel caused formation of non metallic inclusion, which could not eject during the upsetting and remain, associated with the weld. During flaring operation those act as the point of localized stress resulting in necking and cracking. Therefore the materials failed by the ductile fracture mode.

REFERENCES

[1] S. Ray, S. Hadas, A Dey; "Bench Marking: An Innovative Methodology to Achieve Customer Satisfaction"; RAFTS 2012, October 4-6, 2012, Ranchi, India;

[2] http://customwheelsmarket.com/wheeltecin.html

[3] Sandip Bhattacharyya *, M. Adhikary, M.B. Das, Sudipto Sarkar; "Failure analysis of cracking in wheel rims – material and manufacturing aspects"; Engineering Failure Analysis; 15 2008; Page 547–554;

[4] H. K. D. H Bhadeshia, R.W.K Honeycombe; "Steel Microstructure and Properties"; Butterworth-Heinemann Publications; Page 291-293;

[5] J. F. Lancaster; "Metallurgy of Welding"; Abington Publishing; Page 222-223;

[6] A Nayer, m.D. Maheshwari, T. Mukherjee; "Continuous casting of Steel"; Iron and Steel Division, Indian Institute of Metals, Jamshedpur; 1984; Page 36-37;

[7] Yasutomo Ichiyama, Shinji Kodama; "Flush Butt Welding of High Strength Steel"; Nippon Steel Technical Report No. 95; January 2007; Page 81-87.

[8] Gregory L. Dressel; "Sillicon Killed Steel Nozzle Clogging" Report of Dressel Technologies LLC;

[9] J.J. DeLoach, Jr. G.L. Franke, M.G. Vassilaros, R.J. Wong, R. DeNale; "Current Welding Consumables Research in the **U.S.** Navy"; Technical Report Submitted in March 1993;

[10] Widgery, D.J., 1974, *Welding Research International,* Vol. 4, No. 2, pp. 54-80

Improvement in Performance of Bell Annealing Line # 1 of Bokaro Steel Plant through Close Loop Expert System Based Heating Control

S. Mitra[1], S.K. Saha[1], R.S. Kumar[1], M. Gangadaran[1], B.K. Prasad[1], N.K. Ghosh[2] and S. Roy Choudhary[2]

[1]R&D Centre for Iron & Steel, Steel Authority of India Limited, Ranchi, India

[2]Bokaro Steel Plant, Bokaro Steel City, India

Email: smitra@sail-rdcis.com

ABSTRACT

The steel coils in cold rolling mill of an integrated steel plant are annealed to remove the mechanical distortions ensued during hot and cold rolling. It is an extremely crucial process and is an imperative factor upon which final steel quality depends. In annealing process, the coils are heated following a pre-defined thermal regime and then cooled slowly for grain re-crystallization. However in practical no well established co-relations are available, which can determine the thermal regime to be followed for a given set of coil parameters. Such indeterminate applications are the fittest areas, where the theoretical concepts of Artificial Intelligence or expert system in particular can be applied for real time process control. A close loop expert system guided heating control system has been developed for Annealing Furnaces of Line # 1 at Bokaro Steel Plant. The integrated automation is based on the concepts of hierarchical control, wherein a Level-II rule based expert system selects most appropriate thermal regime for a given set of steel coils and Level-I PAC supervises and control the furnace heating operation as per the guidelines of expert system. The application has rendered significant improvement in plants performance in terms of productivity, quality and energy savings.

KEYWORDS: *Annealing, Thermal Regime, Expert System, PAC, Hierarchical Control*

INTRODUCTION

Annealing Furnaces are key constituent of Cold Rolling Mill (CRM) in a steel industry. The process of annealing is extremely significant for achieving desired quality of cold rolled product. During hot and cold rolling of coils in mills, repeated rolling and work hardening causes the original steel grains to become highly distorted, elongated and dislocated. This severely deteriorates its key mechanical characteristics like hardness, ductility, formability, tensile strength etc., which drastically scales downs its commercial applicability. To eliminate these defects, steel coils after cold rolling are usually annealed. The annealing process mainly involves controlled heating of coils as per a pre-defined thermal regime or heating cycle followed by either natural or forced cooling for grain re-crystallization. Annealing softens the steel and makes it susceptible for subsequent forming operations.

In an annealing furnace, generally a group of 3 or 4 coils are stacked over the circular base. Over this base the coils are heated in an inert environment maintained by a mixture of Nitrogen and Hydrogen gas called Protective Gas (PG). To maintain the inert environment the coils are covered by a hood called PG cover. A bell shaped furnace is kept over this PG cover to heat the coils. Due to its bell shaped build these furnaces are generally called Bell or Hood Annealing Furnaces. The coils are heated by burners mounted all along the circumference of furnace. The fuel generally used is mixed gas, which is combination of gas derived from coke oven (called "Coke Oven Gas") and blast furnace (called "Blast Furnace Gas"). These furnaces are of movable type and once heating of a group of coil is completed, they are lifted by overhead cranes and kept over some other base for heating. The heated coils in the previous base are then cooled either naturally or by placing a cooling hood over it. Inert PG environment is maintained during cooling phase also. After the coil temperature is dropped below a certain temperature, PG cover is removed and coils are unloaded from base.

Fig. 1: A typical Annealing Furnace at Bokaro Steel Plant

In batch annealing line, several such bases and furnaces are used to cater the requirement of cold rolling mill. The total annealing operation generally takes around 100 hours. Figure 1 shows a typical Bell Annealing Furnace in Bokaro Steel Plant under heating condition.

One of the key factors responsible for accomplishing optimum annealing is selection of appropriate heating cycle for a set of coils. A typical heating cycle followed in an annealing furnace is indicated in Figure 2.

As evident from Figure 2, heating cycle comprised of four steps – (i) heating from initial temperature to an intermediate temperature 'θ_{S1}' in time 't_{h1}', (ii) maintaining temperature 'θ_{S1}' for time 't_{s1}' called first soaking, (iii) heating from intermediate temperature 'θ_{S1}' to 'θ_{S2}' in time 't_{h2}' called second heating and (iv) maintaining temperature 'θ_{S2}' for time 't_{s2}' called second soaking. These values are generally dependent on coil parameters and steel quality as shown below:

$$\theta_{S1} = f_{n1}(g, w, m, s) \quad (1)$$

$$\theta_{S2} = f_{n2}(g, w, m, s) \quad (2)$$

$$t_{s1} = f_{n3}(g, w, m, s) \quad (3)$$

$$t_{s2} = f_{n4}(g, w, m, s) \quad (4)$$

$$t_{h1} = f_{n5}(g, w, m, s) \quad (5)$$

$$t_{h2} = f_{n5}(g, w, m, s) \quad (6)$$

where, g = coil thickness or gauge, w = coil width, m = coil weight and s = steel grade. However, no well established co-relations are available which can define the exact nature of the functions f_{n1}, f_{n2}, f_{n3}, f_{n4} and f_{n5} as mentioned in equations (1) to (6) above. Further in an annealing furnace, three or

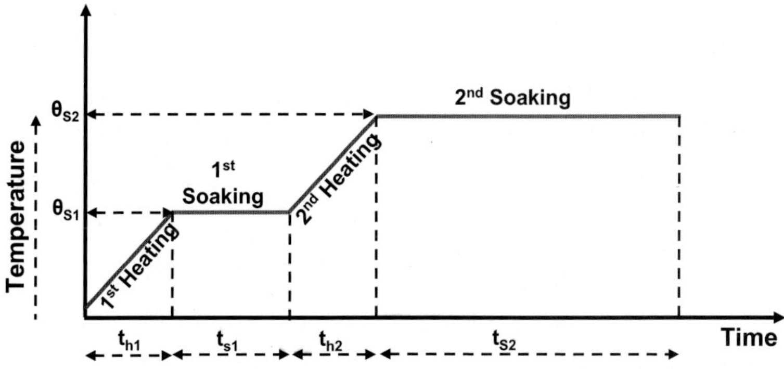

Fig. 2: A typical heating cycle

four coils with different values of 'g', 'w', 'm' and 's' are heated together making the functions even more complicated. However, these are the application areas where the concepts of expert system can be practically implemented. The present paper illustrates, how effectively theoretical concepts of expert system can be implemented in such applications, where exact process behavior is not completely known.

A march forward has been taken by Research & Development Centre for Iron & Steel (RDCIS) and Bokaro Steel Plant (BSL) by successfully implementing a very advanced close loop Expert System based Heating Control System in Bell Annealing line # 1 (BAL1). The expert system selects the optimum heating cycle for a combination of coils as would have been done by a group of experts. The rule based system issues several on-line guidance, warnings and alarms to the furnace heating operators as and when required.

SYSTEM ARCHITECTURE

The integrated system has been designed based on the concepts of hierarchical automation. Rule based expert system comprise the Level-II of hierarchy and is responsible for selection of most appropriate heating cycle for three or four steel coils to be annealed. Expert system also generates guidelines and suggestions as per the prevalent operational practices. Process Automation Controller (PAC) based control system constitutes the Level – I of hierarchy and is responsible for actual furnace temperature control. Field instrumentation comprises the Level-0 of hierarchy.

Expert System

Expert System is a special branch of Artificial Intelligence (AI), where the knowledge or expertise of an expert or a group of expert to resolve a real time problem is transmuted as several rules through software. The system then takes most appropriate decision as would have been taken by an expert, when similar situation actually arise. The availability of an expert cannot be ensured all the times. However his expertise, experience and plant performance analysis data can be embedded together in computer, which takes suitable and appropriate decision as per the situation like an expert. Expert System is one of the most successful applications of AI in industry. The scenario mentioned in aforesaid section, where no exact mathematical models are available to simulate the process behavior, is a fit application area for implementation of an Expert System. The years of experience and expertise of shop floor operation engineers, researchers and quality control agencies framed the knowledge base of expert system. This knowledge base amassed in computer memory combined with operational practices transmuted as hundreds of 'if' and 'then' rules resolved the issue. The expert system selects the optimum heating cycle for a combination of coils as would have been done by a group of experts. The rule based system issues several on-line guidance, warnings and alarms to the furnace heating operators as and when required. Supervisory monitoring of actual heating process and provision of correction / modification in selected heating cycle, if the same is actually being not adhered has been incorporated into the system. The expert system constitutes the Level-II of hierarchical automation.

As mentioned in earlier sections, data congregated from years of expertise, research and quality analysis are imbibed in computer memory. This encompassed with operational practices transmuted in software language through several 'if' and 'then' rules comprise the expert system. The system selects optimum values of θ_{S1}, θ_{S2}, t_{h1}, t_{s1}, t_{h2} and t_{s2} out of hundreds of combination co-related with steel grade, coil width, weight and thickness of three / four coils to be annealed together in the furnace. Expert rules have been framed to tackle the complex situations of different coil combinations like:

i) Comparatively higher and lower thickness coils to be annealed together.

ii) Coils of higher steel grade like "drawing quality" to be annealed together with 'ordinary' quality.

iii) Higher weight coils kept along-with lesser weight.

iv) Different mix in the width.

Expert system renders the furnace operators to enter coil number, weight, width, steel grade etc. through a graphics user interface (GUI). The system issues several warnings and alarms during the data entry to ensure the standard operational practices. For example warnings are issued, if special grade steels are mixed with ordinary grade, very thick coils are mixed with thinner coils, total height exceeds certain pre-defined values etc. Based on the entered data and embedded knowledge base rules, expert system selects the most appropriate and optimum heating cycle for the given combination of coils. A typical heating cycle selected by the expert system is delineated in Figure 3. On confirmation of furnace light-up by operator, expert system configures the attributes of a function generator block of Level- I PAC controller with the selected values of θ_{S1}, θ_{S2}, t_{h1}, t_{s1}, t_{h2} and t_{s2} for further control action.

However, if during the heating process, it is noticed that actual cycle is being deviated from scheduled due to any other reasons like burner mechanical problem etc., the expert system facilitates the furnace operator with provision for modifying the original schedule based on operational rules. At the end of heating cycle the set point is dropped to zero, thus automatically closing the heating gas valve and consequently

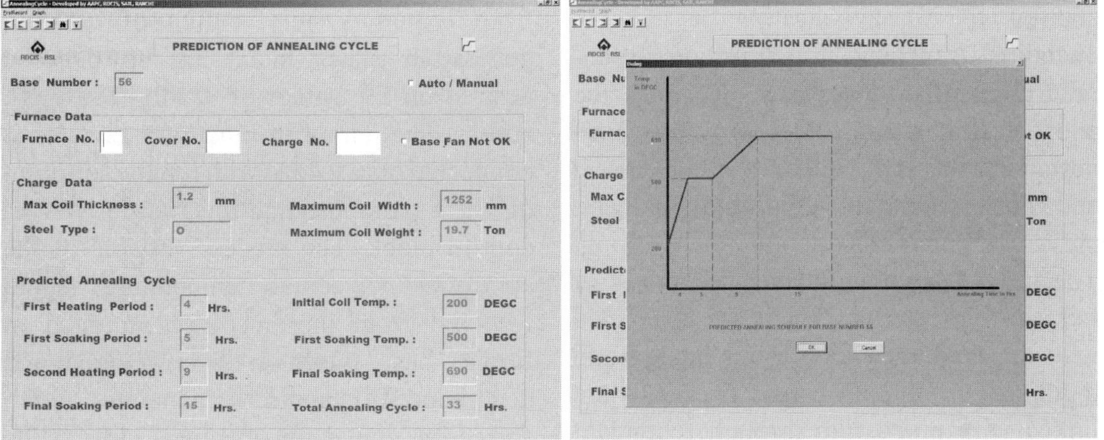

Fig. 3: Heating cycle selected by expert system

stopping the furnace heating. This avoids the unnecessary wastage of scarce energy resource of heating gas. The system keeps on tracking the temperature during cooling cycle and advises the operators to unload the base / furnace, once the coils are cooled down below a certain temperature. All the data pertaining to coils dimensions, selected cycle, actually followed cycle, different events times are stored in large database for future performance and quality analysis purpose. As HMI, expert system generates all operational reports like loading report, production report, unloading report, morning report, monthly report etc.

Expert system is developed using Microsoft VC++ (MFC) while Rockwell Automation make Control-logix is used as Level-I PAC. Both the above applications in different platforms are dovetailed through OPC based interface software. OPC i.e. OLE (Object Linking and Embedding) for Process Control is an open interface for integrating different automation devices and is basically based on Microsoft COM / DCOM and more recently .Net technology. OPC facilitates inter application linkage for data access, historical trends and alarm / event. It is based on Microsoft COM / DCOM base interface IUnknown, which by virtue of inheritance permits read / write data access between two applications written using different language and under different platform. Expert System configures PAC control modules with selected heating cycle using synchronous data access. During furnace heating, Expert System continuously supervises the control action using asynchronous data access. In asynchronous mode PAC intimates expert system as soon as a data value is changed, while in synchronous mode expert system issues exclusive read / write requests to PAC. Expert System also facilitates the shop operators with vital information like current heating / cooling status of the line, base readiness status, unloading status etc. It generates post annealing reports like charge report, morning report, monthly reports etc.

PAC based heating control system

PAC or Process Automation Controller is Level-I tier of automation. As PAC actually controls the furnace heating, its reliable operation is critical for maintaining uninterrupted plant operation. Keeping the above factor in consideration, it has been designed with high availability, fault tolerant and dual redundant configuration w.r.t. processor, communication and networking. The Level-0 automation tier comprises of field instrumentation which are connected with the PAC through various analog and digital input / output cards. The I/O's are linked with processors through redundant deterministic network. Human Machine Interface (HMI) is connected with PAC using redundant industrial Ethernet. The instrumentation comprise of base / furnace temperature, valve position feedback, base fan running status, gas flow / pressure, H_2 % and O_2 PPM in PG etc.

During furnace light-up, Expert System first configures the attributes of a ramp type function generator block in PAC software with the values of θ_{S1}, θ_{S2}, t_{h1}, t_{s1}, t_{h2} & t_{s2} and triggers it for signal generation in auto mode as shown in Fig. 2. With a trigger from Expert System, output of function generation starts increasing as per the desired profile values of θ_{S1}, θ_{S2}, t_{h1}, t_{s1}, t_{h2} and t_{s2}. The output of function generator is connected or cascaded with remote set-point of PID (proportional-integral-derivative) control block. The process variable of the controller is actual temperature signal received from thermocouple mounted in furnace base. Based on the error signal and dead-band, PID controller generates output correction signal for opening / closing the valves. By default expert system configures function generator in auto mode and PID in cascade mode. However, provision of manual intervention has been incorporated to change the modes (cascade, auto or manual) of different modules to meet emergency conditions. Graphics user interface screen for the same is indicated in Figure 4. The different segments of heating cycle can be changed using this GUI. The GUI also indicates the furnace operators about the current status of heating and remaining time left.

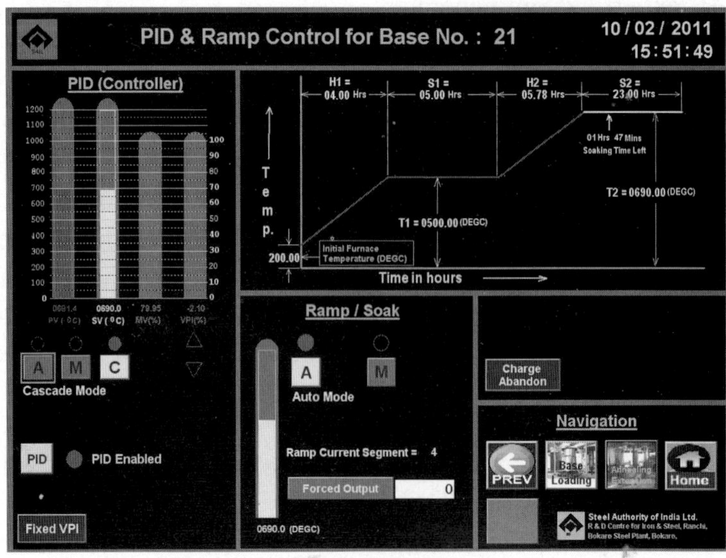

Fig. 4: GUI of Control Operation

At the end of heating cycle the control system forces the set point to drop to zero. This forces the gas valve to close automatically and consequently stopping the furnace heating immediately. Closing of gas valve exactly at the end of heating cycle would not have been possible without the aid of any automatic control system, as manually keeping track of closing time of 100 nos. of base / furnace is extremely difficult. The above action contributes in reducing the unnecessary wastage of heating gas. The control system keeps on tracking the process temperature during cooling phase also. As soon as the coils are cooled below a certain pre-defined temperature, the event is informed to furnace operators by expert system prompting them to remove the inert PG cover and unload the coils.

RESULTS

The close loop expert system is being used for controlling entire 100 bases of Annealing Line # 1 at Cold Rolling Mill of Bokaro Steel Plant. The system exemplifies an industrial application of close loop Expert System. It has demonstrated that expert system can be effectively used in these types of indeterminate situations, where definite mathematical co-relations are unavailable. The rules framed with the knowledge of experts tremendously helped in resolving complex issues like selection of cycle for a divergent combination of coils, actual cycle not being followed due to reasons beyond control etc. The real time close loop control is achieved by dovetailing the decisions of expert system with PAC.

The application has introduced a very efficient temperature control system within ± 5–10 °C accuracy under normal operating conditions The accurate temperature control is depicted in actual historical trends depicted in Figure 5, wherein the actual process temperature has closely followed the heating cycle decided by expert system. Accurate temperature control and

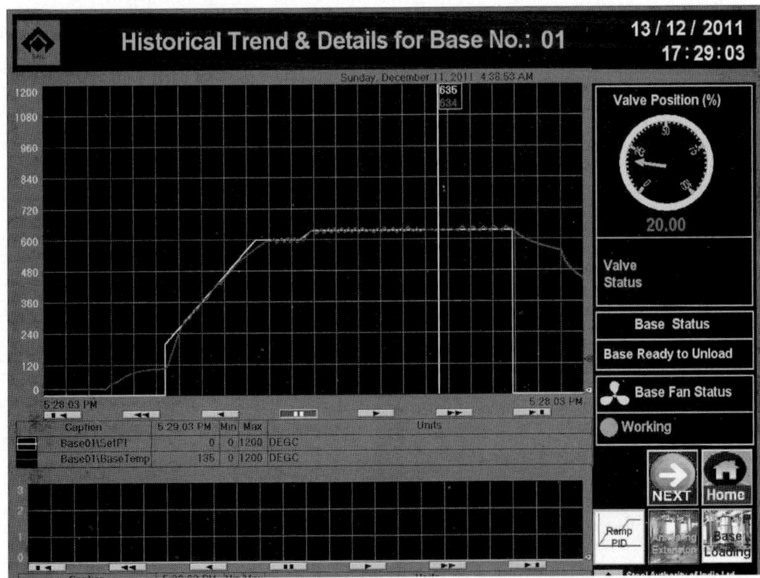

Fig. 5: Trend depicting accurate control

adherence to the guidelines of expert system has significantly improved the product quality in terms of reduction of material downgrading and rejections. The product quality has improved in terms of 62% reduction in sticker formation, which is one of the prime cause of material downgrading and rejections. With the application of Expert System guided control system, plant is able to reduce the average annealing period by around 7.8 hours per charge. This has increased plant productivity by 4%. The same has also resulted in huge savings in consumption of expensive protective gas (PG) and scarce fuel gas by 250 x 10^5 Nm^3 per annum and 435 x 10^3 Nm^3 per annum respectively.

CONCLUSION

The above results are exemplary demonstration of close loop Expert Systems resolving complex industrial process control. It also illustrates that significant benefits can be achieved by introducing appropriate automation like Expert System and Hierarchical Automation by carefully integrating it with actual on-line process control.

REFERENCE

[1] Xiaohui Fan, Xuling Chen and Yi Wang, "Expert System for Sintering Process Control" InTech, January 2010.

[2] Mitra S, Gangadaran M, Bhaskar U, Kumar R. S., "OPC : A COM gateway for Process Control", International Conference on Automation & Information Technology in Iron & Steel Making Processes (AITISM), 2007, pp 26 -28

[3] MA Zhuwu. Automation for Ironmaking, Metallurgy Industry Press, ISBN 7-5024-3639-1, Beijing, 2005.

[4] Chunhua Yanga, Min Wua, Deyao Shena, Geert Deconinck "Hybrid intelligent control of gas collectors of coke ovens", Control Engineering Practice 9 (2001), pp 725–733

[5] Min WU, Michio NAKANO, Jin-Hua SHE, "A Model-Based Expert Control Strategy using Neural Networks for the Coal Blending Process in an Iron and Steel Plant", Expert System with Applications, 1999, Vol. 16, No. 3, pp. 271-281.

[6] Mishra S, Mahajan G. J, Mitra. S, Majumdar. S, Gangadaran. M, Jayaswal N. P., Krishnaswamy. K, Sukhuja S. C., Bedi. B. S. "Control System for hood annealing", Asia Steel, 1997, pp 254 – 256.

[7] Anabuki, Y., Nakazi, S., Iwamura, T., Katae, T., Ohnishi, T. "Application of an expert system to real time coil transportation control of a steel finishing line", Iron and Steel Engineer, January 1996, pp 30-35.

[8] www.opcfoundation.org

Die Stress Optimization Using Quantum Behaved Particle Swarm Optimization

Md. Israr Equbal[1], Anoop Kumar Sood[2] and Aarif Hussain[3]
[1]Department of Forge Technology, National Institute of Foundry and Forge Technology Ranchi-834003, India
[2]Department of Manufacturing Engineering, National Institute of Foundry and Forge Technology, Ranchi-834003, India
[3]Department of Materials and Metallurgical Engineering, National Institute of Foundry and Forge Technology, Ranchi-834003, India
Email: israr_equbal@yahoo.co.in

ABSTRACT

Study proposes a methodology for optimization of die stress via rational design of metal forming process based on three dimensional finite element simulations. For this purpose full factorial experimental design is performed to study the effect of studied factors and their interaction on the die stress. Statistically valid equation is developed and used with quantum behaved particle swarm optimization (QPSO) to determine the optimal values of factors for the minimization of die stress.

KEYWORDS: *DIE Stress, Finite Element Simulation, Design of Experiment, QPSO*

INTRODUCTION

The Forging technology is known for producing parts of superior mechanical properties with minimum waste. In it metals or alloys are plastically deformed to the desired shapes by a compressive force applied with the help of a pair of dies [1]. Die performance and service life decide product quality, time-to-market and production cost. Without suitable die, metal-forming processes are often crippled or rendered totally inefficient [2]. Several tools and approaches have been used such as CAD/CAM, expert system and FEA methods to obtain improvements of die life by proper die design but they rarely include influence of process conditions [3]. Based on this present work concentrates on the study of effect of the process parameters viz. billet temperature, die temperature, friction coefficient and die velocity on the die stress which is a measure of die life. Based on the parametric design approaches the systematic influence of factors and there integrations are considered using three dimensional simulation approach. The approach is statistically validated and finally optimized using quantum behaved particle swarm optimization (QPSO) algorithm [4].

Problem Formulation

For die stress analysis and optimization via rational design of forming system, a part as shown in Fig. 1 (a) is used as a case study. In Fig. 1 (b) also shows the geometry of punch used to get the deformed product. Material of the Product and Die is specified in Table 1 and Table 2 provides the values of operation parameters considered.

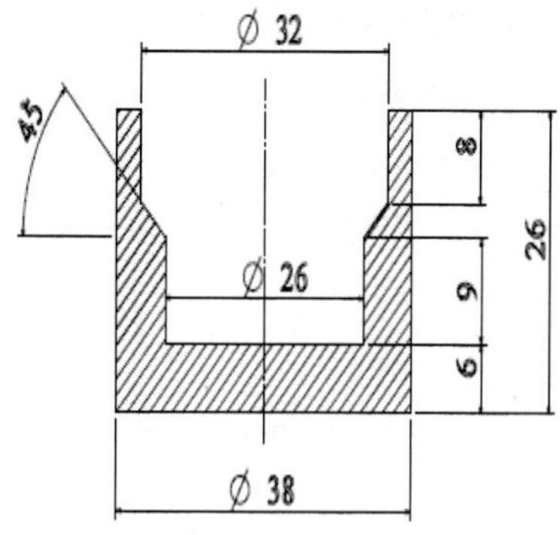

Fig. 1(a): Deformed product

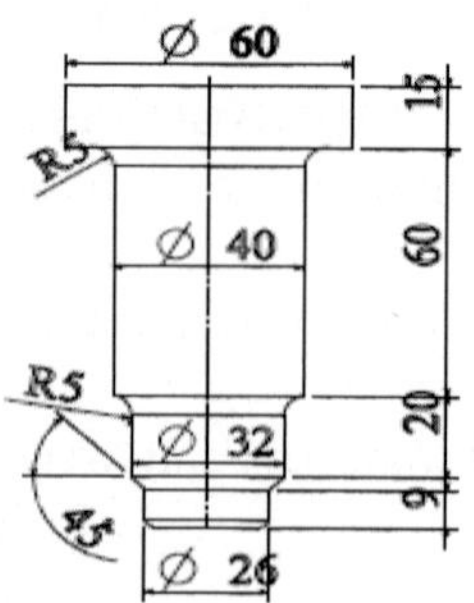

Fig. 1(b): Punch geometry

Table 1: Material properties

	Billet Material	Die Material
Type	AISI 1043	AISI H13
Temperature (^{0}C)	700-1100	800-1000
Thermal Conductivity (W/m-k)	42.7	28.6
Heat Capacity (J/kg.K)	500	460
Density(10^3 kg/m^3)	7.7-8.03	7.76
Tensile Strength, Ultimate(MPa)	1158	640-2000
Tensile Strength, Yield(MPa)	1034	640-2000

Table 2: Process parameters and their levels

Process parameters	Symbol	Level 1	Level 2
Billet temp. (^{0}C)	A	1000	1100
Die temperature (^{0}C)	B	250	300
Friction coefficient	C	0.3	0.7
Die velocity (mm/s)	D	0.5	1

METHODOLOGY

To reveal the die stress in the different process parameters, the simultaneous simulation of billet plastic flow and the elastic deformation of die are conducted. Following are the different steps undertaken:

(i) Calculation of Billet volume

The billet is cylinder and its required dimension is calculated assuming volume constancy of billet and deformed part.

Vol. of Component = 16271.87915 mm^3

Vol. of Billet = Vol. of comp. + Scale loss

Dimension of Billet = 37.6*14.7(mm*mm)

(ii) Preparation of CAD model of the product

A CAD model of the product is prepared in CATIA using the specification provided in Fig. 1. The design is then converted in .STL format so that it can be imported for simulation in DEFORM-3D.

(iii) Design of experiment (DOE)

Design of experiment (DOE) has been a very useful tool to design and analyze complicated industrial design problems. It has been used to systematically determine the optimal process parameters with fewer testing trials and to find out what happens to the output or response when the settings of the input variables in a system are purposely changed.

In many scientific Investigations the interest lies in the study of effects of two or more factors simultaneously. Factorial designs are most commonly used for this type of investigation. Here we consider the important class of factorial designs for k factors each at two levels. Because this class of designs requires 2 x 2 x … x 2 = 2^k observations, it is referred to as 2^k factorial designs. It is also called the class of 2^k full factorial designs. Since each factor only has two levels, they are used to study the linear effect of the response over the range of the factor levels chosen. Factorial effects, which include main effects and interaction effects of different order, are defined. Estimation and testing of factorial effects for location and dispersion models are considered for replicated and un-replicated experiments [5]. An experiment was created and it employed full factorial arrangements, that is, the design comprised all possible combinations of factors considering different levels. The process parameters are selected and examined, each at two levels. The different combinations are shown below in Table 3.

Table 3: Experimental set up

Exp. No.	A	B	C	D
1	1000	250	0.3	0.5
2	1100	250	0.3	0.5
3	1000	300	0.3	0.5
4	1100	300	0.3	0.5
5	1000	250	0.7	0.5
6	1100	250	0.7	0.5
7	1000	300	0.7	0.5
8	1100	300	0.7	0.5
9	1000	250	0.3	1
10	1100	250	0.3	1
11	1000	300	0.3	1
12	1100	300	0.3	1
13	1000	250	0.7	1
14	1100	250	0.7	1
15	1000	300	0.7	1
16	1100	300	0.7	1

On the basis of above experimental set up, total 16 experiments have been conducted by the use of computer simulation technique using Deform 3D software.

(iv) Simulation

Here Simulation has been carried out for various design scenario discussed above. For simulation we used DEFORM-3D software. Following are the simulation properties used as given in Table 4.

Table 4: Operation parameters assigned to complete the simulation

Forging Equipment	Punch Die
Number of elements	45000
Mesh type	Tetrahedral
Die displacement	0.5 mm
Simulation mode	Isothermal
Simulation step	50
Step increment	2
Primary die	Top die
Environment Temperature	20°C

After conducting all the simulation, the different effective stress has been noted for further analysis and optimization.

(v) Analysis of Experiments

Analysis of the experimental data obtained from full factorial design is done on MINITAB R14 software using full quadratic re- sponse surface model as given by eq. (i).

$$y = b_0 + \sum_{i=1}^{k} b_i x_i + \sum_{i=1}^{k} b_{ii} + \sum_{kj} \sum b_{ij} x_i x_j \quad \text{(i)}$$

Where y is the response, x_i is i^th^ factor, k is total number of factors.

The t-test was performed to determine the individual significant term at 95% of confidence level and final response surface equations for the input parameters and the output. To check the feasibility of the experimental value the residuals of each trial are computed. The probability plot of residuals is plotted keeping the 95% confidence level for the fitted distribution.

(vi) Optimization

An optimization consists of maximizing or minimizing a real function by systematically choosing input values from within an allowed set and computing the value of the function. More generally, optimization includes finding "best available" values of some objective function given a defined domain, including a variety of different types of objective functions and different types of domains.

Based on the analysis made, the most significant factors on Die stress are evaluated. The optimized parameter levels that can be established for minimum Die Stress is found using QPSO technique.

In the quantum model of a PSO called here QPSO, the state of a particle is depicted by wave function ψ(x; t), instead of position and velocity. The dynamic behaviour of the particle is widely divergent form that of that the particle in classical PSO systems in that the exact values of xi and vi cannot be determined simultaneously. In this context, the probability of the particle's appearing in position xi from probability density function v(x; t) [6].

Employing the Monte Carlo method, the particles move according to the following iterative eq. (ii) and (iii).

$$X_i\,(t+1) = p + b \times [\text{Mbest} - X_i\,(t)] \times \ln(1/u), \quad \text{if } k \geq 0.5 \quad \text{(ii)}$$

$$X_i\,(t+1) = p - b \times [\text{Mbest} - X_i\,(t)] \times \ln(1/u), \quad \text{if } k < 0.5 \quad \text{(iii)}$$

Where **b** is a design parameter called contraction–expansion coefficient, **u** and **k** are values generated using the uniform probability distribution functions in the range [0,1].

The global point called Mainstream Thought or Mean Best (*Mbest*) of the population is defined as the mean of the *Pbest* positions of all particles and it given by eq. (iv).

$$\text{Mbest} = 1/n \sum_{d=1}^{n} p_{g,d}(t) \sum_{d=1}^{n} p_{g,d}(t) \quad \text{(iv)}$$

where g represents the index of the best particle among all the particles in the swarm. In this case, the local attractor to guarantee convergence of the PSO presents the following coordinates given by eq. v.

$$P = (C_1 P_{i,d} + C_2 P_{g,d})/(C_1 + C_2) \quad \text{(v)}$$

Following are its steps:

Step1. Initialization of swarm positions: Initialize a population (array) of particles with random positions in the n-dimensional problem space using a uniform probability distribution function.

Step2. Evaluation of particle's fitness: Evaluate the fitness value of each particle. We are minimizing, rather than maximizing, the fitness function in this paper.

Step3. Updating of global point: Calculate the ***Mbest***

Step4. Comparison to ***pbest*** (personal best): Compare each particle's fitness with the particle's ***pbest***. If the current value is better than ***pbest***, then set the ***pbest*** value equal to the current value and the pbest location equal to the current location in the n-dimensional space.

Step5. Comparison to ***gbest*** (global best): Compare the fitness with the population's overall previous best. If the current value is better than gbest, then reset gbest to the current particle's array index and value.

Step6. Updating of particles' position: Change the position of the particles where c1 and c2 are two random numbers generated using a uniform probability distribution in the range [0,1].

Step7. Repeating the evolutionary cycle: Loop to Step 2 until a stop criterion is met, usually a sufficiently good fitness or a maximum number of iterations (generations). After utilizing all the steps we find the optimized value of die stress and the factors.

RESULTS

After setting all the parameters in DEFORM pre-processor all 16 simulations are run. The initial Positioning of Dies are shown in Fig.2. Different views of deformed product are shown in Fig.3 and Fig.4. The result of all simulation has been shown in Table 6.

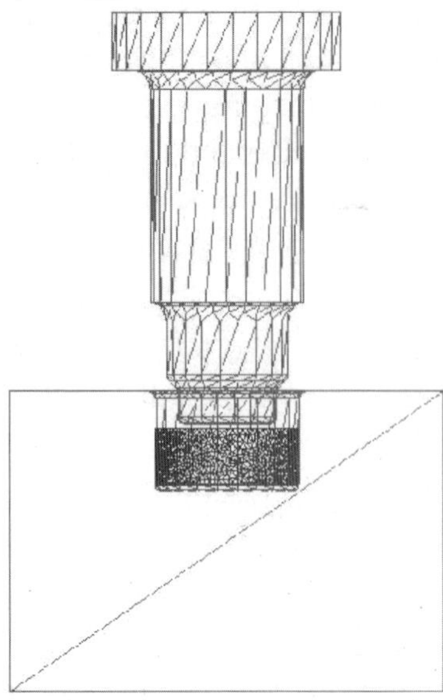

Fig.2: Initial Positioning of Dies

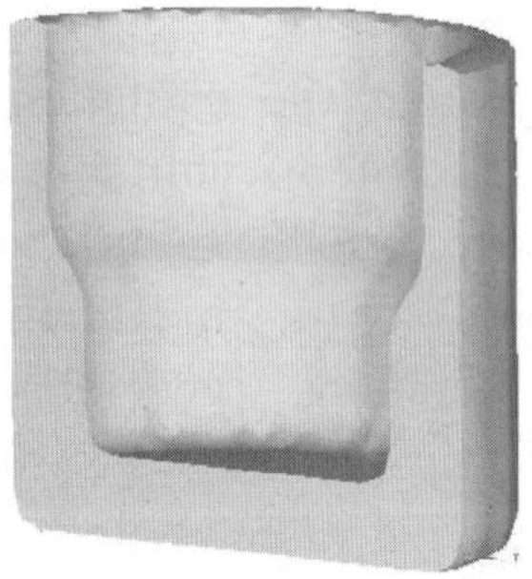

Fig. 3: Solid view of deformed product

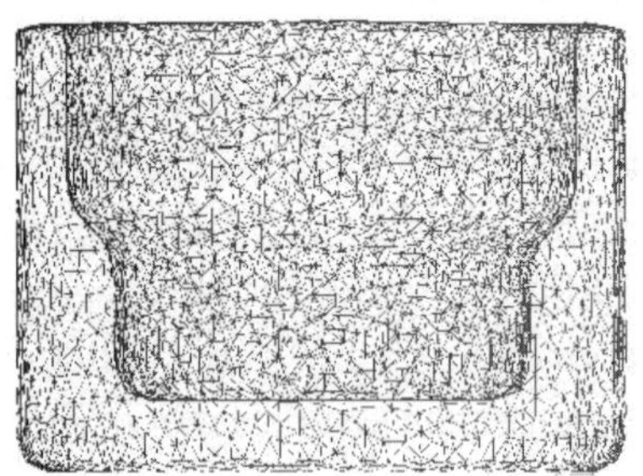

Fig. 4: Meshed view of deformed product

Table 6: Simulation results of Die Stress

Exp. no.	A (^{0}C)	B (^{0}C)	C	D (mm/s)	Die Stress (MPa)
1	1000	250	0.3	0.5	557
2	1100	250	0.3	0.5	467
3	1000	300	0.3	0.5	544
4	1100	300	0.3	0.5	482
5	1000	250	0.7	0.5	835
6	1100	250	0.7	0.5	548
7	1000	300	0.7	0.5	798
8	1100	300	0.7	0.5	566
9	1000	250	0.3	1	523
10	1100	250	0.3	1	476
11	1000	300	0.3	1	625
12	1100	300	0.3	1	473
13	1000	250	0.7	1	773
14	1100	250	0.7	1	439
15	1000	300	0.7	1	835
16	1100	300	0.7	1	546

ANALYSIS OF RESULTS

Relative influence of each factor is determined by analysis of variance method (ANOVA) and results are presented in Table 7. ANOVA results show that both main effect and interaction are important.

Table 7: Analysis of variance for Die Stress

Source	D.O.F	SS	Variance	F- value	P- value
Main Effects	4	273472	68367.9	534.28	0.000
2- Way Interactions	6	44790	7465.0	58.34	0.000
Residual Error	5	640	128.0		
Total	15	318901			
D.O.F. = Degree of Freedom; SS = Sum of Squares					

Final equation is given as:

Stress = *586.69-99.56A+21.94B+80.81C-12.94D+7.69AB-43.19AC-15.69AD-3.19BC+24.06BD-6.31CD* (vi)

Terms in the above equation explain the total variation of 99.80%. The error between the values predicted using developed equation and experimental values shows the normal distribution as in Fig. 5.

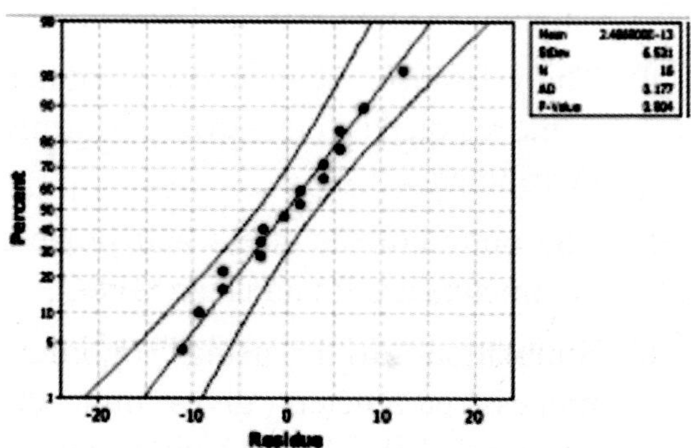

Fig. 5: Probability plot of residue

These results explain that develop model is valid and can be used for optimum value determination. Half normal probability plot is employed to graphically predict the significant factors and terms and results are shown in Fig.6.

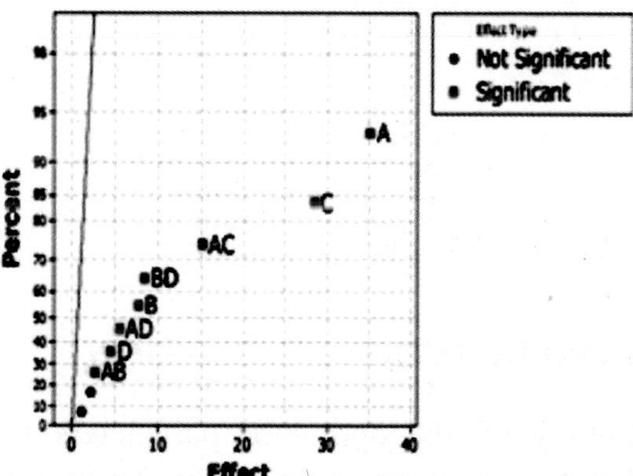

Fig. 6: Half normal plot at 95% of confidence level

OPTIMIZATION

For optimization purpose with population size of of iteration as 50 depicted in Fi

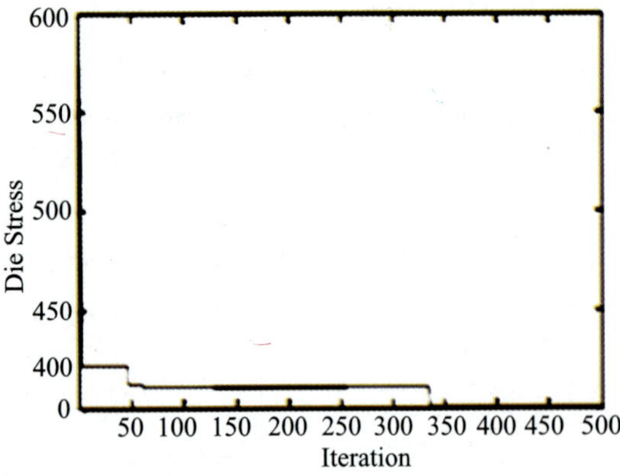

Fig. 7: Convergence curve for QPSO

The minimum die stress value found is **400.57 MPa**. The process parameters at this value are given in Table 8.

Table 8: Optimum process parameter values

Process Parameter	Value
Billet temp.(°C) (A)	1099.51
Die temp.(°C) (B)	252.6375
Friction co. (C)	0.3367
Die vel.(mm/s) (D)	0.904425

CONFIRMATION SIMULATION

Setting all the optimized parameters we get by optimization is set in DEFORM pre-processor and simulation is being run. Die Stress distribution at this value shows that the optimization is achieved Die Stress Distribution is shown in Fig. 8.

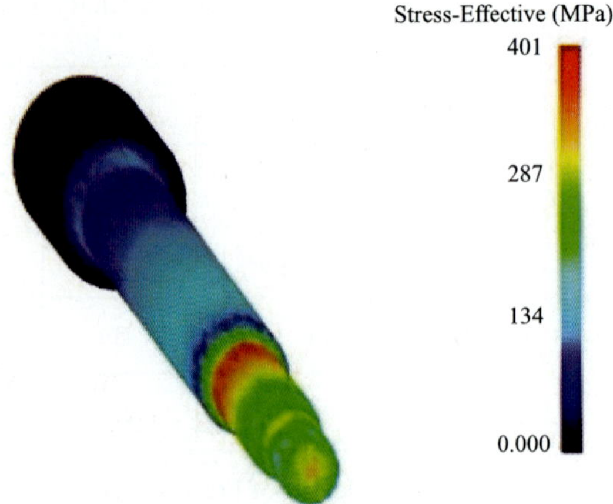

Fig. 8: Stress distribution in the die

CONCLUSIONS

1. The Process parameters taken here are found to be significant in Die Stress.
2. The interaction of process parameters are also found significant in Die Stress.
3. A statistical relation is found using factorial method between the input parameters and the output Die Stress.
4. The relation is statistically validated as the residual is found to be uniformly distributed.
5. The optimum die stress and process parameters are found using QPSO.
6. Simulation trial using optimum process parameters gives min. die stress of 401 MPa which is nearly equal to the optimized die stress found using QPSO.
7. The use of Factorial method and QPSO to optimize the die stress is found to be an effective tool as it saves time and effort both.

REFRENCES

[1] ASM Metals Handbook: Forging and Forming, Vol.14, 9th Edition, ASM Handbook Committee, USA, 1988.

[2] American Society for Metals "Forging Design Handbook, June 1972. pp. 35-60.

[3] M. S. Joun, M. C. Lee, and J. M. Park, Finite element analysis of pre-stressed die set in cold forging, International Journal Machining Dies Manufacturing, Vol. 42, 2002, P.1213–1222.

[4] Leandro dos Santos Coelho, Gaussian quantum-behaved particle swarm optimization approaches for constrained engineering design problems, Expert Systems with Applications, Vol. 37, 2010, P. 1676–1683.

[5] I.N. Vuchkov and L. N. Boyadjieva, Quality Improvement with Design of Experiments: A Response Surface Approach, Kluwer Academic Publishers. Dordrecht, 2001.

[6] Santos Coelho and Leandro dos, Gaussian quantum-behaved particle swarm optimization approaches for constrained engineering design problems, Vol.1155, Curitiba, PR, Brazil, pp.80215-901.

Improvement in Lubrication at Roll-Bite through Heat Balance at Cold Mill

A.K. Marik[1], N. Mondal[1], S. Rath[1], S. Sarkar[2] and Rakesh[2]
[1]Research & Development Centre of Iron & Steel, SAIL, Ranchi-834004, India
[2]Bokaro Steel Limited, Steel Authority of India Limited, Bokaro-834004, India
Email: marik@sail-rdcis.com

ABSTRACT

During Cold rolling process, the desired strip thickness is being achieved through high deformation energy imparted by the rolls. At the time of cold deformation, a huge amount of heat is being generated at the roll-bite due to deformation energy and higher friction of the mating surfaces of the strip and roll. Rolling lubricant is applied at the roll-bite in the form of oil-in-water emulsion to control the friction and wear at the interface during the deformation process. A need for efficient lubrication system is required to control the rolling performance as well as thermal condition at roll-bite of Tandem Cold Mill (TCM). This will address inadequate lubricity at work roll-strip bite, which result in higher roll temperature, roll thermal failure and poor strip surface quality. This paper depicts the method to control of heat balance at roll-bite in an effective way for improved lubricity at roll-bite and emulsion performance in TCM.

A Mathematical Model was developed to determine stand-wise emulsion flow rate considering pass reduction, heat generation due to deformation and friction at roll-bite of TCM. A modified emulsion application system was designed and commissioned based on the predicted emulsion flow rate for achieving maximum heat transfer and optimum lubricity at roll-bite through higher impact density, aligned emulsion spray band on the roll. The total emulsion flow was redistributed for intense cooling of Work Roll with double row spray concept for imparting effective lubricity and controlled friction at interface.

The modified emulsion application system has resulted in reduction of roll surface temperature by 8°C~9°C and thermal crown by 3°C~4° led to reduce specific roll consumption by15% and improved productivity by 5%. It has also helped to reduce the iron pick-up defect leading to elimination.

KEYWORDS: *Deformation, Friction, Emulsion, Lubricity, Roll-Bite*

INTRODUCTION

Friction plays an important role in cold rolling process and ensures traction of strip in the roll-bite. Higher friction, however, leads to increased loss of roll power in the form of thermal energy. Some of the important factors influencing friction at the roll-bite are feed stock quality (material & grade of steel being rolled), surface roughness of mating surfaces(roll & strip), rolling pressure, temperature and its distribution, quality and quantity of rolling lubricant (concentration & distribution), thermal condition, etc[1].

Effective lubrication moderates the friction at the roll-bite, and helps in minimizing the rolling pressure by as much as 20% [2]. The cold rolling lubrication is provided in the form of oil-in-water emulsion; water to keep the rolling temperature under control, and oil to reduce friction by forming a film between the rolls and the strip. The intense heat generated during cold rolling due to cold deformation of stock and higher friction at the roll-bite is largely absorbed by the rolls, the stock and lubricant. As the rolling progresses, heat in the deformation zone builds up rapidly, causing reduction in viscosity and reduction in oil film thickness, hence increases friction co-efficient at the roll-bite [3]. Thicker the oil film hence is applied at the bite of roll / strip in optimum quantity to control friction co-efficient at bite, wear, surface damage of the roll and strip for stable rolling process[4]. Therefore, controlling the heat balance is a very effective way of improving lubricity, which is the theme of this paper.

During cold reduction process at TCM, the free iron powder and metallic fines are generated at the roll-bite due to frictional effect. Higher amount of boundary contacts between asperities of roll and strip leads to increased iron fine generation. These loosen iron particles are being picked up by the rolls during temper rolling and leave pick-up mark on the finished strip [5]. The defect is usually termed as iron pick-up and leads to subsequent finishing coating process interruption. [5]. Iron pick-up, black patch, heat mark, etc are some of the common defects on surfaces of cold rolled strips and are related to tribological parameters. The iron fines amount generated depends mainly on roll surface roughness and the reduction schedule applied (higher roll roughness and more reduction takes place in the first stands and more iron fines generated)[5]. Improved lubricity of cold rolling lubricant retards wear of rolls and strip and thus helps in reducing generation of iron fines [6].

EARLIER SYSTEM

The earlier emulsion application system at TCM-1(4-Stand, 4-Hi Mill) of Bokaro Steel Limited (BSL) was inadequate to control the effective lubricity at the roll-bite due to inadequate emulsion flow-rate, low impinging pressure (2.5~3.0 bar) and improper location of spray application had resulted in higher temperature at the roll-bite as well as roll body which detoriates frictional condition at roll-bite. The total coolant flow rate was about 2800lpm in Stand #1 and hot water was used as lubricant at roll-bite. In Stand # 2, 3 & 4 emulsion of 1.5% ~ 2% concentration of rolling oil were used with a flow rate of 3500lpm. Inadequate emulsion flow resulted in higher roll temperature as well as roll thermal crown of 6°C~8°C. Also due to lack of effective lubrication at roll-bite helps for the generation of higher iron fines and build-up in the emulsion of more than 300ppm and resulted in higher diversion on account of iron pick-up defects to 0.3% of total diversion.

The effective lubrication depends upon lubricant quality and application system. The lubricant characteristics will be controlled by physico-chemical property of lubricant. However, to enhance the efficiency of lubrication and cooling aspect of the rolls, an improvised emulsion application system was designed and commissioned in all the stands at TCM-1 to achieve effective lubricity and controlled friction at roll-bite for improved quality of rolled strips.

EXPERIMENTAL

In order to improve the lubricity and thermal condition at the roll bite, a improvised emulsion application system has been designed, developed and implemented at TCM-1 based on the following activities:

- Estimation of quantity of emulsion flow rate in stands based on friction at roll-bite
- Design and commissioning of modified emulsion application system for achieving higher heat transfer co-efficient (HTC) and effective lubrication at roll-bite.

(a) Estimation of Stand-wise Emulsion Flow-rate based on Friction

A mathematical model has been developed to calculate the emulsion flow rate required for different stands of TCM. The input parameters required for the mathematical model are pass reduction/ draft, rolling speed, dynamic yield strength of material, roll diameter, inter-stand tension and coefficient of friction as shown in the Fig.1.

The model calculates roll force, torque and energy required for deformation using mathematical equations summarized by Roberts [1]. During cold rolling, a large amount of heat is generated due to plastic deformation and friction between roll and coil. The heat generated due to friction is negligible compared to deformation heat. The relationship between the energy of deformation and deformation heat during tandem cold rolling process has been studied in details by Koots [7] and proposed a theory which was later known as Koot's theory. The model uses Koot's theory to calculate the amount of heat required to be transferred by the roll cooling system at different stands of the Tandem Mill. The flow rate of emulsion required by roll cooling system for transferring the heat has been calculated using standard heat transfer equation with established heat transfer coefficient given in literature [1].

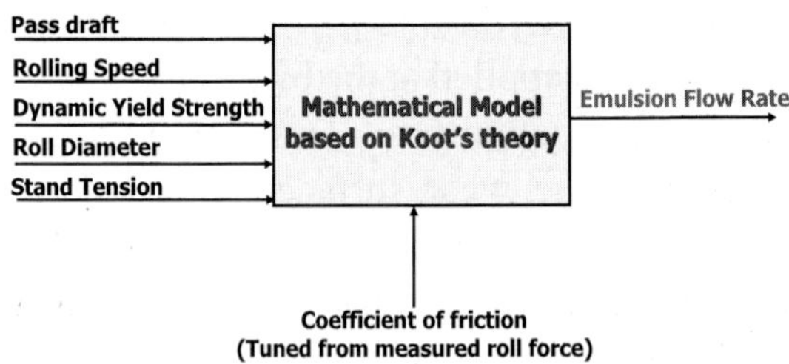

Fig. 1: Conceptual diagram of Mathematical Model for prediction of Flow Rate

All the input parameters except coefficient of friction between roll and the coil are measurable quantities. The coefficient of friction is not measurable and the method of its estimation applied for evaluation. The coefficient of friction varies between 0.01 and 0.07 during the cold rolling process. So, the model has been used to predict the required flow rate for different coefficient of friction. The result of prediction shown in Fig.2, shows the emulsion requirement for all the stands which increases with the increased value of coefficient of friction. If the coefficient of friction is estimated properly and given to model, then the flow rate of emulsion can be known.

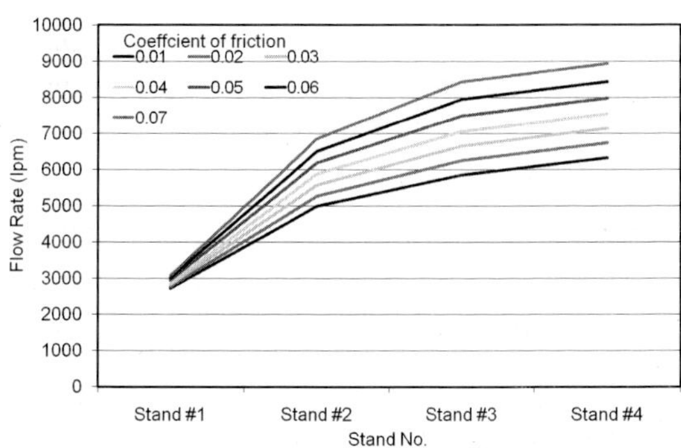

Fig. 2: Spray Header Arrangement at Mill Stand

During the cold rolling process, the coefficient of friction between the roll and strip depends upon process parameters like quality & quantity of rolling oil in emulsion, amount of iron fines in the emulsion, roughness of rolls as well as of strip, composition of roll material, rolling speed and the amount of pass reduction. As most of these parameters are non-measurable, and it is impractical to calculate the coefficient of friction from these parameters directly. Therefore a parameter estimation technique has been applied to estimate coefficient of friction at different stands of Tandem Mill. Coefficient of friction has been used as model calibration parameter in the roll force model. Both input data and roll force data of large number of coils were collected from the TCM–1, BSL. The coefficient of friction has been estimated as that value of calibration factor where the root mean square error between the predicted and measured roll force is minimum. Using this parameter estimation technique, coefficient of friction has been estimated and found that the coefficient of friction for Std #1 is 0.07 whereas for Std # 2 to Std # 4 was 0.05.

(b) Design of Emulsion Application system for achieving maximum HTC

The important criterions for achieving maximum HTC are proper nozzle selection, spray condition, emulsion flow distribution and the spray header configuration.

Nozzle selection: It has been experimented that with the spray-band of flat jet nozzle is having maximum constant HTC over the impact area besides other types of nozzle [8]. The flat jet nozzle is having larger length-width ratio of the impact area in comparison to other types of nozzle. Thus flat jet nozzles are suitable for achieving maximum HTC with relatively low coolant pressure and had been selected for the modified emulsion spray application system.

Spray condition: The spray condition depends upon three basic parameters like emulsion pressure, impinging spray angle and spray height i.e. the distance between nozzle tip to roll surface.

It has been observed that increase in coolant pressure to 15bar results in gradual improvement in HTC, but beyond 15bar pressure no significant improvement in HTC observed (Fig.3). In the model, the emulsion pressure considered to be 6~8bar during estimation of emulsion flow rate. The emulsion pressure has been increased from previous 2.5bar to 6.0bar level, through installation of anew higher capacity pump. As previously discussed, most iron fines are generated in the first two stands of cold rolling mill. Hence a new pump of 5800lpm @12bar capacity installed in Std #1. But due to some consequences, new pump-set could not be installed in the remaining stand # 2, 3 & 4. However, restricting the opening of nozzle led to increased emulsion pressure from 2.5bar to 4.0bar in the spray header design and the connections for pressure loss.

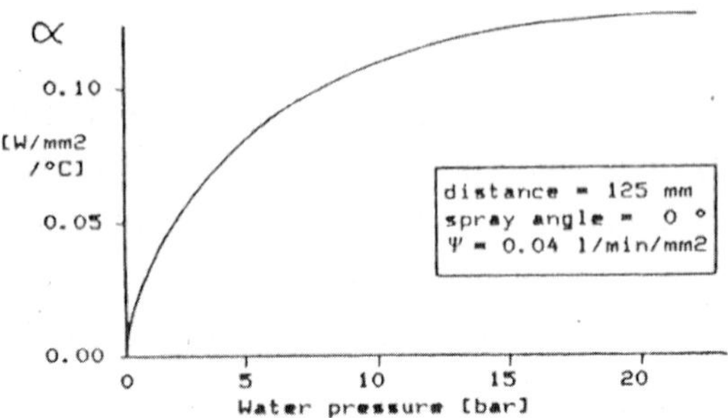

Fig. 3: HTC as a function of Water Pressure

It has been observed that lesser spray distance results in higher HTC[8]. The optimum distance of 300mm is being considered in the model while estimation of the emulsion flow requirement. The spray height of 350 mm for top WR cooling header and 300 mm for bottom WR cooling header had been considered, due to space constraint in the mill fixtures.

The impinging spray angle (β) should be perpendicular to the roll surface for maximum HTC and the cooling efficiency decreases with increased $\beta>25^o$, due to increased disturbance of flow at the impact area as shown in Fig.4. In modified system the impinging spray angle had been kept 20° for top WR and 15° for bottom WR for more roll surface coverage as well as roll-strip bite. The nozzles fitted in the header with an off-set by 20°[8] to avoid interference of adjacent spray.

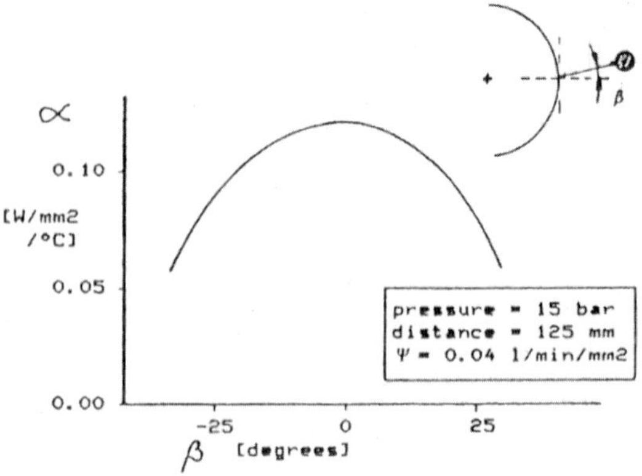

Fig. 4: HTC as a function of Spray Angle

Emulsion flow distribution: The quantity of emulsion flow rate varies, depending on the requirement and type of mill. However, the emulsion pressure ranged between 5bar to 8bar is acceptable for most of the cold rolling mills. The volume and pressure of the emulsion are directly related to the impact density of the spray and in turn to the heat transfer efficiency. Minimum 3~4bar emulsion pressure is needed to achieve any thermal response of spray as shown in Fig.3.

(c) Stand-wise Emulsion flow distribution

The stand-wise emulsion flow distribution was optimized in accordance to the estimated heat generation depending upon the following variables like roll speed, roll force and percentage reduction pattern as output of the Model and thus validation of Model being done. In a TCM, the gradual increment in work hardening on the rolled material takes place from second stand to final stand, and result in increased heat generation due to higher resistance to deformation. Accordingly, the total emulsion flow-rate in the estimation gradually increased from 4000lpm in Std#1 to 8000lpm in Std#4 as given in Table 1. With the installation of new higher capacity pump of 4500lpm @ 8bar in Std#1, the total emulsion quantity increased which in turn improved roll cooling efficiency as well as lubrication aspect at roll-bite than earlier.

Table 1: Recommended Flow Rate & Pressure

Parameters	Std #1	Std #2	Std #3	Std #4
% Reduction	40	30	20	10
Power(kw)	4000	5300	6000	4200
Roll Force(t)	1600	1650	1600	1250
Speed(mps)	4.8	6.8	9.5	12
Flow rate(lpm)	4000	7000	8000	8500
Pressure(bar)	7 ~ 8	5 ~ 6	5 ~ 6	5 ~ 6

The amount of heat generated due to deformation process is higher than friction, therefore the amount of emulsion flow designed higher for intense cooling of roll body than the roll-strip bite in order to arrest heat wave penetration. As the heat flow takes place from WR to BUR during rolling, mild cooling for BUR was kept. The emulsion flow distribution in a stand had been kept in a fashion like 10% for top & bottom BUR each and 40% for top & bottom WR each of total flow.

Spray header configuration

The spray rows for intense cooling of roll body were chambered into three zones - centre, intermediate and edge to accommodate the flow rates without restrictions. In order to control roll thermal crown i.e. the temperature between centre to edges of roll barrel, emulsion flow distribution across roll barrel length varied and was made through zonal emulsion flow control system with three zones. Higher number of spray nozzles had been incorporated in the central zone than intermediate and edge, to control as well as flattened roll thermal crown. Since higher temperature build up takes place at the central portion of the roll barrel due to lower heat dissipation rate, higher coolant flow density was kept at the central zone than the intermediate and edge zones in the design. The flow density varied in different zones by playing with the inter-spacing or number of nozzles in the zone (Fig.6).

The nozzles were equally spaced in the spray headers for BUR cooling as well as lubrication. All these emulsion spray headers are provided with bottom flushing port in each zones for ease of periodic maintenance and made retrofitting.

Fig. 6: Nozzle Arrangement in Spray Header Configuration

Spray header arrangement

As stated throughout, the sprays must impinge as precisely and accurately as possible onto the roll-bite for effective lubricity, which directly depends on the positioning of the headers within the mill fixtures. In the modified system, emulsion flow distribution was designed with two row spray for top and bottom work roll(WR), where one row was oriented towards the roll-strip bite with uniform flow across barrel for controlling friction and other row towards the roll body for extracting heat generated due to deformation process.

As stated earlier, the emulsion flow distributions for each WR were around 40% of total flow in a stand. The major amount of emulsion spray around 25% had been kept for intense cooling of roll body and remaining 15% of emulsion oriented towards roll-bite for basic lubrication with equally spaced nozzles in the row(Fig.7).

The optimal location had been designed taking into account of the existing mill furniture, guides, bridles, ease of maintenance and pipe sizing considering pressure losses. However, the spray header arrangement, location, impinging spray angle, pipe line network for zonal flow, etc. has been shown in Fig.7.

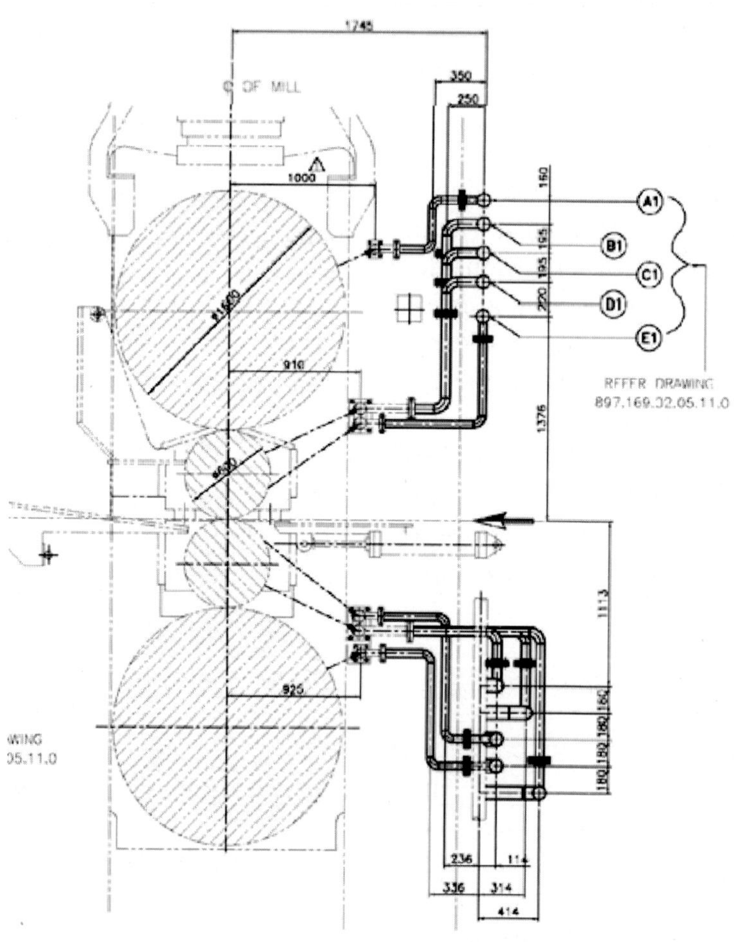

Fig. 7: Spray Header Arrangement at Mill Stand

RESULTS & DISCUSSION

The modified emulsion application system had been commissioned at all stands of TCM-I, and the performance was evaluated and following results had been achieved.

(a) A mathematical model has been developed to calculate the emulsion flow requirement for different stands of TCM-1. Different critical values of input rolling parameters and estimated coefficient of friction have been used in the model to predict required emulsion flow rate at different stands for rolling of thinnest gauge(0.63mm) with rolling speed of 12m/sec(max) at finishing stand. The screenshot of model Graphical User Interface(GUI) as shown in Fig.8. The last row of the figure shows the flow rate of emulsion predicted by the model. Based on the prediction through the model, an estimated flow requirement had been recommended for implementation in the mill as shown in Table 1.

(b) The heat generated due to cold deformation is higher and partly at roll-bite due to friction, are taken-off largely by the work rolls and partially by the strip and emulsion. A nominal part of heat is being conducted to the back-up rolls and the frictional effect between the back-up roll and work roll generates comparatively low heat input to the back-up rolls. Therefore, the intense cooling of top

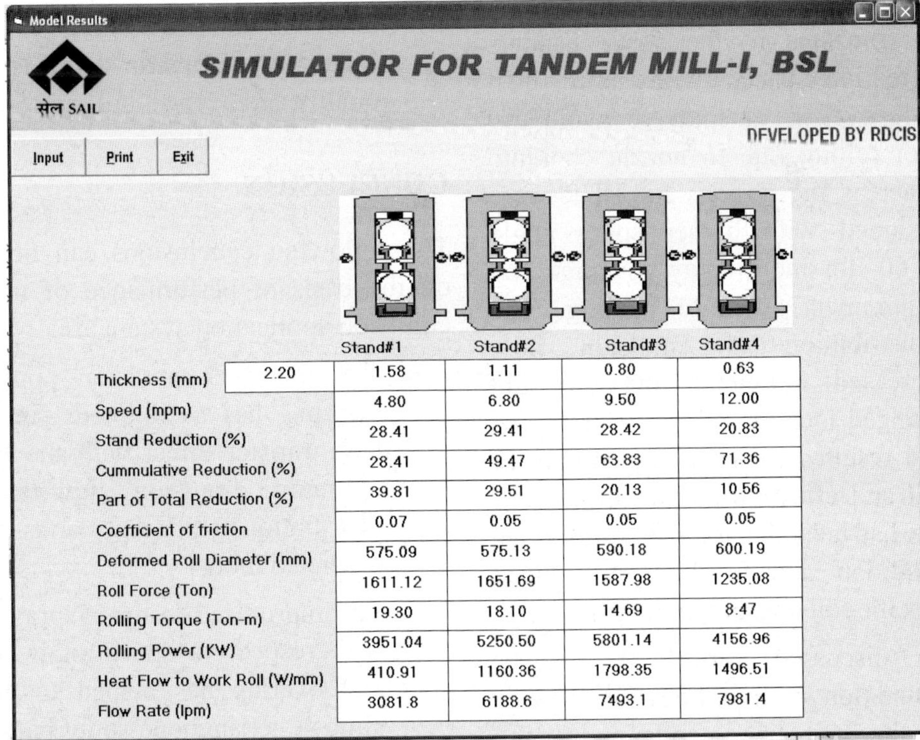

		Stand#1	Stand#2	Stand#3	Stand#4
Thickness (mm)	2.20	1.58	1.11	0.80	0.63
Speed (mpm)		4.80	6.80	9.50	12.00
Stand Reduction (%)		28.41	29.41	28.42	20.83
Cummulative Reduction (%)		28.41	49.47	63.83	71.36
Part of Total Reduction (%)		39.81	29.51	20.13	10.56
Coefficient of friction		0.07	0.05	0.05	0.05
Deformed Roll Diameter (mm)		575.09	575.13	590.18	600.19
Roll Force (Ton)		1611.12	1651.69	1587.98	1235.08
Rolling Torque (Ton-m)		19.30	18.10	14.69	8.47
Rolling Power (KW)		3951.04	5250.50	5801.14	4156.96
Heat Flow to Work Roll (W/mm)		410.91	1160.36	1798.35	1496.51
Flow Rate (lpm)		3081.8	6188.6	7493.1	7981.4

Fig. 8: Screenshot of Model for prediction of flow rate of roll cooling system in TCM

& bottom WR has helped to control and reduce the average roll surface temperature from 64°C to 56°C, and mild cooling of BUR has reduced the BUR temperature by 3°C~4°C.

(c) In the modified system, the emulsion flow was redistributed stand-wise to achieve optimum heat transfer effect and to avoid flooding and wastage of coolant. The modified system had resulted in reduced work roll thermal crown from 6~8°C to 2~3°C which resulted in flattened roll thermal profile and improved strip profile/ shape and observed reduction in diversion of bad shape defect by 40%.

(d) The spray angle of nozzles for work roll cooling header was kept at 20° for top WR and 15° for bottom WR for higher spray coverage of the coolant. The nozzles in the spray headers were interspaced with more than 50% overlap of spray to avoid dry spot rolling due to nozzle choking. Whereas the nozzles for BUR were designed with higher spray angle of 60° for more spray coverage and requirement of mild cooling. The redistribution of emulsion flow increased impinging pressure and changed location of spray impinging had resulted in improved cooling of roll and effective lubrication at roll-bite had helped to improve roll service condition and resulted in reduced specific roll consumption by 10~11%.

(e) As discussed earlier, the higher generation of iron fines takes place in the first two stands and being controlled by introduction of rolling oil concentration of 0.5% in the emulsion of Stand #1 in lieu of hot water and provided better lubrication at the roll-bite. The higher flow rate of emulsion with the help of new pump also helped to reduce roll temperature and controlled thermal condition at the roll-bite. The major quantity of emulsion has been used for effective cooling of roll and oriented towards the WR body with higher impinging pressure at the entry side which helped to wash-off the iron fines picked up by the rolls before getting embedded or carry-over on strip. The total modified emulsion application has helped to reduce the amount of iron fines in the emulsion as well as carryover on the strip surface substantially led to the almost elimination of the iron pick-up defect of total diversion. There was improvement in strip shape as well as quality and over-all mill performance.

CONCLUSION

The following conclusions can be drawn on the basis of performance of modified emulsion application system are:

- The modified emulsion application system has exhibited improved heat transfer effect with the help of optimum emulsion flow-rate and distribution in different stands through predicted model.
- The improvised design of spray header with respect to spray angle, off-set and overlap has helped to achieve aligned herringbone spray pattern for improved HTC.

- Intense cooling of WR with double row spray concept has helped to improve the cooling aspect and reduced roll surface temperature by nearly 8°C, which resulted in reduced specific roll consumption by 10~11%.
- It also helped in improved thermal condition for controlling lubrication at roll-bite and facilitates washing-off the carryover iron fines on strip and almost elimination of iron pick-up defect.
- The emulsion flow distribution across barrel has helped to reduce thermal crown by 2~3°C.
- The modified system also helped to reduce the diversion on account of bad shape defect by 40% and improved mill performance as well as product quality.

ACKNOWLEDGEMENT

The authors offer special thanks to ED I/c RDCIS and Conference Convener for permitting to present the paper in the Conference.

REFERENCES

[1] Roberts, Cold Rolling of Steel, Marcel Decker Inc, pp.384

[2] A Taher, et al, 33rd Conf. Proc MWSP, Cold rollability of Si-steels thin foils with and without using lubricants, ISS-AIME Vol. XXIX, 1992, p.97-101

[3] Oskar Pawelski, et al, Influence of hydrodynamic lubricant entrainment on friction effects in cold rolling; Steel Research, 65(1994) No.1, pp. 488-493

[4] Madhu Ranjan, et al, 2nd ICIT Conf. Proc., Improvement in lubrication during Cold Rolling of Electrical Steel, Dec 1999, pp. 273-279

[5] P Pathak, et al; ISFL Conf. Proc., Influence of Rolling oil quality on generation of Iron Pick-up mark over Cold Rolled Strip surfaces, Oct 2002

[6] SH Wang, et al: Transactions of the ASME, Lubrication with emulsion in cold rolling, Vol.115, 1993, pp. 523-531

[7] L.W. Koot, Proc. ISI meeting on mathematical process models for iron & steel making, Process design criteria for cooling of a cold strip mill, Amsterdam, 1973, pp. 19-21.

[8] G Van Steden and JGM Tellman, 4th Int. Rolling Conference, A new method of designing a work roll cooling system for improved productivity and strip quality.

Subject Index

Author Index